Lees' Process Safety Essentials

Lees' Process Safety Essentials

Hazard Identification, Assessment and Control

Sam Mannan

ELSEVIER

AMSTERDAM • BOSTON • HEIDELBERG • LONDON
NEW YORK • OXFORD • PARIS • SAN DIEGO
SAN FRANCISCO • SINGAPORE • SYDNEY • TOKYO

Butterworth-Heinemann is an imprint of Elsevier

Butterworth-Heinemann is an imprint of Elsevier
The Boulevard, Langford Lane, Kidlington, Oxford, OX5 1GB, UK
225 Wyman Street, Waltham, MA 02451, USA

First edition 2014

Notice
No responsibility is assumed by the publisher for any injury and/or damage to persons or property as a matter of products liability, negligence or otherwise, or from any use or operation of any methods, products, instructions or ideas contained in the material herein. Because of rapid advances in the medical sciences, in particular, independent verification of diagnoses and drug dosages should be made

Library of Congress Cataloging-in-Publication Data
A catalog record for this book is available from the Library of Congress.

British Library Cataloguing-in-Publication Data
A catalogue record for this book is available from the British Library.

ISBN: 978-1-85617-776-4

For information on all Butterworth-Heinemann publications
visit our Web site at www.elsevierdirect.com

Typeset by MPS Limited, Chennai, India
www.adi-mps.com

Working together
to grow libraries in
developing countries

www.elsevier.com • www.bookaid.org

6. Hazard Identification

7. Plant Siting and Layout

Instigated by the tragic industrial accidents of his time, Professor Frank Lees, while writing and editing with passion the first edition of *Loss Prevention in the Process Industries* in the late 1970s, would never have anticipated that his brainchild would almost quadruple in number of pages. Lees was fully committed to process safety. And having the vision that it would become a discipline of its own, he was determined to create a compendium on loss prevention theory, safety concepts, hazard analysis methods, data, and case histories that was complete. But it all further developed, and Lees' renewal in 1996 and Mannan's continuations in 2004 and 2012 have been indispensable, because the field has expanded and intensified greatly.

That growth process still continues, since threats change in nature, challenges posed by technology, scale and intensity of operations, the business, the way we are organized, it all grows and changes. Competition and finite resources force us to scrutinize investments, also those in safety measures, but at the same time the risk of mishap can be disastrous to victims and also to the company that caused it. So, we need better guidance about what to do and more certainty that what we do will be right. In that respect, it is unfortunate that due to economic pressures in quite a number of countries, education in process safety, certainly on the academic level, is diminishing or disappearing. Experience learns that process safety concepts and methods look often rather simple from the outside but appear to require a host of background information and knowledge for successful application. Despite the electronic features of searching and linking in the fourth edition, the nearly 4000 pages may have the effect of missing the wood for the forest, or if on a search may induce the feeling of looking for a needle in a haystack, which both may produce rather frustrating results. Therefore, it has been an insightful initiative of Professor Sam Mannan to work with his crew of Ph.D. students and a team of enthusiastic reviewers on a trimmed down, skinny edition of only 500 pages in 30 chapters!

I hope this book of essentials will contribute much to foster process safety, to guide us to deeper knowledge, and to lay the basis for further developments. Nature is complex, to enable accurate prediction, physics, certainly in its details, is often hard to describe, the spectrum of technology is immensely wide and growing, the result of human action almost impossible to predict reliably, yet we are expected to determine process risks that may occur far less than once in a million years with an accuracy that satisfies CEOs and lawyers. Good success, and don't think this book will be the end!

Hans Pasman
St. Julian's, Malta, Spring 2013

Introduction

Over the last four decades, there has developed in the process industries a distinctive approach to hazards and failures that cause loss of life and property. This approach is commonly called loss prevention. It involves putting much greater emphasis on technological measures to control hazards and on trying to get things right at the beginning. An understanding of loss prevention requires some appreciation of its historical development against a background of heightened public awareness of safety, and environmental problems, of its relation to traditional safety and also to a number of other developments.

1.1. MANAGEMENT LEADERSHIP

By the mid-1960s, it was becoming increasingly clear that there were considerable differences in the performance of companies in terms of occupational safety. These disparities could be attributed only to differences in management. There appeared at this time a number of reports on safety in chemical plants arising from studies by the British chemical industry of the safety performance in the US industry, where certain US companies appeared to have achieved an impressive record. These reports included *Safety and Management* by the Association of British Chemical Manufacturers (ABCM) (1964, p. 3), *Safe and Sound* and *Safety Audits* by the British Chemical Industry Safety Council (BCISC) (1969, p. 9; 1973, p. 12). The companies concerned attributed their success entirely to good management, and this theme was reflected in the reports.

1.2. INDUSTRIAL SAFETY AND LOSS TRENDS

Around 1970, it became increasingly recognized that there was a worldwide trend for losses, due to incidents, to rise more rapidly than gross national product (GNP).

This may be illustrated by the situation in the United Kingdom. The first half of this century saw a falling trend in personal incidents in British factories, but about 1960, this fall bottomed out. Over the next decade, very little progress was made; in fact there was some regression. Figure 1.1 shows the number of fatal incidents and the total number of incidents in factories over the period 1961–1974. The Robens Committee on Health and Safety at Work, commenting on these trends in 1972, suggested that part of the reason was perhaps the increasingly complex technology employed by industry (Robens, 1972).

Another important index is that of fire loss. The estimated fire damage loss in factories and elsewhere in the United Kingdom for the period 1964–1974 is shown in Figure 1.2.

1.3. SAFETY AND ENVIRONMENTAL CONCERNS

There was also at this time growing public awareness and concern regarding the threat to people and to the environment from industrial activities, particularly those in which the process industries are engaged. Taking the United Kingdom as an illustration, the massive vapor cloud explosion at Flixborough in 1974 highlighted the problem

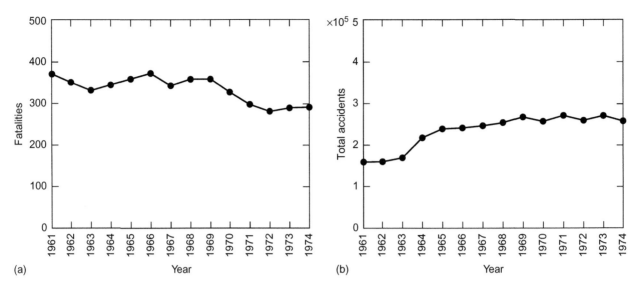

FIGURE 1.1 (a) Fatalities and (b) total accidents in factories in the United Kingdom, 1961–1974. *Source: Courtesy of the Health and Safety Executive; Robens (1972); HM Chief Inspector of Factories (1974).*

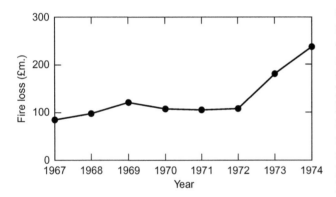

FIGURE 1.2 Total fire losses in the United Kingdom, 1967–1974. *Source: Courtesy of the British Insurance Association; British Insurance Association (1975).*

of major hazards. This led to the setting up of the Advisory Committee on Major Hazards (ACMH) that sat from 1975 to 1983 and to the introduction of legislation to control major hazard installations. Likewise, there was a continuous flow of legislation to tighten up both on emissions from industrial installations and on exposure of workers to noxious substances at those installations.

Similarly, in the United States, the Bhopal incident as well as other highly publicized tragedies (Flixborough, 1974; Seveso, 1976; Three Mile Island, 1979; Cubato, February 1984; Mexico City, November 1984; Houston, 1989) caused widespread public concerns about major incidents in US chemical plants that might disastrously affect the public. Not only was the public's confidence in the chemical industry shaken, but also the chemical industry itself questioned whether its provisions for protection against major accidental releases were

adequate. The recognition of the chemical industry's need for technical advances led to a number of initiatives. For example, in 1985, the Chemical Manufacturers Association (CMA—now known as the ACC, the American Chemistry Council) published its guidelines on Process Safety Management, and the American Institute of Chemical Engineers (AIChE) created the Center for Chemical Process Safety (CCPS) with significant financial support by industry. Over the next several years, many other centers such as the National Institute for Chemical Safety, the National Environmental Law Center, and the Mary Kay O'Connor Process Safety Center also came into existence. During this same period, the United States Environmental Protection Agency (USEPA) and the Occupational Safety and Health Administration (OSHA) of the United States Department of Labor started several technical initiatives aimed at gathering information about major accident risks.

It is against this background, therefore, that the particular problems of the process industries should be viewed. The chemical, oil, and petrochemical industries handle hazardous substances and have always had to devote considerable effort to safety. This effort is directed both to the safe design and operation of the installations and to the personal safety of the people who work on them. However, there was a growing appreciation in these industries that the technological dimension of safety was becoming more important.

1.4. HISTORICAL DEVELOPMENT OF LOSS PREVENTION

The 1960s saw the start of developments that have resulted in great changes in the chemical, oil, and petrochemical industries. A number of factors were involved in these changes. Process operating conditions such as pressure and temperature became more severe. The energy stored in the process increased and represented a greater hazard. Problems in areas such as materials of construction and process control became more taxing. At the same time, plants grew in size, typically by a factor of about 10, and were often single stream. As a result they contained huge items of equipment, such as compressors and distillation columns. Storage, both of raw materials and products and of intermediates, was drastically reduced. There was a high degree of interlinking with other plants through the exchange of by-products.

The operation of such plants is relatively difficult. Whereas previously chemical plants were small and could be started up and shut-down with comparative ease, the start-up and shut-down of a large, single-stream plant on an integrated site is a much more complex and expensive matter. These factors resulted in an increased potential for loss—both human and economic. Such loss may occur in various ways. The most obvious is the major incident, usually arising from loss of containment and taking the form of a serious fire, explosion, or toxic release. But loss due to such situations as delays in commissioning and downtime in operating is also important.

The chemical and oil industries have always paid much attention to safety and have a relatively good record in this respect. In the United Kingdom, for example, the fatal accident rate for the chemical industry has been about equal to that for industry generally, which in view of the nature of the industry may be regarded as reasonable. However, the increasing scale and technology of modern plants caused the chemical industry to re-examine its approach to the problem of safety and loss. If the historical development of this concern in the United Kingdom is considered, there are several problem areas that can be seen, in retrospect, to have given particular impetus to the development of loss prevention.

One of these is the problem of operating a process under extreme conditions and close to the limits of safety. This is usually possible only through the provision of relatively sophisticated instrumentation. About the mid-1960s, several such systems were developed. One of the most sophisticated, influential, and well documented was the high integrity protective system developed by R.M. Stewart (1971) for the ethylene oxide process. Around that same time, many difficulties were being experienced in the commissioning and operation of large, single-stream plants, such as ethylene and ammonia plants, involving severe financial loss. On the design side, too, there was a major problem in getting value for money in expenditure aimed at improving safety and

reducing loss. It was increasingly apparent that a more cost-effective approach was needed.

These developments did not take place in isolation. The social context was also changing and other themes, notably pollution, including effluent and waste disposal and noise, were becoming of increasing concern to the public and the government. In consequence, the industry was obliged to examine the effects of its operations on the public outside the factory fence and, in particular, to analyze more carefully the possible hazards and to reduce emissions and noise. Another matter of concern was the increasing quantities of chemicals transported around the country by road, rail, and pipeline. The industry had to take steps to show that these operations were conducted with due regard to safety. In sum, by the 1970s, these problems became a major preoccupation of senior management. Management's recognition of the problems and its willingness to assign many senior and capable people as well as other resources to their solution has been fundamental in the development of loss prevention.

The historical development of loss prevention is illustrated by some of the milestones listed in Table 1.1. The impact of events has been different in different countries. Within the industry, loss prevention emerged as a theme of technical meetings which indicated an increasingly sophisticated technological approach. The Institution of Chemical Engineers (IChemE) established a Loss Prevention Panel which operates an information exchange scheme and publishes the *Loss Prevention Bulletin*. This growing industrial activity was matched in the regulatory sphere. The Robens Committee (1972) emphasized the need for an approach to industrial safety that is more adapted to modern technology and recommended self-regulation by industry as opposed to regulation from outside. This philosophy is embodied in the Health and Safety at Work Act 1974 (HSWA), which provides the framework for such an approach. However, the Act does more than this. It lays a definite statutory requirement on industry to assess its hazards and demonstrate the effectiveness of its safety systems. It is enforced by the Health and Safety Executive (HSE). The General Duty Clause requirements under US regulations enforced by OSHA and USEPA have been interpreted similarly.

The disastrous explosion at Flixborough in 1974 has proved to be a watershed event. Taken in conjunction with the Act, it has greatly raised the level of concern for SLP in the industries affected. It also led, as mentioned, to the setting up of the ACMH. The incident at Seveso in 1976 has been equally influential. It had a profound impact in Continental Europe and was the stimulus for the development of the EC Directive on Control of Industrial Major Accident Hazards in 1982. Further disasters such as those at San Carlos, Bhopal and Mexico City,

TABLE 1.1 Some Milestones in the Development of Loss Prevention

1960	First UK IChemE symposium on Chemical Process Hazards with Special Reference to Plant Design
1966	Dow Chemical Company's Process Safety Manual
1967	First AIChE symposium on Loss Prevention
1968	First ICI Safety Newsletter; American Insurance Association Hazard Survey of the Chemical and Allied Industries
1971	European Federation of Chemical Engineering symposium on Major Loss Prevention in the Process Industries
1972	UK—Report of Robens Committee on Safety and Health at Work
1973	UK—IChemE Loss Prevention Panel, information exchange scheme and, in 1975, Loss Prevention Bulletin
1974	UK—HSWA 1974; First International Symposium on Loss Prevention and Safety Promotion in the Process Industries; Rasmussen Report; Flixborough disaster
1975	UK—Flixborough Report
1976	First Report of ACMH; Seveso incident
1978	First Canvey Report; San Carlos disaster
1979	Second Report of ACMH; Three Mile Island incident
1981	Norwegian Guidelines for Safety Evaluation of Platform Conceptual Design
1982	EC Directive on Control of Industrial Major Accident Hazards
1984	Third Report of ACMH; Control of Industrial Accident Hazards Regulations 1984; Bhopal disaster; Mexico City disaster
1985	AIChE establishes the CCPS
1986	Russia—Chernobyl disaster; USA, California—Risk Management and Prevention Program; USA, New Jersey—Toxic Catastrophe Prevention Act
1988	Piper Alpha disaster
1990	UK—Piper Alpha Report; USA—•January—API Recommended Practice 750 (Management of Process Hazards); •July—OSHA Process Safety Management Proposed Rule; •October—CMA Responsible Care Code of Management Practices; •November—Clean Air Act Amendments of 1990; Formation of the US Chemical Safety Board
1992	Offshore Safety Act 1992; Offshore Installations (Safety Cases) Regulations 1992
1995	USA—Risk Management Program regulation promulgated by USEPA; Texas A&M University established the Mary Kay O'Connor Process Safety Center
2005	US—Texas City Refinery Explosion, UK-Buncefield Explosion
2007	USA—The report of the BP US Refineries Independent Safety Review Panel, USA Department of Homeland Security-Chemical Facility Anti-Terrorism Standards (CFATS)
2009	UK HSE—Buncefield Explosion Mechanism Report, USA NFPA 59A-Standard for the Production, Storage, and Handling of Liquefied Natural Gas (LNG)
2011	USA-H.R. 501: Implementing the Recommendations of the BP Oil Spill Commission Act of 2011

Buncefield, Texas City, and Deepwater Horizon have reinforced these developments.

1.5. LOSS PREVENTION ESSENTIALS

The area of concern and the type of approach which goes by the name of loss prevention is a development of safety work. But it is a response to a changing situation and need, and it has certain particular characteristics and emphases. The essential problem which loss prevention addresses is the scale, depth, and pace of technology. In fact, control of such hazards is possible only through effective management. The primary emphasis in loss prevention is, therefore, on the management system. This has always been true, of course, with regard to safety. But high technology systems are particularly demanding in terms of formal management organization, competent persons, systems and procedures, and standards and codes of practice. On top of all of this, there is a growing realization that safety performance is quite often influenced by the safety culture of the company.

The 2003 Columbia shuttle disaster and the following investigation report bring to light some of NASA's safety culture issues that may have doomed Columbia as well as Challenger. The 2005 BP Refinery Explosion in Texas City reminded the importance of good safety culture again through the consequence of 15 fatalities and a loss of 1.5 billion dollars. The long list of root causes of this disaster includes lack of management of addressing major hazard control, ineffective role and responsibility of organization, lack of reporting and learning from near-misses, and inadequate resources dedicated to prevent major accidents. Most of the root causes concluded that the lack of safety culture leadership caused the failure of many layers of protection.

According to Mannan (2003a), we should pay more attention to safety culture. One school of thought is that safety culture even though a very important issue is not a specific problem of process safety or loss prevention. However, it seems there is a special set of problems that go along with extreme events and the associated risks. Before the event there is great confidence that such a thing could never happen. Afterward there is denial and no cultural change occurs. This may be true of NASA as well as some process companies. What are the attributes of a good safety culture? How do organizations build a good safety culture and maintain it over the life of the organization? How can the safety culture survive changing leaderships, turnovers, budget pressures, early retirements, and other changes? How can we get organizations that do not have a good safety culture make the necessary changes to move toward a good safety culture? These are questions that should be answered.

Mannan (2003a) goes on to describe the investigation of one incident where everyone involved felt that they had done everything they were supposed to do and the incident was just something that was beyond anyone's control. In fact, a few people in the organization even claimed that if the same set of circumstances were to happen again, the same incident would occur again, possibly with the same consequences. Now, that is a safety culture that needs major overhaul.

Mannan (2003b) states that there are a number of attributes of a good safety culture. It is quite difficult to identify objective characteristics of a good safety culture. However, some known characteristics include:

1. Commitment AND involvement of the highest level personnel.
2. Open communication at all levels of the organization.
3. Everyone's responsibilities and accountabilities regarding safety are clearly defined and understood.
4. Safety is second nature.

5. Zero tolerance for disregard of management systems, procedures, and technology.
6. Information systems allow all parties access to design, operational, and maintenance data.

As part of a multi-layered approach for a good safety culture, organizations use analysis of trends to spot problems. Trend analysis should be focused on leading as well as trailing metrics (indicators). The metrics are utilized to measure the performance of corporations at all levels. Process safety metrics can provide valuable information about plant operations, maintenance, and human behaviors. Establishing uniform metrics will benefit the comparison between different plants, businesses, and companies. However, the selection of indicators for metrics should be sophisticated because they should provide measurable and accurate performance for safety. The CCPS Guidelines for Process Safety Metrics (CCPS, 2009) recommends different metrics, such as absolute metrics, normalized metrics, and near-miss metrics.

A lagging indicator is a downstream measurement of the outcomes of safety and health efforts. These indicators reflect successes or failures of the system to manage hazards. Examples of lagging indicators include fatalities, injuries, and incidents. A leading indicator is an upstream measure that characterizes the level of success in managing safety systems; measurement of activities toward risk reduction prior to occurrence of incidents. While every effort should be made to measure and track lagging indicators, relying on the lagging indicators to assess safety performance is self-defeating. Thus, it is very important to measure and track leading indicators, particularly for high-risk activities such as space flight. Leading indicators, however, are more difficult to define and measure and vary according to the activity and the mission of the organization. Examples of leading indicators might include:

1. the level of near-miss reporting;
2. effectiveness of incident investigation and corrective action;
3. management of change;
4. emphasis on inherently safer design;
5. effective application of risk assessments;
6. level of deferred maintenance;
7. level of repetitive maintenance;
8. number and severity of faults detected by inspection, testing, and audits;
9. number and nature of unresolved safety issues;
10. participation in continuing education and symposia;
11. employee morale, level of expertise.

Thus, loss prevention is characterized by

1. an emphasis on management and management systems, particularly for technology;

2. a concern with hazards arising from technology;
3. a concern with major hazards;
4. a concern for integrity of containment;
5. a systems rather than a trial-and-error approach;
6. techniques for identification of hazards;
7. a quantitative approach to hazards;
8. quantitative assessment of hazards and their evaluation against risk criteria;
9. techniques of reliability engineering;
10. the principle of independence in critical assessments and inspections;
11. planning for emergencies;
12. incident investigation;
13. a critique of traditional practices or existing regulations, standards or codes where these appear outdated by technological change.

The identification of hazards is obviously important, since the battle is often half won if the hazard is recognized. A number of new and effective techniques have been developed for identifying hazards at different stages of a project. These include hazard indices, chemicals screening, hazard and operability studies, and plant safety audits.

Basic to loss prevention is a quantitative approach, which seeks to make a quantitative assessment, however, elementary. This has many parallels with the early development of operational research. This quantitative approach is embodied in the use of quantitative risk assessment (QRA). The assessment produces numerical values of the risk involved. These risks are then evaluated against risk criteria. However, the production of numerical risk values is not the only, or even the most important, aspect. A QRA necessarily involves a thorough examination of the design and operation of the system. It lays bare the underlying assumptions and the conditions that must be met for success and usually reveals possible alternative approaches. It is therefore an aid to decision-making on risk, the value of which goes far beyond the risk numbers obtained.

Reliability engineering is now a well-developed discipline. Loss prevention makes extensive use of the techniques of reliability engineering. It also uses other types of probabilistic calculation that are not usually included in conventional treatments of reliability, such as probabilities of weather conditions or effectiveness of evacuation. Certain aspects of a system may be particularly critical and may require an independent check. Examples are independent assessment of the reliability of protective systems, independent audit of plant safety, and independent inspection of pressure vessels.

Planning for emergencies is a prominent feature of loss prevention work. This includes both works and transport emergencies.

Investigation of incidents plays an important part in loss prevention. Frequently there is some aspect of technology involved. But the recurring theme is the responsibility of management. While a good safety culture varies according to the mission and activities of the organization, one of the attributes of a good safety culture that is a 'must' is 'learning from incidents'. There is no excuse when 'lessons learned' from incidents are ignored or not implemented, particularly 'lessons learned' from incidents that have occurred in one's own organization or incidents that are widely publicized. One of root causes of the 2005 Buncefield tank farm explosion is lack of learning culture. After the incident, people claimed that cold gasoline had never resulted in a vapor cloud explosion. But actually some similar incidents had already been stated in a review of past incidents (Kletz, 1986). The ignorance of past incidents and lack of learning ability can directly lead to the inadequate hazard identification and risk assessment.

Both 2005 Texas City and Buncefield incidents saw the need to perform facility siting studies in permanent and portable occupied buildings. Facility siting evaluates the effects of an incident on the plant area and also the influence to surrounding community. The effects include loss of life, loss of damage, business interruption, and environment as well. Adequate separation between process plants, between equipment and control room, and between tanks and site boundary should be provided by risk assessment. CCPS Guidelines for Facility Siting and layout (CCPS, 2003) gives the most conservative guidance. API 752 and 753 provide guidance to control hazards associated with both permanent and temporary occupied buildings siting problems. In the research area, consequence modeling and optimization methods are currently investigated and have been recommended to find the optimized siting.

The 2010 Deepwater Horizon blowout, which is the largest oil spill incident in the history of America, gained worldwide attention and revealed a series of management and technical gaps in the field of offshore drilling process. The loss of 11 lives and the short-term and long-term environmental impacts have brought the world a big lesson. The trade-off between safety and production, management of change in drilling procedures, planning for emergencies, lessons learned from similar near-misses, and design of blowout preventer system are the main topics that should be improved. The loss prevention approach takes a critical view of existing regulations, standards, rules, or traditional practices where these appear to be outdated by changing technology. Illustrations are criticisms of incident reporting requirements and of requirements for protection of pressure vessels. These developments taken as a whole do constitute a new approach and it is this which characterizes the loss prevention.

Another concept that is gaining increasing prominence is that of safety-critical systems. These are the systems critical to the safe operation of some larger system, whether this is a nuclear power station or a vehicle. In a modern aircraft, particularly of the fly-by-wire type, the computer system is safety critical.

It might perhaps be inferred from the foregoing that the problems which have received special emphasis in loss prevention are regarded somehow as more important than the aspects, particularly personal incidents, with which traditionally safety work in the process industries has been largely concerned. Nothing could be further from the truth. It cannot be too strongly emphasized that mundane incidents are responsible for many more injuries and deaths than those arising from high technology.

1.6. ENVIRONMENT AND SUSTAINABLE DEVELOPMENT

Another major concern of the process industries is the protection of the environment. Developments in environmental protection (EP) have run in parallel with those in SLP. These two aspects of process plant design and operation have much in common and there has been a tendency to assign the same person responsibilities for both. There are some situations where there is a potential conflict between the two, where the safest option has the potential for acute or chronic environmental damage. This can effect decisions regarding equipment venting, flaring, bleeding of dangerous process impurities, leak control of toxic and flammable streams, using halons and other chemicals in fire suppression systems.

Europe and the United States have seen a growing awareness of the need for environmental protection since the 1960s, which led to a growing number of regulations and laws, mostly in the 1970s, 1980s, and early 1990s. In the United States, these laws include the Control of Pollution Act (1974), Air Pollution Control Act (1975), Comprehensive Environmental Resource Conservation and Liability Act (1980), Superfund Amendments and Reauthorization Act (1986), Water Act (1989), Environmental Protection Act (1990), Clean Air Act (1963), and the various amendments to these laws. In Europe, various European Commission Directives play a role in setting pollution control standards for the European Union. There are also advisory bodies on pollution including the Royal Commission on Environmental Pollution (RCEP). The Environmental Protection Agency in the US also provides guidance.

Engineers have the responsibility of adhering to the professional codes and standards, obeying pertinent legislation, fulfilling their job descriptions and terms of employment, and utilizing the information gained through their formal education. All of these responsibilities must be satisfied.

Common elements in an environmental protection plan include use of inherently safer and cleaner design, reduction of intermediate storage, identification, assessment and modeling of hazards, control of fugitive emissions, assessment of environmental impact, and communication with the public. Environmental protection should take a total life cycle approach and consider the final environmental fate of chemicals and wastes produced as well as the chemical capacity of the environment. The different requirements and concerns with regard to liquid, solid, or gaseous pollutants, hazardous vs non-hazardous wastes, reactive vs inert wastes, and different means of disposal, like incineration vs landfill, must all be considered. Environmental monitoring, emergency response planning, spill containment and control, and post-spill clean-up procedures and plans should also be in place, with the facility and responsible parties ready to implement the response and clean-up plans.

The control and management of noxious or unpleasant odors can also be a significant concern depending on odor thresholds, regulations pertaining to odors, and the extent of the impact on the surrounding community. There are a number of methods available for treating air contaminated with odors. Accounts are given by Valentin (1990) and A.M. Martin et al. (1992).

Good environmental protection programs, hand in hand with good safety and loss prevention programs, have been shown to produce cost savings, particularly through reduction in waste generation and clean-up costs. Costs and savings may be computed at a national or company level.

To accomplish the balance between environmental, economic, and social pressures, the concept of sustainable development has been developed in the last two decades. Continuous effort of improving disciplines, technologies, and government strategies toward creating sustainable processes and products has been made, especially for the evaluation of sustainability and life cycle analysis. The tenet of sustainability has been studied a lot in the past decade.

1.7. RESPONSIBLE CARE

In a number of countries, the chemical industry has responded to safety, health, and environmental concerns with the Responsible Care initiative, which was developed in the early 1980s by the Canadian Chemical Producers Association and was then taken up in 1988 in the United States by the ACC, in 1989 in the United Kingdom by the Chemical Industries Association (CIA), and elsewhere.

Companies participating in Responsible Care commit themselves to achieving certain standards in terms of safety, health, and environment. Guidance is given in *Responsible Care* (CIA, 1992 RC53) and *Responsible Care Management Systems* (CIA, 1992 RC51). As stated earlier, in 1973, the IChemE created a Loss Prevention Panel. The Institution itself publishes a range of monographs and books on SLP and the panel publishes the Loss Prevention Bulletin and a range of aids for teaching and training. In 1985, the AIChE formed the CCPS. The Center publishes a series of guidelines on SLP issues.

At the European level, the European Process Safety Centre (EPSC) was set up in 1992 to disseminate information on safety matters, including legislation, research and development, and education and training (EPSC, 1993).

1.8. ACADEMIC AND RESEARCH ACTIVITIES

Over the years, academic and research activities aimed at process SLP have ebbed and flowed. However, lately these activities are becoming more formalized. Engineering departments in various universities throughout the world have begun to realize the importance of the subject and the significant role they can play in educating the students and the solution of industry problems through fundamental research. Many universities in the United States, United Kingdom, elsewhere in Europe, Japan, Korea, China, and other countries now offer specialized courses, certificate programs, and degree programs on process SLP. In some universities, for example Texas A&M University and University of South Carolina, process safety courses instead of being optional electives are now part of the required core curriculum. University professors and researchers are also dedicating extensive efforts toward research topics on process SLP. With regard to academic and research activities, the Mary Kay O'Connor Process Safety Center at Texas A&M University is a classic example of a comprehensive academic and research program dedicated to education, research, and service activities on process SLP.

1.9. OVERVIEW

The modern approach to the avoidance of injury and loss in the process industries is the outcome of the various developments just described. Central to this approach is leadership by management, starting with senior management, and creation of a safety culture that provides the appropriate environment for reduction of incidents and improvement of safety performance. Such leadership and safety culture are indispensable conditions for success. They are not, however, sufficient conditions. Management must also identify the right objectives.

As far as the process industries are concerned, it is the contribution of loss prevention to handle the technological dimension and to provide methods by which failure is eliminated. In the modern approach to SLP, these themes come together. The ends are the safety of personnel and the avoidance of loss. The means to achieve both these aims is leadership by management, informed by an understanding of the technology and directed towards the elimination of failures of all kinds.

REFERENCES

Association of British Chemical Manufacturers (ABCM), 1964. Safety and management. Report number: 3.

British Chemical Industry Safety Council (BCISC), 1969. Safety and sound. Report number: 9.

British Chemical Industry Safety Council (BCISC), 1973. Safety audits. Report number: 12.

British Insurance Association, 1975. Insurance Facts and Figures 1974, London.

Center for Chemical Process Safety (CCPS), 2003. Guidelines for Facility Siting and Layout. Wiley-AIChE.

Center for Chemical Process Safety, 2009. Guidelines for Process Safety Metrics. CCPS, AIChE, New York, NY.

Chemical Industries Association (CIA), 1992. RC53 Responsible care.

Chemical Industries Association (CIA), 1992. RC51 Responsible care management systems.

European Process Safety Centre, 1993. Annual Report, Rugby.

HM Chief Inspector of Factories, 1974. Annual Report of HM Chief Inspector of Factories 1974. HM Stationery Office, London.

Kletz, T.A., 1986. Accident reports and missing recommendations. Loss Prev. Saf. Promotion. 5, 20−21.

Mannan, M.S., 2003a. Director's Corner, Centerline, Mary Kay O'Connor Process Safety Center Newsletter, vol. 7, no. 2, 2003 Summer.

Mannan, M.S., 2003b. Director's Corner, Centerline, Mary Kay O'Connor Process Safety Center Newsletter, vol. 7, no. 1, 2003 Spring.

Martin, A.M., Nolen, S.L., Gess, P.S., Baesen, T.A., 1992. Control odors from CPI facilities. Chem. Eng. Prog. 88 (12), 53.

Robens, L., 1972. Safety and Health at Work. Cmnd 5034, HM Stationery Office, London.

Stewart, R.M., 1971. High integrity protective systems. Major Loss Prev. 99.

Valentin, F.H.H., 1990. Making chemical-process plants odor-free. Chem. Eng. 97 (1), 112.

Incidents and Loss Statistics

It is important in industrial processes to limit (and ideally eliminate) process incidents and their effects. For this to occur it is necessary to analyze hazards and identify root causes of incidents, determine the expected frequency of incidents, and have a financial plan to assure that the company can cover any expenses it incurs. These goals can be reached through proper hazard analysis, statistical information from proper sources, and proper insurance plans. These areas are discussed in greater detail throughout this chapter.

2.1. THE INCIDENT PROCESS

Although in some reporting schemes the investigator is required to determine the cause of the incident, it frequently appears meaningless to assign a single cause as the incident has arisen from a particular combination of circumstances. Second, it is often found that the incident has been preceded by other incidents that have been 'near-misses'. These are cases where most but not all of the conditions for the incident were met. A third characteristic of incidents is that when the critical event has occurred, there are wide variations in the consequences. In one case there may be no injury or damage, while in another case that is similar in most respects, there is some key circumstance that results in severe loss of life or property.

It is helpful to model the incident process in order to understand more clearly the factors that contribute to incidents and the steps that can be taken to avoid them. One type of model, discussed by Houston (1971), is the classical one developed by lawyers and insurers who focuses attention on the 'proximate cause'. It is recognized that many factors contribute to an incident, but for practical, and particularly for legal, purposes, a principal cause is identified. Several incident process models will be discussed to show more details, starting with the Houston model.

2.1.1 The Houston Model

The model given by Houston (1971, 1977) is shown schematically in Figure 2.1. Three input factors are necessary for the incident to occur: (1) target, (2) driving force, and (3) trigger. Principal driving forces are energy and toxins. The target has a threshold intensity θ below which the driving force has no effect. The trigger also has a threshold level θ' below which it does not operate.

The development of the incident is determined by a number of parameters. The contact probability p is the probability that all the necessary input factors are present. The contact efficiency ε defines the fraction of the driving force that actually reaches the target, and the contact effectiveness η is the ratio of damage done to the target under the actual conditions compared to that done under standard conditions. The contact time t is the duration of the process.

The model indicates a number of ways in which the probability or severity of the incident may be reduced. One of the input factors (target, driving force, or trigger) may be removed. The contact probability may be minimized by preventive action. The contact efficiency and contact effectiveness may be reduced by adaptive reaction.

2.1.2 Other Incident Models

A simple fault tree model that can be constructed for an incident is given in Figure 2.2. An initiating event occurs which constitutes a potential incident, but often only if

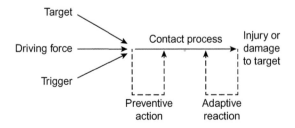

FIGURE 2.1 Houston model of the accident process. *Source: After Houston (1977).*

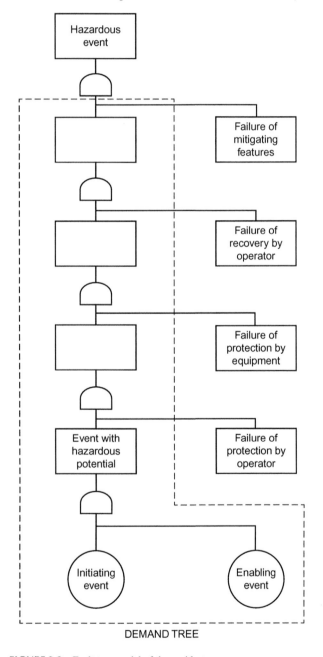

FIGURE 2.2 Fault tree model of the accident process.

some enabling event occurs, or has already occurred. An incident occurs that develops into a more severe incident only if mitigation fails.

A more complex fault tree model is that used in the management oversight and risk tree (MORT) developed by Johnson (1980). This tree is the basis of a complete safety system.

The ACSNI model was proposed by the Advisory Committee on the Safety of Nuclear Installations (ACSNI, 1993). The model provides a general framework that can be used to identify latent failures that are likely to lead to critical errors.

The Bellamy and Geyer model emphasizes the broader, socio-technical background to incidents and was developed by Bellamy and Geyer (1991).

Another approach is that discussed by Kletz (1988), who has developed a model oriented to incident investigation. The model is based essentially on the sequence of decisions and actions that lead up to an incident and shows against each step the recommendations arising from the investigation.

2.2. INJURY STATISTICS

2.2.1 United States of America

In the United States, many federal agencies gather information about the chemical industry. These federal databases, some of which have received information for over three decades, may provide the information needed to develop trends of chemical-related incidents.

Six US federal databases provide information about incidents and incident statistics related to chemical safety at fixed facilities. They are as follows:

1. National Response Center's (NRC) Incident Reporting Information System (IRIS)
2. EPA's Risk Management Program (RMP) Rule's 5-year Accident History Database
3. EPA's Accidental Release Information Program (ARIP) Database
4. Bureau of Labor Statistics' (BLS) Databases for the US Occupational Safety and Health Administration (OSHA)
5. US Centers for Disease Control and Prevention's (CDC) Wide-ranging On-line Data for Epidemiological Reporting (WONDER)
6. US Department of Health and Human Services' Agency for Toxic Substances and Disease Registry's (ATSDR) Hazardous Substances Emergency Events Surveillance (HSEES) Database.

2.2.2 United Kingdom

In the United Kingdom, the definition of a major injury changed with the introduction of the Reporting of

TABLE 2.1 Some of the Worst Non-Industrial and Industrial Disasters Worldwide

Event		Date	Deaths
A Worst World Disasters			
Explosion	Halifax, Nova Scotia, Canada	1917	1963
Industrial	Union Carbide methyl isocyanate plant, Bhopal, India	1984	≈ 2500
Nuclear reactor	Chernobyl Reactor No. 4	1986	31[a]
Explosion	Chilwell, Nottinghamshire (explosives factory)	1918	134
Mining	Universal Colliery, Senghenydd, Mid-Glamorgan	1913	439
Offshore platform	Piper Alpha, North Sea	1988	167
Tornado	Shaturia, Bangladesh	1989	≈ 1300
Nuclear reactor	Windscale (now Sellafield), Cumbria (cancer deaths)	1957	[b]

[a]The Guinness Book of Records states: "Thirty one was the official Soviet total of immediate deaths. On April 25, 1991 Vladimir Shovkoshitny stated in the Ukrainian Parliament that 7000 "clean-up" workers had already died from radiation. The estimate for the eventual death toll has been put as high as 75,000 by Dr Robert Gale, a US bone transplant specialist."
[b]The Guinness Book of Records states: "There were no deaths as a direct result of the fire, but the number of cancer deaths which might be attributed to it was estimated by the National Radiological Protection Board in 1989 to be 100."
Source: Material from Guinness Book of Records, copyright © reproduced by permission of the publishers.

Injuries, Diseases and Dangerous Occurrences Regulations 1985 (RIDDOR). The *Health and Safety Statistics 1990–1991* (HSE, 1992) show that in 1990–1991 there were 572 fatalities reported under RIDDOR, of which 346 were to employees, 87 to the self-employed, and 139 to members of the public. The fatal injury incidence rate for employees was 1.6 per 100,000 workers.

2.3. MAJOR DISASTERS

It is appropriate at this point briefly to consider major disasters. A list of the worst disasters in certain principal categories, both for the world as a whole and for the United Kingdom, is given in Table 2.1.

Those that are of primary concern in the present context are fire, explosion, and toxic release. The explosion at Halifax which killed 1963 people was that of a ship carrying explosives. The Chilwell explosion, in which 134 people died, was in an explosives factory. The toxic gas release at Bhopal, where the death toll was some 2500, was an escape of methyl isocyanate from a storage tank.

2.4. MAJOR PROCESS HAZARDS

The major hazards with which the chemical industry is concerned are fire, explosion, and toxic release. Of these three, fire is the most common but, as shown later, explosion is particularly significant in terms of fatalities and loss. As already mentioned, in the United Kingdom, the explosion at Flixborough killed 28 people, while offshore 167 men died in the explosion and fire on the Piper Alpha oil platform. Toxic release has perhaps the greatest potential to kill a large number of people. Large toxic releases are very rare but, as Bhopal indicates, the death toll can be very high. There have been no major toxic release disasters in the United Kingdom.

The problem of avoiding major hazards is essentially that of avoiding loss of containment. This includes not only preventing an escape of materials from leaks, etc., but also avoidance of an explosion inside the plant vessels and pipework. Some factors that determine the scale of the hazard are as follows:

2.4.1 The Inventory

The most fundamental factor that determines the scale of the hazard is the inventory of the hazardous material. The larger the inventory of material, the greater the potential loss.

2.4.2 The Energy Factor

For an inventory of hazardous material to explode inside the plant or to disperse in the form of a flammable or toxic vapor cloud, there must be energy. In most cases, this energy is stored in the material itself as the energy either of chemical reaction or of material state.

2.4.3 The Time Factor

Another fundamental factor is the development of the hazard in time. The time factor affects both the rate of release and the warning time.

2.4.4 The Intensity–Distance Relationship

An important characteristic of the hazard is the distance over which it may cause injury and/or damage. In general, fire has the shortest potential range, then explosion, and then toxic release, but this statement needs considerable qualification. The range of a fireball is appreciable, and the range of a fire or an explosion from a vapor cloud is much extended if the cloud drifts away from its source.

2.4.5 The Exposure Factor

A factor that can greatly mitigate the potential effects of an incident is the reduction of exposure of the people who are in the affected area. This reduction of exposure may be due to features that apply before the hazard develops or to emergency measures that are taken after the hazard is recognized.

2.4.6 The Intensity–Damage and Intensity–Injury Relationships

The range of the hazard depends also on the relationships between the intensity of the physical effect and the proportion of people who suffer injury at that level of the effect. The annular zone within which injury occurs is determined by the spread of the injury distribution. If the spread is small, the injury zone will be relatively narrow, while if it is large, the zone may extend much further out. Similar considerations apply to damage.

2.5. MAJOR HAZARD CONTROL

Within the last two decades, there has been a growing awareness of the problems associated with industrial growth and technological development. This has manifested itself in debate and conflict on such topics as follows:

1. environment (amenities, countryside, pollution, waste disposal, noise);
2. resources (oil, metals);
3. transport (road building, road accidents, aircraft crashes, tanker shipwrecks);
4. buildings and structures (high rise flats, bridges);
5. oil and chemical industry (pollution, noxious release, noise);
6. nuclear power (major accidents, low-level release, waste disposal);
7. toxic substances (drug and food additive side effects, materials at work).

It is also increasingly recognized that these problems have certain common features and that there is a generic problem in the assessment and control of technological developments. The question is considered in Superstar Technologies (Council for Science and Society, 1976), and the following discussion draws on this work.

The existence of a problem is due to the combination of a number of factors. Human artifacts are now often large and complex. Society also is complex, interdependent, and vulnerable. Technological developments are often rapid and of a large scale, and economic pressures militate against extensive testing. Devices are scaled up by extrapolation into untested areas, and there is no opportunity for gradual evolution and learning by trial and error.

Part of the problem is that high technology projects tend to be vulnerable to failures of all kinds. There is only one way in which they can go right, but many in which they can go wrong. In reliability terms, they are series systems in which all the elements must work if the system is to be a success. Another aspect of the problem is the question of unforeseen dangerous side effects, of which there have been numerous examples. These include the side effects of drugs and of industrial materials such as asbestos and vinyl chloride.

2.5.1 Hazard Monitoring

The Council for Science and Society argues for the creation of mechanisms by which the community can monitor and control hazards. Monitoring is likely to be effective only if it is generally accepted as a necessary and constructive activity that not only protects society but also contributes to the project.

The project needs to be monitored at all stages: conception, decision, and operation. At the conceptual stage, the need is to discover the problems, the hazards, and the disbenefits. The debate should involve the public as well as the experts. Argument continues in the decision stage but now centers on detailed proposals and designs of the project. Here the discussion is mainly between the project team and the other agencies involved. At the operational stage, the emphasis changes to enforcement of operational requirements and monitoring of operating hazards.

2.5.2 Risk Issues

High technology projects involve risks. These risks raise a number of issues including:

1. risk perception;
2. risk criteria;
3. risk management;
4. risk estimation.

There is a growing literature on risk issues, which includes *How Safe is Safe?* (Koshland, 1974), *Of Acceptable Risk* (Lowrance, 1976), *The Acceptability of Risks* (Council for Science and Society, 1977), *An Anatomy of Risk* (Rowe, 1977), *Society, Technology and Risk Assessment* (Conrad, 1980), *Societal Risk Assessment* (Schwing and Albers, 1980), *Science, Technology and the Human Prospect* (Starr and Ritterbush, 1980), *Technological Risk* (Dierkes et al., 1980), *Acceptable Risk* (Fischhof et al., 1981), *The Assessment and Perception of Risk* (Warner and Slater, 1981), *Dealing with Risk* (Griffiths, 1981), *Risk in the Technological Society* (Hohenemser and Kasperson, 1982), *Major Technological Risk* (Lagadec, 1982), *Technological Risk* (Lind, 1982), *Risk in Society* (Jouhar, 1984), *Risk Accessibility According to Social Sciences* (Douglas, 1990), *Behavior Decision Theory and Environmental Risk Management* (Vlek and Keren,

1991), *Risk and Blame* (Douglas, 1992), *Risk, Uncertainty, and Rational Action* (Jaeger et al., 2001), *Social Theories of Risk and Uncertainty: An Introduction* (Zinn, 2008), *Professional Risk and Working with People: Decision-Making in Health, Social Care and Criminal Justice* (Carson and Bain, 2008), and *Risk Analysis: A Quantitative Guide* (Vose, 2008).

The Health and Safety Executive (HSE) has published The Tolerability of Risks from Nuclear Power Stations (HSE, 1988a), Risk Criteria for Land Use Planning in the Vicinity of Major Hazard Installations (HSE, 1989a), and Quantified Risk Assessment: Its Input to Decision Making (HSE, 1989b).

2.5.3 Risk Perception

Work on risk perception shows that people tend not to think in terms of an abstract concept of risk, but rather to evaluate the characteristics of a hazard and to perceive risk in a multi-dimensional way. The idea that risk is an inherent property of the hazard has been criticized by Watson (1981) as the 'phlogiston theory of risk'. Various authors have attempted to define risk. One such attempt is that of Vlek and Keren (1991), who have given 10 different definitions.

2.5.3.1 Acceptable Risk

Some of the considerations affecting judgements about risks given by Lowrance (1976) are shown in Table 2.2.

The Council for Science and Society (1977) takes up the argument advanced by earlier workers that the distinction between risks that are assumed voluntarily and those that

TABLE 2.2 Some Considerations Affecting Judgements on Safety

Risk assumed voluntarily/Risk borne involuntarily
Effect immediate/Effect delayed
No alternative available/Many alternatives available
Risk known with certainty/Risk not known
Exposure is an essential/Exposure is a luxury
Encountered occupationally/Encountered non-occupationally
Common hazard/'Dread' hazard
Affects average people/Affects especially sensitive people
Will be used as intended/Likely to be misused
Consequences reversible/Consequences irreversible

Sources: After Lowrance (1976); reproduced with permission from W.W. Lowrance, *Of Acceptable Risk.* Copyright © 1976, William Kaufman Inc., Los Altos, CA. All rights reserved.

are borne involuntarily is a crucial one. In general, people are prepared to tolerate higher levels of risk for hazards to which they expose themselves voluntarily. The risk to which a member of the public is exposed from an industrial activity is an involuntary one. It is a common view that the risk to which an employee is exposed from this activity is in some degree voluntarily assumed.

When a serious hazard is encountered involuntarily, acceptance may extend only to a much lower level of risk than otherwise. When, in addition, the sufferer feels impotent in the face of danger, tolerance is further reduced.

Perhaps the most difficult class of hazards for judgements of acceptability are those called 'major hazards'. These are defined by a low probability of realization, combined with the likelihood of very great harm if the hazard is realized.

The subjective perceptions of risk can have enormous political importance, possibly to the extent of distorting priorities in programs for coping with the real risks that society encounters. (This problem is most noticeable in medicine.)

There is only one sort of risk 'that is truly "acceptable" in the ethical sense: The risk that is judged worthwhile (in some estimation of costs and benefits), and is incurred by a deliberate choice made by its potential victims in preference to feasible alternatives.' The important question, therefore, is "under what conditions, if any, is someone in society entitled to impose a risk on someone else on behalf of a supposed benefit to yet others?"

2.5.3.2 Acceptable vs Tolerable Risk

In a report by a Royal Society Study Group (RSSG), the account of the acceptability of risk describes some of the approaches taken to the problem of acceptable risk and the distinction between acceptable and tolerable risk. HSE (1988b) gives the following definition:

Tolerability does not mean 'acceptability'. It refers to the willingness to live with a risk to secure certain benefits and in the confidence that it is being properly controlled. To tolerate a risk means that we do not regard it as negligible or something we might ignore, but rather as something we need to keep under review and reduce still further if and as we can.

2.5.3.3 Actual vs Perceived Risk

One line of inquiry draws a distinction between actual and perceived risks and seeks to investigate the extent to which perceived risks differ from actual ones.

The attempt to assess the risks arising from a hazard involves a number of areas of considerable difficulty. The RSSG Report highlights two of these: (1) consequences and (2) uncertainty.

Scenarios for the realization of a hazard generally involve a whole range of consequences.

The assessment of the risk is subject to uncertainty. Traditionally, uncertainty has been discussed in terms of probability, but this does not exhaust the matter.

Expert risk assessments do not provide an unambiguous and reproducible measure of risk.

2.5.3.4 Psychological Issues

There are specific topics bearing on risk where progress has been made using the psychological approach. The RSSG Report describes three such areas: (1) estimation of fatality, (2) characterization of hazards, and (3) differences between individual and group perceptions.

Lichtenstein et al. (1978) investigated the estimation by laymen of the annual number of fatalities in the United States from 40 different hazards. The estimates made by the laymen showed two biases. One was that they overestimated the number of deaths due to infrequent causes such as tornadoes, but underestimated those due to frequent causes such as diabetes. The other was that the infrequent events for which overestimates were highest tended to be those of which laymen have a vivid picture and which are salient in their memory.

A study of the perception of the qualitative characteristics of hazards was made by Slovic et al. (1980). In this study, respondents were asked to rate 90 hazards in respect of 18 qualitative characteristics. The latter were then reduced to three dominant factors of which the first two were 'dread risk' and 'unknown risk.' The third factor was the number of people exposed to the hazard.

An individual's perception of risk is likely to be affected by his personality but also by his membership of a group. A number of studies have been done of the effect of membership of groups differentiated by culture, gender, or age. Work on groups also includes investigation of the attitudes of individuals within groups with different stances on a particular issue such as pro- and anti-nuclear groups.

2.5.3.5 Social Science Issues

Cultural theory, developed by Douglas (1966, 1982, 1985, 1990, 1992), holds that attitudes such as that to risks vary systematically with cultural bias. Four principal biases are recognized: hierarchists (high grid/high group), egalitarians (low grid/high group), fatalists (high grid/low group), and individualists (low grid/low group).

The concept of social amplification is that risk perceptions are amplified or attenuated by social processes. This occurs because the knowledge of risk that the individual possesses is largely second-hand.

2.5.3.6 Risk Communication

Risk communication occurs at national and company levels. At the latter level a company needs to communicate about risks both with its employees and with the public. In particular, risk communication is an essential part of emergency planning. More recently, legislative requirements for the provision of information to the public have been introduced. Risk communication at the national level is seen most clearly when opposition arises to developments such as hazardous installations or waste disposal sites.

Risk communication in such a context is now recognized as a discipline, but its study is not well advanced.

The credence placed in a communication about risk depends crucially on the trust reposed in the communicator.

Trust in a company depends on many factors and may be put at risk in a number of ways. Its general reputation may be undermined by persistent practice or by particular incidents. Distrust of the political process may rub off on the regulatory body. Risk assessment involves the treatment of uncertainties, and mishandling can result in loss of trust.

2.5.4 Risk Management

'Risk management' is the term generally used to cover the whole process of identifying and assessing risks and setting goals and creating and operating systems for their control. An important input to risk management is risk assessment.

The hazards with which risk management is concerned include, in general, those from natural events and man-made systems that give rise to a range of physical, financial, legal, and social risks.

Ultimately it is up to the government to decide which activities can be allowed and which must be banned and how any disbenefits from the former are handled. Thus, the government is involved in risk management, but the extent to, and manner in, which it should be is a matter of debate.

Many activities involve external 'costs' that are imposed on other parties. These externalities include danger and environmental pollution. Another market approach is the use of insurance to pool risks.

Insurance also has its problems. Insurers may be reluctant to provide cover for perils that are poorly defined such as pollution. Even where insurance is available, there are recognized problems. Adverse selection occurs where parties know their own risks better than the insurers and insure only the high risks. Moral hazard occurs where the possession of insurance lessens the incentive to reduce the risk.

The RSSG Report identifies, but does not attempt to resolve, seven areas of debate in risk management. These issues are (1) anticipation, (2) liability and blame, (3) quantitative risk assessment (QRA), (4) institutional design, (5) risk reduction costs, (6) participation, and (7) regulatory goals. For each issue, the Report presents both sides of the argument, but with caveats that the issues

are not usually so starkly posed, that opposing views are not necessarily irreconcilable and that one of the two viewpoints may be much more accepted than the other.

The RSSG Report contains an appendix on cost−benefit analysis in relation to risks by Marin (1992), with particular emphasis on risks to people. The application of cost−benefit analysis to such risks is based on the principle of reducing the risks until the marginal cost equals the marginal benefit. In order to do this, costs and benefits have to be expressed in the same units, which means in money terms.

2.5.5 Hazard Control Policy

It is appropriate at this point to consider some of the elements that go to make up a hazard control policy. An account of this has been given in Hazard Control Policy in Britain (Chicken, 1975). Chicken considers five fields that illustrate hazard control policy. These are (1) road transport, (2) air transport, (3) factories, (4) nuclear power, and (5) air pollution. These fields represent very different hazard control situations.

Later studies contain further issues of hazard control policies related to chemical and biological attacks and the reopening/recovery of facilities due to terrorism issues. Under the United States Department of Homeland Security's Chemical Facility Anti-Terrorism Standards (CFATS April 2007), facilities may be required to comply with at least some provisions of the CFATS regulation: chemical manufacturing, storage, and distribution; energy utilities; mining; agriculture and food; plastics; health care; electronics; painting and coating. Also, from the National Research Council Committee (National Research Council, 2005), the Standards and Policies for Decontaminating Public Facilities Affected by Exposure to Harmful Biological Agents Report states that in biological attack cases, the hazard control major issue is clarified as the evaluation for determining when a facility is safe for reopening.

2.5.6 Process Hazard Control: Advisory Committee on Major Hazards

The ACMH under the chairmanship of Professor B.H. Harvey, a former Deputy Director-General of the HSE, was set up soon after Flixborough in 1974.

The work of the Committee, which ended in 1983, is described in the three Reports of the ACMH (Harvey, 1976, 1979, 1984). The first task of the Committee was to define, or rather identify, major hazards. It concluded that, with few exceptions, a major hazard arises only if there is a release from containment. The hazard that then arises is a release that is flammable, toxic, or both. A flammable release may give rise to a jet or pool fire or a fireball, or the vapor cloud may burn as a flash fire, or it may give a vapor cloud explosion.

The first step in the control of major hazards was clearly that they should be notified, and an outline notification scheme was presented in the First Report. The principle adopted was notification of hazardous substances above a certain level of inventory. It was recognized that inventory is by no means the only factor that determines the hazard, but it was nevertheless adopted as the basis because it is simple to administer.

At this basic level of inventory, only notification was proposed, but it was envisaged that for a level of inventory 10 times the notification level, a hazard survey would be required and that the HSE would also have the further power to call for a detailed assessment.

The Committee strongly endorsed the principle of self-regulation by industry that it considered particularly appropriate to major hazard plants with their high level of technology. Self-regulation is a constant theme in the reports.

The First Report also began consideration of the measures necessary to prevent major accidents, against the background of Flixborough. Here the factor on which the Committee placed most emphasis was the management. Unless the management is competent to operate a major hazard plant, other measures are likely to be rendered ineffective.

The Second Report gave revised proposals for notification. This revision took into account the problem of ultratoxic materials, which had been highlighted by the Seveso disaster. Various models of regulatory control were considered, including those for the nuclear industry, mines, and pharmaceuticals.

Much of the work described in the report is concerned with comparison of theoretical estimates of major hazard scenarios with historical experience and with putting some bounds to the effects to be expected from major releases.

The Third Report opens with a discussion of risk. It does not give any specific numerical criteria but states certain principles which may assist in deriving such criteria. The report brings together the overall system of hazard control proposed by the Committee. The essential elements are identification, avoidance, and mitigation of the hazard and planning for it.

The report gives support to the avoidance of hazard by inherently safer design, a generalization of the earlier theme of limitation of inventory.

Measures to mitigate consequences which are discussed in the report are separation distances between the hazard and the public and emergency planning.

The Committee took the view that, although a major hazard plant should be designed to high standards, it is nevertheless prudent to seek to have some separation between the plant and the public as a further line of defense.

2.5.7 Process Hazard Control: Major Hazards Arrangements

The first legislative initiative in response to the ACMH's recommendations was proposals for the Hazardous Installations (Notification and Survey) Regulations 1978. These draft regulations contained requirements for the notification of installations holding a specified inventory of listed hazardous materials and for a hazard survey for installations containing 10 times the notifiable level.

These legislative proposals were overtaken by the EC Directive on the Major-Accident Hazards of Certain Industrial Activities 1982 (82/501/EEC) (the Major Accident Hazards Directive) and were never implemented. They were replaced by two sets of regulations, the Notification of Installations Handling Hazardous Substances Regulations 1982 (NIHHS), and the Control of Industrial Major Accident Hazards Regulations 1984 (CIMAH).

2.5.7.1 NIHHS Regulations 1982–2002

The NIHHS Regulations implement a notification scheme based on the ACMH proposals and similar to that of the original Hazardous Installations Regulations proposals. They are confined essentially to notification and do not contain requirements for a hazard survey, which is now covered by the CIMAH Regulations. The NIHHS Regulations provide for the notification of the installations which the authorities in Britain wish to see notified and provide the basis for planning controls over these installations.

2.5.7.2 CIMAH Regulations 1984

The EC Major Accident Hazards Directive 1982 has been amended twice, in 1987 (87/216/EEC) and 1988 (88/610/EEC). The first amendment, prompted by the disastrous toxic release at Bhopal, made a revision of some of the threshold inventories. The second, following the pollution of the Rhine by chemicals from the Sandoz warehouse fire, modified the controls on storage. These amendments of the Directive were implemented by amendments to the CIMAH Regulations in 1988 and 1990.

There are certain activities to which the regulations do not apply, notably defense, explosives, and nuclear installations (Regulation 3). Two levels of activity are defined. For the first level, the requirements are to take the precautions necessary to prevent a major accident and to limit the consequences and generally to demonstrate safe operation (Regulation 4) and to report any major accident which does occur (Regulation 5). For the higher level of activity (defined in Regulation 6), more extensive requirements apply (Regulations 7–12). These include the requirements to submit a safety report (Regulation 7), to update the safety report (Regulation 8), to provide on request additional information (Regulation 9), to prepare an on-site emergency plan (Regulation 10), and to provide information to the public (Regulation 12). The local authority is required to prepare an off-site emergency plan (Regulation 11).

2.5.7.3 CIMAH Safety Case

A central feature of the regulations is the safety report, commonly called the 'safety case'. Two aspects of particular interest are the requirements concerning the management system and the quantification of the hazards. With regard to the former, Schedule 6 of the original CIMAH Regulations 1984 requires that the report provide information on the management system and specifically on the staffing arrangements, including: for certain responsibilities, the names of the persons assigned; the arrangements for safe operation; and the arrangements for training. The wording of the 1990 Regulations is identical. However, the Guide to the latter (HSE, 1990 HS(R) 21 rev.) is much more explicit on the matters to be covered and is in effect a description of good practice in respect of a safety management system.

2.5.7.4 COMAH Regulations 1999

The COMAH Regulations implement the Seveso II Directive except for the land-use planning requirements. They replaced the Control of Industrial Major Accident Hazards Regulations 1984 (CIMAH) and was assigned to come into force on 1st April 1999. The Regulations have amendment at June 30, 2005.

For those who shall be responsible for COMAH Regulations, the competent authority comprises three organizations: the Health Safety Executive (HSE), the Environment Agency (EA—for England and Wales), and the Scottish Environmental Protection Agency (SEPA). They are responsible for the enforcement COMAH Regulations. The Competent Authority Strategic Management Group (CASMG) is responsible for setting a strategic direction and plan for safety works. They are also responsible for reporting back to administration organizations on progress, both with the United Kingdom and the EU.

2.5.8 Process Hazard Control: Planning

Some appreciation of the planning system is necessary for the understanding of overall arrangements for the control of major hazards. The treatment given here is limited to providing this essential background.

2.5.8.1 Planning System

The essential function of planning is the control of land use. This control may be used to prevent incompatible uses of adjacent pieces of land, but such control is much easier to exercise before development has taken place than after it. The control of land use is affected through the structure plans of the counties, which set the overall framework, and through the local plans of the LPAs, which deal in specific developments. New developments are required to relate to these plans. Structure and local plans are essentially instruments for forward planning. Difficulties can arise when the problem is one not of a greenfield development but of an existing installation.

Since the definition of 'development' is so wide, it has been necessary in order not to overload the planning system to grant certain general permissions. Under the Town and Country Planning General Development Order (GDO) 1977, a number of changes of land use are classed as 'permitted' developments, which do not require planning permission.

2.5.8.2 Planning and Major Hazards

The problem of developments involving major hazards was recognized well before the Flixborough disaster. In 1972, the Department of Environment (DOE) issued the first of a number of circulars to Local Planning Authorities (LPAs) on hazardous installations. This circular, DoE 72/1, gave a list of hazardous inventories for which it was recommended that the LPA should seek advice.

Following Flixborough, the HSE experienced an increased number of inquiries from LPAs concerning developments involving hazardous installations and set up a risk appraisal group (RAG) to give advice in response to these queries. In 1974, the ACMH was set up with a membership which included planners. Planning was one of the principal topics considered by the Committee which made a number of recommendations in this area.

On siting policy, the committee stated:

The overall objective should always be to reduce the number of people at risk, and in the case of people who unavoidably remain at risk, to reduce the likelihood and the extent of harm if loss of control or containment occurs. (Third Report, paragraph 109)

With regard to development in the vicinity of a hazardous installation, the Committee endorsed the aim of first stabilizing and then reducing the number of people exposed to the hazard. It recommended that, for proposed development, the LPA should consult the HSE. It also drew attention to the power of the LPA to enter into a voluntary agreement with the owners of land to restrict the use to which the land is put.

The Committee recognized that, in a limited number of cases, the LPA might consider it necessary to revoke or discontinue planning permission and recommended that the government review its discretionary powers to make payments to local authorities to meet compensation liabilities.

2.5.8.3 Planning Reforms

The Housing and Planning Act 1986 creates a requirement for a written consent from the LPA for a new notifiable inventory. This consent is separate from planning permission.

The Town and Country Planning (Assessment of Environmental Effects) Regulations 1988 require that an environmental impact assessment be submitted to the LPA along with the planning application.

The loophole associated with the concept of development is closed by the provisions of the Planning (Hazardous Substances) Act 1990 and the Planning (Hazardous Substances) Regulations 1992.

2.5.8.4 HSE Consultation and Advice

As stated above, the policy of the HSE is to stabilize and then reduce the number of people at risk from a hazardous installation. The HSE implements this policy primarily through its advice to LPAs.

The vast majority of planning decisions are concerned not with development on industrial sites, but with development in the vicinity of such sites. The HSE issues to LPAs guidelines on the size of 'consultation zones' around these sites and asks to be consulted on developments within these zones. Consultation with the HSE is not appropriate for minor planning applications. Guidance on the type of application for which consultation is required is given in DoE Circular 9/84.

Initially, the HSE set a general consultation distance of 2 km around a hazardous site. This proved to be excessive and the HSE has since refined its assessment methods and now specifies consultation distances which are related to the particular hazard.

The HSE distinguishes three types of development: A, B, and S (special). Type A developments are those where control may well be appropriate. They are situations where people would be present most of the time or where large numbers might be present quite often and the people are not highly protected. Type B developments are those where the risk would not normally be great enough to warrant control. Type S is intended to cover developments which are significantly more vulnerable than type A. Examples are hospitals, old people's homes, and schools.

The replies from the HSE following consultation by the LPA are generally in terms of the risk categories given in DoE 9/84: 'negligible', 'marginal', and 'substantial'. Negligible risk means that the HSE assessment of the consequences of realization of the hazard is that it is unlikely, if not inconceivable, that people in the development would be killed, and that if someone was killed, this would be a 'freak' effect. Marginal risk means that the probability of people being affected by a major release is remote, that in such a case they might be seriously affected but that death is unlikely. Safety reasons would not in themselves justify a refusal of planning permission, but might justifiably be among those contributing to it. Substantial risk means that people might be seriously injured by a major release. The risk could in itself justify refusal of planning permission, and safety should be a major factor in the decision.

2.5.8.5 Emergency Planning

There is a statutory requirement for counties and equivalent authorities to undertake emergency planning for civil defense, and the tendency has been to extend these arrangements to cover process site and transport emergencies.

2.5.8.6 Information to Public

The CIMAH Regulations 1984 contain a requirement that the public be given information about the hazards to which they are exposed and the action which they should take in an emergency. It is intended that the body which informs the public is the local authority. The manufacturer is required to try to reach with the local council an agreement on the information to be provided and then to furnish this to the council, which then issues it to the public. If, however, an agreement cannot be reached, the manufacturer is responsible for informing the public directly.

2.5.8.7 Public Inquiries

A planning application may become the subject of a public inquiry. This may occur if the developer is refused planning permission and appeals or if the application is called in by the Secretary of State. In some cases, this call-in may be at the behest of the HSE.

It is only relatively rarely that a planning application for a process plant becomes the subject of a formal planning or public inquiry.

2.5.8.8 Planner's Viewpoint

The viewpoint of the LPA is given in Major Hazard Installations: Planning and Assessment (Petts, 1984b) and Major Hazard Installations: Planning Implications and Problems (Petts, 1984c).

Certain LPAs have been conscious from an early date of the need to provide for major hazard installations in their area. Accounts of the measures taken following Flixborough by Halton Borough Council have been given by Brough (1981) and Payne (1981).

2.5.9 Process Hazard Control: European Community

The arrangements for the control of major hazards in the EC and in certain other European countries are now briefly reviewed. Comparative accounts are given in Risk Assessment for Hazardous Installations (Q.C. Consultancy Ltd, 1986) and by Beveridge and Waite (1985) and Milburn and Cameron (1992).

2.5.9.1 European Community

Controls over major hazards in the EC are established by Directive 82/501/EEC, the Major Accident Hazards Directive, amended by 87/216/EEC and 88/610/EEC, as described in Section 4.10. The Directive contains requirements for notification of installations, for a safety report, and for emergency planning.

A fundamental revision of the major hazard arrangements is in prospect, as outlined in Proposal for a Council Directive on the Control of Major-Accident Hazards Involving Dangerous Substances (COMAH) (Commission of the European Communities, 1994). Features emphasized in the document are land use planning, management and human factors, and the safety report.

2.5.9.2 Germany

In Germany, basic safety requirements are embodied in a number of measures which include the Industrial Code 1869–1978, the Imperial Insurance Code 1911, the rules for Prevention of Accidents to Man and the Technical Means of Work Act 1968.

Control of major hazards is effected, still under the FECA, through the Hazardous Incident Ordinance 1980 (HIO) (*Stb'rfallverordnung*). The ordinance gives lists of processes and chemicals. The licensing requirements apply to installations which carry out these processes and use these chemicals.

Exemption from the HIO is allowed if the inventory of the hazardous substance is very small. The levels of inventory below which exemption is given are specified in the First General Administrative Regulation 1981 to the ordinance.

The HIO requires the preparation of a safety analysis. The requirements for this are specified in greater detail in the Second General Administrative Regulation 1982.

2.5.9.3 France

In France, safety legislation is based on the Imperial Decree of 1810, which has been subject to amending Laws in 1917, 1932, and 1961. This legislation was extensively modified in the Law of July 19, 1976 on Registered Works for Environment Protection (Installations Classes) and the associated Decree of September 10, 1977.

A safety study is required for an authorized installation. The study must include a justification of the measures taken to reduce the probability of realization of the hazard and its effects.

2.5.9.4 The Netherlands

In the Netherlands, there are separate systems for the assurance of safety of employees at the workplace and of the public. The first is the responsibility of the Ministry of Social Affairs, and specifically the Labor Inspectorate, and the second that of the Ministry of the Environment.

Safety legislation in the workplace is based on the employee safety law of 1934. The Statute of November 23, 1977 creates a requirement for hazardous installations to submit a safety report. This occupational safety report statute is the basic legislation on major hazards at the workplace. Accounts of the safety report have been given by Meppelder (1977), van de Putte and Meppelder (1980), van de Putte (1981, 1983), Husmann and Ens (1989), Oh and Albers (1989), and Bottelberghs (1995).

The installations for which a safety report is required are defined not by inventory but on a 'threshold value', which is defined as the quantity of material which would present a serious danger to human life at a distance of over 100 m from the point of release.

The report must contain an assessment of the hazards in the form of a specified fire and explosion index and a toxicity index, derived from the Dow Index method.

2.5.10 Process Hazard Control: USA

The arrangements for the control of major hazards in the United States are now briefly described. The account given here is confined to accidental releases, which include major hazards.

2.5.10.1 Plant Siting

The development of these controls needs to be seen against the background of the growing problems experienced in the siting of new hazardous installations since the late 1970s.

A case in point at this time was the siting of LNG facilities, as described in the Transportation of Liquefied Natural Gas by the Office of Technology Assessment (OTA) of the

US Congress (1977). The Federal Power Commission (FPC) was the lead agency in determining whether an individual LNG project should be allowed. It was the practice of the FPC to make decisions on siting on a case-by-case basis. The OTA identified several problems in this approach. The FPC was a regulatory rather than a policy-making body, and there was no national siting policy to which it could refer. The criteria on which the FPC made its decisions on siting were not known to the industry, since no guidelines were issued. There had been pressure from the state legal authorities and from industry for the issue of uniform siting criteria. In some cases, the FPC had made its approval contingent on the receipt of state and local approval, so that the criteria that the industry had to satisfy had become even more obscure.

There were several areas of overlapping jurisdiction. The Office of Pipeline Safety Operations (OPSO) was involved through the Natural Gas Pipeline Safety Act 1968, although there appeared to be a statutory provision against OPSO standards prescribing the location of LNG facilities. In the past, the two agencies had clashed directly on the standards required. The FPC required a temporary shut-down of an LNG facility which OPSO had inspected and approved. The US Coast Guard (USCG) was also involved through the Coastal Zone Management Act 1972. This Act required an applicant for a Federal license or permit for an activity in a coastal zone of a state to certify the project to be consistent with the state's program.

Siting was also affected by other legislation. The National Environmental Policy Act 1969 required an environmental impact statement.

2.5.10.2 Accidental Releases

The principal Federal legislation is the Emergency Planning and Community Right-to-Know Act 1986 (EPCRA), which is Title III of the Superfund Amendments and Reauthorization Act 1986 (SARA), generally known as SARA Title III, and is enforced by the Environmental Protection Agency (EPA). Other relevant federal legislation is the Occupational Health and Safety Administration (OSHA) rule for Process Safety Management of Highly Hazardous Chemicals 1990. There are also the state laws in New Jersey and California. One focus of concern is the control of highly toxic hazardous materials (HTHMs).

2.5.10.3 SARA Title III

SARA Title III is the Federal legislation that creates controls on accidental releases of hazardous substances. The legislation followed soon after Bhopal. Accounts are given by Brooks et al. (1988), Bowman (1989), Horner (1989), Burk (1990), and Fillo and Keyworth (1992). SARA Title III has four main parts dealing with

emergency planning, emergency notification, community right-to-know, and toxic chemicals inventory.

For each state the governor appoints an Emergency Response Commission (ERC) that in turn designates local emergency planning committees charged with the preparation of an emergency plan. The facilities that attract such an emergency plan are those containing certain chemicals above specified inventory levels. These chemicals and inventories are given in the List of Extremely Hazardous Substances and Their Threshold Planning Quantities (TPQ) developed by the EPA. Management is required to provide any information needed by the local emergency planning committee to implement the emergency plan. The wording has deliberately been left broad enough for the local emergency planning committee to request a QRA (Florio, 1987).

The emergency notification arrangements require the management of the facility to report immediately to the local planning committee any accidental release and the area likely to be affected.

With regard to the right-to-know requirements, for any hazardous chemical that attracts the requirement for a Material Safety Data Sheet (MSDS) under the OSHA 1970, the facility has to submit either the data sheets or a list of the chemicals to the local emergency planning committee, the state ERC and the responsible fire department. Section 313 contains the requirement to submit information on the 'toxic chemical release emission inventory', or Toxic Release Inventory (TRI), on the Toxic Chemical Release Form. The subjects of the principal sections of SARA Title III may be summarized as follows: Section 301, state ERC; Section 302, notification requirements; Section 304, accidental release reporting; Section 305, study of safety capabilities; Sections 311 and 312, information for the public; and Section 313, toxic emissions.

2.5.10.4 Process Safety Management Rule

The OSHA has further extended its controls to prevent accidental releases with the rule for Process Safety Management of Highly Hazardous Chemicals (the Process Safety Management Rule). The governing document (Federal Register July 17, 1990, 29 CFR Part 1901) refers to major accidents such as Flixborough, Seveso, and Bhopal, and to EC controls on major hazard installations in the 'Seveso' Directive and to the need to strengthen such controls in the United States, which otherwise largely rely on the OSHA 1970 and on standards.

2.5.10.5 New Jersey Toxic Catastrophe Prevention Act

The New Jersey Toxic Catastrophe Prevention Act 1985 (TCPA) is also addressed to the identification, evaluation, and control of hazards for facilities handling extraordinarily hazardous substances (EHSs), there being 11 such chemicals on the initial list. It creates a requirement for a risk management plan (RMP). The RMP covers risk management, safety review, risk assessment, and emergency planning.

2.5.10.6 California Hazardous Materials Planning Program

The California Hazardous Materials Planning Program draws together four complementary laws that are codified in Chapter 6.95 of the state Health and Safety Code.

2.5.10.7 EPA's RMP Rule

According to the Clean Air Act, facilities that are involved with certain chemicals must create a Risk Management Program and Risk Management Plan (RMP). After completion, the RMP must be sent to the EPA. A full list of affected chemicals can be found in Section 112(r) of the Clean Air Act.

2.5.10.8 Regulatory Agencies

As just described, both the EPA and the OSHA are involved in the regulation of accidental releases, primarily in the case of the former through SARA Title III and of the latter through the OSHA 1970 and, more recently, the Process Safety Management Rule and the Risk Management Program.

2.5.10.9 Voluntary Initiatives

Mention should also be made of some of the voluntary initiatives that have been taken in the United States in this area. Since 1985, the Chemical Manufacturers Association (CMA) has recommended to its members the Community Awareness and Emergency Response (CAER) program.

2.6. FIRE AND EXPLOSION LOSS

The loss statistics of interest are primarily those for fire loss. In the United Kingdom, principal sources of statistics on such losses are the Home Office, the Fire Protection Association (FPA), the Loss Prevention Council, and the insurance companies. These organizations produce annual statistics for fire losses. Loss due to explosions is generally included in that for fire loss. There are no regular loss statistics on toxic release, since this is a rare event and usually causes minimal damage to property.

The chemical industry had the largest number of fires, but the oil industry the most expensive. The origin of the fires is predominantly in storage and in leakages. The sources of ignition are fairly evenly spread. There is only one fire attributed to static electricity and two to arson.

But in 35 cases, i.e., 44%, the ignition source was unknown. The material phase first ignited also shows a balanced spread, with the solid phase actually being predominant. There is no marked trend in the time of day of the fires, although the number is somewhat higher during the day shift.

So far no distinction has been made between fire and explosion losses. The latter are normally included in the overall fire statistics. In fact it is explosions that cause the most serious losses. Some two-thirds of the loss is attributable to explosions.

2.7. CAUSES OF LOSS

There are almost as many analyses of the causes of loss as there are investigators. Unfortunately, there is no accepted taxonomy, so that it is often difficult to reconcile different analyses. Two typical breakdowns of the causes of loss are given in Table 2.3 (Doyle, 1969) and in Table 2.4 (American Insurance Association (AIA), 1979), the latter being an up-date of an earlier table (Spiegelman, 1969).

The losses considered so far are insured losses arising from fire and explosion. Another important, but often uninsured, loss arises from plant shut-down and down-time. Once again there are many different analyses of shut-down and down-time and its causes at different types of plants.

2.8. TREND OF INJURIES AND LOSSES

The long-term trend of injury rates in the process industry is downwards. The trend in the United Kingdom shows that between 1970−1974 and 1987−1990, the fatal accident rate for the chemical and allied industries fell from 4.3 to 1.2.

TABLE 2.3 Large Losses in the Chemical Industry Insured by the Factory Insurance Association: Causes of Loss (Doyle, 1969)

	No.	Loss (%)
Incomplete knowledge of the properties of a specific chemical	6	11.2
Incomplete knowledge of the chemical system or process	6	3.5
Poor design or layout of equipment	13	20.5
Maintenance failure	14	31.0
Operator error	5	6.9
Total	44	73.1

Source: Courtesy of the American Institute of Chemical Engineers.

TABLE 2.4 Hazard Factors for 465 Fires and Explosions in the Chemical Industry 1960−1977 (American Insurance Association, 1979)

	No. of Times Assigned	Proportion (%)
Equipment failure	223	29.2
Operational failure	160	20.9
Inadequate material evaluation	120	15.7
Chemical process problems	83	10.9
Material movement problems	69	9.0
Ineffective loss prevention program	47	6.2
Plant site problems	27	3.5
Inadequate plant layout	18	2.4
Structures not in conformity with use requirements	17	2.2
Total	764	100.0

TABLE 2.5 The 100 Largest Losses in the Chemical and Oil Industries Worldwide

Period	No. of Losses	Average Loss
1957−1966	15	28.5
1967−1976	29	38.2
1977−1986	56	36.6

Source: After Marsh and MacLennan (1987).

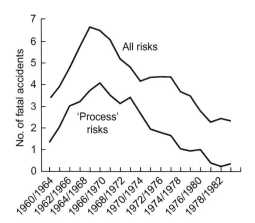

FIGURE 2.3 Trend of the fatal accident rate in ICI, 1960−1982 (Hawksley, 1984). The number of fatal accidents is given as the number in 10^8 working hours or the number in 1000 men in a working lifetime expressed as a 5-year moving average.

Evidence is available that the efforts devoted to safety and loss prevention by some companies have borne fruit. Figure 2.3 (Hawksley, 1984) shows the trend of the fatal accident rate in ICI over the period 1960−1982. The data represent the 5-year moving average.

There have been a number of studies addressing the question of whether the number of major incidents in the chemical and oil industries is increasing. Early work on these lines was done by Marshall (1975) and Kletz and Turner (1979). The insurers Marsh and McLennan (M&M) publish periodically a list of the 100 largest losses in the chemical and oil industries worldwide over a running 30-year period (Garrison, 1988; Mahoney, 1990; Marsh and McLennan, 1992). The 1987 edition contains the analysis shown in Table 2.5. This indicates a trend that is rising up to 1986.

2.9. ECONOMICS OF LOSS PREVENTION

Loss prevention is concerned with the avoidance of personal injury, environmental damage, and economic loss. In all of these spheres, there is an economic balance to be struck. It also represents an important factor that will be used in Cost−Benefit analysis, which is necessary to ensure appropriate decision making.

In today's environment, many loss prevention measures that were optional in the past are now mandated by government regulation. Others are voluntarily provided in some countries as they have become 'industry practice'. Still others are provided due to court rulings in liability cases. In the above situations, traditional cost−benefit analysis studies may not be relevant.

2.9.1 Cost of Losses

The result of not taking adequate loss prevention measures is to give rise to losses and costs such as those of: (1) accidents, (2) property damage, (3) plant design delays, (4) plant commissioning delays, (5) plant downtime, restricted output, (6) equipment repairs, (7) loss of markets, (8) public reaction, (9) insurance, and (10) social costs. More recently, costs of natural environmental damage and recovery, as well as personal injuries and human lives losses, have also become very important factors in considering the cost of losses estimations. A rational approach to loss prevention requires some understanding of the economic importance of these factors and of the means of estimating them.

2.9.1.1 National Level

Accident statistics are fairly readily available from the US Department of Labor, Bureau of Labor Statistics and from the National Safety Council in the United States for injuries and from the National Fire Protection Association (NFPA) for fires in the United States. Accident statistics in the United Kingdom can be obtained from the *Annual Report* of HM Chief Inspector of Factories and the annual *Health and Safety Statistics*. By making certain assumptions, the national resource cost, or cost to the nation, of these accidents may be estimated.

In the United Kingdom, statistics for fires are published by the Home Office in the annual *Fire Statistics United Kingdom* and by the Fire Protection Association (FPA) (the FPA defines a large fire as one costing £50,000 or more and the Home Office uses this definition). The *Fire Statistics United Kingdom 1990* (Home Office, 1992) gives the annual loss due to large fires over the 4-year period 1986–1989 as some £290 million.

In previous editions of this guide, the expenditure on insurance was used as a measure of the cost of incidents. With the continued growth of multinational corporations, the globalization of the world economy and the emergence of super majors in the energy business, insurance costs can no longer be used to effectively estimate the cost of incidents. A large number of events are no longer reported to insurers, and the costs of these are absorbed fully by the corporation.

2.9.1.2 Company and Works Level

If the methods of determining these costs are considered at the level of the individual company or works, estimates may again be based on the same types of information as before. There are some important differences, however, between the company and the national level. The two principal points are that the company generally has access to higher quality information than is available at national level. The other is that the company is much more vulnerable to a single major incident.

Company-level fire loss statistics may be used, but the effect of a possible single large loss fire needs to be borne in mind.

For companies that purchase insurance with relatively low deductible or self-retention layers, insurance costs may be viewed. The insurance costs incurred by the company are another measure of the cost of losses.

Costs which the company is especially well placed to assess are those of commissioning delays, downtime, and maintenance in its own plants. The effect of downtime on plant profitability is another standard topic in process economics (Allen, 1973).

2.9.2 Cost of Prevention

Some of the areas in which costs tend to be incurred to prevent loss are (1) management effort, (2) research effort,

(3) design effort, (4) process route, (5) operational constraints, (6) plant siting, (7) plant layout and construction, (8) plant equipment (safety margins, materials, duplication), (9) process instrumentation (trip systems), (10) fire protection, (11) inspection effort, (12) maintenance operations, (13) costs during downtime, and (14) emergency planning. Since all aspects of a project have safety implications, any such list is somewhat arbitrary. Others are given in the literature (Pratt, 1974). The areas listed are those in which additional costs attributable to loss prevention are particularly likely to occur.

The standards of safety which are already obligatory are high and have tended to rise. Since in the chemical industry safety is both relatively good and relatively expensive to improve, the high minimum standard may well tend to become the norm. This is not unreasonable provided, but only provided, that additional expenditure is incurred where it is cost-effective in terms of human life.

It can be misleading to assume that increased safety necessarily involves increased expenditure. It is undoubtedly true that safety expenditure which is based on reaction to incidents can be very expensive. But it is general experience that good practice usually costs no more than bad and gives both improvements in safety and reductions in costs. Thus, given good loss prevention practice, large improvements in safety can be achieved relative to poor practice at no additional cost.

The various techniques of economic analysis such as net present value (NPV) and discounted Cash Flow (DCF) may also be used in support of assessment of the economic level of loss prevention expenditure (Allen, 1973; Park, 2002).

Some decisions on the level of loss prevention expenditure are amenable to treatment by other formal decision-making techniques, in particular decision analysis. Current techniques and models for economic decision analysis can be found in Contemporary Engineering Economics by Park (2002) and Capital Investment Analysis for Engineering and Management by Canada et al. (1995).

Accounts of decision analysis are given by Mitchell (1972) and Hastings and Mellow (1978). In a typical decision analysis, there is a set of possible outcomes, or states of nature, and a set of possible options for action, or decisions. The aim is to choose the decision(s) which maximize the gain, or utility, or minimize the loss, or regret. A common form of the problem is that there is a possibility of obtaining better information, which will improve the decision but will itself involve a cost.

2.10. INSURANCE OF PROCESS PLANT

Insurance has always been based on the principle of solidarity, and the more balanced, or homogeneous, the risk structure is within an insured community, the better this principle works.

The insurer cannot be expected to pass judgement on the safety culture of individual companies as such. His only responsibility is to set up insured communities. However, this involves evening out recognizable inhomogeneity either through selection, or the pricing and design of cover, or by supporting the insureds' risk management efforts.

Assessing individual industrial risks requires more than experience and technical know-how; it also requires special tools for the systematic appraisal of the factors relevant to safety and risk.

The company operating a hazardous plant must either itself bear the whole financial risk or seek outside insurance for this risk. The decision is an important but difficult one and it needs to be related to the overall approach taken by the company to prevent loss. Insurance practices vary, of course, from country to country and it is essentially the practice in the United Kingdom which is described here.

2.10.1 The Insurance Process

The procedure adopted by the insurer in response to an insurance inquiry is to do a breakdown of insurance values by site and installation and to obtain a report from the insurance surveyor. Inspection is a normal preliminary to insurance. In general, the industry operates on the principle of 'no insurance without inspection'. Frequently, however, insurers will accept reports or an inspection or a survey conducted by an insurance broker or an engineering consultant retained by the insurer or the broker.

2.10.2 Insurance Policies

In the past, an insurance policy covered a named peril for a specified period. The perils may be standard or special perils. Some of the types of insurance available in a multi-line policy in the United States are given by Spiegelman (1987): (1) public liability, (2) product liability, (3) workers' compensation, (4) motor vehicle, (5) business interruption, (6) professional liability, (7) boiler and machinery, and (8) property damage.

More recently, insurers have provided policies covering 'all risks of direct physical loss' except for perils or events specifically excluded by the policy language.

2.10.3 Loss Measures

The insurance policy may provide cover for the full sum at risk, but, as just described, insurers generally find it convenient to work in terms of a more limited sum. There are various measures which are used, notably normal maximum loss (NML), probable maximum loss (PML), and estimated maximum loss (EML), but the definitions

of the terms tend to vary. These terms are discussed further below.

2.10.4 Insurance Surveyors

It is the function of the insurance surveyor to provide the insurer with a report on the risk to be insured. Insurance surveyors are employees of the insurance companies or insurance brokers, and, in the United Kingdom, are typically experienced people but are not necessarily qualified engineers.

The insurance surveyor's report includes an assessment of the loss measure used such as the PML and serves as a guide to the underwriter in setting the premium.

2.10.5 Tariff and Non-Tariff Systems

The traditional approach to setting premium rates for industrial fire damage risks has been classification by trade or occupation. Classifications run to as many as 300 trades. For many trades, there is a detailed tariff which consists of a basic rate together with increments or decrements for factors which increase or decrease the risk, respectively. However, for some trades there is no tariff.

2.10.6 Fire Insurance in the United Kingdom

Throughout the 1960s, there was a history of growing fire loss in many trades in the United Kingdom. Many of these fires were in industries where fire losses have always been high. Insurers dealt with this by applying to each trade appropriate premium weighting factors.

2.10.7 Business Interruption Insurance

Business interruption (BI) insurance covers the consequential loss from an insured peril that interrupts or diminishes production. Following a waiting period stated in the policy, BI insurance covers lost profit and continuing or non-avoidable expenses during the time required to restore production. The need for this is due mainly to the very heavy losses which can occur if a large, single-stream plant is out of action for more than a short period. The costs of losses due to, and of insurance for, BI now tend to exceed several times those of damage.

2.10.8 Large, Single-Train Plants

Large plants pose several problems to insurers. One is the magnitude of the PML for both property and BI. Another is the high premium rates required to provide cover for BI. Another is the difficulty of assessing the adequacy of the management and of the technology.

Insurers have been virtually unanimous in expressing their dislike of large plant risks and in suggesting to the industry that it may have gone too far in putting its eggs in one basket. On the other hand, while the early generations of large plants gave rise to many problems, it is arguable that the management and technology of such plants are now better understood, not least due to the development of loss prevention.

2.10.9 Insurance Market

As just indicated, the loss experience of the insurance industry is variable, and this has an impact on both capacity and premium rates.

In the United Kingdom, the late 1960s and the 1970s saw loss on an increasingly serious scale in the chemical industry, which up to this time had a good record.

Likewise in the United States, the insurance industry experienced large losses around 1967 with the vapor cloud explosion at Lake Charles in 1967 and other major incidents. The period 1974−1976 also saw heavy losses. This trend culminated in the late 1980s with vapor cloud explosion losses at Pampa, Texas (1987), Norco, Louisiana (1988), and Pasadena, Texas (1989).

2.10.10 Insurance Capacity

The capacity of the insurance market is usually given in terms of the extent of coverage which can be obtained for a single potential loss. As just described, this capacity fluctuates. Although generally the market is able to provide insurance even for very large loss potentials, there is some degree of restriction.

The vapor cloud explosion at Flixborough had an influence on the insurance market. Some insurers recognized the possibility of such an explosion, either by allowing for it in their assessment of PML or by buying specific re-insurance. A loss of US$100 million from damage and BI due to a major vapor cloud explosion was considered within the realm of possibility.

Redmond (1990) describes the capacity available through the Oil Insurance Limited (OIL) mutual together with the London Master Energy Line Slip, which is a facility whereby many insurers commit their capacity automatically. However, disasters such as Piper Alpha, Pasadena, and the Exxon Valdez have caused reduction of capacity for periods of time called 'hard markets'. A similar hard market condition developed after the terrorist attacks of September 11, 2001. Market capacity decreased dramatically and has only recently started to improve. Worldwide capacity is now in excess of US$1 billion for certain risks.

2.10.11 Insurance Restrictions

In periods when the capacity of the insurance market is restricted, there may be pressure on the insured to bear part of the risk. There are various ways in which this may be affected. The restrictions may take the form of exclusions, limited coverage for particular activities, or the application of deductibles. If deductibles, or excesses, are applied, the insured company itself is required to meet a fixed degree of loss.

2.10.12 Self-Insurance

A company may often choose not to seek full insurance cover but to bear some or all of the losses itself. Indeed, the demands placed on the world insurance market by the chemical industry have tended to leave companies with little choice but to carry some of the risk themselves. Policy on the purchase of insurance is a complex matter and depends very much on the circumstances of the individual firm, so that only a few general points are made here. Perhaps, the most obvious and important is that the scale of potential loss is such that it would be severely damaging to even the largest firm, so that some degree of outside insurance is advisable.

2.10.13 Vapor Cloud Explosions

A significant proportion of the loss potential on process plants is attributable to vapor cloud explosions. Alexander (1990) quotes his own estimate made in about 1980 that this proportion was in the range 5−10%. A discussion of insurance with special reference to vapor cloud explosions is given by Davenport (1987). The results have been less than desirable since a point source explosion overpredicts blast loads close to the epicenter and underpredicts blast loads in the far field. Extensive research has been conducted during the past 15 years to analyze vapor cloud explosions of various purity and mixture materials to determine what factors influence the energy released in an explosion.

2.10.14 Major Disasters

Disasters such as Piper Alpha in 1988 and Pasadena in 1989 have a significant impact on the insurance world.

The losses incurred in these two accidents have been discussed by Redmond (1990). The interim estimates of loss given are for physical damage, US$680 million; BI, US$275 million; fatalities and injuries, US$160 million; removal of wreckage, US$100 million; and other items, US$155 million. These items total US$1370 million. The interim estimates for the loss at Pasadena quoted by Redmond are very similar, about US$1400 million, divided almost equally between property damage and BI. The HSE (1993) has estimated the loss due to Piper Alpha as amounting to some £2 billion, including about £750 million in direct insured losses.

2.11. PROPERTY INSURANCE

There are a variety of factors which are taken into account by insurance surveyors to assess property damage risk for the purpose of premium rating. Some of the features which are frequently mentioned by insurers are (1) size of plant, (2) novelty of technology, (3) process materials (flammable liquids at high pressures and temperatures), (4) process features and controls, (5) plant layout, (6) building design, (7) fire protection of structures, (8) fire fighting arrangements, (9) management of hazards, and (10) natural perils such as earthquakes, floods, and hurricanes.

2.11.1 Loss Measures

As already mentioned, a number of measures of loss are utilized, but different meanings are attached to the terms used. Alexander (1990) utilizes the following terms: average loss; NML or PML; EML or possible maximum loss; and maximum credible loss (MCL) or maximum foreseeable loss. The average loss is self-explanatory. NML is the maximum loss which would be expected if all protective functions work, EML is the maximum loss if one critical item of protection does not work, and MCL is the maximum loss in circumstances where a number of critical items of protection do not work or where a catastrophic event occurs which is just credible. He gives the following approximate frequencies:

$$NML \geq 10^{-3}/year$$
$$EML \; 10^{-4} - 10^{-3}/year$$
$$MCL \; 10^{-5} - 10^{-4}/year$$

2.11.2 Risk Assessment Methods

Insurance assessors use a variety of methods to assess the risk and to arrive at a premium rating. These methods include the use of checklists, of various types of hazard index, and of formal premium rating plans. The last two are closely related. Increasingly, such hazard indices and premium rating plans emphasize management aspects. A number of specific management audit methods have been developed.

2.11.2.1 Management Audits

The increasing emphasis on management is reflected in the use of audit methods which concentrate particularly on management. Redmond (1984) gives a checklist by which the management may be assessed. A more detailed discussion of the assessment of management for insurance purposes is given by Instone (1990) and in the Swiss Re publication entitled *Safety Culture—a reflection of risk awareness* (2003).

2.11.2.2 Risk Profiles

It is also possible, and instructive, to construct the loss profile for a particular plant. The construction and use of such profiles is discussed by Alexander (1990). The method which he describes is the use of what may be termed a frequency—loss (FL) plot.

2.11.2.3 Technical Services

In the early 1970s, the British insurance industry recognized the need for technological backup and supported the formation of the Insurance Technical Bureau (ITB) (Chilver, 1973). The ITB was subsequently dissolved and technical support is now provided by in-house engineers or outside consultancies.

2.11.3 Checklists

The insurance surveyor tends particularly to use checklists of features to be assessed. A series of such checklists is given by Spiegelman (1969). Later versions of such checklists are given by Norstrom (1982) and Spiegelman (1987), the latter in the context of gas plants. The original checklists are much more detailed. Further checklists are given by Ashe (1971) and Ackroyd (1974).

2.11.4 Hazard Indices

From such checklists, a number of hazard indices and premium rating plans have evolved. The best known hazard index is probably the Dow Index, given in the so-called Dow Guide, which has gone through a number of versions. The most recent is the *Dow Fire and Explosion Index Hazard Classification Guide* (Dow Chemical, 1994). Another hazard index is the Mond Index, which is an adaptation of the Dow Index (D.J. Lewis, 1979). A third index is the instantaneous fractional annual loss (IFAL).

2.11.5 Premium Rating Plans

Premium rating plans have evolved over the years to incorporate many other risk management factors than previously considered. The plans are no longer used to develop a premium rate for insurance but rather are used to judge the quality of a risk. With this information, the underwriter can determine if he wishes to participate in an insurance program, at what level he will participate and how much of the risk he is willing to take.

Another scheme which is relevant to premium rating for damage insurance is the Dow Index (Dow Chemical, 1994). The scheme involves the determination of the Fire and Explosion Index (F&EI) and then of the Maximum Probable Property Damage (MPPD).

2.11.6 Estimation of EML

Some of the problems of assessing the EML are discussed by Redmond (1984). In general, this assessment is relatively easy for indoor plant with subdivisions each of which is provided with fire protection. It is more difficult to determine for outdoor plant, where there are no such physical barriers.

Improvements in vapor cloud explosion EML loss estimating techniques have been made during the 1990s, but these have not been adopted by all insurers.

2.11.7 Risk Assessment Approaches

There are variations of practice in the methods used to determine the premium rating. As indicated, some insurers have developed formal quantitative methods. Others consider that such quantitative methods offer little advantage over qualitative approaches. This is the view argued by Hallam (1990).

By the 1980s, the setting of insurance premiums had become much more market-driven and rating-schedule or similar analytical approaches fell by the wayside. Insurers continued to consider a prospective insured's loss record and the industry's loss record over the previous 5- or 10-year period as well as the state of the prospective insured's hazard-control efforts and results as documented in survey reports. However, supply and demand forces were paramount.

2.12. INDIVIDUAL INSURANCE

Besides process and property insurances, loss prevention also involves the use of individual insurance. It represents a pool that provides coverage for worker medical expenses and/or legal claims related to medical conditions.

In many cases, employers are required to provide health insurance to their workers. It represents an agreement between a beneficiary, or sponsor (i.e., employer), and the insurance company.

Employers assumed different obligations regarding this insurance, it depends on many factors such as the company size (number of workers), type of company, federal and state insurance laws, etc. Some obligations include the payment of the total or partial cost of the insurance, maternity coverage, participation in premiums, disability, etc.

2.13. BUSINESS INTERRUPTION INSURANCE

The need for BI insurance did not really arise until the mid-1960s, when the first large, single-train plants began to make their appearance. It then became apparent that consequential losses could be as costly as, or more costly than, property damage losses. Accounts of BI insurance are given by Neil (1971), Cloughton (1981), and Di Gesso (1989).

2.13.1.1 Large, Single-Train Plants

As just indicated, the advent of the large, single-train plant changed the relationship between property damage and BI insurance. The consequential losses which may occur with such plants apply not only to the plant itself but also to the suppliers of the plant's inputs and the customers for its outputs.

2.13.1.2 Minor Incidents

Serious BI can be caused by incidents in which the property damage is relatively minor. Typical incidents which may involve only minor damage but which can cause BI for a number of weeks include damage to control rooms, instrument cables, or individual items of equipment, especially large compressors, reactors, and fractionation towers.

2.13.1.3 BI Insurance

The initial approach to BI loss was to assess it as a proportion of damage loss. There is in fact only a weak connection between the two, the BI loss depending rather on the extent to which the loss interrupts a complex business chain. Thus, BI insurance merits separate treatment.

BI insurance is for a specified maximum indemnity period. Typical periods are 12 or 18 months. There is a tendency for insurers to require the insured to bear, as an excess, the losses sustained over some initial period such as 5, 10, or 20 days. Insurance may be extended to cover also the contingent liability for losses due to incidents at supplier and customer plants. The cover generally applies provided the losses are due to damage from a peril for which the insured's plant is covered.

2.13.1.4 BI Survey

In assessing the vulnerability of a single-train operation, Cloughton somewhat discounts in the UK context problems of availability of services and of fuels but emphasizes weak points in single-train operation. In particular, he highlights vulnerability to loss of particular pieces of equipment.

Generally speaking, the existence or lack of inventory of intermediate or finished products will not significantly affect BI analysis. Inventory costs money to finance and management is not likely to maintain more inventory than that which is necessary to sustain normal operations.

2.13.1.5 Estimation of EML

It will be apparent that the assessment of the EML for BI is not straightforward. It involves first determining the downtime of the plant, or sections of it, and then converting this loss of production to loss of financial contribution to fixed costs and profit. This last step may be particularly problematic. The assessment of EML for BI is discussed by Di Gesso (1989), who gives an illustrative example.

2.13.1.6 Property Damage and BI

The worldwide data from Munich Re (1991) quoted above show that over the 5-year period 1985−1989 for property material damage, the annual average number of incidents was 144 and the average annual loss was US $754 million. The corresponding figures for BI are annual number of incidents 39 and annual loss US$572 million.

At the level of the individual plant, some examples of claims for property damage and BI have been given by Di Gesso (1989).

2.13.1.7 BI Insurance Capacity

During the 1960s, it rapidly became apparent, however, that the premiums asked for this class of business had been set at an unrealistically low level. The nature of the problem can be seen from the fact that Neil (1971) estimated the total BI insurance premium income at £35 million. Such a sum could easily be swallowed up by one or two disasters. As already described, since then the market has adapted and has been able, for example, to cover the large BI risks of offshore oil platforms. For Piper Alpha, one estimate of the BI loss, given above, is US$275 million.

Because of the greater uncertainty in assessing BI loss exposure as opposed to PD loss exposure, many insurers will not underwrite BI unless the insured also purchases insurance for PD. Such insurers view an insured's attempt to purchase insurance for BI as a form of adverse selection.

2.14. OTHER INSURANCE ASPECTS

2.14.1 Insurance Credit

The extent to which insurers should encourage and give credit for good practice in general and for loss prevention measures is a long-standing question.

Insurers usually emphasize that they prefer to keep both risks and premiums down. Buyers of insurance are equally keen to hold premiums down and seek credit for loss prevention measures. This is well illustrated by the growth of the Factory Mutual Insurance Corporation. The loss prevention measures which constitute the Factory Mutual System are described in the *Handbook of Industrial Loss Prevention* (Factory Mutual Engineering Corporation (FMEC), 1967).

2.14.2 Insurance in Design

The degree to which reduction of insurance costs is a design objective is discussed by Alexander (1990), who suggests that it is not a major aim. The factors governing the safety and loss prevention features of the design, he argues, are, in order: (1) the attitude of the planning and licensing authorities, (2) the policy of the company, (3) the BI assessment, (4) the acceptability as an insurance risk, and (5) the cost of insurance.

2.14.3 Insurers' Advice

Spiegelman (1987) emphasizes that the primary and only real function of the insurance surveyor is to make a report to the underwriter. Although it is often assumed that he is also there to offer a free consulting service, this is a misconception. However, as described by Redmond (1984), an insurance survey report generally contains recommendations for improvements. Hallam (1990) enumerates some of the recommendations typically made in such reports.

2.14.4 Loss Adjusters

When an incident involving loss has occurred, it is the job of the loss adjuster to assess the financial value of that loss. The loss adjuster is usually employed by a firm specializing in this business and is independent of the insurer and insured. The report of the loss adjuster is primarily a financial rather than a technical document.

2.14.5 Loss Data and Analysis

The insurance industry is uniquely placed to provide information on loss and is in fact a prime source of loss data. Data sources include the annual report of bodies such as the Association of British Insurers and the Loss Prevention Council in the United Kingdom, the corresponding bodies in other countries, the periodic reports *Large Property Damage Losses in the Hydrocarbon−Chemical Industries: A Thirty Year Review* (Marsh and McClennan, 1992), earlier versions being those by Garrison (1988) and Mahoney (1990), as well as occasional publications such as *Losses in the Oil, Petrochemical and Chemical Industries* (Munich Re, 1991). Insurers are also well placed to assess the significant sources of loss in the process industries.

REFERENCES

Ackroyd, G.C., 1974. Insurance aspects of loss prevention. Loss Prev. Saf. Promotion. 1, 139.

Advisory Committee on the Safety of Nuclear Installations (ACSNI), 1993. Study Group on Human Factors. Third Report: Organising for Safety. HM Stationery Office, London.

Alexander, J.M., 1990. Chemical plant underwriting and risk assessment—a personal view. Saf. Loss Prev. Chem. Oil Process. Ind.75.

Allen, D.H., 1973. Economic aspects of plant reliability. Chem. Eng. (London). 278, 467, Department of Chemical Engineering, Teesside Polytechnic, Middlesbrough.

American Insurance Association, 1979. Hazard Survey of the Chemical and Allied Industries, Technical Survey No 3, Engineering and Safety Service. New York, NY, 1979, p. 65.

Ashe, J.B., 1971. The fire insurance surveyor's view. Major Loss Prev. 9.

Bellamy, L.J., Geyer, T.A.W., 1991. Incorporating human factors into formal safety assessments: the offshore safety case. In: BHR Group Ltd, Management and Engineering of Fire Safety and Loss Prevention. Elsevier, London.

Beveridge, H., 1985. Improving the efficiency of rupture-disk systems. Process Eng. 66 (1), 35.

Bowman, J.A., 1989. The consequences of Section 313. Chem. Eng. Prog. 85 (6), 48.

Brooks, J.K., Haney, R.W., Kaiser, G.D., Leyva, A.J., Mckelvey, T.C., 1988. A survey of recent major accident legislation in the USA. Prev. Major Chem. Relat. Process Accid. 425. See also Lees, F.P., Ang, M.L. (1989), op. cit., p. 45.

Brough, C.W., 1981. Dealing With Hazard and Risk In Planning (2). In: Griffiths, R.F. op. cit., p. 89.

Burk, A.F., 1990. Regulatory initiatives and related activities in the United States. Plant/Operations Prog. 9, 254.

Canada, J.R., Sullivan, W.G., White, J.A., 1995. Capital Investment Analysis for Engineering and Management. Prentice Hall, Upper Saddle River, NJ.

Carson, D., Bain, A., 2008. Professional Risk and Working with People: Decision-Making in Health, Social Care and Criminal Justice. Jessica Kingsley Publishers, London; Philadelphia.

Chicken, J.C., 1975. Hazard Control Policy in Britain. Pergamon Press, Oxford.

Chilver, R.C., 1973. The Insurance Technical Bureau. The Post Magazine, August 9, 2196.

Cloughton, D.W., 1981. Aspects of interruption surveys. Fire Surveyor. 10 (3), 35.

Commission of the European Communities, 1994. Proposal for a Council Directive on the Control of Major-Accident Hazards involving Dangerous Substances (COMAH), (Brussels).

Conrad, J., 1980. Society, Technology and Risk Assessment. Academic Press, London.

Council for Science and Society, 1976. Superstar Technologies. Barry Rose, London.

Council for Science and Society, 1977. The Acceptability of Risks. Barry Rose, London.

Davenport, J.A., 1987. Gas plant and fuel handling facilities: an insurer's view. Plant/Operations Prog. 6, 199.

Di Gesso, J., 1989. Business interruption insurance. Chem. Eng. (London). 465, 52.

Dierkes, M., Edwards, S., Coppock, R., 1980. Technological Risk. Oelgeschlager, Gunn & Hain, Cambridge, MA.

Douglas, M., 1966. Purity and Danger. Routledge & Kegan Paul, London.

Douglas, M., 1982. Cultural bias. In: The Active Voice. Routledge & Kegan Paul, London.

Douglas, M., 1985. Risk Acceptability According to the Social Sciences. Routledge & Kegan Paul, London.

Douglas, M., 1990. Risk as a forensic resource. Daedalus. 119 (4), 1.

Douglas, M., 1992. Risk and Blame. Routledge, London.

Dow Chemical Company, 1994. Dow's Fire and Explosion Index Hazard Classification Guide. Seventh ed. American Institute of Chemical Engineers, New York.

Doyle, W.H., 1969. Industrial explosions and insurance. Loss Prev. 3, 11.

Factory Mutual Engineering Corporation, 1967. Handbook of Industrial Loss Prevention. McGraw-Hill, New York.

Fillo, J.P., Keyworth, C.J., 1992. SARA Title III—a new era of corporate responsibility and accountability. J. Hazard. Mater. 31, 219.

Florio, J., 1987. New right-to-know radically changes approach to toxic threats by providing information officials need for emergency planning. New Jersey's First Congressional District, News Release, March 13.

Garrison, W.G., 1988. 100 large losses, A Thirty-Year Review of Property Damage Losses in the Hydrocarbon-Chemical Industries, eleventh ed. M&M Protection Consultants, New York, NY.

Griffiths, R.F., 1981. Dealing with risk. The Planning, Management and Acceptability of Technological Risk. Manchester University Press, Manchester.

Hallam, M.J., 1990. How the insurance industry views risk and some common risk improvement recommendations. Safety and Loss Prevention in the Chemical and Oil Processing Industries. p. 87.

Harvey, B.H., 1976. First Report of the Advisory Committee on Major Hazards. HM Stationery Office, London.

Harvey, B.H., 1979. Second Report of the Advisory Committee on Major Hazards. HM Stationery Office, London.

Harvey, B.H., 1984. Third Report of the Advisory Committee on Major Hazards. HM Stationery Office, London.

Hastings, N.A.J., Mellow, J.M.C., 1978. Decision Networks. Wiley, New York.

Hawksley, J.L., 1984. Some Social, Technical and Economic Aspects of the Risks of Large Chemical Plants. Chemrawn III. World Conference on Resource Material Conversion, The Hague, June 25–29.

Health and Safety Executive, 1988a. The Tolerability of Risks from Nuclear Power Stations. HM Stationery Office, London.

Health and Safety Executive, 1988b. COSHH Assessments. HM Stationery Office, London.

Health and Safety Executive, 1989a. Risk Criteria for Land Use Planning in the Vicinity of Major Hazard Installations. HM Stationery Office, London.

Health and Safety Executive, 1989b. Quantified Risk Assessment: Its Input to Decision Making. HM Stationery Office, London.

Health and Safety Executive, 1990. A Guide to the Health and Safety at Work, etc. Act 1974. HM Stationery Office, London.

Health and Safety Executive, 1992. Health and Safety Statistics 1990/91 (London). Issue Employ. Gazette. 110, 9.

Health and Safety Executive, 1993. The Costs of Accidents at Work. HM Stationery Office, London.

Hohenemser, C., Kasperson, J.X. (Eds.), 1982. Risk in the technological society. Westview, Boulder, CO.

Home Office, 1992. Fire Statistics United Kingdom 1990, London.

Horner, R.A., 1989. Direction of process plant safety regulation in the United States. J. Loss Prev. Process Ind. 2 (3), 123.

Houston, D.E.L., 1971. New approaches to the safety problem. Major Loss Prev. 210.

Houston, D.E.L., 1977. Private communication.

Husmann, C.A.W.A., Ens, H., 1989. The Dutch Occupational Safety Report, In Lees, F.P. and Ang, M.L. (1989), op. cit., p. 21.

Instone, B.T., 1990. The Management Factor: Property Insurance Hazard Assessment In Hydrocarbon Processing And Petrochemical Plants. Safety and Loss Prevention in the Chemical and Oil Processing Industries, p. 113.

Jaeger, C., 2001. Risk, Uncertainty, and Rational Action. Earthscan, Oxford.

Johnson, W.G., 1980. MORT Safety Assurance Systems. Dekker, Basel.

Jouhar, A.J., 1984. Risk in Society. Libbey.

Kletz, T.A., 1988. Learning from Accidents in Industry. Butterworths, London.

Kletz, T.A., Turner, E., 1979. Is the number of serious accidents in the oil and chemical industries increasing? Imperial Chemical Industries Paper. Chemical Industries Association, London.

Koshland, D.E. (Ed.), 1974. How safe is safe? National Academy of Science, Washington, DC.

Lagadec, P., 1982. Major Technological Risk. Oxford: Pergamon Press, Oxford.

Lewis, D.J.,1979. The Mond Fire, Explosion and Toxicity Index — a development of the Dow Index. Am. Inst. Chem. Engrs, Symp. on Loss Prevention, New York.

Lichtenstein, S., Slovic, P., Fischhoff, B., Layman, M., Combs, B., 1978. Judged frequency of lethal events. J. Exp. Psychol. 4, 551.

Lind, N.C. (Ed.), 1982. Technological Risk. Univ. of Waterloo Press, Waterloo, Ontario.

Lowrance, W.W., 1976. Of Acceptable Risk. William Kaufmann, Los Altos, CA.

Mahoney, D.G., 1990. Mahoney, Large Property Damage Losses in the Hydrocarbon-Chemical Industries: A Thirty-Year Review, thirteenth ed. M&M Protection Consultants, New York, NY.

Marsh and, McClennan, 1992. Large Property Damage Losses in the Hydrocarbon-Chemical Industries. A Thirty-Year Review, fourteenth ed. M&M Protection Consultants, New York, NY.

Marshall, V.C., 1975. Process-plant safety—a strategic approach. Chem. Eng. 82 (December), 58.

Meppelder, F.M., 1977. The introduction of safety analysis reports for the chemical process industries in the Netherlands. Loss Prevention and Safety Promotion. 2, p. 43.

Milburn, M.W., Cameron, I.T., 1992. Planning for Hazardous Industrial Activities in Queensland (report). Australian Ins. for Urban Studies, St Lucia, Queensland.

Mitchell, G.H., 1972. Operational Research. English Univ. Press, London.

Munich Re, 1991. Losses in the Oil, Petrochemical and Chemical Industries — A Report, Munich.

National Research Council (U.S.). Committee on Standards and Policies for Decontaminating Public Facilities Affected by Exposure to Harmful Biological Agents: How Clean Is Safe?, 2005. Reopening Public Facilities After a Biological Attack: a Decision Making Framework. National Academies Press, Washington D.C.

Neil, G.D., 1971. The practicalities of insuring a petrochemical complex in the UK. Major Loss Prevention. p. 13.

Norstrom, G.P., 1982. An insurer's perspective of the chemical industry. In: Fawcett, H.H., Wood, W.S. (Eds.), op. cit., p. 509.

Oh, J.I.H., Albers, H.J., 1989. Procedure aspects of safety in the process industry. Loss Prevention and Safety Promotion. 6, p. 2.1.

Park, C.S., 2002. Contemporary Engineering Economics. Prentice Hall, Upper Saddle River, NJ.

Payne, B.J., 1981. Dealing with Hazard and Risk in Planning (1). In Griffiths, R.F. (1981), op. cit., 77.

Petts, J.I. (Ed.), 1984a. Major Hazard Installations: Planning and Assessment (seminar proc.). Dept of Chem. Engng. Loughborough Univ. of Technol.

Petts, J.I. (Ed.), 1984b. Major Hazard Installations: Planning Implications and Problems (seminar proc.). Dept of Chem. Engng. Loughborough Univ. of Technol.

Pratt, M.W.T., 1974. The Cost of Safety. Course of Process Safety — Theory and Practice. Dept of Chem Engng, Teesside Polytechnic, Middlesbrough.

Redmond, T., 1984. Hazard assessment for insurance purposes. Chem. Engr.

Spiegelman, A., 1969. Risk evaluation of chemical plants. Loss Prev. 3, 1.

Vlek, C.J.H., Keren, G., 1991. Behavioural decision theory and environmental risk management: what we have learned and what has been neglected. Thirteenth Resolution Conference on Subjective Probability. Utility and Decision-Making, Fribourg, Switzerland.

Legislation, Law, and Standards

3.1. US LEGISLATION

In the United States, the Federal Government is a source of legislation. In recent years, the involvement of the Federal Government in health and safety legislation has greatly increased. Major categories of safety legislation impacting the process industry can be categorized as follows:

1. occupational health and safety;
2. environment;
3. toxic substances;
4. accidental releases;
5. security.

Laws are created by the legislative process in several steps. Once the bill is debated and approved by the committee, it goes before the floor of the House or Senate for further debate. The debate, revision, and amendment process is repeated in committee and then on the floor. The House and Senate versions of the bill must match exactly. If there are discrepancies, a joint committee is formed to create identical versions. The revised bill is again voted on by both houses. If signed, the bill becomes law. Congress can override a veto by a two-thirds vote from each house.

Once a law has been created, there exists a rulemaking process to create and enact federal regulations to ensure the provisions of the law are implemented. There are many regulatory agencies with the authority to create regulations as well as penalties for violating these regulations. The Administration Procedure Act (APA) outlines how these agencies may create regulations. The regulation then becomes a final rule and is published in the Federal Register and the Code of Federal Regulations (CFR). Additionally, many agencies post regulations on their web sites.

Occupational Safety and Health Act 1970

Hodgson (1971); Ross (1971, 1972); Vervalin (1971, 1972a,b); Nilsen (1972); White (1972, 1973, 1974); Anon. (1973a,b); Ludwig (1973a,b); Anon. (1974); Peck (1974); Stender (1974); Hopf (1975); Petersen (1975); Corn (1976); Anon. (1977); Demery (1977); Morgan (1979); Ladou (1981); Foster and Burd (1993)

Occupational Safety and Health Administration (OSHA)

Shaw (1978, 1979); Morgan (1979); Cahan (1982); Bradford (1986); Spiegelman (1987); Burk (1990); Seymour (1992); Donnelly (1994)

Environmental Protection Agency (EPA)

Shaw (1978); Porter (1988); Friedman (1989a,b); Stephan and Atcheson (1989); Matthiessen (1994); Rosenthal and Lewis (1994)

National Standards

van Atta (1982); Bradford (1985); Henry (1985); Johnson (1985)

Process Safety Management

Rosenthal and Lewis (1994)

Emergency Planning and Community Right-to-Know Act 1986 (EPCRA)

Makris (1988); Bowman (1989); Bisio (1992b)

New Jersey Toxic Catastrophe Prevention Act 1986 (TCPA)

Florio (1987); Somerville (1990)

Clean Air Act (CAA)

Harrison (1978); Siegel et al. (1979); Brown et al. (1988); Zahodiakin (1990); Matthiessen (1992); Kaiser (1993)

Resource Conservation and Recovery Act (RCRA)

Glaubinger (1979); Sobel (1979); Basta (1981); Fawcett (1982); Hoppe (1982)

Superfund

Shaw (1980); Resen (1984); Casler (1985); Sidley and Austin (1987); Habicht (1988); Nott (1988); Bowman (1989); Hirschorn and Oldenburg (1989); Melamed (1989); Rhein (1989); Bisio (1992a)

SARA

Friedman (1989a,b); Anon. (1990); Curran and Kizior (1992); Fillo and Keyworth (1992); Heinold (1992); Keyworth et al. (1992)Chipperfield (1979); Cussell (1979); Malle (1979); Otter (1979); Furlong (1991)

Occupational Safety History

Occupational safety in the United States began with the passage of the Occupational Safety and Health Act of 1970. Some significant events in occupational safety in the United States include:

a. 1864, The Pennsylvania Mine Safety Act (PMSA) was passed into law.

b. 1878, the first recorded call by a labor organization for a federal occupational safety and health law is heard. A professional, technical organization responsible for developing safety codes for boilers and elevators.

c. 1911, the American Society of Safety Engineers (ASSE) was founded. (Today, the NSC carries on major safety campaigns for the general public, as well as assists industry in the development of safety promotion programs.)

d. 1916, the Supreme Court upheld the constitutionality of state workers' compensation laws.

e. 1918, the American Standards Association was founded. Responsible for the development of many voluntary safety standards, some of which are referenced into laws. Today, it is now called the American National Standards Institute (ANSI).

f. 1936, Frances Perkins, Secretary of Labor, called for a federal occupational safety and health law. 1948, all states (48 at the time) now had workers' compensation laws.

g. 1952, Coal Mine Safety Act (CMSA) was passed into law.

h. 1966, the US Department of Transportation (DOT) and its sections, the National Highway Traffic Safety Administration (NHTSA) and the National Transportation Safety Board (NTSB), were established.

i. 1968, President Lyndon Johnson called for a federal occupational safety and health law.

j. 1969, the Construction Safety Act (CSA) was passed.

k. 1970, President Richard Nixon signed into law the Occupational Safety and Health Act, thus creating the OSHA administration and the National Institute for Occupational Safety and Health (NIOSH).

l. 1990, CAAA authorized establishment of the CSB and the Risk Management Program requirements of the EPA.

May 1992, The Process Safety Management rule was adopted by OSHA. June 1996, The Risk Management Plan was published by The EPA. The EPA Risk Management Program (both the Program and Plan are referred to as RMP) purposely mirrors the OSHA PSM standard. EPA's targeted population is the general public and environment, while OSHA's targeted population is workers.

3.1.1 Local and State Regulation

By law, states have the option of upholding the standards of most federal safety regulations themselves. States may request representation from the federal government by submitting a state implementation program. The state must prove that its program is at least as strict in both the content of the program and the enforcement of the rules.

Many states have OSHA approval and operate their own occupational and health safety programs. Although the programs are OSHA approved, they vary based on the specific needs of the state, as long as they maintain the effectiveness of the federal OSHA program. Most of the states that operate their own programs do so to address issues that are not part of the federal OSHA program. For this reason, they elect to use many of the federal standards, but add standards of their own to reflect state-specific issues.

When a state runs its own program, it must fill all of the roles that OSHA provides. These roles include inspections to enforce standards, covering public employees, and operating occupational safety and health training programs. Additionally, most states provide employers with on-site consultation to recognize and correct workplace hazards. This consultation can be provided through the plan or under Section 21(d) of the act.

Before OSHA approves a state program, the state must submit a developmental plan, which includes appropriate legislation, regulations, and procedures for standards setting, enforcement, appeal of citations and penalties, and a sufficient number of qualified enforcement personnel. A state is eligible for certification when it has completed all of its developmental steps.

When a state is ready to be certified, the last step is called the 'final approval'. Under Section 18(e), when OSHA grants a state final approval, it loses all authority to handle matters of occupational health and safety now covered by the state.

3.2. US REGULATORY AGENCIES

The OSHA was created by the Occupational Safety and Health Act of 1970 and has responsibility for enforcing that Act. OSHA comes under the Secretary of Labor and is headed by an Assistant Secretary of Labor. Its activities include the adoption of standards and making of rules, the inspection of workplaces, and the investigation of workplace accidents.

The EPA, created by the National Environmental Policy Act (NEPA) of 1969, is responsible for environmental legislation, including that on air pollution, water pollution, and hazardous wastes. The areas of responsibility between the OSHA and the EPA are not clear-cut. Although OSHA is concerned with the workplace and the EPA with the environment, there are some areas where both agencies are involved. One of these is accidental releases and another is toxic substances, as described below. The relation between the two agencies is discussed by Spiegelman (1987) and Burk (1990).

BSEE works to promote safety, protect the environment, and conserve resources offshore through vigorous regulatory oversight and enforcement.

Mining and mineral processing (such as Alumina refineries and lime processing) are the responsibility of the Mine Safety and Health Administration (MSHA).

Offshore production safety is the responsibility of the Mineral Management Service (MMS), which took over from the original US Coast Guard duties in the mid-1980s. However, the Coast Guard now has responsibility for deepwater facilities, in addition to its traditional role of marine safety.

Pipeline safety, rail transport safety, and motor carrier safety issues are the responsibility of the Department of Transportation (DOT), the Pipeline and Hazardous Materials Safety Administration (PHMSA), and the Federal Energy Regulatory Commission (FERC). Moreover, FERC regulates natural gas, oil, and electricity transportation.

The Department of Homeland Security (DHS) was created by the Homeland Security Act of 2002 to reduce the vulnerability of United States by prevention of terrorism attacks, law enforcement, and emergency preparedness in the eventually of a disaster. In 2003, the Federal Emergency Management Agency (FEMA) became part of the DHS.

Cooperation between federal agencies is common when their respective jurisdictions overlap. A formal agreement between the agencies called a 'Memorandum of Understanding' provides the guidelines under which each agency will address the specific jurisdictional issues.

3.3. CODES AND STANDARDS

Codes, on the other hand, are legal requirements, which are incorporated in the US Code of Federal Regulations or the equivalent state or local code document. Hence, code and regulation are synonymous, but standards may or may not be regulations. Then Government agencies (such as OSHA and EPA) are authorized to create regulations, which are eventually codified in the US Code of Federal Regulations. Announcement of new regulations or changes to existing regulations must be published in the US Federal Register.

Many industrial standards are incorporated directly by reference. For example, the OSHA flammable liquid code 29 CFR 1910.106 directly references the ASME standard on pressure vessels. OSHA eventually updated their regulations by publishing an announcement in the Federal Register.

For example, the OSHA regulation on gas turbine facilities 29 CFR 1910.110(b)(20) directly references the 1970 version of NFPA 37 (standard for the installation and use of stationary combustion engines and gas turbines) even though the latest update of this particular NFPA standard is 1998.

3.4. OCCUPATIONAL SAFETY AND HEALTH ACT 1970

The Occupational Safety and Health Act (OSHA) 1970 is the framework act. It provides that the Secretary of Labor should promulgate safety standards, inspect workplaces, and assess penalties. There is a separate commission called the Occupational Safety and Health Commission, which deals with violations of the OSHA. OSHA regulations do not override any other equivalent federal safety regulation.

Regulations are issued under the Act by OSHA. The OSHA also established the National Institute of Occupational Safety & Health to develop and establish recommended occupational safety and health standards.

3.5. US ENVIRONMENTAL LEGISLATION

Legislation of particular importance in this field includes: the Water Quality Act 1965, the NEPA 1969, the CAA 1970, the Federal Environmental Pollution Control Act 1970, the Federal Water Pollution Control Act (FWPCA) 1972, the Safe Drinking Water Act 1974 and the Noise Control Act 1972, and the CAAA of 1990.

The CAA 1970 gives the EPA power to adopt and enforce air pollution standards. The EPA has promulgated air quality standards that set the maximum concentrations for various gaseous pollutants. It also mandated the Chemical Safety and Hazards Investigation Board or CSB, which was established in 1998.

The FWPCA 1972 gives the EPA power to adopt and enforce water pollution standards.

The Comprehensive Environmental Response, Compensation and Liability Act 1980 (CERCLA or Superfund) strengthens the regulation of hazardous wastes by the EPA. It creates controls on hazardous waste sites and enforces clean-up of existing and abandoned sites.

3.6. US TOXIC SUBSTANCES LEGISLATION

The Toxic Substances Control Act 1976 (TSCA) is a framework Act, which creates a comprehensive system of controls over toxic substances. It empowers the EPA to regulate the manufacture, processing, distribution, use and disposal of existing and new chemicals in order to avoid unreasonable risk to health or the environment and to delay or ban manufacture or marketing. It provides for the creation of an inventory of existing chemicals and for notification by a manufacturer of a new chemical or of new uses for a chemical. There are requirements for the toxicity testing of chemicals and for Premanufacture Notices (PMNs). The EPA has an Office of Toxic Substances (OTS) dealing with the Act.

Section 313 of SARA Title III also deals with toxic substances, but essentially in relation to high-level emissions. It requires the EPA to compile a toxic chemical inventory database, the Toxic Release Inventory (TRI). Section 313 is discussed by Bowman (1989). The control of toxic substances in the workplace and atmosphere is also covered by the OSHA 1970 and pollution legislation.

3.7. US ACCIDENTAL CHEMICAL RELEASE LEGISLATION

Controls on accidental releases have also been introduced by the OSHA in the form of a rule for the Process Safety Management of Highly Hazardous Chemicals. This introduces a requirement for a process safety management system to identify, evaluate, and control such hazards. An account of OSHA's process safety management rule is

given by Donnelly (1994). On June 20, 1996 the Risk Management Plan was also a reactive program created in response to the Bhopal accident. It includes five elements aimed to reduce the number of accidents and magnitude as: hazard assessment, prevention program, emergency response program, and required documentation. Individual states have also brought in their own legislation. The New Jersey TCPA 1985 creates a requirement for a risk management program for facilities handling certain toxic chemicals above a given inventory. The California Hazardous Materials Planning Program is based on a codification of four laws within the California Health and Safety Program.

3.8. US TRANSPORT LEGISLATION

Another major concern in the United States in recent years has been the problem of the transport of hazardous materials, and this has given rise to a large volume of legislation. Some principal items in the field include:

1. Rivers and Harbors Act 1899
2. Explosives Transportation Acts 1908, 1909, and 1921
3. Tank Vessels Act 1936
4. Natural Gas Act 1938
5. Federal Aviation Act 1958
6. Natural Gas Pipeline Safety Act 1968
7. Dangerous Cargo Act 1970
8. Hazardous Materials Transportation Act 1970
9. Coastal Zone Management Act 1972
10. Ports and Waterways Safety Act 1972
11. Deep Water Port Act 1974
12. Maritime Transportation Security Act 2002
13. 49 CFR Part 192: Pipeline Safety: Integrity Management Program for Gas Distribution Pipelines.

3.9. US SECURITY LEGISLATION

DHS issued the Chemical Facility Anti-Terrorism Standards (CFATS) on April 9, 2007 aiming *'to identify, assess, and ensure effective security at high-risk chemical facilities'*.

The Public Health Security and Bioterrorism Preparedness and Response Act of 2002 (the Bioterrorism Act) entered into law on June 12, 2002 to improve the security of the United States. It is divided in five sections, which encompass security and safety of food, drug supply, water systems, emergency preparedness for bioterrorist events, and controls for biological agents and toxins.

Other related regulations are as follows:

33 CFR Part 165: Security Zone; Freeport LNG Basin, Freeport, TX

49 CFR Parts 1520 and 1580: Rail Transportation Security
6 CFR Part 27:Appendix to Chemical Facility Anti-Terrorism Standards.

In addition to federal regulations, states such as New Jersey (NJ) and California (CA) have issued and strengthened regulation aiming to enhance the chemical industry's security.

In 1985, New Jersey created the Toxic Catastrophe Prevention Act (TCPA), which covers facilities handling designated substances above established thresholds. In November of 2005, at TCPA/DPCC Chemical Sector Facilities, Best Practices Standards was created by the state. This regulation requires facilities covered by TCPA or DPCC to perform hazard assessments and to establish risk management programs for those facilities that might be targets for terrorism acts. Main differences are the larger number of chemicals covered, smaller thresholds, and requirement of external analysis events.

3.10. US DEVELOPING LEGISLATION

After the events of April 20, 2010 on the gulf coast, a new bill called 'Blowout Prevention Act of 2010' has been proposed as a response to the largest oil disaster in the history of United States. The act is intended to control drilling operations of high-risk wells and assure the safety of the public and the environment, and so far it has passed committed and it is on its way to be brought to the house for a full vote. This bill addresses topics such as:

- Permitting approval on drilling operations after the applicant has demonstrated its capabilities to contain leaks and prevent them;
- Requirements for a safe design and operation of a blowout preventer devices;
- Standards for the safe design of wells, from the production to the completion stage;
- Stop-work requirements when a potential endangering situation arises;
- Requirement of certification and inspections by a third party.

The Safe Chemicals Act of 2010 was introduced on April 2010 to ensure consumer safety. The proposed bill is intended to update the TSCA of 1976. Highlights of this reform include:

- Expansion on the number of chemicals (currently covered by TSCA in addition to new products targeting to enter the market) that are required to undergo safety testing before reaching consumer hands.

- Use of a risk-ranking system based on the preliminary safety information provided by the manufacturer. EPA will require the producer of high-risk substances to provide further information on and proof of chemical safety.
- Encouragement of the use of inherently safer alternatives by providing monetary incentives for research toward less harmful substances and technology for the environment, personnel, and the consumer.
- Requirement for safety information of each individual chemical ingredient existing in the composition of a final product. Currently, EPA requires MSDS (material safety data sheet) for listed hazardous substances; however, once they are combined into a solid product this requirement no longer applies.

In 2009, OSHA proposed to modify the Hazard Communication Standard (HCS) to be consistent with the United Nations Globally Harmonized System of Classification and Labeling of Chemicals (GHS). Currently, the revision is expected to formalize with the promulgation of a final rule in the CFR. The proposed changes by GHS encompass:

- Labeling: will include harmonized signal word, pictogram, hazard statement, and precautionary statement.
- Hazard classification: no longer a performance-oriented standard. It will include criteria for evaluation of health and physical hazards and classification of mixtures.
- Safety data sheets: current MSDS will become safety data sheet, with 16 pre-specified sections.
- Training and information: training only addressed the promulgation of this rule. Training is mandatory within the two years after adoption of the standard.

3.11. EU LEGISLATIONS

The Seveso Directive came about as a response to the accident that occurred in Seveso, Italy in 1976. Adopted in 1982 and subsequently amendment in 1987 and 1988, its objective was to improve the safety of the facilities that deal with hazardous substances. Formally known as the Control of Industrial Major Accidents Hazards Regulations (CIMAH), this regulation was later replaced by the Control of Major Accident Hazards Regulations (COMAH) in 1999. This new regulation includes the implementation of management systems, accident reporting, inspections, and emergency planning.

On June 2007, the Registration, Evaluation, Authorization, and Restriction of Chemical Regulation (REACH) entered into force in the European Union (EU). According to the European Commission, REACH's primary objective is '*to ensure a high level of protection of human health and the environment'*. Manufacturers of hazardous chemicals are required to provide a comprehensive characterization of the associated hazards and to communicate this safety information to the central database managed by the European Chemicals Agency (ECHA). It also provides guidelines of chemicals' decommission and guidelines on the implementation of inherently safer alternatives.

In 2009, a new regulation complemented REACH regulation by providing unified guidelines for labeling and characterization of chemicals. The new regulation, Classification, Labeling, and Packaging of Substances and Mixtures (CLP), incorporates the Globally Harmonized System of Classification and Labeling of Chemicals (GHS) rules recognized by the United Nations (UN).

3.12. US CHEMICAL SAFETY BOARD

In 1990, the US Chemical Safety and Hazard Investigation Board (CSB) was created as an independent board as mandated by the amendments to the Clean Air Act. Modeled after the National Transportation Safety Board (NTSB), the CSB was directed by Congress to conduct investigations and report on findings regarding the causes of any accidental chemical releases resulting in a fatality, serious injury, or substantial property damages. In October 1997, Congress authorized initial funding for the CSB. "The principal role of the new chemical safety board is to investigate accidents to determine the conditions and circumstances which led up to the event and to identify the cause or causes so that similar events might be prevented." Congress gave the CSB a unique statutory mission and provided in law that no other agency or executive branch official may direct the activities of the Board. Following the successful model of the National Transportation Safety Board and the Department of Transportation, Congress directed that the CSB's investigative function be completely independent of the rulemaking, inspection, and enforcement authorities of EPA and OSHA. Congress recognized that Board investigations would identify chemical hazards that were not addressed by other agencies. CSB has conducted over 60 investigations and published a comprehensive report for each. All CSB reports are available to the public at their website, while enforcement agency reports such as OSHA and EPA are confidential and subject to the Freedom of Information Act.

3.13. THE RISK MANAGEMENT PROGRAM

In 1996, EPA promulgated the regulation for *Risk Management Programs for Chemical Accident Release Prevention* (40 CFR 68). This federal regulation was mandated by Section 112(r) of the Clean Air Act Amendments of 1990. The regulation requires regulated facilities to develop and implement appropriate Risk Management Programs to minimize the frequency and severity of chemical plant accidents. In keeping with regulatory trends, EPA required a performance-based approach toward compliance with the Risk Management Program regulation. The eligibility criteria and requirements for the three different program levels are given in Table 3.1.

The Risk Management Program regulation defines the worst-case release as the release of the largest quantity of a regulated substance from a vessel or process line failure, including administrative controls and passive mitigation that limit the total quantity involved or release rate. For gases, the worst-case release scenario assumes the quantity is released in 10 min. For liquids, the scenario assumes an instantaneous spill and that the release rate to the air is the volatilization rate from a pool 1 cm deep unless passive mitigation systems contain the substance in a smaller area. For flammables, the scenario assumes an instantaneous release and a vapor cloud explosion using a 10% yield factor. For alternative scenarios (note: EPA used the term *alternative scenario* as compared to the term *more-likely scenario* used earlier in the proposed regulation), facilities may take credit for both passive and active mitigation systems.

3.14. THE PROCESS SAFETY MANAGEMENT PROGRAM

The process safety management regulation applies to processes that involve certain specified chemicals at or above threshold quantities, processes that involve flammable liquids or gases on-site in one location, in quantities of 10,000 lb or more (subject to few exceptions), and processes which involve the manufacture of explosives and pyrotechnics. Hydrocarbon fuels, which may be excluded if used solely as a fuel, are included if the fuel is part of a process covered by this regulation. In addition, the regulation does not apply to retail facilities, oil or gas well drilling or servicing operations, or normally unoccupied remote facilities.

3.14.1 Process Safety Information

This element of the PSM regulation requires employers to develop and maintain important information about the different processes involved. This information is intended to provide a foundation for identifying and understanding potential hazards involved in the process.

The process safety information covers three different areas, that is, chemicals, technology, and equipment. A complete listing of the process safety information that must be compiled in these three areas is shown in Table 3.2 This information is intended to provide a foundation for identifying and understanding potential hazards involved in the process.

TABLE 3.1 Eligibility Criteria and Compliance Requirements for Different Program Levels—EPA's Risk Management Program Regulation

Program 1	Program 2	Program 3
Program Eligibility Criteria		
No offsite accident history	Process not eligible for Program 1 or 3	Process is subject to OSHA PSM (29 CFR 1910.119)
No public receptors in worst-case circle		Process is SIC code 2611, 2812, 2819, 2821, 2865, 2869, 2873, 2879, or 2911
Emergency response coordinated with local responders		
Program Requirements		
Hazard assessment	*Hazard assessment*	*Hazard assessment*
Worst-case analysis	Worst-case analysis	Worst-case analysis
5-year accident history	Alternative releases	Alternative releases
Certify no additional steps needed	5-year accident history	5-year accident history
	Management program	*Management program*
	Document management system	Document management system
	Prevention program	*Prevention program*
	Safety information	Process safety information
	Hazard review	Process hazard analysis
	Operating procedures	Operating procedures
	Training	Training
	Maintenance	Mechanical integrity
	Incident investigation	Incident investigation
	Compliance audit	Compliance audit
		Management of change
		Pre-startup safety review
		Contractors
		Employee participation
		Hot work permits
	Emergency response program	*Emergency response program*
	Develop plan and program	Develop plan and program

TABLE 3.2 Process Safety Information

Chemicals	Technology	Equipment
Toxicity	Block flow diagram or process flow diagram	Design codes employed
Permissible exposure limit	Process chemistry	Materials of construction
Physical data	Maximum intended inventory	Piping and instrumentation diagrams
Reactivity data	Safe limits for process parameters	Electrical classification
Thermal and chemical stability data	Consequence of deviations	Ventilation system design
Effects of mixing		Material and energy balances
		Safety systems
		Relief system design and design basis

The information in Table 3.2 is essential for developing and implementing an effective process safety management program. The fundamental concept is that complete, accurate, and up-to-date process knowledge is essential for safe and profitable operations.

3.14.2 Process Hazards Analysis

This element of the PSM regulation requires facilities to perform a PHA. The PHA must address the hazards of the process, previous hazardous incidents, engineering and administrative controls, the consequences of the failure of engineering and administrative controls, human factors, and an evaluation of effects of failure of controls on employees. This element requires that the PHA be performed by one or more of the following methods or any other equivalent method:

1. what-if;
2. checklist;
3. what-if/checklist;
4. hazard and operability (HAZOP) studies;
5. failure modes and effects analysis (FMEA);
6. fault tree analysis.

3.14.3 Operating Procedures

The operating procedures must be in writing and provide clear instructions for safely operating processes; must include steps for each operating phase, operating limits, safety and health considerations, and safety systems. Procedures must be readily accessible to employees, must be reviewed as often as necessary to assure they are up-to-date, and must cover special circumstances such as lockout/tagout and confined space entry. The employer must certify annually that the operating procedures are current and accurate.

3.14.4 Training

The regulation requires that facilities certify that employees responsible for operating the facility have successfully completed (including means to verify understanding) the required training. The training must cover specific safety and health hazards, emergency operations and safe work practices. Initial training must occur before assignment. Refresher training must be provided at least every 3 years.

3.14.5 Pre-startup Safety Review

This element of the PSM regulation requires a pre-startup safety review of all new and modified facilities to confirm integrity of equipment; to assure that appropriate safety, operating, maintenance and emergency procedures are in place; and to verify that a PHA has been performed.

Modified facilities for this purpose are defined as those for which the modification required a change in the process safety information.

3.14.6 Management of Change

This element of the regulation specifies a written program to manage changes in chemicals, technology, equipment, and procedures, which addresses the technical basis for the change, impact of the change on safety and health, modification to operating procedures, time period necessary for the change, and authorization requirements for the change. The regulation requires employers to notify and train affected employees and update process safety information and operating procedures as necessary.

REFERENCES

Anon., 1973a. Occupational Safety and Health Act. Chem. Eng. 80 (February), 13.

Anon., 1973b. The Occupational Safety and Health Act. Chem. Eng. 80 (June), 3.

Anon., 1974. OSHA citations—an analysis. Chem. Eng. 81 (November), 44.

Anon., 1977. OSHA: moving to balance. Chem. Eng. 84 (April), 108.

Anon., 1990. SARA goes public. Chem. Eng. 97 (3), 30.

Basta, N., 1981. The RCRA law—when the states take over. Chem. Eng. 88 (June), 24.

Bisio, A., 1992a. The front end of superfund. Chem. Eng. (London). 529, s24.

Bisio, A., 1992b. The right to know. Chem. Eng. (London). 529, s14.

Bowman, J.A., 1989. The consequences of Section 313. Chem. Eng. Prog. 85 (6), 48.

Bradford, H., 1986. EPA's risk-assessment guidelines take shape despite OMB opposition. Chem. Eng. 93 (19), 19.

Bradford, W.J., 1985. Standards on safety: how helpful are they for practicing engineers? Chem. Eng. Prog. 81 (8), 16.

Brown, G.A., Cramer, J.J., Samela, D., 1988. The impact of the proposed Clean Air Act amendments. Chem. Eng. Prog. 84 (12), 41.

Burk, A.F., 1990. Regulatory initiatives and related activities in the United States. Plant/Operations Prog. 9, 254.

Cahan, V., 1982. OSHA: is it going back instead of forward? Chem. Eng. 89 (January), 47.

Casler, J., 1985. Superfund Handbook. Sidley & Austin and Concorde, Chicago, IL.

Chipperfield, P.N.J., 1979. EEC 'dangerous substances' directive. The UK industry view.

Corn, M., 1976. OSHA's Morton Corn talks to CE. Chem. Eng. 83 (August 30), 57.

Curran, L.M., Kizior, G.J., 1992. A computerized method for reporting SARA Title III Section 313 emissions from a petroleum refinery. J. Hazard. Mater. 31, 255.

Cussell, F., 1979. EEC 'dangerous substances' directive. Fixed emission standards with reference to the EEC Directive. Chem. Ind. (May), 306.

Demery, W.P., 1977. OSHA—where it stands, where it's going. Chem. Eng. 84 (April), 110.

Donnelly, R.E., 1994. An overview of OSHA's process safety management standards (USA). Process Saf. Prog. 13 (2), 53–58.

Fawcett, H.H., 1982. Chemical wastes: new frontiers for the chemist and engineers—RCRA and Superfund. In: Fawcett, H.H., Wood, W.S. (Eds.), op. cit., p. 597.

Fillo, J.P., Keyworth, C.J., 1992. SARA Title III—a new era of corporate responsibility and accountability. J. Hazard. Mater. 31, 219.

Florio, J., 1987. New right-to-know radically changes approach to toxic threats by providing information officials need for emergency planning. New Jersey's First Congressional District, News Release, March 13.

Foster, A., Burd, J.E., 1993. Hold the line on OSHA compliance costs. Chem. Eng. Prog. 89 (6), 71.

Friedman, K.A., 1989a. Can the EPA meet the challenges of the 1990s? Chem. Eng. Prog. 85 (8), 47.

Friedman, K.A., 1989b. EPA and SARA title III. Chem. Eng. Prog. 85 (6), 18.

Furlong, J., 1991. The EC dangerous substances directive: a seventh amendment. Chem. Ind.(June), 418.

Glaubinger, R.S., 1979. A guide to the Resource Conservation and Recovery Act. Chem. Eng. 86 (January), 79.

Habicht, F.H., 1988. The role of the agency for toxic substances and disease registry under the Superfund Amendments and Reauthorization Act of 1986. J. Hazard. Mater. 18 (3), 219.

Harrison, E.B., 1978. What the new Clean Water Act means to HPI plant managers. Hydrocarbon Process. 57 (2), 165.

Heinold, D.W., 1992. Quantifying potential off-site impacts of SARA Title III releases. J. Hazard. Mater. 31, 297.

Henry, M.F., 1985. NFPA's consensus standards at work. Chem. Eng. Prog. 81 (8), 20.

Hirschorn, J.S., Oldenburg, K.U., 1989. Are we cleaned up? An assessment of superfund. Chem. Eng. Prog. 96, 55.

Hodgson, J.D., 1971. The new Occupational Safety and Health Act. Chem. Eng. 78 (June), 108.

Hopf, P.S., 1975. Designer's Guide to OSHA. McGraw-Hill, New York, NY.

Hoppe, R.M., 1982. Widespread dissension over RCRA amendments. Chem. Eng. 89 (August), 31.

Johnson, R.K., 1985. Legal implications of compliance and noncompliance with standards and internal corporate guidelines. Chem. Eng. Prog. 81 (8), 9.

Kaiser, G.D., 1993. Accident prevention and the Clean Air Act Amendments of 1990 with particular reference to anhydrous hydrogen fluoride. Process Saf. Prog. 12, 176.

Keyworth, C.J., Smith, D.G., Archer, H.E., 1992. Emergency notification under SARA Title III: impacts on facility emergency planning. J. Hazard. Mater. 31, 241.

Ladou, J., 1981. Occupational Health Law. Dekker, Basel.

Ludwig, E.E., 1973a. Designing process plants to meet OSHA standards. Chem. Eng. 80 (September), 88.

Ludwig, E.E., 1973b. Project managers should know OSHA. Hydrocarbon Process. 52 (6), 135.

Makris, J., 1988. Comments on the Emergency Planning and Community Right-to-Know Act of 1986. Preventing Major Chemical and Related Process Accidents, p. 729.

Malle, K.G., 1979. EEC 'dangerous substances' directive. The German chemical industry view. Chem. Ind. 5 (May), 308.

Matthiessen, C., 1994. An overview of EPA's chemical Accident Release Prevention programs. Process Saf. Prog. 13, 61.

Matthiessen, R.C., 1992. Chemical accident prevention under the Clean Air Act Amendments. Plant/Operations Prog. 11, 99.

Melamed, D., 1989. Fixing superfund. Chem. Eng. 96 (11), 30.

Morgan, R.D., 1979. Prepare for an OSHA inspection. Hydrocarbon Process. 58 (10), 185.

Nilsen, J.M., 1972. OSHA: acronym for trouble. Chem. Eng. 79 (March), 58.

Nott, S.L., 1988. Implementing the new superfund. J. Hazard. Mater. 18 (3), 229.

Otter, R.J., 1979. EEC 'environmental substances' directive. Environmental quality objectives. Chem. Ind. (May), 302.

Peck, T.P., 1974. Occupational Safety and Health—Guide to Information Sources. Gale Research Company, Detroit, MI.

Petersen, D., 1975. OSHA Compliance Manual. McGraw-Hill, New York, NY.

Porter, J.W., 1988. The major concerns of the Environmental Protection Agency. Preventing Major Chemical and Related Process Accidents, p. 765.

Resen, L., 1984. Superfund a headache for the US chemical industry. Chem. Eng. (London). 409, 11.

Rhein, R., 1989. Report lambasts superfund work. Chem. Eng. 96 (3), 58.

Rosenthal, J., Lewis, P.G., 1994. The real costs of risk analysis. Process Saf. Prog. 13, 92.

Ross, S., 1971. Current legislation. Chem. Eng. 78 (June), 9.

Ross, S.S., 1972. Federal laws and regulations. Chem. Eng. 79 (May), 9.

Seymour, T., 1992. OSHA's interest in plant safety. Plant/Operations Prog. 11, 164.

Shaw, J., 1978. EPA and OSHA: gearing for change. Chem. Eng. 85 (October), 70.

Shaw, J., 1979. OSHA performance faulted as accident rate rises. Chem. Eng. 86 (February), 56.

Shaw, J.S., 1980. Superfund proves, if not painless, at least bearable. Chem. Eng. 87 (December), 24.

Sidley and Austin, 1987. Superfund Handbook, second ed. ERT Inc., Concorde, MA. See also Casler, J. (1985), ERT Inc., Concorde, MA.

Siegel, R.D., Foye, C.M., Petrillo, J.L., Rosenfeld, M.F., 1979. The impact of the Clean Air Act amendments on new and expanded plants. Chem. Eng. Prog. 75 (8), 13.

Sobel, R., 1979. How industry can prepare for RCRA. Chem. Eng. 86 (January), 82.

Somerville, R.L., 1990. Reduce risks of handling liquified toxic gas. Chem Eng. Prog. 86 (12), 64.

Spiegelman, A., 1987. Gas plant problems and the insurance industry. Plant/Operations Prog. 6, 190.

Stender, J.H., 1974. What the HPI can expect from OSHA. Hydrocarbon Process. 53 (7), 190.

Stephan, D.G., Atcheson, J., 1989. The EPA's approach to pollution prevention. Chem. Eng. Prog. 85 (6), 53.

Van Atta, F.A., 1982. Federal standards on occupational safety and health. In: Fawcett, H.H., Wood, W.S. (Eds.), op. cit., p. 29.

Vervalin, C.H., 1971. Keep up with OSHA. Hydrocarbon Process. 107.

Vervalin, C.H., 1972a. What governments expect. Hydrocarbon Process. 51 (10), 85.

Vervalin, C.H., 1972b. Know where OSHA is taking you. Hydrocarbon Process. 51 (12), 61.

White, L.J., 1972. State laws and enforcement. Chem. Eng. 79 (May), 13.

White, L.J., 1973. Occupational Safety and Health Act. Chem. Eng. 80 (February), 13.

White, L.J., 1974. The laws and their enforcement. Progress amid problems. Chem. Eng. 81 (October), 7.

Zahodiakin, P., 1990. Puzzling out the new Clean Air Act. Chem. Eng. 97 (12), 24.

Management Systems

Chapter Outline

The starting point for developing and implementing a safety and loss prevention (SLP) program is the management and the management system.

4.1. MANAGEMENT ATTITUDE

Safety and loss prevention in an organization stands or falls with the attitude of senior management. This fact is simply stated, but it is difficult to overemphasize and it has far-reaching implications.

It is the duty of senior management to ensure that this attitude to SLP is realized throughout the company by the creation of a safety culture in which the company's way of doing things is also the safe way of doing things.

4.2. MANAGEMENT COMMITMENT AND LEADERSHIP

The creation and maintenance of a safety culture requires strong commitment and leadership by senior management. This means that the attitude of senior management must be demonstrated in practical ways so that all concerned are convinced of its commitment. Without management commitment and leadership of management, SLP programs are doomed to fail. An account of some of the ways in which management commitment and leadership is demonstrated was given by McKee (1990) of Conoco, part of Dupont, in evidence to the Piper Alpha Inquiry. They include giving safety a high profile, giving managers safety objectives, backing managers who give priority to safety in their decisions, operating an active audit system, and responding to deficiencies and incidents. McKee believes, "The fastest way to fail in our company is to do something unsafe, illegal or environmentally unsound." The attitude and leadership of senior management, then, are vital, but they are not in themselves sufficient. Appropriate organization, competent people, and effective systems are equally necessary.

4.3. MANAGEMENT ORGANIZATION AND COMPETENT PEOPLE

The discharge of senior management's duty to exercise due care for the safety of its employees, of other people on the site, and of the public requires that it create a rather comprehensive and formal system and that it be active in creating, operating, maintaining, auditing, and adapting the system. It is necessary to define clearly the management structure. Responsibility needs to be assigned unambiguously. There should be a job description for each of the jobs shown in the management structure. Once a job is defined, a competent person should be selected to fill it.

In addition to strong management systems, the design, operation, and maintenance of hazardous processes require competent people. It is generally accepted that academic qualifications, practical training, recent relevant experience, and personal qualities are all important.

4.4. SYSTEMS AND PROCEDURES

It is fundamental that responsibility for SLP should be shared by all concerned in the project. It cannot be delegated to a separate safety function. This does not mean, however, that reliance should be placed simply on individual competence and conscientiousness. It is essential to support the competent people with appropriate systems of work. Experience indicates that effective systems require quite a high degree of formality.

The purpose of these systems of work is to ensure a personal and collective discipline, to exploit the experience gained by the organization, and to provide checks to minimize problems and errors. Some key systems are those which are concerned with (1) management leadership, commitment, and accountability, (2) risk analysis, assessment, and management, (3) facilities design and construction, (4) process and facilities information and documentation, (5) operation of plant (normal, emergency), (6) control of access to plant, (7) control of plant maintenance (permits-to-work), (8) management of change, (9) inspection and maintenance of plant equipment, (10) emergency preparedness, (11) third party services, and (12) incident reporting.

It is important that the various activities be properly phased and matched to the project stages.

4.5. PROJECT SAFETY REVIEWS

The management system should include a formal system of project safety reviews for the identification of hazards, evaluation of risk, and an assessment of the adequacy of controls. The system of reviews should be comprehensive in that it covers all aspects of the project and does so over the whole life cycle. A typical set of project safety reviews includes: (1) Inherent safety/health/environment review, (2) Hazard and Operability (HAZOP) Review, (3) Detailed design review, (4) Construction review, (5) Pre-commissioning review, and (6) Post-commissioning review.

4.6. MANAGEMENT OF CHANGE

Modification control systems are required to detect the intent to modify, to refer the modification to the appropriate function for checking, to record a modification authorized, to inform others of the modification, and to follow up any implications such as a need for training. Modifications may be proposed during the design of a new plant or for an existing plant. The latter also embraces major plant extensions. In the chemical and petroleum industry, the change procedure is known as management of change (MOC).

The need for change has to be formally initiated. Once the change is initiated, the next step is to perform an initial hazard analysis and, based on the results of the hazard analysis, a risk analysis to understand the likelihood and the severity of potential consequences for different scenarios which could arise because of the proposed changes. After this, the formal request for change and the risk analysis should be reviewed and authorized by appropriate persons in charge and either approved or declined. If the change is approved, then it has to be implemented as per the review and any recommendations made after the risk analysis. An audit of the approved changes must be performed to ensure that the approved changes are accurately implemented and any suggestions based on risk analysis are completed. If personnel operate a particular piece of equipment after changes have been implemented, proper initial and periodic training must be provided to company employees and contractors as appropriately needed. Finally, all changes must be documented for knowledge retention and formal completion of the MOC process.

4.7. STANDARDS AND CODES OF PRACTICE

An important aspect of the procedures is the use of standards and codes of practice, both external and internal. They represent a distillation of industry's experience and are not to be disregarded lightly. Although the majority of standards and codes relate to design, there are also many which are concerned with operation.

4.8. PRESSURE SYSTEMS

Central to a successful SLP program is the effectiveness of loss of containment program. The management system for the design and operation of pressure systems is therefore of crucial importance. Major failure of a properly designed, fabricated, constructed, tested, inspected, and operated pressure vessel is very rare. But failures do occur in pressure systems. They tend to be failures of other pressure system components such as pipework and fittings, pumps and heat exchangers, or failures due to maloperation of the system. The management systems for the control and monitoring of a pressure system should be in two parts, covering the two broad areas of design and operation and administered by two separate authorities. The design authority should be responsible for systems for control of design, fabrication, testing, and inspection, and the operating authority for those for control of commissioning, operation, maintenance, and modification.

4.9. MAJOR HAZARDS

There are some particular considerations which need to be borne in mind in dealing with major hazards. According to Challis (1979) when there is a combination on major hazard plants of high technology and people, to operate safely, it is necessary that there be strong leadership by management. The level of management which is particularly crucial is the first level of executive technical management. The workforce needs to have an understanding of the process and the plant, of the hazards involved, and of the actions required. It should be well trained. It should be provided with clear instructions, both for normal operation and for emergencies. On such an installation, the manager needs to be 'out and about' rather than in the office. There should be a strong executive atmosphere and discipline on the plant.

4.10. TOTAL QUALITY MANAGEMENT

Increasingly, SLP is subject to the influence of quality assurance (QA) and total quality management (TQM). TQM is unlikely to be introduced for the sole purpose of SLP, but if adopted it will have an impact. The concept underlying loss prevention is that the problem of failures and their effects has an influence on company performance which goes far beyond safety of personnel. Essentially the same concept underlies the TQM approach. Thus, TQM and SLP have much in common.

Stemming from the idea of TQM, more emphasis is placed on the six sigma process and certification and training organizations for black belt. Six sigma has emerged as a very disciplined and quantitative approach for improving operations for different types of functions. There are five basic steps in the six sigma process, which are define, measure, analyze, improve, and control. Implementation of the six sigma process (SSP) rests on the shoulders of the highest authorities of a company and then tapers down to all other levels of management in an organization, with very clear-cut responsibilities for each level of management (Hoerl, 1998).

4.11. SAFETY MANAGEMENT AND SAFETY POLICY

The key elements and structure of a safety management system (SMS) as given in *Successful Health and Safety Management* (HSE, 1991) consider policy, organization, planning, measurement, control, and audit.

Policy on safety should aim to set appropriate goals and objectives, to organize and plan to achieve these objectives in a cost-effective way, and to ensure by systems of measurement and control and of audit that the plan is implemented.

4.12. ORGANIZATION

Organization for safety is considered by the HSE under the following headings: (1) control, (2) cooperation, (3) communication, and (4) competence.

The safety goal is to ensure that activities take place in a controlled, and therefore safe, manner. This means putting in place systems of control. Such a control system is analogous to the typical feedback control system used in process control. A safety culture necessarily involves the cooperation of the whole workforce. Specific measures have to be taken to obtain this.

Effective operation of the control systems described involves large flows of information. The forms of communication distinguished by the HSE are (1) information inputs, (2) internal information flows, (3) visible behavior, (4) written communications, (5) face-to-face discussion, and (6) information outputs. The competence of personnel, particularly those in key positions, is crucial. Competence depends partly on education and training, and partly on experience.

4.13. PLANNING

Broad safety goals need to be translated into specific objectives. The objectives which it is practical to set are essentially determined by the variables which can be measured.

In management theory, a distinction is commonly made between the following three stages of an activity to which controls may be applied: (1) input, (2) process, and (3) output. Again, on the analogy of process control, the first two may be likened to open loop control in that the required output is obtained if the process gain remains constant and the input is maintained at an appropriate value. Alternatively, the output can be measured, and closed loop (or feedback) control applied. The application of controls to all three stages therefore involves a degree of redundancy. As redundancy enhances reliability, this is a desirable feature. The systems required for hazards include systems for (1) hazard identification, (2) evaluation of risk, and (3) an assessment of the adequacy of hazard controls.

Once objectives have been defined, suitable measurements, or metrics, have been selected and control systems based on those devised, the performance standards which these control systems should meet can be set. Performance standards should be set for the systems for hazard identification, hazard assessment, and hazard control just mentioned.

4.14. MEASUREMENT

The account just given brings out the crucial role played by measurement. Effective systems of control can be devised only if suitable measures, or metrics, can be found. For every serious or disabling injury incident, there are a number of minor injury incidents; that for every minor injury incident, there are a number of damage-only incidents; and that for every damage-only incident, there are a number of incidents with no visible injury or damage. It follows that if incidents at this latter, and lowest, level can be kept under control, the incidence of the former will be much reduced.

There are various metrics for accidents. One of the best known is the lost time accident (LTA). Statistics of such accidents can be compiled and monitored. If the accidents are occurring with a sufficiently high frequency, the accident statistics provide a suitable measure of performance.

Turning then to specific measures, some items which may be made the subject of proactive monitoring include (1) achievement of objectives, (2) adherence to systems and procedures, (3) conduct of auditing, (4) state of plant and equipment, and (5) state of documentation. Some items which may be made the subject of reactive monitoring include (1) injuries, (2) damage, (3) other losses, (4) incidents, and (5) workplace deteriorations.

4.15. CONTROL

The purpose of the measurement activities just described is to provide the basis for control actions to correct deviations and for review of the performance of the system. If a deviation is detected, control action is required to correct it. In some cases, the corrective action needed is obvious and it can then be taken. In other cases, it is necessary to carry out an investigation. The other aspect of the control system is the review of performance to ensure that the performance standards are being met.

4.16. AUDIT SYSTEM AND AUDIT

It is essential that there is a mechanism to monitor the system as a whole and to make sure that it is working correctly and is not falling into decay; in other words, an audit system.

The whole management system just described needs to be examined periodically to ensure that its goals and objectives remain appropriate and that the control systems to achieve these objectives are working. This is the function of the audit system. Continuing the analogy of process control, the audit system constitutes a set of outer, higher-level loops around the control systems. The purpose of an audit is to detect degradations and defects in a management system.

4.17. SAFETY MANAGEMENT SYSTEMS

The system that delivers the approach to process SLP described above is in effect an SMS. The 10 components of a typical management system are (1) a charter defining the system's purpose, responsible resources, and time expectations; (2) the rationale of why the system's purpose is important, its expected outcome, and how the output will be used; (3) the scope, clearly defining boundaries, constraints, specifications of input/output, resources required, and excluded areas; (4) an administrator, defined as a single function or an individual responsible for ensuring that the system is effective and provides continuity over time; (5) tools/procedures/resources/schedules, the documentation of what, when, how, and by whom tasks must be done, including scheduling and measuring or results and periodic assessment of competency of resources; (6) communication of results and plans to whom, by whom, how often, by what method; (7) management sponsor to facilitate, monitor, recognize, provide commitment and sustainability; (8) verification, which includes a measurement process to determine if desired results are being achieved and communicated to customers, sponsors, and members of the system; (9) a continuous improvement mechanism to facilitate improvements over time; (10) a document defining how the system works.

4.18. PROCESS SAFETY MANAGEMENT

In the United States, certain hazardous chemicals attract the statutory requirement for a PSM system, introduced by the Occupational Safety and Health Administration (OSHA) Rule Process Safety Management of Highly Hazardous Chemicals (Federal Register 29 CFR 1910.119). The PSM rule specifies performance standards in 14 elements: (1) employee participation, (2) process safety information, (3) process hazard analysis, (4) operating procedures, (5) training, (6) contractors, (7) pre-start-up safety review, (8) mechanical integrity, (9) hot work permit, (10) management of change, (11) incident investigation, (12) emergency planning and response, (13) compliance audits, and (14) trade secrets.

In addition to the PSM system given in the OSHA Rule, there are several other American systems. One is the EPA Risk Management Program (Federal Register 40 CFR Part 68), the components of which are as follows: (1) hazard assessment, (2) prevention program,

(3) emergency response program, and (4) risk management plan. The prevention program covers: (1) management system, (2) process hazard analysis, (3) process safety information, (4) standard operating procedures, (5) training, (6) maintenance of mechanical integrity, (7) pre-start-up review, (8) management of change, (9) safety audits, and (10) accident investigation.

Another process safety system is that given in API RP 750: 1990 *Management of Process Hazards*. The elements of the American Petroleum Institute (API) system are (1) process safety information, (2) process hazards analysis, (3) management of change, (4) operating procedures, (5) safe work practices, (6) training, (7) assurance of the quality and mechanical integrity of critical equipment, (8) pre-start-up safety review, (9) emergency response and control, (10) investigation of process-related incidents, and (11) audit of process hazards management system.

The process safety system developed by the American Chemistry Council (ACC) has four main parts: (1) management leadership in process safety, (2) process safety management of technology, (3) process safety management of facilities, and (4) managing personnel for safety.

4.19. CCPS MANAGEMENT GUIDELINES

The CCPS has published several sets of guidance on management. The first of these was A Challenge to Commitment (CCPS, 1985) addressed to senior management in the industry. The *Guidelines for Technical Management of Chemical Process Safety* (the *Technical Management Guidelines*) by the CCPS (1989/7) give a comprehensive treatment of the management of SLP.

The *Guidelines* are concerned with PSM. Process safety is defined as 'the operation of facilities that handle, use, process, or store hazardous materials in a manner free from episodic or catastrophic incidents' and PSM as 'the application of management systems to the identification, understanding, and control of process hazards to prevent process-related injuries and accidents.'

The Plant Guidelines for Technical Management of Chemical Process Safety (the Plant Technical Management Guidelines) by the CCPS (1992/11) have the same basic structure as the Technical Guidelines but emphasize concrete examples.

The implementation of a PSM system is covered in *Guidelines for Implementing Process Safety Management Systems* (CCPS 1994/13). The auditing of a PSM system is dealt with in *Guidelines for Auditing Process Safety Management Systems* (CCPS, 1993/12). The book by the Center for Chemical Process Safety (CCPS), entitled Guidelines for Risk Based Process Safety (2007), provides the guidelines for operations involving the manufacture, use, storage, and handling of hazardous substances

by focusing on new design paths for improving process safety management practices. The new framework provided by CCPS for encouraging risk-based process safety builds upon the original process safety management ideas published in the early 1990s.

When firmly committing to process safety, facilities should focus on developing and sustaining a strong safety culture that promotes the understanding of process safety as a core value and as a means to go beyond compliance with standards and regulations. Facilities should focus on gathering, documenting, and retaining process knowledge along with hazard identification and risk analysis studies. Facilities should have developed written procedures describing safety measures during plant start up, normal operation, and unintended shut down processes. Facilities and organizations as a whole should focus on investigating incidents and near-misses alike that occur in the facilities and identify and address the root causes.

4.20. SAFETY CULTURE

It is crucial that senior management should give appropriate priority to safety and loss prevention. It is equally important that this attitude be shared by middle and junior management and by the workforce. A positive attitude to safety, however, is not in itself sufficient to create a safety culture. Senior management needs to give leadership in quite specific ways.

Organizations find safety culture a critical factor that sets the tone for implementation of safety within their workplaces. It is generally accepted that an organization that develops and maintains a well-built safety culture becomes more effective at preventing accidents. It is important to emphasize management involvement and behavior-based interventions that reflect the real commitment to safety at all levels in the organization.

4.21. SAFETY ORGANIZATION

It is normal for there to be a separate department responsible for safety and loss prevention. Titles vary, but generally include some combination of those two phrases. The safety department needs to be, and be seen to be, independent, and the organizational structure should reflect this. In particular, the department should be independent of the production function.

In broad terms, it is the responsibility of the safety organization to participate in the formulation of safety policy; ensure that safety systems are created, maintained, and adapted; ensure compliance with the regulatory requirements; review plants and procedures; identify, assess, and monitor hazards; educate and train in safety; assist communication and promote feedback in safety matters; and contribute to technical developments in safety. Thus, safety personnel are concerned with the development and maintenance of the overall safety system and with its application to particular works and to plants.

The safety department has a responsibility to ensure that there exists a safety management system which is appropriate and comprehensive, to monitor its implementation and operation, to undertake periodic audits and reviews, and to make proposals for revision as necessary. On safety matters, the department should be the company's window to the outside world, monitoring developments, whether in legislation, technology, or good practice. The safety department will have a substantial commitment to education and training.

Personnel involved in work on safety and loss prevention tend to come from a variety of backgrounds and have a variety of qualifications and experience. The role of the safety officer is in most respects advisory. It is essential, however, for the safety officer to be influential and to have the technical competence and experience to be accepted by line management. The safety officer should have direct access to a senior manager. Much of his work is concerned with systems and procedures, with hazards and with technical matters.

Safety is commonly associated with one or both of two additional areas: (1) occupational health and (2) environmental protection. The association of safety with occupational health is a long-standing one. In some cases, the two have operated as separate, parallel functions, while in others one has been subsumed into the other. As environmental protection has grown in prominence, it has become increasingly linked with safety.

According to Dawson et al. (1985), the three main activities of safety personnel are as follows: (1) processing and generating information, (2) giving advice and participating in problem solving, and (3) taking direct action.

In respect of the dependence of senior management on safety specialists, they see the relationship extending from one in which the safety function is required to assure compliance and to deal with external agencies to one in which it has increasing relevance to core managerial objectives and activities.

4.22. SAFETY POLICY STATEMENT

In both Europe and the United States, legislation has decreed that the company must protect workers from hazards of the plant. The US Occupational Safety and Health Administration has a general duty clause. The general duty clause, Section 5(a)(1) of the Occupational Safety and Health Act of 1970, applies to all employers and requires each employer to provide employees with a place of employment which is free of recognized hazards that may cause death or serious physical harm.

4.23. SAFETY REPRESENTATIVES AND SAFETY COMMITTEES

A system of safety representatives with responsibility for their fellow workers at a particular workplace has long existed in chemical works. A safety representative is appointed by a trade union. He represents in the first instance, the employees who are members of that union, but may by agreement represent other employees also. The functions of the safety representative are to represent the employees in consultation with the employer. The safety representative should keep informed on the hazards of the workplace, the relevant legislation, and the employer's safety policy, organization, and arrangements. The safety representative should act as a channel of communication between the management and the workforce on safety matters.

The most common practice requires the employer to establish a safety committee if requested to do so by at least two safety representatives. The guidance notes suggest that the objectives of a safety committee are to promote cooperation between employers and employees and to provide a forum for participation by employees in matters of health and safety.

The existence of a safety committee does not in any way relieve management of its responsibility for safety, but such a committee can fulfill a valuable function in complementing the work of the professional safety personnel.

4.24. SAFETY ADVISER

The safety department is led by a safety adviser. The approach taken by the safety adviser may well have a crucial influence on safety and loss prevention in the company. In the process industries, the work of the safety adviser is likely to have a large technical content. As loss prevention has grown in importance, so has the function of the safety adviser.

4.25. SAFETY TRAINING

Training in safety is important for both management and workers. Guidance on safety training includes the many books issued by the AIChE Center for Chemical Process Safety and the Safety series of the IChemE. Managers require training, particularly in technical aspects of safety and in the loss prevention approach, in company systems and procedures, in the division of labor between the company's specialist safety personnel and themselves, and in training workers.

4.26. SAFETY COMMUNICATION

A major aspect of the activity of safety personnel is communication in its various forms. This follows from the advisory role of the safety function. For the most part, communication takes place as a result of the participation of safety personnel in the various activities such as process safety reviews and hazard studies, emergency planning, safety training, and safety cases. Such participation can be an effective means of creating and maintaining the safety culture.

4.27. SAFETY AUDITING

A major aspect of the work of the safety department is auditing. Audits are the periodic examination of the functioning of the safety system.

The overall system of control and auditing needs to operate at three levels, which correspond to increasingly long timescales. Subjects of audit at both Levels 2 and 3 should be (1) the safety management system of the company, (2) the safety management system and operation of the plant, and (3) special features of the plant.

The overall safety management system of the company should be subject to audit. At Level 2, the audit should check that the systems established are being operated and at Level 3, that these systems are still appropriate. At plant level, there should be a post-commissioning audit carried out soon after a plant has been commissioned. Audits should then be repeated at intervals throughout the life of the plant. Plant audits should cover the safety management system of the plant, including operation, maintenance, and control of change.

These general audits may be supplemented by audits in which some particular feature is examined in greater depth. Typical topics for such special feature audits might be the fire protection system or the permit-to-work system.

The conduct of a plant audit is assisted by the use of aids such as checklists. A structured set of checklists related to the root causes of accidents is typically used. The effectiveness of an audit is highly dependent on the technique used. There is some guidance available on these practical aspects of auditing.

4.28. MANAGEMENT PROCEDURE TO IMPLEMENT REQUIRED CHANGES TO ESTABLISH PROPER SAFETY

The implementation of a Safety Management System (SMS) is a complex task because it takes into account several functional areas of a company's organization. For this reason, practical assistance on how to implement the ideas and concepts needs to be provided, and there are some CCPS publications related with this topic.

While the process industry has been collectively responsible for establishing and encouraging the adoption of different process safety management elements under

various regulations and directives for maintaining the integrity and optimum operability of all assets of the plant, major incidents still do occur. This is mainly because of the lack of accurate measurements of operational systems in the facilities. Many important indicators that would provide valuable information for improving safety are not measured. So, what is not measured cannot be used to predict the safety level or the trends in the process industry for improvements. Thus, there is a need to utilize metrics (or indicators) for measuring the performance of different sectors of corporations at all levels.

4.29. NEED FOR PROCESS SAFETY METRICS

Organizations seeking to understand operational performances by monitoring operational activities and behavioral trends are better capable of preventing failures by using proper layers of protection and thus avoiding major incidents. Critical incident information such as the contributing factors and root causes for the incidents are valuable information that can be data mined by tracking organizational performances. Therein lies the importance of generating metrics based on all aspects of plant performance qualitatively or quantitatively for the purpose of generating important performance trends in the corporation and the process industry.

Process Safety Metrics are valuable tools that provide important information about plant operations, maintenance, and personnel behavior to effectively develop and use appropriate safety measures for hazard elimination or risk mitigation.

4.30. DIFFERENT TYPES OF METRICS

Organizations must make conscious effort in determining different management systems and indicators for the metrics to be analyzed so that they are efficient and completely capture the safety culture at various levels of the organization. The CCPS Guideline for Process Safety Metrics (CCPS, 2009) recommends the following metrics.

Absolute metrics provide measure of performance by way of counting the number of recorded events based on reported events. The drawback of this metrics is that it heavily relies on the accuracy of the reporting. Normalized metrics provide the measure of performance by generating ratios of operational activities and deviations across various sites within an organization or different entities within the process industry. However, using only normalized data could take away from lessons learned from particular incidents. Lagging metrics provide retrospective information based on incidents that have already occurred. These metrics are effective in providing a glimpse of potential deviations or incidents which could lead to more severe consequences such as fire, explosions, and fatalities. Lagging metrics are a useful subjective measure which is usually classified based on severity of consequences. Near-miss metrics include actual process safety incidents that could have led to an incident with greater operating deviations and consequences. If near-misses are not tracked and analyzed, they lead to further degradation of process safety programs leading to catastrophic incidents. Leading metrics provide information about incidents or events which could potentially occur in the future. Hence, these are extremely valuable metrics.

4.31. CHOOSING USEFUL METRICS

In order to improve the efforts made for improving process safety systems, choosing effective metrics is vital. The objective for adopting particular metrics (or indicators) and their scopes have to be defined clearly. The main aim of adopting metrics must be to obtain useful information about the operations of the corporation at different levels and different divisions. Hence, the metrics chosen must reflect any changes or deviations from the normal or established practice.

4.32. IMPLEMENTING THE SELECTED METRICS

After the selection of metrics, the collection of relevant data must be undertaken for many years. Only then can the trends of the systems be decipherable and more distinct. Management support and sufficient resources to sustain the continued efforts of data accumulation are critical. Insufficient and incomplete data could result in skewed results (CCPS, 2009). Once the inventory of data is collected, the information must be data mined and appropriate statistical analysis performed in order to understand the results from the gathered information. Even if fewer data is collected, it should be such that those metrics are efficient and must be collected for longer periods of time.

4.33. FUTURE EFFORTS FOR GENERATING INDUSTRY-WIDE METRICS

With the emergence of newer more complex processes, newer risks could surface, making study of performance indicators invaluable to learn more about industry incidents and seek incentives to better understand the profile of incidents. Along with lagging indicators for metrics to measure safety performance, leading indicators must also be used (Prem, 2010). Quantitative risk analysis is one of the most efficient methods for generating leading metrics for outlining future expected losses. If the losses are monetized then, it would prove to be of value to management

and regulators for risk-informed decision making (Prem et al., 2009).

Probabilistic methods are more cost-effective methods to analyze risks as they give results that are easier to communicate to decision makers. Therefore, the different concepts of probability theory must be used to estimate the portfolio risks. Metrics should be established such that they help track leading and lagging indicators by way of monetizing the asset loss and include societal consequences for potential accidents (Prem et al., 2009).

In chemical engineering, when metrics are chosen and decisions are made for safety, the stakes at risk are always measured. This includes making hard decisions on optimizing safety and performance (Clemen and Reilly, 2001). Expected utility principles could be most effectively utilized to decide which performance indicators are most preferred in the presence of competing design alternatives. For each of the decisions, the different criteria have to be evaluated and estimates can then be benchmarked. For understanding how the different criteria interact to provide optimality, game theory concepts can be utilized by choosing dominant criteria among the multi-criteria (Dixit and Skeath, 2004; Prem, 2010). In other words, the gain will benefit the overall system performance.

4.34. CONCLUSION

Establishing an all encompassing consistent and common metrics will aid in performance comparison between different companies and the different sectors of the chemical process industry. Such consensus metrics are now gaining interest among corporations globally. Both leading and lagging indicators must be chosen as metrics to be consistently monitored and analyzed for predicting future industry trends to improve safety. Continuous improvement of different metrics would also aid in fine-tuning better management systems. This must be aggressively sought and common knowledge actively shared across all levels of corporation, to various industry sectors including going beyond geographical borders.

As published in the CCPS book, *Guidelines for Risk Based Process Safety* (CCPS, 2007), organizations must also move toward quantitative risk analysis as opposed only a qualitative analysis of deviations from normal operations. Monetization of assets is also one very effective method to bridge the gap between engineers and decision makers (Prem et al., 2009).

A greater emphasis is now being placed on process safety, and there is a growing awareness that improving the safety culture boosts the production and revenue. As organizations become more global in the nature of businesses conducted, the importance of emphasizing process safety culture and use of consistent metrics will only be magnified. Continued management commitment is extremely important to achieve this feat within an organization. Management support is crucial for sharing of lessons learned and knowledge about useful metrics among different organizations both nationally and beyond geographical borders. Metrics must be used and data collected from all constituent states within a country to completely understand the overall process safety performance. Only by consistently using metrics will the government and the industry as a whole have the accurate measure of how safe the chemical process industry is and undertake suitable efforts for improvement.

REFERENCES

Center for Chemical Process Safety (CCPS), 1985. A Challenge to Commitment. American Institute of Chemical Engineers, New York, NY.

Center for Chemical Process Safety (CCPS), 1989. Guidelines for Technical Management of Chemical Process Safety. AIChE, New York, NY.

Center for Chemical Process Safety (CCPS), 1992. Plant Guidelines for Technical Management of Chemical Process Safety. AlChE, New York, NY.

Center for Chemical Process Safety (CCPS), 1993. Guidelines for Auditing Process Safety Management Systems. AlChE, New York, NY.

Center for Chemical Process Safety (CCPS), 1994. Guidelines for Implementing Process Safety Management Systems. AIChE, New York, NY.

Center for Chemical Process Safety (CCPS), 2007. Guidelines for Risk Based Process Safety. Wiley-AIChE, New York, NY.

Center for Chemical Process Safety (CCPS), 2009. Guidelines for Process Safety Metrics. CCPS, AIChE, New York, NY.

Challis, E.J., 1979. Management of major hazard plant. J. Hazard. Mater. 3 (1), 49.

Clemen, R.T., Reilly, T., 2001. Making Hard Decision with Decision Tools. Duxbury Thompson Learning, New Jersey.

Dawson, S., Poynter, P., Stevens, D., 1985. Safety specialists in industry: roles, constraints and opportunities. J. Occup. Behav. 5, 253.

Dixit, A., Skeath, S., 2004. Games of Strategy, second ed. W. W. Norton & Company, Inc, New York, NY.

Health and Safety Executive, 1991. Successful Health and Safety Management. HM Stationery Office, London.

Hoerl, R.W., 1998. Six sigma and the future of quality profession. Quality Progress. 31 (6), 35–42.

McKee, R.E., 1990. Piper Alpha Public Inquiry. Transcript, Days 171, 172.

Prem, K.P., 2010. Risk Measures Constituting Risk Metrics for Decision Making in the Chemical Process Industry, Texas A&M University, College Station. Retrieved March 15, 2010 from Dissertation and Thesis Database.

Prem, K.P., Ng, D., Sawyer, M., Guo, Y., Pasman, H.J., Mannan, M.S., 2009. Risk measures constituting a risk metrics which enables improved decision making: value-at-risk. J. Loss Prev. Process Ind. 23, 211–219.

Reliability

The ability of equipment to perform its expected or designed functions throughout a period of time is important in any kind of process. The performance of equipment may decline or even fail over the period of its lifetime due to various factors such as aging. As a result, reliability analysis of equipment with proper maintenance and modifications has become an important area of process safety. The first section of this chapter gives an introduction to reliability analysis. Several aspects of equipment maintenance and modifications along with the management of change are discussed in the rest of the sections.

5.1. RELIABILITY ENGINEERING

The discipline which is concerned with the probabilistic treatment of failure in systems in general is reliability engineering. From these early beginnings, the study of reliability has become a fully developed discipline. It has received particular impetus from the reliability requirements in the fields of defense and aerospace, and electronics and computers. This section gives an account of reliability engineering and of some reliability techniques.

The earliest documented developments in reliability engineering began during the Second World War. The Germans had problems with the reliability of the V-1 missile (Bazovsky, 1961). The project team leader, Lusser, described how the first approach taken to the problem was based on the argument that a chain is no stronger than its weakest link. This concentrated attention on the small number of low reliability components. But this approach was not successful. It was then pointed out by a mathematician, Pieruschka, that the probability of success p in a system in which all the components must work if the system is to work is the product of the individual probabilities of success p_i:

$$p = \prod_{i-1}^{n} p_i \qquad (5.1)$$

This drew attention to the need to improve the reliability of the many medium reliability components. This approach was much more successful in improving missile reliability. Equation 5.1 is known as Lusser's product law of reliabilities.

On the other side of the English Channel, Blackett (1962) was drawing attention to the significance of Equation 5.1 to military operations in general:

In the simplest case of air attack on a ship, the four main probabilities are (1) the chance of a sighting, (2) the chance the aircraft gets in an attack, (3) the chance of a hit on a ship, and (4) the chance that the hit causes the ship to sink.

The US armed forces also had serious reliability problems, particularly with vacuum tubes used in electronic equipment. Studies of electronic equipment reliability at the end of the War showed some startling situations (Shooman, 1968). In the Navy, the number of vacuum tubes in a destroyer had risen from 60 in 1937 to 3200 in 1952. A study conducted during maneuvers revealed that equipment was operational only 30% of the time. An Army study showed that equipment was broken down between two-thirds and three-quarters of the time. The Air Force found that over a 5-year period, maintenance and repair costs of equipment exceeded the initial cost by a factor of 10. It was also discovered that for every vacuum tube in use there was one held as spare and seven in transit, and that one electronics technician was needed for every 250 vacuum tubes. These studies illustrate well the typical problems in reliability engineering, which is concerned not only with reliability but also with availability, maintenance, and so on.

One of the main fields of application of reliability engineering has been in electronic equipment. Such equipment typically has a large number of components. Initially, the reliability of electronic equipment was much less than that of mechanical equipment. But the application of reliability engineering to electronic equipment has now made it generally as reliable. Another area in which reliability engineering has been widely used is nuclear energy. Methods have had to be developed to assess the hazards of nuclear reactors and to design instrument trip systems to shut them down safely.

5.1.1 Reliability Engineering in the Process Industries

Reliability engineering involves an iterative process of reliability assessment and improvement. In some cases, the assessment shows that the system is sufficiently reliable. In other cases, the reliability is found to be inadequate, but the assessment work reveals ways in which the reliability can be improved. It is generally agreed that the value of reliability assessment lies not in the figure obtained for system reliability, but in the discovery of the ways in which reliability can be improved.

The reliability engineer, however, cannot wait until his fellow engineers have solved all their reliability problems. It is his job to identify the areas where improvements are essential for success. But for the rest, he is obliged to accept, as given, the levels of reliability currently being achieved. This is not the case, however, with other engineers. It is they who are in a position to reduce the number of failures. They too should use reliability techniques. But it would be disastrous if they were to take the existing level of reliability as unalterable.

There is a close link between reliability and quality control of equipment. But the two are not identical and the distinction between them is important. Equipment is likely to be unreliable unless there is good quality control over its manufacture. In general, quality control is a necessary condition for reliability. It is not, however, a sufficient condition. Deficiencies in specification, design, or application are also causes of unreliability. A badly designed piece of equipment may be manufactured with good quality control, but it will remain unreliable.

5.1.2 Definition of Reliability

Definitions of reliability are given in various British Standards dealing with terminology on quality and reliability. BS 4778: Part 1: 1987 defines reliability as:

The ability of an item to perform a required function under stated conditions for a stated period of time.

Later definitions of reliability states that the word 'time' may be replaced by 'distance', 'cycles', or other quantities or units, as appropriate.

An alternative definition of reliability is:

The probability that an item will perform a required function under stated conditions for a stated period of time.

This definition brings out several important points about reliability: (1) it is a probability, (2) it is a function of time, (3) it is a function of defined conditions, and (4) it is a function of the definition of failure.

Failure, in turn, is defined as: (1) failure in operation, (2) failure to operate on demand, (3) operation before demand, and (4) operation after demand to cease. The first is applicable to a piece of equipment which operates continuously, while the other definitions are applicable to one which operates on demand.

5.1.3 Some Probability Definitions and Relationships

It is appropriate to give, at this point, a brief treatment of some basic probability relationships. These are important in the present context not only because they are the basis of reliability expressions, but also because they are needed for work in areas such as fault trees.

- *Equal likelihood*: One definition of probability derives from the principle of equal likelihood. If a situation has n equally likely and mutually exclusive outcomes, and if n_A of these outcomes are event A, then the probability $P(A)$ of event A is:

$$P(A) = \frac{n_A}{n} \qquad (5.2)$$

This probability can be calculated *a priori* and without doing experiments.

The example usually given is the throw of an unbiased die, which has six equally likely outcomes: the probability of throwing a one is 1/6. Another example is the withdrawal of a ball from a bag containing four white balls and two red ones: the probability of withdrawing a red one is 1/3. The principle of equal likelihood applies to the second case also, because, although the likelihood of withdrawing a red ball and a white one is unequal, the likelihood of withdrawing any individual ball is equal.

- *Relative frequency*: The second definition of probability is based on the concept of relative frequency. If an experiment is performed n times and if the event A occurs on n_A of these occasions, then the probability $P(A)$ of event A is:

$$P(A) = \lim_{n \to \infty} \frac{n_A}{n} \qquad (5.3)$$

This probability can only be determined by experiment. This definition of reliability is the one that is most widely used in engineering. In particular, it is this definition that is implied in the estimation of probability from field failure data.

- *Probability of unions*: The probability of an event X which occurs if any of the events A_i occur and is thus the union of those events is:

$$P(X) = P\left(\bigcup_{i=1}^{n} A_i\right) \qquad (5.4)$$

If the events are mutually exclusive, Equation 5.4 simplifies to:

$$P(X) = \sum_{i=1}^{n} p(A_i) \qquad (5.5)$$

For events not mutually exclusive but of low probability the error in using Equation 5.4 instead of Equation 5.5 is small. Equation 5.5 is sometimes called the low probability, or rare event, approximation. The estimate of probability given by Equation 5.5 errs on the high side and hence is conservative in calculating failure probabilities, but is not conservative in calculating success probabilities or reliabilities.

- *Joint and marginal probability*: So far the events considered are the outcomes of a single experiment. Consideration is now given to events which are the outcome of several sub-experiments.

The probability of an event X which occurs only if all the n events A_i, occur and is thus the intersection of these events is:

$$P(X) = P(A_1, \ldots, A_n) = P\left(\bigcap_{i=1}^{n} A_i\right) \qquad (5.6)$$

$P(A_1, \ldots, A_n)$ is the joint probability of the event.

The probability of an event X which occurs if any of the mutually exclusive and exhaustive n events A_i in one sub-experiment occurs and the event B in a second sub-experiment occurs is:

$$P(X) = P\left(\bigcup_{i=1}^{n} A_i B\right) \qquad (5.7)$$

$$= \sum_{i=1}^{n} P(A_i)P(B) \qquad (5.8)$$

$$= P(B) \qquad (5.9)$$

$P(B)$ is the marginal probability of the event.

- *Conditional probability*: The probability of an event X which occurs if the event A occurs in one sub-experiment and the event B occurs in a second sub-experiment where the event A depends on the event B is:

$$P(X) = P(AB) = P(A|B)P(B) \qquad (5.10)$$

$P(A|B)$ is the conditional probability of A, given B.

The probability obtained in Equation 5.10 is a joint probability. Marginal probabilities may also be obtained from conditional probabilities:

$$P(X) = P(B) = \sum_{i=1}^{n} P(B|A_i) \qquad (5.11)$$

- *Independence and conditional independence*:
  If events A and B are independent:

$$P(AB) = P(A|B)P(B) = P(A)P(B) \qquad (5.12)$$

The probability of an event X, which occurs only if all n events A_i occur is given by Equation 5.6. If all n events are independent:

$$P(X) = P(A_1, \ldots, A_n) = \prod_{i=1}^{n} p(A_i) \qquad (5.13)$$

Two events A and B are conditionally dependent if their relationship with a third event C is:

$$P(AB|C) = P(A|C)P(B|C) \qquad (5.14)$$

Conditional independence does not imply independence.

— *Bayes' theorem*: The relationship given in Equation 5.10 is a form of Bayes' theorem. This theorem is extremely important in probability work. It appears in various forms, some of which are:

$$P(AB) = P(A|B)P(B) = P(B|A)P(A) \qquad (5.15)$$

$$P(A|B) = \frac{P(AB)}{P(B)} = \frac{P(B|A)P(A)}{P(B)} \qquad (5.16)$$

On rewriting the denominator as a marginal probability, and A as A_k.

$$P(A_k|B) = \frac{P(B|A_k)P(A_k)}{\sum_{i=1}^{n} P(B|A_i)P(A_i)} \qquad (5.17)$$

where $P(A_k|B)$ is the posterior probability, $P(B|A_k)$ is the likelihood, and $P(A_k)$ is the prior probability.

5.1.4 Reliability Relationships

The fundamental relationships for reliability analysis are as follows:

— *Reliability function and hazard rate*: If n equipments operate without replacement, then after time t the numbers which have survived and failed are $n_s(t)$ and $n_f(t)$, respectively, and the probability of survival, or reliability, $R(t)$ is:

$$R(t) = 1 - \frac{n_f(t)}{n} \qquad (5.18)$$

The instantaneous failure rate, or failure rate expressed as a function of the number of equipments surviving, $z(t)$ is:

$$z(t) = \frac{1}{n - n_f} \frac{dn_f(t)}{dt} = -\frac{1}{R(t)} \cdot \frac{dr(t)}{dt} = -\frac{d[\ln R(t)]}{dt} \qquad (5.19)$$

$z(t)$ is also called the 'hazard rate' or just the 'failure rate'.

The cumulative hazard function $H(t)$ is:

$$H(t) = \int_0^t z(t)\, dt \qquad (5.20)$$

Then, by integration of Equation 5.3 the reliability $R(t)$ is

$$R(t) = \exp\left(-\int_0^t z(t)dt\right) = \exp[-H(t)] \qquad (5.21)$$

$R(t)$ is also called the 'reliability function'.

— *Failure density and failure distribution functions*:
The overall failure rate, or failure rate expressed as a function of the original number of equipments, $f(t)$ is:

$$f(t) = \frac{1}{n} \frac{dn_f(t)}{dt} = -\frac{dR(t)}{dt} \qquad (5.22)$$

$f(t)$ is also called the 'failure density function'.

The complement of the reliability, or the unreliability, $Q(t)$ is:

$$Q(t) = 1 - R(t) \qquad (5.23)$$

$Q(t)$ is also called the 'failure distribution function' and is then commonly written as $F(t)$.

The failure density function and failure distribution function are often referred to, respectively, as the 'failure density' or the 'density function' and the 'failure distribution', the 'distribution function' or the 'cumulative distribution function'.

— *Relationships between basic functions*: The following relationships can readily be derived from Equations 5.18 and 5.23 and are particularly useful:

$$z(t) = \frac{f(t)}{R(t)} \qquad (5.24a)$$

$$= \frac{f(t)}{1 - Q(t)} = \frac{f(t)}{1 - F(t)} \qquad (5.24b)$$

$$R(t) = \int_t^{\infty} f(t)dt \qquad (5.25)$$

$$Q(t) = F(t) = \int_0^t f(t)dt \qquad (5.26)$$

— *Exponential distribution*: An important special case is that in which the hazard rate $z(t)$ is constant:

$$z(t) = \lambda \qquad (5.27)$$

Then,

$$R(t) = \exp(-\lambda t) \qquad (5.28)$$

$$f(t) = \lambda \exp(-\lambda t) \qquad (5.29)$$

$$Q(t) = 1 - \exp(-\lambda t) \qquad (5.30)$$

These four quantities are shown in Figure 5.1.

In Figure 5.1, the vertical axes for the reliability $R(t)$ and the unreliability $Q(t)$ have the range $0-1$, but those for the hazard rate $z(t)$ and the failure density function $f(t)$ are proportional to λ. At low values of λt:

$$R(t) = 1 - \lambda t \quad \lambda t \ll 1 \qquad (5.31)$$

$$Q(t) = \lambda t \quad \lambda t \ll 1 \qquad (5.32)$$

Equation 5.31 is useful in obtaining accuracy in numerical computation of high values of the reliability $R(t)$. Equation 5.32 is useful in making simple computations of the unreliability $Q(t)$.

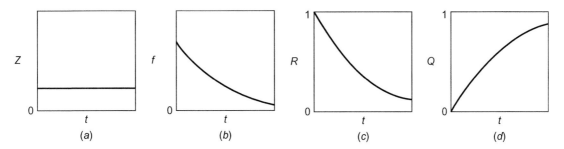

FIGURE 5.1 The exponential distribution: (a) hazard rate, (b) failure density, (c) reliability, (d) unreliability.

The assumption of constant hazard rate is that normally made in the absence of other information. This is therefore a special case of particular importance.

The failure distribution with constant hazard rate is called the 'exponential distribution' or, more accurately but less commonly, the 'negative exponential distribution'. It is also referred to as the 'random failure distribution', although in most cases it is a moot point whether the failures are appropriately called random.

— *Unreliability and failure rate*: There is sometimes confusion between unreliability $Q(t)$ and failure rate λ. This confusion arises because, given certain assumptions, the two are numerically identical. This occurs if in Equation 5.32 time t is equal to unity. Then,

$$Q(t) = \lambda \quad \lambda \ll 1 \qquad (5.33)$$

— *Bathtub curve*: In general, the failure behavior of equipment exhibits three stages: initially during commissioning the rate is high, then it declines during normal operation, and finally it rises again as deterioration sets in. For many kinds of equipment, particularly electronics, the rate has been found to form a bathtub curve, as shown in Figure 5.2(a) (Carhart, 1953). This curve has three regimes: (1) early failure, (2) constant failure, and (3) wearout failure.

Early failure, or infant mortality, is usually due to such factors as defective equipment, incorrect installation. It also tends to reflect the learning curve of the equipment user. Constant failure, or the so-called 'random failure', is often caused by random fluctuations of load which exceed the design strength of the equipment. A constant failure characteristic is also shown by a piece of equipment which has a number of components that individually exhibit different failure distributions. Wear-out failure is self-explanatory. The corresponding curve for the failure density function is shown in Figure 5.2(b). The bathtub curve is widely quoted in the reliability literature, but it should be emphasized that its

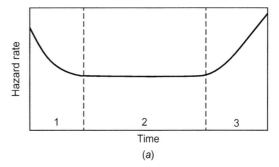

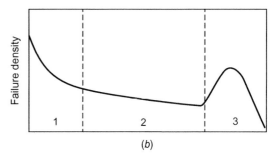

FIGURE 5.2 The bathtub curve: (a) hazard rate and (b) failure density.

applicability to all types of equipment, particularly mechanical equipment, is not established.

— *Mean life*: The mean life m is defined as the first moment of the failure density function

$$m = \int_0^\infty tf \, \mathrm{d}t \qquad (5.34)$$

Alternatively, from Equations 5.22 and 5.34

$$m = -\int_0^\infty t \frac{\mathrm{d}R}{\mathrm{d}t} \, \mathrm{d}t \qquad (5.35a)$$

$$= -[tR]_0^\infty + \int_0^\infty R \, \mathrm{d}t \qquad (5.35b)$$

Since

$$[tR]_0^\infty = 0$$

provided that

$$\lim_{t \to \infty} (tR) = 0$$

then

$$m = \int_0^\infty R \, dt \qquad (5.36)$$

Alternatively, use may be made of the relationships between the moments μ_i and the Laplace transform $\bar{g}(s)$ of a function $g(t)$:

$$\mathcal{L}[g(t)] = \bar{g}(s) = \int_0^\infty g(t)\exp(-st)dt \qquad (5.37)$$

Hence:

$$\frac{d^i \bar{g}(s)}{ds^i} = (-1)^i \int_0^\infty t^i g(t)\exp(-st)dt \qquad (5.38)$$

and

$$\left(\frac{d^i \bar{g}(s)}{ds^i} \right)_{s=0} = (-1)^i \int_0^\infty t^i g(t)dt \qquad (5.39a)$$

$$= (-1)^i \mu_i \qquad (5.39b)$$

Applying these relationships to the present case by substituting dR/dt for $g(t)$ in Equation 5.39a,

$$m = \left(\frac{d[sR - R(0)]}{ds} \right)_{s=0} = (R)_{S=0} \qquad (5.40)$$

For the exponential distribution

$$m = \int_0^\infty t\lambda \exp(-\lambda t)dt \qquad (5.41a)$$

$$= \frac{1}{\lambda} \qquad (5.41b)$$

Other terms used in addition to 'mean life' are the Mean Time Between Failures (MTBF), the Mean Time To Failure (MTTF), and the Mean Time To First Failure (MTTFF). These times are sometimes used interchangeably, but they are not identical.

The most widely used is probably MTBF. MTBF has meaning only when applied to a population of components, equipments, or systems in which there is repair. It is the total operating time of the items divided by the total number of failures. It is also the mean of the failure distribution, regardless of its form.

MTTF is applied to items without repair and is the mean of the distribution of times to failure. MTTFF is applied to items with repair and is the mean of the distribution of times to first failure.

MTTF and MTTFF are applied particularly to systems. For an n-component parallel system with an exponential failure distribution of the individual components and without repair,

$$\text{MTTF} = \sum_{i=1}^n \frac{1}{i\lambda} \qquad (5.42)$$

and for an n-component parallel system with repair

$$\text{MTTF} = \frac{1}{\lambda} \cdot \sum_{i=0}^{n-1} \frac{(1+\mu/\lambda)^i}{i+1} \qquad (5.43)$$

where λ is the failure rate of equipment, and μ is the repair rate of equipment. If the repair rate μ is zero, Equation 5.43 reduces to Equation 5.42.

- _Expected value_: Use is frequently made in reliability engineering of the concept of the expected value. The expected value of a distribution is its mean value. The use of the expected value concept allows convenient manipulation, as the following treatment demonstrates. For a variable t, the expected value, or mean value, is:

$$m = E[t] \qquad (5.44)$$

The variance is:

$$\sigma^2 = E[(t-m)^2] = E[t^2] - E[t]^2 \qquad (5.45)$$

5.1.5 Failure Distributions

There are several statistical distributions which are fundamental in work on reliability. Important discrete distributions are (1) binomial distribution, (2) multinomial distribution, and (3) Poisson distribution. On the other hand, several important continuous distributions are (1) exponential distribution, (2) normal distribution, (3) log−normal distribution, (4) Weibull distribution, (4) rectangular distribution, (5) gamma distribution, (6) Pareto distribution, and (7) extreme value distribution.

Some properties of these distributions are given in Table 5.1.

5.1.6 Reliability of Some Standard Systems

The reliability of some standard systems is now considered. It is assumed that the exponential failure distribution applies, unless otherwise stated.

- _Series systems_: A series system is one which operates only if all its components operate. It is not implied that the components are necessarily laid out physically in series configuration.

For a series system, the reliability R is the product of the reliabilities R_i of the components:

$$R = \prod_{i=1}^n R_i \qquad (5.46)$$

This follows directly from Equation 5.13.

TABLE 5.1 Some Properties of Failure Distributions

Distribution	Failure density function $f(t)$	Mean $m = \int_{-\infty}^{\infty} tf(t)dt$	Median $= t$ where $\int_{-\infty}^{t} f(r)dr = 0.5$	Mode $= t$ where $\dfrac{df}{dt} = 0$	Variance $\sigma^2 = \int_{-\infty}^{\infty} (t-m)^2 f(t)dt$
Exponential	$\lambda\exp(-\lambda t)$	$\dfrac{1}{\lambda}$	$\dfrac{1}{\lambda}\ln 2$	0	$\dfrac{1}{\lambda^2}$
Normal	$\dfrac{1}{\sigma(2\pi)^{1/2}}\exp\left[-\dfrac{(t-m^2)}{2\sigma^2}\right]$	M	m	m	σ^2
Log-normal	$\dfrac{1}{\sigma t(2\pi)^{1/2}}\exp\left[-\dfrac{[\ln(t)-m^*]^2}{2\sigma^2}\right]$	$\exp\left(m^* + \dfrac{\sigma^2}{2}\right)$	$\exp(m^*)$	$\exp(m^* - \sigma^2)$	$\exp(2m^* + \sigma^2)[\exp(\sigma^2) - 1]$
Weibull (two-parameter)	$\dfrac{\beta}{\eta}\left(\dfrac{t}{\eta}\right)^{\beta-1}\exp\left[-\left(\dfrac{t}{\eta}\right)^{\beta}\right]$	$\eta\Gamma\left(1+\dfrac{1}{\beta}\right)$	$\eta(\ln 2)^{1/\beta}$	$\eta\left(1-\dfrac{1}{\beta}\right)^{1/\beta}$, $\quad \beta > 1$ $0 \qquad\qquad\quad \beta \le 1$	$\eta^2\left\{\Gamma\left(1+\dfrac{2}{\beta}\right) - \left[\Gamma\left(1+\dfrac{1}{\beta}\right)\right]^2\right\}$
Rectangular	$\dfrac{1}{b}$	$a + \dfrac{b}{2}$	$a + \dfrac{b}{2}$	—	$\dfrac{b^2}{12}$
Gamma	$\dfrac{1}{b\Gamma(a)}\left(\dfrac{t}{b}\right)^{a-1}\exp\left(-\dfrac{t}{b}\right)$	B^a	$-a$	$b(a-1)$, $\quad a \ge 1$	$b^2 a$
Pareto	$at^{-(a+1)}$	$\dfrac{a}{a-1}$, $\quad a > 1$	$2^{1/a}$	0	$\dfrac{a}{a-2} - \left(\dfrac{a}{a-1}\right)^2$, $\quad a > 2$
Extreme value	$\dfrac{1}{b}\exp\left(\dfrac{t-a}{b}\right)\exp\left[-\exp\left(\dfrac{t-a}{b}\right)\right]$	$a + b\Gamma'(1)^b$	$a + b\ln 2$	a	$\dfrac{b^2\pi^2}{6}$

[a]No simple expression is available.
[b]$\Gamma'(1) = -0.57721$ is the first derivative of the gamma function.

For the exponential distribution, if the failure rates of the components are constants λ_i,

$$R_i = \exp(-\lambda_i) \tag{5.47}$$

and hence

$$R = \prod_{i=1}^{n} \exp(-\lambda_i t) = \exp\left(-\sum_{i=1}^{n} \lambda_i t\right) \tag{5.48}$$

If the overall failure rate of the system is a constant λ,

$$R = \exp(-\lambda t) \tag{5.49}$$

and hence

$$\lambda = \sum_{i=1}^{n} \lambda_i \tag{5.50}$$

– *Parallel systems*: A parallel system is one which fails to operate only if all its components fail to operate. Again it is not implied that the components are necessarily laid out physically in a parallel configuration. For a parallel system, the unreliability Q is the product of the unreliabilities Q_i of the components:

$$Q = \prod_{i=1}^{n} Q_i \tag{5.51}$$

Again this follows directly from Equation 5.13. The reliability R of the system is:

$$R = 1 - \prod_{i=1}^{n}(1 - R_i) \tag{5.52}$$

Since parallel configurations incorporate redundancy, they are also referred to as 'parallel redundant systems'. For the exponential distribution, Equation 5.47 applies, and hence:

$$R = 1 - \prod_{i=1}^{n}[1 - \exp(-\lambda_i t)] \tag{5.53}$$

If the overall failure rate of the system is a constant λ, Equation 5.49 is applicable. In this case, there is no simple general relationship between the system and the component failure rates. But, if the component failure rates are all the same,

$$\frac{1}{\lambda} = \sum_{i=1}^{n} \frac{1}{i\lambda_i} \tag{5.54}$$

– *r-out-of-n parallel systems*: For a system which has n components in parallel and operates as long as r components survive, the binomial distribution is applicable. Thus, for an r-out-of-n system the reliability R is obtained from the component reliability R_x:

$$R(r \leq k \leq n) = \sum_{k=r}^{n} \binom{n}{k} R_x^k Q_x^{n-k} \tag{5.55}$$

As an illustration, consider the reliability of a 2-out-of-4 system. Expanding

$$(R_x + Q_x)^n = 1 \tag{5.56}$$

for the terms between n and r to get the individual terms of Equation 5.55 gives

Probability of 0 failure $= R_x^4$

Probability of 1 failure $= 4R_x^3 Q_x$

Probability of 2 failures $= 6R_x^2 Q_x^2$

The coefficients of the terms may be obtained from Pascal's triangle. The system survives provided no more than two failures occur, and hence

$$R = R_x^4 + 4R_x^3 Q_x + 6R_x^2 Q_x^2 \tag{5.57}$$

An r-out-of-n system reduces in the limiting cases to a series or a parallel system. If all the components must operate for the system to survive, it becomes a series system

$$R = R_x^n \tag{5.58}$$

while if it is sufficient for one component to operate for the system to survive, it becomes a parallel system

$$Q = Q_x^n \tag{5.59}$$

– *Standby systems*: The simplest standby system is one in which there is one component operating and one or more on standby, all components have the same failure rate in the operational mode, the standby components have zero failure rate in the standby mode, and there is perfect switchover. For this case, the Poisson distribution can be used. Thus, for such a standby system, the failure rate is λ, the number of failures which the system can withstand is r, and the system reliability R is

$$R(0 \leq k \leq r) = \exp(-\lambda t) \sum_{k=0}^{r} \frac{(\lambda t)^k}{k!} \tag{5.60}$$

As an illustration, consider the reliability of a standby system with one component operating and one on standby, so that the system can survive one failure. Then

$$R = \exp(-\lambda t)(1 + \lambda t) \tag{5.61}$$

It is relatively easy in this case to take into account imperfect switchover. If the reliability of switchover is R_{SW}, the system reliability R is:

$$R = \exp(-\lambda t)(1 + R_{SW}\lambda t) \qquad (5.62)$$

Standby systems may be more complex than this simple case in a number of ways. The standby component may be different from the component normally operating and may have a different failure rate in the operational mode. It may also have a finite failure rate in the standby mode, and there may be imperfect switchover.

— *Systems with repair*: If it is possible to carry out repair on a system, a much higher reliability can be achieved. The repair time which is used here is the total time from initial failure to final repair, and therefore includes any time required for detection of the failure and organization of the repair. Obviously, the determination of the reliability of systems with repair requires data on repair times as well as failure rates.

Repair times, like failure times, may fit various distributions. It is sometimes assumed that the repair time is a constant τ, but a more common assumption is that the repair times τ_r have an exponential distribution with constant repair rate μ:

$$f_r(t_r) = \mu \exp(-\mu t_r) \qquad (5.63)$$

For the exponential distribution the mean repair time m_r is

$$m_r = \frac{1}{\mu} \qquad (5.64)$$

The reliability of some standard systems with repair is now considered. It is assumed that the exponential repair time distribution applies, unless otherwise stated.

— *Parallel systems with repair*: For a system which has n components in parallel, and which survives provided that at least one component operates, and for which the failure and repair rates are λ and μ, respectively, expressions for the reliability R can be derived using Markov models. For a parallel system with two components, it can be shown that the reliability R is:

$$R = \frac{1}{r_1 - r_2}[(3\lambda + \mu + r_1)\exp(r_1 t) \\ - (3\lambda + \mu + r_2)\ \exp(r_2 t)] \qquad (5.65)$$

with

$$r_1, r_2 = \frac{-(3\lambda + \mu) \pm [(3\lambda + \mu)^2 - 8\lambda^2]^{1/2}}{2} \qquad (5.66)$$

— *Standby systems with repair*: For a standby system with two components in which there is one component operating and one is on standby, both components have the same failure rate in the operational

mode, the standby component has zero failure rate in the standby mode, the switchover is perfect, and there is repair, it can be shown using Markov models that the reliability R is:

$$R = \frac{1}{r_1 - r_2}[(2\lambda + \mu + r_1)\exp(r_1 t)] \\ - (2\lambda + \mu + r_2)\exp(r_2 t)] \qquad (5.67)$$

with

$$r_1, r_2 = \frac{-(2\lambda + \mu) \pm [(2\lambda + \mu)^2 - 4\lambda^2]^{1/2}}{2} \qquad (5.68)$$

— *Constant repair time*: As just described, the usual assumption for repair times is a constant repair rate μ, which implies a distribution of repair times. An alternative assumption is a single constant repair time τ.

For a parallel system of n components with repair, the probability of the system being in the 'as-new' condition and then suffering an initial failure a given number of times, is given by the Poisson distribution. The probability of the system failing completely after an initial failure and before repair to the as-new condition is:

$$Q(\tau) = [1 - \exp(-\lambda t)]^{n-1} \quad \lambda t \ll 1 \qquad (5.69)$$

Then the probability of the system suffering one initial failure followed by system failure is:

$$P(t) = \exp(-n\lambda t)n\lambda t Q(\tau) \qquad (5.70)$$

The probabilities of the system suffering high numbers of initial failures followed by system failure may be obtained in a similar manner. Hence, it can be shown that the reliability of the system is:

$$R = \exp(-\lambda' t) \qquad (5.71)$$

with

$$\lambda' = n\lambda Q(\tau) \quad \lambda t \ll 1 \qquad (5.72)$$

5.1.7 Reliability of Complex Systems

— *System reliability analysis*: The reliability analysis of systems is carried out first to obtain an assessment of system reliability and to identify critical features, and then to achieve necessary improvements. A full analysis is likely to require considerable effort. It is important, therefore, to identify those subsystems which particularly affect the overall system reliability. In other words, the direction of effort is assisted by a sensitivity analysis. If the reliability of a particular subsystem does not greatly affect that of the system, it is not necessary to analyze it in great detail.

Reliability calculations require data not only on failure rates but also on repair times. The sensitivity of the system reliability determines the accuracy required in the data in a given case. In some instances it is necessary to obtain a rather accurate estimate from field data. In others an engineering estimate made without any field data may be quite sufficient. The matching of the data to the application is an important element in the art of the reliability engineer.

There are a number of methods which can be used to analyze more complex systems and to decompose them into their subsystems. These techniques may be illustrated by reference to Figure 5.3. Figure 5.3(a) illustrates a system consisting of two intermediate storage tanks and three pumps. Figure 5.3(b) shows the three pumps all in parallel, with complete interchangeability between them. Figure 5.3(c) shows two separate streams, with one tank and one pump in each, and with the third pump available to either stream. The system is defined as successful provided flow is maintained by pumping with at least one pump from one tank.

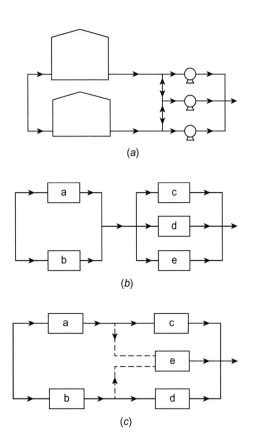

FIGURE 5.3 A system to illustrate methods of analyzing complex systems: (a) storage tank pump system, (b) system with three pumps in parallel, (c) system with separate streams, but with a common spare pump.

- *Reliability graphs*: The structure of a system may be represented in terms of simple reliability graphs. Each branch of the graph represents a component. Some examples are shown in Figure 5.4. Figure 5.4(a) shows the reliability graph for three components in series, and Figure 5.4(b) shows that for three components in parallel. Figure 5.4(c) shows the reliability graph for two components in parallel, in series with three components in parallel, which corresponds to the storage tank pump system shown in Figure 5.3(a), assuming the configuration shown in Figure 5.3(b). The system reliability may then be determined by computing the reliability of the subsystems and then that of the system itself.

- *Logic flow diagrams*: The most widely used method of graphical representation is the logic flow diagram, also called the 'logic sequence diagram' or simply the 'logic diagram'. Figure 5.5 shows a logic flow diagram for the storage tank pump system with the configuration shown in Figure 5.3(b). In this case the top event considered is a success, but in other cases it may be a failure. Some typical applications of logic flow diagrams are given in Table 5.2. Logic flow diagrams can be used for systems of considerable complexity.

- *Event space method*: The event space method involves making a comprehensive list of all the possible states of the system and determining which of these events correspond to success. The events E_i, so listed are mutually exclusive and system reliability R is:

$$R = \sum_{i=1}^{n} P(E_i) \qquad (5.73)$$

- *Tree diagrams*: Another graphical representation is the tree diagram. Such diagrams can be used for simple cases, but rapidly become rather cumbersome as systems become more complex.

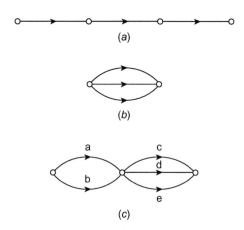

FIGURE 5.4 Reliability graphs: (a) for a three-component series system, (b) for a three-component parallel system, (c) for a storage tank pump system (configuration as in Figure 5.3(b)).

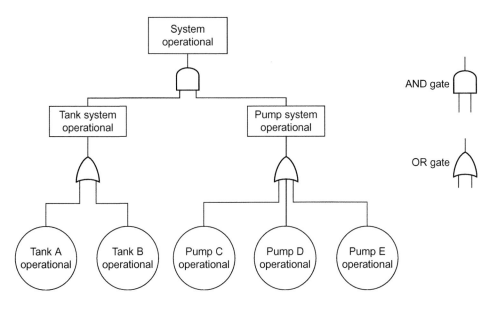

FIGURE 5.5 Logic flow diagram for storage tank pump system (configuration as in Figure 5.3(b)).

TABLE 5.2 Some Typical Applications of Logic Flow Diagrams

Event Investigated	Type of Diagram
Hazardous occurrences	Failure diagram (fault tree)
Trip system operation	Success diagram
Plant availability	Success diagram

— *Path tracing and tie sets*: The path tracing method consists of tracing through the system the paths which constitute success. These paths are known in graph theory as 'tie sets'. System reliability is obtained from the minimum tie sets T_i. These sets of events are not mutually exclusive and the system reliability is given by:

$$R = P\left(\bigcup_{i=1}^{n} T_i \right) \qquad (5.74)$$

— *Path breaks and cut sets*: An alternative approach is to determine the sets of events which break all the paths and thus ensure failure. These sets are known in graph theory as 'cut sets'. System unreliability is obtained from the minimum cut sets C_i. Again these sets of events are not mutually exclusive and the system reliability is given by:

$$R = 1 - \left(\bigcup_{i=1}^{n} C_i \right) \qquad (5.75)$$

— *System decomposition*: It is often possible to simplify the analysis of a system by selecting a

key component which makes it possible to decompose the system into subsystems. In such a case, one approach is the use of Bayes' theorem. Here it is sufficient to note that the theorem may be written in the form:

$$P(A) = P(A|B)P(B) + P(A|\overline{B})P(\overline{B}) \qquad (5.76)$$

Thus, the reliability $R(S)$ of the system is related to the reliability $R(X)$ of the key component as follows:

$$R(S) = R(S|X)R(X) + R(S|\overline{X})Q(\overline{X}) \qquad (5.77)$$

For a storage tank pump system having the configuration shown in Figure 5.3(c) the system unreliability Q is:

$$Q = Q_a Q_b R_c + Q_{ac} Q_{bd} Q_c \qquad (5.78)$$

with

$$R_{ac} = R_a R_c \qquad (5.79a)$$

$$R_{bd} = R_b R_d \qquad (5.79b)$$

5.1.8 Availability

— *System availability analysis*: The availability analysis of systems, like reliability analysis, has as its objectives: (1) to assess system availability and to identify critical aspects and (2) to effect the required improvements.

It is usually necessary to consider a number of availability characteristics. For process plants these are likely to include: (1) the probability of obtaining nominal output and down-time, (2) the probability of obtaining other outputs and down-times, and (3) the probability of infrequent but very long down-times. Long outages are particularly important. It is usually appropriate to deal separately with these. Thus, for example, the frequency of a long outage of 1 year might be about 10^{-4}/year. It is not very meaningful to express this as an additional loss of mean availability of 0.04 days/year.

A full analysis may well involve much effort. Therefore, it is important to identify those subsystems which have a particularly strong influence on the overall system availability. Those subsystems which do not greatly affect system availability can be treated in less detail.

There are two main types of availability data which may be used. Sometimes data on the availability of whole subsystems can be obtained. In other cases it is necessary to calculate system availability from failure rate and repair time data. In either case it is important that the accuracy of the data be matched to the sensitivity of the system availability to errors in the data.

As far as possible data on the availability of a process unit should describe the performance of that unit independent of other units. Downtime of the unit due to unavailability of other units or to other extraneous causes should not be included.

Particular attention should be paid to the availability of process raw materials and services. It is also often necessary to consider the availability of other plants which receive from the plant under consideration outputs such as by-products, electricity, or steam.

There are a number of methods available for the analysis of more complex systems, and some of these are described below.

— *The availability function*: In general, availability $A(t)$ is a function of time. It may be expressed in terms of the up-time $u(t)$ and down-time $d(t)$:

$$A(t) = \frac{u(t)}{u(t) + d(t)} \qquad (5.80)$$

The unavailability $V(t)$ is

$$V(t) = 1 - A(t) \qquad (5.81)$$

$$= \frac{d(t)}{u(t) + d(t)} \qquad (5.82)$$

The point availability $A(t)$ can be integrated over the time period T to give the mean availability over that period:

$$A(T) = \frac{1}{T} \int_0^T A(t) dt \qquad (5.83)$$

Often, however, it is the long-term, or steady-state, availability $A(\infty)$ which is of most interest:

$$A(\infty) = \frac{u(\infty)}{u(\infty) + d(\infty)} \qquad (5.84)$$

The long-term availability may also be expressed in terms of the mean up-time m_u and the mean down-time m_d:

$$A(\infty) = \frac{m_u}{m_u + m_d} \qquad (5.85)$$

For a single equipment or system with exponential failure and repair distributions, failure rate λ and repair rate μ:

$$m_u = \frac{1}{\lambda} \qquad (5.86)$$

$$m_d = \frac{1}{\mu} \qquad (5.87)$$

$$= m_r \qquad (5.88)$$

Then, from Equations 5.85−5.88:

$$A(\infty) = \frac{\mu}{\lambda + \mu} \qquad (5.89a)$$

$$= \frac{1}{1 + \lambda m_r} \qquad (5.89b)$$

$$= 1 - \lambda m_r \quad \lambda m_r \ll 1 \qquad (5.89c)$$

The long-term unavailability is:

$$V(\infty) = \frac{\lambda}{\lambda + \mu} \qquad (5.90a)$$

$$= \frac{1}{1 + \lambda m_r} \qquad (5.90b)$$

$$= \lambda m_r \quad \lambda m_r \ll 1 \qquad (5.90c)$$

— *The repair time density function*: It is appropriate at this point to consider the distribution of repair times. The mean repair time m_r is defined as the first moment of the repair time density function $f_r(t_r)$:

$$m_r = \int_0^\infty t_r f_r(t_r) dt_r \qquad (5.91)$$

So far, the repair time distribution which has been considered is the exponential. For this distribution the repair time density function $f_r(t_r)$ is

$$f_r(t_r) = \mu \exp(-\mu t_r) \qquad (5.92)$$

The repair time distribution function $F_r(t_r)$ is:

$$F_r(t_r) = \int_0^{t_r} f_r(t_r) dt_r \qquad (5.93a)$$

$$= 1 - \exp(-\mu t_r) \qquad (5.93b)$$

The mean repair time is:

$$m_r = \int_0^\infty t_r \mu \exp(-\mu t_r) dt_r \qquad (5.94a)$$

$$= \frac{1}{\mu} \qquad (5.94b)$$

For the normal distribution:

$$f_r(t_r) = \frac{1}{\sigma_r(2\pi)^{1/2}} \exp\left(-\frac{(t_r - m_r)^2}{2\sigma_r^2}\right) \qquad (5.95)$$

where the mean repair time is m_r.

For the log−normal distribution:

$$f_r(t_r) = \frac{1}{\sigma_r t_r (2\pi)^{1/2}} \exp\left(-\frac{[\ln(t_r) - m_r^*]^2}{2\sigma_r^2}\right) \qquad (5.96)$$

Repair time data often fit a log−normal distribution.

— *Logic flow diagrams*: Logic flow diagrams are often used to represent system availability. Thus, the logic flow diagram shown in Figure 5.5 for the storage tank pump system with the configuration shown in Figure 5.3(b) is in fact an availability diagram. In order to use the logic flow diagram method it is necessary to represent the plant flow diagram in logical form. It is necessary, therefore, to convert the plant flow diagram into a network of relatively simple blocks such as series or parallel units.

The plant availability determined using a logic flow diagram may be the mean availability or the availability density function. Usually the calculation is limited to the former. There are two main methods of determining the mean availability of the plant. In many cases the availability data are the mean availabilities of the units which constitute the plant. Typical units might be a tank pump system, a complete distillation column or a set of centrifuges. The mean unavailability of a unit is effectively a fractional dead time or probability of being unavailable. Then the mean availability of the plant may be calculated from the mean unavailabilities of these units.

In other cases, the availability data obtainable are the failure rates and repair times of the equipments which constitute the units. Typical equipment is a pump, a heat exchanger, or a drier. The mean unavailability, or fractional dead time, of a piece of equipment may be calculated from the failure rate and repair time data using Equation 5.54. The fractional dead time of the unit may be calculated from those of the constituent equipment. Then the mean unavailability of the plant may again be calculated from the unavailabilities of the units. The mean repair time of the plant may be determined by calculating the failure rate and the mean unavailability.

System availabilities can also be obtained by other methods such as throughput capability method, flow sheeting, and simulations.

5.1.9 Failure and Event Data and Sources

Many of the methods described in this chapter require the use of data for failure and other events. This section redirects to a selection of references, which have been published in the open literature.

The data are given in summary form and primarily for illustrative purposes only. It is emphasized that there are many factors that determine the failure rates of equipment and the range of failure rates observed can be quite wide. For industrial reliability work, therefore, it is necessary to consult the original literature and to make appropriate use of data banks and plant data.

5.1.9.1 Type of Data

There are several types of data that may be required. For equipment, these data include: (1) failure frequency, probability; (2) repair time; (3) unavailability; and for events: (4) event frequency.

Failure rate data are usually given as frequencies or mean time to failure (MTTF). Two types of failure rate data are typically used: time-related and demand-related. Time-related failure rates are established for equipment that is continuously in use while demand-related failure rates are for equipment that is normally dormant and required to operate on demand.

5.1.9.2 Definition and Regimes of Failure

The failure rate recorded for equipment depends on the definition of failure. An equipment failure can be catastrophic, degraded, or incipient. Bathtub and Weibull distribution have been used to represent a wide range of failure data. The hazard rate may be decreasing, constant, or increasing, corresponding to the regimes of early failure, constant failure, or wearout failure and characterized by values of the Weibull shape parameter, as mentioned in Section 5.1.9.

Equipment performance can be evaluated by observing and analyzing failure data over a period of time. The measure of performance can be in the form of reliability or availability. Reliability $R(t)$ is defined as the probability that a replaceable component or hazard barrier will function at time t without failure. The exponential reliability distribution function also known as the survivor function is given by the following equation:

$$R(t) = 1 - F(t) \qquad (5.97)$$

where $R(t)$ is reliability, λ is the rate of occurrence of failure (ROCOF), and t is the working time period of the component. Failure distribution function:

$$F(t) = 1 - e^{-\lambda t} \qquad (5.98)$$

For active components and system that undergo maintenance and experience degradation (such as pump, valves, switches), availability is used as a measure of performance because it takes into account test and maintenance downtime. Per definition, availability is the probability that a repairable item will function at specified time when called upon demand. It can be expressed as

$$A(t) = \frac{u(t)}{u(t) + d(t)} \qquad (5.99)$$

where $A(t)$ is availability, $u(t)$ is up-time, and $d(t)$ is down-time of the component.

- *Reliability in terms of distribution*: The reliability function is usually described by probability distributions. Among them are the exponential distribution, the Weibull distribution, the Binomial distribution, and other. An extensive account of the failure distributions is given in Sections 5.1.3 through 5.1.5.
- *Frequent testing of a component*: As the time between test increases, the probability of protective system failure approaches 1. If the time between the test decreases, $F(t) \sim \lambda t$ and the $F(t^*)_{test} = 0$. The mean time to failure (MTTF) is used to express the expected time during which a non-repairable component will perform its function successfully.

$$\text{MTTF} = \mu = \int_0^\infty e^{-\lambda t}\, dt = \frac{1}{\lambda} \qquad (5.100)$$

On the other hand, mean time between failures (MTBF) is associated with repairable component.

5.1.9.3 Collection of Data

Collection of data is receiving more and more attention from engineers. Failure rate data are characterized by complex interactions of components, specifications, maintenance procedures, operating environments, etc. Each company should examine the sources of data carefully and take full advantage of the available information, internal and external.

The most desirable information for reliability and risk analysis is to have sufficient plant-specific data. In general, failure of plant equipment needs to be recorded and investigated, both in order to identify types of failure so that failure rates can be reduced by better engineering and to obtain failure data for reliability calculations.

It is common practice for production management to record the downtime of the plant as well as the cause of the downtime. Frequently the cause assigned is the failure of a particular piece of equipment. This type of data collection may yield useful data, particularly on plant's availability data. Likewise, it is common practice for maintenance management to record failures of plant equipment. Useful data on failure rates may be generated through this data gathering.

If data collection is to be instituted for reliability work, it is necessary to define the system carefully. This includes (1) data requirement, (2) data capture, (3) data classification, (4) data retrieval, and (5) data utilization. The system should be appropriate in scale and duration. In some cases the aim is to monitor continuously features such as equipment failure and/or unit availability. In this case continuous collection is clearly necessary. On the other hand, if the aim is simply to obtain data for reliability engineering or hazard assessment purposes, it may be sufficient to collect certain data over a limited period on a 'campaign' basis. Further discussions of data collection are given by Smith and Babb (1973), Wingender (1991) and Davidson (1994).

5.1.9.4 Sources of Data

Failure data may be obtained from external sources such as the literature and data banks or internal sources, where the data is collected as part of routine maintenance or inspection. The sources available depend on the user. A company operating plant has access to data from its own works that are not available to other parties, unless supplied to an accessible database. The possible sources may include warranty claims, previous experience with similar or identical equipment, repair facility records, factory acceptance testing, records generated during the development phase, customer reports on equipment failure, test, and inspection records generated by quality control/manufacturing group.

The ideal situation is to have sufficient in-house data from identical equipment in the same application. On the other hand, internal collection may not be appropriate, or for new plant designs there is apparently no in-house failure data. Generic data are often used in reliability work, especially at the design stage. However, there is available in the literature a number of collections of data, and knowledge of the background and origin of the data is necessary to choose the appropriate generic failure rate data.

5.1.9.5 Status, Processing, and Uncertainty of Data

In hazard assessment, it is a good practice to indicate the status of the data. Some relevant distinctions are (1) value based on historical data including value based on large number of events (narrow confidence limits), value based on small number of events (wide confidence limits), and value based on number of event-free years or occasions; (2) value based on judgment of a number of experts; (3) value synthesized using fault tree methods; (4) value based partly on data and partly on judgment of analyst; and (5) value entirely on judgment of analyst.

Using failure rate data requires thorough understanding and judicious judgment. The quality of the data from different sources will vary. For a particular reliability study, it is essential to consult the original data sources and identify appropriate data sources to reach a better estimation of the equipment failure rates. Reliability data can deviate by a factor of three or four, and a factor of 10 is not unusual, as suggested by Trevor Kletz (1999). However, not much information exists in the open literature regarding the uncertainty of data. Since the uncertainty is quite high, it is inappropriate to report failure rate data or calculations using failure rate data to more than two significant figures.

5.1.9.6 Databases

A database is a collection of data managed through computer programs for specified applications while a data bank is a collection of data in a specific area of knowledge, such as a handbook. The two main kinds of databases are incident databases and reliability databases. An incident database does not have an inventory of items at risk but concentrates on the attributes and development of the incidents. The information available is often limited to whatever was recorded at the time. A variety of databases have been established to store and share this type of information. Among them are FACTS, WOAD (Worldwide Offshore Accident Database), MARS (the European Union Major Accident Reporting System). A reliability database may well record incidents but treats them primarily as events from which statistical information on reliability, availability, and maintainability are to be derived. Reliability data can be overall failure rates or values separated into failure modes and causes.

High-quality reliability data is expensive to collect and often require a long time or a large sample to be statistically viable. It is essential to define clearly the data to be held, the means by which they will be acquired, and the uses to which they will be put. There are a variety of reliability databases, such as the reliability data bank operated by AEA Technology, ERDS, and OREDA.

CCPS (1989) provides a review of the available databases.

Six of the principal databases are (1) the NCSR Reliability Data Bank is a major database system which has been in operation for over 20 years serving the nuclear, aerospace, electronics, oil, chemical, and other industries; (2) the European Reliability Data System (ERDS) is an EC database operated by the JRC at Ispra; (3) the Failure and Accident Technical Information System (FACTS) is an incident data bank operated by TNO in the Netherlands; (4) the Offshore Reliability Data (OREDA) project is an offshore reliability database described in the handbook by OREDA (2009). It gives failure rates, failure mode distribution, and repair data for offshore installations, mainly in the United Kingdom and Norwegian sectors of the North Sea; (5) the EIReDA (European Industry Reliability Data Bank) is operated by ESReDA, which was established in 1992 by EuReDATA (European Reliability Data Bank Association) and ESReDA (European Safety and Reliability Research and Development Association) jointly; and (6) GIDEP (Government Industry Data Exchange Program) was established in 1959 as the Inter-service Data Exchange Program (IDEP) by the US government to reduce expenditures of resources in the life cycle of systems facilities and equipment through sharing of data among government agencies and industry organizations. The Hazardous Substances Emergency Events Surveillance (HSEES) system was maintained by the Agency for Toxic Substances and Disease Registry (ATSDR) of the Centers for Disease Control (CDC). HSEES stopped in 2010 due to funding-related issue and was replaced by another chemical incident database called the National Toxic Substances Incident Program (NTSIP) in 2010.

5.1.9.6.1 Design of Database

As to its definition, a database is first of all a computer system, which may have billions of data collections. The construction of a database involves the translation of the logical data model, giving the relationships between the data field into the physical database embodied in the computer. This is ensured by the database management system (DBMS). Many software companies have developed very effective database management programs that can be tailored for any specific needs. Four main architecture models of the database management system are the hierarchical, network, relational, and object models.

Information about components of the database is the component inventory and the component history. The component inventory contains a description of each component on which information is held. The component history gives the detailed failure and maintenance history of each component. For inventory it is first necessary to define the component boundary. It is then necessary to specify the attributes of the components. For component history, information is typically recorded on failure mode and failure cause. It is also desirable to have a plain language fault descriptor. In general, it is desirable to store the data in their "original" form rather than in an abstracted form, which involves loss of information. For this reason, the data in a reliability data bank may be stored as a set of events from which statistical data may then be obtained.

Table 5.3 is the cross-reference between principal data sources prepared by the Data Acquisition Working Party of the Mechanical Reliability Committee of the Institution of Mechanical Engineering (Davidson, 1994), and Table 5.4 is the equipment category associated with Table 5.3.

A selection of data on equipment failure rates obtained by the UKAEA and given by Green and Bourne are given in Table 5.5. These data derive from the work of the UKAEA initially in the nuclear field but subsequently in non-nuclear applications also.

The rest of the databases present information more or less similar of that in Table 5.5, with aid from Tables 5.3 and 5.4 is possible to find which databases can be of help regarding a particular equipment.

TABLE 5.3 Summary of Principal Reliability Data Sources

Reference	Application	Equipment					
		Rotation	Static	Instrumentation	Safety	Process	Electrical
CCPS (1992)	I	Y	Y	Y	Y	Y	Y
Davenport and Warwick	N, I	–	Y	–	–	–	–
DEF STAN 00-41 pt 3 Issue 1	M	Y	Y	Y	Y	Y	Y
Dela Mare	O	–	Y	–	–	–	–
Dexter and Perkins DP-1633	N, I	Y	Y	Y	Y	Y	Y
EIReDA	N, I	Y	Y	Y	Y	Y	Y
ENI	I	Y	Y	Y	Y	Y	Y
Green and Bourne	N, I	Y	Y	Y	Y	Y	Y
IAEA-TECDOC	N	Y	Y	Y	Y	Y	Y
IEEE 500	I	Y	–	Y	Y	–	Y
Lees	N, I	Y	Y	Y	Y	Y	Y
MIL-HDBK-217F	M	Y	Y	–	Y	–	Y
NPRD-91	N, I	Y	Y	Y	Y	–	Y
OREDA 84	O	Y	Y	Y	Y	Y	Y
OREDA 92	O	Y	Y	Y	Y	Y	Y
RKS/SKI 85-25	N, I	Y	Y	Y	Y	–	Y
Rothbart	I	Y	Y	Y	Y	–	Y
Smith	N, I	Y	Y	Y	Y	Y	Y
Smith and Warwick	N, I	–	Y	–	–	–	–
Wash 1400	N	Y	Y	Y	Y	–	Y
RMC HARIS	O, N, I	Y	Y	Y	Y	Y	Y
SRD Data Center	O, N, I, M	Y	Y	Y	Y	Y	Y

O, offshore; N, nuclear; I, industrial; M, military.
Source: Reproduced from The Reliability of Mechanical Systems edited by Davidson (technical editor Hunsley), 1994, with permission from Professional Engineering Publishing Ltd.

TABLE 5.4 Equipment Category Code for Table 5.3

Equipment Category	Typical Component
1 Rotating equipment	Pumps, compressors, turbines, motors
2 Static equipment	Pipes, pipelines, flowlines, valves, vessels
3 Instrumentation	Sensor (temperature, pressure, level, etc.) and controllers
4 Safety	Fire pumps, safety valves, fire/gas/smoke detectors
5 Process	Pumps, compressors, valves vessels, piping
6 Electrical	Cables, motors, circuit boards lamps

Note that the categories are not intended to be mutually exclusive. For example, pumps can be categorized as both 'rotating' and 'process'.
Source: Reproduced from The Reliability of Mechanical Systems edited by Davidson (technical editor Hunsley), 1994, with the permission from Profession Engineering Publishing Ltd.

5.1.9.6.2 Operation of Database

The operator of a reliability database has to maintain a sufficient flow of data from the data suppliers, synthesize the data from different sources, and satisfy the needs of the data users in a manner which is cost-effective. This is no easy task.

Fresh data is the lifeblood of a reliability data bank. There must be sufficient incentive for an organization to supply them, either by collecting them itself or by allowing access for their collection. Typically, it will be one that is also a user of the database.

The data user will have a set of requirements, which often can be met only in part. Ideally, the user would like data from which to obtain failure and repair time distributions, and so on. Often these will simply not be available, the best on offer being perhaps failure or repair rates based on a constant rate assumption. There may also be problems of sample size. The data bank operator has to balance the often-conflicting requirements of user demand, data supply, and economy in handling.

5.1.9.6.3 Data Acquisition

Information on equipment for the component inventory is obtained from the plant documentation, which includes equipment records and equipment and plant drawings. Acquisition of the inventory data for a plant may not be straightforward. Records and drawings may be incomplete from the outset, and it is even more common that

modifications made are not entered. There can be confusing identifications, with the same item allocated several quite different code numbers.

There can be a corruption of the component design parameters, with differences between the item as designed and as purchased and installed being quite common. As Cross and Stevens comment, "It is incredible how poor the average plant inventory records are."

For information on equipment failure and repair a basic source of information is the job card or ticket. Other documentation such as logbooks, permits-to-work, and stores requisitions are generally useful as cross-checks rather than prime sources. In general, the failure data should be such as to permit analysis to obtain failure distributions as well as average values.

5.2. EQUIPMENT MAINTENANCE

5.2.1 Management of Maintenance

The need for maintenance of equipment implies some abnormality and often some change or increase in hazard. The conduct of the maintenance work may also introduce its own hazards to the equipment. In addition, the maintenance activities may be a hazard for the personnel involved in them. Careful control of maintenance work must be exercised in order to eliminate hazards to the equipment and to the personnel. The consequences of failure to exercise this control can be serious, examples where maintenance played a significant role are the Piper Alpha and the Flixborough disaster.

Maintenance activities are regulated by numerous codes, standards, and guidelines. Some of which are Occupational Safety and Health Administration (OSHA), Environmental Protection Agency (EPA), Department of Transportation (DOT), Department of Energy (DOE), Nuclear Regulatory Commission (NRC), National Institute for Occupational Safety and Health (NIOSH), National Fire Protection Association (NFPA), American Society of Mechanical Engineers (ASME), American Institute of Chemical Engineers (AIChE), National Association of Corrosion Engineers (NACE), Institute of Electrical and Electronics Engineers (IEEE), International Electrotechnical Commission (IEC), American Society of Testing and Materials (ASTM), American National Standards Institute (ANSI), National Board Inspection Code (NBIC), and American Petroleum Institute (API).

Important factors of a safe maintenance management should consider: Safe system of work, equipment documentation, maintenance procedures, permit system, handover system, control of workforce, contractors, and supervision and training.

TABLE 5.5 Some Data on Equipment Failure Rates Published by the UKAEA

	Failure Rate (Failures/10^6h)
Electric motors (general)	10.0
Transformers (<15 kV)	0.6
(132–400 kV)	7.0
Circuit breakers (general, <33 kV)	2.0
(400 kV)	10.0
Pressure vessels (general)	3.0
(high standard)	0.3
Pipes	0.2
Pipe joints	0.5
Ducts	1.0
Gaskets	0.5
Bellows	5.0
Diaphragms (metal)	5.0
(rubber)	8.0
Unions and junctions	0.4
Hoses (heavily stressed)	40.0
(lightly stressed)	4.0
Ball bearings (heavy duty)	20.0
(light duty)	10.0
Roller bearings	5.0
Sleeve bearings	5.0
Shafts (heavily stressed)	0.2
(lightly stressed)	0.02
Relief valves: leakage	2.0
blockage	0.5
Hand-operated valves	15.0
Control valves	30.0
Ball valves	0.5
Solenoid valves	30.0
Rotating seals	7.0
Sliding seals	3.0
'O' ring seals	0.2
Couplings	5.0
Belt drives	40.0
Spur gears	10.0
Helical gears	1.0
Friction clutches	3.0
Magnetic clutches	6.0

(Continued)

TABLE 5.5 (Continued)

	Failure Rate (Failures/10^6h)
Fixed orifices	1.0
Variable orifices	5.0
Nozzle and flapper assemblies: blockage	6.0
breakage	0.2
Filters: blockage	1.0
leakage	1.0
Rack-and-pinion assemblies	2.0
Knife-edge fulcrum: wear	10.0
Springs (heavily stressed)	1.0
(lightly stressed)	0.2
Hair springs	1.0
Calibration springs: creep	2.0
Breakage	0.2
Vibration mounts	9.0
Mechanical joints	0.2
Grub screws	0.5
Pins	15.0
Pivots	1.0
Nuts	0.02
Bolts	0.02
Boilers (all types)	1.1
Boiler feed pumps	1012.5
Cranes	7.8

Note: Further failure data on electronic, mechanical, pneumatic, and hydraulic components are given by Green and Bourne (1972).
Sources: Reproduced with permission; Farmer (1971); Green and Bourne (1972).

5.2.1.1 Maintenance Information Systems

Information on equipment failures, repairs, and incidents is needed both for maintenance and safety and loss prevention (SLP) purposes. It should be a specific objective of the maintenance system to generate such information.

The principal data required are those for failure, repair, and availability. Data on failure and related aspects are essential to the loss prevention approach. The maintenance engineer, therefore, has a crucial role to play here. The requirement for failure data is one aspect of the more general need for feedback of information from the maintenance to the design function.

The maintenance policies which are followed in a process plant can have a marked influence on its reliability, and availability, and hence its safety and economics. Traditional policies have been based on breakdown maintenance and/or scheduled maintenance, and also opportunity maintenance, which are supplemented increasingly by on-condition maintenance and reliability-centered maintenance. The effectiveness of the maintenance should be monitored using measures suitable for the policy adopted.

5.2.1.2 Maintenance of Protective Devices

Maintenance of the protective devices on the equipment is particularly important and should be covered by a formal system with full documentation.

Some of the principal protective devices are as follows:

1. pressure relief valves;
2. bursting discs;
3. tank vents and filters;
4. other pressure relief devices;
5. non-return valves;
6. mechanical trips and governors;
7. instrument trips;
8. other instrumentation;
9. alarm systems;
10. fire and gas detection systems;
11. sprinkler systems;
12. fire water systems.

Fire protection equipment should be well maintained so that it is available when required. Maintenance of such equipment is dealt with in NFPA 25: *Inspection, Testing and Maintenance of Water-Based Fire Protection Systems*.

5.2.1.3 Maintenance Optimization

The objective is to conduct predictive and preventive maintenance to reduce major repairs and eliminate potential catastrophic accidents by correcting minor deviations as soon as they became evident. Operators usually recognize when a component or piece of equipment is making "strange noise" or there are other irregularities. Maintaining equipment in a productive state will also increase productivity and decrease interruptions. Any lost and worn parts must be repaired and replaced immediately. Performing predictive and preventive maintenance can eventually result in a reduction of overall operating costs. Preventive maintenance is more desirable than corrective maintenance.

Maintenance optimization can be used to estimate maintenance schedules, requirements and resource needs, including personnel, spares, equipment, etc. The optimization process includes selection of maintenance techniques, periodicity to accomplish maintenance requirements, and maintenance targets, while taking into account safety, regulatory requirements, reliability, availability, and cost. Maintenance optimization is a useful tool to achieve systematic maintenance programs in an effective manner to establish plant reliability, which includes safety and cost-effective approaches. Based on how the maintenance is performed there are two basic types, (1) time-based maintenance and (2) condition-based maintenance. Time-based maintenance can be conducted using an interval approach or a running-time approach. The interval approach is usually used for components with a high safety significance. The selection of the interval is based on the function, availability, and historical data on failure or degradation. The running-time approach is generally performed on critical components.

5.2.2 Preparation for Maintenance

There are a number of preparatory measures which may need to be taken before maintenance work is started. The principal means of control of such work is the permit system. The preparatory measures to be taken should be specified on the permit.

Many of these measures are aimed at allowing work inside equipment to be done safely. If work is to be done internally on an item of equipment, it should be prepared by the following operations: (1) depressurization, (2) cooling down, (3) isolation, (4) removal of contents (gas, liquid, solids), and (5) cleaning.

As a preparation for maintenance the following points should be considered: Identification of equipment, depressurization, cooling, isolation, emptying of liquids, gas freeing, removal of solids, and cleaning.

5.2.2.1 Maintenance Personnel Training

There are different kinds of maintenance work that is generally conducted in industry, such as repair and maintenance of mechanical equipment, electrical equipment, software and networks, buildings, and general maintenance. It is important for maintenance work personnel to recognize the hazards of maintenance work and modifications for any unit equipment. Education, training, and certification may result in better advancement and accomplishment of tasks. Depending on the level of skill required, the length of training for maintenance personnel can range from several days to months to become fully qualified. There may also be some prerequisites prior to the training such as P&ID reading, mechanical drawing, electricity, basic computer skill, and others; technical education is important part of the training.

Documents and guidelines for personnel training requirements are provided by Department of Energy (DOE) under 10 CFR 830.122, Criteria 2-Management/ Personnel Training Qualification. The documents are consistent and applicable to the current industry standards of American National Standard Institute. For example, the training documentation for nuclear power plant gives in DOE training requirements which are consistent to ANSI/ ANS 3.1-1993, American National Standard, Selection, Qualification and Training of Personnel for nuclear power plant, ANSI/ANS 15.4-2007, American National Standard, Selection and Training of Personnel for Research Reactors and 10 CFR 55, Operators' Licenses.

5.2.3 Isolation, Purging and Cleaning

Isolation is governed by OSHA standards 1910.147 *The Control of Hazardous Energy* (Lockout/tagout) and 1910.146 *Permit Required Confined Spaces*. Isolation may be required prior to installation, inspection, repair, cleaning, or dismantling. Where a job involves isolation of equipment, there should always be a permit for the job. The methods available for isolating a vessel or pipe are, in ascending order of effectiveness, the use of (1) a closed and locked valve, or valves, (2) a double block and bleed valve system, (3) a blind, and (4) physical disconnection.

Where work is to be done on powered machinery, the source of power should first be isolated. Sources of power include electrical, hydraulic, and pneumatic power and engines. Isolation of an engine-driven system should be done by shutting off the engine fuel supply and then isolating and disconnecting all starting systems. It may sometimes be possible for the machinery to move even though it is disconnected from its power source, and in such cases it should be secured to prevent such movement. A lockout/tagout system should be used based on principles similar to that for the isolation of equipment.

Electrical isolation may be required either to immobilize machinery or to protect personnel working on electrical equipment. Where isolation of fluids is required, electrical isolation should be a complement to, but not a substitute for, mechanical isolation.

OSHA standards 1910.147 and Subpart S govern maintenance work on electrical systems. The isolation of electrical equipment is governed by the UK *Electricity at Work Regulations* 1989. Electrical isolation should be performed only by an electrically competent person.

Two methods of isolating electrical equipment are to withdraw the fuses and to lock-out the breaker. A lockout device is a mechanism or an arrangement which allows the use of key operated padlocks to hold a switch lever or handle in the 'off' position. The lock-out procedure is to switch off or de-energize electric power, lockout, tagout and confirm lock-out by checking that the

equipment will not restart. The lockout should be applied to the breaker or disconnect switch itself and not to some remote stop/start button, selector switch or interlock. Where there is more than one power source, all should be locked out and tagged. All items of electrical equipment should have permanent labels, and their separate parts should be identifiable. For a prime mover, the IChemE *Guide* gives these as the drive unit, the breaker/disconnect, and the stop/start button. A system of warning tags should also be used.

Purging involves replacing one gas or vapor with another. It is performed for a variety of purposes and using a number of different purging media. Guidance on purging is given in *Purging Principles and Practice* (AGA, 2001) (the AGA *Purging Guide*). Further accounts of purging are given in the IP *Refining Safety Code*, the IP *LPG Code*, the IChemE *Maintenance Guide*, and by Kletz (1982). Purging may be used to take a unit out of service by replacing flammable or toxic process gas with an inert medium and then with air. It may be used to bring a unit back into service by replacing air with an inert medium and then with the process gas.

There is a wide variety of methods of cleaning equipment. They include: water washing, chemical cleaning, steaming, water jetting, solvent jetting, shot blasting, and manual cleaning, cleaning in place, line clearing. Cleaning operations generate an appreciable fraction of the liquid effluents from process equipment and measures to reduce effluents from this source can contribute significantly to waste minimization. Certain types of cleaning methods have a risk of their own, such as the use of solvents in solvent jetting; appropriate protective equipment and other considerations must be addressed for each type of cleaning. The choice of cleaning method depends on the type of equipment and on the nature of the material to be cleaned out. Most of these methods are described in CS 15 and/or the IChemE *Maintenance Guide*.

5.2.4 Confined Spaces

Accidents associated with entry into and work in confined spaces and vessels have occurred with depressing regularity. This activity is recognized as presenting particular potential hazards and is governed by a statutory requirement for a permit. Permits as such are considered in Section 5.2.5. The account given here is confined to the hazards of confined spaces and to precautions which should be taken. Accounts of these hazards and precautions are given in the IChemE *Maintenance Guide*.

Requirements for confined spaces are governed by OSHA 1910.146 *Permit Required Confined Space*, *NIOSH Recommended Standard for Working in Confined Spaces*, API 2217A *Guidelines for Confined Space Work in the Petroleum Industry*, and ANSI Z117.1 *Safety Requirements for Confined Spaces*.

A closed equipment with restricted access may be the obvious example of a confined space, but it also includes open manholes, trenches, pipes, flues, ducts, ceiling voids, enclosed rooms such as basements, and other places where there is inadequate natural ventilation. Open top tanks, furnaces and ovens, even those with a large aperture open to the atmosphere, are confined spaces.

Some of the specific hazards of confined space entry are flammable materials, toxic substances, hazardous atmospheres, oxygen-deficient atmospheres, oxygen-enriched atmospheres, and noxious fumes. Prior to entry into equipment, internal atmospheres should be tested for oxygen content, flammable gases, and vapors and for potential toxic air contaminants. The gas tests which are commonly done are for hydrocarbons, carbon monoxide and hydrogen sulfide, and oxygen, together with tests for any toxic substance which has been contained in the equipment.

5.2.5 Permit Systems

Maintenance work on process equipments should be controlled by a formal permit system. US companies use a work permit system to control maintenance activities in process units and entry into equipment. The United Kingdom uses a similar system of permits-to-work (PTWs).

In essence, the objectives of the permit system are to exercise control over the maintenance activities by assigning responsibilities, ensuring communication between interested functions, and requiring that proper consideration be given to the job, its hazards, and the precautions required.

In the normal system, there is an issuing authority and a performing authority. The issuing authority is the operations supervisor. The performing authority is usually the maintenance craftsman who is to do the work, but may sometimes be the maintenance supervisor. It is the responsibility of the issuing authority to ensure that the equipment is safe for the work to proceed. The performing authority is responsible for ensuring that the further working precautions are taken. It is the responsibility of these same two authorities to terminate the permit. The normal system is that on completion of the work, the performing authority signs that the work is complete and the issuing authority, after inspecting the work site, signs that the permit is complete. There may be a requirement that in certain defined cases, where work at a particular unit may affect an adjacent unit, the operations supervisor on the latter should countersign the permit.

5.2.5.1 Types of Permit

There is a variety of types of permit, some of which have special names, for example clearance certificates and fire permits. Permits may be classified by reference to the operation to be performed, the equipment to be worked on, the classification of the areas where the work is to be done, the special hazards which may be encountered, the equipment to be used or the time of day specified for the work. A list of typical permits, which illustrates all these categories, is as follows:

Operations	Area Classification
Equipment removal	Flammable area
Excavation	Special hazards
Hot work	Corrosive substances
Leak sealing	Fire
Line breaking	Toxic substances
Vessel entry	Ionizing radiations
Waste disposal	Equipment used
Equipment worked on	Mobile crane
Electrical equipment	Time of day
Interplant pipelines	After-hours work
Sprinkler system	

A common basic set of permits covers entry, cold work, hot work, and electrical work. Some companies use a single permit system to cover all work activities. The IP *Refining Safety Code* gives model permit forms for the following: general work, electrical work, hot work and work involving ionizing radiations; line disconnecting and vessel opening; entry; and excavation.

The contents of a permit need to be carefully defined. Accounts of permit contents are given in the OIAC *Permit Systems Guide*, the IChemE *Maintenance Guide* and by S. Scott (1992). Such accounts usually distinguish between features which are essential and those which are desirable.

Particular types of permit require additional features. For example, a hot work permit should contain an entry for the hazardous area classification of the location of the work and gas test results.

Sample permit forms are given in a number of publications, including the IP *Refining Safety Code*, the IP *LPG Code*, and the IChemE *Maintenance Guide*.

5.2.5.2 Entry Permits

Entry into vessels and other confined spaces has resulted in numerous accidents. As a result this operation has had a unique status. OSHA 1910.146 defines entry permit requirements. The entry permit identifies:

1. permit space to be entered;
2. purpose of entry;
3. date and authorized duration of the entry permit;
4. authorized entrants by name;

5. personnel, by name, serving as attendants;
6. individual, by name, currently serving as entry supervisor;
7. hazards of permit space to be entered;
8. measures used to isolate the permit space and to eliminate or control permit space hazards before entry;
9. acceptable entry conditions;
10. results of initial and periodic gas tests;
11. rescue and emergency services that can be summoned and the means for summoning those services;
12. communication procedures used by authorized entrants and attendants to maintain contact during the entry;
13. equipment such as personal protective equipment, testing equipment, communications equipment, alarm systems, and rescue equipment to be provided;
14. any other information necessary to ensure employee safety;
15. any additional permits such as hot work that have been issued to authorize work in permit spaces.

5.2.5.3 Design of Permit Systems and Auditing of Permit System

Aspects of the design of a permit system include (1) standardization, (2) the personnel involved, (3) scope, (4) cross-referencing, (5) display, (6) multiple jobs, (7) change of intent, (8) suspension, (9) handback, (10) training, and (11) monitoring, auditing, and review.

It is desirable that the permit system used within a company be as uniform as is practicable. This applies not just to the design of permit forms but to the whole system. A standard system eases the problem of training and reduces the probability of confusion and hence error.

The design of the permit form should allow for the necessary cross-referencing. There should be cross-references to other jobs, by description of job as well as permit number; isolations and the associated permit numbers; and equipment identification tags. In particular, the cross-referencing should cover the case where isolation is common to more than one job.

It is not enough to create a permit system to control maintenance work. There should also be a routine audit of the system to ensure that it is operating properly. These may consist of a specific instruction to the plant manager to check each week a portion of the permits issued. The OIAC *Permit Systems Guide* gives a checklist for permit systems.

5.2.5.4 Operation of Permit Systems

If the permit has been well designed, the operation of the system is largely a matter of compliance. If this is not the case, the operations function is obliged to develop solutions to problems as they arise.

As just stated, personnel should be fully trained so that they have an understanding of the reasons for, as well as the application of the system. It is the responsibility of management to ensure that the conditions exist for the permit system to be operated properly. An excessive workload on the plant, with numerous modifications or extensions being made simultaneously, can overload the system. The issuing authority must have the time necessary to discharge his responsibilities for each permit. In particular, he has a responsibility to ensure that it is safe for maintenance to begin and to visit the work site on completion to ensure that it is safe to restart operation.

Where the workload is heavy, the policy is sometimes adopted of assigning an additional supervisor to deal with some of the permits. However, a permit system is in large part a communication system, and this practice introduces into the system an additional interface. The communications in the permit system should be verbal as well as written. The issuing authority should discuss, and should be given the opportunity to discuss, the work. It is bad practice to leave a permit to be picked up by the performing authority without discussion.

The issuing authority has the responsibility of enforcing compliance with the permit system. He needs to be watchful for violations such as extensions of work beyond the original scope.

5.2.6 Maintenance Equipment

Many of the equipment used to perform maintenance may present hazards if safety precautions are not considered. Tools such as hammers and wrenches can be a source of ignition by causing sparks. If there is presence of flammable materials, non-sparking tools should be used. In the case of lifting equipment, incidents sometimes occur in which a lifting lug gives way. Lifting equipment is governed by OSHA 1910.184 Slings and 1926.251 Construction Rigging Equipment. UK requirements are given in the Factories Act 1961, Sections 22−27. Mobile cranes present several different types of hazard, one is that of collision with process equipment, particularly pipe racks, when the crane is on the move. A minimum measure to prevent this is clear identification of pipe racks including labeling with clearance height. Additional guidance is provided in OSHA 1910.333. Lift trucks, including forklifts, present another set of hazards, one is the hazard due to any moving vehicle, especially in a relatively congested plant area. Guidance on lift trucks is given in OSHA 1910.178 *Powered Industrial Trucks, Safety in Working with Lift Trucks* (HSE, 2000). Manlifts are used to hoist personnel. They provide quick access for work at elevated heights. Personnel operating the manlift must exercise care so as not to become entangled in the work area. They should never exit the manlift platform; OSHA 1910.68 governs manlifts. Scaffolding is used as a temporary platform for work. Before initial use, scaffolding should be inspected by a competent person and certified as safe for use. Scaffolding is governed by OSHA 1910.28 *Safety Requirements for Scaffolding.*

A less conventional form of equipment is a robot. The use of robots in relation to process equipment is most developed in the nuclear industry and in offshore activities, but a gradual growth of their use in special applications on conventional process plants can be envisaged. The hazards posed by a fixed robot are those of a machine, essentially striking and trapping and entanglement. The characteristic problem with a robot is the realization of these hazards due to aberrant behavior. Two potential causes of such behavior are power transients and program errors. HSE guidance on robots is given in *Industrial Robot Safety* (HSE, 1988).

5.2.7 Flanged Joints

Flanged joints maintenance is a special process operation that requires considerable review before and after working on the joints. The flanged joint is deceptively simple yet, in common with the welded joint, its integrity relies on a number of parameters. Hence, the assembly and disassembly of flanged joints are treated in this section.

5.2.7.1 Making Joints

The normal pipe joint on a process plant is the flanged joint, as described in Chapter 8. The two main varieties are the gasketed joint and the ring-type joint. For high pressures, the ring-type joint is used. For the installation of both types, the surfaces of the flanges should be free of deterioration, clean, and the gasket used should be of the correct type. When a flanged joint is opened, the gasket should not be reused. The proper procedure to tighten the bolts in a joint is sequence tightening, which involves gradual tightening of the bolts following a criss-cross pattern.

There are a variety of tightening tools available. They include: traditional wrenches and torque wrenches, impact wrenches, pneumatic torque multipliers, and hydraulic torque wrenches. There are various bolt stretching devices, notably direct-acting bolt tensioners. Likewise, there are a variety of techniques for measuring bolt stress. One of the simplest is the micrometer, which measures strain, from which stress may then be calculated. Lubrication may be used to reduce the friction and to give more uniform tensioning. The lubricant should be matched to the application, selection being a specialist

matter. It is not unusual for a flanged joint to leak. A common response is to tighten the bolts. This may be effective if the bolts were loose in the first place. However, further tightening of bolts which were correctly tensioned originally is bad practice.

5.2.7.2 Breaking Joints

Breaking into a line can involve a number of hazards and it is necessary to exercise considerable care. OSHA Process Safety Management 1910.119(o) requires safe work practices for opening process equipment or piping. Safe methods for the breaking of pipelines are described in *Safety in Inspection and Maintenance of Chemical Plant* (BCISC, 1959).

For the breaking of a line there should be clear instructions on the work and on the area in which it is to be done. Information should be given on the material in the line and its hazards and on the precautions to be taken. The joint to be broken should be indicated by an identification tag. The work is normally covered by a permit.

Some preliminary measures and precautions include:

1. isolation of the working area and posting of warning notices;
2. provision of safe access;
3. support of the pipe on either side of the joint;
4. isolation of the pipe section;
5. release of pressure from the pipe section;
6. draining of the pipe section and disposal of the fluid drained;
7. precautions against fire;
8. protection of personnel.

A particular hazard occurs on certain types of valve, where a joint containing the process fluid is liable to be broken in error. Whenever a joint is to be placed adjacent to a valve special precautions must be taken to not damage the valve.

5.2.8 Hot Work

Hot work such as welding and cutting is a common activity on process equipment, but involves potential hazards both to equipment and personnel, and needs to be closely controlled. OSHA 1910.119(k) requires a hot work permit and 1910 Subpart Q provides general requirements for welding, cutting, and brazing operations. Additional guidance is provided by API RP 2009: *Safe Welding, Cutting and other Hot Work Practices in the Petroleum and Petrochemical Industries*, API Publ. 2201: *Procedures for Welding or Hot Tapping on Equipment in Service*, API RP 1107: 1991 *Pipeline Maintenance Welding Practices* and NFPA 51B: 1999 *Cutting and Welding Processes*. Also relevant is

HS(G) 5 *Hot Work* (HSE, 1979) (the HSE *Hot Work Guide*).

Welding operations are covered by Section 31(4) of the Factories Act 1961, which states:

No plant, tank or vessel which contains or has contained any explosive or inflammable substance shall be subjected

a. to any welding, brazing or soldering operation;
b. to any cutting operation which involves the application of heat;
c. to any operation involving the application of heat for the purpose of taking apart or removing the plant, tank or vessel or any part of it; until all practicable steps have been taken to remove the substance and any fumes arising from it, or to render them non-explosive or non-inflammable; and if any plant, tank or vessel has been subjected to any such operation, no explosive or inflammable substance shall be allowed to enter the plant, tank or vessel until the metal has cooled sufficiently to prevent any risk of igniting the substance.

Section 31(5) provides that, where he is satisfied that compliance is unnecessary or impracticable, the Chief Inspector may grant exemption from any of the above requirements by the issue of a certificate, which may contain conditions.

The welding operations should be controlled by a permit system. The general arrangements of a permit system are described in Section 5.2.5, but certain points of particular relevance to welding merit mention.

A principal hazard of welding is that it introduces a source of ignition. The equipment to be worked on should normally be isolated, emptied and cleaned, tested for flammables, and inspected before welding is undertaken. The surfaces of the surrounding equipment should be free of flammable deposits.

Special consideration must be given when welding on equipment which contains flammable fluids, the conditions under which welding may be carried out on equipment containing flammable fluids are defined in API 2201 and HSE *Hot Work Guide*.

Some welding operations require temporary isolation of a section of a pipe, in some cases it will be necessary to use seals. The most common seals are atmospheric seal plugs, high pressure stopple or seal plug, frozen product seal plug, and bag seal. An example of operation that requires temporary isolation is when welding on operating pipes or pipelines. The method to be followed in welding operations on pipelines carrying flammable fluids is described in the API 2201, RP 1107, and the HSE *Hot Work Guide*.

Before performing a welding job there are a number of limitations when working with equipment containing flammable fluids. For example, welding should not generally be carried out on equipment where the metal thickness is less than 0.2 in. (5 mm); on the external surface of a tank at a point less than 3 ft (1 m) below the liquid surface, or on equipment at a temperature below 45°F (7°C), unless it can be shown to be safe to do so.

The HSE *Hot Work Guide* gives the procedures to be followed in welding operations on a pipeline in the form of a checklist, under the headings of: (1) initial planning of operations, (2) site preparations, (3) standby services, (4) pipeline preparation, (5) fitting alignment, (6) welding procedure, (7) action immediately prior to welding, (8) action during welding, (9) integrity of the completed weld, and (10) final completion.

The procedure of fitting a branch onto a pipeline which is still operational is known as hot tapping. As already stated, guidance is available in API RP 2201, RP 1107, and the HSE *Hot Work Guide*.

5.2.9 Tank Cleaning, Repair, and Demolition

A set of operations of some importance in the maintenance of equipment is the cleaning, repair, and demolition of tanks for storage of flammable liquids. Guidance is given in OSHA 1910.146: *Permit Required Confined Space*, API Std 2015: *Requirements for Safe Entry and Cleaning of Petroleum Storage Tanks*, API Std 653: *Tank Inspection, Repair, Alteration, and Reconstruction*, NFPA 326: *Safeguarding of Tanks and Containers for Entry, Cleaning and Repair*, CS 15: *The Cleaning and Gas Freeing of Tanks Containing Flammable Residues* (HSE, 1985).

The basic considerations when cleaning, repairing, or demolishing a tank include: preparation for entry or hot work; gas freeing of tanks; emptying of liquid from equipment; precautions for ignition at equipment; gas freeing and cleaning of small equipment, mobile tanks and large equipments, relevant advice is given in API Publ. 2207: 1999 *Preparing Tank Bottoms for Hot Work*; hot work on outside of equipment; inerting of equipment; demolition of equipment, methods for the demolition of equipment are given in API Std 653: *Tank Inspection, Repair, Alteration and Reconstruction*.

5.2.10 On-Line Repairs

There has been a steady growth in the development of techniques for on-line, or on-steam, repair and in their application. The growth of such techniques has been driven by a number of factors, including general economic pressures and the need to accommodate the lengthening of times between scheduled shut-downs, or turnarounds; to keep major pipelines operating; to economize on energy and reduce steam leaks; and to reduce hydrocarbon emissions.

One of the most common types of on-line repair is the sealing of a leak. A variety of simple methods are used to seal leaks such as steam leaks, including clamps and various forms of enclosure or box. These may often be applied without the need to weld on an operating line. A specialized leak repair technique is that of on-line compound leak sealing. One of the main points where a leak is liable to occur is at a flanged joint. If the leak is severe, there may be no choice but to shut-down and repair it, unless it can be repaired online with the equipment still operating. Compound leak sealing of a flanged joint involves putting containment around the joint and injecting a sealant under pressure. If leak sealing is to be undertaken, it should be assessed by a competent engineer and a formal work plan agreed, covered by a permit. The assessment should cover: the hazard of a major release; the residual thickness of the pipe, including the method of, and apparatus for, its measurement, the bolts, including materials, deterioration, initial and final stresses; the external loads on the pipe and flange; the record of past injections; the maximum injection pressure to be used; the design codes and procedures for the boxes and clamps; the possibility of internal blockage of the pipe by the sealing compound; the access to the site; the personal protection required; and the expected life of the temporary repair.

5.2.11 Maintenance of Particular Equipment

The deterioration and failure of equipment, which is dealt with by maintenance, has causes in design and installation as well as operation, and has first to be detected by inspection. OSHA PSM standard 1910.119 (j) Mechanical Integrity requires programs to maintain the integrity of various types of process equipment. The process equipment along with general guidance for maintenance can be, for example:

1. Pressure vessels. The maintenance function should maintain a general oversight of pressure vessels, focusing on external corrosion, support defects, stress inducing features, disabled relief protection, and disabled instrumentation. Pressure vessel maintenance and inspection guidance is provided in API RP 510: Pressure Vessel Inspection Code, RP 572: Inspection of Pressure Vessels, RP 579: Fitness-for-Service, RP 580: Risk Based Inspection, RP 573: Inspection of Fired Boilers and Heaters, RP 570: Inspection, Repair, Alteration, and Rerating of In-Service Piping Systems.
2. Pressure relief devices. They provide essential protection for pressure systems and much effort goes into their design. This is negated if the relief system is not properly maintained. Guidance on inspection of pressure relief systems is provided in API RP 576: Inspection of Pressure Relief Devices.
3. Storage tanks. The maintenance of atmospheric storage tanks presents a quite different set of problems. Guidance on inspection of storage tanks is provided in

API RP 575: *Inspection of Atmospheric and Low-Pressure Storage Tanks*, RP 653: *Tank Inspection, Repair, Alteration, and Reconstruction.*

4. Ammonia storage spheres. The occurrence of stress corrosion cracking in ammonia storage spheres has created a considerable problem in the maintenance of such storages. This work is discussed further in Chapter 16.
5. Heat exchangers. Where a heat exchanger is delivered to a construction site and is left in the open for some time before it is installed, it may well suffer deterioration even before it begins operation. In order to avoid this, it should be stored carefully before installation and its surfaces cleaned periodically.
6. Equipment supports, foundations, and dikes. They provide another illustration of the interaction between design, inspection, and repair. The general problems for equipment support include corrosion, overstressing, frost damage and also fire or explosion.

5.2.12 Deteriorated Equipment

The equipment on the plant starts to deteriorate from the first day of operation, and sometimes earlier if it has been neglected prior to installation. As the plant ages, it becomes necessary to make a decision on the continued operation of some equipment. Tools that are useful when considering working with deteriorated equipment is the remaining life assessment, API RP 579 provides guidance on evaluating the present integrity of the equipment and the projected remaining life. Likewise, special attention should be given to used equipment, the situation can arise where it is necessary to consider either the purchase or the sale of used equipment. For purchase, three approaches may be taken: purchase 'as is', on approval, and rebuilt and guaranteed. Another special case for deteriorated equipment should be given in the aftermath of a large fire, there may be equipment which has been in the fire. The question then arises as to the continued safe use of the equipment. In all such cases an inspection should be made.

5.2.13 Major Shut-Downs

Two particular kinds of shut-down merit specific mention. These are major scheduled shut-downs and mothballed shut-downs.

5.2.13.1 Scheduled Shut-Downs

Most units require a periodic major scheduled shut-down, or turnaround, in order to carry out maintenance and other jobs which cannot be done while the unit is operating.

The turnaround arrangements described are broadly as follows. The responsibility for the turnaround is assigned to a turnaround manager. The turnaround passes through four stages: (1) formulation, (2) planning, (3) implementation or execution, and (4) startup.

In the formulation stage, the turnaround manager solicits job requests from the interested parties. In the planning stage, the overall critical path network is developed together with a bar chart showing the start and end time of each job and the personnel required. The management of the implementation stage needs to be sufficiently flexible to accommodate the facts that some jobs may reveal problems, that some may take longer than planned, or require parts other than those expected, or that some completely new jobs may need to be done. The purpose of the turnaround is to get the work completed and the unit started up again. The emphasis, therefore, should be on the startup, rather than simply completion of the turnaround. When the startup is over, the personnel involved should be debriefed and a report written while matters are still fresh in the mind. In particular, note should be made of times actually taken and the resources used for various jobs and activities, of problems encountered, equipment required, techniques developed or lessons learned, and of jobs which will need to be done in the next turnaround.

5.2.13.2 Mothballed Shut-Downs

During an economic recession, the decision may be made to take a unit out of commission and to 'mothball' it, with the intent to recommission it when times are better.

There are a number of factors that affect the way in which the mothballing is done. The length of the shutdown is relevant, the location, etc. It may not be appropriate to mothball every item. Some items of equipment and much instrumentation are particularly prone to obsolescence. For other items, such as small bore piping and valves, it may be more economic to let them go and replace them when the unit is recommissioned.

Many problems arise during a mothball shutdown. Equipment of the process should be guarded from corrosion. Two widely used methods of reducing corrosion are the use of dry inert atmospheres in vessels and the coating of surfaces with oil. Corrosion is reduced by the exclusion of oxygen, and filling vessels with nitrogen is an effective anticorrosion measure. The gas should be dry, with a dew point at least 40°F (5°C) below that of the lowest expected ambient temperature.

The mothballing of equipment does not remove the need to inspect and maintain, or even to operate parts of it. Inspection is still required, but the techniques involved are somewhat different. There is also a need for some degree of maintenance.

A full set of documentation on the plant, and on its operating and maintenance procedures, needs to be compiled. Such documentation is necessary for any equipment, but it is crucial for one which is to be recommissioned at some date in the future by personnel who are likely to be unfamiliar with it. The methods of protection used in mothballing may also be applicable where equipment arrives on site but cannot be installed immediately, for example heat exchangers.

5.2.14 Spares Inventory and Computer Systems

The control of the spares inventory, or stock control, is of particular importance on process equipment, where the cost of downtime is generally very high.

Policy on the spare parts to be held in stock has to be a compromise between holding a large number of parts in an attempt to provide for virtually every contingency and holding only a small number in order to keep holding costs down. The methods of reliability engineering may in certain cases be used to determine the spares holdings. The level of spares will be affected by the degree of standardization. Indeed, it is a principal aim of standardization to reduce spares holdings. Another feature affecting spares holdings is the least replaceable assembly. The spares holdings will be strongly affected by the extent to which a 'just-in-time' policy is practiced.

Features of the spares organization are the stores layout, the stores documentation, the coding system, policies on aspects such as used parts and cannibalization of parts and access to the store. An explicit policy should be formulated for the stores and their organization. Access to the store should be controlled, but in some cases it is policy to provide an open store with free access for minor items, where the cost of wastage is less than that of the control paperwork. Materials for a major project should be treated separately from those for normal maintenance. Failure to do this can cause considerable disruption to the maintenance spares inventory. In this context a turnaround may count as a major project requiring its own dedicated store, as already described.

The use of some form of computer system is virtually universal for maintenance systems on process equipment. Some functions which a computer-based maintenance information system may perform include: (1) work orders, (2) planning and scheduling, (3) failure analysis, (4) cost, and (5) budget.

5.3. MANAGEMENT OF CHANGES AND MODIFICATIONS

5.3.1 Modifications to Equipment

Some work goes beyond mere maintenance and constitutes modification or change. Such modification involves a change in the equipment and/or process and can introduce a hazard. The outstanding example of this is the Flixborough disaster. The Flixborough Report (R.J. Parker, 1975, para. 209) states, "The disaster was caused by the introduction into a well designed and constructed plant of a modification, which destroyed its integrity."

It is essential for there to be a system of identifying and controlling changes. Changes may be made to the equipment or the process, or both. It is primarily equipment changes which are discussed here, but some consideration is given to the latter.

OSHA PSM 1910.119 (l) requires a written program to manage changes to process chemicals, technology, equipment, procedures, and facilities. OSHA PSM 1910.119 (i) also requires a pre-start-up safety review. The control of plant expansions is dealt with in *Memorandum of Guidance on Extensions to Existing Chemical Plant Introducing a Major Hazard* (BCISC, 1972). Changes may be classified according to the stage at which they are made: design, commissioning, operating. They may also be distinguished by the degree of permanence: temporary or permanent.

5.3.1.1 Pressure Relief and Blowdown Hazards

A major hazard which is liable to arise as a result of change and one which is particularly dangerous is the pressure relief and blowdown invalidation.

Another type of change is the de-rating of equipment to a lower operating pressure. In this case, while the need to alter the pressure relief valve setting is usually appreciated, it is sometimes forgotten that it is necessary also to check the relief valve capacity, which is reduced by the de-rating. Similarly, an increase in equipment throughput may require a change in pressure relief valve capacity.

The ease with which hazards associated with pressure relief arise mean that the review of pressure relief is particularly important.

5.3.1.2 Other Hazards of Changes

A common change is the temporary replacement or bypassing of equipment such as a reactor or heat exchanger with a length of pipe. This appears to be a simple matter, but it is still necessary to design the pipe

properly and to provide supports. The hazard is shown by the Flixborough disaster.

An alteration of a device which in some way limits flow or pressure can create a hazard. Examples include the removal of a regulator, orifice, etc. installed specifically to restrict flow, and the increase in the size of a valve trim or installation of a pump impeller capable of a greater head.

5.3.1.3 Changes to the Process

The lesson commonly drawn from the Flixborough disaster is the importance of maintaining the integrity of the equipment and of avoiding degradation due to an equipment change. The equipment can also be put at risk, however, by the operation of the process outside the envelope of operating conditions for which the pressure system is designed. It is essential, therefore, to control such process changes as well.

Moreover, even if the envelope of operating conditions remains the same, a change in operating practice may affect features such as inspection and test intervals, which tend to be based on historical equipment experience.

5.3.2 Software and Network Maintenance

Industrial communications networks play a vital role in plant control systems; therefore, it is necessary to have very reliable software and network systems. If networks do not meet the expected degree of reliability, the entire process will not be considered reliable. The reliability of a network system must meet two main criteria, integrity and availability. Failure of software and networks can significantly impact the productivity and the reliability of the process plant. It is important to have a group of personnel who has the right knowledge and tools to maintain the facility's networks and software. Problems in network communications generally relate to three main areas, (1) noise problems, (2) loading problems, and (3) device/software incompatibilities. Once the problem is identified, it requires the utilization of different tools to determine the cause of the problem and how to solve it. Security is very important in the network communications which include confidentiality, integrity, and availability. Company must update their security technology and practices to ensure that they have a safe and reliable network and communication system.

5.3.3 Managing Change

A simple definition of change is any addition, modification, or replacement that is not a replacement-in-kind.

The elements of a system for managing change are (1) procedures, (2) assessment, (3) inspection, (4) documentation, and (5) training.

There should be a formal procedure which requires all modifications to be authorized by competent people and a standard method of making safety assessments; there should be a system of inspection of changes by a competent person to make sure the work has been done as intended and is complete; there should be a system of documentation to record the change; and there should be adequate training so that all personnel concerned understand the system of control.

OSHA PSM Standard requires a written procedure to manage changes. Several considerations must be addressed:

1. technical basis;
2. impact of change on safety and health;
3. modifications to operating procedures;
4. necessary time period for the change;
5. authorization requirements.

Additional management of change (MOC) requirements are as follows:

1. employees involved in operating and maintaining the equipment shall be informed of, and trained in, the change prior to implementation of the change;
2. process safety information shall be updated;
3. operating procedures or practices shall be updated.

Systems for managing change and the levels at which particular types of change can be authorized vary somewhat, but the basic principles are quite clear. There should be a well defined and understood system of authorization. The system described by Henderson and Kletz (1976) is that any change to equipment or process must be authorized in writing by a competent manager and engineer.

5.3.3.1 Procedure for Changes

The procedure given by Henderson and Kletz requires that before authorizing a modification particular attention should be paid to ensuring that:

1. The number and size of relief valves required are not changed (or any necessary changes are specified).
2. The electrical area classification is not changed (or any necessary changes to the electrical equipment are specified).
3. There are no effects on trips or alarms (or any necessary changes are specified).
4. There are no other effects which might reduce the standard of safety.
5. The appropriate engineering standards are followed.

6. The right materials of construction and fabrication standards are used.
7. Existing equipment is not subjected to conditions beyond the design basis without checking that it can withstand the new conditions.
8. Any necessary changes in operating conditions are made.
9. Adequate instruction and training are provided to operating and maintenance teams.

5.3.3.2 Safety Assessment of Modifications

The above procedures need to be supplemented by a system for the identification of hazards in the modification. A checklist and guidance for safety assessment of modification is provided in CCPS (1992), *Guidelines for Hazard Evaluation Procedures*. Any modification should include the inspection of modifications; documentation of modifications; training on modified systems; commissioning of modifications and take into account the variations of control for the modifications.

5.3.4 Some Modification Problems

5.3.4.1 Design and Repair Codes and Standards

Like equipment design, change should be done in accordance with applicable standards or codes. Design standards or codes are not applied retrospectively to existing equipment and its changes.

There is a problem, however, which arises from the nature of the advanced codes. The design criteria in such codes are quite complex, may require analysis of stress concentration sites, of fatigue and creep and assume finite equipment life The main guide for equipment repairs and modifications is API RP 510: *Inspection, Rating and Repair of Pressure Vessels in Petroleum Refinery Service*.

5.3.4.2 Materials Aspects

The materials for changes or repair should be 'suitable' and should have properties at least equal to that of the parent material originally used. If the codes apply, the materials should have guaranteed minimum properties and should be so certified. It is also necessary that both parent and replacement materials should be capable either of withstanding the fabrication processes without losing their required properties or of having these properties restored.

The availability of suitable materials may be a problem, either because a material is no longer made or because delivery times are long.

The parent material is sometimes degraded and may require treatment before welding can be done on it. Thus, for example, if there is surface sulfur contamination, it may be necessary to remove the contamination by grinding and then to preheat to allow welding. In some cases, the parent metal may have been rendered unweldable by high temperature exposure or hydrogen attack.

The application of quality control procedures can give rise to difficulties. There has been continuous progress in the measurement of defects in materials and in acceptance standards. The situation can easily arise, therefore, where the quality demanded for the replacement materials far exceeds that of the original equipment material.

The properties of both parent and replacement materials may be affected by activities such as cold working, preheating, or welding. Material properties may be restored to some extent by suitable heat treatment, but the heat treatment operations which can be carried out are strictly limited.

5.3.5 Major Plant Expansions

A major expansion involves changes on a greater scale than a normal equipment modification. Major plant expansions are in large part covered by the procedures for design, on one hand, and for change on the other. It is important to address the following aspects when major plant expansions are planned to take place: Design of changes, this means that it is particularly necessary to check on the various facilities which may become inadequate or overloaded as a result of the expansion; layout for expansions, pipelines, utilities, etc. must be considered during the time of expansion and after; safe-working procedures during the expansion, this will help warrant the safety of employees and contractors participating during the construction of the expansion.

ACRONYMS

MTTF	Mean Time to Failure
MTTFF	Mean Time to first failure
ROCOF	Rate of occurrence of failure
MTBF	Mean time between failures
FACTS	Failure and Accident Technical Information System
WOAD	Worldwide Offshore Accident Database
MARS	European Union Major Accident Reporting System
ERDS	European Reliability Data System
OREDA	Offshore Reliability Data
EIReDA	European Industry Reliability Data Bank
EuReDATA	European Reliability Data Bank Association
ESReDA	European Safety and Reliability Research and Development Association
GIDEP	Government Industry Data Exchange Program
HSEES	Hazardous Substances Emergency Events Surveillance
NTSIP	National Toxic Substance Incident Program
CCPS	Center for Chemical Process Safety
DOE	Department of Energy
DBMS	Database management system

PTW	Permit to Work
NCSR	National Center for Safety and Reliability
OSHA	Occupational Safety and Health Administration
EPA	Environmental Protection Agency
DOT	Department of Transportation
NCR	Nuclear Regulatory Commission
NIOSH	National Institute for Occupational Safety and Health
NFPA	National Fire Protection Association
ASME	American Society of Mechanical Engineers
AIChE	American Institute of Chemical Engineers
NACE	National Association of Corrosion Engineers
IEEE	Institute of Electrical and Electronics Engineers
IEC	International Electrotechnical Commission
ASTM	American Society of Testing and Materials
ANSI	American National Standards Institute
NBIC	National Board Inspection Code
API	American Petroleum Institute

NOTATION

$\mathcal{L}(\ )$	Laplace transform
$\bar{g}(s)$	Laplace transform of $g(t)$
a	constant; parameter in rectangular, gamma, Pareto, and extreme value distributions
A	event A
$A(t)$	availability
a, b	constants
A_i	ith individual sub events of event A
B	event B
C	event C; minimum cut set
E	event E
$E(\ )$	expected value
$f(t)$	failure density function
$F(t)$	failure distribution function
$H(t)$	cumulative hazard function
$H(t)$	cumulative hazard function
k	counter; state counter; parameter in Weibull distribution (alternative form)
k	parameter in Weibull distribution (alternative form)
m	mean life
m	mean life; parameter in Weibull distribution (alternative form); mean of normal distribution; number of repairmen
m	parameter in Weibull distribution (alternative form)
m^*	location parameter in log−normal distribution
m_d	mean down-time
m_r	mean repair time; mean of normal distribution of repair times
m_u	mean up-time
n	total number of outcomes; counter; number of items; number of states; number or trials; number of repairmen; number of applications of load
n_A	number of outcomes of event A
$n_f(t)$	number of items failed
n_i	number of times outcome i occurs
$n_s(t)$	number of items surviving
P	probability
p	probability; probability of success

P	probability; probability of success
$P(\mid)$	conditional probability
p_i	individual probability of success
P_k	probability of being in state k
q	probability of failure
$Q(t)$	probability of failure, unreliability, failure distribution function
r	counter
$R(t)$	probability of success or survival, reliability
$R(t)$	probability of success or survival, reliability, reliability function
$r_1 r_2$	equation roots defined by Equations 5.65 and 5.68
R_i	reliability of individual component
R_x	reliability of component
s	Laplace operator
t	time
T	time period; minimum tie set
t	time; failure time
t_i	time to failure of component i
t_r	repair time
$u(t)$	up-time
X	event X
$z(t)$	hazard rate
α	parameter in Weibull distribution (alternative form)
β	shape factor in Weibull distribution
γ	location parameter in Weibull distribution; number of standard deviations
η	characteristic life in Weibull distribution
λ	failure rate; transition rate
λ'	failure rate
μ	moment; mean number of events; repair rate
μ_i	ith moment (theoretical)
σ	standard deviation of normal distribution; shape factor in log−normal distribution
τ	time; repair time

REFERENCES

AGA, 2001. Purging Principles and Practice. third ed. American Gas Association, Washington, DC.

Bazovsky, I., 1961. Reliability Theory and Practice. Prentice-Hall, Englewood Cliffs, NJ.

BCISC, 1959. Safety in Inspection and Maintenance of Chemical Plant. British Chemical Industry and Safety Council, London.

BCISC, 1972. Memorandum of Guidance on Extensions to Existing Chemical Plant Introducing a major Hazards. British Chemical Industry and Safety Council, London.

Blackett, P.M.S., 1962. Studies of War. Oliver & Boyd, Edinburgh.

CCPS, 1989. Guidelines for Process Equipment Reliability Data, with Data Tables. AIChE, New York, NY.

CCPS, 1992. Guidelines for Hazard Evaluation Procedures, second ed. with Worked Examples. AIChE, New York, NY.

Carhart, R.R., 1953. A Survey of the Current Status of the Reliability Problem. Res. Memo RM-1131. Rand Corp, Santa Monica, CA.

Davidson, 1994. UKAEA, (data from Reliability Technology by A.E. Green and J.R. Bourne, 1972).

Farmer, F.R., 1971. Experience in the reduction of risk. Major Loss Prevention. 82.

Green, A.E., Bourne,A.J., 1972. Reliability Technology, New York: Wiley.

Henderson, J.M., Kletz, T.A., 1976. Must plant modifications lead to accidents. Process Industry Hazards, pp. 53.

HSE, 1979. Hot work, welding and cutting in plant containing flammable materials. HS(G) 5, London.

HSE, 1985. The Cleaning and Gas Freeing of Tanks Containing Flammable Residues. Health Safety Executive, UK.

HSE, 1988. Industrial Robot Safety. HS(G)43. HMSO, UK.

HSE, 2000. Safety in Working with Lift Trucks. Health Safety Executive, UK.

Kletz, T.A., 1982. Hazards in Chemical System Maintenance: Permits. In Fawcett, H.H., and Wood, W.S., (1982), op. cit.p. 807.

Kletz, T.A., 1999. The origins and history of loss prevention. Trans IChemE. 77 (Part B (May)), 109−116.

OREDA, 2009. Offshore Reliability Data Handbook, fivth ed., Norway.

Scott, S., 1992. Management of safety—permit-to-work systems. Major Hazards Onshore and Offshore, pp. 171.

Shooman, M.L., 1968. Reliability physics models. IEEE Trans. Reliab. R-17, 14.

Smith, D.J., Babb, A.H., 1973. Maintainability Engineering. Pitman, London.

Wingender, H.J., 1991. Reliability data collection in process plants. In: Cannon, A.G., Bendell, A. (Eds.), op. cit., p. 81.

Hazard Identification

The identification of areas of vulnerability and of specific hazards is of fundamental importance in loss prevention. Once these have been identified, the battle is more than half won. Such identification is not a simple matter, however.

In many ways, it has become more difficult as the depth of technology has increased. Loss prevention tends, increasingly, to depend on the management system, and it is not always easy to discover the weaknesses in this. The physical hazards also no longer lie on the surface, accessible to simple visual inspection. On the other hand, there is now available a whole battery of hazard identification methods which may be used to solve these problems. Selected references on hazard identification are given in Table 8.1 of full version Lees' book (fourth edition).

Different methods are required at different stages of a project. In the full version Lees' book (fourth edition), Table 8.2 lists some of these stages and the corresponding hazard identification techniques. The list is illustrative and, in particular, a technique quoted for one stage may also be applicable to another stage. There is no single ideal system of hazard identification procedures. The most appropriate system varies to some extent with the type of industry and process. Thus, for example, a firm involved in the batch manufacture of a large number of organic chemicals is likely to be much more interested in techniques of screening and testing chemicals and reactions than one operating ethylene plants.

6.1. SAFETY AUDITS

One of the first systematic methods of hazard identification used in the chemical industry was the safety audit. Audits of various types are a normal management tool and are of considerable importance in safety.

An early account of safety audits in the United Kingdom was the BCISC report *Safe and Sound* (1969/9), which drew on the experience of the US chemical industry, where the safety audit was established as a prime means of ensuring safety. A full description of such an audit is given in the BCISC's *Safety Audits* (1973/12). A safety audit subjects every area of the organization's activity to a systematic critical examination. It aims, like other audits, to reveal the strengths and weaknesses and the areas of vulnerability. It is carried out by professionals and results in a formal report and action plan.

A safety audit examines and assesses in detail the standards of all facets of a particular activity. It extends from complex technical operations and emergency procedures to clearance certificates, job descriptions, housekeeping, and attitudes. It involves labor relations since people are asked about their training, their understanding of works policy, and whether they think they are making their contribution in the right way.

An audit might cover a company-wide problem or a total works situation (say, its emergency procedures or effluent systems) or simply a single plant activity. At any level an audit is carried out thoroughly by a team made up of people immediately concerned, assisted by experts in various fields, and experienced people not connected with the area of audit, to ensure that a fresh and unbiased look is injected into the inspection process.

Safety audits thus cover the following main areas:

1. Site level
   a. Management system;
   b. Specific technical features.
2. Plant level
   a. Management system.

The audit has five main elements:

1. Identification of possible hazards;
2. Assessment of potential consequences of realization of these hazards;
3. Selection of measures to minimize frequency and/or consequences of realization;
4. Implementation of these measures within the organization;
5. Monitoring of the changes.

The audit is conducted in a fairly formal way, using a checklist, although experience-based approaches are used with appropriate protocols. In full version of Lees' book (fourth edition), Table 8.3, taken from *Safety Audits*, gives a list of activities and activity standards. These standards can be used as the basis of a numerical rating scheme. If a safety audit is conducted, it is essential that the management act on it. This does not necessarily mean that every recommendation must be implemented, but there should be a reasoned response.

6.2. MANAGEMENT SYSTEM AUDITS

The management system is crucial to loss prevention and it is essential that this system itself be monitored. Management audit is so important that a brief, summary account is in order at this point.

The management system may be audited in several ways. These include (1) self-checking procedures, (2) internal audit, and (3) external audit. An illustration of a specific check built into the system is a formal requirement that a plant manager audit a proportion of the permits-to-work each week.

The safety audits described above include an internal audit of the overall system. In particular, there are checks on:

1. Overall management attitude, policies, systems and procedures, and personnel selection;
2. Plant level management, attitude, systems, training, and feedback;
3. Incident reporting, investigation, and statistics.

An external audit may be carried out by an outside body such as a firm of consultants.

There are now available a number of audit systems, including proprietary systems. Some of these yield quantitative factors applicable in quantitative risk assessment (QRA).

The HSE is also concerned with auditing the management system. Indeed, it is fair to say that this is its most important function. When visiting a factory, an inspector is very much concerned with the way the system works in practice. Audit by the HSE is another aspect discussed in Chapter 4.

6.3. CHECKLISTS

One of the most simplistic tools of hazard identification is the checklist. Like a standard or a code of practice, a checklist is a means of passing on lessons learned from experience. It is impossible to envisage high standards in hazard control unless this experience is effectively utilized. The checklist is one of the main tools available to assist in this.

Checklists are applicable to management systems in general and to a project throughout all its stages. Obviously the checklist must be appropriate to the stage of the project, starting with checklists of basic material properties and process features, continuing on to checklists for detailed design and terminating with operation audit checklists.

There are a large number of checklists given in literature—indeed a paper on practical engineering is quite likely to include a checklist. Selected references on checklists are given in Table 8.4 of full version Lees' book (fourth edition), but these are only a tiny fraction of those available. A number of checklists are given in full version Lees' book (fourth edition). Checklist should be used for just one purpose only—as a final check that nothing has been neglected. Also, it is more effective if the questions cannot be answered by a simple 'yes' or 'no' but require some thought in formulating an answer.

6.4. MATERIALS PROPERTIES

6.4.1 Physical and Chemical Properties

It is obviously necessary to have comprehensive information on all the chemicals in the process: raw materials, intermediates, and final products. A list of some important physical and chemical properties of any chemical is given in Table 8.5 of full version Lees' book (fourth edition). The significance of most of the items in the list is self-evident.

Compilations of properties of hazardous substances are typically found in accompanying Material Safety Data Sheets (MSDSs). Other sources include those by the NFPA (NFPA 704) and the NIOSH.

6.4.2 Material Safety Data Sheets

Information on the physical and chemical properties of chemicals and on the associated hazards and the appropriate precautions cast in standard format is available in the form of MSDSs.

Safety data sheet compilations are prepared by various organizations and are available on the Internet.

An EC Directive (91/155/EEC) lays down the contents of a safety data sheet. In the United Kingdom, this is implemented by the Chemical Hazard Information and Packaging (CHIP) Regulations. HSE guidance: *Safety Data Sheets for Substances and Preparations Dangerous for Supply* (1993) has been amended on a regular basis since its inception, and further guidance is available in *The EC Safety Data Sheet Directive* by CONCAWE (1992 92/55).

The Directive requires that the contents of a safety data sheet cover:

1. identification of the substance/preparation and company;
2. composition/information on ingredients;
3. hazard identification;
4. first-aid measures;
5. firefighting measures;
6. accidental release measures;
7. handling and storage;
8. exposure controls/personal protection;
9. physical and chemical properties;
10. stability and reactivity;
11. toxicological information;
12. ecological information;
13. disposal considerations;
14. transport information;
15. regulatory information;
16. other information.

6.4.3 Impurities

Impurities, foreseen or unforeseen, can have serious adverse effects on many aspects of process operation, including safety and loss prevention. Here consideration is confined to the identification of possible problems with impurities.

There is no established methodology of identifying impurities which may arise in the process that is comparable with that for checking the reactivity of a chemical. Reliance usually has to be placed on the literature and on experience. As discussed below, however, indication of possible impurities and of their adverse effects is one of the particularly valuable features of a pilot plant.

6.5. PILOT PLANTS

Scale-up from laboratory to full-size plant has always been a difficult problem in the chemical industry. Indeed it is this problem which in large part is responsible for the profession of chemical engineering. An important tool for tackling the problem is the pilot plant. In the words of L.H. Baekeland, the philosophy of the pilot plant is: 'Make your mistakes on a small scale and your profits on a large scale.'

A pilot plant is used principally to assist in the scale-up of the process design rather than the mechanical design, for example, reaction conditions rather than pump specifications.

Some types of information relevant to safety which a pilot plant may yield include:

1. operating conditions (e.g., pressure, temperature);
2. design parameters (e.g., reaction rates, heat transfer coefficients);
3. reactor problems;
4. unit operation problems;
5. material handling problems (e.g., foaming, solids handling, catalyst handling);
6. decomposition;
7. impurities;
8. corrosion;
9. fouling;
10. analytical problems;
11. operating problems;
12. working environment problems (e.g., toxics);
13. effluent and waste disposal problems.

If a pilot plant is used, it is essential, of course, that due attention is paid to safety in its design and operation. This rather different aspect is dealt with in Appendix 10 of full version Lees' book (fourth edition).

6.6. HAZARD INDICES

There are a number of hazard indices which have been developed for various purposes. Mention has already been made at the level of the material processed in the indices of energy hazard potential. There are other indices which are applicable to the process and plant as a whole. Some principal indices of process and plant hazard are the Dow Index, the Mond Index, and the IFAL Index. Another technique of ranking hazards is that of rapid ranking.

6.6.1 Dow Index

The Dow *Guide* was originally published in 1964 and has gone through seven editions (Dow Chemical Company, 1964, 1966; Dow Chemical Company, 1976, 1980). Descriptions of the first and third editions have been given by Pratt (1965) and van Gaalen (1974), respectively.

The procedure is to calculate the fire and explosion (F&EI) and to use this to determine fire protection measures and, in combination with a Damage Factor, to derive the base Maximum Probable Property Damage (MPPD). This is then used, in combination with the loss control credits, to determine the actual MPPD, the Maximum Probable Property Damage (MPPD), and the Business Interruption (BI) loss.

The procedure may be summarized as follows. The basic information required for the F&EI is the plot plan and flow sheet of the plant. For the economic losses, it is also necessary to know the value of the plant per unit area, or the 'capital density', and the monthly value of production.

The F&EI is calculated as follows. First a material factor (MF) is obtained. Then two penalty factors (F_1 and F_2), one for general process hazards (GPHs) and one for special process hazards (SPHs), respectively, are determined, and the process unit hazards factor (PUHF) (F_3), which is the product of these, is calculated. The product of the MF and PUHF is the F&EI.

6.6.2 Mond Index

The Mond Fire, Explosion, and Toxicity Index are an extension of the Dow Index. The index was developed at the Mond Division of ICI.

The principal modifications to the Dow method made in the Mond Index are as follows:

1. To enable a wider range of processes and storage installations to be studied;
2. To cover the processing of chemicals which are recognized as having explosive properties;
3. To offset difficulties raised by high heat of combustion per unit mass of hydrogen and to enable distinctions to be made between processes where a given fuel is reacted with different reactants;
4. To include a number of additional special process type of hazard considerations that have been shown by a study of incidents to affect the level of hazard significantly;
5. To allow aspects of toxicity to be included in the assessment;
6. To include a range of offsetting factors for good design of plant and control/safety instrumentation systems to enable realistic hazard levels to be assessed for plant units under varying levels of safety features;
7. To indicate how the results of using the method can be applied logically in the design of plants having a greater degree of 'inherent safety'.

The Mond Index was developed from the 1973 version of the Dow Index and the comparisons between the two given here relate to that version. Since then, there have been a number of developments in the Dow Index itself, in particular, the introduction of the loss control credit factors.

The Mond method involves making an initial assessment of hazard in a manner similar to that used in the Dow Index, but taking into account additional hazard considerations. The potential hazard is expressed in terms of the initial value of a set of indices for fire, explosion, and toxicity. A hazard factor review is then carried out to see if there is scope to reduce the hazard by making design changes, and intermediate values of the indices are determined. Offsetting factors for preventive and protective features are applied, and the final values of the indices, or offset indices, are calculated.

In outline, the method of determining the Mond Index is as follows. The plant is divided into units, the demarcation between units being based on the feasibility of locating a separating barrier (open space, wall or floor) between the unit and its neighbors.

The material factor is determined as in the Dow method, but in addition, special material hazard factors are introduced. Again as in the Dow method, use is made of factors for GPHs and SPHs although the particular factors are different. A quantity factor, based on the inventory of material, and layout hazard factors are also introduced. There are also factors for toxicity hazard.

The features to be taken into account in calculating the various factors are specified in detail in the *Technical Manual*, which contains a wealth of practical information, particularly in respect of the offsetting features.

6.6.3 IFAL Index

The IFAL index is a separate index developed by the Insurance Technical Bureau primarily for insurance assessment purposes. Accounts have been given by J. Singh and Munday (1979), Munday et al. (1980), and H.B. Whitehouse (1985). Calculation of the index is described in the *IFAL p Factor Workbook* (Insurance Technical Bureau, 1981).

The index was developed in order to provide a means of assessment that was more scientific-based and satisfactory than the historical loss record, which tends to be subject to chance fluctuations.

The method involves dividing the plant into blocks and examining each major item of process equipment in turn to assess its contribution to the index. The main hazards which contribute to the index are as follows:

1. pool fires;
2. vapor fires;
3. unconfined vapor cloud explosions;
4. confined vapor cloud explosions;
5. internal explosions.

For hazards (2)–(5), emission frequency and hole size distributions are used; and for each case, emission, ignition, fire, and explosion are modeled and the damage effects are estimated.

The IFAL Index is the product of the process factor p and of two modifying factors, the engineering factor e and the management factor m. For a process engineered and managed to 'standard good practice', the last two factors are unity and the IFAL Index equals the p factor.

In contrast to the Dow and Mond Index methods, the IFAL Index method is too complex for manual calculation and is carried out using a computer program.

6.6.4 Dow Chemical Exposure Index

As stated earlier, the main Dow Index contained at one time a Toxicity Index. Dow now uses a separate Chemical Exposure Index (CEI).

The CEI is a measure of the relative acute toxicity risk. It is used for initial process hazard analysis (PHA), in the calculation of its Distribution Ranking Index (DRI) and in emergency response planning.

The information needed for the calculation of the CEI is the physical and chemical properties of the material; a simplified process flow sheet, showing vessels and major pipework with inventories; and an accurate plot plan of the plant and the surrounding area.

It is also necessary to have toxicity limits in the form of the American Industrial Hygiene Association (AIHA) Emergency Response Planning Guidelines (ERPGs) for the material. There are three levels of ERPG (ERPG-1, ERPG-2, and ERPG-3), which are essentially the maximum airborne concentrations below which it is believed that nearly all individuals could be exposed without experiencing some defined effect, the effects being: for level 1, experience of something more than a mild transient health effect or odor; for level 2, serious or irreversible health effects; and for level 3, life-threatening effects. ERPGs are considered further in Chapter 18 of full version Lees' book (fourth edition).

The procedure for the calculation of the CEI is to determine (1) the release scenarios, (2) the ERPG-2, (3) the airborne quantities, (4) the CEI, and (5) the hazard distance. The CEI is based on the scenario giving the largest airborne quantity (AQ).

The Dow *CEI Guide* gives a set of rules for defining release scenarios and of equations for estimating the AQ based on gas or liquid flow from vessels and pipes, liquid flash-off, and pool formation and evaporation. These, in effect, constitute a release model.

6.6.5 Mortality Index

Another index which should be mentioned here is the Mortality Index. This is an index of the lethality of materials with major hazard potential and is defined as the number of deaths per tonne of material involved. The index is discussed in Chapters 16–18 of full version Lees' book (fourth edition).

6.7. HAZARD STUDIES

One of the widely accepted definitions for hazard and risk is that, hazard is the potential to cause harm, while risk is the likelihood of harm. For a long time, people have cared about risk rather than hazard. Policies, standards, and techniques have been implemented to reduce risk. However, the truth is that risk can be eliminated by making events less hazardous. The first and most critical step to achieve a lower risk is through hazard identification.

There are now a number of methods that have been developed for the identification of hazards. A brief overview of these is given in this section, and a more detailed account of each method is then presented in the succeeding sections.

Overviews of hazard identification methods are available from numerous sources. One primary information source is the *Guidelines for Hazard Evaluation Procedures* (CCPS, 1992).

Some of the methods described are complete methods. Methods in this category are What-If analysis, Preliminary Hazard Analysis (PHA), screening analysis techniques, HAZOP studies, and FMECA. Of these, PHA and screening analysis are designed for use at an early stage in the design.

Other methods, such as event tree, fault tree analysis, sneak analysis, computer control, and human error analysis (known as CHAZOP), are specialist techniques used to complement or support more comprehensive methods. Scenario development and consequence analysis are part of hazard identification but are also major features in hazard assessment.

6.8. WHAT-IF ANALYSIS

The What-If method involves asking a series of questions beginning with this phrase as a means of identifying hazards. Apart from checklists, What-If analysis is possibly the oldest method of hazard identification. The method is to ask questions such as:

What if the pump stops?
What if the temperature sensor fails?
The questions posed need not necessarily all start with What-If; other phrases may be used.

The method involves review of the whole design by a team using questions of this type, often using a list of pre-determined questions. A checklist for use in What-If analysis is given by Burk (1992).

6.9. EVENT TREE AND FAULT TREE ANALYSIS

The event tree and fault tree methods may be used either qualitatively or quantitatively. Here it is sufficient to draw attention to the value of tree methods, particularly fault trees, in hazard identification.

An event tree involves the development of the consequences (outcomes) of an event. The overall approach is similar to that adopted in FMEA, which is described below.

A fault tree involves the development of the contributing causes of an undesirable event, often a hazard. The possibility of this event must be foreseen before the fault tree can be constructed. What the fault tree helps to reveal are the possible causes of the hazard, some of which may not have been foreseen.

Fault trees are used extensively in hazard assessment, but they are also of great value in hazard identification. In many cases, it is sufficient to be able to identify the fault paths and the base events, which can give rise to the top event, it being unnecessary to quantify the frequency of occurrence of these events.

In an FMEA or an event tree, the approach is 'bottom up', while in a fault tree, it is 'top down' (deductive). The HAZOP method involves both approaches, starting with the deviations and tracing down to the causes and up to the consequences.

6.10. BOW-TIE METHOD

The bow-tie diagram is a risk assessment method that is used to identify critical events, build accident scenarios, revise causes of accidents, and study the effectiveness and influence of safety barriers in the diagram (ARAMIS Project: Effect of safety systems on the definition of reference accident scenarios on SEVESO establishments). The bow-tie method can be described as the combination of a fault tree and an event tree, connected by a critical event (CE) which lies in the center of the diagram (its name comes from the shape the combination of both diagrams make). The bow-tie diagrams may also show the safety barriers of safeguards to prevent the critical event or mitigate the Major Event (ME) or consequence.

The fault tree identifies the plausible causes for a critical event to happen. Generic fault trees could be used (ARAMIS final user guide provides a few in the appendix), but it is recommended that they are modified.

Undesirable Events (UE) and Current Events (CE) may combine to produce a Detailed Direct Cause (DDC) that may lead to a Direct Causes (DC) leading to Necessary and Sufficient Conditions (NSC), and finally to the Critical Event (*ARAMIS Project: Effect of safety systems on the definition of reference accident scenarios on SEVESO establishments*). The right side of the bow-tie diagram is represented by an event tree. After the Critical Event has occurred, the Secondary Critical Event (SCE) and Tertiary Critical Event (TCE) may occur leading to a Dangerous Phenomena (DP) and finally a Major Event (ME).

The critical event lies in the center of a bow-tie diagram. The most common critical events are associated with loss of containment (LOC) or a loss of physical integrity (LPI).

The first step in the development of a bow-tie diagram is the identification of the critical events. This is done by associating critical events to each hazardous equipment. For this first step, two matrices may be used: equipment type—critical events matrix and substance state—critical event matrix. After all the critical events have been identified, a fault tree has to be built for each.

6.11. PRELIMINARY HAZARD ANALYSIS

PHA is a method for the identification of hazards at an early stage in the design process. Accounts of PHA are given in the CCPS *Hazard Evaluation Guidelines* and by Kavianian et al. (1992). PHA is a requirement of the MILSTD-882D Standard Practice for System Safety.

Since the early identification of hazards is of prime importance, many companies have some sort of technique for this purpose, a portion of which are called PHA; the use of the term tends to be fairly loose.

The CCPS *Guidelines* state that PHA is intended for use only in the preliminary stage of plant development, in cases where past experience provides little insight into the potential hazards, as with a new process. The information required for the study is the design criteria, the material and equipment specification, and so on. The *Guidelines* list the entities examined for hazards as: (1) raw materials, intermediates, and final products; (2) plant equipment; (3) facilities; (4) safety equipment; (5) interfaces between system components; (6) operating environment; and (7) operations (maintenance, testing, etc.).

6.12. SCREENING ANALYSIS TECHNIQUES

One of the principal methods of identifying process hazards is the HAZOP study. Before describing this,

however, it is appropriate to consider first the method developed for use in conjunction with, but earlier than the HAZOP.

The screen analysis is done at an early stage in the design where there is a block layout of plant items. The objective is to determine whether there are problems in areas such as data on the chemicals, information about the hazards, basic features of the process design, or layout and siting of the plant.

There are a number of advantages to be gained from conducting screening prior to the HAZOP study itself. It assists in the identification of those hazards and other problems which are quite basic and which are therefore capable, in principle, of being identified at this earlier stage. It reveals deficiencies in the design information. And it exposes hazards due to interactions between the plant and other plants or the environment, at which the HAZOP study is less effective. These features contribute to removing potential delays on the critical path of the project.

The conduct of a screening analysis involves compilation of two lists: (1) a database of hazards and nuisance properties of the material and (2) a list of potential hazards, nuisances, and other matters of concern. The study therefore covers, albeit in a different way from a HAZOP study, the two essential features of the design intent: (1) materials and (2) equipment and activities.

Use is made of a list of suitable guidewords. Guidewords may be generated by considering the below:

1. Energy sources;
2. Chemical interactions;
3. Environment.

Each of these headings may be broken down further. A suitable energy set is a list of the main forms of energy.

1. People
2. Materials
3. Equipment.

The study results in a set of work assignments to follow-up the queries raised. As an illustration, the block layout might show a storage tank containing flammable liquids. The application of the guideword 'fire' to this tank might raise a number of queries. Among those that may be considered are (1) frequency of fire, (2) detection of fire, (3) access for firefighting, (4) method of firefighting, (5) source and disposal of fire water, (6) use of alternative firefighting agents, (7) nature of combustion products, (8) effects of radiant heat, and (9) need for a disaster plan.

6.13. HAZARD AND OPERABILITY STUDIES

The HAZOP study is carried out when the ELD (piping and instrument diagram) of the plant becomes available. Accounts of HAZOP studies are given in HAZOP and HAZAN (Kletz, 1983, 1986, 1992), Guidelines for Hazard Evaluation Procedures (CCPS, 1992) (the CCPS *Hazard Evaluation Guidelines*), and *A Manual of Hazard and Operability Studies* (Knowlton, 1992), and by Kletz (1972), S.B. Gibson (1974, 1976a,b), Lawley (1974a,b), and Kavianian et al. (1992).

The basis of such a study may, in principle, be a process flow diagram (PFD), piping and instrumentation diagram (P&ID), or other information, which reveals the design intent. The basis of a HAZOP study is generally the ELD (P&ID) supplemented, as appropriate, by other information such as operating instructions, MSDSs, details of metallurgy, process control information, etc.

The HAZOP study technique is not a substitute for good design. There is something fundamentally wrong if the application of the method consistently reveals too many basic design faults.

The HAZOP technique has been widely adopted and is the centerpiece of the hazard identification system in many companies. It does, however, have some limitations.

These limitations are of two kinds. The first type arises from the assumptions underlying the method and is an intended limitation of scope. In its original form, the method assumes that the design has been carried out in accordance with the appropriate codes. Thus, for example, it is assumed that the design caters for the pressures at normal operating conditions and intended relief conditions. It is then the function of HAZOP to identify pressure deviations which may not have been foreseen.

The other type of limitation is that which is not intended, or desirable, but is simply inherent in the method. HAZOP is not, for example, particularly well suited to deal with spatial features associated with plant layout and their resultant effects.

There are a large number of variants on the original HAZOP study. Many, perhaps most, companies have adapted the technique to their own needs. Many of these adaptations extend the method in respect of non-normal operation or of other activities such as maintenance, or to take account of other concerns such as environmental effects. Another type of variant is the application of the basic technique of HAZOP to systems other than process vessels and pipework. Some of these are described in Section 8.12.29. In this regard, it should be remembered that HAZOP itself was an application of method study, and it may often be better to step back to this origin and then forward again from there rather than to try to apply HAZOP *per se*.

The methodology of HAZOP, or its parent critical examination, has been applied to a number of other situations, many of which are of interest in the process industries. These include (1) plant modification, (2) plant commissioning, (3) plant maintenance, (4) emergency shutdown and emergency systems, (5) mechanical handling, (6) tanker loading and unloading, (7) works traffic, (8) construction and demolition, and (9) buildings and building services.

6.14. FAILURE MODES, EFFECTS AND CRITICALITY ANALYSIS

Another method of hazard identification is FMEA and its extension FMECA.

Accounts of FMEA and of FMECA are given in the CCPS *Hazard Evaluation Guidelines* (1985/1, 1992/9) and by Recht (1966), J.R. Taylor (1973, 1974, 1975), Himmelblau (1978), Lambert (1978), Flothmann and Mjaavatten (1986), Moubray (1991), Kavianian et al. (1992), Scott and Crawley (1992), and Goyal (1993). A relevant code is BS 5760, which is described below.

FMEA involves reviewing systems to discover the mode of failure, which may occur, and the effects of each failure mode. The technique is oriented toward equipment rather than process parameters.

6.14.1 BS 5760

Guidance on FMEA is given in BS 5760 *Reliability of Systems, Equipment and Components*, Part 5: 1991 *Guide to Failure Modes, Effects and Criticality Analysis* (FMEA and FMECA). BS 5760: Part 5: 1991 deals with the purposes, principles, procedure, and applications of FMEA, with its limitations and its relationship to other methods of hazard identification, and gives examples.

FMECA is an enhancement of FMEA in which a criticality analysis is performed. Criticality is a function of the severity of the effect and the frequency with which it is expected to occur. The criticality analysis involves assigning to each failure mode a frequency and to each failure effect a severity.

The purpose of an FMEA is to identify the failures which have undesired effects on system operation. Its objectives include: (1) identification of each failure mode, of the sequence of events associated with it and of its causes and effects; (2) a classification of each failure mode by relevant characteristics, including detectability, diagnosability, testability, item replaceability, compensating and operating provisions; and, for FMECA, (3) an assessment of the criticality of each failure mode.

FMEA is an efficient method of analyzing elements which can cause failure of the whole, or of a large part, of a system. It works best where the failure logic is essentially a series one. It is much less suitable where complex logic is required to describe system failure.

FMEA is an inductive method. A complementary deductive method is provided by fault tree analysis, which is the more suitable, where analysis of complex failure logic is required.

6.14.2 Application of FMEA

The range of applications of FMEA is very wide. At one end of the spectrum, It has been applied to a pneumatic differential pressure transmitter, while at the other, It was also applied to a liquefied natural gas terminal.

6.15. SNEAK ANALYSIS

In contrast to general methods such as HAZOP and FMECA, there are also niche methods. One of these is sneak analysis. Accounts of sneak analysis are given by E.J. Hill and Bose (1975), J.R. Taylor (1979, 1992), Rankin (1984), Dore (1991), Hahn et al. (1991), Hokstad et al. (1991), and Whetton (1993b).

Sneak analysis originated in sneak circuit analysis (SCA), a method of identifying design errors in electronic circuits (E.J. Hill and Bose, 1975). The technique was developed within a particular company and was slow to spread. An early use in the process industry was in the work of J.R. Taylor (1979) on sneak path analysis. A lead in encouraging the wider use of sneak analysis has been taken by the European Space Agency (ESA), as described by Dore (1991).

A review of sneak analysis is given by Whetton (1993b), who gives as a rough definition: 'A sneak is an undesired condition which occurs as a consequence of a design error, sometimes, but not necessarily, in conjunction with a failure.'

In sneak analysis generally, four recognized categories of sneak have emerged: (1) path, (2) indication, (3) label, and (4) timing. Whetton refers to the following categories, which have developed from process work: (1) flow, (2) indication, (3) energy, (4) procedure, (5) reaction, and (6) label.

6.15.1 Types of Sneak

The different types of sneak which can occur are discussed by Whetton.

- Sneak flow
- Sneak indication
- Sneak label

- Sneak energy
- Sneak reaction
- Sneak procedure or sequence.

6.15.2 Sneak Analysis Methods

Whetton distinguishes three basic methods of sneak analysis: (1) topological, (2) path, and (3) clue.

The original sneak analysis method was based on decomposing electrical circuits into standard subnetworks. According to Whetton, the method has proved difficult to adapt to process plants in that successful application depends on the ability of the analyst to cast process phenomena in the form of electrical analogues.

Sneak path analysis, developed by J.R. Taylor (1979, 1992), is the investigation of unintended flow between a source and a target. It is performed in a systematic manner by decomposing the P&ID into functionally independent sections and identifying sources and targets, source−target pairs and paths between sources and targets. One method of doing this is the use of colored markers to trace the various paths, sometimes known as the 'rainbow' technique.

The clue method involves the use of a structured checklist. It is applicable to all types of sneak and examines each type by means of a question set.

6.15.3 Sneak-Augmented HAZOP

Sneak analysis may be used as an enhancement of HAZOP. The techniques used are the path and clue methods. Whetton gives a detailed procedure for the conduct of a sneak-augmented HAZOP. All drawings used should be 'as built'. Information on labels should cover the as-built condition. Essentially, the sneak analysis aspect of the procedure involves an examination of the various types of sneak against the appropriate clue list. Among the other points made are the following:

- Sneak flow
- Sneak indications
- Sneak labels
- Sneak procedures
- Sneak energy.

6.16. COMPUTER HAZOP

The HAZOP study method just described was developed for plants in which predominantly the control system was based on analogue controllers. The advent of computer control has created the need for some method, which addresses the specific problems of this form of control.

In recent years, process engineers have increasingly chosen to use Programmable Logic Controllers (PLCs), DCSs, and computers for process control. While these systems provide flexibility and close control of the process, they introduce an additional mode of failure to the plant. Modern HAZOP techniques including CHAZOP are used to consider such systems. This is quite distinct from computer aiding of HAZOP.

6.16.1 Checklist and Guideword Methods

A method for CHAZOP that has the same general approach as conventional HAZOP has been described by Andow (1991). Like HAZOP, it is intended to be carried out by a multi-disciplinary team following a systematic methodology and using standard guidewords and questions.

Andow describes a two-part study format, comprising a preliminary study and then the main study. The purpose of the preliminary study is to identify critical features early in the design. The study covers (1) system architecture, (2) safety-related features, (3) system failure, and (4) utilities failure. In the full study, attention is directed to (1) the computer-system environment, (2) input/output signals, and (3) complex control schemes.

6.16.2 Task Analysis Method

A quite different approach is discussed below. The methodology is based on an analysis of some 300 incidents in two organizations—one a process industry and one an avionics company. As an example of a process industry incident, they quote the premature start of a control sequence due to inadequate specification of the preconditions for initiation.

The underlying premise of the method is that most incidents are the result of deficiencies in the software–hardware interface. The authors state that attempts to conduct separate analyses of the hardware and the software in order to simplify the task, actually complicates it.

This interface is represented by four concentric functional levels. Each level is associated with certain system components. Each component has a defined task. Starting at the periphery and working in to the center.

6.17. HUMAN ERROR ANALYSIS

An important source of hazards and losses is maloperation of the plant, and other forms of operator error. There are a number of methods for addressing this problem. One of these is the HAZOP study which, as just described, utilizes aids such as the sequence diagram to discover potential for maloperation. Other methods which may be mentioned here are task analysis and action error analysis (AEA).

6.17.1 Task Analysis

Task analysis is a technique, which was originally developed as a training tool. As applied to the process operator, it involves breaking down the task of running the plant into separate operations carried out according to a plan. The objective is to discover potential difficulties, and hence errors, in executing the individual operations or the overall plan. A detailed account is given in Chapter 9.

Clearly such task analysis also ranks as a hazard identification technique. Task analysis has been applied to process plants with good results, but is not in widespread use. It may also be noted in passing that the writing of operating instructions constitutes a less formalized type of task analysis, and may likewise serve to identify hazards.

6.17.2 Action Error Analysis

Another technique of identifying operating errors is AEA, developed by J.R. Taylor (1979). This is a method of analyzing the operating procedures to discover possible errors in carrying them out.

The actions to be carried out on the process interface are listed in turn, each action being followed by its effects on the plant, so that a sequence is obtained of the form:

Action—Effect on plant—Action—Effect on plant...

The actions are interventions on the plant such as pushing buttons, moving valves.

The errors handled in the method are given in Table 6.1. The effects of possible errors are then

TABLE 6.1 Action Error Analysis: Errors Handled (J.R. Taylor, 1979)

Cessation of a procedure
Excessive delay in carrying out an action or omission of an action
Premature execution of an action—timing error
Premature execution of an action—preconditions not fulfilled
Execution on wrong object of action
Single extraneous action
In making a decision explicitly included in a procedure, taking the wrong alternative
In making an adjustment or an instrument reading, making an error outside tolerance limits

examined using guidewords for action. The main guide-words are as follows:

TOO EARLY
TOO LATE
TOO MUCH
TOO LITTLE
TOO LONG
TOO SHORT
WRONG DIRECTION
ON WRONG OBJECT
WRONG ACTION

The technique of AEA appears to have been quite widely used in Nordic countries, but much less so elsewhere.

6.18. SCENARIO DEVELOPMENT

6.18.1 Release Scenarios

The methods just described are concerned with HAZOP problems, which occur on most plants. A rather different question is the identification of potential sources of major release of hazardous materials, which may pose a risk to the workforce and/or the plant, and in unfavorable circumstances may even affect the public.

Information on release sources is required for hazard assessment and for emergency planning. In order to identify such sources, it is necessary to carry out a review.

In principle, virtually all elements of the pressure system (vessels, pumps, pipework) are points at which a release may occur. Table 6.2 gives a list of some of these release sources. A more detailed checklist of release sources is given in the CCPS *QRA Guidelines* (1989/5). The identification of release sources in a hazard assessment is illustrated by the studies given in the *Rijnmond Report* described in Appendix 8 of full version Lees' book (fourth edition).

The role of the HAZOP study in identifying such sources merits mention. A HAZOP study starts from the assumption that the plant is basically well designed mechanically and concentrates primarily on process parameter deviations. The type of release which a HAZOP study might identify would be overfilling of a tank as a result of maloperation, or brittle fracture of a pipe due to contact with cold liquefied gas, but the study is not normally concerned with mechanical failures which occur while the process parameters are within their design range. The points on the plant at which a release may occur can generally be identified by the release source review.

What the HAZOP study does is to identify specific credible realizations of release from these points. For a potential release, it is necessary not only to identify a source but also to decide on the nature of the release,

TABLE 6.2 Some Sources of Release on a Plant
Pipework
Pipe rupture
Pipe flanges
Valves
Hoses
Compressors
Pumps
Agitators
Vessels and tanks
Vessel/tank rupture
Vessel/tank overfilling
Reactors
Drain points
Sample points
Drains
Relief devices
Pressure relief valves
Bursting discs
Vents
Flares
Operational release
Maintenance release

which could occur. The tendency is to assume that the fluid released is at its design conditions. A HAZOP study may reveal that this is not so. For example, materials may be released from a storage tank as a result of a chemical reaction in the tank, which causes the liquid to heat up considerably above its storage temperature, and hence when released to give a much larger fraction of vapor.

6.18.2 Escalation Scenarios

The identification process should not stop at the point where a release occurs, but in principle should be continued to embrace the consequences of the release and the failures and other events, which may allow these consequences to escalate. In practice, this aspect is usually treated as part of the hazard assessment. While this is a reasonable approach, there is a danger that, unless the identification of the escalation modes is treated with a thoroughness matching that applied to the identification of the release modes, features which permit escalation and against which measures might be taken will be missed. The identification of the modes of escalation appears to be a rather neglected topic.

6.19. CONSEQUENCE MODELING

For major hazard plants, it is common practice, following the development of the scenarios just described, to carry out some degree of quantitative modeling of the consequences. Here it is sufficient to note its effect in giving a much better understanding of the likely development of the release and its potential for the identification of both hazards and mitigatory measures.

6.20. PROCESS SAFETY REVIEW SYSTEM

It is convenient at this point to consider under the heading of process safety reviews the family of techniques for hazard identification, hazard assessment, audit, and accident investigation developed by Wells and co-workers.

6.20.1 General Incident Scenario

The next element is a method of defining incident scenarios, which is a necessary part of both hazard assessment and accident investigation. An account of this is given by Wells et al. (1992). The general incident scenario (GIS) described by these authors is a structure intended to be applicable to any incident.

6.20.2 Preliminary Safety Analysis

The GIS provides the basis for a set of techniques for PSA. Accounts of PSA are given by Wells et al. (1993, 1994). The PSA methods described are:

1. concept safety review (CSR);
2. concept hazard analysis (CHA);
3. critical examination (CE);
4. preliminary consequence analysis (PCA);
5. preliminary hazard analysis (PHA).

These methods are now considered in turn, based on the account by Wells et al. (1993).

6.20.3 Concept Safety Review (CSR)

The CSR is carried out as early as practical, possibly during the process development stage. Features of the CSR are review of the process and the alternatives, the safety, health, and environment (SHE) information requirements, any previous incidents within the company and elsewhere, movement of raw materials and products, and organizational aspects.

6.20.4 Concept Hazard Analysis (CHA)

The CHA is an examination of the hazards of the process and is commenced as soon as the preliminary PFD becomes available. Wells et al. (1993) distinguish between a hazard and a hazardous condition. The former is purely qualitative, but the latter has a quantitative element. For example, a hazardous material constitutes a hazard.

The hazardous condition is determined by the quantity of the material. The CHA is akin to a coarse hazard study.

6.20.5 Critical Examination (CE)

The CE of system safety is in large part a critical examination of the inherent safety of the system and is therefore different from the critical examination involved in conventional HAZOP. The CE examines how the proposal is to be achieved and, in particular, (1) the materials, (2) the method, and (3) the equipment. It also examines any dangerous condition and its cause. A checklist of keywords for the CE is given in Table 8.35 of full version Lees' book (fourth edition).

6.20.6 Preliminary Consequence Analysis (PCA)

The PCA, which may be started when the preliminary P&ID becomes available, is a hazard assessment kept at a fairly simple level. The main hazardous events considered are (1) fire, (2) explosion, including missiles, and (3) toxic release. The analysis covers the effects on the environment as well as on humans.

The analysis seeks to identify worst-case scenarios, to establish whether these could have impact outside the site and to define the types of emergency which might arise.

6.20.7 Preliminary Hazard Analysis (PHA)

The PHA is an analysis, which centers on the dangerous disturbance entry in the GIS. It involves estimating the frequency/probability of certain events and thus, in contrast to some other PHAs described in Section 8.10, it is a simplified form of conventional hazard analysis.

The results of a PHA are presented in tabular form, the core of which is the events:

1. significant event to be prevented;
2. failure to recover the situation;
3. dangerous disturbance;
4. inadequate emergency control or action;
5. hazardous disturbances.

Together with expansions at both ends, into causes of the set of hazardous disturbances and into escalation of the significant event. For each event, an estimate is given of the frequency/probability.

6.20.8 Short-Cut Risk Assessment (SCRAM)

The SCRAM is essentially an extension of the PHA. It is described by Allum and Wells (1993). The SCRAM generates a set of significant events to be prevented, or top events, each of which is described in a risk evaluation sheet, which is essentially the tabulation of the PHA as just outlined. The risk assessment is developed using:

1. target risk;
2. severity categories;
3. priority categories.

6.20.9 Goal-Oriented Failure Analysis (GOFA)

The GOFA technique is another based on the CIS checklists. An account is given by Reeves et al. (1989). The method is described as a combination of a fault tree analysis and checklists. The example given by the authors is loss of containment, the causes of which are traced through the checklist structure.

6.20.10 Socio-technical Systems Analysis

A socio-technical systems analysis is used to examine the preconditions for failure. An account is given by Wells et al. (1994). These authors describe the procedure and give the primary checklist. The main headings of this list are as follows:

1. system climate;
2. communication and information systems;
3. working environment;
4. organization and management;
5. management control;
6. operator performance;
7. external systems;
8. procedures and practice;
9. site and plant facilities;
10. equipment integrity.

They also give the subsidiary checklist for external systems.

6.20.11 Layers of Protection Analysis (LOPA)

LOPA (adapted from Primatec Inc) is a simplified assessment approach to evaluate the risk of hazard scenarios, which can be treated as an extension of PHA. The information acquired from the PHA is the basis for LOPA. With comparison to the risk criteria, the existing safeguards can be evaluated. So far, there are several available LOPA methods. LOPA can assist to choose between safeguards rather than suggest additional ones.

The development of LOPA is mainly because it is more rational and objective method, which can avoid the disagreements of additional safeguards and the implementation of inappropriate methods caused by subjective engineering judgment.

6.21. CHOICE OF METHOD

A variety of methods of hazard identification applicable at the different stages of the project have been described. In many cases, the aspect of hazard identification covered by a particular method is self-evident, but in others it is not. Some methods are alternatives or are complementary. In full version Lees' book (fourth edition), Table 8.44 lists some of the principal methods as a guide to selection. The table gives the prime purpose(s) of each method. With regard to checklists, the applications given in the table are limited, but some of the other methods such as HAZOP may also make use of a checklist specific to that method.

6.22. FILTERING AND FOLLOW-UP

Although it is convenient for purposes of exposition to treat hazard identification and assessment separately, the two are often interwoven. A process of filtering of the hazards takes place. Two forms of filtering, which occur with many of the identification methods described, may, be mentioned. The first involves discarding a hazard, which is considered to be unrealizable.

In the second form of filtering, a hazard is discarded if it is considered not to be sufficiently important. A measure of the need for action on an identified hazard which is widely used, albeit often not explicitly, is the product of the magnitude and frequency of the event. It is also used in task analysis as a stopping rule to decide how far to carry the process of subdivision of the task.

Once a hazard has been identified and accepted, as worth further consideration, it is necessary to decide what to do about it. In many cases, once the problem has been identified, the solution is obvious or is known from experience. In many others it is taken care of by codes of practice. In a small proportion of cases, it may be appropriate to carry out a hazard analysis.

6.23. SAFETY REVIEW SYSTEMS

Formal systems for the conduct of safety reviews appropriate to the stages of the project are now widespread in the process industries. These systems go by different names.

6.23.1 Hazard Study Systems

The system used in Mond Division of ICI has been described by several authors (N.C. Harris, 1975; S.B. Gibson, 1976c; Hawksley, 1984). The system involves six hazard studies.

Hazard Study I is the first formal study, carried out at the project exploration stage. Its purpose is to ensure that there is full understanding of the materials involved in the process and their potential interactions, and also the constraints on the project due to the particular site, not only in terms of safety, but also in terms of environmental factors, including pollution and noise. This type of study was referred to above as a 'coarse hazard study'.

Hazard Study II is performed using the PFDs. Each section such as reaction, distillation is studied in turn and potential hazardous events such as fire, explosion are identified. At this stage it is often possible to eliminate such a hazard by means of design changes or to protect against it by the use of protective systems. In some instances, it may be necessary to study an event further. In this case, use may be made of a fault tree analysis, with the hazardous event as the top event in the tree.

Hazard Study III is the HAZOP study, which has already been described.

Hazard Study IV involves a review of the actions generated in Hazard Studies I–III by the plant or commissioning manager, who also ensures that operating procedures, safe systems of work, and emergency procedures are available.

Hazard Study V is the plant inspection carried out to ensure that the plant complies with the legal requirements on such matters as access and escape, guarding, emergency equipment, so on.

Hazard Study VI is undertaken during the commissioning and early operation of the plant and involves a review of changes made to the plant and its operating procedures, and of any differences between design intent and actual operation.

Once in operation, the plant is subject to further audits, but these fall outside the system as described.

6.23.2 Project Safety Review System

A broadly similar system is the project safety review system operated by BP. The reviews conducted in this system are as follows:

1. pre-project review;
2. design proposal review;
3. detailed design review;
4. construction review;
5. pre-commissioning review;
6. post-commissioning review.

The two sets of safety review systems described above typify those in use.

6.24. HAZARD RANKING METHODS

There are available a number of hazard ranking methods for assigning priorities among hazards. Accounts include those of Gillett (1985), Ashmore and Sharma (1988), Rooney et al. (1988), Casada et al. (1990), and Keey (1991).

6.24.1 Rapid Ranking

Typical of these methods is that of rapid ranking described by Gillett (1985). The basis of the method is a trade-off between the consequence and the frequency of the event considered. The more serious the consequence, the lower the frequency, which is tolerable for the event. In principle, therefore, the criterion for action may be taken, as a first approximation, as some threshold level of the product of consequence and frequency.

The rapid ranking method actually utilizes a mixture of quantitative and qualitative measures. The cost of damage to the plant and the frequency of occurrence are expressed quantitatively and the other consequences qualitatively.

6.25. HAZARD WARNING ANALYSIS

In some cases, analyses are carried out to identify the lesser events, which may serve as precursors to, or warnings of, more serious events. Such analyses clearly rank as hazard identification techniques. One such method is the technique of hazard warning analysis described by Lees (1982). Like fault tree analysis, on which it is based, this technique may be used to obtain quantitative results.

6.26. PLANT SAFETY AUDITS

The conduct of some form of plant safety review is a normal part of plant commissioning. Traditionally this is a joint inspection by the plant manager and the plant safety officer and is concerned mainly with checking that the company complies with the legal and company safety requirements. Attention is directed to features such as: access and means of escape; walkways, stairs, and floors; and firefighting and protective equipment.

There has developed from this, however, the more comprehensive plant safety audit. There is no uniform system for such audits. The BCISC publications *Safe and Sound* (1969/9) and *Safety Audits* (1973/12) give information on several systems.

The general approach is on the following lines. The plant safety audit is conducted by a small, interdisciplinary team which is not connected with the plant. The members may come from the same works or from other works. The team may be led by a quite senior manager. The audit is carried out first during the plant commissioning but is also repeated later at intervals. Typical intervals are a year after initial start-up and every 5 years thereafter.

The topics covered by the audit are illustrated by the checklists given in Tables 8.48 and 8.49 of full version Lees' book (fourth edition). The former gives topics listed in *Safety Audits*. The latter is a checklist for both audits and surveys given. As already emphasized, it is essential that management act on any audit made.

6.27. OTHER METHODS

6.27.1 Feedback from Workforce

It is the responsibility of the management to create systems and attitudes which lead to the effective identification of hazards, but it is the responsibility of everyone involved in the design and operation of the process to play his role in identifying and reporting any hazard of which he becomes aware.

This aspect has already been considered in the context of the safety system in chapter 4.

6.27.2 Process Design Checks

There is an enormous amount of experience on the hazards associated with unit processes, unit operations, plant equipment, pressure systems, instrumentation, etc., and it is essential that this be utilized. It is not enough, therefore, to rely only on generalized methods of hazard identification. Full use needs to be made of what is already known. This may be done by methods such as the use of checklists for such processes, operations. This aspect is discussed in more detail in Chapter 8.

6.27.3 Plant Equipment Checks

There are numerous types of check which are carried out on the plant equipment at various stages in the project. There are checks done before and during plant commissioning, including testing in manufacturers' works, checking the correct installation of the equipment, testing *in situ*, etc. There is a whole range of inspection and examination techniques, including non-destructive testing and condition monitoring, which are applied both during commissioning and in operation. Many items are checked or tested on a scheduled basis. While the results of some tests are unambiguous, others require careful interpretation and the application of sophisticated inspection standards.

6.27.4 Emergency Planning

If a works contains hazardous plant, it is essential to plan for possible emergency situations. The first step in emergency planning is identification of those hazards, which are sufficiently serious to warrant such planning. Such a study is not necessarily an exercise separate from the other hazard identification procedures, but it does emphasize particular aspects, especially the development and scale of the emergency, the effects inside and outside the works boundary, the parties involved in dealing with the situation, etc. Emergency planning is treated in Chapter 17.

6.28. QUALITY ASSURANCE

The quality of the hazard identification system should be assured by a suitable system of quality assurance. Normally this will be part of an accredited system covering the quality assurance of all aspects of company operation. Principles of quality assurance are discussed in Chapters 1 and 6 of full version Lees' book (fourth edition) and are therefore not repeated here, but it is appropriate to consider some aspects which apply to hazard identification.

There should be a formal system for hazard identification. For each stage of a project, a set of hazard identification techniques should be selected. The system should then ensure that these techniques are used.

Quality assurance may be based on inputs and process and/or outputs. It is generally prudent to utilize both. Inputs and process for hazard identification include: the qualifications, training, experience, and plant knowledge of the team performing the task; the methods of hazard identification used and their suitability; the project documentation available, its comprehensiveness and accuracy; and the time available for the task.

Outputs include the whole range of information generated by the team. They are therefore not confined just to items such as release scenarios or deviations with associated causes and consequences, but extend also to the statements of scope, the report on the analysis and the follow-up measures.

Quality assurance of hazard identification requires documentation both of the hazard identification system and of the hazard identification activities conducted in a particular project. It should leave an audit trail which can be followed later if necessary.

6.29. QUALITY ASSURANCE: COMPLETENESS

The quality of hazard identification virtually parallels the completeness of identification. It is of interest, therefore, to consider attempts made to assess the completeness of hazard identification. Some identified hazards are discarded as unrealizable. In discarding such hazards, discrimination is exercised. It is desirable to have some measure of this also.

A study of completeness and discrimination in hazard identification has been made by J.R. Taylor (1979). It is possible to specify a list of hazards which is complete. According to Taylor, process plant hazards are covered by the following list: fires, explosions, toxic releases, asphyxiations, drownings, and mechanical impact and cutting accidents. However, the price paid for completeness is a list so general that it is of little practical use. As soon as the list is developed into more specific categories, the degree of completeness is liable to decrease.

6.30. QUALITY ASSURANCE: QUASA

A formal method for quality assessment assurance of safety analyses (QUASA) has been developed by Rouhiainen (1990). The method covers both qualitative and quantitative aspects, but with emphasis on hazard identification, and it is therefore appropriate to consider it at this point.

QUASA is based on quality theory. In accordance with this theory, methods for the assessment of the quality of a safety analysis may be based essentially on inputs and process and/or on outputs. The features assessed in QUASA are given in Table 8.51 of full version Lees' book (fourth edition). As stated earlier, they include not just the immediate results of the hazard identification and risk assessment, but other features such as the statements of scope, the report on the analysis, and the follow-up measures. The emphasis in the method is thus on outputs, broadly defined.

6.31. STANDARDS

6.31.1 PSM—OSHA

Unexpected releases of hazardous chemicals, such as flammable, reactive, or toxic liquids and gases, can cause serious incidents in relevant industries, which are mainly attributed to the improper control and use of them, and have the potential to harm the surroundings. The Occupational Safety & Health Administration (OSHA) has issued the standards of Process Safety Management (PSM) of highly hazardous chemicals (29 CFR 1910.119), to make the workplace safer. PSM requirements in processes dealing with highly hazardous chemicals have been extensively discussed in the standards pertaining to general and construction industries. The standards of OSHA provide a fully developed method, with a combination of technologies, procedures, and practices to manage the hazards related to hazardous chemicals. The following issues are addressed as PSM requirements.

6.31.2 RMP

Risk Management Program (RMP) is regulation issued by the Environmental Protection Agency (EPA), which was implemented to manage about 70,000 facilities associated with regulated chemicals in the United States. As discussed by Air Products, the purpose of RMP is to mitigate the potential risks of these chemicals to the public through advocating the emergency preparedness in facilities and communities around them. The three steps to achieve this goal are hazard evaluation, emergency planning, and sharing information.

6.31.3 ARAMIS

Accidental Risk Assessment Methodology for Industries (ARAMIS) is a tool to assess the risk level of industrial establishments in Europe by considering the prevention methods used by the operators. The risk level is characterized by three factors, which should be taken into account in the reference accident scenarios. The factors include consequence severity evaluation, prevention management efficiency, and environment vulnerability estimation. ARAMIS aims to combine the strong points of various methods in European countries to develop a novel risk assessment methodology, which can promote process safety in industry.

NOTATION

Section 6.6

AQ	Airborne quantity (kg/s)
BI	Business interruption loss
CEI	Chemical exposure index
F_1	GPH factor
F_2	SPH factor
F_3	PUHF factor
F&EI	Fire and explosion index
GPH	General process hazard
MF	Material factor
MPDO	Maximum probable days outage
MPPD	Maximum probable property damage
PUHF	Process unit hazards factor
SPH	Special process hazard

REFERENCES

Andow, P.K., 1991. Guidance on HAZOP Procedures for Computer-Controlled Plants. HM Stationery Office, London.

Allum, S., Wells, G.L., 1993. Short-cut risk assessment. Process Saf. Environ. 71B, 161.

Ashmore, F.S., Shama, Y., 1988. Decisions on risk. Fire Saf. J. 13, 125.

Burk, A.F., 1992. Strengthen process hazards reviews. Chem. Eng. Prog. 88 (6), 90.

Casada, M.L., Kirkman, J.Q., Paula, H.M., 1990. Facility risk review as an approach to prioritizing loss prevention efforts. Plant/Operations Prog. 9, 213.

CCPS, 1992. Guidelines for Hazard Evaluation Procedures, second edition With Worked Examples. AICHE, New York, NY.

Dow Chemical Company, 1964. Dow's Process Safety Guide, first ed. Midland, MI.

Dow Chemical Company, 1966. Dow's Process Safety Guide, second ed. Midland, MI.

Dow Chemical Company, 1976. Fire and Explosion Index. Hazard Classification Guide, fourth ed. Midland, MI.

Dow Chemical Company, 1980. Fire and Explosion Index. Hazard Classification Guide, fifth ed. Midland, MI.

Flothmann, D., Mjaavatten, A., 1986. Qualitative methods for hazard identification. In: Hartwig, S. (Ed.), op. cit., p. 329.

Gibson, S.B., 1974. Reliability engineering applied to the safety of new projects. Course on Process Safety—Theory and Practice. Department of Chemical Engineering, Teesside Polytechnic, Middlesbrough.

Gibson, S.B., 1976a. How we fixed the flowsheet—safely. Process Eng. 119.

Gibson, S.B., 1976b. Safety in the design of new chemical plants (hazard and operability studies). Course on Loss Prevention in the Process Industries. Department of Chemical Engineering, Loughborough University of technology.

Gibson, S.B., 1976c. The design of new chemical plants using hazard analysis. Process Industry Hazards, pp. 135.

Gillett, J., 1985. Rapid ranking of hazards. Process Eng. February.

Goyal, R.K., 1993. FMEA, the alternative process hazard method. Hydrocarbon Process. 72 (5), 95.

Harris, N.C., 1975. Risk assessment. Electr. Saf. 2, 132.

Hawksley, J.L., 1984. Some social, technical and economic aspects of the risks of large plants. CHEMRAWN III.

Himmelblau, D.M., 1978. Fault Detection and Diagnosis in Chemical and Petrochemical Processes. Elsevier, Amsterdam.

Hill, E.J., Bose, L.J., 1975. Sneak circuit analysis of military systems. Proc. Second Int. Systems Safety Conf., p. 351.

Hokstad, P., Aro, R., Taylor, J.R., 1991. Integration of SA and RAMS Techniques. Rep. 757302. SINTEF, Trondheim.

Insurance Technical Bureau, 1981. IFAL p Factor Workbook. London.

Kavianian, H.R., Rao, J.K., Brown, G.V., 1992. Application of Hazard Evaluation Techniques to the Design of Potentially Hazardous Industrial Chemical Processes. Report. Division of Training and Manpower Development, National Institute of Occupational Safety and Health, Cincinnati, OH.

Keey, R.B., 1991. A rapid hazard-assessment method for smaller-scale industries. Process Saf. Environ. 69B, 85.

Kletz, T.A., 1972. Specifying and designing protective systems. Loss Prev. 6, 15.

Kletz, T.A., 1983. HAZOP and HAZAN. Institution of Chemical Engineers, Rugby.

Kletz, T.A., 1986. HAZOP and HAZAN, second ed. Institution of Chemical Engineers, Rugby.

Kletz, T.A., 1992. HAZOP and HAZAN, third ed. Institution of Chemical Engineers, Rugby.

Knowlton, R.E.,1992. A Manual of Hazard and Operability Studies. Chemetics International Ltd, Vancouver, BC.

Lambert, H.E., 1978. Failure Modes and Effects Analysis. NATO Advanced Study Institute on Synthesis and Analysis Methods for Safety and Reliability Studies, Urbino, Italy.

Lawley, H.G., 1974a. Operability studies. Course of Process Safety—Theory and Practice. Department of Chemical Engineering, Teesside Polytechnic, Middlesbrough.

Lawley, H.G., 1974b. Operability studies and hazard analysis. Loss Prev. 8, 105.

Lees, F.P., 1982. The hazard warning structure of major hazards. Trans. Inst. Chem. Eng. 60, 211.

Moubray, J., 1991. Reliability-Centred Maintenance. Butterworth-Heinemann, Oxford.

Munday, G., Phillips, H.C.D., Singh, J., Windebank, C.S., 1980. Instantaneous fractional annual loss—a measure of the hazard of an industrial operation. Loss Prev. Saf. Promotion. 3, 1460.

Pratt, N.R., 1965. From laboratory to production. Chem. Eng. Prog. 61 (2), 41.

Recht, J.L., 1966. Systems safety analysis: failure mode and effect. Nat. Saf. News February.24.

Reeves, A.B., Wells, G.L., Linkens, D.A., 1989. A description of GOFA. Loss Prev. Saf. Promotion. 6, 28.1.

Rooney, J.J., Turner, J.H., Arendt, J.S., 1988. A preliminary hazards analysis of a fluid catalytic cracking unit complex. J. Loss Prev. Process Ind. 1 (2), 96.

Rouhiainen, V., 1990. The quality assessment of safety analysis. Doctoral Thesis. Technical Research Centre of Finland (VTT), Espoo, Finland.

Scott, D., Crawley, F., 1992. Process Plant Design and Operation. Institution of Chemical Engineers, Rugby.

Singh, J., Munday, G., 1979. IFAL: a model for the evaluation of chemical process losses. In: Design 79. Institution of Chemical Engineers, London.

Taylor, J.R., 1973. A Formalisation of Failure Mode Analysis of Control Systems. Rep. Riso -M-1654. Atomic Energy Commission, Research Establishment, Riso, Denmark.

Taylor, J.R., 1974. A Semiautomatic Method for Qualitative Failure Mode Analysis. Rep. Riso -M-1707. Atomic Energy Commission, Research Establishment, Riso, Denmark.

Taylor, J.R., 1975. Sequential effects in failure mode analysis. In: Barlow, R.E., Fussell, J.B., Singpurwalla, N.D. (Eds.), op. cit., p. 881.

Taylor, J.R., 1979. A Background to Risk Analysis, vols. 1−4 (Report). Riso National Laboratory, Riso, Denmark.

Taylor, J.R., 1992. A comparative evaluation of safety features based on risk analysis for 25 plants. Loss Prevention and Safety Promotion. 1 (7), 9−1.

Van Gaalen, A.S., 1974. Loss prevention and property protection in the chemical industry. Loss Prev. Saf. Promotion. 1, 35.

Wells, G., Phang, C., Wardman, M., 1994. Improvements in process safety reviews prior to HAZOP. Hazards XII.301.

Wells, G., Wardman, M., Whetton, C., 1993. Preliminary safety analysis. J. Loss Prev. Process Ind. 6, 47.

Wells, G.L., Phang, C., Wardman, M., Whetton, C., 1992. Incident scenarios: their identification and evaluation. Process Saf. Environ. 70B, 179.

Whitehouse, H.B., 1985. IFAL—a new risk analysis tool. In: The Assessment and Control of Major Hazards. Institution of Chemical Engineers, Rugby.

Plant Siting and Layout

7.1. PLANT SITING

Safety is a prime consideration in plant siting. As far as the safety of the public is concerned, the most important feature of siting is the distance between the site and built-up areas. Separation between a hazard and the public is beneficial in mitigating the effects of a major accident. The physical effects of a major accident tend to decay quite rapidly with distance. Models for fire dispersion display decay using the inverse square law, as do many of the simpler models for explosion and toxic release. Information on the potential effects of a major accident on the surrounding area is one of the main results obtained from a hazard assessment, and such an assessment is of assistance in making decisions on plant siting. In selecting a site, allowance should be made for on-site emergencies. Emergency utilities such as electrical power and water should always be available. Another is the availability and experience of outside emergency services, particularly the fire service. A discussion of siting for highly toxic hazardous materials (HTHMs) is given in the CCPS HTHM *Storage Guidelines* (1988/2).

7.2. PLANT LAYOUT

Plant layout is a crucial factor in the economics and safety of process plants. Some of the ways in which plant layout contributes to safety and loss prevention (SLP) are as follows:

1. segregation of different risks;
2. minimization of vulnerable piping;
3. containment of accidents;
4. limitation of exposure;
5. efficient and safe construction;
6. efficient and safe operation;
7. efficient and safe maintenance;
8. safe control room design;
9. emergency control facilities;
10. fire fighting facilities;
11. access for emergency services.

Plant layout can have a large impact on plant economics. Additional space tends to increase safety but is expensive in terms of land, additional piping, and operating costs. Space needs to be provided where it is necessary for safety but not wasted.

A general guide to the subject is given in *Process Plant Layout* (Mecklenburgh, 1985). This is based on the work of an Institution of Chemical Engineers (IChemE) working party and expands upon an earlier guide *Plant Layout* (Mecklenburgh, 1973). The treatment of hazard assessment in particular is much expanded in this later volume. The loss prevention aspects of plant layout have also been considered specifically by Mecklenburgh (1976).

Other work on plant layout, and in particular SLP, includes that of Armistead (1959), R. Kern (1977a–f, 1978a–f), and Bausbacher and Hunt (1993), on general aspects and spacing recommendations; Simpson (1971) and R.B. Robertson (1974a, 1976b), on fire protection; Fowler and Spiegelman (1968), Balemans et al. (1974) and Drewitt (1975), on checklists; and Madden (1993), on synthesis techniques.

Plant layout is one of the principal aspects treated in various versions of the Dow *Guide* by the Dow Chemical Company (1994). It is also dealt with in the *Engineering Design Guidelines* of the Center for Chemical Process Safety (CCPS, 2003). The treatment given here follows that of Mecklenburgh, except where otherwise indicated.

7.3. LAYOUT GENERATION

The design of a process plant layout involves synthesis and then analysis. One excellent reference for process plant siting and layout is the CCPS Guidelines for Facility Siting and Layout (2003). Other sources include property insurance guidelines such as GE GAP and Factory Mutual. For process plants, one method is to lay the plant out so that the material flow follows the process flow diagrams. This is the process flow method.

Another method is the Process plant layout whose principles have been given by Madden (1993). He describes a structured approach to the generation of the layout which has four stages: (1) three-dimensional (3D) model, (2) flow, (3) relationships, and (4) groups. In 3D model, the first step is to produce a 3D model of the space occupied by each item of the equipment. This 3D envelope should include space for (1) operations access, (2) maintenance access, and (3) piping connections. The effect of allowing for these aspects is generally to increase the volume of the envelope by several-fold. In the concept of 'flow' Madden gave two meanings: (1) progression of materials toward a higher degree of completion and (2) mass flows of process or utility materials.

Third, a relationship exists between two items when they share some common factor. Relationships may be identified by considering the plant from the viewpoint of each discipline in turn. Broad classes of relationships are (1) process, (2) operational, (3) mechanical, (4) electrical, (5) structural, and (6) safety. From the relationships identified it is then necessary to select those which are to be given priority. It is finally then possible to arrange the items into groups. In the segregation technique, a relationship of particular importance in plant layout is that between a hazard and a potential target of that hazard. The minimization of the risk to the target is effected by segregating the hazard from the target. The requirement for segregation therefore places constraints on the layout.

7.4. LAYOUT TECHNIQUES AND AIDS

There are a number of methods available for layout design. They include:

1. classification, rating, and ranking;
2. critical examination;
3. hazard assessment;
4. consequence modeling;
5. economic/risk optimization.

There are also various aids, including:

1. visualization aids;
2. computer aids.

7.4.1 Classification, Rating, and Ranking

There are several methods of classification, rating, and ranking which are used in layout design. The main techniques are those used for the classification of (1) fire hazard areas, (2) storage, (3) fire fighting facilities, (4) access zones, and (5) electrical area classification together with methods based on hazard indices.

7.4.2 Critical Examination

Critical examination, which is part of the technique of method study (Currie, 1960), may be applied to plant layout. This application has been described by Elliott and Owen (1968). In critical examination of plant layout, typical questions asked are: Where is the plant equipment placed? Why is it placed there? Where else could it go?

7.4.3 Hazard Assessment

Hazard assessment of plant layout is practiced with respect to both major hazards which affect the whole site, and lesser hazards, notably leaks and their escalation. The traditional method of dealing with the latter has been the use of minimum safe separation distances, but there has been an increasing trend to supplement the latter with hazard assessment.

7.4.4 Economic/Risk Optimization

The process of layout development generates alternative candidate layouts. Economic optimization is a principal method of selection from among these. Some factors which are of importance for the cost of a plant layout include foundations, structures, piping, major equipment, and power consumption.

A grid-based method to map expected explosion risk from BLEVE and VCE of a process unit is given done by Jung (2011). In addition to it optimization of the layout of new plant units around it which utilizes a mixed integer linear program to solve a global optimum solution can be performed. The BLEVE and VCE risk is usually mapped using integral models in which weather conditions were taken into account using Monte Carlo simulations. The objective function balances land cost, piping cost, and of course risk cost associated with building a unit on a certain grid point.

In the same way, a continuous method can be formulated using disjunctive programming that takes into account the fire and explosion risks associated with process plants.

7.4.5 Computer Aids

One form of computer-generated visualization is a 2D layout equivalent to a cutout. The 3D layout, equivalent to either a block model or a piping model, is also produced to give a powerful way of describing the plant siting.

A particular application of 3D modeling is that model dynamics can be input into other computer programs such as computational fluid dynamics programs for explosion simulation. The 3D layout required for the latter is provided by the 3D visualization code, which then forms the front end of the total package.

Another type of code tackles the synthesis of layouts. The general approach is to first define a priority sequence for locating items of equipment inside a block and then for the actual location of the block itself.

A third type of code deals with the analysis of layouts to obtain an economic optimum. Typical factors taken into account in such programs include the costs of piping, space, and buildings.

7.5. LAYOUT PLANNING AND DEVELOPMENT

7.5.1 Layout Activities and Stages

Plant layout is usually divided into the following activities:

1. Site layout
2. Plot layout
3. Equipment layout.

The layout developed typically goes through three stages:

1. Stage One layout
2. Stage Two layout
3. Final layout.

The sequence of layout development described by Mecklenburgh is as follows:

1. Stage One plot layout
2. Stage One site layout
3. Stage Two site layout
4. Stage Two plot layout.

7.5.2 Stage One Plot Layout

In the Stage One plot layout, the information available should include preliminary flow sheets showing the major items of equipment and major piping, with an indication of equipment elevations and process engineering designs for the equipment. The outcome of this process for each plot is a set of candidate layouts. These are then presented for review as layout models in block model or computer graphics form. The different disciplines can then be invited to comment. A revised cost estimate is then generated for these plot layouts and a short list is selected.

The plot layouts are then subjected to hazard assessment. This assessment is concerned largely with the smaller, more frequent leaks which may occur as well as with sources of ignition for possible leaks. The process of hazardous area classification is also performed.

7.5.3 Stage One Site Layout

The Stage One plot layouts provide the information necessary for the Stage One site layout. These include the size and shape of each plot, the desirable separation distances, the access requirements, and traffic characteristics. The flow method is followed in laying out the plots but may need to be modified to meet constraints. Guidance is available on separation distances for this preliminary site layout. Hazard assessment is then performed on the site layout with particular reference to escalation of incidents and to vulnerable features such as service buildings and buildings just over the site boundary.

7.5.4 Stage Two Site Layout

Stage Two layout is the secondary, intermediate, or sanction layout. As the latter term implies, it is carried out to provide a layout which is sufficiently detailed for sanction purposes. It starts with the site layout and then proceeds to the layout. At this stage, information on the specific characteristics of the site is brought to bear, such as the legal requirements, the soil and drainage, the meteorological conditions, the environs, the environmental aspects, and the services. Site standards are set for building lines and finishes, service corridors, pipe racks, and roads.

Stage Two layout involves reworking the Stage One site layout in more detail and for the specific site and repeating the hazard assessment, economic optimization, and critical examination.

Features of the specific site which may well influence this stage are planning matters; environmental aspects; neighboring plants, which may constitute hazards and/or targets; other targets such as public buildings; and road, rail, and service access points.

At this stage, there should be full consultation with the various regulatory authorities, insurers and emergency services, including the police and fire services.

A final site plan is drawn up in the form of drawings and models, both physical and computer-based ones, showing in particular the layout of the plots within the site, the main buildings and roads, railways, service corridors, pipe racks, and drainage.

7.5.5 Stage Two Plot Layout

There then follows the Stage Two plot layout. The information available for this phase includes (1) standards, (2) site data, (3) Stage Two site layout, (4) process engineering design, and (5) Stage One plot layout. The Stage Two plot layout involves reworking in more detail, and subject to the site constraints the plot plans and layouts and repeating the hazard assessment, piping layout, and critical examination.

7.6. SITE LAYOUT FEATURES

Once a site has been selected, the next step is to establish the site constraints and standards. The constraints include:

1. topography and geology;
2. weather;
3. environment;
4. transport;
5. services;
6. legal constraints.

The site central services, such as the boiler house, power station, switch station, and pumping stations, should be placed in suitable locations.

Electrical substations, pumping stations, etc., should be located in areas where non-flameproof equipment can be used, except where they are an integral part of the plant.

Some or part of the plants may need to be located inside a building, but the use of a building is always expensive and it can create hazards and needs to be justified. Buildings which are the work base for a number of people should be located so as to limit their exposure to hazards.

An aspect of segregation which is of particular importance is the limitation of exposure of people to the hazards. The measures required to effect such limitation are location of the workbase outside and control of entry to the high hazard zone. The contribution of plant layout to limitation of exposure, therefore, lies largely in workbase location.

7.7. PLOT LAYOUT CONSIDERATIONS

Some considerations which bear upon plot layout are as follows:

1. process considerations;
2. economic considerations;
3. construction;
4. operations;
5. maintenance;
6. hazards;
7. fire fighting;
8. escape.

Process Considerations
Process considerations include some of the relationships, such as gravity flow and availability of head for pump suctions, control valves, and reflux returns.
Economic Considerations
Some features which have particularly strong influence on costs are foundations, structures, piping, and electrical cabling. This creates the incentive to locate items on the ground, to group items so that they can

share a foundation or a structure, and to keep pipe and cable runs to a minimum.

Construction

The installation of large and heavy plant items requires space and perhaps access for cranes. Such items tend to have long delivery times and may arrive late; the layout may need to take this into account.

Operation

Access and operability are important to plant operation. The routine activities performed by the operator should be studied with a view to providing the shortest and most direct routes from the control room to items requiring most frequent attention.

Maintenance

Plant items from which the internals need to be removed for maintenance should have the necessary space and lifting arrangements.

Hazards

The hazards on the plant should be identified and allowed for in the plot layout. Plot layout should be designed and checked with intent to reduce the magnitude and frequency of the hazards and assisting preventive measures. The principle of segregation of hazards applies also to plot layout.

Fire Fighting

Access is essential for fire fighting. This is provided by the suggested plot size of 100 m × 200 m with approaches preferably on all four sides and by spacing between plots and buildings of 15 m.

Fire water should be available from hydrants on a main between the road and the plant. Hydrant points should be positioned so that any fire on the plot can be reached by the hoses. Hydrant spacings of 48, 65, and 95 m are suitable for high, medium, and low-risk plots, respectively. Plants over 18 m high should be provided with dry riser mains and those over 60 m high or of high risk should be equipped with wet riser mains.

7.8. EQUIPMENT LAYOUT

Furnaces and fired heaters are very important. Furnace location is governed by a number of factors, including the location of other furnaces, the use of common facilities such as stacks, the minimization of the length of transfer lines, the disposal of the gaseous and liquid effluents, the potential of the furnace as an ignition source, and the fire fighting arrangements.

No trenches or pits which might hold flammables should extend under a furnace, and connections with underground drains should be sealed over an area of 12 m from the furnace wall. On wall-fired furnaces, there should be an escape route at least 1 m wide at each end. The top-fired incinerators and burning areas for waste disposal should be treated as fired equipment. Chemical

reactors, in which a violent reaction can occur, may need to be segregated by firebreaks or even enclosed behind blast walls.

Heat exchangers should have connecting piping kept to a minimum, consistent with provision of pipe lengths and bends to allow for pipe stresses and with access for maintenance.

If the process materials are corrosive, this aspect should be taken into account in the plant layout. The layout of plants for corrosive materials is discussed in Safety and Management by the Association of British Chemical Manufacturers.

7.9. SEPARATION DISTANCES

For hazards, there are basically three approaches to determine a suitable separation distance. The first and most traditional one is to use standard distances developed by the industry. The second is to apply a ranking method to decide the separation required. The third is to estimate a suitable separation based on an engineering calculation for the particular case.

7.9.1 Types of Separation

The types of separation which need to be taken into consideration are illustrated by the set of tables of separation distances given by Mecklenburgh (1985) and include:

1. site areas and sizes;
2. preliminary spacing for equipment
   a. spacing between equipment,
   b. access requirements at equipment,
   c. minimum clearances at equipment;
3. preliminary spacings for storage layout
   a. tank farms,
   b. petroleum products,
   c. liquefied flammable gas,
   d. liquid oxygen;
4. preliminary distances for electrical area classification;
5. Size of storage piles.

7.9.2 Standard Distances

There are a large number of standards, codes of practice, and other publications which give minimum safe separation distances. The guidance available relates mainly to separation distances for storage, either of petroleum products, of flammable liquids, of liquefied petroleum gas (LPG), or of liquefied flammable gas (LFG). Typical spacing for plant equipment under fire circumstances given by CCPS Guidelines for Facility Siting and Layout (2003) is shown in Figure 7.1.

Table A
TYPICAL SPACING FOR PLANT EQUIPMENT FOR FIRE CONSEQUENCES
Explosion and toxic concerns may require greater spacing
Horizontal distance (ft)

	Text references	Process unit battery limits	Property	ESD valves - manual	Fire pumps	Hydrants, monitors	Water spray and ESD activation switches	Equipment handling nonflammables, noncombustibles, nontoxics	Reactors and desalters	Towers, drums, knock out pots, on-site storage tanks	Air cooled heat exchangers - process	Boilers, air compressors, power generation (utility area)	Cooling towers	Exchangers (< autoignition or non-self-igniting)	Exchangers (> autoignition or self-igniting)	Fired heaters	Gas compressor, expanders	Pumps handling flammables > autoignition or self-igniting	Pumps handling flammables < autoignition or non-self-igniting	Central loading racks for trucks and rail cars (except LFG)	Any liquefied flammable gas loading racks (trucks and rail cars)	Main pipe racks (piping not associated with unit)	Process pipe racks
Boundaries																							
Process unit battery limits	5.7.3 6.8.1	100																					
Property	5.2.5	200	NM																				
Emergency																							
ESD valves - manual	6.8.15	50	NM	NM																			
Fire pumps	5.8.2	200	NM	NM	NM																		
Hydrants, monitors	6.8.18	NM	NM	NM	NM	NM																	
Water spray and ESD activation switches	6.8.14	50	NM	NM	NM	NM	NM																
Process vessels																							
Equipment handling nonflammables, noncombustibles, nontoxics	6.8.2	NA	NM	NM	NM	NM	NM	NM															
Reactors and desalters	6.8.4	NA	200	50	200	50	50	NM	NM														
Towers, drums, knock out pots, on-site storage tanks	6.8.3 6.8.5	NA	200	50	200	50	50	NM	15	15													
Heat transfer equipment																							
Air-cooled heat exchangers - process	6.8.7	NA	200	50	200	50	50	NM	15	15	NM												
Boilers, air compressors, power generation (utility area)	5.5	100	100	50	200	50	50	NM	100	100	100	NM											
Cooling towers	5.5.6	100	100	50	100	50	50	NM	100	100	100	100	25										
Exchangers (< autoignition or non-self-igniting)	6.8.6	NA	200	50	200	50	50	NM	15	15	15	100	100	NM									
Exchangers (> autoignition or self-igniting)	6.8.6	NA	200	50	200	50	50	NM	15	15	15	100	100	15	NM								
Fired heaters	6.8.6	NA	200	50	200	50	50	NM	50	50	50	100	100	50	50	NM							
Rotating equipment																							
Gas compressor, expanders	6.8.10	NA	200	50	200	50	50	NM	15	15	15	100	100	15	50	50	NM						
Pumps handling flammables > autoignition or self-igniting	5.8.5 6.8.11	NA	200	50	200	50	50	NM	15	15	15	100	100	15	50	50	150	NM					
Pumps handling flammables < autoignition or non-self-igniting	5.8.5 6.8.11	NA	200	50	200	50	50	NM	15	15	15	100	100	15	50	50	50	NM	NM				
Transfer equipment																							
Central loading racks for trucks and rail cars (except LFG)	5.8.8	200	100	50	200	50	50	NM	200	200	200	200	150	100	200	200	200	200	50	NM			
Any liquefied flammable gas loading racks (trucks and rail cars)	5.8.8	250	350	50	250	50	50	NM	250	250	250	250	250	250	250	250	250	250	50	150	NM		
Main pipe racks (piping not associated with unit)	5.8.8	NM	100	NM	50	NM	NM	NM	50	50	NM	50	50	15	15	50	50	50	50	50	50	NM	
Process pipe racks	6.8.19	NM	200	NM	200	NM	NM	NM	15	15	NM	100	100	15	15	50	15	15	15	200	200	NM	NM

FIGURE 7.1 Typical spacing for plant equipment under fire circumstances, in feet. *Source: Adapted from CCPS Guidelines for Facility Siting and Layout (2003).*

7.9.3 Rating and Ranking Methods

The most widely applied method of this kind is that used in hazardous area classification. This method ranks items by their leak potential.

7.9.4 Mond Index

In the Mond Index method, two values are calculated for the overall risk rating (ORR), those before and after allowance is made for offsetting factors. It is the latter rating R_2 which is used in applying the technique to plant layout. The ORR assigns categories ranging from mild to very extreme.

1. Flare spacing should be based on heat intensity with a minimum space of 60 m from equipment containing hydrocarbons.
2. The minimum spacing can be down to one-quarter of these typical values when properly assessed.

The objectives of layout are to minimize risk to personnel; to minimize escalation, both within the plant and to adjacent plants; to ensure adequate access for fire fighting and rescue; and to allow flexibility in combining together units of similar hazard potential.

7.9.5 Hazard Models

Hazard models determine the separation distance at which the concentration from a vapor escape or the thermal radiation from a fire falls to an acceptable level. This is the other side of the coin to hazard assessment of a proposed layout.

Two principal factors considered as determining separation are (1) heat from burning liquid and (2) ignition of a vapor escape.

7.9.6 Liquefied Flammable Gas

A separation distance of 15 m frequently occurs in codes for the storage of petroleum products, excluding LPG. The general approach taken is that there is significant risk of failure for a refrigerated storage but negligible risk for a pressure storage vessel.

7.10. HAZARDOUS AREA CLASSIFICATION

Plant layout has a major role to play in preventing the ignition of any flammable release which may occur. This aspect of layout is known as 'area classification'.

Hazardous area classification is dealt with in BS 5345: 1977 Code of Practice for the Selection, Installation and Maintenance of Electrical Apparatus for Use in Potentially Explosive Atmospheres (Other than Mining Applications or Explosive Processing and Manufacture), and in a number of industry codes, including the Area Classification Code for Petroleum Installations (IP, 1990).

7.11. HAZARD ASSESSMENT

In the methodology for plant layout described by Mecklenburgh (1985), hazard assessment is used at several points in the development of the layout. The nature of the hazard assessment will vary depending on whether it is done in support of site location, site layout, Stage One plot layout, or Stage Two plot layout. Hazard assessment in support of site location is essentially some form of quantitative risk assessment.

Hazard assessment for site layout concentrates on major events. It provides guidance on the separation distances required to minimize fire, explosion and toxic effects and on the location of features such as utilities and office buildings.

Hazard assessment for plot layout deals with lesser events and with avoidance of the escalation of such events. It is used as part of the hazardous area classification process and it provides guidance on separation distances to prevent fire spread and for control building location.

7.12. HAZARD MODELS

7.12.1 Mecklenburgh System

A set of hazard models specifically for use in plant layout has been given by Mecklenburgh (1985). A summary of the models in this hazard model system is given in Table 7.1. Although the modeling of some of the phenomena has undergone further development, this hazard model system remains one of the most comprehensive available for its purpose.

7.12.2 IP System

Another, more limited, set of hazard models for plant layout is that given in *Liquefied Petroleum Gas* (IP, 1987 MCSP Pt 9). The models cover:

1. emission
   a. pressurized liquid,
   b. refrigerated liquid;
2. pool fire;
3. jet flame.

The models include view factors for thermal radiation from cylinders at a range of angles to the vertical and of positions of the target.

7.13. FIRE PROTECTION

Plant layout for fire protection is covered in BS 5908: 1990 *Fire Precautions in the Chemical and Allied Industries*. Also relevant are BS 5306: 1976 *Fire*

TABLE 7.1 Hazard Assessment in Support of Plant Layout: Mecklenburgh Hazard Model System

Table No.	
B1	Source term: instantaneous release from storage of flashing liquid (catastrophic failure of vessel) Flash fraction Mass in, and volume of, vapor cloud
B2	Dispersion of flammable vapor from instantaneous release Distance to lower flammability limit (LFL)
B3	Explosion of flammable vapor cloud from instantaneous release Explosion overpressure Damage (as function of overpressure)
B4	Fireball of flammable vapor cloud from instantaneous release Fireball diameter, duration, thermal radiation
B5	Dispersion of toxic vapor from instantaneous release Peak concentration, time of passage Distance to safe concentration, outdoors and indoors
B6	Source term: continuous release of fluid a. Gas (subsonic) b. Gas (sonic) c. Flashing liquid (not choked) d. Flashing liquid (choked)
B7	Dispersion of flammable vapor jet Jet length, diameter (to LFL)
B8	Jet flame from flammable vapor jet Flame length, diameter Flame temperature, surface heat flux, distance to given heat flux
B9	Dispersion of toxic vapor plume Distance to given concentration Distance to safe concentration, outdoors and indoors
B10	Growth of, and evaporation from, a pool Pool diameter Evaporation rate
B11	Pool or tank fire Flame height Regression rate, surface heat flux View factor
B12	Effect of heat flux on targets Tolerable heat fluxes
B13	Risk criteria Individual risk to employees (as a range) Individual risk to public (as a range) Multiple fatality accident
B14	Explosion overpressure Damage (as function of overpressure) (see also B3)

(Continued)

TABLE 7.1 (Continued)

Table No.	
B15	Dispersion of flammable vapor from small continuous release Jet dispersion: distance to given concentration, to LFL (see B7) Passive dispersion: distance to given concentration (see B9) Jet flame: distance to given heat flux (see B8)
B16	Evaporation and dispersion from small liquid pool Distance to given concentration
B17	Dispersion of flammable vapor from small continuous release in a building Jet dispersion: distance to given concentration Passive dispersion: distance to given concentration Jet flame: distance to given heat flux (see B8)
B18	Evaporation and dispersion from small liquid pool in a building Evaporation rate Mean space concentration Other parameters for (a) horizontal air flow and (b) vertical air flow

Extinguishing Installations and Equipment on Premises, particularly Part 1 on fire hydrants, and BS 5041: 1987 *Fire Hydrant Systems Equipment*. An important earlier code, BS CP 3013: 1974 *Fire Precautions in Chemical Plant*, is now withdrawn.

7.13.1 Fire Water

In a fire, water is required for extinguishing the fire, for cooling tanks and vessels, and for foam blanketing systems. The quantities of water required can be large, both in terms of the instantaneous values involved and of the duration for which they may be needed.

The design of a fire water system requires the determination of the maximum fire water flow which the system should deliver.

7.13.2 Fire Protection Equipment

The other aspects of fire protection of plant, include the fire containment by layout, gas, smoke and fire detection, passive fire protection such as fire insulation, and active fire protection such as the use of fixed, mobile, and portable fire fighting equipment.

7.14. EFFLUENTS

General arrangements for dealing with effluents are discussed by Mecklenburgh (1973, 1985). Pollution of any kind is a sensitive issue and attracts a growing degree of public attention.

Hazard identification methods should be used to identify situations which may give rise to acute pollution incidents and measures similar to those used to control other hazards should be used to ensure that this type of hazard is also under control.

7.14.1 Liquid Effluents

Liquid effluents include soil effluents, domestic and process effluents, and cooling, storm and fire water. Harmless aqueous effluents and clean storm water may be run away in open sewers, but obnoxious effluents require a closed sewer. One arrangement is to have three separate systems: an open sewer system for clean storm water and two closed sewer systems, one for domestic sewage and one for aqueous effluent from the plant and for contaminated storm water.

7.14.2 Gaseous Effluents

Gaseous effluents should be burned or discharged from a tall stack so that the fumes are not obnoxious to the site or the public. The local Industrial Pollution Inspector is able to advise on suitable stack heights and should be consulted. It is also necessary to check whether a high stack constitutes an aerial hazard and needs to be fitted with warning lights.

7.14.3 Solid Wastes

Solid waste should preferably be transferred directly from the process to transport. If intermediate storage is unavoidable, care should be taken that it does not constitute a hazard or a nuisance. If combustible solid and solvent wastes are burnt, the incinerator should be convenient to the process.

7.15. BLAST-RESISTANT STRUCTURES

It is sometimes necessary in the design of structures such as plant and buildings to allow for the effect of shocks from explosions and/or earthquakes. In both these cases, there is a strong probabilistic element in the design in that it is not possible to define the precise load to which the plant structure may be subjected. The starting point is therefore the definition of the design load in terms of the relation between the magnitude of the load and the frequency of occurrence.

7.16. CONTROL BUILDINGS

Until the mid-1970s, there were few generally accepted principles, and many variations in practice, in the design of control buildings. Frequently the control buildings constructed were rather vulnerable, being in or close to the plant and built of brick with large picture windows.

7.16.1 Flixborough

The Flixborough disaster, in which 18 of the 28 deaths occurred in the control building, caused the Court of Inquiry to call for a fundamental reassessment of practice in this area.

The control building at Flixborough has been described by V.C. Marshall (1976). It was constructed with a reinforced concrete frame, brick panels, and considerable window area. It was 2½ storeys high in its middle section, the 1½ storeys over the control room consisting of a half-story cable duct, and a full-storey electrical switchgear room. The control room was part of a complex of buildings some 160 m long, which also housed managers' offices, a model room, the control laboratory, an amenities building, and a production block.

This building complex was 100 m from the assumed epicenter of the explosion and was subjected to an estimated overpressure of 0.7 bar. The complex lays with its long axis at right angles to the direction of the blast. It was completely demolished by the blast and at the control room the roof fell in. The occupants of the control room were presumably killed mainly by the collapse of the roof, but some had been severely injured by window glass or wired glass from the internal doors. It took mine rescue teams 19 days to complete the recovery of the bodies.

The main office block, which was a three-storey building, again constructed with a reinforced concrete frame, brick panels, and windows, was only 40 m from the assumed epicenter and was also totally demolished.

The implications of the Flixborough disaster for control building location and design have been discussed by V.C. Marshall (1974) and Kletz (1975).

7.16.2 Building Function

The control building should protect its occupants against the hazards of fire, explosion, and toxic release. Much of the most common hazard is fire, and this should receive particular attention.

There are several reasons for seeking to make control buildings safer. One is to reduce to a minimum level the risk to which operators and other personnel are exposed. Another is to allow control to be maintained in the early stages of an incident and so reduce the probability of escalation into a disaster. A third is to protect plant records, including those of the period immediately before an accident.

7.16.3 ACMH Recommendations

Control building location and design is one of the topics raised by the Court of Inquiry on the Flixborough disaster and considered in the First and Second *Reports* of the Advisory Committee on Major Hazards (ACMH) (Harvey, 1976, 1979).

7.16.4 Control Facilities

The proper policy is to build a secure control room in which the functions performed are limited to those essential for the control of the plant and to remove all other functions to a distance where a less elaborate construction is permissible. The essential functions which are required in the control room are those of process control. There are other types of control which are required for the operation of the plant, such as analytical control and management control, but they need not be exercised from the control room. Thus, other facilities such as analytical laboratories and amenities rooms should be located separately from the control room.

7.16.5 Location

It is good practice to lay plant out in blocks with a standard separation distance. The control building should be situated on the edge of the plant to allow an escape route. Recommended minimum distances between the plant and the control building tend to lie in the range 20−30 m.

A control building should not be sited in a hazardous area as defined in BS 5345: 1977. Further guidance on location of the control building is given below.

7.16.6 Basic Principles

Arising from the experience of Flixborough, V.C. Marshall (1976) has suggested certain principles for control building design which may be summarized as follows:

1. The control room should contain only the essential process control functions.
2. There should be only one storey above ground.
3. There should be only the roof above the operator's head. The roof should not carry machinery or cabling.
4. The building should have cellars built to withstand earthshock and to exclude process leaks and should have ventilation from an uncontaminated intake.
5. The building should be oriented to present minimum area to probable centers of explosion.

6. There should be no structures which can fall on the building.

7. Windows should be minimal or non-existent and glass in internal doors should be avoided.

8. Construction should be strong enough to avoid spalling of the concrete, but it is acceptable that, if necessary, the building be written off after a major explosion.

7.17. TOXICS PROTECTION

In general, ordinary buildings off site and even onsite can afford an appreciable degree of protection against a transient toxic gas release, but for certain functions enhanced protection is required. Buildings of particular interest here are (1) the control building, (2) the emergency control center, and (3) any temporary refuges.

7.17.1 Control Room

The design of a control room for protection against toxic release is discussed in the CIA *Control Building Guide* (Chemical Industries Association, 1979). The design should start by identifying the release scenarios against which protection is required and by making some quantitative assessment of the dispersion of the gas.

The control building should be located at the edge of the plant, and its siting should take into account both fire/explosion and toxic gas hazards.

7.17.2 Emergency Control Center

Protection of the emergency control center from a toxic release will normally be in large part by location. It is necessary for the emergency controllers to gain access to the center at the start of the emergency and it is undesirable that they should have to pass through a toxic gas cloud.

7.17.3 Refuges

A temporary refuge, or haven, has the quite different function of providing temporary shelter for personnel. The design of such havens is described in the *Vapor Release Mitigation Guidelines* (CCPS, 1988/3). The *Guidelines* distinguish between temporary and permanent havens, or more effective temporary havens.

7.18. MODULAR PLANTS

Accounts of modular plants have been given by Glaser et al. (1979), Zambon and Hull (1982), Glaser and Kramer (1983), Hulme and La Trobe-Bateman (1983), Kliewer (1983), Clement (1989), and Shelley (1990).

Modular plants were seen as offering benefits where site construction was unusually difficult, particularly on remote sites. Factors favoring modular plants include problems associated with (1) access difficulties, (2) severe weather, and (3) the labor force.

REFERENCES

Balemans, A.W.M., 1974. Check-list. Loss Prev. Saf. Promotion 1, 7.

Bausbacher, E., Hunt, R., 1993. Process Plant Layout and Piping Design. PTR Prentice Hall, Englewood Cliffs, NJ.

BS 5041, 1987. Fire Hydrant Systems Equipment Specification for Landing Valves for Wet Risers. British Standard Institute, United Kingdom.

BS 5306, 1976. Fire Extinguishing Installations and Equipment on Premises. British Standards Institute, United Kingdom.

BS 5345, 1977. Code of Practice for the Selection, Installation and Maintenance of Electrical Apparatus for Use in Potentially Explosive Atmospheres (Other Than Mining Applications or Explosive Processing and Manufacture). British Standards Institute, United Kingdom.

BS 5908, 1990. Fire Precautions in the Chemical and Allied Industries. British Standards Institute, United Kingdom.

BS CP 3013, 1974. Fire Precautions in Chemical Plants. British Standards Institute, United Kingdom.

CCPS, 1988. Guidelines for Safe Storage and Handling of High Toxic Hazard Material. AIChE, New York, NY.

Center for Chemical Process Safety (CCPS), 2003. Guidelines for Facility Siting and Layout. Wiley-AIChE, New York, NY.

Chemical Industries Association, 1979. An Approach to the Categorisation of Process Plant Hazard and Control Building Design. Chemical Industries Association, London.

Clement, R., 1989. Deciding when to use modular construction. Chem. Eng. 96 (8), 169.

Currie, R.M., 1960. Work Study. Pitman, London.

Dow, 1994. Dow's Fire and Explosion Index Hazard Classification Guide. Dow Chemical Company, Midland, MI.

Drewitt, D.F., 1975. The insurance of chemical plants. In: Course on Process Safety—Theory and Practice. Department of Chemical Engineering Teesside Polytechnic, Middlesbrough.

Elliott, D.M., Owen, J.M., 1968. Critical examination in process design. Chem. Eng. London. 223, CE377.

Fowler, E.W., Spiegelman, A., 1968. Hazard Survey of the Chemical and Allied Industries. American Insurance Association, New York, NY.

Glaser, L.B., Kramer, J., 1983. Does modularization reduce plant investment? Chem. Eng. Prog. 79 (10), 63.

Glaser, L.B., Kramer, J., Causey, E.D., 1979. Practical aspects of modular and barge-mounted plants. Chem. Eng. Prog. 75 (10), 49.

Harvey, B.H., 1976. First Report of the Advisory Committee on Major Hazards. HM Stationery Office, London.

Harvey, B.H., 1979. Second Report of the Advisory Committee on Major Hazards. HM Stationery Office, London.

Hulme, D., La Trobe-Bateman, J., 1983. Take a closer look at modular construction. Hydrocarbon Process. 82 (1), 34-C.

IP, 1987. MCSP Pt 9: Large Bulk Pressure Storage and Refrigerated LPG. Energy Institute, London.

IP, 1990. MCSP Pt 15: Area Classification Code for Installations Handling Flammable Fluids. Energy Institute, London.

Jung, S., Ng, D., Vazquez-Roman, R., Diaz-Ovalle, C., Mannan, S., 2011. New approach to optimizing the facility siting and layout for fire and explosion scenarios. Ind. Eng. Chem. Res. 50 (7), 3928–3937.

Kern, R., 1977a. Pressure relief valves for process plants. Chem. Eng. 84 (February), 187.

Kern, R., 1977b. Specifications are the key to successful plant design. Chem. Eng. 84 (July), 123.

Kern, R., 1977c. Arrangements of process and storage vessels. Chem. Eng. 84 (November), 93.

Kern, R., 1977d. How to find optimum layout for heat exchangers. Chem. Eng. 84 (September), 169.

Kern, R., 1977e. How to get the best process plant layouts for pumps and compressors. Chem. Eng. 84 (December), 131.

Kern, R., 1977f. How the manage plant design to obtain minimum cost. Chem. Eng. 84 (May), 130.

Kern, R., 1978a. Arranging the housed chemical process plant. Chem. Eng. 85 (July), 123.

Kern, R., 1978b. Controlling the cost factors in plant design. Chem. Eng. 85 (August), 141.

Kern, R., 1978c. How to arrange the plot plan for process plants. Chem. Eng. 85 (May), 191.

Kern, R., 1978d. Instrument arrangements for ease of maintenance and convenient operation. Chem. Eng. 85 (April), 127.

Kern, R., 1978e. Piperack design for process plants. Chem. Eng. 85 (January), 105.

Kern, R., 1978f. Space requirements and layout for process furnaces. Chem. Eng. 85 (February), 117.

Kletz, T.A., 1975. Some of the wider questions raised by Flixborough. Symposium on Technical Lessons of Flixbrough. Institution of Chemical Engineers, London.

Kliewer, V.D., 1983. Benefits of modular plant design. Chem. Eng. Prog. 79 (10), 58.

Madden, J., 1993. Safety Aspects of Plant Layout. Course on Hazard Control and Major Hazards. Department of Chemical Engineering, Loughborough University of Technology, United Kingdom.

Marshall, V.C., 1974. Flixborough (letter). Chem. Eng. London. 287/288, 483.

Marshall, V.C., 1976. Seveso and Flixborough. Chem. Eng. London. 314, 697.

Mecklenburgh, J.C., 1973. Plant Layout. Leonard Hill, London.

Mecklenburgh, J.C., 1976. Plant Layout and Loss Prevention. Course on Loss Prevention in the Process Industries. Department of Chemical Engineering, Loughborough University of Technology, United Kingdom.

Mecklenburgh, J.C., 1985. Process Plant Layout. Godwin, London.

Robertson, R.B., 1974. Fire engineering in relation to process plant design. Fire Int. 45, 69.

Robertson, R.B., 1976. Spacing in chemical plant design against loss by fire. Process Ind. Hazards 157.

Simmonds, John G., 1959. Safety in Petrsoleum Refinery and Related Industries. George Armistead, Washington, D.C.

Simpson, H.G., 1971. Design for loss prevention – plant layout. Major Loss Prevention 105.

Shelley, S., 1990. Making inroads with modular construction. Chem. Eng. 97 (8), 30.

Zambon, D.M., Hull, G.B., 1982. A look at five modular projects. Chem. Eng. Prog. 78 (11), 53.

Chapter 8

Process Design

Chapter Outline

8.1. PROCESS DESIGN

Process design designates the Chemical Engineering development aspects of a process plant. This is augmented by the classical related engineering disciplines: Mechanical, Electrical, Civil Engineering along with others, which provide constraints as to the availability and feasibility of industrial equipment, electrical hazards, and siting issues. The design of a large process plant, or plant modification or extension is a complex activity involving engineers of many disciplines. The design process normally involves engineering organizations and expertise other than the operating company. The project evolves under the influence of: research and development; safety, health, and environmental studies; economic studies; and the financial approvals. The political, social, and environmental issues at the location must be recognized. An overall risk of the plant to the community must be developed and agreed with the appropriate local, state, and federal authorities before further process design can proceed. This will largely comply with regulations and avoid a Bhopal-type catastrophe. The infrastructure available will dictate whether the design is a self-supporting grassroots or an integrated facility. Of course, the additional constraint of the limited space and congestion affect the design of offshore facilities.

The decisions made in the early stages, particularly concerning the process route, the plant capacity and their locations, are crucial. Thereafter, many options are foreclosed such that further fundamental changes are difficult, if not simply impractical. From the Safety and Loss Prevention (SLP) viewpoint, it is essential to get the process fundamentally right from the start, as illustrated above. The design should be to eliminate hazard rather than devise measures to control it. The safety of the plant is determined primarily by the quality of the basic design rather than by the addition of special safety features. It is necessary, however, to incorporate specific systematic safety checks throughout the design process at each level, generally termed hazard identification and assessment studies.

8.1.1 The Design Process

Process risk management strategies are important for the process design. Risk has been defined as a measure of economic loss or human injury in terms of both the incident likelihood and the magnitude of the loss or injury (CCPS, 1989). Thus, any effort to reduce the risk arising from the operation of a chemical processing facility can be directed toward reducing the likelihood of incidents (incident frequency), reducing the magnitude of the loss or injury should an incident occur (incident consequences), or some combination of both. In general, the strategy for reducing risk, whether directed toward reducing frequency or consequence of potential accidents, falls into one of the following categories:

1. *Inherent or Intrinsic*: Eliminating the hazard by using materials and process conditions that are non-hazardous.
2. *Passive*: Eliminating or minimizing the hazard by process and equipment design features that do not eliminate the hazard, but do reduce either the frequency or consequence of realization of the hazard without the need for any device to function actively.
3. *Active*: Using controls, safety interlocks, and emergency shut-down systems to detect potentially hazardous process deviations and take corrective action. These are commonly referred to as engineering controls.
4. *Procedural*: Using operating procedures, administrative checks, emergency response, and other management approaches to prevent incidents or to minimize the effects of an incident. These are commonly referred to as administrative controls.

Some important stages of the design of a process plant are (1) research and development—pilot plant, (2) process design, (3) front-end conceptual engineering, (4) engineering design and equipment specification—selection.

Process design can be properly done and executed only with adequate and correct design data. This should include as a minimum: (1) the physical and chemical properties of the chemicals; (2) the reaction characteristics; (3) fire, explosion, and toxic hazards; (4) the effect of trace impurities which may have a major impact on metallurgy. It is useful to consider the forms in which process design experience and know-how are available. The information may relate to: chemical reactors, unit processes, unit operations and equipment, operating conditions, utilities, particular chemicals, particular processes, and plants.

Effective communication is also essential in the design process. Much of this communication is done on an individual basis or in design committees. The most important channel of communication, however, is the documentation. Of particular importance are the process flow diagram, process heat and mass balances, and the engineering flow sheets (Piping and Instrumentation Diagram, P&ID). At a critical point in the design, a freeze is instituted to complete the design effectively. Inevitably, changes occur subsequently, usually due to external factors which must be accommodated. A Management of Change (MOC) procedure is instituted to ensure all aspects of the change are properly reviewed, accredited and authorized, and communicated to all parties involved.

Overdesign may exist in engineering which is often equivalent to the incorporation of an extra factor of safety, but this is by no means always so. In some cases such overdesign can reduce safety. There is an inherent tendency to overdesign in a project as the various individuals in the chain introduce factors of safety. In this context, overdesign is taken to cover the purchasing as well as the design decisions. What matters is the item which is finally installed.

Modern process design involves considerable use of Computer-Aided Design (CAD) techniques, particularly for flow sheeting, equipment design, plant layout, piping, and instrument diagrams. It is sometimes appropriate to check the error associated with a design calculation, particularly if this has important safety aspects. A discussion of error propagation is given by Park and Himmelblau (1980).

8.1.2 Conceptual—Front-End Design

At the conceptual design stage, the process concept is developed, its implications are explored, and potential problems are identified. Design in general and conceptual design in particular is generally regarded as an art. Much work is going on, however, to put it on a more systematic basis. Early work is described in *Strategy of Process Design* (Rudd and Watson, 1968). A systematic approach is stated in *The Conceptual Design of Chemical Processes* (Douglas, 1988).

In most cases, the process is an established one, so that the conceptual design stage may be quite short. The nature of this stage is best appreciated, however, by considering the conceptual design of a new process. Elements of the process and plant considered in conceptual design are (1) process materials, (2) chemical reaction, (3) overall process, (4) effluents, (5) storage, (6) transport, (7) utilities, (8) siting and layout.

The topics addressed in the conceptual design are principally: (1) process design; (2) mechanical design; (3) pressure relief, blow down, venting, disposal, and drains; (4) control and instrumentation; (5) plant construction and commissioning; (6) plant operation; (7) plant maintenance; (8) health and safety; (9) environment; (10) costing.

8.1.3 Detailed Engineering

The detailed design stage involves the detailed process and mechanical design together with detailed design from a large number of supporting disciplines. The treatment given here is confined to a broad outline, with particular reference to features bearing on SLP.

Elements of the detailed design of the process and plant include: (1) process design; (2) mechanical design; (3) storage; (4) transport; (5) utilities; (6) layout; (7) pressure relief, blow down, venting and disposal and drains; (8) control and instrumentation; (9) fire protection; (10) explosion protection; (11) toxic emission protection; (12) personnel protection; (13) plant failures; (14) plant operation; (15) plant maintenance; (16) plant reliability, availability, and maintainability; (17) equipment specification, selection, and procurement; (18) health and safety; (19) environment.

8.1.4 Design Assessments

As the design progresses, it is subject to various assessments, which draw on a number of specific techniques. Those considered here are (1) critical examination, which is an effective method for fundamental review, exposing assumptions and generating alternative options, (2) value engineering assessment, (3) energy efficiency assessment, (4) reliability and availability assessment, (5) hazard identification and assessment, (6) occupational health assessment, and (7) environmental assessment.

Value Improving Practices (VIPs) are formal structured practices applied to the front-end stages of any project to improve profitability above that attained through good engineering and project management practices. VIPs are also intended to challenge existing project and design premises to ensure the most robust and profitable facility results. Well-recognized VIPs include: (1) classes of facility quality, (2) technology selection, (3) process simplification, (4) constructability, (5) customization of standards and specifications, (6) predictive maintenance, (7) process reliability simulation, (8) value engineering, (9) design to capacity, (10) energy optimization, (11) waste minimization, (12) three-dimensional CAD.

The process simplification VIP and the value engineering VIP require the use of the value methodology. The value methodology follows an established set of procedures that include: (1) pre-study planning and information gathering, (2) information analysis, (3) function identification and analysis, (4) identification of creative alternatives for key functions via brainstorming, (5) evaluation of selected alternatives, (6) development of selected alternatives to estimate costs, benefits, risks, and other impacts to the project, (7) recommendation of the best alternatives, (8) implementation or approved recommendations by the project team, and (9) post-study follow-up.

It consistently yields the greatest value improvements of the VIPs. It is a formal, rigorous process to search for opportunities to eliminate or combine process and utility system steps or equipment ultimately resulting in the reduction of investment and operating costs.

It is the facilitated systematic implementation of the latest engineering, procurement and construction concepts, and lessons learned consistent with the facility's operations and maintenance requirements to enhance construction safety, scope, cost, schedule, and quality.

8.1.5 Licensors, Vendors, and Contractors

It should take the appropriate steps to ensure that the processes designed and the equipment supplied by the other parties are safe.

Some design responsibility resides, however, with other parties. These may be (1) licensor, (2) vendor, and (3) contractor. A treatment of licensing is given in *Licensing Technology and Patents* (Parker, 1991). Contractual arrangements are the subject of the series *Model Form of Conditions of Contract for Process Plants* in three volumes—(1) subcontracts, (2) lump sum contract, and (3) reimbursable contracts—by the IChemE (1992a−c) (the 'Yellow Book', 'Red Book', and 'Green Book', respectively) with a guide by Wright (1993) (the 'Purple Book').

If the process has been bought under license, then the licensor has some design responsibility. There is a responsibility on the vendor to supply equipment which is safe. The responsibilities of the contractor depend on whether he is responsible for the whole process and engineering design on a turnkey basis or whether he is acting as an extension of the client's own organization and is thus undertaking detailed engineering only. It may happen in some cases that the client proposes a feature which the contractor considers unsafe. If agreement cannot be reached, the contractor should not undertake this feature. Another point of difficulty can arise if the client wishes to make a modification. The contractor should make all reasonable checks that the modification does not introduce a hazard.

8.1.6 Second Chance Design

Another relevant concept is that of 'second chance design'. This means the provision of a second line of defense to guard against an initial hazard or a failure. Some features which are prominent in second chance design are (1) plant layout, (2) pressure system design, (3) materials of construction, (4) isolation arrangements, (5) alarms and trips, (6) operating and maintenance procedures.

8.1.7 Unit Processes

A particular unit process tends to have certain characteristic features. These often relate to: (1) reactant and product; (2) reactor phase; (3) main reaction thermodynamic equilibrium as well as the heat of reaction, velocity constant, activation energy; (4) side reactions; (5) materials of construction.

Descriptions of unit processes and of their thermodynamics and kinetics are given in *Unit Processes in Organic Synthesis* (Groggins, 1952), *Industrial Chemicals* (Faith et al., 1965). Further information is given in other standard texts, including encyclopedias such as those by Kirk and Othmer (1963, 1978, 1991) and Ullman (1969, 1985). Unit processes are listed by Fowler and Spiegelman (1968) as follows: acylation, alkaline fusion, alkylation, animation, aromatization, calcination, carboxylation, causticization, combustion, condensation, coupling, cracking, diazotization, double decomposition, electrolysis, esterification, fermentation, halogenation, hydration, hydroforming, hydrogenation, hydrolysis, ion exchange, isomerization, neutralization, nitration, oxidation/reduction, polymerization, pyrolysis, and sulfonation. The unit processes which are generally more hazardous are given by Fowler and Spiegelman as alkylation, animation, aromatization, combustion, condensation, diazotization, halogenation, hydrogenation, nitration, oxidation, and polymerization. The unit processes considered here by way of illustration are (1) oxidation, which are those of an organic compound with molecular oxygen; (2) hydrogenation, which are those of an organic compound with molecular hydrogen in the presence of a catalyst; (3) chlorination, which are those of an organic compound with molecular chlorine; and (4) nitration, which are those of an organic compound with a nitrating agent such as nitric acid or mixed acid.

8.1.8 Unit Operations and Equipments

Unit operations include: mixing, dispersion; distillation, gas absorption, liquid—liquid extraction; leaching, ion exchange; precipitation, crystallization; centrifugation, filtration, sedimentation; classification, screening, sieving; crushing, grinding; compacting, granulation, pelletizing; gas cleaning; heat transfer; and drying and dehydration. A further listing is given by Fowler and Spiegelman (1968).

The unit operations and equipments considered here by way of illustration are (1) mixers, (2) centrifuges, (3) driers, (4) distillation columns, and (5) activated carbon adsorbers. All five feature in the CCPS *Engineering Design Guidelines* (1993) and the first three are the subject of IChemE guides.

Mixers present a variety of hazards. An account of hazards of, and precautions for, mixers is given in the *Guide to Safety in Mixing* (Schofield, 1982) (the IChemE *Mixer Guide*). The body of the *Mixer Guide* is in the form of a checklist of hazards and precautions under the headings: (1) equipment, (2) operations, (3) substances, and (4) plant.

Centrifuges likewise present a variety of hazards. The hazards of, and the precautions for, centrifuges are considered in the *User Guide for the Safe Operation of Centrifuges with Particular Reference to Hazardous Atmospheres* (Butterwick, 1976) and the later publication *User Guide for the Safe Operation of Centrifuges* (Lindley, 1987) (the IChemE *Centrifuge Guide*).

The principal hazards presented by driers are those of fire and explosion. These hazards and the corresponding precautions are discussed in *Prevention of Fires and Explosions in Dryers* (Abbott, 1990) (the IChemE *Drier Guide*). This is the second edition of, and thus supersedes, the *User Guide to Fire and Explosion Hazards in the Drying of Particulate Materials* (Reay, 1977).

Distillation columns present a hazard in that they contain large inventories of flammable boiling liquid, usually under pressure. Another hazard is overpressure due to heat radiation from fire; again pressure relief devices are required to provide protection. One quite different hazard in a distillation column is the ingress of water; the rapid expansion of the water as it flashes to steam can create very damaging overpressures.

8.1.9 Operating Conditions

Operation at extremes of pressure and temperature has its own characteristic problems and hazards. Extreme conditions may occur at: (1) high pressure, (2) low pressure, (3) high temperature, and (4) low temperature.

Some general characteristics of these operating conditions are described below, together with some of the associated literature. In relation to extreme operating conditions in general, mention may be made of *Chemical Engineering under Extreme Conditions* (IChemE, 1965/40) and of the series *Safety in Air and Ammonia Plants* (1960−1969/17−26), *Operating Practices in Air and Ammonia Plants* (1961−1962/27−28), and *Ammonia Plant Safety and Related Facilities* (1970−1994/31−53) by the American Institute of Chemical Engineers (AIChE).

The range of pressures handled in high pressure technology is wide. Three broad ranges of pressure may be distinguished: (1) pressures up to about 250 bar, (2) pressures of several thousand bar, and (3) pressures greater than 8000 bar.

The degrees of vacuum handled in low pressure or vacuum technology span many orders of magnitude. In the present context, it is relatively low vacuum from 760 (atmospheric pressure) to 1 torr and medium vacuum from 1 to 10^{-3} torr which are the ranges of prime interest.

The literature on high temperature technology is relatively diffuse and deals mainly with particular types of plant (e.g., ammonia plants) or particular types of equipment (e.g., fired heaters). Mention may be made, however, of *High Temperature Materials and Technology* (Campbell and Sherwood, 1967).

The temperatures handled in low temperature technology are those below 0°C. There is a somewhat distinct technology dealing with ultra-low temperatures below 20 K, but this is not of prime interest in the present context.

8.1.10 Utilities

The process plant is dependent on its utilities. These include in particular: (1) electricity includes general, uninterrupted power supplies, and electrical heating; (2) fuels; (3) steam; (4) compressed air like plant air, instrument air, process air, and breathing air; (5) inert gas; (6) water which contains cooling water, process water, hot water, and fire water; (7) heat transfer media like hot fluids and refrigerants.

Important features of a utility are its: (1) security, (2) quality, (3) economy, (4) safety, and (5) environmental effects.

A treatment of the principal utilities is given in *Process Utility Systems* (Broughton, 1993) (*The* IChemE *Utilities Guide*). The Utilities Guide deals with the following topics: (1) efficient use of utilities, (2) fuel, (3) compressed air, (4) inert gases, (5) thermal fluid systems, (6) water preparation, (7) the boiler house, (8) steam distribution, (9) electricity use and distribution, (10) air and water cooling, (11) refrigeration, (12) fire protection system, (13) building services, and (14) pipework and safety.

8.1.11 Operational Deviations

An important aspect of process design is consideration of possible deviations of operating parameters from their design values. The causes of deviation of a process variable are mostly specific to that variable, but there are some general causes of deviation. One is the deterioration or failure of equipment. Another is the maloperation of the control system, including the process operator.

The effects of deviations and the measures which may be taken to prevent them are also specific. But there are certain devices which are generally provided to give protection against extreme deviations, particularly of pressure and temperature. A further discussion of operating deviations is given by Wells et al. (1976).

Effects of pressure deviations include overpressure and underpressure of equipment and changes of temperature, flow, and level. Measures should be taken as appropriate to reduce the pressure deviations. In addition, it is essential to provide protective devices which prevent overpressure and underpressure.

Effects of temperature deviations include over temperature and under temperature of equipment, changes of pressure and flow, and runaway reactions. The appropriate measures should be taken to reduce the deviations. In particular, attention should be paid to the heat input and output from the plant and to the control of reactors.

Flow and level deviations may be too high, too low, zero, reverse, or fluctuating. There are many ways in which inhomogeneities of conditions and accumulations of material can arise. Poor mixing in reactors or other equipment can result in: side reactions and reaction runaway; hot spots, overheating, and thermal degradation; and fouling. The accumulation of substances which are relatively innocuous at high dilution but which when concentrated in some way, such as by vaporization with an insufficient bleed-off or by recycling of material, may present a hazard. Material which is held up in the system in a dead leg or elsewhere, for an abnormally long time, may undergo degradation and become hazardous.

The control system, including the process operator, may be a cause of operating deviations. One cause of this is errors, faults, or lags in measurement. Deviations may also be caused by the action of the automatic control system. It is important for the process design to provide sufficient potential correction for control.

8.1.12 Guidelines for Engineering Design for Process Safety

Process design is one of the principal topics covered in the *Guidelines for Engineering Design for Process Safety* (CCPS, 1993) (the CCPS *Engineering Design Guidelines*).

As far as process design is concerned, the *Guidelines* deal in particular with: (1) inherently safer design, (2) process equipment, (3) utilities, (4) heat transfer fluids, (5) effluent disposal, and (6) documentation. The treatment of inherently safer design is organized around the themes of: (1) intensification, (2) substitution, (3) attenuation, (4) limitation of effects, and (5) simplification and error tolerance and includes an inherent safety checklist.

The items of process equipment treated in the *Guidelines* are (1) chemical reactors, (2) columns, (3) heat exchangers, (4) furnaces and boilers, (5) filters, (6) centrifuges, (7) process vessels, (8) gas/liquid separators, (9) driers, (10) solids handling equipment, (11) pumps and compressors, (12) vacuum equipment, and (13) activated carbon adsorbers.

The account of heat transfer fluid systems describes the fluids available and their applications, discusses the relative merits of steam, liquid, and vapor−liquid systems, outlines the system design considerations and describes the system components, and considers the safety issues.

With regard to effluent disposal, the systems considered are (1) flares, (2) blowdown systems, (3) incineration systems, and (4) vapor control systems. The treatment does not extend to disposal of materials discharged in reactor venting.

The section on documentation covers: (1) design documentation, (2) operations documentation, (3) maintenance documentation, and (4) record keeping.

8.2. INTEGRATION OF SAFETY INTO THE PROCESS DESIGN

Integrating safety assessment into process design software, such as Aspen Plus, allows the synchronism of design and safety evaluation.

Several studies have stressed the importance of establishing a new methodology or using existing ones to evaluate the inherent safety of a chemical .Some have tried to integrate different hazard assessment tools into simulation software to build an inherent safer model at the design stage.

Controllability is another factor usually considered in safety evaluation. A systematic approach to evaluate the resilience of chemical process in design aspect has been studied (Dinh, 2011). The evaluation framework to identify resilience design index is developed by means of the multi-attribute model approach, including design, detection potential, emergency response, human, and safety management.

8.2.1 Inherently Safer Design

The best way of dealing with a hazard is to remove it completely if possible. The provision of means to control the risk associated with a hazard is very much the second best solution. This has been succinctly stated by Kletz (1978, 2003, 2009) as 'What you don't have, can't leak'. This, of course, immediately eliminates fire and explosive hazards. A corollary to this is the importance of limiting inventory of hazardous materials, which is one of the lessons learned in the Flixborough disaster (Parker, 1975; Turney, 1994). This principle has been incorporated in the systematic application of inherent safety.

One of the principal ways in which a process may be made inherently safer is to limit the inventory of hazardous material. It is better to have only a small inventory of hazardous material than a large one which can be rendered relatively safe only by highly engineered safety systems. In general, storage capacity is of great assistance to plant management in the operation of the plant, but this has to be balanced against the cost and the hazard of storage. Application of inherently safer design to storage may involve: (1) elimination of intermediate storage, (2) reduction in storage inventory, (3) storage under less hazardous conditions.

Some basic principles of inherently safer design Kletz (1984a−m, 1991a) are (1) intensification, (2) substitution, (3) attenuation, (4) simplicity, (5) operability, (6) fail-safe design, (7) second chance design.

A plant which embodies these principles is described by Kletz (1989, 1990) as a 'friendly plant'. These basic principles have been distilled down to the following systemized approaches (CCPS, 1996a 'Gold Book'): (1) minimize, which is the principle to limit the inventory of the process; (2) substitute, in which a hazardous feature is replaced by a less hazardous one; (3) moderate, that of attenuation, involves the use of less hazardous process conditions; (4) simplify, which means the selection of the process with a view to eliminate particularly hazardous chemicals and/or to operate under less hazardous conditions.

As far as operating conditions are concerned, a compromise solution may turn out to be the most hazardous. The compromise solution of moderate inventory at moderate pressure and temperature may actually be the most hazardous, if the conditions are such that on release a large fraction of the material will flash off and the inventory is such that the quantity escaping is likely to be large.

The fundamental principle of inherently safer design is simplicity. Many actual plant designs are extremely complex. Aspects of simpler design instanced by Kletz (1984d) include: (1) design for full overpressure, (2) design modification to avoid instrumentation, (3) use of resistant materials of construction, (4) use of simple alternatives to instrumentation.

Some other things should be taken into account as well as inherent safety. Flexibility is one of them, which is a valuable feature on a plant, but it can be taken too far. Not only can the interconnections needed to give such flexibility become very complex and costly but also they introduce further potential sources of leaks and human errors. General multiple single trains are employed rather than complex crossovers.

Another source of complexity is the modification chain: an initial modification is made which leads to others. This is a problem not only in design, but perhaps even more on existing plant, where a single *ad hoc* modification can lead to a whole series.

Some processes are inherently more operable than others. This aspect should be borne in mind in the selection of the process. The operability of process and plant is one of the two main aspects which are examined in HAZOP studies.

A rather different aspect of inherently safer design is the limitation of exposure of personnel. This subject is covered in detail with a quantitative analysis in the CCPS *Guidelines for Evaluating Process Plant Buildings for External Explosion and Fires* (CCPS, 1996b).

Progress in the realization of inherently safer designs has not been as rapid as its advocates would wish. Some reasons for this have been examined by Kletz (1991b, 1992) and Mansfield and Cassidy (1994).

The Health and Safety Executive (HSE) now has an explicit policy of encouraging the application of inherently safer design concepts. Its philosophy and activity in this area have been described by Barrell (1988) and Jones (1992). The work described by Mansfield and Cassidy (1994) is one aspect of this program.

8.2.2 Inherently Safer Design Methodology

The work of Mansfield and Cassidy is aimed at furthering the practice of inherently safer design by the development of suitable tools. Such a tool needs four broad categories for inherently safer design: (1) brainstorming with a degree of structure, (2) a more highly structured HAZOP-style examination of flow sheets and process diagrams, (3) examination, using checklists, of the plant layout, and (4) indices of inherent safety.

8.2.3 Inherently Safer Design Index

The development of an index for inherently safer design has been described by Edwards and Lawrence (1993). They state that one of their aims is to test the hypothesis that inherently safer design means cheaper plants, and give as an illustrative example a comparison of six routes to methylmethacrylate in respect of: (1) the estimated costs and (2) the inherent safety index. The *Dow Fire and Explosion Index—Hazard Classification Guide* (Dow chemical company, 1976, 1980, 1987, 1994a–b) which has been available for over 20 years, now in the eighth edition, provides also quantitative results based on the same issues previously described.

8.2.4 Safety Instrumented Systems

The most effective way to prevent incidents is use of inherently safer design to eliminate or reduce the potential hazard. In many processes, technical or manufacturing issues limit the engineer's capability to design an inherently safer process. Further, there is generally a point where the required capital investment is disproportional to the additional risk reduction provided by the process modification. In other words, the derived safety benefit is too low relative to the economic investment. When this occurs, protection layers or safeguards must be provided to prevent or mitigate the process risk. A Safety Instrumented System (SIS) is a protection layer, which shuts down the plant, or part of it, if a hazardous condition is detected.

The essential characteristic of SIS is that it is composed of instruments, which detect that process variables are exceeding preset limits, a logic solver, which processes this information and makes decisions, and final control elements, which take necessary action on the process to achieve a safe state.

For start-up, the various instrumented permissives and operator actions must be defined. These permissives may require that certain process conditions exist prior to allowing start-up. They may also involve the use of temporary bypass of the SIS for start-up, such as the bypass of a low flow shutdown of a pump. From the case histories, it is apparent that the permissive and operator action requirements have often been ill-defined, allowing start-up under unsafe operating conditions. The required performance or integrity of the SIS must also be defined. Not all SIS provide the same performance.

The risk reduction that the SIS must provide is defined by process hazards analysis, supplemented with semi-quantitative or quantitative evaluation. Qualitative process hazards analysis techniques, such as what-if analysis, checklists, and Hazard & Operability (HAZOPs) analysis, are used to identify the causes and consequences of potential process hazards. Semi-quantitative or quantitative techniques are then used to evaluate the risk reduction requirements. The semi-quantitative evaluation is typically performed using Layers of Protection Analysis (LOPA). For quantitative risk assessment, the most commonly used technique is fault tree analysis.

The hazards analysis allocates risk reduction to the SIS. This risk reduction defines the integrity requirement for the SIS. Once the SIS functionality and integrity requirements are known, it is possible to design an SIS that will mitigate the process risk. However, the design simply represents the intended operation of the SIS. How the SIS is actually maintained, tested, and managed determines the risk reduction that the SIS will achieve over its lifetime.

8.2.5 Layers of Protection Analysis (LOPA)

The intent of LOPA, according to *Layers of Protection Analysis* (CCPS, 2001) (the CCPS *LOPA book*), is to provide a rational, objective, risk-based approach to identifying and specifying the protection layers that are used to prevent or mitigate process risk. Process risk is defined by the frequency and severity of potential hazards. The severity may be assessed in terms of human impact, such as injuries and fatalities, environmental impact, such as noise, releases or spills, or financial losses, such as production loss or equipment damage.

LOPA builds upon traditional Process Hazards Analysis (PHA) techniques to identify the frequency of process deviations that potentially propagate into hazardous events. LOPA determines the frequency of each initiating cause using either an order of magnitude assessment or estimated failure rate. The consequence of the hazardous event is defined by: (1) the team using qualitative judgment, (2) look-up tables, or (3) consequence modeling for specific scenarios. The method chosen is typically dependent on the nature of the hazards present in a facility. If the hazards are easily understood by the team, the assessment may be completely qualitative with the team instructed to use past experience and process knowledge in its assessment.

With the initiating cause frequency and consequence, the unmitigated process risk is the product of initiating cause frequency and consequence. This unmitigated risk is compared to the risk tolerance criteria to determine whether protection layers are required. The risk tolerance can be defined as a specific, numerical target, such as 1 fatality in 10,000 years, or in general categories using a risk matrix. The overall risk reduction required to reduce the unmitigated risk to the tolerable risk may be provided by one protection layer or multiple protection layers. It is always important to evaluate the claimed protection layers against the independence criterion. When the protection layer is an SIS, the risk reduction allocated to the protection layer is the SIL. This establishes the performance requirement for SIS, which must be met by the design, operation, maintenance, and testing philosophy.

After allocating the appropriate risk reduction to the existing or planned protection layers, the mitigated risk is again compared to the tolerable risk. If the tolerable risk is not satisfied, recommendations for additional risk mitigation layers or design modifications are made. LOPA is a tool used to ensure that process risk is successfully mitigated to the tolerable level. It can be used at any point in the lifecycle of a project or process, but it is most cost effective when implemented during front-end loading, as soon as process flow diagrams are complete and the Piping & Instrumentation Diagrams (P&IDs) are under development. For existing processes, LOPA should be used during or after the HAZOP review. LOPA is typically applied after a qualitative hazards analysis has been completed, which provides the LOPA team with a listing of hazard scenarios with associated initiating causes, consequence ratings, and potential safeguards for consideration as IPLs.

8.2.6 Design of SIS

Appropriate SIS design requires that the engineer understands the underlying purpose of the SIS. This entails reviewing the Process Hazards Analysis (PHA) documentation to better define the required SIS functionality and integrity. These basic requirements should then be developed into complete specifications.

The functional specification should include the definition of the process safe state; SIS inputs, voting configuration, and trip points; SIS outputs, redundancy requirements, and safe state actions; functional relationship between inputs and outputs; automatic diagnostics for inputs and outputs; provisions for bypassing, testing, and maintenance; provisions for manual shutdown; required reset actions; response time requirements for SIS actions; and means by which alarm, trip, and diagnostic information are to be communicated to plant operator.

The integrity specification should include the integrity level assigned to the SIS, test interval requirements, and limitations on acceptable spurious trip rate.

8.2.7 Separation of BPCS and SIS

Most process control systems involve some sequential control even if it is largely limited to start-up and shutdown. The sequential logic should operate in such a way as not to cause any safety problems. Its operation should be tested against the SIS logic to ensure that normal operation of the sequential control does not trigger unnecessary SIS action or inhibit any required SIS action. BPCS provides protection against hazardous events by maintaining the process within control tolerances. Process alarms are issued to the operator through the BPCS human—machine interface, notifying the operator that the process has exceeded preset process conditions. This allows the operator the opportunity to bring the process back into control, preventing a unit shutdown.

8.2.8 Response Time Requirement

The dependability of an SIS is based on its functionality and integrity. An important aspect of functionality is the dynamic response. There will normally be delays due to: (1) sampling, (2) dynamic response of the measuring instrument, and (3) normal measuring instrument error. Even when the measuring instrument has responded, there will be delays in the safety circuitry and the closure of the shutdown valve. There will be a further delay in the process itself, due to the amount of time required for the valve closure to affect the process variable. All of these factors, delays, and errors erode the available safety margin and should be considered carefully. The maximum rate of rise of the process variable is also critical. If the process variable rises very rapidly, SIS valve closure speed may become a major design factor.

Further reduction of the nominal trip point may be appropriate, but the setting should not be put so low that noise in the process variable measurement activates the trip at normal operational conditions. After all, a spurious trip can arise from an incorrect trip set point, as well as from instrument unreliability.

The dynamic response of the hazard scenario against which the SIS is designed to protect may be modeled using standard methods. If the time interval from the initial process deviation to the hazard point exceeds 2 min, there is normally no problem in designing the SIS. On the other hand, if the time interval is less than this, there is a potential problem. If the time interval is only a few seconds, special SIS design is required.

The signal transmission lags to and from the logic system, and the logic system delays itself are normally negligible. The more significant lags are likely to be in the sampling and the sensor, in the final control element, and in the process itself. For analyzers, sampling lags may amount to a dead time of the order of seconds to minutes. For smart transmitters, the sensor lag may be of the order of $100-250$ ms. Transfer lags in sensors vary, with temperature measurement lags often being large due to the thermal inertia of the measuring pocket. The lag at the control valve can vary from a fraction of a second up to several minutes, depending on the valve size. The lag in the process itself is also highly variable.

8.2.9 Power Supply

Loss of power to the process unit results in the failure of pumps and other electrical driven equipment. During the flare load analysis, as discussed in this volume, the loss of power case was cited as often presenting the largest flare load. How long one can run safely without power is dependent on the process operation. In designing the SIS, the choice must be made between Energize-to-Trip (ETT) and De-energize-to-Trip (DTT).

In a DTT design, loss of power causes the final elements to take their fail-safe action. To prevent brief power interruptions from causing process shutdowns,

Uninterruptible Power Supplies (UPS) are used. Power is supplied to the SIS instrumentation by the UPS when the power is lost. The UPS can be sized to supply power for any length of time, but for practical reasons, generally do not supply power for periods in excess of $30-45$ min.

In DTT design, a power outage results in the final elements going to the safe state. Due to historical plant power supply problems, some companies migrated to an ETT design to allow the plant to continue to operate during short power outages. An ETT design requires that power be applied in order for the final elements to achieve their safe state. This migration to ETT improved process reliability, but placed the process at greater risk of an incident during a loss of power condition. ETT SISs are generally limited to SIL 1 applications, because it is difficult to implement back-up power generation and/or alternative power supplies (e.g., UPS) to support SIL 2 and 3. If ETT circuits are used, line monitoring must be performed, because any loose wire or open circuit will inhibit the safety action. But even line monitoring does not guarantee that the SIS will work when a process demand occurs, because fuses or coils may burn-out when power is supplied.

8.2.10 Integration of SISs

As already described, an SIS is normally dormant and operates in a demand mode. An SIS device such as a sensor or a valve may fail in such a way that its failure will remain unrevealed until detected by proof testing or some other means. By contrast, similar devices in a control system are exercised continuously, and these failures are more likely to cause an operational excursion. The failure in the control system is often a revealed one. Yet the actual physical fault in both cases may be identical. A sensor may fail giving a low/zero or high reading, or a valve may stick open or closed. The concept of trip integration is based on this contrast between a fault that lies unrevealed in an SIS but is revealed in a measurement and control system. The principle applies to any system that has a protective function. The system is regarded as integrated provided it is regularly exercised, which generally means that it is in use during the normal operation of the plant.

8.2.11 Field Design

The SISs most commonly used are the 1oo1, 1oo2, 2oo2, and 2oo3 systems.

The use of parallel redundant, or 1oo$n$ (1-out-of-n), systems gives an increase in functional integrity but a decrease in operational reliability compared with a 1oo1 (1-out-of-1) system. Better overall reliability characteristics can be obtained by the use of a majority voting system, of which the 2oo3 system is the simplest. A comparison of a 2oo3 system with a 1oo1 system shows that the 2oo3 system has a higher functional integrity and operational reliability, while a comparison with a 1oo2 system shows that the 2oo3 system has a slightly lower functional integrity, but a much higher operational reliability. The 2oo2 system has some interesting characteristics. It has a lower functional integrity than a 1oo1 system, but its operational reliability exceeds not only that of the 1oo1 but also that of the 2oo3 system. When 2oo2 systems are used to improve operational reliability, the functional integrity can be improved by increased proof testing or enhanced diagnostics. The requirement for functional integrity is rarely such as to justify a 1oo3 system and that for operational reliability rarely such as to justify a 3oo3 system.

The SIS field sensors should be independent from the initiating cause for the potential hazardous event and from any other protection layers used to reduce the process risk to the risk tolerance. Independence is necessary to minimize common cause failure. For example, if the initiating cause is the process control transmitter, the control transmitter cannot be used as the sole means for SIS initiation. For enhanced diagnostics, the control transmitter signal can be compared to the SIS transmitter signal to detect faults. Any sharing of the control transmitter with the SIS should be carefully assessed to ensure that potential common cause failures are adequately addressed. Further, the design should ensure that no failure of the control system can disable the SIS.

Field sensor redundancy is related to the required integrity. For SIL 1, redundancy is usually not necessary, though it may be appropriate for a lower integrity (i.e., high failure rate) field sensor or for improving operational reliability. For SIL 2, redundancy is often employed to reduce the required proof test interval and improve the operational reliability. For SIL 3 or 4, there should be full redundancy, such that there are no single points of failure in the SIS design. IEC 61511 requires the demonstration of fault tolerance for SIL 3 or 4. Depending on the voting configuration selection, the incorporation of fault tolerance may also improve operational reliability.

The SIS technologies given in the CCPS *Safe Automation Guidelines* include: (1) fluid logic (pneumatic, hydraulic); (2) electrical logic, including direct-wired systems, electromechanical devices (relays, timers), solid-state relays, solid-state logic, and motor-driven timers; (3) Programmable Electronic Systems (PES) technology, involving programmable logic controllers (PLCs); and (4) hybrid systems. The technologies are detailed in Appendix A of the CCPS *Safe Automation Guidelines*.

The hardware of a typical PES consists of a chassis, input modules receiving sensor signals, main processors, output modules sending out signals to final control elements, terminations, power supplies, a BPCS interface, a human—machine interface, and an engineering interface. The software of a typical PES consists of embedded software and utility software, which perform the system-level fault detection and execute basic PES functions. In addition, an application program within the PES executes the required user functions. The application program is constructed using software that is typically provided by the PES vendor.

The SIS logic solver must provide performance that meets the required SIL. It was general practice for many years to design SISs as separate, hardwired systems. Hardwired systems have highly predictable failure modes and exhibit low failure rates. In contrast, a PES is composed of hardware and software components with complex failure modes and substantially higher failure rates. The major advantages of PES are the ease of logic sequencing implementation, the ability to perform complex SIS logic, and enhanced diagnostics of field components. Over the last 15 years, the practice of using PES has gained wider acceptance.

A PES uses a large number of automatic diagnostic circuits and functions that are implemented in hardware and software. The purpose of the diagnostics is to detect failures, particularly dangerous failures, which would inhibit the operation of the PES. Upon detection of failure, the PES may perform one of the following actions: (1) for a simplex PES, the SIS fail-safe action is initiated; (2) for 1oo$n$ PES, the SIS fail-safe action is initiated; (3) for 1oo$n$ PES, if the process safety time is sufficiently long, the operational state is continued under alternative operating procedures or by other compensating measures that ensure that safe operation is maintained; (4) for 2oo$n$ PES, the failed circuit is voted toward the trip state. The operational state is continued under alternative operating procedures or by other compensating measures that ensure that safe operation is maintained. In the event of any detected failure, an alarm is generated by the PES indicating the nature of the failure and that maintenance is required.

The final elements used for shutdown action should be independent of the initiating cause of the incident and any other protection layer used to mitigate the incident. The final elements can be valves, circuit breakers, motor starters, protective relays, etc.

Final element redundancy is related to the required integrity. For SIL 1, redundancy is usually not necessary, though it may be appropriate for a lower integrity (i.e., high failure rate) final element. For SIL 2, redundancy is often employed to reduce the required proof test interval. For SIL 3 or SIL 4, there should be full redundancy, such that there are no single points of failure in the SIS design.

8.2.12 Verification of SIS

Since an SIS protects against a hazardous condition, it is essential for the system itself to be dependable. The dependability of an SIS is related to its: (1) functionality and (2) integrity. It is necessary for the system to have the capability of carrying out its function in terms of features, such as accuracy, dynamic response, and so on, and for it to have a high probability of doing so.

SIS functionality is defined by identifying what must be detected and the response that must be taken to achieve the safe state. The SIS functionality is, therefore, related to the input and output architecture.

The SIS integrity can be improved by the use of: (1) redundancy, (2) diversity, and (3) diagnostics. Redundancy involves the use of multiple instruments to detect the potentially hazardous event. With voting, multiple failures are required for the SIS to fail to function. This yields a lower probability of failure than that of a single instrument. But, redundancy is not always the full answer, because there are some dependent failures which may disable the whole set of redundant instruments. This difficulty can be minimized by implementing diversity, which often involves using different process variables to detect the same hazard, or different types of instrumentation to measure the same process variable. On-line diagnostics can also be used to detect component failure including dependent failures. Diagnostics allow the operator to be notified that an SIS fault has occurred, significantly improving SIS integrity.

Most SISs consist of a single channel comprising a sensor, a logic processor, and a final control element, but where the integrity required is higher than that which can be obtained from a single channel, redundancy is generally used.

In the process industry hazardous conditions are seldom present, so most SISs operate in a demand mode, that is, the SIS is dormant until a process excursion or demand occurs. Thus, functional failures of the system are generally unrevealed failures. Device failure that results in the SIS failing to function when required by a process demand can be predicted using probabilistic techniques. A dependable SIS should have a low PFD.

The SIS should also be reliable and not result in spurious trips, that is, failure that causes the SIS to shut the plant down when no hazardous condition exists. Thus, operational failures of the system are generally revealed failures due to direct process impact or system diagnostics.

It is the objective of SIS design and operation to avoid both loss of protection against the hazardous condition, due to functional failures, and plant shutdown, due to operational failure or spurious trip. Since functional failure of the system is generally unrevealed, it is necessary to carry out periodic proof testing to detect such failure. The simpler theoretical treatments of SISs usually assume that the functional failures are unrevealed, operational failures are revealed, and failure rates are constant.

A method of determining the frequency of events of interest has been described by Kumamoto et al. (1981). These events are the process demand rate, the functional failure rate, and the operational failure rate. The procedure is to designate each of these events in turn as the top event of a fault tree, to create the fault tree, and to determine its cut-sets. These cut-sets together with the proof test interval for the SIS are the inputs used by the model to provide estimates of the frequency of the events mentioned.

8.2.13 Operation of SIS

To ensure the control system integrity, the design process must be supported by administrative procedures and actions. The SIS consists of instrumentation and logic processors that can and do fail. When these failures are detected, immediate operator response is essential to ensure safe operation. This response includes issuing a request for repair of the faulted device, monitoring other process variables associated with the unit, and, if necessary, initiating a manual shutdown action.

Bypasses (or overrides) may be required for start-up, maintenance, or testing activities. When a bypass is placed in service, the bypass should be alarmed in the control room to notify operations staff that an SIS is inoperable. Then, compensating measures should be implemented to maintain safe operation, while any portion of the SIS is bypassed. Compensating measures may include, for example, increased process monitoring, increased supervision, revised operational procedures, or reduction in production rates.

Bypasses are typically only used during maintenance or testing activities. Occasionally, a start-up bypass is required for start-up, and these bypasses must be carefully evaluated to ensure that potentially unsafe process conditions cannot occur during this bypass condition. Timers, operation sequences, or process condition-based permissives may be used to automatically bring the SIS out of bypass during start-up to prevent prolonged operation in the bypass condition. When SISs are bypassed during normal operation, it is generally due to poor SIS reliability. If an SIS proves troublesome, it is prone to be bypassed without appropriate authorization. This is particularly likely to occur if there are frequent spurious trips due to sensor failure or other causes.

8.2.14 Maintenance and Testing of SIS

The importance of proof testing cannot be overemphasized. Testing and, more generally, SIS maintenance should be of high quality, if the functional and integrity requirements are to be achieved.

Testing plays an important role in achieving functional integrity. All devices eventually fail. The most hazardous failure is a dangerous, undetected failure, which results in the device not functioning when a process demand occurs. This type of failure results in the need to periodically test SIS devices to determine if they are functional. Thus, the required proof test interval is related to the device failure rate.

Perhaps the most important functional test is the validation of the SIS prior to introduction of hazardous chemicals into the process equipment. This validation is sometimes referred to as the pre-start-up acceptance test (PSAT) or the site acceptance test (SAT). Validation is a fully functional test of the SIS performed to determine whether the installed SIS is functioning as intended from a specification, design, and installation standpoint. Throughout the lifecycle, verification activities are performed to maintain a quality design process and to ensure that the design remains in agreement with the required functionality and integrity. Validation is the ultimate test of the SIS functionality. Since the majority of testing after start-up will consist of testing individual devices, it is extremely important that complete input to output testing is performed to identify any system-level problems.

The test should be conducted using a written procedure, which details the actions to be taken and requires the technician's signature at key steps. Changes to the procedure should be managed through a management of change (MOC) process.

The test should cover the entire SIS from sensor to final element. From the point of view of testing, the preferred method is to take the process variable to the trip point using test equipment and to verify the trip action. This method requires a shutdown or provisions for bypass of the final element in order to perform the test. For inputs, the test should include process variable indication leading to alarm and trip initiation. For input devices requiring field calibration, the sensor calibration range should be checked to ensure that it matches the specification and that the trip setpoint is achieved at the corresponding process condition. This is accomplished by inducing the process condition at the sensor and monitoring the corresponding output of the input device. It should be demonstrated that the input device generates the correct signal at the alarm and trip setpoints. It should also be shown that if there is a bad sensor output, there is appropriate indication of this fault to the control room.

Logic solvers should demonstrate that on receipt of an input trip condition that it generates the correct output response. At the simplest level, this involves validating that the logic performed in the logic solver matches the specification.

The test for final elements should determine that the final element can achieve the safe-state condition. The *final element* boundary includes all of the devices from the logic solver output to the control element that takes action on the process. While testing the functionality of the field devices and logic solver is an obvious requirement, it is also necessary to test auxiliary systems and communications. Auxiliary systems may include electrical power and air supplies. Any communication between the SIS and the basic process control system should be tested to ensure proper transfer of the data between the two systems. The test should demonstrate that no communication failure can occur that would prevent proper functioning of the SIS.

8.3. PRESSURE SYSTEMS

Any system in which the pressure departs significantly from atmospheric, and which is therefore of rigid construction, needs to be considered as a pressure system. In the United Kingdom, pressure systems are covered by the Pressure Systems and Transportable Gas Containers Regulations 1989 (the Pressure Systems Regulations). For fixed plant, the associated code is ACOP L122 *Safety of Pressure Systems* (HSE, 2000).

The main components of pressure systems have been described by Dickenson (1976). They are as follows:

1. pressure vessels (reactors, distillation columns, storage drums, and vessels);
2. piping system components (pipes, bends, tees, reducers, flanges, valves, nozzles, nipples);
3. means of adding, controlling, or removing heat (fired heaters, reboilers, vaporizers, condensers, coolers, heat exchangers generally);
4. means of increasing, controlling, or reducing pressure (pumps, compressors, fans, letdown turbines, control valves);
5. means of adding or removing fluids or solids to or from the process system (pumps, compressors, dump valves);
6. measurement and control devices and systems (instrumentation);
7. utilities and services (electricity, steam, water, air).

8.3.1 Steels and Their Properties

The main material of construction used in pressure systems is steel, and it is necessary to consider some of the properties of steel which are particularly important in relation to pressure systems and which are the basis of pressure systems standards and codes of practice.

8.3.2 Types of Steel

The main types of steel which are used in process plants are (1) carbon steel, including mild steel, (2) low alloy steel, and (3) stainless steel, including ferritic and austenitic types. Of prime importance is the temperature at which the steel begins to undergo creep. At room temperature, the strength of the steel is limited by the yield and/ or tensile strengths. As the operating temperature increases, however, the creep strength becomes the limiting factor. Thus, the operating temperatures may be divided into: (1) temperatures below the creep range and (2) temperatures above the creep range. The basis of design is different for these two cases.

8.3.3 Pressure Vessel Design

Pressure vessels are subject to a variety of loads and other conditions which stress them and, in certain cases, may cause serious failure.

In general, structures are subject to two types of loading: (1) static loading and (2) dynamic loading. For pressure vessels, this loading is normally due to pressure. The load caused by the pressure creates stresses in the vessel. There may also be other stresses, which include: (1) residual stress, (2) local stress, and (3) thermal stress.

The basis of the design of pressure vessels is the use of appropriate formulas for vessel dimensions in conjunction with suitable values of design strength. In determining the strength of materials, a basic distinction is drawn between (1) temperatures below the creep range and (2) temperatures inside the creep range.

For temperatures below the creep range, two important properties are (1) tensile strength, R and (2) yield strength, E. Different design methods may use these properties as measured at room temperature (generally 20°C) or at the operating temperature (f°C). Use has been made, therefore, of all the four properties: (1) tensile strength at room temperature, R_{20}; (2) tensile strength at operating temperature, R_t; (3) yield strength at room temperature, E_{20}; and (4) yield strength at operating temperature, E_t.

For temperatures inside the creep range, two further important properties are (5) stress for 1% extension in 100,000 h at the operating temperature, S_c and (6) stress for rupture in 100,000 h at the operating temperature, S_r. These material strengths are divided by a factor of safety to obtain the design strengths for use in the design.

8.3.4 Joining, Fastening, and Welding

There are two main methods by which materials may be joined: (1) mechanical fastening and (2) physical bonding. For pressure systems, the main fastening methods are riveting and bolts and nuts, while the methods of bonding are soldering and brazing and fusion welding.

Soldering and brazing are methods of bonding in which the joint is made by introducing molten metal between the two parts to be joined without deliberate fusion of the parent metal. With fusion welding, by contrast, the parts of the parent metal adjacent to the joint are brought to molten temperature and caused to fuse together, generally with the addition of molten filler material. Fusion welding is the principal method of making permanent joints in pressure vessels and pipework.

8.3.5 Pipework and Valves

Loss of containment from a pressure system generally occurs not from pressure vessels but from pipework and associated fittings. At least as much attention should be paid to the pipework and fittings as to the vessels.

A large proportion of failures of containment in process plants occur on the pipework and fittings. Some suggestions for reducing pipework failures have been given by Kletz (1984k) as part of a survey of such failures. The design of pipework should be done by a fully integrated design organization working in a structured manner. There should be a relatively small number of designers of high quality, making full use of computer aids. Similar principles should apply to the fabrication and construction stages.

Kletz recommends detailed design of even small bore pipework, though he recognizes that some organizations consider this impractical. He states that efforts should be made to reduce the number of grades of steel required so as to reduce the chance of installation of the incorrect grade, and instances restriction of steam temperatures so as to avoid the need for creep-resistant steel. His survey highlights the high proportion of failures which are attributable to the construction phase, and he makes suggestions for improved inspection during and after construction. Kletz states that the incidents which he lists suggest that of all the measures proposed this would be the most effective.

The pipework should be designed for ease of maintenance. If a joint may have to be broken, there should be adequate access and sufficient 'spring' in the pipework. If the insertion of a slip plate into a joint is likely to be a frequent operation, consideration should be given to installing a slip ring or spectacle plate.

Concentrated loads, direction changes, and pipe joints require special attention. Increase in pipe diameter is sometimes used to obtain a greater distance between supports. Some devices used to support horizontal pipes include pipe hangers, slider supports, and roller supports. Spring and turnbuckle hangers are used for suspending hot insulated pipes. For vertical pipes, methods of support include trunnions supported on guide lugs and sling rods supported by pipe hangers.

Pipe clips or hangers should be in direct contact with the pipe or insulation but should not be too tight around them. Distance pieces are generally used to prevent this. The insulating material should be capable of bearing the compression load imposed. Pipe guides are used to restrain sideways movement of pipes. Anchors are used where it is necessary to provide fixed points for pipe bends and loops.

Where the consequences of pipework failure could be especially serious, the pipework should be of high integrity. Measures should be taken to reduce the stresses to which the pipework is liable to be subjected and to give it robustness in the face of such stresses. Sources of vibration and of severe cycling should be minimized and, if necessary, countermeasures taken. Joints on pipework should be welded or flanged, and threaded joints should be avoided. Features which may prove to be weak points should be minimized. Gaskets should be suitable for the fluid handled and the temperatures.

8.3.6 Process Equipment

8.3.6.1 Heat Exchangers

Heat exchangers are components of importance in pressure systems. They include: (1) vaporizers, (2) reboilers, (3) condensers, and (4) waste heat boilers, as well as heat exchangers on general heat interchange duties. Some problems that occur in heat exchangers and which may affect safe operation are (1) fouling, (2) tube vibration, and (3) tube rupture.

If fouling occurs in a heat exchanger, it may affect (1) heat transfer and (2) pressure drop. Reduction of heat transfer capacity may be particularly serious if the heat exchanger is a cooler on a critical duty such as removal of heat from a reactor in which an exothermic reaction is carried out. Fouling may be reduced by correct selection of tube and shell side fluids and by the use of appropriate fluid velocities. An allowance may be made for fouling by the use of a fouling factor which is equivalent to a heat transfer coefficient. This may be a general value taken from the literature or a specific value determined for the plant.

The use of a large fouling factor is not necessarily conservative. It may give a degree of overdesign which results in higher than expected temperatures and greater corrosion or coke deposition in the equipment. The use of

very compact equipment can give rise to fouling due to lack of turbulence. Fouling can also occur due to operational factors such as stoppage of a pump, omission of a filter, and transfer of debris from other equipment. If this can have a serious effect, it may be necessary to take appropriate precautions, but the incorporation of a fouling factor in the design is of little use in this situation.

The use of high capacity, compact heat exchangers with long tubes and high fluid velocities has intensified the problem of tube vibration. Tube vibration can cause damage to the tubes themselves and possibly to the exchanger shell and its supports or to the pipework connected to it, can transmit destructive pressure fluctuations from the exchanger, and can generate considerable noise.

In a heat exchanger with high pressure gas or vapor in the tubes and low pressure liquid in the shell, tube rupture can lead to overpressure of the shell. Although the localized overpressure is less than the overall peak pressure rise in the shell which occurs later, the latter is normally catered for by a pressure relief device, while the former is not. The effect of local overpressure on the shell and its implications for the positioning of tubes relative to the shell should be considered as a separate matter.

8.3.6.2 Fired Heaters and Furnaces

Fired heaters are another important component in pressure systems. They include pipe stills on crude oil units, furnaces on olefins plants, and heaters for reactors and for heat transfer media. Fired heaters are a prime source of hazards. These hazards are principally (1) explosion in firing space and (2) rupture of tubes. Explosion in the firing space occurs mainly either during lighting up or as the result of flame failure. Rupture of tubes is usually caused by loss of feed or by overheating. In addition, heaters are a source of ignition for escapes of flammable materials from other parts of the plant.

Many explosions in the firing space take place during start-up when an attempt is made to light a burner. There are a number of measures which should be taken to prevent this hazard. The most important is to eliminate the admission of unburnt fuel into the firing space. For isolation it is not sufficient to rely on a single valve. Use should be made either of double block and bleed valves or of slip plates. Before a burner is started up, the firing space should be purged with air. The atmosphere in the space should then be sampled to confirm that it is not flammable. The purging of the firing space is essential.

It is essential that if the flame fails, the burner should be shut-down immediately. A flame failure detector is used to monitor the flame and to initiate the trip. Loss of combustion air also necessitates immediate shut-down. Methods of detecting this condition are measurement of static pressure or of air flow.

The sequence of operations for the start-up of a burner should be carefully specified. In particular, the following operations should be carried out in sequence and the start-up should not proceed unless this has been done: establishment of air flow; purging of firing space; testing of atmosphere in space; and establishment of flame.

Many cases of tube rupture in fired heaters occur because the flow of process fluid through the tubes either falls too low or ceases altogether. Overheating of the tubes is another cause of rupture. It is necessary, therefore, to have measurements of tube temperature.

Instrumentation should be provided to measure and control the main operating parameters and, if necessary, to trip the furnace.

There should be trip systems which shut off the fuel flow to the furnace and inject snuffing steam: (1) if the electrical power fails, (2) if the fuel pressure and/or flow is low, (3) if the flame goes out, (4) if the combustion air pressure and/or flow is low, or (5) if the process fluid pressure and/or flow is low or its outlet temperature high. Difficulties are sometimes experienced with the flame failure device. One problem is instrument failure. Flame failure devices are available with some degree of self-checking. Another is sighting of a pilot burner flame instead of the main flame.

8.3.6.3 Process Machinery

Process machines such as compressors and pumps are particularly important items in pressure systems. Not only are they themselves potential sources of loss of containment, but they also affect the rest of the plant by imposing pressure and/or flow fluctuations and by causing vibrations. Several important pieces of process machinery are (1) Process Compressors, (2) Reciprocating Compressors, (3) Centrifugal Compressors, (4) Screw Compressors, (5) Gas Engines, and (6) Process Pumps. These are described in the following paragraphs.

1. *Process Compressors*: The process industries use both positive displacement and centrifugal compressors, some of which are very large machines with high throughputs and energy. Compressors are complex machines and their reliability is crucial. Some general features of compressors which are important are (a) lubrication, (b) protection, (c) isolation, (d) purging, (e) liquid slugs, (f) housekeeping, and (g) observation.

2. *Reciprocating Compressors*: Reciprocating compressors are utilized for high compressions. They can be provided with capacity control, or turndown, by the use of volume pockets and by drive speed control. Some principal malfunctions on reciprocating compressors are (a) valve leakage, (b) cylinder/piston scoring, (c) piston ring leakage, (d) gasket failure, (e) tail rod failure, and (f) vibration, as well as the general compressor failures such as those caused by liquid slugs or loss of lubricating oil or cooling water.

3. *Centrifugal Compressors*: Centrifugal compressors are the main workhorse machines in the process industries. They can be built for very high throughputs. Although the compression obtained has been lower than that given by reciprocating machines, the range of pressures attainable has gradually been extended. As mentioned, centrifugal compressors are used for the main duties on ethylene plants, for both process gas and refrigeration. Centrifugal compressors have relatively limited turndown. On centrifugal compressors, some of the main malfunctions are (a) rotor or shaft failure, (b) bearing failure, (c) vibration, and (d) surge, as well as the general compressor failures mentioned earlier.

4. *Screw Compressors*: Screw compressors are positive displacement machines with rotary motion and are also known as 'helical screw' or 'spiral lobe' compressors. They are relatively simple and low in capital cost. Capacity control on screw compressors is effected by the use of a slide valve which moves axially along the housing. They have good turndown, being able to operate at loads as low as 10% of the normal throughput. Another feature of screw compressors is that they can tolerate relatively large changes in suction pressure, a characteristic useful in refrigeration duties.

5. *Gas Engines*: Gas engines are used to drive compressors. Some principal failures which occur are explosions of: (a) starting air line, (b) fuel line, and (c) crankcase. When the engine is shut-down, it is desirable that the fuel be shut off also. If this is not done, fuel may collect in the engine and the exhaust system and may explode when the engine is restarted.

6. *Process Pumps*: Most process pumps are centrifugal machines, although reciprocating machines are used in some cases. Where the application is severe and/or critical, a degree of redundancy may be provided, using one or more standby pumps or using several pumps operating at less than full capacity. Thus, for example, the equipment to perform the duty of a single pump may be one pump operating at 100% throughput and a similar pump on stand-by or two pumps each rated for 100% capacity, but both operating at 50% throughput. The most common pump faults are failures of bearings or of glands or seals. A common cause of failure is shaft misalignment. Also, a bearing failure can induce a gland or seal failure. If seal failure is particularly undesirable, a type of pump may be used which has a more reliable sealing arrangement.

8.3.7 Insulation

Another element of the pressure system on process plants is the insulation. Insulation is employed to control heat transfer in both normal operation and fire conditions. There are two basic types of insulation: thermal insulation and fire insulation. Thermal insulation is used to: (1) reduce heat loss from plant operating at temperatures above ambient, (2) reduce heat gain from plant at temperatures below ambient, (3) protect personnel from hot or cold surfaces on the plant, and (4) attenuate sound from the plant.

Some of the principal safety aspects of insulation are (1) corrosion beneath insulation, (2) self-heating in insulation, (3) insulation against fire, and (4) effect on the process of inadequacies, defects, or failures of insulation.

The thermal conductivity is by no means the sole criterion in selecting an insulation. Materials with similar thermal conductivities may differ widely in their other relevant properties. Categories of non-combustible insulation material are (1) calcium silicate, (2) expanded perlite, (3) expanded vermiculite, (4) mineral fiber, and (5) cellular glass.

The insulation should provide complete cover of the areas of the equipment which it is intended to insulate. Properties which bear on this are dimensional stability and shrinkage. Furthermore, it should be sufficiently easy to fabricate that gaps do not occur as a result of fabrication difficulties. Insulation is often subject to conditions which lead to its being crushed or torn, damage from feet being quite common.

An important property of an insulation is the extent to which it retains liquid that it already contains or that leaks into it. This property is relevant to the insulation in respect of its (1) thermal performance, (2) mechanical performance, and (3) weight, and to (4) corrosion beneath the insulation, and (5) self-heating in the insulation.

With some insulations the moisture can be driven out and thermal performance restored, but with others this is not so. Calcium silicate, which has a high propensity to absorb water, falls in this latter group.

Process liquids may enter from pipework as may heat transfer fluids. Possible sources of in-leak are flanges and drain and sample points. Some liquids react with the binder in the insulation and promote its disintegration.

Corrosion under insulation, fire insulation, and personnel protection by insulation are described below.

1. *Corrosion Under Insulation*: Factors which determine the extent of external corrosion under insulation include: (a) the insulation material, (b) the equipment material, (c) the equipment configuration, (d) the equipment coating, (e) the equipment stress, (f) the service temperature, (g) any temperature transients, and (h) the climate and location. External corrosion under insulation occurs mainly on equipment operating in the range $-5°C$ to $105°C$ and especially in the range $60-80°C$. At lower temperatures the reaction rate is slower, while at higher temperatures water tends to be driven off. Much equipment is subject to temperature transients, which can increase external corrosion. These may be major changes occurring during the start-up and shut-down of high or low temperature plants, or they may take the form of temperature cycling. High plant temperatures can cause concentration of salts which then cause severe corrosion when rewetted. Corrosion also occurs in low temperature plants at thawing zones which tend to remain wet and corrosive.

2. *Fire Insulation*: Fire insulation proper is provided on equipment which requires fire protection but not thermal insulation. For fire insulation use is made of material which is generally cementitious, such as vermiculite cement. From the fire protection viewpoint, the ideal thermal insulation is one which does not burn or melt and is non-absorbent. To the extent that it does undergo combustion, it should not give off toxic fumes. Some insulating materials such as polyisocyanurate foam are not suitable. The insulation needs to be protected against the weather. Methods of weatherproofing include the use of caulking, of mastic, or of metal jacketing. Metal jacketing is to be preferred. Stainless steel provides a quality jacket, with galvanized steel being a less expensive alternative.

3. *Personnel Protection by Insulation*: One function of thermal insulation is to prevent injury to personnel. Hazard exists not only on high temperature plants but also on those operating at low temperatures which can cause 'cold burns'. The usual method of protection is to make the insulation sufficiently thick to prevent injury. For high temperatures, a criterion commonly used is that the external surface of the insulation should not exceed $60°C$. As already emphasized, design must address the total insulation system rather than just the insulation itself. It needs to have regard for the several goals of thermal insulation, fire protection, and personnel protection, and to the avoidance of corrosion beneath the insulation, self-heating, and hazardous effects on the process.

8.3.8 Overpressure Protection

Pressure systems need to be provided with protection against failure, particularly from overpressure. The overpressure protection considered here applies to situations in which the pressure rise is relatively gradual.

API RP 521 gives detailed guidance on causes of overpressure. The principal events may be summarized as follows: (1) connection to a high pressure source, (2) disconnection from a low pressure sink, (3) increased heat input, (4) decreased heat output, (5) vapor evolution, (6) absorbent failure, (7) heat exchanger tube failure, (8) expansion of blocked-in liquid, (9) reverse flow, (10) fluid transients, and (11) plant fire.

Connection to a high-pressure source can occur if a valve is opened in error. Likewise, loss of connection to a low-pressure sink can occur if a valve is erroneously closed. Increased heat input can occur due to malfunction of heating equipment or chemical reaction. A heat exchanger such as a reboiler when just cleaned may temporarily have a high heat transfer capacity.

Decreased heat output, or loss of cooling, can occur in numerous ways. Malfunction of cooling equipment such as a heat exchanger is an obvious case. Distillation columns particularly can lose cooling in a number of ways. In a single column, loss of cooling can occur not only due to malfunction of the condenser but also due to loss of reflux or of subcooled feed. Where there is a set of columns in series, loss of heat input to one column can cause overpressure of the next column, due to carry forward of light ends in the bottoms from the first column which then overload the second column.

Overpressure can be caused by admission of water or light hydrocarbons to hot oil, resulting in rapid evolution of vapor. Failure to remove sufficient gas due to loss of flow of absorbent can also cause overpressure. Failure of a tube in a heat exchanger so that the low-pressure shell is exposed to the high-pressure tube fluid can cause overpressure. Overpressure can be caused by expansion of liquid in a section of line between two closed block valves.

A particular case of increased heat input is a fire on the plant. Failure of a pressure raiser and reverse flow from the high-pressure discharge to the low-pressure suction can cause overpressure of the latter.

8.3.8.1 Identification of Relief Requirements

Hazard identification is an essential part of the design of protection for pressure systems. The technique of hazard and operability (HAZOP) studies can be helpful in ensuring that all potential overpressure scenarios have been considered.

Once the hazards have been identified, it is appropriate to consider whether measures other than overpressure protection are more suitable. Some alternatives are to make the plant inherently safer by means such as increasing vessel strength or limiting the delivery pressure of pumps. Full consideration should be given to the action of the process operator. He or she is an integral part of the hazard situation. It is usually reasonable to place some reliance on operator action in averting a hazard, but the extent of such reliance depends on such factors as

ease of recognition of the existence of the hazard, instructions and training for dealing with it, and difficulty of and time available for this task.

Pressure relief becomes necessary when plant conditions are abnormal. Failure of pressure let-down devices can cause a large and sudden rise in downstream pressure, particularly where a liquid line is blown down by gas. On a continuous distillation column, overpressure may be created by an increase in heat input. Some causes of increased heat input include increase in temperature difference in the reboiler, loss of cooling, loss of reflux, and loss of subcooled feed.

Pressure relief should be provided between isolations if the equipment is subject to pressure from a source of high pressure or of process heat and, generally, if the equipment can be isolated when the plant is operating and it is in a fire zone. Failure in a control loop may occur in the measuring element, in the controller, in the control valve or its actuator, in the transmission lines or by operator action. The system downstream of a power turbine or let-down engine is liable to be subjected to the upstream pressure. Such machines often have a low resistance to flow when stopped. For the bursting of a high-pressure tube in a heat exchanger, the pressure relief valve on the low-pressure side should be sized to handle the flow from twice the cross-sectional area of the tube. Batch operations require special consideration, because the conditions change throughout the cycle.

8.3.8.2 Overpressure Protection: Pressure Relief Devices

Valves for the relief of pressure are referred to by a number of names and there is no universally agreed terminology. Two names commonly used as generic terms are 'pressure relief valve' and 'safety valve'.

There are a number of different types of pressure relief valve. Three broad categories of valve are (1) the conventional, direct-loaded valve, (2) the balanced valve, and (3) the pilot-operated, indirect-loaded valve. A conventional pressure relief valve, the simplest type, is a spring-loaded valve. The pressure at which the valve relieves is affected by the back pressure.

The direct-loaded pressure relief valve has a number of variants. One is the assisted-opening valve, which has power-assisted opening. One use of this type of valve is to give depressurization down to a predetermined level. In a supplementary-loaded pressure relief valve, an external power source is used to impose an additional sealing force, which is released automatically when the set pressure is reached. This arrangement gives an improved degree of leak-tightness.

A third variant is the pilot-assisted valve, which is a different type from the pilot-operated valve. A pilot-assisted valve is a direct-loaded valve in which some three-quarters of the load is due to the spring and the rest to the pressure of the fluid from the pilot valve. When the set pressure is reached, the pilot opens and that part of the load contributed by it is removed.

A balanced pressure relief valve is one that incorporates means of minimizing the effect of back pressure on performance. With this type of valve, the back pressure has little effect on the pressure at which the valve relieves. A balanced relief valve is used to accommodate the back pressure in a relief header.

A pilot-operated, indirect-loaded pressure relief valve is one in which the main valve is combined with and controlled by an auxiliary direct-loaded pressure relief valve. The whole load on the main valve is provided by the fluid pressure from the pilot valve. When the set pressure is reached, the pilot valve opens and releases the loading pressure.

For reclosing pressure relief devices, there is no universally agreed terminology for non-reclosing diaphragm-type devices. In API and ASME parlance, such devices are 'rupture disks'. According to the BSI, they are 'bursting discs', which is the term used here.

A conventional bursting disc should be carefully installed. Manufacturers provide disc mountings that have a number of fool proofing features of which the following are typical. The ring holding the disc on the vent side is made thicker than the dome in order to protect the latter. The disc has an identification tag with full details stamped on. The tag neck serves to center the disc in place and is notched on one side so that, if correctly installed, the disc does not seat properly and, if not readjusted, will vent at a low pressure. Discs with different pressures have pegs located at different points so that a disc for one pressure will not fit into a holder for a different pressure. Conventional bursting discs are normally made of metal, but graphite discs are also used.

Some other types of disc which are used in general applications include the composite slotted disc and the reverse buckling disc.

In addition to pressure relief valves and bursting discs, there are a number of other devices that may be used to relieve pressure. One such device is the buckling pin device, which is classified by ASME Section VIII—along with the bursting disc—as a 'non-reclosing' pressure relief device.

A variety of devices are used to provide overpressure protection of low pressure and atmospheric storage tanks. The simplest of these devices is the atmospheric vent, typically a short pipe. An atmospheric vent is used on atmospheric storage tanks where there is minimal hazard from the ingress of air or egress of vapor.

On an atmospheric storage tank where the liquid held is more volatile, use is generally made of pressure-vacuum valves, which allow the tank to 'breathe'. Another device that can be used to provide pressure relief on low-pressure tanks is the liquid seal. Fire relief for an atmospheric storage tank may be provided in the form of a weak roof-to-shell seam or rupture seam. For a pressure vessel a device that can provide fire relief is the fusible plug, a non-reclosing device actuated by high temperature.

8.3.8.3 Overpressure Protection: Relief System Design

The only logical method for determining the required location of pressure relief devices is by starting with the identification of potential sources of overpressure. This is best done for each equipment item in the process unit. When this task has been completed, relief devices are then 'located' on equipment or associated piping in such a way that they will be accessible to provide relief for each potential overpressure scenario identified. The relief devices should be installed not simply to protect the particular points at which they are located but to provide protection for all vulnerable points.

The first decision is whether to make use of a bursting disc, either alone or in combination with a pressure relief valve. It advises the use of a bursting disc if there is a completely free choice and if this is the more economic option or if either (1) the pressure rise is too rapid for a pressure relief valve or (2) the process fluid properties make a pressure relief valve unsuitable (toxicity, corrosiveness, blockage-forming components, aggressiveness).

If a system based on pressure relief valves alone is indicated, this may be a single valve or a set of multiple valves in parallel, depending on whether a single valve can provide the necessary capacity. One fundamental factor that affects the choice of relief valve type is the sink to which the valve will discharge. If this is a relief header and therefore exerts a back pressure, a balanced valve may be indicated. On the other hand, if discharge is to atmosphere, a pilot-operated valve may be selected for its high discharge velocity. Other factors that affect the choice are (1) the margin between the operating and the set pressure, (2) the required speed of opening, and (3) the required valve tightness. The first and the third factors favor a pilot-operated valve.

For a bursting disc/pressure relief valve system, the valve may have a bursting disc: (1) upstream, (2) downstream, or (3) on both sides, depending on whether it is the process-side fluid, the discharge-side fluid, or both fluids that are aggressive.

The principal types of bursting disc are (1) conventional domed disc, (2) reverse domed disc, and (3) composite slotted disc. Use of the conventional type is suitable, provided that there is a wide margin, say 30%, between the operating and the design pressures and that the pressure does not pulsate. If these conditions are not met, if a long life is required, or if the disc is to be used in series with a safety valve, consideration should be given to the other two types.

The approach taken is to describe the scenario and to define a basis for the determination of the relief capacity, indicating whether credits can be taken for particular mitigating factors and discussing features to be taken into account. An essentially similar philosophy is applicable to the sizing of bursting discs.

8.3.8.4 Overpressure Protection: Fire Relief

The pressure relief requirements just discussed are those for relief of conditions that may arise during operation. In addition to such operational relief, there is a requirement for fire relief. Except in storage areas, operational upsets are more likely to give rise to a requirement for pressure relief than are fires, but in cases where fire can affect a number of vessels, the requirement for fire relief is much greater than that for operational relief. Equipment for which fire relief may be required includes atmospheric storage tanks, pressure storage vessels, and process systems. For pressure vessels, there are two distinct cases: vessels containing gas, vapor, or supercritical fluids only and vessels containing liquids. See API RP 521 for details on overpressure protection in the case of fire relief.

The first step in design for fire relief is to define the scenarios for which such relief is to be provided. The amount of fire relief required depends on the surfaces which may be exposed to fire. The requirement is greatest for groups of process vessels and storage vessels. If the scenario indicates that the fire would develop slowly and would cause overpressure only after a long period, say several hours, it is sometimes assumed that the fire would be brought under control by fire fighting before overpressure occurs.

Some methods of protection against fire in addition to pressure relief are (1) inventory reduction, (2) fire insulation, and (3) water sprays. Consideration needs to be given to measures for limitation of developed pressure and for reduction of inventory in fire conditions. One method of limiting the overpressure which can develop is vapor depressurization or blowdown. Vapor is blown down to a suitable disposal system through a depressurization valve which is separate from any pressure relief device and which is normally operated before the pressure reaches the set pressure of the relief device.

Provision may also be made for the removal of the liquid inventory. One reason for doing this is to limit the amount of vapor which has to be handled. Another is to limit the release of material which could then feed the fire if the vessel should fail. The preferred means of removal is the normal liquid withdrawal system, but a separate liquid pulldown system may be used.

It is also necessary to consider the effect of fire on pressure relief devices. On a pressure relief valve, there are two main effects of fire. One is to cause thermal expansion of the spring, which reduces the force on the valve and is a deviation in the safe direction. The other is to cause thermal expansion of the valve spindle which can lead to jamming.

Provision of fire relief does not give full protection against fire. Unwetted surfaces which are exposed to fire may experience overtemperature and may rupture, even though overpressure of the vessel has not occurred.

8.3.8.5 Overpressure Protection: Vacuum and Thermal Relief

Vacuum collapse of equipment is destructive and hazardous, despite the fact that in some instances the pressure differential may appear relatively small. There is a wide variety of situations that can lead to at least a partial vacuum. They include: (1) pumping out with inadequate vent opening, (2) condensation of a vapor, (3) absorption of a vapor, (4) cooling of a volatile liquid, (5) connection to a source of vacuum or suction, (6) depletion of oxygen in air by rusting, and (7) sudden arrest of a moving column of liquid.

Vacuum relief devices include vacuum valves and bursting discs. A vacuum valve is a direct-loaded valve which opens to admit air. An alternative is a vacuum bursting disc, which may be preferred where the fluid is corrosive or liable to create blockage, or where it is sufficiently toxic that even a small leak is unacceptable. Bursting disc arrangements are also available which provide both overpressure and vacuum protection.

Another form of pressure relief which may be required is thermal hydraulic relief. This is the relief of pressure due to expansion of a liquid that is blocked in and then heated. The pressure generated by a blocked-in and heated liquid is independent of the volume, but the latter does affect the consequences of any escape. Separate thermal relief may not be needed if other means of relief exist. These may include a small hole in the valve or a small bore bypass around it, bellows, or relief provided for other purposes.

8.3.8.6 Overpressure Protection: Bursting Discs

Like a relief valve, a bursting disc may suffer functional or operational failures. In other words, it may fail to burst at the set burst pressure and thus fail to danger, or it may burst below that pressure and thus fail prematurely but, generally, be safe. Some of the principal modes of failure of a conventional bursting disc are fatigue, creep, and corrosion. Pulsation of the fluid pressure induces fatigue, while high temperature causes creep and contributes to fatigue.

Another type of failure is caused by blockage. Blockage may occur on the process side and may be caused by corrosion, crystallization, or polymerization. It can also occur on the vent side.

Mal-installation constitutes a third type of failure. The disc may be installed upside down. Another example of mal-installation is putting more than one disc in the holder. Discs are sometimes supplied in stacks and in this case duplication may be a simple error. In other cases, the use of more than one disc is done deliberately to avoid frequent bursting, particularly during commissioning.

8.3.9 Flare and Vent Systems

A principal method of disposal of gases and vapors discharged from the process is flaring. The function of a flare system is generally to handle materials vented during: (1) normal operations, particularly start-up, and (2) emergency conditions. There are three main types of flare system: (1) elevated flare, (2) ground flare, and (3) low-pressure flare.

A ground flare generally consists of a battery of tubular burners contained in a short refractory-lined stack, which serves as combustion chamber and windshield. The bottom of the stack is above grade so that the combustion air can enter. The elements of a ground flare system are knockout drum, seal drum, burner system, ignition system, stack, and windshield. A ground flare may be favored where there is a limitation on the height of the flare which can be installed.

There are in process plants various low pressure units which may produce an off-gas stream. These include storage tanks, oil−water separator, and wastewater treatment units. These off-gases may be routed to a low-pressure flare. Where the discharge source pressure is insufficient, use may be made of a blower installed in the off-gas line. The use of an off-gas blower, however, is not an ideal arrangement in that it introduces rotating equipment which may fail and which may have the potential to act as a source of ignition.

Elevated flares are used to handle both normal and emergency loads and are generally the system of choice except where environmental considerations prevent their use. An elevated flare is a normal feature of a refinery or a petrochemical plant. A flare system consists of a flare stack and of gathering pipes which collect the gases to be vented. Other features include the flare tip, which typically has steam nozzles to assist entrainment of air into the flare, seals installed on the stack to prevent flashback of the flame, and a knockout drum at the base of the stack to remove the liquid from the gases passing to the flare.

There are a number of hazards associated with a flare system. These include: (1) failure of the collection system pipework due to low-temperature embrittlement or corrosion, (2) obstruction in the flare system, (3) explosion in the flare system, (4) heat radiation from the flare, (5) liquid carryover from the flare, and (6) emission of toxic materials from the flare.

8.3.10 Blowdown and Depressuring Systems

The response to a plant emergency may require the disposal of hazardous inventories. This is affected by the emergency blowdown and depressuring systems.

Disposal of liquid streams in an emergency is through the blowdown system. An emergency blowdown system has two basic features: treatment and disposal. Blowdown streams may be unsuitable for immediate disposal and may therefore require treatment. Principal types of treatment are disengagement of gases and vapors from liquid, and quenching to cool and partially condense hot vapors.

Disposal of vapors from equipment at pressures below those of the pressure relief devices is by means of the emergency depressuring system. Depressuring is used to reduce hazard by removing vapor or gas from an equipment. This may be necessary where the equipment may fail at a pressure below the set pressure of the pressure relief valve, as in fire engulfment of a vessel which is filled with gas or one which contains some liquid but in which part of the walls is unwetted.

8.3.11 Containment of Toxic Materials

Some plants handle substances which rank as high toxic hazard materials (HTHMs) and the pressure system must ensure their containment. Containments for HTHMs should be designed and constructed to standards which are higher than average. This applies to, among other things, the vessel wall thickness and connections, the materials of construction, the welds, and the quality control.

8.3.11.1 Storage Tanks and Vessels

US standards and codes applicable to the storage of HTHMs include API Std 620: 1990, API Std 650: 1988, and the ASME *Boiler and Pressure Vessel Code* 1992. Also relevant is the ASME B31.3: 1990 *Chemical Plant and Petroleum Refinery Piping Code*.

API Std 650 is the standard for large oil storage tanks and has limited relevance to HTHMs. The safety factors and welding standards are less stringent and the tank relief is to atmosphere. The more relevant standard is API Std 620. For storage of larger quantities of HTHMs the preferred method will often be the use of refrigerated storage. This is dealt with in the standard in appendices R and Q which treat, respectively, tanks for refrigerated liquids stored in the temperature range 40°F to −60°F and those stored at temperatures in the range down to −270°F.

Tank designs for refrigerated storage given in API Std 620 include both single and double wall metal tanks, but in the latter the outer wall is not intended to withstand major failure of the inner tank, either in respect of the hydrostatic head or of the temperature of the liquid stored. A more suitable design is one incorporating a high bund close to the metal tank. The CCPS *Guidelines* refer to that given the CIA *Refrigerated Ammonia Storage Code*. Materials of construction for refrigerated storage of HTHMs need to possess both the necessary impact toughness and corrosion resistance.

In design of storage tanks for HTHMs, particular attention should be given to features such as the foundations, the weld radiography, and the pressure testing. API Std 620 contains stress relief requirements which are given in Appendices H and I.

With a storage tank designed to API Std 620, the pressure relief is generally designed for normal 'breathing', and for operational variations. These arrangements are less suitable for HTHMs, for which there are more severe restrictions on release to atmosphere. It may be necessary to use a closed relief and vent system. If direct relief to atmosphere is permissible, a rupture disc should not be used alone, as it does not reclose after opening.

API Std 620 contains provisions by which fire relief may be provided by a rupture seam. The CCPS *Guidelines* state that a rupture seam is not normally acceptable for HTHMs. The implication of this is that there need to be specific alternative arrangements for fire relief. Whatever arrangements are adopted, it is essential to ensure that the tank does not rupture at the wall−floor seam, which would allow discharge of its entire contents.

For storage vessels, the ASME Boiler and Pressure Vessel Code recognizes lethal service. It states that 'by lethal substances are meant poisonous gases or liquids of such a nature that a very small amount of gas or of the vapor of the liquid, mixed or unmixed with air, is dangerous to life when inhaled … this class includes substances of this nature which are stored under pressure or which may generate a pressure if stored in a closed vessel.'

The minimum wall thicknesses given in pressure vessel codes may yield for small vessels at low pressure relatively thin walls, and it may be prudent for certain HTHMs to use a thicker walled container. A suggestion made in the CCPS *Guidelines* is the use of compatible DOT cylinders.

The nozzles are frequently the weak points on a vessel and for containment of HTHMs they need to be of high integrity. Measures to ensure this include high nozzle specifications, use of sufficient thickness and of suitable welds, and avoidance of small nozzles and of excessive projection. The CCPS *Guidelines* recommend that: nozzles should preferably be Class 300 flanged welding necks; all nozzle-to-shell welds be full penetration welds; nozzles of less than 1−1.5 in. nominal bore (NB) be avoided or at least protected; and nozzles less than 2 in. NB do not project more than 6 in. or be provided with additional strength or protection.

Since for HTHMs avoidance of catastrophic failure is of particular importance, use should be made of fracture mechanics so that, where practical, the vessel operates in the lower risk 'leak-before-break' regime.

8.3.11.2 Pipework

As for tanks and vessels, so for piping—design and construction for HTHM service should be of high standards. ASME B31.3 recognizes a Category M Fluid Service in which the potential for personnel exposure is judged to be significant, and in which a single exposure to a very small quantity of a toxic fluid, caused by leakage, can produce serious irreversible harm to persons upon breathing or bodily contact, even when prompt restorative measures are taken. Chapter VIII of the code deals with Category M Fluid Service.

It is preferable that filling and discharge lines enter a storage tank through the top, thus avoiding piping penetration below the liquid level. However, for discharge this requires the use of submerged pumps, which may constitute a maintenance problem.

The pipework should have an adequate degree of fire protection. Whereas, in general, the emphasis in fire protection is avoidance of further releases which could feed the fire, in the case of HTHMs it shifts rather to the avoidance of serious toxic release. It follows that for HTHMs it may be necessary to provide pipework with protection which would otherwise be considered unnecessary. The CCPS Guidelines recommend a fire resistance of at least 30 min at 1100°F.

8.3.11.3 Limitation of Release

There are a number of devices which may be used to shut off or reduce releases of HTHMs. These include EIVs, NRVs, excess flow valves, restrictor orifices, and pump trips.

Storage tanks and vessels for HTHMs should be provided with emergency isolation valves or control valves with separate shut-off facilities. Other situations which may merit an emergency isolation valve are a large vapor line on a pressurized storage vessel and a long pipeline.

NRVs, or check valves, can be used with the purpose of preventing back flow out of a tank in the event of rupture of a liquid line on it, but are notoriously unreliable and are not suitable as the sole device in a critical duty. The flow from a line may be reduced by the use of an excess flow valve or a restrictor orifice. Another measure is remote shut-off of pumps providing the pressure behind the release.

8.3.11.4 Secondary Containment

There are a number of arrangements which can be used to contain a release and prevent its escape to atmosphere. Such secondary containments include double-walled equipment, enclosures, and bunds. Pressure relief and vent systems are also sometimes so classified. Common to all these arrangements is the need to provide in addition some form of disposal system.

Double-walled equipment is used particularly for storage tanks. In particular, the principle is exploited in the design of large refrigerated ammonia storage tanks.

The double-wall principle is also used to a limited extent for pipework. One system involves the enclosure of the pipe carrying the HTHM in a second, concentric pipe. The annular space between the two pipes is filled with inert gas either at a pressure higher than that in the inner pipe or at a lower pressure with monitoring to detect leakage of the toxic fluid. An alternative system involves the use of double seals on flanges and of bellows seals on valves, both purged with monitored inert gas.

The use of an enclosure building provides another form of secondary containment but allows accumulation of small leaks. It therefore requires the provision of toxic detectors and alarms and of ventilation and disposal systems. Where toxic liquids are handled, it is also necessary to provide for their retention, collection, and disposal. The sump should be of small cross-sectional area but deep, in order to minimize vaporization. If there is a possibility of an internal overpressure sufficient to cause structural damage, consideration should be given to this.

Material from the pressure relief and depressuring, or blowdown, devices should be collected in a closed relief or vent header system. This is usually a common header, though in some instances this approach may need to be modified in order to avoid a problem arising from the interconnection of systems. The header typically passes to a knockout drum or catchpot and then to the disposal system, usually a flare or scrubber.

8.3.12 Failure in Pressure Systems

Maintenance of the integrity of the pressure system and avoidance of loss of containment are the essence of the loss prevention problem. It is necessary, therefore, to consider the failures which occur in pressure systems. Of particular importance are catastrophic failures in service.

Catastrophic failure of a properly designed, constructed, and operated pressure vessel is comparatively rare. The most common cause of such failure is inadequate operational procedures. Most failures in pressure systems occur, however, not in pressure vessels but in the rest of the system, which includes pipework, valves and fittings, and equipment such as heat exchangers and pumps.

The problem of failure in pressure systems is treated here by first describing some principal causes of failure, particularly sudden failures, and then giving some historical data on failure. In the present context, it is service failures which are of prime importance. Service failures of pressure systems are generally caused by exposure to operating conditions more severe than those for which the system was designed.

It is usual to classify service failures as (1) mechanical failure, through stress and fatigue and (2) corrosion failure, although many failures have an element of both of these. Accounts of causes of failure in pressure systems are given in *Defects and Failures in Pressure Vessels and Piping* (Thielsch, 1965) and *Inspection of Chemical Plant* (Pilborough, 1971, 1989). The following are common causes of failure in pressure systems:

1. *Materials Identification Errors*: Mistake in identifying materials of construction which result in the construction of plant using the wrong materials is a significant problem. Such errors are particularly likely to occur where the materials have a somewhat similar appearance, e.g., low alloy steel and mild steel, or stainless steel, and aluminum-painted mild steel. It is necessary, therefore, to exercise careful control of materials. Methods of reducing errors involve marking, segregation, and instrument spot checks. For critical applications, it may be necessary to carry out a full 100% *in situ* check on materials using an instrument such as a Metascope.

2. *Mechanical Failure*: Some common causes of mechanical failure in process plant are (a) excessive stress, (b) external loading, (c) overpressure, (d) overheating, (e) mechanical fatigue and shock, (f) thermal fatigue and shock, (g) brittle fracture, (h) creep, and (i) hydrogen attack.

3. *Mechanical Fatigue and Shock*: Conditions which give rise to mechanical fatigue include: (a) pressure variations, (b) flow variations, (c) expansion effects, and (d) imposed vibrations. Stress cycling can be caused by normal changes of pressure in the process. Normal flow fluctuations or effects such as hammer blow or cavitation also give rise to stress cycles. So do differential expansions and contractions of the process plant. Other

stress cycling is imposed by vibrations from such equipment as compressors, pumps, or valves.

4. *Thermal Fatigue and Shock*: Similarly, the distinction between thermal fatigue and thermal shock is that in the latter the applied temperature difference and the rate of change of temperature are greater and cause failure in one, or a few, cycles. A given temperature cycle may give rise to thermal fatigue or shock, depending on the material. It may constitute thermal fatigue for a ductile material but shock for a brittle one. Thermal cycling is caused by: (a) intermittent operation and (b) particular equipment conditions. There are many types of intermittent operation ranging from batch processes to throughput changes and including forced shut-downs. These all create thermal stress cycles. This source of thermal fatigue is much reduced if the plant is operated continuously, even though the operating temperatures may be relatively high.

5. *Brittle Fracture*: The materials used in plants handling low temperature fluids should have a ductile−brittle transition temperature below not only the normal operating temperature but also the minimum temperature which may be expected to occur under abnormal conditions. If this requirement is not observed and cold fluid contacts metal below the transition temperature, brittle fracture may occur. Brittle fracture is catastrophic, since the fracture can propagate at a velocity close to that of sound.

6. *Creep*: Creep involves plastic deformation and, eventually, ruptures. However, some metals have a very low ductility in the normal creep range, although they are more ductile at low and at very high temperatures. Low ductility fractures with an elongation of less than 1% are frequently associated with the formation of cavities at the grain boundaries. The development of cavities can occur at low stress over long time periods. The application of higher temperatures and higher strain rates, as in a fire, can accelerate cavitation and cause low ductility failure. Since design for high temperatures is based on creep strength, failure due to creep is usually unlikely under normal operating conditions and is generally caused by abnormal conditions such as maloperation or fire. It is appropriate to emphasize here the very rapid increase in creep rate with temperature. For mild steel in the creep range, a temperature increase of about $10-15°C$ is sufficient to double the creep rate. A stress which gives an 11-year life at $500°C$ can result in rupture in 1 h at $700°C$.

7. *Hydrogen Attack*: Mild steel is subject to the following types of hydrogen attack: (a) hydrogen blistering and (b) hydrogen embrittlement and damage. Hydrogen blistering can occur when atomic hydrogen diffuses into steel. Normally, the atomic hydrogen diffuses right through, but if it encounters voids it forms molecular hydrogen, which can generate pressures of several hundred bar. Internal splitting of the steel occurs and gives surface blisters. At high temperatures and high hydrogen partial pressures, steels can suffer hydrogen embrittlement and damage. The steel suffers decarburization by the hydrogen, which causes a loss of ductility, and development of microfissures at the grain boundaries, which leads to loss of strength.

8. *Corrosion Failure*: Corrosion occurs as: (a) general corrosion, (b) local corrosion, and (c) erosion. In general corrosion, there is a fairly uniform deterioration of the overall surface. In contrast, localized corrosion involves little generalized corrosion but severe local attack, often at points of surface defects or stress. Erosion is also localized at points of high velocity or impact.

8.3.13 Corrosion Testing

Corrosion is generally a function of a number of variables and can be very sensitive to the particular process conditions. It is not always easy to predict, and in some cases corrosion testing is necessary.

Since in a candidate design corrosion is likely to proceed relatively slowly, testing can take a long time. It is therefore not uncommon to use some form of accelerated test. As with any form of accelerated testing, this can be misleading. An alternative is the use of corrosion rate monitoring for which instrumentation now exists. This allows information to be obtained in tests which simulate the plant conditions. However, since the corrosion rate can vary with time, if this method is used it is necessary to ensure that the steady-state rate has been reached.

8.3.14 Safety Factors

It is important that a safety factor be applied to take account of both uncertainty in the data and inaccuracy in the calculation methods. Among other reasons it is useful (1) to take account of uncertainty in the data, (2) to allow for the inaccuracy in the fluid flow calculation methods which is present even for incompressible liquid and/or ideal gas, (3) to allow for the effects of gas/vapor-phase nonideality if these have not been taken into account in the mass flow capacity calculation, and (4) to take account of errors in the flow calculation for two-phase gas/vapor−liquid flow resulting from neglecting the variation in liquid density, liquid specific heat, and latent heat along the pipe. The overall safety factor can be the product of these sub-factors and more.

8.4. CONTROL SYSTEM DESIGN

8.4.1 Control System Design

The operation of the plant according to specified conditions is an important aspect of loss prevention. This is very largely a matter of keeping the system under control and preventing deviations. The control system, which includes both the process instrumentation and the process operator, therefore has a crucial role to play. Traditionally, control systems have tended to grow by a process of accretion as further functions are added. One of the thrusts of current work is to move toward a more systematic design approach in which there is a more formal statement of the control objectives, hierarchy, systems, and subsystems. Once the objectives have been defined, the functions of the systems and subsystems can be specified.

It is convenient to distinguish several broad categories of function that the control system has to perform: these are (1) information collection, (2) normal control, and (3) fault administration. A control system is usually also an information collection system. In addition to that required for immediate control of the process, other information is collected and transmitted. Much of this is used in the longer term control of the process. Another category which is somewhat distinct from normal control is the administration of fault conditions which represent disturbances more severe than the control loops can handle.

8.4.2 Process Characteristics

The control system required depends very much on the process characteristics (Edwards and Lees, 1973). Important characteristics include those relating to the disturbances and the feedback and sequential features. A review of the process characteristics under these headings assists in understanding the nature of the control problem in a particular process and of the control system required to handle it.

Processes are subject to disturbances due to unavoidable fluctuations and management decisions. The disturbances include:(1) raw materials quality and availability, (2) services quality and availability, (3) product quality and throughput, (4) plant equipment availability, (5) environmental conditions, and due to (6) links with other plants, (7) drifting and decaying factors, (8) process materials behavior, (9) plant equipment malfunction, (10) control system malfunction.

Some process characteristics which tend to make feedback control more difficult include: (1) measurement problems, (2) instability, (3) very short time constants, (4) very long time constants, (5) recycle, (6) non-linearity, (7) inherent limit cycles, (8) dead time, (9) strong interactions, (10) high sensitivity, (11) high penalties, (12) parameter changes, (13) constraint changes.

The sequential control characteristics of a process include: (1) plant start-up, (2) plant shut-down, (3) batch operation, (4) equipment changeover, (5) product quality changes, (6) product throughput changes, (7) equipment availability changes, (8) mechanical handling operations.

Some other process characteristics which may be significant include requirements for: (1) monitoring, (2) feedforward control, (3) optimization, (4) scheduling, (5) process investigation, (6) plant commissioning.

8.4.3 Control System Characteristics

The characteristics of process control systems have passed through three broad phases: (1) manual control, (2) analogue control, and (3) computer control (covering all forms of programmable electronic system). However, such a classification can be misleading because it does not bring out the importance of measuring instrumentation and displays, because neither analogue nor computer control is homogeneous stage, and because it says very little about the quality of control engineering and reliability engineering and the human factors involved.

The sophistication of the measuring instrumentation greatly affects the nature of the control system even at the manual control stage. This covers instruments for measuring the whole range of chemical and physical properties. The displays provided can also vary widely.

The stage of analogue control implies the use of simple analogue controllers but may also involve the use of other special purpose equipment. Most of this equipment serves to facilitate one of the following functions: (1) measurement, (2) information reduction, and (3) sequential control.

Another crucial distinction is in the provision of protective or trip systems. In some cases, the safety shutdown function is assigned primarily to automatic systems; in others it is left to the operator. Similarly, computer control is not a homogeneous stage of development. In some early systems, the function of the computer was limited to the execution of Direct Digital Control (DDC). The real control of the plant was then carried out by the operator with the computer as a rather powerful tool at his disposal. In other systems, the computer had a complex supervisory program which took most of the control decisions and altered the control loop set points, leaving the operator a largely monitoring function. The two types of system are very different.

The quality of the theoretical control engineering is another factor which distinguishes a system and largely determines its effectiveness in coping with problems such as throughput changes, dead time, and loop interactions.

Equally important is reliability engineering. Unless good reliability is achieved nominally automated functions will be degraded so that they have to be done manually or not at all. Control loops on manual setting are the typical result.

8.4.4 Instrument System Design

The design of process instrument systems, like most kinds of design, is largely based on previous practice. The control panel instrumentation and the control systems on particular operations tend to become fairly standardized. Some points should be taken into account like:

1. *Some design principles*: There are some basic principles which are important for control and instrument systems on hazardous processes. It is also necessary to pay careful attention to the details of the individual instruments used.
2. *Instrument distribution*: A feel for the distribution of types of instrument on a process plant is given by Tayler (1987).
3. *Instrument accuracy*: Most process plant instrumentation is quite accurate provided they are working properly. Information on the expected error limits of commercially available instrumentation has been given by Andrew and Williams (1980), who list limits for over 100 generic types of instrument.
4. *Instrument signal transmission*: Pneumatic instrument signals are transmitted by tubing, but several means are available for the transmission of electrical signals.
5. *Instrument utilities*: Instrument systems require high quality and high reliability utilities. As far as quality is concerned, pneumatic systems require instrument air which is free of dirt and oil. Many electronic instrument systems can operate from an electrical feed which does not constitute an Uninterruptible Power Supply (UPS).
6. *Valve leak-tightness*: In many situations on process plants, the leak-tightness of a valve is of some importance. The leak-tightness of valves is discussed by Hutchison (1976) in the *ISA Handbook of Control Valves*.
7. *Hazardous area compatibility*: The instrument system, including the links to the control computers, should be compatible with the hazardous area classification. Hazardous area classification involves first zoning the plant and then installing in each zone instrumentation with a degree of safeguarding appropriate to that zone.
8. *Multi-functional vs dedicated systems*: An aspect of basic design philosophy which occurs repeatedly in different guises is the choice which has to be made between a multi-functional and a dedicated system. A particular but common example of the multi-functional vs dedicated system problem is the choice between a computer-based and a hardwired trip system.

8.4.5 Control of Batch Processes

The control of batch processes involves a considerable technology over and above that required for the control of continuous processes. Batch processes constitute a large proportion of those in the process industries. Many batch plants are multi-purpose and can make multiple products. Their outstanding characteristic is their flexibility. They differ from continuous plants in that: the operations are sequential rather than continuous; the environment in which they operate is often subject to major variability; and the intervention of the operator is to a much greater extent part of their normal operation rather than a response to abnormal conditions.

1. *Models of batch processing*: There are a number of models which have been developed to represent batch processing. Three described by Fisher (1990) are (a) the recipe model, which centers on the recipe required to make a particular product, and its elements are the procedures, the formula, the equipment requirements, and the 'header'; (b) the procedure model, which has the form: Procedure → Operation → Phase → Control step; (c) the unit model, which is equipment-oriented and has the form: Unit → Equipment module → Device/loop → Element.
2. *Representation of sequential operations*: The control of a batch process is a form of sequential control. Various methods are available for the specification of sequences. They include (1) flowcharts, (2) sequential function charts, and (3) structured plain language.
3. *Structure of batch processing*: The overall structure of batch processing is commonly represented as a hierarchy.
4. *Batch control systems*: Batch processing may be controlled by the process operator, a system of single controllers or a Programmable Logic Control (PLC) system, a distributed Control Logic System (DCL) or a Centralized Control System (CCS). The selection of the system architecture and hardware is discussed by Sawyer (1993).

8.4.6 Control of Particular Units

The safe operation of process units is critically dependent on their control systems. Two particularly important features of control in process plant are (1) compressor control and (2) chemical reactor control. These are now considered in turn.

1. *Compressor control*: Centrifugal and axial compressors are subject to the phenomenon of surging.

Surging occurs when flow through the compressor falls to a critical value so that a momentary reversal of flow occurs. This reversal of flow tends to lower the discharge pressure and normal flow resumes. The surge cycle is then repeated. Severe surging causes violent mechanical shock and noise and can result in complete destruction of parts of the compressor such as the rotor blades.

2. *Chemical reactor control*: A continuous stirred tank reactor is generally stable under open-loop conditions, but in some cases, a reactor may be unstable under open-loop but stable under closed-loop conditions. Some polymerization reactors and some fluidized bed reactors may be open-loop unstable under certain conditions. The reactor should be designed so that it is open-loop stable unless there is good reason to the contrary. One method of achieving this is to use jacket cooling with a large heat transfer area. Another is to cool by vaporization of the liquid in the reactor. This latter method gives a virtually isothermal reactor.

8.4.7 Computer Integrated Manufacturing

There is now a strong trend in the process industries to integrate the business and plant control functions in a total system of Computer Integrated Manufacturing (CIM). The aim of CIM is essentially to obtain a flexible and optimal response to changes in market demand, on the one hand, and to plant capabilities on the other. It has been common practice for many years for production plans to be formulated and production schedules to be produced by computer and for these schedules to be passed down to the plant. In refineries, use of large scheduling programs is widespread. In addition to flexibility, other benefits claimed are improved product quality, higher throughputs, lower costs, and greater safety.

A characteristic feature of CIM is that information also flows the other way, that is, up from the plant to the planning function. This provides the latter with a continuous flow of up-to-date information on the capability of the plant so that the schedule can be modified to produce the optimal solution. A CIM system may therefore carry out not only the process control and quality control but also scheduling, inventory control, customer order processing, and accounting functions.

The architecture of a CIM system is generally hierarchical and distributed. Treatments of such architecture are given in Controlling Automated Manufacturing Systems (O'Grady, 1986) and by Dempster et al. (1981).

For such a system to be effective, it is necessary that the data passing up from the plant be of high quality. The system needs to have a full model of the plant, including the mass and energy balances and the states and capabilities of the equipment. This involves various forms of model-based control, which is of such prominence in CIM that the two are sometimes treated as if they are equivalent.

Plant data are corrupted by noise and errors of various kinds, and in order to obtain a consistent data set, it is necessary to perform data reconciliation. Methods based on estimation theory and other techniques are used to achieve this. Complete and Rigorous Model-Based Reconciliation (CRMR) is therefore a feature of CIM. One implication of CIM is that the plant is run under much tighter control, which should be beneficial to safety.

8.4.8 Instrument Failure

Process plants are dependent on complex control systems, and instrument failures may have serious effects. It is helpful to consider first the ways in which instruments are used. Measuring instruments are taken to include digital as well as analogue outputs. Control elements are normally control valves but can include power cylinders, motors, etc.

The important point is that some of these applications constitute a more severe test of the instrumentation than others. The accuracy of a flowmeter may be sufficient for flow control, but it may not be good enough for an input to a mass balance model in a computer. The dynamic response of a thermocouple may be adequate for a panel display, but it may be quite unacceptable in a trip system.

This leads directly, of course, to the question of the definition of failure. It is sufficient here to emphasize that the reliability of an instrument depends on the definition of failure and may vary depending on the application.

8.4.9 Trip Systems

It is increasingly the practice in situations where a hazardous condition may arise in the plant to provide some form of automatic protective system. One of the principal types of protective system is the trip system, which shuts down the plant, or part of it, if a hazardous condition is detected. Another important type of protective system is the interlock system, which prevents the operator or the automatic control system from following a hazardous sequence of control actions.

The existence of a hazard which may require a protective system is usually revealed either during the design process, which includes, as routine, consideration of protective features, or by hazard identification techniques such as HAZOP studies. Some operator calculates Safety Integrity Level (SIL) for having specification of trip systems. Oil and Gas UK has a guideline for it but it is not mandatory for all operators to follow it. Risk graphs and Layer of Protection Analysis (LOPA) are also used for specifying and designing of trip systems for offshore, but it is complicated to use these methods as there are arguments about the required of degree of independence.

The decision as to whether a trip system is necessary in a given case depends on the design philosophy. There are quite wide variations in practice on the use of trip systems. There is no doubt, however, about the general trend, which is toward the provision of a more comprehensive coverage by trip systems. The decision as to whether to install a trip system can be put on a less subjective basis by making a quantitative assessment of the hazard and of the reliability of the operator in preventing it.

Since a trip system is used to protect against a hazardous condition, it is essential for the system itself to be dependable. The dependability of a trip system depends on (1) capability and (2) reliability. Thus, it is necessary both for the system to have the capability of carrying out its function in terms of features such as accuracy, dynamic response, etc., and for it to be reliable in doing so. The reliability of the trip system may be improved by the use of (1) redundancy and (2) diversity.

Most trip systems consist of a single channel comprising of a sensor, a switch, and a shut-off valve, but where the integrity required is higher than that which can be obtained from a single channel, redundancy is generally used.

A trip system is normally dormant and comes to life only when a demand occurs. An element of the trip system such as a sensor or a valve may experience failure, and such a failure will lie unrevealed unless detected by proof testing or some other means. By contrast, equivalent elements in a control system are exercised continuously, and failure in such an element is liable to cause an operational excursion of some kind.

It should be an aim of trip system design to convert unrevealed failures into revealed failures, and hence to enhance reliability, by the judicious exploitation of benign integration.

8.4.10 Interlock Systems

Interlocks are another important type of protective device. They are used to control operations which must take place in a specified sequence and equipments which must have specified relations between their states. This definition of an interlock differs from that often used in the American literature, where the term 'interlock' tends to be applied to both trip and interlock systems (as defined here).

There are various kinds of interlocks. The original type is a mechanical device such as a padlock and chain on a hand valve. Another common type is the key interlock. Increasing use is made of software interlocks based on process computers.

Some typical applications of interlocks are in such areas as: (1) electrical switchgear, (2) test cubicles, (3) machinery guards, (4) vehicle loading, (5) conveyor systems, (6) machine start-up and shut-down, (7) valve systems, (8) instrument systems, (9) fire protection systems, (10) plant maintenance.

An interlock is often used to prevent access as long as a piece of equipment is operating. Thus, electrical switchgear may be installed in a room where an interlock prevents the door opening until there is electrical isolation. Similarly, an interlock prevents access to a test cubicle for operations involving high pressure or explosive materials until safe conditions pertain. An interlock may be used to stop access to a machine or entry into a vessel unless the associated machinery cannot move. In vehicle loading, interlocks are used to prevent a tanker moving away while it is still connected to the discharge point.

Pressure relief valves have interlocks to prevent all the valves being shut off simultaneously. There may be interlocks on other critical valve systems. Interlocks are also a part of instrument systems. An interlock may be used to prevent the disarming of a trip system unless certain conditions are met. Fire protection systems are provided with interlocks as a safeguard against leaving the system disabled, particularly after testing or maintenance. Plant maintenance operations make much use of interlocks to prevent valves being opened or machinery started up while work is in progress.

Some features of a good hardware interlock are that it (1) controls operations positively, (2) is incapable of defeat, (3) is simple, robust, and inexpensive, (4) is readily and securely attachable to engineering devices, and (5) is regularly tested and maintained.

8.4.11 Programmable Logic Systems

As already indicated, increasing use is made in process control systems of PLCs. An account of the application of PLCs to functions such as pump change over, fire and gas detection, and ESD has been given by Margetts (1986a,b). He describes the planning of an operation such as pump change over using hierarchical task analysis, in which the change over task is successively re-described until it has been broken down into executable elements, and the application of the hazard and operability (HAZOP) method to assess the adequacy of the resultant design.

He also deals with the reliability of the PLC system. For the system which he considers, the MTBFs of the input device, the control logic, and the output device are 100,000, 10,000, and 50,000 h, respectively, giving an overall system MTBF of 7690 h. Use of as many as four control logic units in parallel would raise the system MTBF to 14,480 h, but this is not the complete answer. The method described by the author for the further enhancement of reliability is the exploitation of the ability of the PLC to test the input and output devices and also itself.

8.4.12 Programmable Electronic Systems

Increasingly, the concept of computer control has become subsumed in the broader one of the PES. The account given here is confined to the safety aspects of PESs and is based on the *HSE PES Guide*.

HSE PES Guide: An account of PESs and their safety implications is given in *Programmable Electronic Systems in Safety Related Applications* (HSE, 1987) (the *HSE PES Guide*), of which Part 1 is an *Introductory Guide* (PES 1) and Part 2 the *General Technical Guidelines* (PES 2). Whereas in a safety-related system the use of conventional hardwired equipment is routine, the use of a PES in such an application has been relatively unknown territory. The approach taken, therefore, has been to assess the level of integrity required in the PES by reference to that obtained with a conventional system based on good practice. This level of integrity is referred to as 'conventional safety integrity'. PES 2 gives three system elements which should be taken into account in the design and analysis of safety-related systems: (1) configuration; (2) reliability; (3) overall quality. Safety integrity criteria for the system should be specified which cover all three of these system elements.

a. *Configuration*: The configuration of the system should be such as to protect against failures, both random and systematic. The former are associated particularly with hardware and the latter with software.
b. *Reliability*: The governing principle for reliability of the hardware is that the overall failure rate in a dangerous mode of failure, or, for a protection system, the probability of failure to operate on demand, should meet the standard of conventional safety integrity. Essentially, the level of reliability should be governed by the conventional safety integrity principle. Where the acceptable level of reliability is relatively low, the first method may suffice, but where a higher reliability is required, the second and third methods will be appropriate.
c. *Overall quality*: It is concerned essentially with high-quality procedures and engineering. These should cover the quality of the specification, design, construction, testing, commissioning, operation, maintenance, and modification of the hardware and software.
d. *Design considerations*: PES 2 describes a number of design considerations which are particularly relevant to the safety integrity of PESs. The replacement of a control chain in which the sensor sends a signal directly to the actuator by one which involves analogue-to-digital (A/D) converters and PES may reduce the safety integrity.
e. *Software considerations*: The software for use in safety-related applications needs to be of high quality, and PES 2 gives an account of some of the measures which may be taken to achieve this.

8.4.13 Emergency Shut-Down Systems

In quite a large proportion of cases, the plant is provided not just with individual trips but with a complete automatic emergency shut-down (ESD) system. There is relatively little written about ESD systems. The following is an account and explanation of the steps that are taken in the synthesis of an emergency shutdown system.

1. *Conceptual design of ESD*: The function of an ESD system is to detect a condition or an event sufficiently hazardous or undesirable as to require shut-down and then to effect transition to a safe state. The potential hazards are determined by a method of hazard identification such as HAZOP, Layer of Protection Analysis, Fault tree techniques. Estimates are then made of the frequency and consequences of these hazards. The hazards against which the ESD system is to protect are then defined.
2. *Initiation of ESD*: The arrangements for initiation of the ESD are critical. If these are defective, so that the system is not activated when it should be, all the rest of the design goes for nothing. There is a balance to be struck between the functional and the operational reliability of the ESD system. It should act when a hazard arises, but should not cause unnecessary shut-downs or other hazards.
3. *Action on ESD*: There are a variety of actions which an ESD system may take. Three principal types are (a) flow shut-off, (b) energy reduction, (c) material transfer. Flow shut-off includes shut-off of feed and other flows. It often involves shut-down of machinery and may include isolation of units. Energy reduction covers shut-off of heat input and initiation of additional cooling. Material transfer refers to pressure reduction, venting, and blow-down.
4. *Detail design of ESD system*: It is a fundamental principle that protective systems be independent of the rest of the instrument and control system, and this applies equally to an ESD system. The design of the ESD system should follow the principles which apply to trip systems generally. There should be a balance between functional and operational reliability. Dependent failures should be considered. The reliability may be assessed using fault tree and other methods. The techniques of diversity and redundancy

should be used as appropriate. Use may be made of majority voting systems.

5. *Operation of ESD system*: The status of the ESD system should be clear at all times. There should be a separate display showing this status in the control center. This display should give the status of any part of the ESD system which is under test or maintenance and of any part which is disarmed. Initiation of ESD should activate audible and visual alarms in the control center. There should be an indication of the source of the initiation, whether manual or instrument. ESD should also be signaled by an alarm which is part of the general alarm system.

6. *Testing and maintenance of ESD system*: The ESD system should be subject to periodic proof testing, and such testing should be governed by a formal system. As far as is practical, the test should cover the complete system from initiation to shut-down condition. The need for proof testing and, more generally, for the detection of unrevealed failure, should be taken into account in the design. The equipment should be designed for ease of testing. It should be segregated and clearly identified. Techniques for detection of instrument malfunction should be exploited. In voting systems, the failure of a single channel should be signaled.

7. *Documentation of an ESD system*: The ESD system should be fully documented. *The HSE Design, Construction and Certification Guidance Notes* give details of recommended documentation.

8. *ESD of a gas terminal*: The design of systems for ESD and (Emergency Depressurization) EDP of a gas terminal has been described by Valk and Sylvester-Evans (1985). The design philosophy described is that the ESD system should operate only in an extreme emergency, that the ESD and EDP systems are separate from the control, trip, and relief systems, and that the systems should be simple and reliable.

ACRONYMS

SLP	Safety and Loss Prevention
CCPS	Center for Chemical Process Safety
P&ID	Piping and Instrument Diagram
MOC	Management of Change
CAD	Computer-Aided Design
VIP	Value Improving Practice
IChemE	Institution of Chemical Engineers
AIChE	American Institute of Chemical Engineers
HAZOP	Hazard and Operability (study)
HSE	Health and Safety Executive
SIS	Safety Instrumented System
LOPA	Layers of Protection Analysis
PHA	Process Hazards Analysis
IPL	Independent Layer of Protection

BPCS	Basic Process Control System
ETT	Energize-to-trip
DTT	De-energize-to-trip
UPS	Uninterruptible Power Supply
PLC	Programmable Logic Control
PFD	Process Flow Diagram
PSAT	Pre-start-up Acceptance Test
SAT	Site Acceptance Test
API	American Petroleum Institute
ASME	American Society of Mechanical Engineers
BSI	British Standards Institution
HTHM	Highly Toxic Hazardous Material
DOT	Department of Transportation
NB	Nominal Bore
NRV	Non-Return Valve (Check Valve)
EIV	Emergency Isolation Valve
PLC	Programmable Logic Control
DCL	Distributed Control Logic
CCS	Centralized Control System
CIM	Computer Integrated Manufacturing
CRMR	Complete and Rigorous Model-Based Reconciliation
SIL	Safety Integrity Level
MTBF	Mean Time Between Failures
PES	Programmable Electronic System
ESD	Emergency Shut Down
EDP	Emergency Depressurization

REFERENCES

Abbott, J.A., 1990. Prevention of Fires and Explosions in Dryers, second ed. Institution of Chemical Engineers, Rugby.

Andrew, W.G., Williams, H.B., 1980. Applied Instrumentation in the Process Industries, vol. 2: Practical Guidelines. Gulf Publishing Company, Houston, TX.

Barrell, A., 1988. Inherent safety—only by design? (editorial). Chem. Eng. (London). 451, 3.

Broughton, J., 1993. Process Utility Systems. Institution of Chemical Engineers, Rugby.

Butterwick, B., 1976. User Guide for the Safe Operation of Centrifuges. Institution of Chemical Engineers, Rugby.

Campbell, I.E., Sherwood, E.M., 1967. High Temperature Materials and Technology. Wiley, New York, NY.

CCPS (Center for Chemical Process Safety), 1989. Guidelines for Chemical Process Quantitative Risk Analysis. American Institute of Chemical Engineers, New York, NY, ISBN 0-8169-0402-2, p. 585.

CCPS (Center for Chemical Process Safety), 1993. Guidelines for Engineering Design for Process Safety. Wiley—American Institute of Chemical Engineers, New York, NY.

CCPS, 1996a. Inherently Safer Chemical Processes, A Life Cycle Approach, 'Gold Book'. American Institute of Chemical Engineers, New York, NY, ISBN 0-8169-0703-X.

CCPS (Center for Chemical Process Safety), 1996b. Guidelines for Evaluating Process Plant Buildings for External Explosions and Fires. American Institute of Chemical Engineers, New York, NY, ISBN 0-8169-0646-7, p. 208.

CCPS (Center for Chemical Process Safety), 2001. Layer of Protection Analysis. American Institute of Chemical Engineers, New York, NY.

Dempster, M.A.H., Fisher, M.L., Jansen, L., Lageweg, B.J., Lenstra, J.K., Rinnooy Kan, A.H.G., 1981. Analytical evaluation of hierarchical planning systems. Oper. Res. 29, 707.

Dickenson, S.W., 1976. Pressure systems: general considerations. In: Course on Loss Prevention in the Process Industries. Department of Chemical Engineering, Loughborough University of Technology, Loughborough, Leicestershire, England.

Dinh, L., 2011. Safety-oriented Resilience Evaluation in Chemical Processes. Doctoral dissertation, Texas A&M University, College Station, TX.

Douglas, J.M., 1988. The Conceptual Design of Chemical Processes. McGraw-Hill, New York, NY.

Dow Chemical Company, 1976. Fire and Explosion Index. Hazard Classification Guide, fourth ed. Midland, MI.

Dow Chemical Company, 1980. Fire and Explosion Index. Hazard Classification Guide, fifth ed. Midland, MI.

Dow Chemical Company, 1987. Dow's Fire and Explosion Index. Hazard Classification Guide, sixth ed. Midland, MI.

Dow Chemical Company, 1994a. Dow's Fire and Explosion Index Hazard Classification Guide, seventh ed. American Institute of Chemical Engineers, New York, NY, ISBN 0-8169-0623-8.

Dow Chemical Company, 1994b. Dow's Chemical Exposure Index. American Institute of Chemical Engineers, New York, NY.

Edwards, D.W., Lawrence, D., 1993. Assessing the inherent safety of chemical process routes: is there a relation between plant costs and inherent safety? Trans. IChemE. 71 (Part B), 252–258.

Edwards, E., Lees, F.P., 1973. Man and Computer in Process Control. Institution of Chemical Engineers, London.

Faith, W.L., Keyes, D.B., Clark, R.L., 1965. Industrial Chemicals, second ed. Wiley, London.

Fisher, T.G., 1990. Batch Control Systems. Instrument Society of America, Pittsburgh, PA.

Fowler, E.W., Spiegelman, A., 1968. Hazard Survey of the Chemical and Allied Industries. American Insurance Association, New York, NY.

Groggins, P.H., 1952. Unit Processes in Organic Synthesis. McGraw-Hill, New York, NY.

HSE (Health and Safety Executive), 1987. Programmable Electronic Systems in Safety Related Applications: Part 1, an Introductory Guide. HM Stationery Office, London.

HSE (Health and Safety Executive), 2000. Pressure Systems Safety Regulation. HSE Books. Health and Safety Executive), 2000, London.

HSE (Health and Safety Executive), 2000. Safety of pressure systems: Pressure Systems Safety Regulation 2000. HSE Books. Health and Safety Executive), 2000, London.

Hutchison, J.W., 1976. Valve Trim. ISA Handbook of Control Valves, Pittsburgh, PA.

Institution of Chemical Engineers, 1992a. Model Form of Conditions of Contract for Process Plants: Subcontracts (Rugby) (the 'Yellow Book').

Institution of Chemical Engineers, 1992b. Model Form of Conditions of Contract for Process Plants: Suitable for Lump Sum Contracts (Rugby) (the 'Red Book').

Institution of Chemical Engineers, 1992c. Model Form of Conditions of Contract for Process Plants: Suitable for Reimbursable Contracts (Rugby) (the 'Green Book').

Jones, P., 1992. Inherent safety: action by HSE. Chem. Eng. (London). 526, 29.

Kirk, R.E., Othmer, D.F., 1963. Encyclopaedia of Chemical Technology. Wiley-Interscience, New York, NY.

Kirk, R.E., Othmer, D.F., 1978. Encyclopaedia of Chemical Technology. Wiley-Interscience, New York, NY.

Kirk, R.E., Othmer, D.F., 1991. Encyclopaedia of Chemical Technology. Wiley, New York, NY.

Kletz, T.A., 1978. What you don't have, can't leak. Chem. Ind. 9124 (May), 287–292.

Kletz, T.A., 1984a. Accident investigation: how far should we go? Plant/Operations Prog. 3, 1.

Kletz, T.A., 1984b. The analysis of hazards and the economy of risk prevention. CHEMRAWN III.

Kletz, T.A., 1984c. Anglo-American glossary. Chem. Eng. (London). 410, 33.

Kletz, T.A., 1984d. Cheaper, Safer Plants. Institution of Chemical Engineers, Rugby.

Kletz, T.A., 1984e. Flixborough. The view from ten years on. Chem. Eng. (London). 404, 28.

Kletz, T.A., 1984f. The Flixborough explosion—ten years later. Plant/Operations Prog. 3, 133.

Kletz, T.A., 1984g. Grey hairs cost nothing. Plant/Operations Prog. 3, 210.

Kletz, T.A., 1984h. Hazardous consequences—prediction and measurement. In: Petts, J.I. (1984), op. cit.

Kletz, T.A., 1984i. The industrial approach to hazard identification and analysis and setting criteria for risk. In: Petts, J.I. (1984), op. cit.

Kletz, T.A., 1984j. Myths of the Chemical Industry. Institution of Chemical Engineers, Rugby.

Kletz, T.A., 1984k. The prevention of major leaks—better inspection after construction? Plant/Operations Prog. 3, 19.

Kletz, T.A., 1984l. Talking about safety. Chem. Eng. (London). 399, 33.

Kletz, T.A., 1984m. Talking about safety. Chem. Eng. (London). 406, 29.

Kletz, T.A., 1989. Friendly plants. Chem. Eng. Prog. 85 (7), 18.

Kletz, T.A., 1990. The need for friendly plants. J. Occup. Accid. 13 (1–2), 3.

Kletz, T.A., 1991a. Are managers above the law? Chem. Eng. (London). 507, 38.

Kletz, T.A., 1991b. Inherently safer plants—recent progress. Hazards. XI, 225.

Kletz, T.A., 1992. Inherently safer design—a review. Loss Prev. Saf. Promotion. 7, 1.

Kletz, T.A., 2003. Still Going Wrong!: Case Histories of Process Plant Disasters and How They Could Have Been Avoided. Gulf Publishing Company, Burlington, MA.

Kletz, T.A., 2009. What Went Wrong: Case Histories of Process Plant Disasters and How They Could Have Been Avoided, fifth ed. Butterworth-Heinemann.

Kumamoto, H., Inoue, K., Henley, E.J., 1981. Computer-aided protective system hazard analysis. Comput. Chem. Eng. 5, 93.

Lindley, J., 1987. User Guide for the Safe Operation of Centrifuges, second ed. Institution of Chemical Engineers, Rugby.

Mansfield, D., Cassidy, K., 1994. Inherently safer approaches to plant design: the benefits of an inherently safer approach and how this can be built into the design process. Hazards. XII, 285.

Margetts, A., 1986a. Design and test your PLC control logic this way. Process Eng. (March), 41.

Margetts, A., 1986b. PLCs for alarm and shutdown systems. Chem. Eng. (London). 427, 35.

O'grady, P.J., 1986. Controlling Automated Manufacturing Systems. Kogan Page, London.

Park, S.W., Himmelblau, D.M., 1980. Error in the propagation of error formula. AIChE J. 26, 168.

Parker, R.J., 1975. The Flixborough Disaster. Report of the Court of Inquiry. HM Stationery Office, London.

Parker, V., 1991. Licensing Technology and Patents. Institution of Chemical Engineers, Rugby.

Pilborough, L., 1971. Inspection of Chemical Plant. Leonard Hill, London.

Pilborough, L., 1989. Inspection of Industrial Plant, second ed. Gower Technical, Aldershot, Hampshire.

Reay, D., 1977. User Guide to Fire and Explosion Hazards in Drying of Particulate Materials. Institution of Chemical Engineers, Rugby.

Rudd, D.F., Watson, C.C., 1968. Strategy of Process Design. Wiley, New York, NY.

Safety in Air and Ammonia Operations Progress, 1960. Vol. 2–(Ammonia Plant Safety and Related Facilities), 1970, vol. 12–. American Institute of Chemical Engineers, New York, NY.

Sawyer, P., 1993. Computer-Controlled Batch Processing. Institution of Chemical Engineering, Rugby.

Schofield, C., 1982. Guide to Safety in Mixing. Institution of Chemical Engineers, Rugby.

Tayler, C., 1987. Reduce instrument costs: go talk to your piping engineer. Process Eng. 33 (July).

Thielsch, H., 1965. Defects and Failures in Pressure Vessels and Piping. Reinhold, New York, NY.

Turney, R.D., 1994. Flixborough: 20 years on. Loss Prev. Bull. 117, 1.

Ullman, 1969. Ullmanns Encyklopädie der technischen Chemie. Urban & Schwarzenberg, Munich.

Ullman, 1985. Ullmanns Encyclopedia of Chemical Technology. VCH, Weinheim, Germany.

Valk, A., Sylvester-Evans, R., 1985. Safety in the design of gas terminals. Gastech. 41 (84), 157.

Wells, G.L., Seagrave, C.J., Whiteway, R.N.C., 1976. Flowsheeting for Safety. Institution of Chemical Engineers, London.

Wright, D., 1993. A Guide to the IChemE's Model Forms of Conditions of Contract. Institution of Chemical Engineers, Rugby.

Human Factors and Human Error

9.1. CONCEPT OF HUMAN FACTORS

Human factor is the discipline that considers the interaction of the human component with the remainder of a system (the environment, organization, equipment, etc.). A key aim of human factors is to identify and mitigate the sources and consequences of human error.

Human factors have three main considerations:

1. Ergonomics: matching the design of equipment and procedures to the physical capabilities of the human operator.
2. Cognitive Human Factors: matching the design of equipment and procedures to the cognitive capabilities and characteristics of the human operator.
3. Safety Culture: consideration of how the organization's culture and attitude influence the way in which operations are carried out.

The purpose of applying human factors is to maximize human performance within a system in order to maintain and improve health, safety, and overall system performance. This is achieved by matching the design of equipment and processes to human capabilities and limitations.

9.2. ROLE OF THE PROCESS OPERATOR

Much of the early work on human factors was concerned with physical tasks: performing a task efficiently, in the correct body position, and avoiding long-term injuries. In more recent years, the focus is increasing on mental tasks (i.e., cognitive processes).

The model of humans as the information processors provides a constructive perspective and has increasing value within assessment and design. Understanding human capabilities is essential in determining how to optimally implement automation, i.e., to give the operator those tasks that are most effectively carried out by humans, and automate tasks more effectively carried out by technology.

The human operator is utilized as a positive rather than a negative factor in process plants. The operator gives the system a much enhanced ability to deal with abnormal situations. But it is necessary to design systems which protect against, and are tolerant of, human error and to train operators to improve their decision making in these situations.

Although modern control systems achieve a high degree of automation, the process operator still has the overall immediate responsibility for safe and economic operation of the process. There are different philosophies on the extent to which the function of safety shut-down should be removed from the operator and assigned to automatic trip system. The operator has the vital function of running the plant so that shut-down conditions are avoided.

The job of the process operator has developed over the years from one based largely on manual work to one consisting primarily of decision making. Nowadays, the process operator is an integral part of the control system. The nature of the control system influences strongly the operator's functions. The job of the process operator, at least in the control room, is essentially decision making in a rather artificial situation, involving the manipulation of symbolic displays.

If the process control task as a whole is considered, a number of distinct operator functions may be identified (Edwards and Lees, 1973, 1974): (1) goal formulation, (2) measurement, (3) data processing and handling, (4) monitoring, (5) single variable control, (6) sequential control, (7) other control, (8) optimization, (9) communication, (10) scheduling, (11) manual operations.

The state of the process also affects the operator's task. If process conditions are abnormal, they have the crucial function of fault administration. In simple terms, this may be regarded as having three stages: (1) fault detection, (2) fault diagnosis, and (3) fault correction.

9.3. ALLOCATION OF FUNCTION

As already emphasized, the role of human factors in matters of system design, such as allocation of function, is just as important as in those of detailed design. The classic approach to allocation of function is to list the functions which machines perform well and those which humans perform well, and to use this as a guide. One of the original lists was compiled by Fitts (1962), and such a list is often referred to as a 'Fitts' list'.

Rasmussen (1986) discusses the behavior of humans as a component in a control loop, and his needs for aiding, in relation to the successively more difficult tasks of (1) direct manipulation, (2) indirect manipulation, (3) remote manipulation, and finally, (4) remote process control.

The automation of functions previously performed by the human operator can give rise to certain characteristic problems. Automated systems should be designed with the consideration of the capacities of both the machines and the humans. Human factors have a significant role in automation system. Human factors relevant to computerized and automated systems are discussed by Tzafestas (2010).

9.4. HUMAN INFORMATION PROCESSING

The task of the process operator is largely one of processing information. An understanding of the characteristics of human information processing in such a task is therefore crucial to the design of his work situation. In particular, it can contribute to the design of the interface and to the assessment of human error. The process operator as an information processor is described in *Information Processing and Human—Machine Interaction* (Rasmussen, 1986).

The operator may think about the system at several different levels of abstraction: (1) physical form, (2) physical function, (3) generic function, (4) abstract function, and (5) functional purpose.

One of the principal functions of the process operator is diagnosis. The identification of the fault is typically not the first priority. The priority is generally to evaluate the situation, decide whether the ultimate goal is to be modified, decide the new target state of the plant and define the tasks, and formulate and execute the procedures required to bring the plant to that state.

There are various models used in cognitive tasks. At the level of knowledge-based behavior, the operator is required to effect transformation between different types of model. Rasmussen identifies three strategies which he uses to do this: (1) aggregation, (2) abstraction, and (3) analogy.

If a system is viewed with a high degree of resolution, it may appear complex. If the elements of the system are aggregated into a larger whole by reducing the degree of resolution, a simplification may be achieved. These considerations have implications for the decision support systems provided for the operator. There is need for displays to support decision making at the different levels.

9.5. CASE STUDIES IN HUMAN ERROR

There are various accidents caused mainly by human error. The most important and severe ones are Bhopal, Chernobyl, Three-Mile Island, and BP Texas City incidents. These accidents are explained briefly in this section.

9.5.1 Bhopal

Early in the morning of December 3, 1984, a relief valve lifted on a storage tank containing highly toxic Methyl IsoCyanate (MIC) at the Union Carbide India Ltd (UCIL) works at Bhopal, India. A cloud of MIC gas was released onto housing, including shantytowns, adjoining the site. More than 2000 people died within a short period and tens of thousands were injured. The accident at Bhopal is by far the worst disaster which has ever occurred in the chemical industry.

The Bhopal incident was a wakeup call for the chemical industry around the world. There are numerous lessons learned from the Bhopal incident. At the time of the accident, the plant was losing money and listed a number of measures which had been taken, apparently to cut costs.

9.5.2 Chernobyl

On Saturday April 26, 1986, an accident occurred in Unit 4 of commercial nuclear power generation at Chernobyl in Ukraine. On that day, an experiment to check the use of the turbine during rundown as an emergency power supply for the reactor went catastrophically wrong. There was a power surge in the reactor, and the coolant tubes burst. There were two explosions, the second within some 3 s of the first, and debris and sparks shot into the air above the reactor. A plume of radioactive material was sent out. Some 30 fires broke out. Emergency measures to put out the fire and stop the release were not effective until 6 May.

The Chernobyl disaster was caused by a series of actions by the operators of the plant. It appears to be a case of human error which is virtually impossible to foresee and prevent. The Chernobyl disaster is reported in *The Chernobyl Disaster* by Haynes and Bojcum (1988).

9.5.3 Three-Mile Island

At 4.00 a.m. on March 28, 1979, a transient occurred on Reactor No. 2 at Three Mile Island, near Harrisburg, Pennsylvania. A turbine tripped and caused a plant upset. The operators tried to restore conditions, but, misinterpreting the instrument signals, misjudged the situation and took actions which resulted in the loss of much of the water in the reactor and the partial uncovering of the core. Radioactivity escaped into the containment building. Site and general emergencies were declared.

In this incident, there were a number of human errors which were made by the operators in the TMI-2 control room. Some of these were failures to make a correct diagnosis of the situation; others were undesirable acts of intervention.

9.5.4 BP Texas City Refinery Explosion

At approximately 1.20 p.m. on March 23, 2005, a massive explosion occurred at BP Texas City Refinery during the restarting of a hydrocarbon isomerization unit after turnaround maintenance. This accident resulted in 15 fatalities, and 66 serious injuries, while 114 others required medical attention. The damage caused by this explosion also impacted the area up to a distance of three-quarters of a mile away from the ISOM (ISOMerization) unit. This incident was considered one of the most catastrophic industrial accidents in US history. The financial losses from this accident exceed $1.5 billion.

The lack of considerations of human factors, e.g., training, staffing, scheduling, and operator fatigue, can be considered a negative factor in the incident. Adequate staffing requirements for the startup and obligations to correct unsafe conditions were not fulfilled. Based on analysis, operator fatigue might have contributed to the incident by impairing operator performance. The already shrinking operations personnel were not adequately trained. Furthermore, miscommunication was another issue contributing to the incident.

Texas City Disaster is reported in the Investigation Report Refinery Explosion and Fire at BP Texas City by US Chemical Safety and Hazard Investigation Board (2007).

9.6. DEFINITION OF HUMAN ERROR

Engineering interest in human error derives from two principal sources. One is accident investigation, where the apparent dominance of human errors gives rise to concern. The other is hazard assessment, where the requirement to quantify the effect of human error on system performance creates a demand for a methodology capable of doing this.

Human error is classified under the headings of slips and lapses, mistakes, misperceptions, mistaken priorities and violations.

A systematic approach to human error must involve the classification of errors and must, therefore, be based on appropriate models, either explicitly or implicitly. The classification is not necessarily along a single dimension. Most workers in the field have found it necessary to classify in terms of at least two dimensions: (1) human behavior and (2) task characteristics.

Human error in cognitive tasks is another important issue. Rasmussen distinguishes three types or levels of human approach to execute a task such as process control. These are skill-based, rule-based, and knowledge-based.

This description of operator behavior by Rasmussen is frequently referred to as the *Skill-Rule-Knowledge* or SRK-model. Skill-based, or skilled, behavior occurs without conscious attention and is data-driven. Rule-based behavior is consciously controlled and goal-oriented. Knowledge-based behavior is also conscious and involves reasoning.

There are other human error models in the literature. These are demand-capacity mismatch model, tolerance-variability model, time availability model, skills-rules-knowledge model, absentmindedness model, organizational model, violations.

9.7. HUMAN FACTORS APPROACHES TO ASSESSING HUMAN ERROR

The ergonomist's approach to assessing a task tends to be to enquire into the skill involved, including the nature of the skill, its acquisition through the learning process, and its disintegration under stress. An important finding is that skilled performance tends to improve with moderate stress, but that beyond a certain threshold, which varies greatly with the individual, it deteriorates rapidly.

9.7.1 Human Error Data, Framework, and Approaches

The various methods described for quantitative assessment of human reliability give rise to a demand for data on human error. The account given in *Guidelines for Preventing Human Error in Process Safety* by the CCPS (1994/17) deals with the essentials.

Data on human error may be acquired by a number of methods. These methods are study of the task from documentation, based on some form of direct observation, based on debriefing, the elicitation of information from experts.

The acquisition of high-quality data on human error is clearly of central importance. Human error data collection and data collection systems are also treated by the CCPS. The account deals with (1) types of data collection system, (2) design principles for data collection systems, (3) organizational and cultural aspects of data collection, types of data collected, (5) methods of data collection and storage, and (6) data interpretation.

Some types of data collection system that are described in the CCPS *Guidelines* are (1) the Incident Reporting and Investigation System (IRIS), (2) the Root Cause Analysis System (RCAS), (3) the Near Miss Reporting System (NMRS), and (4) the Quantitative Human Reliability Data Collection System (QHRDCS).

Human error data banks were a relatively early development. The data entered into these data banks were of the type necessary to support techniques such as THERP. One such data bank is the American Institute for Research (AIR) data bank developed by Altman (1964).

The Human Error Database (HED), described by Kirwan (1988), is based on the HEP data given in the *Rasmussen Report*, tempered by expert judgment.

Framework for human error models by means of task analysis is an important issue. The general framework in which most models and classifications of human error are applied is that of task analysis. That is to say, the task is decomposed into elements such as plans and actions and the errors associated with these are modeled and classified.

In addition to being a general framework within which particular models are applied, task analysis, particularly hierarchical task analysis, may be regarded as a model in its own right. One common classification of human error is in terms of actions. This type of classification refers to acts of omission, acts of commission, delays in taking action, and so on.

Task Analysis is an approach used in human error assessment. Task analysis has developed into a family of techniques, undertaken for a variety of purposes and employing different methodologies. For example, task analysis may be used as an aid to identifying information requirements, writing operating procedures, defining training needs, specifying manning levels, estimating human reliability in probabilistic safety assessment, and investigating problems. There are some 25 or more major techniques available. In many areas of human factors in the process industries, task analysis has become an indispensable tool.

Early work was shown in *Task Analysis* (Annett et al., 1971). The method described by these authors is Hierarchical Task Analysis (HTA), in which the task is broken down into a hierarchy of task elements.

The method adopted is to break the task down into a hierarchy. The elements of the hierarchy are a goal, plans, and operations. The task involves a goal and this is then re-described in terms of the plans and operations necessary to achieve it. Operations are units of behavior, typically with an action–information feedback structure.

9.7.2 Assessment of Human Error

Assessment of Human Error in view of QRA applications is mostly in the fault tree part of bow-ties, but human action can also appear in an event tree, in barriers or protective measures. Human activity can be broken down in tasks which can in turn be broken down further into its elements. Error rates can be assigned to each of these elements from the reliability of performance of the task. The reliability of the overall system in which the human is a part can then be found by the linear combination of the reliabilities of the various components.

Assessment of Human Error in Process Operations is another important aspect. A methodology based on the task analytical approach, the Technique for Human Error Rate Prediction (THERP), developed by Swain and co-workers, is described below. An overview of this type of approach is given here based on an early account by Swain (1972).

Furthermore, there are various qualitative methods of human error assessment. These methods may be used in their own right to reduce error or as a stage in a human reliability assessment method. Their significance in the latter application is that they provide a structured approach to gaining understanding of the problem. In the absence of a high-quality technique for this essential preliminary stage, quantification is premature.

Task analysis may be regarded as the prime technique, or rather family of techniques, but there are also a number of others. Early work in this area was that of Taylor (1979), who described a variety of approaches.

A method based on hierarchical task analysis is predictive human error analysis. Another method is the work analysis technique described by Pedersen (1985).

There are various error analysis strategies in the literature. Against the background of a long-term program of work on human error, Taylor (1979) has developed a set of four error analysis strategies. These are action error method, pattern search method, sneak path analysis, and Predictive Human Error Analysis (PHEA).

Another qualitative method is System for Predictive Error Analysis and Reduction (SPEAR). The SPEAR is a set of qualitative techniques, of which PHEA is one. It is described by the CCPS (1994/17). SPEAR comprises the following techniques: (1) task analysis, (2) Performance Influencing Factor (PIF) analysis, (3) PHEA, (4) consequence analysis, and (5) error reduction analysis.

9.8. QUANTITATIVE HUMAN RELIABILITY ANALYSIS (HRA)

The first systematic approach to the treatment of human error within a PRA was the THERP. An early account of THERP was given by Swain (1972).

The accident at Three Mile Island gave impetus to work in this area and led to the publication of the *Handbook of Human Reliability Analysis with Emphasis on Nuclear Power Plant Applications, Final Report* (the *HRA Handbook*) by Swain and Guttmann (1983).

The overall approach used in the *Handbook* is shown in Figure 9.1. The tasks to be performed are identified as part of the main PRA. The HRA involves the assessment of the reliability of performance of these tasks.

A quantitative human reliability analysis technique is the THERP. The starting point is a task analysis for each of the tasks to be performed. The method is based on the original THERP technique and uses a task analytic

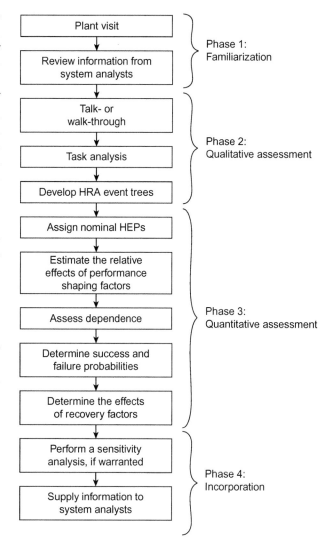

FIGURE 9.1 HRA Handbook: methodology for human reliability analysis (Swain and Guttmann, 1983).

approach in which the task is broken down into its constituent elements along the general lines described above. The basic assumption is that the task being performed is a planned one.

Human Error Probability (HEP) is another quantitative risk analysis method. Several different human error probabilities are distinguished: nominal, basic, conditional, and joint. A nominal HEP is a generic value before application of any performance shaping factors. A basic HEP (BHEP) is the basic unconditional HEP after application of performance shaping factors. A conditional HEP (CHEP) is a BHEP adjusted to take account of dependency. A joint HEP (JHEP) is the HEP for the overall task.

Estimates of HEP are given in a number of fault trees. Operator error occurs in the tree as errors that either initiate or enable the fault sequence and as errors that constitute failures of protection. Errors that result in failure of protection (expressed as probabilities) predominate over errors that initiate or enable the fault sequence (expressed as frequencies). Initiating and enabling errors tend to be associated with an item of equipment, and protection errors with a process variable. Such a variable may have (1) no measurement, (2) measurement only, or (3) measurement and alarm, and this is an important feature influencing the error rate.

As already mentioned, there has been an increasing tendency, associated mainly with experimental work with operators on simulators, to correlate the probability of operator failure with time. In particular, use is made of the Time–Reliability Correlation (TRC) to obtain human error probabilities for complex, or non-routine, tasks, including handling an emergency. The assumption underlying such a TRC is that, although there are in principle other factors that affect operator performance in such tasks, time is the dominant one. Early work on this was done by R.E. Hall (1982). A number of TRC models have since been produced based on simulator results.

The task analysis approach on which THERP is based is not well adapted to handling the behavior of the operators in an abnormal situation. For this, use is made of the TRC approach. Several TRCs are given in the *Handbook*. Two are used for screening: one for diagnosis and one for post-diagnosis performance.

Some errors are not recoverable, but many are and the recovery model is therefore another important feature of the methodology. Recovery is treated with human redundancy, annunciated indications, active inspections, and passive inspections. Human redundancy is essentially the checking of one person's work by another person. For checking, the *Handbook* gives the following HEP values (see Chapter 20).

There are various factors affecting the performance of the operators. The performance shaping factors (PSFs) are divided into several classes. These are (1) external factors: (a) situational characteristics, (b) task and equipment characteristics, (c) job and task instructions; (2) internal factors; and (3) stressors.

9.9. HUMAN RELIABILITY ASSESSMENT METHODS

Success Likelihood Index Method (SLIM): A method of obtaining HEP estimates based on PSFs is the Success Likelihood Index (SLI) which is incorporated in the SLI method (SLIM). Accounts are given in *SLIM-MAUD: An Approach to Assessing Human Error Probabilities using Structured Judgement* by Embrey et al. (1984 NUREG/CR-3518).

SLIM treats not only the quality of the individual PSFs but also the weighting of these in the task. It is thus a complete method for assessing of human error, and not merely a technique for determining values of the PSFs. The basic premise of SLIM is that the HEP depends on the combined effects of the PSFs. A systematic approach is used to obtain the quality weightings and relevancy factors for the PSFs, utilizing structured expert judgment. From these PSFs, the SLI for the task is obtained.

Human Error Assessment and Reduction Technique (HEART): In the HEART, described by Williams (1986), the HEP of the task is treated as a function of the type of task and of associated Error Producing Conditions (EPCs), effectively PSFs. The method is based on a classification of tasks into the generic types, which also gives the proposed nominal human unreliabilities for execution of the tasks. There is an associated set of EPCs, for each of which is given an estimate of the maximum predicted normal amount by which the unreliability might change going from 'good' to 'bad'. HEART has been designed as a practical method and is easy to understand and use.

Dougherty and Fragola Method (D&F): The deficiencies of THERP in respect of non-routine behavior have led to the development of alternative methods. One of these methods is that described in *Human Reliability Analysis* by Dougherty and Fragola (1988). Like THERP, this method has been developed essentially as an adjunct to fault tree analysis, but the approach taken is rather different.

CCPS Method: Another methodology for human reliability assessment is that described in the CCPS *Guidelines for Preventing Human Error in Process Safety* (CCPS, 1994/17). The core is the four stages of (1) critical human interaction identification and screening, (2) qualitative prediction of human error, (3) representation of event development, and (4) quantification of significant human errors.

Other Methods: The following methods also merit mention. A brief account of each is given in the Second Report of the Study Group on human factors of the ACSNI (1993). These methods are discussed below.

TESEO method by Bello and Colombari (1980), absolute probability judgement (APJ) method by Seaver and Sitwell (1983), method of paired comparisons (PCs) by Blanchard (1966) and Hunns (1980, 1982), systematic human error reduction and prediction approach (SHERPA) by Murgatroyd (1986).

9.10. HUMAN FACTORS APPROACHES TO MITIGATING HUMAN ERROR

Mitigation can be achieved by two means: (1) reducing or eliminating the likelihood of an error occurring in the first place and (2) reducing, eliminating, or protecting against the consequences of an error once it has been made.

There exist a number of strategies for prevention and mitigation of human error. Essentially these aim to: (1) reduce frequency, (2) improve observability, (3) improve recoverability, (4) reduce impact.

The *Guide to Reducing Human Error in Process Operation* compiled by the human factors Reliability Group (HFRG) (1985 SRD R347) gives guidance in checklist form. The checklists cover (1) operator-process interface, (2) procedures, (3) workplace and working environment, (4) training, and (5) task design and job organization.

There is a significant relationship between human error and plant design. Turning to the design of the plant, design offers wide scope for reduction of both the incidence and the consequences of human error. It goes without saying that the plant should be designed in accordance with good process and mechanical engineering practice. In addition, however, the designer should seek to envisage errors that may occur and to guard against them.

9.10.1 Displays

Much work has been done on displays, in terms of both the detailed design of dials, etc., and the display layout, and it is perhaps this aspect which the engineer most readily identifies as human factors.

There are very different uses and a display (Indicating, Quantitative reading, Check reading, Setting, and Tracking (Murrell, 1965)) which is optimal for one is not necessarily so for another.

Control−display relations can be important. In a given culture, there tend to be expectations of particular relations between control movements and display readings. Although an operator can be trained to use equipment that embodies faulty control−display relations in its design, he may tend under stress to revert to the expected relation.

In designing the work situation, the aim should be to prevent the occurrence of error, to provide opportunities to observe and recover from error, and to reduce the consequences of error.

The acquisition of information from large display layouts is another important problem. Work in this area tends to emphasize the value to the operator of being familiar with the position of dials which give particular readings.

Monitoring, signal detection and vigilance, is another related and important area on which a large amount of work has been done. Some of this has been concerned with the falloff in attention over the watch-keeping

period, i.e., the vigilance effect. More recently the concept of situation awareness is getting a lot of attention.

9.10.2 Design of Information Display

Information display is an important problem, which is intensified by the increasing density of information in modern control rooms. The traditional display is the conventional control panel. Computer graphics now present the engineer with a more versatile display facility, offering scope for all kinds of display for the operator.

The first thing that should be emphasized is that a display is only a means to an end, the end being improved performance by the operator in executing some control function. The proper design of this function in its human factors aspects is more important than the details of the display itself.

The provision of displays that the operator deliberately samples with a specific object in view is only part of the problem. It is important also for the display system to cater both for his characteristic of acquiring information 'at a glance' and for his requirement for information redundancy. There is a need, therefore, for the development of displays that allow the operator to make a quick and effortless survey of the state of the system.

9.10.3 Regular Instrumentation

The conventional panel does constitute a survey display, in which the instruments have spatial coding, from which the operator can obtain information at a glance and on which he can recognize patterns. These are solid advantages not to be discarded lightly. This is only true; however, if the density of information in the panel is not allowed to become too great.

Traditionally, an important individual display is the chart recorder. A trend record has many advantages over an instantaneous display. As the work of Crossman (1964) shows, it assists the operator to learn the signal characteristics and facilitates his information sampling. Attwood (1970) found that recorders are useful to the operator in making coarse adjustments of operating point, while the latter authors also noted the operator's use of recorders in handling fault conditions.

9.10.4 Computer Consoles

The computer console presents a marked contrast to conventional instrumentation. Some of the features of ergonomic importance in conventional panel are (1) specific action is required to obtain a display; (2) there is no spatial coding and the coding of the information required has to be remembered or looked up; (3) only one variable is displayed at a time; (4) only the instantaneous value of the variable is displayed; and (5) the presentation is digital rather than analogue.

9.11. ALARM SYSTEMS

An alarm system is a normal feature of conventional control systems. If a process variable exceeds specified limits or if an equipment is not in a specified state, an alarm is signaled. Both audible and visual signals are used. Reliability should be high because false alarms undermine severely reaction alertness.

Accounts of alarm systems include those by E. Edwards and Lees (1973), Swain and Guttmann (1983), and the Center for Chemical Process Safety (CCPS, 1993/14). Other good practices for Design, Management, and Procurement of Alarm systems are described in EEMUA 191 (Engineering Equipment and Materials Users Association, 2007), and further in ISA SP-18 (International Society of Automation, 2009), NAMUR NA 102 in 2003 , API 1167 (Dunn and Sands, 2005).

9.11.1 Basic Alarm Systems

The traditional equipment used for process alarms is a lightbox annunciator. When a new alarm occurs, the hooter sounds and the fascia light flashes until the operator acknowledges receipt by pressing a button. The panel then remains lit until the alarm condition is cleared by operator action or otherwise, as described below.

In modern computer-controlled systems, there are numerous display options, some of which are detailed by Kortlandt and Kragt (1980b). One is the use of a dedicated Visual Display Unit (VDU) in which the alarms come up in time sequence, as on the printer. Another is a group display in which all the alarms on an item of equipment are shown with the active alarms(s) highlighted.

In a conventional instrument system, the hardware used to generate an alarm consists of a sensor, a logic module, and the visual display. In a computer-based system, there are two approaches that may be used. In one, the computer receives from the sensor an analogue signal to which the program applies logic to generate the alarm. In the other, the signal enters in digital form.

9.11.2 Alarm System Features

The process computer has enormous potential for the development of improved alarm systems but also brings with it the danger of excess. There is first the choice of variables which are to be monitored. It is no longer necessary for these to be confined to the process variables measured by the plant sensors. In addition, 'indirect' or 'inferred' measurements calculated from one or more process measurements may be utilized. This considerably increases the power of the alarm system.

The conventional alarm system, therefore, is severely limited by hardware considerations and is relatively inflexible. The type of alarm is usually restricted to an absolute alarm. The computer-based alarm system is potentially much more versatile.

The alarm system, however, is frequently one of the least satisfactory features of the control system. The most common defect is that there are too many alarms and that they stay active for too long. Therefore, alarm management (rationalization and prioritization) techniques are an essential part of control system design.

The process computer provides the basis for better alarm systems. It makes it possible to monitor indirect measurements and to generate different types and degrees of alarm, to distinguish between alarms and statuses and to adapt the alarms to the process state. The abundance of instrumentation and data has over the years worsened the situation. Alarm management is therefore today a must.

9.11.3 Alarm Management

Adopting a consistent alarm management philosophy is essential for abnormal situation management. If no alarm management is in place, abnormal situations can 'inundate' operators with alarms even to such an extent that override becomes general practice and higher level or super alarms become necessary to draw immediate attention and action. The dynamics of a process can demand most of the operators to maintain good control. A good alarm system can generate appropriate information to operators guiding them to the cause of the upset and helping to restore the plant to normal operations.

Alarm management is discussed by the CCPS (1993/14). Approaches to the problem include (1) alarm prioritization and segregation, (2) alarm suppression, and (3) alarm handling in sequential operations. Alarms may be ranked in priority.

9.11.4 Alarm System Operation

Studies of the operation of process alarm systems have been conducted by Kragt and co-workers (Kortlandt and Kragt, 1980a,1980b; Kragt, 1984). Kortland and Kragt (1980a,1980b) studied five different control room situations, the authors identified two confusing features of the alarms. One was the occurrence of oscillations in which the measured values moved back and forth across the alarm limit. The other was the occurrence of clusters of alarms. Kragt (1984) has described an investigation of the operator's use of alarms in a computer-controlled plant. The main finding was that sequential information presentation is markedly inferior to simultaneous presentation.

9.12. FAULT ADMINISTRATION

Fault administration can be divided into three stages: fault detection, fault diagnosis, and fault correction or shut-down. For the first of these functions, the operator has a job aid in the form of the alarm system, while fault correction in the form of shut-down is also frequently automatic, but for the other two functions, fault diagnosis and fault correction less drastic than shut-down, they are largely on their own.

Fault detection: The alarm system represents a partial automation of fault detection. The operator still has much to do, however, in detecting faults. This is partly a matter of the additional sensory inputs such as vision, sound, and vibration that the operator possesses. But it is also partly due to his ability to interpret information, to recognize patterns, and to detect instrument errors.

Fault diagnosis: There are various ways in which the operator may approach fault diagnosis. Several workers have observed that an operator frequently seems to respond only to the first alarm which comes up. This is an incomplete strategy, although it may be successful in quite a high proportion of cases. An alternative approach is pattern recognition from the displays on the control panel. The pattern may be static or dynamic. The static pattern is obtained by instantaneous observation of the displays, like a still photograph. The operator then tries to match this pattern with model patterns or templates for different faults. Duncan and Shepherd (1975) have developed a technique for training in fault diagnosis in which some operators use this method. The alternative, and more complex, dynamic pattern recognition involves matching the development of the fault over a period of time.

Another approach is the use of some kind of mental decision tree in which the operator works down the paths of the tree, taking particular branches, depending on the instrument readings. Duncan and Gray (1975) have used this as the basis of an alternative training technique.

Fault diagnosis is not an easy task for the operator. There is scope, therefore, for computer aids, if these can be devised. Some developments on these lines are described below.

Fault correction and shut-down: When, or possibly before, a fault has been diagnosed, it is usually possible to take some corrective action which does not involve shutting the plant down. In some cases, the fault correction is trivial, but in others, such as operating a complex sequence of valves, it is not. Operating instructions are written for many of these activities, but otherwise this is a relatively unexplored area.

Some fault conditions, however, require plant shut-down. Although fault administration has been described in terms of successive stages of detection, diagnosis, and correction, in emergency shut-down usually little diagnosis is involved. The shut-down action is triggered directly when it is detected that a critical process limit has been passed.

In a plant without protective systems, the operator is effectively given the duty of keeping the plant running if he can, but shutting it down if he must. This tends to create in his mind a conflict of priorities. Usually, he will try to keep the plant running if he possibly can and, if shut-down becomes necessary, he may tend to take action too late. Mention has already been made of the human tendency to gamble on beating the odds. There are numerous case histories which show the dangers inherent in this situation (Lees, 1976).

Although the use of protective systems is rapidly increasing, the process operator usually retains some responsibility for safe plant shut-down. There are a number of reasons for this. In the first place, although high integrity protective systems with 2/3 voting are used on particularly hazardous processes (Stewart, 1971), the majority of trip systems do not have this degree of integrity. The failure rate of this simple 1/1 trip system has been quoted as 0.67 faults/year (Kletz, 1974).

9.13. MALFUNCTION DETECTION

Another aspect of the administration of fault conditions by the control system is the detection of malfunctions, particularly incipient malfunctions in plant equipment and instruments. These malfunctions are distinguished from alarms in that although, they constitute a fault condition of some kind, they have not as yet given rise to a formal alarm.

Detection of instrument malfunction by the operator has been investigated by Anyakora and Lees (1972). In general, malfunction may be detected either from the condition of an instrument or its performance.

Some of the ways in which the operator detects malfunction in instruments are illustrated by considering the way in which he uses for this purpose one of his principal detection aids—the chart recorder. The operator detects error in such signals by utilizing some form of redundant information and making a comparison. Some types of redundant information are (1) *a priori* expectations, (2) past signals of instrument, (3) duplicate instruments, (4) other instruments, and (5) control valve position.

The detection of instrument malfunction by the process operator is important for a number of reasons. Instrument malfunctions tend to degrade the alarm system and introduce difficulties into loop control and fault diagnosis. Their detection is usually left to the operator and it is essential for him to have the facilities to do this. This includes appropriate displays and may extend to computer aids.

9.14. TRAINING

Many human errors in process plants are due to poor training and instructions. In terms of the categories of skill, rule and knowledge-based behavior, instructions provide the basis of the second, while training is an aid to the first and the third, and should also provide a motivation for the second.

Certain features of the operation and maintenance of process plants should be covered in elementary training. This elementary training is discussed in more detail in Chapter 4.

9.14.1 Training and Education

There is a distinction to be made between training and education: the former is specific to a particular task or job, the latter is more general.

As already mentioned, training on a simulator helps. It can also be used to simulate abnormal situations and to exercise complex procedures. However, hands-on experience on hardware is even more valuable and where equipment becomes more automated and the operator is further away of the plant proper, such experience is more needed. At several places, mini-plants offer the possibility to make errors, observe the effects, and correct.

Courses on various topics and at different levels are being offered nowadays in many places. However, process safety education which is also very necessary in chemical engineering and chemistry education at university level, is provided only at a relative small number of places. CCPS SACHE (Safety and Chemical Engineering Education, www.sache.org) provides workshops and organizes other activities. The Mary Kay O'Connor Process Safety Center at the Chemical Engineering Department of the Texas A&M University in College Station, Texas, is fully dedicated to process safety education up to the highest level.

9.14.2 Training Principles

The job of the trainer is to observe and analyze, and to arrange to supply the right amount of the right kind of information to the learner at the right time. He must know a great deal about the task being trained, but the kind of knowledge needed is what comes from careful, objective analysis of the job and of the necessary skill rather than from the experience of becoming personally proficient.

This account by Holding (1965) highlights some of the important features in training. It clearly brings out the importance of the prior task analysis to determine where the operator may have difficulty and where training may be necessary. The content of training should be appropriate, which in effect means it should be related to these difficulties. The training should be at a suitable pace. It should provide feedback of results, since this is essential for learning.

The process operator needs to have the safety training received by other employees also, but in addition has certain specific needs of his own. In this section, consideration is given to the content of training for process operators. Safety training is considered in Chapter 4. Designing training programs for complex cognitive skills were discussed in detail by Van Merriënboer (1997).

9.14.3 Job Design

Job design involves the arrangement of the individual tasks which a person is capable of doing and from which they obtain satisfaction. An account of job design is given by Davis and Wacker (1987), (Meshkati, 1991), (Morgeson et al., 2006).

An approach to job design starting from this fact is to develop explicitly the functions of the operator that are concerned with handling faults and keeping the plant running, notably fault administration and malfunction detection.

9.14.4 Personnel Selection

One of the first studies of the process operator was by Hiscock (1938), who described the development of a set of selection tests, but subsequent work by Davies (1967) showed little correlation between performance assessed by a selection test battery and by the judgment of supervisors.

The abilities an operator needs to understand include signal detection, signal filtering, probability estimation, system state evaluation, manual control, and fault diagnosis. Tests may be devised to measure these abilities, using perhaps a computer-based simulator.

It is also important, however, to take into account in selection personal qualities. Crossman (1960) suggested that it is desirable that a process operator should be responsible, conscientious, reliable, and trustworthy. Other important factors are temperament, motivation, and social skills. A further account of personnel selection is given by (Schmidt et al., 1992).

9.15. CCPS GUIDELINES FOR PREVENTING HUMAN ERROR IN PROCESS SAFETY

The prevention of human error in process plants is addressed in the *Guidelines for Preventing Human Error in Process Safety* edited by Embrey for the CCPS (1994/ 17) (the CCPS *Guidelines for Preventing Human Error in Process Safety*).

The *Guidelines* distinguish four basic perspectives on human error, which they term: (1) the traditional safety engineering approach, (2) the human factors engineering and ergonomics (HF/E) approach, (3) the cognitive engineering approach, and (4) the socio-technical systems approach.

The factors shaping human performance are referred to in the *Guidelines* as PIFs. They give a PIF classification structure and a detailed commentary on each factor. Furthermore, The *Guidelines* deal with methods for predicting and reducing human error in terms of (1) data acquisition techniques, (2) task analysis, (3) human error analysis, and (4) ergonomics checklists.

The *Guidelines* describe a methodology for HRA, as part of quantitative risk assessment, utilizing both qualitative and quantitative methods for predicting human error.

The core of the method is the four stages of (1) critical human interaction identification and screening, (2) qualitative prediction of human error, (3) representation of event development, and (4) quantification of significant human errors.

The *Guidelines* give a number of methods of incident analysis. They are (1) the causal tree/variation diagram, (2) the Management Oversight and Risk Tree (MORT), (3) the Sequentially Timed Events Plotting procedure (STEP), (4) root cause coding, (5) the Human Performance Investigation Process (HPIP), and (6) change analysis.

A feature of the *Guidelines for Preventing Human Error in Process Safety* is the number of case histories given. The section on case studies gives five such studies: (1) incident analysis of a hydrocarbon leak from a pipe (Piper Alpha); (2) incident investigation of mischarging of solvent in a batch plant; (3) design of standard operating procedures for the task in Case Study 2; (4) design of VDUs for a computer-controlled plant; and (5) audit of offshore emergency blow-down operations.

The *Guidelines* provide guidance on the implementation of an error reduction program in a process plant. A necessary precondition for such a program is a management culture that provides the background and support for such initiatives.

NOTATION

CCPS	Center for Chemical Process Safety
EPCs	Error Producing Conditions
HEART	Human Error Assessment and Reduction Technique
HEP	Human Error Probability
HRA	Human Reliability Analysis
PHEA	Predictive Human Error Analysis
PRA	Probabilistic Risk Assessment (In nuclear engineering the term for QRA)
PSF	Performance Shaping Factor

QRA	Quantitative Risk Assessment (In chemical engineering the term for PRA)
SLIM	Success Likelihood Index Method
SPEAR	System for Predictive Error Analysis and Reduction
THERP	Technique for Human Error Rate Prediction
TRC	Time−Reliability Correlation
UCB	Uncertainty Bound
VDU	Visual Display Unit

REFERENCES

Advisory Committee on the Safety of Nuclear Installations (ACSNI), 1993. Study Group on Human Factors. Third Report: Organising for Safety. HM Stationery Office, London.

Altman, J.W., 1964. A central store of human performance data. Symposium on Quantification of Human Performance. Sandia Laboratories, Albuquerque, NM.

Annett, J., Duncan, K.D., Stammers, R.B., Gray, M.J., et al., 1971. Task Analysis. Department of Employment, Training Information. Paper 6. HM Stationery Office, London.

Anyakora, S.N., Lees, F.P., 1972. Detection of instrument malfunction by the process operator. Chem. Eng. (London). 264, 304.

Attwood, D., 1970. The interaction between human and automatic control. In: Bolam, F. (Ed.), Paper-Making Systems and Their Control. The British Paper and Board Makers Association, London, p. 69.

Bello, G.C., Colombari, V., 1980. The human factors in risk analyses of process plants: the control room operator model 'TESEO'. Reliab. Eng. 1 (1), 3.

Blanchard, P.C., 1966. Likelihood-of-Accomplishment Scale for a Sample of Man−Machine Activities. Dunlop and Associates Inc., Santa Monica, CA.

CCPS, 1994. Guidelines for Preventing Human Error in Process Safety. American Institute of Chemical Engineers, New York, NY.

Crossman, E.R.F.W., 1960. Automation and Skill. HM Stationery Office, London.

Crossman, E.R.F.W., Cooke, J.E., Beishon, R.J., 1964. Visual Attention and the Sampling of Displayed Information in Process Control. Rep. HFT 64−11−7. Human Factors in Technology Research Group, University of California, Berkeley, CA.

Davies, D.G., 1967. A Psycho-Physiological Investigation of Process Control Skill (M.Sc. Thesis). University of Aston in Birmingham.

Davis, L.E., Wacker, G.J., 1987. Job design. In: Salvendy, G. (Ed.), Handbook of Human Factors. Wiley, New York, p. 431., op. cit.

Dougherty, E.M., Fragola, J.R., 1988. Human Reliability Analysis. Wiley, New York, NY.

Duncan, K.D., Gray, M.J., 1975. An evaluation of a fault-finding course for refinery process operators. J. Occup. Psychol. 48, 199.

Duncan, K.D., Shepherd, A., 1975. A simulator and training technique for diagnosing plant failures from control panels. Ergonomics. 18, 627.

Dunn, C.L., 1952. The Emergency Medical Services, vol. 1. HM Stationery Office, London.

Dunn, D.G., Sands, N.P., 2005. ISA-SP18-Alarm Systems Management and Design Guide, ISA EXPO 2005, Chicago, IL, United States.

Edwards, E., Lees, F.P., 1973. Man and Computer in Process Control. Institution of Chemical Engineers, London.

Edwards, E., Lees, F.P. (Eds.), 1974. The Human Operator in Process Control. Taylor & Francis, London.

Embrey, D., Humphreys, P.C., Rose, E.A., Kirwan, B., Rea, K., 1984. SLIM-MAUD: An Approach to Assessing Human Error Probabilities Using Structured Expert Judgment. Rep. NUREG/CR-3518. Nuclear Regulatory Commission, Washington, DC.

Fitts, P.M., 1962. Functions of man in complex systems. Aerospace Eng. 21 (1), 34.

Hall, R.E., Fragola, J.R., Wreathall, J., 1982. Post Event Human Decision Errors: Operator Action Trees/Time Reliability Correlation. Rep. NUREG/CR-3010. Nuclear Regulatory Commission, Washington, DC.

Haynes, V., Bojcum, M., 1988. The Chernobyl Disaster. Hogarth Press, London.

Hiscock, W.G., 1938. Selection tests for chemical process workers. Occup. Psychol. 12, 178.

Holding, D.H., 1965. Principles of Training. Pergamon Press, Oxford.

Hunns, D.M., 1980. Discussions Around a Human Factors Data-Base: An Interim Solution: The Method of Paired Comparisons. National Centre for Systems Reliability, Culcheth, Warrington.

Hunns, D.M., 1982. Discussion around a human factors data-base. An Interim Solution: The Method of Paired Comparisons. In: Green, A. E. (Ed.), op. cit., p. 181.

Kirwan, B., 1988. A comparative evaluation of five human reliability assessment techniques. In: Sayers, B.A. (Ed.), Human Factors and Decision Making, op. cit., p. 87.

Kletz, T.A., 1974. Over-pressure and other risks—some myths of the chemical industry. Loss Prev. Saf. Promotion. 1, 309.

Kortlandt, D., Kragt, H., 1980a. Ergonomics in the struggle against 'alarm inflation' in process control systems. In: Aune, A.B., Vlietstia, J. (Eds.), Automation for Safety in Shipping and Offshore Petroleum Operations. North Holland, New York, NY, 231.

Kortlandt, D., Kragt, H., 1980b. Process alarm systems as a monitoring tool for the operator. Loss Prev. Saf. Promotion. 3 (2), 804.

Kragt, H., 1984. A comparative simulation study of annunciator systems. Ergonomics. 27, 927–945.

Lees, F.P., 1976. Visual displays and the man-machine interface. Acta IMEKO..

Merrifield, R., 1985. The development of ball valves. Chem. Eng. (London). 418, 36.

Meshkati, N., 1991. Integration of workstation, job, and team structure design in complex human-machine systems: a framework. Int. J. Ind. Ergon. 7 (2), 111–122.

Morgeson, F., Medsker, G., Campion, M., 2006. Job and team design. Handbook of Human Factors and Ergonomics. Wiley, Hoboken, NJ.

Murgatroyd, R.A., 1986. A Systematic Approach to Optimizing Human Performance in Ultrasonic Inspection (Report). Safety and Reliability Directorate, Culcheth, Warrington.

Murrell, K.F.H., 1965. Ergonomics: Man in His Working Environment. Chapman & Hall, London.

Oser, B.L., 1981. The rat as a model for human toxicological evaluation. J. Toxicol. Environ. Health. 8, 521.

Pedersen, O.M., 1985. Human Risk Contributions in the Process Industry: Guides for Their Pre-identification in Well-structured Activities and for Post-incident Analysis. Rep. Riso-M-2513. Riso National Laboratory, Riso, Denmark.

Rasmussen, J., 1986. Information Processing and Human−Machine Interaction. Elsevier, Amsterdam.

Safety and Reliability Directorate, 1985. Guide to reducing human error in process operation: short version, UK atomic energy authority, SR347, February 1985.

Schmidt, F., Ones, D., Hunter, J., 1992. Personnel selection. Ann. Rev. Psychol. 43 (1), 627−670.

Schmidt, F.H., 1965. On the rise of hot plumes in the atmosphere. Int. J. Air Water Pollut. 9, 175.

Seaver, D.A., Sitwell, W.G., 1983. Procedures for Using Expert Judgement to Estimate Human Error Probabilistics in Nuclear Power Plant Operations. Rep. NUREG/CR-2743. Nuclear Regulatory Commission, Washington, DC.

Semonelli, C.T., 1990. Secondary containment of underground storage tanks. Chem. Eng. Prog. 86 (6), 78.

Stewart, R.M., 1971. High integrity protective systems. Major Loss Prev. 99.

Swain, A.D., 1972. Design Techniques for Improving Human Performance in Production. Industrial & Commercial Techniques Ltd, London.

Swain, A.D., Guttmann, H.E., 1983. Handbook of human-reliability analysis with emphasis on nuclear power plant applications. Final report (No. NUREG/CR-1278; SAND-80-0200). Sandia National Labs, Albuquerque, NM (USA).

Taylor, A., 1979. Comparison of predicted and actual hazard rates. Second National Reliability Conference, paper 4B/1.

Tzafestas, S., 2010. Human factors in automation (II): psychological, physical strength, human error and human values factors. Human Nat. Minding Autom. 1, 35−46.

US Chemical Safety and Hazard Investigation Board, 2005. Investigation report on refinery explosion and fire at BP Texas City, Texas on March 23, 2005. Report No. 2005-04-I-TX.

Van Merriënboer, J.J.G., 1997. Training Complex Cognitive Skills: A Four-Component Instructional Design Model for Technical Training. Educational Technology Publications, Englewood Cliffs, NJ.

Whalley, S., 1987. What can cause human error? Chem. Eng. (London). 433, 37.

White, A.T., 1984. Graphs, Groups and Surfaces. revised ed. North Holland, Amsterdam.

Williams, J.C., 1986. HEART—a proposed method for assessing and reducing human error. Ninth Advances in Reliability Technology Symposium, NEC, Birmingham, June, AEA Technology, Culcheth, Warrington.

Safety Culture

10.1. INTRODUCTION OF SAFETY CULTURE

The term 'safety culture' arose after its first official use in an initial report (INSAG, 1986) about the Chernobyl accident by International Nuclear Safety Advisory Group (INSAG). It was then defined as '...assembly of characteristics and attitudes in organizations and individuals with established that, as an overriding priority, nuclear plant safety issues receive the attention warranted by their significance' (INSAG, 1991). Many implicit or explicit definitions of safety culture have been given in literature (Cox and Cox, 1991; International Safety Advisory Group, 1991; Pidgon, 1991; Ostrom et al., 1993; Geller, 1994; Berends, 1996). The definition given by the Advisory Committee on the Safety of Nuclear Installations (ACSNI) (1993) and adopted by HSE is the most explicit:

The safety culture of an organization is the product of individual and group values, attitudes, perceptions, competencies and patterns of behavior that determine the commitment to, and the style and proficiency of, an organization's health and safety management.

Besides the definition of safety culture, it is also important to distinguish two concepts, 'safety climate' and 'safety culture'. Based on the work done by Cooper (2000), the term safety culture can be used to refer to the behavioral aspects (i.e., 'what people do') and the situational aspects of the company (i.e., 'what the organization has'). The term safety climate should be used to refer to psychological characteristics of employees (i.e., 'how people feel'), corresponding to the values, attitudes, and perceptions of employees with regard to safety within an organization. Culture can be viewed as the background

influence on the organization, while climate is the foreground (BST, 2004). Therefore, safety climate changes more quickly and more readily than safety culture (Olive et al., 2006).

10.2. DEVELOPMENTS IN SAFETY CULTURE

10.2.1 HSE Safety Culture Model

The safety model developed by the HSE is demonstrated as an example of where to start in the long process of developing a world class safety culture.

Five organizational levels, as shown in Figure 10.1, included in the HSE model, are as follows.

In the first level, safety is not a very important priority. Only the minimum effort is put into safety, and the company only wishes to avoid penalties from failing to meet regulations.

In the second level, safety is more important; however, the focus of management and the company, in regard to safety, is reactive rather than preventative. The majority of safety problems are believed by management to be the fault of employees who do not strictly follow company rules, procedures, and engineering controls.

In the third level, the accident rate is relatively low but has reached an area of little variation. The management of the company realizes that not all accidents are a result of the frontline employees making mistakes. The frontline employees must be involved for future improvement of safety.

In the fourth level, almost every member of the company recognizes the importance of safety. In this level, frontline employees make an effort to be personally

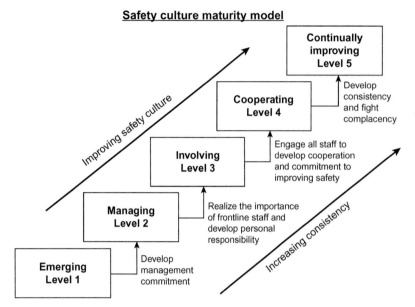

Safety culture maturity model

FIGURE 10.1 HSE SCMM. *Source: Fleiming (2001). ©The Keil Centre, 1999.*

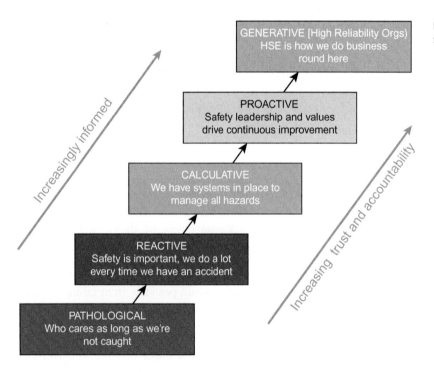

FIGURE 10.2 Hearts and minds safety culture mode. *Source: Energy Institute, Shell E&P.*

responsible for the safety of the company and themselves in order to improve safety further. To make further improvements, the company realizes that employees and management must work together.

In the fifth level, the company has matured to such a degree, that there is a zero incident goal. A company at this level of safety culture maturity has had no major accidents in years. Performance and non-work related injuries are still monitored, and a great effort is invested in maintaining employee's health and safety away from work.

10.2.2 Shell's Heart and Mind Safety Culture Model

The safety culture model of Shell's Hearts and Minds Program is shown in Figure 10.2. The five organizational categories are briefly summarized as follows (Hudson et al., 2002).

In the pathological category, safety issues are blamed on workers by management, who are often afraid of punishment for failure to follow regulations. Management in

FIGURE 10.3 DuPont 22 elements of safety. *Source: Hewitt (2008).*

this category does not care about safety at all unless they are caught.

In the reactive category, safety is taken more seriously, but safety precautions are mainly taken to prevent a repeat incident. Safety is important to management, but it is still at a crude level of competence since it only becomes important after an incident.

In the calculative category, safety is data driven. Quantitative risk assessment techniques and overt cost−benefit analyses are blindly followed. Management uses these tools to justify safety and to make decisions. The organization forces safety upon itself, but the individual members of the company are often not yet committed to safety.

In the proactive category, safety is taken seriously by both management and frontline employees. The organization realizes that with ever increasing performance, the unexpected incidents are the challenge to future safety improvement.

In the generative category, the safety culture of an organization has improved even further. Safety is integrated into the very business of the company. The organization is characterized by a nearly constant fear of the next big incident and is always trying to reduce potential risks.

10.2.3 DuPont's Safety Culture Model

The DuPont model contains 22 elements (Figure 10.3) that are necessary in order to achieve both safety leadership and operational excellence. The DuPont Bradley Curve is shown in Figure 10.4.

The following is a description of the four major categories of the curve as it approaches a zero incidence of injuries within a company (Holstvoogd and Van der Graaf, 2006).

In the reactive category, there is a lack of focus in the area of safety. Management is rarely involved, with responsibility for safety resting on a Safety Manager. Companies at this stage are indeed reactive to problems instead of proactive.

In the dependent category, there is a small increase in management commitment. Safety becomes an issue of employment, with a focus on rules and procedures. Companies in this stage will begin to provide safety training.

In the independent category, personal knowledge and commitment to standards is stressed. Safety management is internalized by the company with an emphasis on personal value. For improvement or achievements in safety,

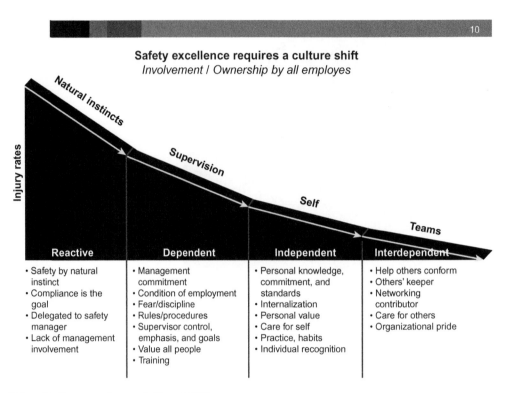

FIGURE 10.4 DuPont Bradley curve. *Source: Griffioen (2008).*

there is often individual recognition within the company to set a standard for safety excellence.

In the interdependent category, companies begin to help others conform to safety initiatives. The company will become a major safety network contributor out of genuine care for others. A company at the interdependent level has great organizational pride.

10.3. EVALUATING AND IMPLEMENTING SAFETY CULTURE

Good safety culture is of several characteristics, which are easily to be observed (Olive et al., 2006):

- A commitment to the improvement of safety behaviors and attitudes at all organizational levels;
- An organizational structure and atmosphere that promotes open and clear communication where people feel free from intimidation or retribution in raising issues and are encouraged to ask questions;
- A propensity for resilience and flexibility to adapt effectively and safety to new situations; a prevailing attitude of constant vigilance.

There are many ways to evaluate if a company has a poor, good, or world class safety culture. The manifestation of safety culture can be observed through the analysis

of leading and lagging indicators. Many companies use performance indicators to evaluate their current level of safety culture, for example, the lagging indicator of fatalities per annum for Shell Oil and Gas and British Petroleum (BP). A good safety culture will be reflected through a reduction in occupational injuries and chemical incidents, increase in profits from a lack of incidents and lawsuits, and compliance with codes, standards, and best practices. A poor safety culture will be reflected in lagging indicators such as an increase in occupational injuries.

Methods such as interviews, audits, and questionnaires can be used to aid in implementing a safety culture within an organization. However, it is important to note that changing a safety culture requires changing the behavior and mentality of every person related to a company, which is not an easy or quick fix (Step Change, 2002). Specific safety culture improvement tools, with their appropriate safety culture level, are shown in Figure 10.5. These tools are matched with the current level of safety culture maturity in which they are most effective.

For multi-national companies, in implementing a good safety culture (Hudson, 2007), they should make sure the program follows the strategy by avoiding facilitators cascading their skills more than one level further down. They should make the tools fit the local environment as well. It is not just translating into local languages to penetrate effectively, but also taking care of the differences

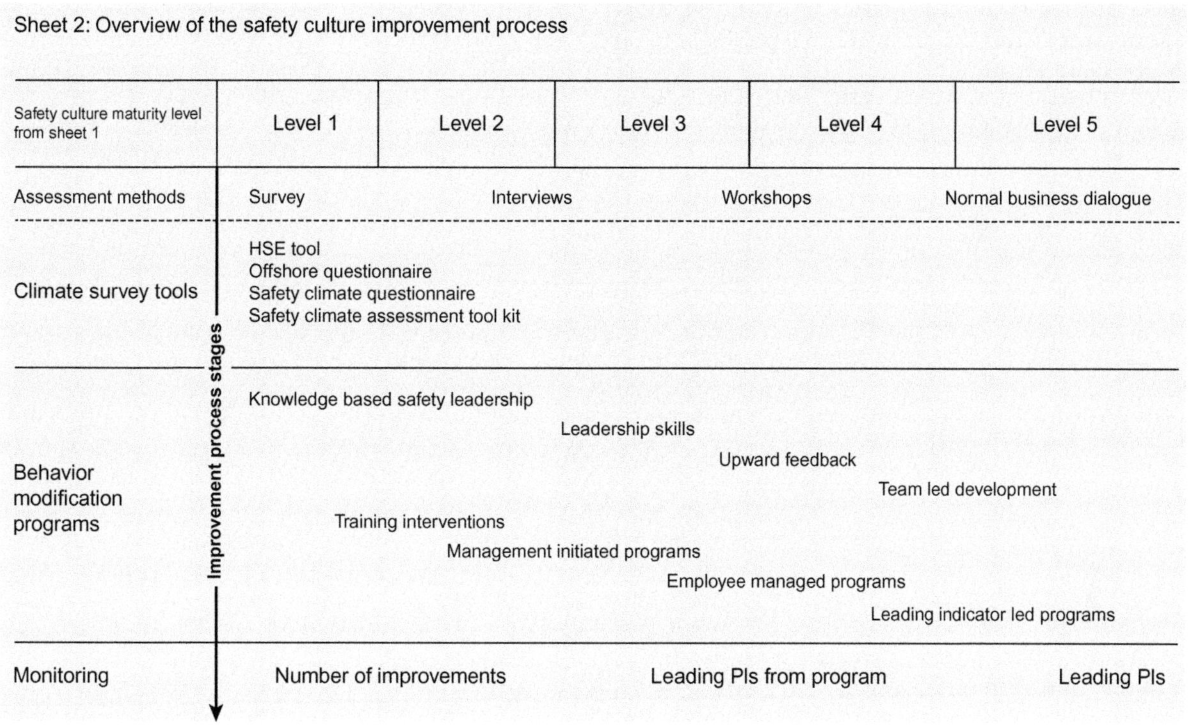

Sheet 2: Overview of the safety culture improvement process

Safety culture maturity level from sheet 1	Level 1	Level 2	Level 3	Level 4	Level 5
Assessment methods	Survey	Interviews		Workshops	Normal business dialogue
Climate survey tools	HSE tool Offshore questionnaire Safety climate questionnaire Safety climate assessment tool kit				
Behavior modification programs	Knowledge based safety leadership Leadership skills Upward feedback Team led development Training interventions Management initiated programs Employee managed programs Leading indicator led programs				
Monitoring	Number of improvements		Leading PIs from program		Leading PIs

(vertical label: Improvement process stages)

FIGURE 10.5   Step Change overview of the safety culture improvement process. *Source: Step Change (2002).*

between collectivist cultures and individualist societies. It is also important to realize that both top-down and bottom-up approaches are necessary and to treat cognitive dissonance as a mechanism to induce change.

10.4. CONCLUSION

In order to improve the safety culture of a company, the most important aspect is proper assessment of the current safety culture. Choosing a proper model is necessary. After a model has been chosen, it is necessary to evaluate the current level of safety within the company through leading and lagging indicators. The next step in assessing the current culture is through interviews, audits, and questionnaires. Development of a plan to improve the safety culture will follow, based on this company and organization specific data. Progress in improving the culture can then be assessed through periodic/annual questionnaires.

REFERENCES

Advisory Committee on the Safety of Nuclear Installations (ACSNI), 1993. Study Group on Human Factors. Third Report: Organising for Safety. HM Stationery Office, London.

Behavioral Science Technology (BST), 2004. Assessment and Plan for Organizational Culture Change at NASA. BST, Ojai, CA.

Berends, J.J., 1996. On the Measurement of Safety Culture. Unpublished Graduation Report). Eindhoven University of Technology, Eindhoven.

Cox, S., Cox, T., 1991. The structure of employee attitudes to safety: an European example. Work Stress. 5 (2), 93–106.

Cooper, M.D., 2000. Towards a model of safety culture. Saf. Sci. 36 (2), 111–136.

Fleming, M., 2001. Safety Culture Maturity Model. Health and Safety Executive, London.

Geller, E.S., 1994. Ten principles for achieving a total safety culture. Prof. Saf. 39 (9), 18–24.

Griffioen, G., 2008, Oct 24. Culture Change. Paper presented at the Mine Metallurgical Managers' Association of South Africa, Johannesburg.

Hewitt, M., 2008. Strategies for Creating and Sustaining a Culture of Safety. Wilmington, DE, DuPont.

Holstvoogd, R., Van der Graaf, G., 2006. Hearts and Minds Programmes the Road Map to Improved HSE Culture. Institute of Chemical Engineers, London.

Hudson, P., 2007. Implementing a safety culture in a major multinational. Saf. Sci.

Hudson, P., Parker, D., van der Graaf, G.C., et al., 2002. The Hearts and Minds Program: Understanding HSE Culture. SPE International Conference on Health, Safety and Environment in Oil and Gas Exploration and Production. Society of Petroleum Engineers Inc., Kuala Lumpur, Malaysia, pp. 20–22.

International Nuclear Safety Advisory Group (INSAG), 1986. Summary Report on the Post Accident Review Meeting on the Chernobyl Accident, INSAG-1, International Atomic Energy Agency, Vienna.

International Nuclear Safety Advisory Group (INSAG), 1991. Safety Culture, Safety Series No. 75-INSAG-4, International Atomic Energy Agency, Vienna.

Olive, C., O'Connor, T.M., Mannan, M.S., 2006. Relationship of safety culture and process safety. J. Hazard. Mater. 130 (1-2), 133–140.

Ostrom, L., Wilhelmsen, C., Kaplan, B., 1993. Assessing safety culture. Nucl. Saf. 34 (2), 163–172.

Pidgeon, N.F., 1991. Safety culture and risk management in organizations. J. Cross Cult. Psychol. 22 (1), 129–140.

Step Change, 2002. Changing Minds: A Practical Guide for Behavioural Change in the Oil and Gas Industry. second ed. BIOCC.

Chapter 11

Emission, Dispersion, and Toxic Release

Chapter Outline

The three major hazards—fire, explosion, and toxic release—usually involve the emission of material from containment followed by vaporization and dispersion of the material. The treatment given here is relevant to all three hazards; in particular, the development of the emission and dispersion phases is relevant to situations such as:

1. Escape of flammable material, mixing of the material with air, formation of a flammable cloud, drifting of the cloud, and finding of a source of ignition, leading to (a) a fire and/or (b) a vapor cloud explosion, affecting the site and possibly populated areas.
2. Escape of toxic material, formation of a toxic gas cloud, and drifting of the cloud affecting the site and possibly populated areas.

11.1. EMISSION

There is a wide range of circumstances which can give rise to emission. The situation which appears to be most frequently discussed is a failure of plant integrity, but it is important to consider other occurrences, including escape from valves which have been deliberately opened and forced venting in emergencies. Thus, on a low-pressure refrigerated liquefied petroleum gas (LPG) installation, for example, a large quantity of flammable vapor may be released by failure of a pressure vessel or of pipework. But loss of refrigeration with resultant forced venting could also give a large release of vapor. Finally, emissions caused by terrorist acts are of growing concern.

In view of the many different situations which can give rise to emission, it is peculiarly difficult to obtain meaningful estimates of the quantity and duration of emissions. The emission stage is therefore subject to great uncertainty. It is nevertheless very important, because the way in which emission occurs can greatly influence the nature and effect of the release, and particularly of any vapor cloud which is formed.

Emission flows may be determined by the basic relations of fluid mechanics. Fundamentals of fluid flow are treated in such texts as *Hydraulics* (Lewitt, 1952), *Thermodynamics Applied to Heat Engines* (Lewitt, 1953),

The Dynamics and Thermodynamics of Compressible Flow (Shapiro, 1953), *Chemical Engineering* (Coulson and Richardson, 1955, 1977), *Mechanics of Fluids* (Massey, 1968), and *Chemical Engineers Handbook* (Perry and Green, 1984).

11.1.1 Emission Situations

Emission situations may be classified as follows:

1. Fluid
   a. gas/vapor,
   b. liquid,
   c. vapor—liquid mixture;
2. Plant
   a. vessel,
   b. other equipment,
   c. pipework;
3. Aperture
   a. complete rupture,
   b. limited aperture;
2. Enclosure
   a. in building,
   b. in open air;
3. Height
   a. below ground level,
   b. at ground level,
   c. above ground level;
4. Fluid momentum
   a. low momentum,
   b. high momentum.

Some typical emission situations are shown in Figure 11.1.

The fluid released may be gas, vapor, liquid, or a two-phase vapor—liquid mixture, as shown in Figure 11.1(a). If the escape is from a container holding liquid under pressure, it will normally be liquid if the aperture is below the liquid level and vapor or vapor—liquid mixture if it is above the liquid level. For a given pressure difference, the mass of release is usually much greater for a liquid or vapor—liquid mixture than for a gas or vapor.

The plant in which the release occurs may be a vessel, other equipment such as a heat exchanger or a pump, or pipework, as shown in Figure 11.1(b). The maximum amount which can escape depends on the inventory of the container and on the isolation arrangements.

The aperture through which the release occurs may range from a large fraction of the envelope of the container in the case of complete rupture of a vessel to a limited aperture such as a hole, as shown in Figure 11.1(c). An aperture on a vessel may be (1) a sharp-edged orifice; (2) conventional pipe branch; (3) a rounded nozzle branch; or (4) a crack. The flow through a rounded nozzle is greater than through a conventional pipe branch, but it

is the latter which is generally used. Other apertures on vessels, equipment, and pipework include (5) drain and sample points; (6) pressure relief devices—(a) pressure relief valves, (b) bursting disks, and (c) liquid relief valves; (7) seals; (8) flanges; and (9) pipe ends.

The relief may take place from the plant in a building or in the open air, as shown in Figure 11.1(d). This greatly affects the dispersion of the material. A large proportion of escapes to the open air are dispersed without incident.

The height at which the release occurs also has a strong influence on the dispersion, as shown in Figure 11.1 (e). A liquid escape from plant below ground level may be completely contained. An escape of gas or vapor from above ground level may be dispersed over a considerable distance.

Dispersion is further affected by the momentum of the escaping fluid, as shown in Figure 11.1(f). Low and high momentum releases of gas or vapor form a plume and a turbulent momentum jet, respectively. Low and high momentum releases of liquid form a liquid stream and a high 'throw' liquid jet, respectively, both of which then form a liquid pool.

A generalized chart for the calculation of the discharge of liquids and gases under pressure has been given by Pilz and van Herck (1976) and is reproduced in Figure 11.2.

11.1.2 Elementary Relations

For flow of a single fluid

$$W = Q_p \tag{11.1}$$

$$Q = uA \tag{11.2}$$

$$G = \frac{W}{A} \tag{11.3}$$

$$G = u_p \tag{11.4a}$$

$$G = \frac{u}{\nu} \tag{11.4b}$$

where A is the cross-sectional area of the pipe, G is the mass flow per unit area, Q is the total volumetric flow, u is the velocity, v is the specific volume of fluid, W is the total mass flow, and ρ is the density of the fluid.

The general differential form of the energy balance for the flow of unit mass of fluid is

$$-dq + dW_s + dH + d\left(\frac{u^2}{2}\right) + g\,dz = 0 \tag{11.5}$$

where g is the acceleration due to gravity, H is the enthalpy, q is the heat absorbed from the surroundings, W_s is the work done on the surroundings (shaft work), and z is the height.

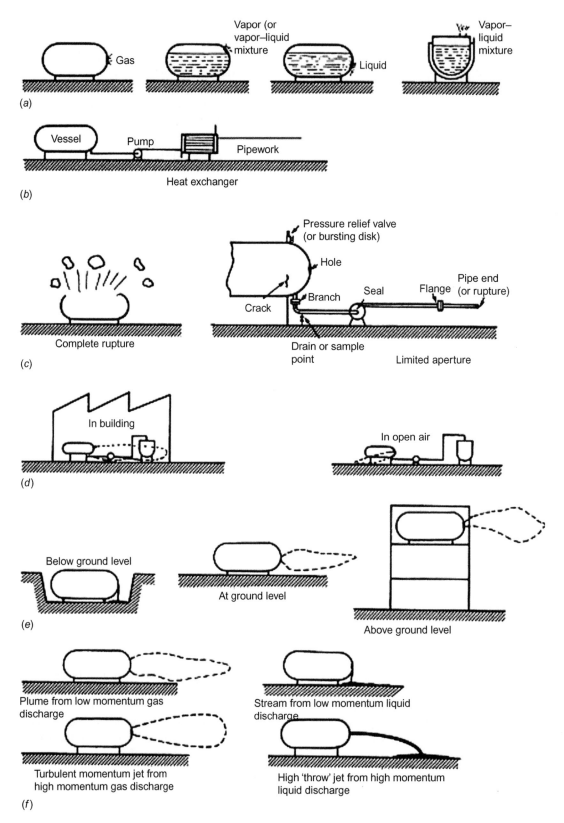

FIGURE 11.1 Some emission situations.

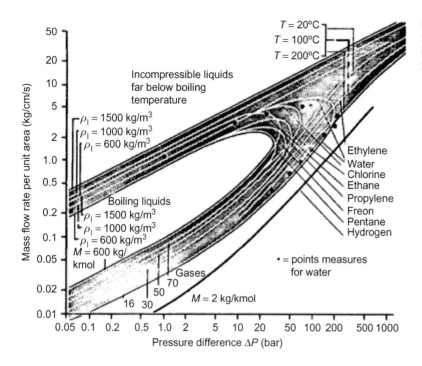

FIGURE 11.2 Mass flow vs pressure difference for flow of gases, vapors, and liquids through an orifice. Source: Pitz and van Herck (1976); reproduced by permission.

It is often convenient to eliminate H from Equation 11.5 using the relation

$$dH = T\, dS + \nu\, dp = dq + dF + \nu\, dP \qquad (11.6)$$

where F is the mechanical energy irreversibly converted to heat (frictional loss) and P is the absolute pressure.

Then, combining Equations 11.5 and 11.6 gives

$$dW_s + \nu\, dP + dF + d\left(\frac{u^2}{2}\right) + g\, dz = 0 \qquad (11.7)$$

For the discharge processes considered here, the work done on the surroundings W_s is zero. Then Equation 11.7 becomes

$$\nu\, dP + dF + d\left(\frac{u^2}{2}\right) + g\, dz + 0 \qquad (11.8a)$$

or, alternatively,

$$\nu\, dP + dF + u\, du + g\, dz = 0 \qquad (11.8b)$$

In some applications, the friction loss and potential energy terms may be neglected, in which case Equation 11.8b reduces to

$$\nu\, dP + u\, du = 0 \qquad (11.9)$$

Equation 11.8b may also be expressed in the form of a pressure drop relation

$$\frac{dP}{dl} + \frac{1}{\nu}\frac{dF}{dl} + G\frac{du}{dl} + \frac{g}{\nu}\frac{dz}{dl} = 0 \qquad (11.10a)$$

or

$$\frac{dP}{dl} + \rho\frac{dF}{dl} + G\frac{du}{dl} + \rho g\frac{dz}{dl} = 0 \qquad (11.10b)$$

or

$$\frac{dP}{dl} + \frac{1}{\nu}\frac{dF}{dl} + \frac{u}{\nu}\frac{du}{dl} + \frac{g}{\nu}\frac{dz}{dl} = 0 \qquad (11.10c)$$

where l is the distance along the pipe.

A relation for pressure drop may also be obtained from the momentum balance

$$\frac{dP}{dl} + \frac{S}{A}R_0 + G\frac{dz}{dl} + \rho g\frac{dz}{dl} = 0 \qquad (11.11)$$

where R_0 is the shear stress at the pipe wall and S is the perimeter of pipe. The term dz/dl in Equations 11.10 and 11.11 is often replaced by the term $\sin\theta$, where θ is the angle between pipe and horizontal.

Comparing Equations 11.10b and 11.11, which each give the pressure drop, it follows that

$$\frac{S}{A}R_0 = \rho\frac{dF}{dl} \qquad (11.12)$$

The pressure drop dP/dl may also be expressed in terms of its constituent elements

$$-\frac{dP}{dl} = -\frac{dP_f}{dl} - \frac{dP_a}{dl} - \frac{dP_g}{dl} \qquad (11.13)$$

where dP_f/dl is the frictional pressure change, dP_a/dl is the acceleration pressure change, and dP_g/dl is the gravitational pressure change.

For flow in pipes, the friction term dF is

$$dF = 4\phi \frac{dl}{d} u^2 \qquad (11.14a)$$

$$= 8\phi \frac{dl}{d} \frac{u^2}{2} \qquad (11.14b)$$

with

$$\phi = \frac{R}{\rho u^2} \qquad (11.15)$$

where d is the diameter of the pipe, R is the shear stress at the pipe wall, and ϕ is the friction factor.

The most commonly used friction factor f is twice the friction factor ϕ

$$f = 2\phi \qquad (11.16)$$

Hence from Equations 11.14b and 11.16,

$$dF = 4f \frac{dl}{d} \frac{u^2}{2} \qquad (11.17)$$

Another form of the pressure drop equation which is derived from Equation 11.10b together with Equations 11.14 and 11.17 is

$$\frac{dP}{dl} + G^2\left(\frac{d\nu}{dl} + \frac{4f\nu}{2d}\right) + \frac{g\,dz}{\nu\,dl} = 0 \qquad (11.18)$$

The equations just given are in differential form. In order to obtain relations useful in engineering, they need to be integrated. Integrating the energy balance in Equation 11.8a,

$$\int_1^2 \nu\,dP + F + g\,\Delta z + \Delta\left(\frac{u^2}{2}\right) = 0 \qquad (11.19)$$

where limit 1 is the initial state and limit 2 is the final state.

Integration of the first term in Equation 11.19 requires a relation between pressure P and specific volume v. Three principal relations are those for incompressible, isothermal, and isentropic flow

$$\nu = \text{Constant incompressible flow} \qquad (11.20a)$$

$$P\nu = \text{Constant isothermal flow} \qquad (11.20b)$$

$$P\gamma = \text{Constant isentropic flow} \qquad (11.20c)$$

where γ is the ratio of the gas specific heats. Also, for an adiabatic expansion,

$$P_k = \text{Constant adiabatic flow} \qquad (11.20d)$$

where k is the expansion index.

Then it can be shown that for these flow regimes,

$$\int_1^2 \nu\,dP = \nu(P_2 - P_1)\text{Incompressible flow} \qquad (11.21a)$$

$$= P_1\nu_1 \ln\left(\frac{P_2}{P_1}\right)\text{Isothermal flow} \qquad (11.21b)$$

$$= \frac{\gamma}{\gamma-1}P_1\nu_1\left[\left(\frac{P_2}{P_1}\right)^{(\gamma-1)/\gamma} - 1\right]\text{Isentropic flow} \qquad (11.21c)$$

$$= \frac{k}{k-1}p_1\nu 1\left[\left(\frac{P_2}{P_1}\right)^{(k-1)/k} - 1\right]\text{Adiabatic flow} \qquad (11.21d)$$

It is frequently convenient to work in terms of the head h rather than the pressure P of the fluid. The relation between the two is

$$p = bpg \qquad (11.22)$$

It is generally necessary to define the flow regime. The principal criterion of similarity is the Reynolds number Re which is

$$Re = \frac{Gd}{u} \qquad (11.23)$$

where d is the diameter of the orifice or pipe and μ is the viscosity of the fluid.

The discharge of liquids and of gases and vapors is now considered.

11.2. TWO-PHASE FLOW

In some cases, the fluid is neither a pure gas (or vapor) nor a pure liquid, but is a vapor–liquid mixture, and it is then necessary to use a correlation for two-phase flow.

The behavior of fluids in two-phase flow is complex and by no means understood. Of the large literature on the topic mentioned may be made of the accounts in *One Dimensional Two-Phase Flow* (Wallis, 1969), *Two-Phase Flow and Heat Transfer* (Butterworth and Hewitt, 1977), *Two-Phase Flow in Pipelines and Heat Exchangers* (Chisholm, 1983), and *Emergency Relief Systems for Runaway Reactions and Storage Vessels: A Summary of Multiphase Flow Methods* (AIChE, 1992/149), and of the work of Benjaminsen and Miller (1941), Burnell (1947), Lockhart and Martinelli (1949), Pasqua (1953), Schweppe and Foust (1953), O. Baker (1954, 1958), Isbin et al. (1957), Zaloudek (1961), Fauske (1962, 1963, 1964), Isbin et al. (1962), Fauske and Min (1963), Dukler et al. (1964), Levy (1965), F.J. Moody (1965), Baroczy (1966), Chisholm and Watson (1966), Chisholm and Sutherland (1969–1970), R.E. Henry and Fauske (1971), M.R.O. Jones and Underwood (1983, 1984), van den Akker et al. (1983), Nyren and Winter (1983), B. Fletcher (1984a,b), B. Fletcher and Johnson (1984), Leung (1986, 1990a,b, 1992), and S.D. Morris (1988a,b, 1990a,b).

Work on two-phase flow, particularly in pipelines, has been carried out by the American Institute of Chemical Engineers (AIChE) and Design Institute for Multiphase Processing (DIMP). An important distinction in two-phase flow is that between one-component systems, such as steam and water, and two-component systems, such as air and water. It is the former which is of prime concern here and, unless otherwise stated, it is to these systems that the account given here refers. Two situations which are of particular interest in the present context are the escape of a superheated, flashing liquid, and the venting of a chemical reactor.

11.2.1 Modeling of Two-Phase Flow

Two-phase flow is a complex phenomenon and has given rise to a large number of models. Of particular interest here are models for critical two-phase flow. A review of such models has been given by Wallis (1980). Wallis distinguishes essentially three types of models for critical two-phase flow:

1. Homogeneous equilibrium model
2. Models incorporating limiting assumptions
   a. frozen flow models,
   b. slip flow models,
   c. isentropic stream tube models;
3. Equilibrium models
   a. empirical models,
   b. physically based models for thermal equilibrium,
   c. two fluid models.

The homogeneous equilibrium model (HEM) is based on the two limiting assumptions that the flow is homogeneous, that is, there is no slip between the two phases; and that the flow is in equilibrium, that is, there is effectively perfect transfer between the two phases.

This is a well-established and widely used model; it is used in a number of computer codes. Wallis states that the model gives reasonably good predictions for the mass velocity under conditions where the pipe is long enough for equilibrium to be established and the flow pattern is such as to suppress relative motion. For short pipes where there is insufficient relaxation time, the flow can be in error by a factor of about 5. For longer pipes with a flow pattern which allows large relative velocity, such as annular flow, the error factor tends to be somewhat less than 2.

The homogeneous equilibrium model is based on limiting assumptions. The second group of models is based on alternative limiting assumptions. The limiting assumption made in the frozen flow model is that there is no transfer between phases so that flow is effectively frozen and quality remains constant. This assumption is most nearly met if the outlet is short. While no interphase

transfer is the key assumption in this model, it may be combined with other assumptions such as those of no slip or isentropic expansion.

The limiting assumption made in the slip equilibrium model is that there is equilibrium between the phases, which is the opposite of the previous case. Two principal slip equilibrium models are those by Fauske (1962, 1963) and F.J. Moody (1965). These two models differ in terms of the way in which the exit condition is determined. In the Moody model, the exit conditions are determined by an energy balance and the maximum flow occurs at a slip ratio $k = (\rho_1/\rho_g)^{1/3}$, while in the Fauske model the exit conditions are determined by a momentum balance and the maximum flow occurs at $k = (\rho_1/\rho_g)^{1/3}$.

The slip equilibrium models just described treat the velocity ratio k as a parameter to be adjusted to obtain the maximum flow, without indicating how this condition is actually achieved. Wallis and Richter (1978) have described an isentropic stream tube model, based on isentropic expansion of individual stream tubes originating from the vapor–liquid interface, which attempts to address this point.

The third group of models incorporates non-equilibrium effects. One approach used to handle such effects is the use of empirical correction factors. This is the basis of the model by R.E. Henry and Fauske (1971).

Other non-equilibrium models are based on physical descriptions of the processes of bubble nucleation and vapor generation. Wallis concludes that the development of such models is not such as to offer much improvement on the purely empirical models.

Another type of non-equilibrium model is the two fluid, or separated flow, model, in which effects such as interphase drag are taken into account, but again Wallis concludes that the stage of development is not such as to make these models more useful than the empirical ones.

11.2.2 Elementary Relations

For a two-phase system, the following equations apply for continuity and other relations:

$$W_g = Wx \tag{11.24}$$

$$W_1 = W(1 - x) \tag{11.25}$$

$$G_g = u_g \rho_g \tag{11.26}$$

$$G_g = \frac{Gx}{\alpha} \tag{11.27}$$

$$G_1 = \frac{G(1 - x)}{1 - \alpha} \tag{11.28}$$

$$Q_g = \frac{W_x}{\rho_g} \tag{11.29}$$

$$Q_1 = \frac{W(1-x)}{\rho_1} \qquad (11.30)$$

$$u_g = \frac{Gx}{\alpha \rho_g} \qquad (11.31)$$

$$u_1 = \frac{G(1-x)}{(1-\alpha)\rho_1} \qquad (11.32)$$

$$k = \frac{u_g}{ul} \qquad (11.33)$$

$$k = \frac{x}{1-x}\frac{1-\alpha}{\alpha}\frac{\rho l}{\rho_g} \qquad (11.34)$$

$$\alpha = \left(1 + \frac{1-x}{x}k\frac{\rho g}{\rho l}\right) \qquad (11.35a)$$

$$\alpha = \left(1 + \frac{1-x}{x}k\frac{vl}{vg}\right)^{-1} \qquad (11.35b)$$

$$\beta = \frac{Q_g}{Q_1 + Q_g} \qquad (11.36)$$

$$\beta = \left(1 + \frac{1-x}{x}\frac{\rho_g}{\rho_1}\right)^{-1} \qquad (11.37)$$

where Q is the volumetric flow, v is the specific volume, W is the mass flow, ρ is the fraction of volume occupied by vapor, or void fraction, and β is the vapor phase volumetric flow fraction.

The pressure drop derived from the energy balance for two-phase flow, analogous to Equation 11.10b for single phase flow, is

$$\frac{dP}{dl} + \rho_h\frac{dF}{dl} + G\frac{du}{dl} + \rho_h g\frac{dz}{dl} = 0 \qquad (11.38)$$

with

$$\rho_h = \left[\frac{x}{\rho_g} + \frac{1-x}{\rho_1}\right]^{-1} \qquad (11.39)$$

where F is the friction loss, g is the acceleration due to gravity, l is the distance along the pipe, P is the absolute pressure, z is the vertical distance, and subscript h denotes homogeneous flow.

The pressure drop derived from the momentum balance for two-phase flow, analogous to Equation 11.11 for single phase flow, may be written as

$$\frac{dP}{dl} + \frac{S}{A}R_0 + \frac{d[G_g\alpha u_g + G_1(1-\alpha)u_1]}{dl} \qquad (11.40)$$
$$+ g\left[\alpha\rho_g + (1-\alpha)\rho_1\right]\frac{dz}{dl} = 0$$

where A is the cross-sectional area of the pipe, R_0 is the shear stress at the pipe wall, and S is the perimeter of the pipe.

Comparing Equations 11.37 and 11.39, each of which gives the pressure drop, it follows that, in a manner analogous to the derivation of Equation 11.17,

$$\frac{S}{A}R_0 = \rho_h\frac{dF}{dl} \qquad (11.41)$$

As with Equation 11.14a,b for single-phase flow, the pressure drop given in Equation 11.37 may also be expressed in terms of its constituent elements, as given in Equation 11.13.

11.3. VESSEL DEPRESSURIZATION

Some of the most complex fluid flow problems arise in the depressurization or venting of vessels, particularly reactors and storage vessels. In order to specify the venting arrangements for a vessel, it is necessary to determine the flow, the phase condition of fluid, and the vent area required for this flow. The venting of vessels is complicated by the fact that in many situations the flow is likely to be two-phase. It is necessary, therefore, to be able to estimate both the vapor mass fraction, or quality, of the fluid entering the vent and the flow through the vent.

The need to determine the quality of the fluid vented has been highlighted during the work of the DIERS project. The quality of the fluid entering the vent is determined by liquid swell and vapor disengagement. There are two principal regimes that are recognized as occurring when a vessel is depressurized. If the liquid is non-foaming, the regime tends to be churn turbulent, whereas if it is foaming the regime is bubbly.

For a non-foaming liquid, methods have been developed to predict the onset of two-phase flow. In the depressurization of a vessel containing such a liquid, there will, in general, be a region in which the flow is all vapor and another region in which it is two-phase. A large initial vapor space, or freeboard, will favor all vapor flow. For a foaming liquid, it is necessary to assume two-phase flow. Broadly, a pure liquid held in a storage vessel may well be non-foaming. It is to this situation that the methods developed principally apply.

An otherwise non-foaming liquid may be rendered foaming by the presence of impurities. Quite small quantities of a surface active agent may suffice to effect this. Since impurities are likely to be present in the liquid in a reactor, it is usual to treat a reaction mass as foaming.

In this section, an account is given of models of liquid swell and vapor disengagement. Models have been given in the DIERS Technical Summary and by Fauske et al. (1983), Swift (1984), and Grolmes and Epstein (1985). The application of these models to venting, particularly of storage vessels, is described in Chapter 13.

11.4. PRESSURE RELIEF VALVES

As components of a pressure system, pressure relief valves have been treated in Chapter 8. In this section, some further consideration is given in respect of the flow through such devices. Dispersion of the discharge is considered in Section 11.20 and for dense gases in Section 11.43.

11.4.1 Single-Phase Flow

Relations for single-phase flow of gases and liquids have been presented in Section 11.1. Formulae derived from these are given in the codes for pressure relief. The API and BS formulae are stated in Chapter 8.

11.5. VESSEL RUPTURE

In certain circumstances, a vessel may rupture completely. If vessel rupture occurs, a large vapor cloud can be formed very rapidly. Accounts of vessel rupture include those by Hardee and Lee (1974, 1975), Hess et al. (1974), J.D. Reed (1974), Maurer et al. (1977), A.F. Roberts (1981/1982), and B. Fletcher (1982).

11.5.1 Vaporization

If a vessel containing a superheated liquid under pressure ruptures, a proportion of the liquid vaporizes. This initial flash fraction is determined by the heat balance, the latent heat of vaporization being supplied by the fall in the sensible heat of the liquid. The rapid formation of vapor bubbles also generates a spray of liquid drops so that typically most or all of the remaining liquid becomes airborne, leaving little or no residue in the vessel.

This effect has been demonstrated by J.D. Reed (1974), who carried out experiments on sudden vessel depressurization. In one series of experiments, 3.5 kg of liquid ammonia contained at an absolute pressure of 3 bar and a temperature of $-9°C$ in a vessel 15 cm diameter and 45 cm high was released using a quick release lid. One of the experiments is shown in Figure 11.3, in which the time interval between the first and last frame is one-sixth of a second. In all the experiments, at least 90% of the liquid ammonia was vaporized.

Similarly, Maurer et al. (1977) have carried out experiments (described below) on the rapid release of propylene held at pressures of 22−39 bar and temperatures of 50−80°C. The flash fraction of vapor was 50−65% and the remaining liquid formed spray.

An investigation of the extent of vapor and spray formation and of retained liquid has been made by B. Fletcher (1982), who carried out experiments, principally in a vertical vessel of 127 mm × 47 mm cross-section, in

FIGURE 11.3 Sudden depressurization of a vessel containing liquid ammonia under pressure (J.D. Reed, 1974) *Source: Courtesy of Elsevier Publishing Company.*

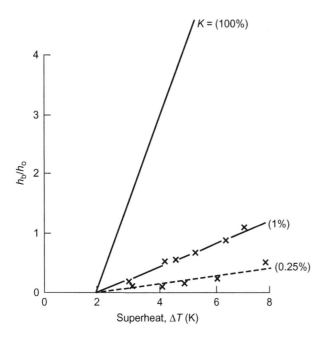

FIGURE 11.4 Sudden depressurization of a vessel containing super-heated liquid: effect of a restriction on the height to which the liquid rises (B. Fletcher, 1982). *Source: Courtesy of the Institution of Chemical Engineers.*

which a charge of superheated Refrigerant 11 was depressurized and the liquid residue was determined. The results are correlated in terms of the ratio (h_b/h_o) of the height reached by the liquid on depressurization h_b to the initial height of the liquid h_o, as shown in Figure 11.4. The parameter in the figure is the ratio K of the vent area to the vessel cross-sectional area. The mass of retained liquid is given by the relation

$$\frac{m_r}{m_o} = \frac{1}{K}\left[\frac{m_v\rho_1}{m_o\rho_v} + \left(1 - \frac{m_v}{m_o}\right)\right]^{-1} \tag{11.42}$$

where m is the mass of fluid, ρ is the density, and subscripts l, o, r, and v denote the liquid, initial, retained, and vapor, respectively. For a full bore release, the value of the constant K is unity.

Further correlation of the results is given in terms of the superheat ΔT_{onb} required for vapor nucleation, and hence nucleate boiling. The expression for superheat is

$$\Delta T \approx \frac{4RT_{sat}\sigma}{\Delta H_v M\delta P_1}\rho_1 \gg \rho_v; \quad 4\sigma/P_1\delta \ll 1 \tag{11.43}$$

where ΔH_v is the latent heat of vaporization, M is the molecular weight, P is the absolute pressure, R is the universal gas constant, T is the absolute temperature, ΔT is the superheat, Δ is the diameter of the vapor nucleus, σ is the surface tension, and subscripts onb and sat denote onset of nucleate boiling and saturation, respectively.

The proportion of retained liquid was expressed in terms of the ratio (h/h_v) of the height h of the liquid

residue to the height h_v of the vessel. The limiting value of the superheat, or value at the onset of nucleate boiling (ΔT_{onb}), corresponds to the situation where all the liquid is retained and thus $h/h_v = 1$.

For Refrigerant 11, the value of ΔT_{onb} is about 1.9 K. Using this value in Equation 11.43 with the appropriate physical value for Refrigerant 11, a value of δ can be obtained. Then, utilizing this value, the superheat ΔT_{onb} for other substances can be obtained.

The correlation given by Fletcher for the superheat effect is shown in Figure 11.5. The proportion of retained liquid h/h_v falls off rapidly with increase in the ratio $\Delta T/\Delta T_{onb}$ and for values in excess of about 12 falls to less than 5%.

A comparison of the proportion of liquid retained in the tank cars in several transport incidents with that estimated from Equation 11.42 is given in Table 11.1.

Further work on vaporization following vessel rupture has been described by Schmidli, Bannerjee, and Yadigaroglu (1990).

11.6. PIPELINE RUPTURE

Another situation that can lead to a large release of gas or vapor is rupture of a pipeline. Pipelines that may give rise to such a release are principally those carrying either high-pressure gas or liquefied gas. In each case, while the determination of the initial emission rate is relatively straightforward, the situation then becomes rather more complex.

11.6.1 Gas Pipelines

Accounts of emission from pipelines containing high-pressure gas have been given by R.P. Bell (1978), D.J. Wilson (1979), Picard and Bishnoi (1988, 1989), and J.R. Chen et al. (1992).

An empirical model for flow from a pipeline rupture has been given by R.P. Bell (1978). This model may be written as

$$m = \frac{m_o}{m_o} + m_r\left[m_o\exp(-t/\mathscr{T}_2) + m_r\exp(-t/\mathscr{T}_1)\right] \tag{11.44}$$

with

$$m_r = A\left(\frac{2P\rho d}{flN}\right)^{1/2} \tag{11.45}$$

$$\mathscr{T}_1 = \frac{W_o}{m_r} \tag{11.46}$$

$$\mathscr{T}_2 = \frac{W_o m_r}{m_o^2} \tag{11.47}$$

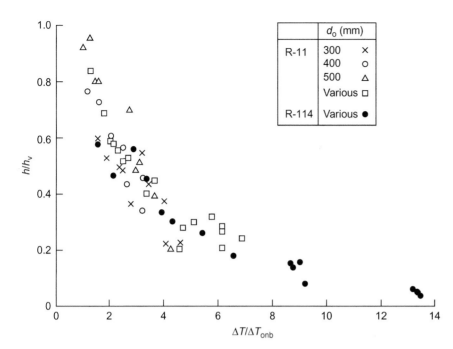

FIGURE 11.5 Sudden depressurization of a vessel containing superheated liquid: depth of liquid remaining (B. Fletcher, 1982). *Source: Courtesy of the Institution of Chemical Engineers.*

TABLE 11.1 Comparison of the Proportion of Liquid Retained in the Tank Cars in Several Transport Incidents

Incident	Liquid	Mass Fraction Retained (%)	
Reported	Estimated		
Pensacola	Ammonia	50	40
Mississauga	Chlorine	10	13
Youngstown	Chlorine	44	37

where A is the cross-sectional area of the pipeline, d is the diameter of the pipeline, f is the friction factor, l is the length of the pipeline, m_o is the initial mass flow from the pipeline, m_r is a steady-state, or reference, flow from the pipeline defined by Equation 11.45, N is a correction factor, P is the absolute pressure, W_o is the initial mass holdup in the pipeline, ρ is the density, and τ_1 and τ_2 are time constants. The friction factor f is evidently the Darcy friction factor f_D ($= 8\phi$).

For the correction factor N, Bell gives the empirical formula

$$N = 8[1 - \exp(-26,400)d/l] \qquad (11.48)$$

He also states that he used a value of 0.02 for the friction factor f.

Bell also discusses the dispersion from the pipeline.

D.J. Wilson (1979) has derived a rather more complex model but also states that Bell's model compares quite well.

A formulation of the Bell model has been given by the CCPS (1987/2) and this may be put in the following form:

$$m = \frac{m_o}{1 + \psi} \left[\exp(-t/\mathscr{T}_2) + \psi \exp(-t/\mathscr{T}_1) \right] \qquad (11.49)$$

with

$$\psi = m_r/m_o \qquad (11.50a)$$

$$\psi = W_o/m_o \mathscr{T}_l \qquad (11.50b)$$

$$\mathscr{T}_1 = W_o/m_r \qquad (11.51)$$

$$\mathscr{T}_2 = \psi^2 \mathscr{T}_1 \qquad (11.52)$$

where ψ is a parameter. Furthermore,

$$\mathscr{T}_1 = 0.67 \left(\frac{\gamma f l}{d} \right)^{1/2} \frac{1}{u_s} \qquad (11.53)$$

with

$$u_s = \left(\frac{\gamma R T}{M} \right)^{1/2} \qquad (11.54)$$

where M is the molecular weight, R is the universal gas constant, T is the absolute temperature, u_s is the velocity of sound, and γ is the ratio of the gas specific heats.

More complex treatments are given by Picard and Bishnoi (1988, 1989) and J.R. Chen et al. (1992).

11.6.2 Liquefied Gas Pipelines

Accounts of emission from pipelines containing liquefied gas are given by Inkofer (1969), Westbrook (1974), and T.B. Morrow et al. (1982).

Inkofer (1969) discusses the factors determining the rate of emission in a liquid ammonia pipeline rupture. He envisages an initial spurt of liquid followed by a period of prolonged and spasmodic ejection of liquid and vapor due to the effect of vapor locks at humps along the pipeline.

Estimates of the rate of emission from a rupture in a chlorine pipeline have been given by Westbrook (1974). These estimates are an initial escape rate of 60.3 ton/h from each of two 4 in orifices and a total escape of 28.4 ton in 24 min.

T.B. Morrow et al. (1982) have given a method of estimating the flow from a pipeline containing liquefied gas. They consider two cases: complete rupture and partial rupture. For the two-phase critical flow at the rupture point, the method utilizes Fauske's slip equilibrium model. Upstream of the rupture point, it is assumed that there is a transition, or interface, point at which the flow changes from liquid flow to two-phase flow and that this point is that at which the pressure corresponds to the bubble point of the liquid. The basic equation for the two-phase pressure drop is

$$\frac{dP}{dz} = \frac{\phi_g^2 4 f u_{fs}}{2 \, d\nu_f} \tag{11.55}$$

where f is the friction factor, u is the velocity, v is the specific volume, z is the distance along the pipe from the rupture point, ϕ_g is a parameter, and subscripts f and fs denote liquid and superficial value for liquid, respectively. The friction factor f is the Fanning friction factor $(= 2\phi)$.

The actual liquid velocity u_f is

$$u_f = \frac{u_{fs}}{1 - y} \tag{11.56}$$

where y is the void fraction. The parameter ϕ_g is

$$\phi_g^2 = \frac{1}{(1-y)^2} \tag{11.57}$$

Then from Equations 11.55–11.57

$$\frac{dp}{dz} = \frac{2 f u_f^2}{d\nu_f} \tag{11.58}$$

Using the same assumption for slip as made by Fauske in the derivation of the slip equilibrium model, namely,

$$\frac{u_g}{u_f} = \left(\frac{\nu_g}{u_f}\right)^{1/2} \tag{11.59}$$

where subscript g denotes vapor. Expressing the quality in terms of the fluid enthalpies, and hence of pressure, the authors obtain

$$\left(\frac{m\nu_f}{A u_f}\right)^2 = f(P) \tag{11.60}$$

Hence

$$\frac{dP}{dz} = \frac{2 f m^2 \nu_f}{A^2 \, df(P)} \tag{11.61}$$

To integrate Equation 11.61, the mass flow m is expressed as a function of the distance z from the rupture point

$$m = m_e \left(\frac{m_e - m_i}{Z_i}\right) \tag{11.62}$$

where subscript e denotes the rupture point, or exit, and i denotes the transition, or interface, point. Expressions are also given for the volume of the vapor space, and hence of the liquid removed, between the transition and rupture points. The model allows the mass flow from the rupture point and the position of the transition point to be determined as a function of time.

The authors have used the model to study ruptures in the propane pipeline system shown in Figure 11.6. They investigated different pumping station distances, valve

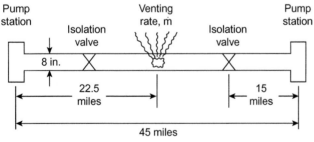

- Pipeline flow rate
- Isolation valve spacing
- Pump stations
- Pump shut-off time
- Valve closure time
- Break sizes

40 000 bbl/day 45, 15 and 5 miles
45 miles apart 5 min and
20 min after rupture 5 min after
pump shut-off time complete
break (venting from both ends),
and partial break (initial venting
rate 10 000 bbl/day)

FIGURE 11.6 Emission from an LPG pipeline: configuration of pipeline studied (T.B. Morrow, Bass, and Lock, 1982). *Source: Courtesy of the American Society of Mechanical Engineers.*

spacings, and shut-down times of the upstream pump, with valves shut down 5 min after the pump. Typical results for the flow profiles given by ruptures are shown in Figure 11.7. Curve A shows the flow for a complete rupture. This flow is for some time unaffected by pump shut-down and valve closure, since the two-phase interface moves relatively slowly. For example, for a spacing between pumps of 15 miles, the interface is estimated to reach the isolation valves at 7½ miles only after some 25 min. Curve B in Figure 11.7 shows the flow for a partial rupture. In this case, the pressure is maintained at the value before rupture until the pump shuts off, then falls, and finally reaches a new steady value corresponding to the bubble point. The fall may approximate to a ramp or may change to a steeper slope as the isolation valve shuts.

A description of the actual pipeline break at Port Hudson has been given by the NTSB (1972 PAR-72-01) and Burgess and Zabetakis (1973 BM RI 7752). The event is described in Case History A52.

11.7. VAPORIZATION

If the fluid that escapes from containment is a liquid, then vaporization must occur before a vapor cloud is formed. The process of vaporization determines the rate at which

material enters the cloud. It also determines the amount of air entrained into the cloud. Both aspects are important for the subsequent dispersion.

11.7.1 Vaporization Situations

In considering the generation of a vapor cloud from the liquid spillage, the following situations can be distinguished:

1. A volatile liquid at atmospheric temperature and pressure, for example, acetone.
2. A superheated liquid.
   a. at ambient temperature and under pressure, for example, butane;
   b. at high temperature and under pressure, for example, hot cyclohexane.
3. A refrigerated liquefied gas at low temperature but at atmospheric pressure, for example, cold methane.

The vaporization of the liquid is different for these three cases. In the first case, the liquid after spillage is approximately at equilibrium and evaporates relatively slowly. In the second case, the liquid is superheated and flashes off when spilt and then undergoes slower evaporation. The first category of a superheated liquid, at ambient temperature but under pressure, is that of a liquefied gas, while the second, at high temperature and under pressure, is that of a liquid heated above its normal boiling point. The third case is that of a refrigerated liquefied gas that on spillage evaporates rapidly at first and then more slowly.

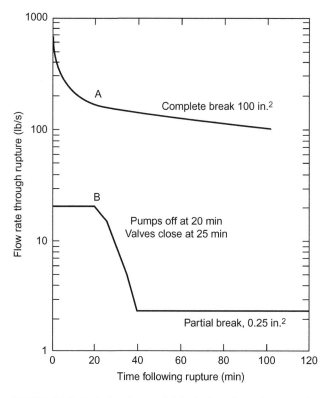

FIGURE 11.7 Emission from an LPG pipeline: flow of propane for partial and complete ruptures. Isolation value spacing = 45 miles (T.B. Morrow, Bass, and Lock, 1982). *Source: Courtesy of the American Society of Mechanical Engineers.*

11.7.2 Vaporization of a Superheated Liquid

If the liquid released from containment is superheated, a proportion flashes off as vapor. The remaining liquid is cooled by the removal of the latent heat of vaporization and falls to its atmospheric boiling point.

The theoretical adiabatic flash fraction (TAFF) of vapor so formed is usually determined by the simple heat balance

$$\phi = \frac{c_p l}{\Delta H_v}(T_i - T_b), \quad T_i > T_b \qquad (11.63)$$

where c_p is the specific heat, T is the absolute temperature, ΔH_v is the latent heat of vaporization, ϕ is the fraction of liquid vaporized, and subscripts b, I, and l denote boiling point, initial, and liquid, respectively.

An alternative expression that takes account of the differential nature of the vaporization is

$$(1 - \phi)c_{pl}(-dT) = \Delta H_v \, d\phi \qquad (11.64)$$

and hence

$$\phi = 1 - \exp\left[-\frac{c_{pl}}{\Delta H_v}(T_i - T_b)\right] \quad T_i > T_b \qquad (11.65)$$

Equations 11.63 and 11.65 give the theoretical fraction of vapor formed. The sudden growth and release of vapor bubbles also results in the formation of liquid droplets, or spray. The mass of liquid in spray form is generally of the same order as that in the initial vapor flash and may exceed it.

This spray then either vaporizes, increasing the vapor cloud, or rains out as liquid, forming a pool on the ground. The total amount of vapor formed, both that from the initial flash and that from the evaporation of spray, constitutes the ultimate flash.

For the fraction of liquid which forms spray a rule-of-thumb frequently used is that it is equal to the initial vapor flash (Kletz, 1977J). Another rule-of-thumb is that if the fraction flashing off is small, it may be appropriate to assume that the spray fraction is two or three times the initial vapor flash-off (W.G. High, 1976).

A discussion of spray and rainout following the discharge of a flashing liquid, with particular reference to liquid ammonia, has been given by Wheatley (1986). Two discharge situations may be distinguished: metastable flow of a superheated liquid and choked two-phase flow. These two cases have been modeled by Wheatley. As given in Table 11.2, the flow characteristics obtained are significantly different.

In the non-choked liquid flow case, there is a large increase in diameter of the flow at the outlet. Such large increases have been observed in the Desert Tortoise tests (Koopman et al., 1986). In the choked two-phase flow case, the flow velocity is much higher. The flash fraction is similar in the two cases.

The drops formed are subject to shear stress and there is a maximum size of drop given by the drop Weber number

$$W_{ed} = \frac{u_d^2 d\rho d}{\sigma_d} \qquad (11.66)$$

where d is the diameter of the drop, u is the velocity, ρ is the density, σ is the surface tension, and subscript d denotes drop. Table 11.2 gives the drop sizes obtained for the two cases.

The extent to which rainout of the drops occurs depends on a balance between the inertial and gravity forces. The drops will have higher inertia than the vapor but will be more affected by gravity. Wheatley treats rainout in terms of the inclination of the bounding drop trajectory, which is obtained from the ratio of the settling velocity to the horizontal velocity of the drops. Following flash-off, the residual liquid is at its normal boiling point. Vaporization then continues as a rate-limited process. This secondary stage of rate-limited vaporization is usually regarded as relatively less important compared with the initial flash-off, particularly with respect to the formation of flammable gas clouds.

11.8. DISPERSION

Emission and vaporization are followed by dispersion of the vapor to form a vapor cloud.

Work on dispersion is primarily concerned with the dispersion of pollutants from industrial chimney stacks. Most of the fundamental work on dispersion relates to this problem. There is, however, an increasing amount of work on dispersion of hazardous releases from process plant.

11.8.1 Dispersion Situations

Dispersion situations may be classified as follows. The fluid and the source may be classified as

1. Fluid buoyancy
   a. neutral buoyancy,
   b. positive buoyancy,
   c. negative buoyancy;
2. Momentum
   a. low momentum,
   b. high momentum;

TABLE 11.2 Flow Characteristics of Discharge of Liquid Ammonia

Storage Temperature (°C)	Flow Type	Axial Velocity (m/s)	Flow Diameter (cm)	Flash Fraction (%)	Spray Characteristics	Inclination (°)	
Drop Size (μm)	Settling Velocity (m/s)						
20	Liquid	154	13	17.6	25	0.013	0.0048
20	Two phase	37.7	61	18.7	420	1.4	2.1
34?	Liquid				1900	5.2	16

Source: After Wheatley (1986). Courtesy of the Institution of Chemical Engineers.

3. Source geometry
   a. point source,
   b. line source,
   c. area source.
4. Source duration
   a. instantaneous,
   b. continuous,
   c. intermediate;
5. Source elevation
   a. ground level source,
   b. elevated source.

The dispersion takes place under particular meteorological and topographical conditions. Some principal features of these are as follows:

2. Meteorology
   a. wind,
   b. stability;
3. Topography
   a. surface roughness,
   b. near buildings and obstructions,
   c. over urban areas,
   d. over coastal zones and sea,
   e. over complex terrain.

These aspects of the dispersion situation are now considered.

11.8.2 Buoyancy Effects

The fluid may have neutral, positive, or negative buoyancy. Neutral density is generally the default assumption and applies where the density of the gas−air mixture is close to that of air. This is the case where the density of the gas released is close to that of air or where the concentration of the gas is low. In determining the density of the gas, it is necessary to consider not only molecular weight but also the temperature and liquid droplets. Gases with positive buoyancy include those with low molecular weight and hot gases. Many hazardous materials, however, form negatively buoyant gases or heavy gases.

Much of the fundamental work on dispersion, and the models derived from this work, relates to the dispersion of gas of neutral density or neutral buoyancy. This work is relevant to dispersion from stacks once buoyancy effects have decayed. There are separate models that treat gases of positive buoyancy, which apply to releases close to stacks and gases of negative buoyancy. Dispersion of gases that do not exhibit positive or negative buoyancy is generally referred to as 'passive dispersion.'

11.8.3 Momentum Effects

A continuous release of material with low kinetic energy forms a plume that tends to billow. If the kinetic energy

is high; however, a momentum jet is formed which has a well-defined shape.

The momentum of the release has a marked effect on the extent of air entrainment. If the kinetic energy is high, large quantities of air are entrained. The degree of air entrainment affects the density of the cloud and is important in its further dispersion.

11.8.4 Source Terms

The principal types of source used in idealized treatments of dispersion are the point source, the line source, and the area source. An escape from a pipe is normally treated as a point source, while vaporization from a pool may be treated as an area source. There may also be some situations that may be modeled as an infinite or semi-infinite line source.

A very short and a prolonged escape may approximate to an instantaneous release and to a continuous release, respectively. An escape of intermediate duration, however, may need to be treated as a quasi-instantaneous or, alternatively, quasi-continuous release.

The most common scenarios considered are an instantaneous release from a point source, or 'puff,' and a continuous release from a point source, or 'plume.'

It will be apparent that the source terms described are idealizations of the actual situation.

11.8.5 Source Elevation

Another distinction is the elevation of the source. Sources are classed as ground level or elevated. Most hazardous escapes are treated as ground level sources. Stacks are the principal elevated sources.

11.8.6 Models for Passive Gas Dispersion

The modeling of dispersion, particularly that of neutral density gas, and passive gas dispersion models are discussed further in Sections 11.15 and 11.16, respectively. Passive gas dispersion over particular surfaces and in particular conditions is considered in Sections 11.17 and 11.18, while dispersion parameters for passive gas dispersion models are described in Section 11.19.

11.8.7 Models for Dense Gas Dispersion

For high concentration releases of many of the hazardous materials of interest in process plants, the assumption of neutral density gas behavior is not valid. In particular, the gas cloud is often heavier than air. In this situation, the common neutral density gas models are not applicable. However, the behavior of dense gases has been the subject of much work in recent years and dense gas dispersion models have been developed. Dispersion of dense gas is discussed further in Section 11.22 and succeeding sections.

11.9. DISPERSION MODELING

There are a number of different approaches to the modeling of dispersion. These include

1. gradient transfer models;
2. statistical models;
3. similarity models;
4. top hat, box, and slab models.

These are now described in turn.

11.9.1 Diffusion Equation

The fundamental equation for diffusion of a gas, in rectangular coordinates, is

$$\frac{d\chi}{dt} + u\frac{d\chi}{dx} + v\frac{d\chi}{dy} + \omega\frac{d\chi}{dz} = K_x\frac{d^2\chi}{dx^2} + K_y\frac{d^2\chi}{dy^2} + K_z\frac{d^2\chi}{dz^2}$$

(11.67)

where x, y, z are the rectangular coordinates (m), K_x, K_y, K_z are the diffusion coefficients in the x, y, z directions (m^2/s), t is the time (s), u, v, w are the mean wind speeds in the x, y, z directions (m/s), and χ is the concentration (kg/m^3). The coordinate system is shown in Figure 11.8(a).

If the wind speed in the y and z directions is zero ($v = w = 0$) and the diffusion coefficients are the same in each direction ($K_x = K_y = K_z = K$), Equation 11.67 becomes

$$\frac{d\chi}{dt} + u\frac{d\chi}{dx} = K\left(\frac{d^2\chi}{dx^2} + \frac{d^2\chi}{dy^2} + \frac{d^2\chi}{dz^2}\right)$$

(11.68)

where x, y, z are the distances in the downwind, crosswind, and vertical directions (m), respectively, and K is the diffusion coefficient (m^2/s).

If the wind speed in the x direction is also zero ($u = 0$), Equation 11.68 reduces to

$$\frac{d\chi}{dt} = K\left(\frac{d^2\chi}{dx^2} + \frac{d^2\chi}{dy^2} + \frac{d^2\chi}{dz^2}\right)$$

(11.69)

The corresponding equation for a symmetrical spherical system is

$$\frac{d\chi}{dt} = \frac{K}{r^2}\frac{d}{dr}\left(r^2\frac{d\chi}{dr}\right)$$

(11.70)

with

$$r^2 = x^2 + y^2 + z^2$$

(11.71)

where r is the radial coordinate (m).

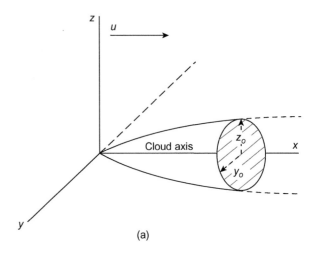

(a)

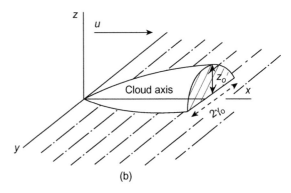

(b)

FIGURE 11.8 Coordinates for dispersion equations: (a) elevated source and (b) ground level source

Analytical solutions of the above equations with a constant value of the diffusion coefficient K have been given by O.F.T. Roberts (1923) as described below.

11.9.2 Gradient Transfer Models

Gradient transfer models, or K models, are solutions of the diffusion equation. Although the assumption of a constant diffusion, or turbulent exchange, coefficient was made in the early work, it is known that this is an oversimplification that yields unsatisfactory results. The approach now adopted is to solve the diffusion equation using relationships for the variation of the individual exchange coefficients K_x, K_y, and K_z and for the wind speed u. If the form of these relations is amenable, an analytical solution may be obtained, but usually it is necessary to resort to numerical solution.

11.10. PASSIVE DISPERSION

The dispersion of gases with neutral buoyancy, or passive dispersion, has been the subject of a very large volume of work. Some of the early work was oriented to gas warfare, but later air pollution has been the principal concern.

Neutral buoyancy is commonly due to the low concentration of the contaminant gas released, although it may also occur if the density of the gas is close to that of air. The neutral buoyancy condition may be negated if the gas release causes a large change in the temperature of the resultant cloud.

Hazard assessment utilizes, but has not greatly contributed to, work on neutral density gas dispersion. Most of the work has therefore been concerned with dispersion of continuous releases from an elevated point source, as represented by an industrial chimney stack. The two other main types of release studied that are of industrial relevance are continuous and instantaneous point source releases at ground level. There is some work on continuous line sources at ground level, relevant to gas warfare, but also to industrial area releases.

Earlier work on the subject includes that of G.I. Taylor (1915) and O.F.T. Roberts (1923). The models that are most widely used, however, are those of O.G. Sutton (1953) and of Pasquill (1961, 1962a) and the Pasquill−Gifford model (Pasquill, 1961; Gifford, 1961).

In this section an account is given of experimental studies and empirical features of passive dispersion. Section 11.16 describes passive dispersion models and Sections 11.17−11.19 describe dispersion over particular surfaces, dispersion in particular conditions, and dispersion parameters, respectively.

11.10.1 Empirical Features

Experiments on passive gas dispersion indicate several important empirical features. The fundamental features have been described by O.G. Sutton (1953).

One of the most important features is that, for both continuous and instantaneous releases from a point source at ground level, the concentration profiles are Gaussian. Another basic feature is that for both types of release, the spread of the measured concentration increases as the sampling period increases. It is observed that the plume from a continuous point source release tends to meander and that the dispersion due to turbulence is augmented by that due to this meandering.

The concentration downwind of a continuous or instantaneous point source at ground level is found to vary according to the strength of the source, provided that the latter does not itself cause appreciable convection. For a continuous point source, the concentration is also inversely proportional to the mean wind speed.

The concentration on the centerline of a continuous point source is

$$\chi \propto x^{-1.76} \qquad (11.72)$$

and that on the center plane of a continuous infinite line crosswind source is

$$\chi \propto x^{-1.09} \qquad (11.73)$$

This information on the variation of concentration with distance has played an important role in guiding the development of dispersion models.

11.11. PASSIVE DISPERSION: MODELS

Some principal models for passive dispersion are

1. Roberts model
2. Sutton model
3. Pasquill model
4. Pasquill−Gifford model.

An account is now given of each of these models in turn.

11.12. DISPERSION OF JETS AND PLUMES

The dispersion of material issuing as a leak from a plant is determined by its momentum and buoyancy. If momentum forces predominate, the fluid forms a jet, while if buoyancy forces predominate, it forms a plume. Such dispersion contrasts with the dispersion by atmospheric turbulence considered so far. However, once the momentum or buoyancy decays, dispersion by atmospheric turbulence becomes the predominant factor for leaks also.

Emission situations were classified in Section 11.1. In respect of jets and plumes, the following distinctions may be made:

1. Fluid
   a. gas,
   b. liquid,
   c. two-phase vapor−liquid mixture;
2. Fluid momentum
   a. low momentum,
   b. high momentum;
3. Fluid buoyancy
   a. positive buoyancy,
   b. neutral buoyancy,
   c. negative buoyancy;
2. Atmospheric conditions
   a. low turbulence,
   b. high turbulence.

If the momentum of the material issuing from an orifice on a plant is high, the dispersion in the initial phase is at least due to the momentum, and the emission is described as a momentum jet. If the momentum is low, either because the initial momentum is low or because it has decayed, the dispersion is due to buoyancy and atmospheric turbulence, and, if buoyancy is involved, the emission is described as a buoyant plume; the buoyancy may be positive or negative.

An account is now given of dispersion by momentum jets and buoyant plumes. Both types of emission are sometimes described as plumes, the jet being a forced plume. The forces in a buoyant plume are often of the same order as those in a momentum jet. Some jet and plume dispersion situations are illustrated in Figure 11.9. The figure shows schematically instances of releases issuing as jets and becoming plumes as the influence of buoyancy takes over from that of momentum.

11.12.1 Dispersion of Two-Phase Flashing Jets

The two-phase jets of main interest here are those of superheated liquids, which give flashing jets. A review of such jets has been given by Appleton (1984 SRD R303).

Work by Bushnell and Gooderum (1968) has shown that there is a critical degree of superheat above that the jet will disintegrate due to vapor bubble formation.

11.13. DENSE GAS DISPERSION

11.13.1 Dense Gas

A gas cloud that is denser than air depends on a number of factors. These are:

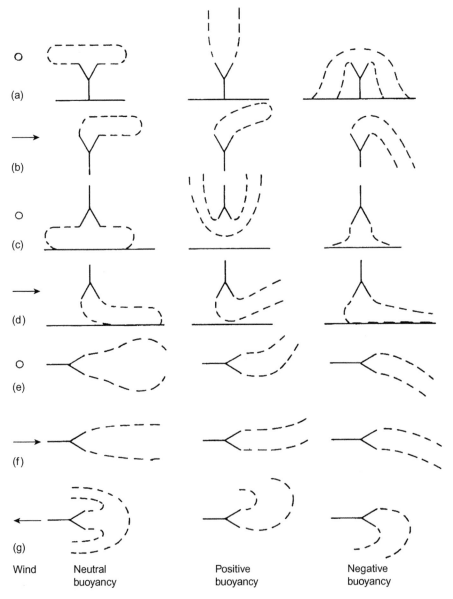

FIGURE 11.9 Some jet and plume dispersion situations (after TNO, 1979): (a) upward vertical release, zero wind speed; (b) upward vertical release, finite wind speed; (c) downward vertical release, zero wind speed; (d) upward vertical release, finite wind speed; (e) horizontal release, zero wind speed; (f) horizontal release, wind speed in direction of release; and (g) horizontal release, wind speed in direction opposed to release.

Wind　　　　Neutral　　　　　　　　Positive　　　　　　　Negative
　　　　　　buoyancy　　　　　　　buoyancy　　　　　　buoyancy

1. the molecular weight of the gas;
2. the temperature of the gas;
3. the presence of liquid spray;
4. the temperature and humidity of the air.

11.13.2 Dense Gas Behavior

The behavior of a dense gas may be described by considering the development of the gas cloud from the bursting of a vessel. The source term for this is usually modeled by assuming that air is entrained in a ratio of some 10−20:1 and that a cloud of unit aspect ratio is formed.

The behavior of the cloud is influenced by gravity. Three stages may be distinguished:

1. Gravity slumping
2. Gravity spreading
3. Passive dispersion.

11.13.3 LNG Spill Study

Another piece of work which was influential in the development of the topic was the study for the US Coast Guard by Havens (1978) of the hazard range from a spill of 25,000 m^3 of LNG onto water. The work was a comparative study of the predictions of the distance to the lower flammability limit given by seven gas dispersion models. These predictions ranged from 0.75 up to 50 km.

11.14. DISPERSION OF DENSE GAS: SOURCE TERMS

The source term can be critical for the modeling of the subsequent dispersion of a dense gas. It is therefore important that the source model should be realistic and complete.

11.14.1 Release Scenarios

The modeling of release scenarios is assisted by an appropriate classification.

The main concern is the modeling of the source term for a release of liquefied gas. The classification given for liquefied gases is

1. Pressurized release
    a. small hole in vapor space;
    b. large hole in vapor space—catastrophic failure;
    c. intermediate-sized hole in vapor space;
    d. hole in liquid space;
2. Refrigerated release
    a. spill onto land;
    b. spill onto water;
3. Jet release.

These cases are considered in turn for a dense gas.

11.14.2 Release from Pressurized Containment

For a pressurized liquefied gas, a release from a small hole in the vapor space gives rise to a vapor jet which loses momentum and turns into a dense gas plume.

A catastrophic failure of the vessel results in the flash-off of a vapor fraction with the liquid falling to its normal boiling point. A release from a hole of intermediate size may be presumed to give as spray a liquid fraction intermediate between the above two cases. A hole in the liquid space gives a two-phase flashing jet, as considered below.

11.14.3 Release from Refrigerated Containment

For a fully refrigerated liquefied gas, a spill onto land results in a spreading, vaporizing pool.

The vaporization from the pool is a function of the area of the pool and of the rate of vaporization per unit area. The vaporization process is governed by the heat transfer from the surface to the liquid, which is initially high but rapidly falls off as the surface chills.

A spill onto water again results in a pool spreading until the liquid depth becomes small and break-up occurs.

11.14.4 Release from Semi-Refrigerated Containment

For a semi-refrigerated liquefied gas, the scenarios are basically similar to those for the fully pressurized case. The liquid is, however, at a lower pressure and a lower temperature.

11.14.5 Two-Phase Jets

For a pressurized liquefied gas, a hole in the liquid space of a vessel or in a pipe gives a two-phase flashing jet. The development of such a jet proceeds through three zones. The first is the depressurization, or flashing, zone in which the vapor and liquid droplets form. This extends several hole diameters and increases rapidly in diameter. The second zone is that of the two-phase momentum jet at atmospheric pressure into which air is entrained. The third zone starts where the jet loses momentum and other dispersion mechanisms take over.

11.15. DISPERSION OF DENSE GAS: SLAB AND FEM3

11.15.1 SLAB

SLAB was originally formulated by Zeman (1982a,b) and its development has been described by Ermak, Chan et al. (1982), D.L. Morgan et al. (1984), and Ermak and Chan (1986, 1988).

The SLAB model is for a continuous release. It is a slab model with properties averaged in the horizontal and vertical directions and is thus one-dimensional. The model is based on a set of six simultaneous differential equations, for the conservation of total mass, conservation of the material released, conservation of momentum, and conservation of energy.

11.15.2 FEM3

Accounts of FEM3 have been given by Ermak, Chan et al. (1982), S.T. Chan and Ermak (1984), S.T. Chan et al. (1984), and Ermak and Chan (1986, 1988).

The FEM3 model is also for a continuous release. It is a full three-dimensional model. The model is based on a set of four simultaneous differential equations, in terms of tensor quantities, for the conservation of total mass, conservation of the material released, conservation of momentum, and conservation of energy.

11.16. DISPERSION OF DENSE GAS: DEGADIS

11.16.1 DEGADIS

DEGADIS has been developed by Havens and co-workers (Havens, 1985, 1986; Spicer and Havens, 1986, 1987). DEGADIS is based on HEGADAS but differs in several respects and, in particular, it incorporates its own source, vertical entrainments, and heat transfer models. DEGADIS is for a continuous release but may be adapted to time-varying releases, and an instantaneous release, by treating these as a series of quasi-steady-state releases.

11.17. DISPERSION OF DENSE GAS: FIELD TRIALS

As already indicated, there have been a large number of field trials conducted on the dispersion of dense gases. Accounts of such field trials have been given by Puttock et al. (1982), McQuaid (1984), Koopman et al. (1989), and Havens (1992), and also by Blackmore et al. (1982) and Koopman et al. (1986).

11.17.1 Bureau of Mines Trials

The Bureau of Mines (BM) performed a series of trials in 1970−1972 involving the vaporization and dispersion of LNG following spillage on water, as reported by Burgess and co-workers (Burgess, Murphy, and Zabetakis, 1970 BM RI 7448, S 4105; Burgess, Biordi, and Murphy, 1972 BM PMSRS 4177). In the 1970 work, some 51 trials were conducted in which LNG was spilled instantaneously on water and the pool spread and vaporization were measured; measurements of gas dispersion were not taken. Four trials were also carried out with continuous spillage of LNG and in these the gas dispersion was measured. The Pasquill−Gifford relations were used to model the dispersion, the data being correlated by taking the ratio σ_y/σ_z as 5, indicating a relatively flat cloud. A further 13 continuous spillage experiments were done in 1972, of which 6 were usable.

11.17.2 Esso/API Trials

In 1971, a series of spillages of LNG onto the sea were carried out in Matagordo Bay, Texas, by Esso under the auspices of the AGA (Feldbauer et al., 1972). The concentration of methane in the cloud was measured. There were 10 smaller experiments of about 0.9 m^3 and 7 larger ones of between 2.5 and 10.2 m^3; the former were virtually instantaneous and the latter nearly so.

11.17.3 Gaz de France Trials

A series of 40 experiments involving the instantaneous spillage of LNG into a bund were performed in 1972 at Nantes by Gaz de France to investigate vaporization, dispersion, and combustion (Humbert Basset and Montet, 1972). Most of the tests involved the spillage of 3 m^3 into bunds with areas 9−200 m^3. The vaporization rate decreased with time. There was also a test with a continuous release of 10 m^3/h.

11.17.4 SS Gadila Trials

In 1973, larger scale trials of spillage of LNG onto the sea were performed by Shell from the SS Gadila, an LNG carrier (Kneebone and Prew, 1974; Te Riele, 1977). The vessel was being commissioned and the purpose of the trials was to confirm recommended procedures for jettisoning cargo in an emergency. There were six tests lasting some 10 min and involving releases of up to 200 m^3.

11.17.5 China Lake Trials: Avocet, Burro Series

The Naval Weapons Center (NWC) test facility at China Lake, California, has been the site of several series of dense gas dispersion trials. Accounts of the facility have been given by Ermak et al. (1983).

The Avocet series was a set of four trials involving spillage onto water of some $4.5\,m^3$ of LNG over about 1 min. The main purpose of these trials, described by Koopman et al. (1980), was the development of the test facility.

In 1980, the Burro series of trials was performed involving the continuous release onto water of up to $40\,m^3$ of LNG over periods up to 3.5 min. These tests are therefore classed as continuous releases.

The experiments were well instrumented and obtained a fairly comprehensive set of measurements, including turbulence and concentration fluctuations in the plume.

11.17.6 Maplin Sands Trials

In 1980, a series of trials were carried out by Shell at Maplin Sands. Accounts of the work have been given by Puttock and co-workers (Puttock et al., 1982, 1983, 1984; Colenbrander and Puttock, 1984).

The liquefied gases were released onto water from a pipeline terminating in a downwards-pointing pipe and measurements were taken by instruments located on pontoons placed in arcs around the release point.

Two series of trials were carried out, one with propane and one with LNG. The experiments were mainly continuous spills but, for both gases, some trials with instantaneous spills were conducted.

11.18. DISPERSION OF DENSE GAS: PARTICULAR GASES

It is now necessary to consider some gases of industrial interest for which the dense gas behavior is governed by the specific characteristics of the gas. The gases considered are:

1. propane;
2. LNG;
3. ammonia;

The account in this section deals with the effect on the density of the gas cloud of the following factors:

1. molecular weight;
2. boiling point;
3. chemical behavior;
4. atmospheric humidity;
5. surface heat transfer;
6. liquid spray.

11.18.1. PROPANE

The normal boiling point of propane is $-42°C$. If the release is one of liquefied propane, the low temperature

makes the gas cloud initially somewhat more dense. The initial temperature difference between the gas at its boiling point and the environmental conditions is not great and heat transfer to the gas does not play a major role Any density changes due to temperature changes tend to be small.

11.18.2 LNG

The characteristics of the gas cloud from a release of LNG, mainly methane, have been described by Puttock (1987). Broadly, LNG releases normally give rise to dense gas clouds. However, under conditions of very high atmospheric humidity, the initial cloud may be buoyant, while in other cases the cloud, initially dense, may become buoyant with time due to surface heat transfer.

LNG is mainly liquefied methane and, as such, has a molecular weight of 16 and a normal boiling point of $-161°C$. The gas is sufficiently cold that on release it is dense.

11.18.3 Ammonia

A release of liquid anhydrous ammonia normally gives rise to a dense gas cloud.

The molecular weight of ammonia is 17 and its normal boiling point is $-33°C$. Thus the molecular weight is not responsible for any dense gas behavior. A systematic investigation of the factors governing the density of clouds of ammonia in air has been made by Haddock and Williams (1978 SRD R103, 1979). They consider chemical interactions, atmospheric humidity, and liquid spray.

11.19. DISPERSION OF DENSE GAS: PLUMES FROM ELEVATED SOURCES

The treatment of dense gas dispersion given so far has been largely confined to releases from sources at ground level. It is now necessary to consider elevated releases. A review of work on the dispersion of an elevated plume of dense gas has been given by Ooms and Duijm (1984).

11.19.1 Dispersion of Dense Gas: Plumes from Elevated Sources—PLUME

A model for dispersion from a jet or plume has been developed by Shell as part of the HGSYSTEM package. This model, PLUME, is one of the front-end models for the main dispersion model, HEGADAS. It has been described by McFarlane (1991).

11.19.2 Plume Development

The scenario model is that of a jet or plume issuing along the direction of the wind at an inclination to the horizontal. The plume is treated as passing through three stages:

1. Airborne plume
2. Touchdown plume
3. Slumping plume.

The cross-section of the airborne plume is assumed to be circular. The touchdown regime starts when the circumference of the airborne plume touches the ground and it ends when the cross-section of the plume has become semicircular. The slumping regime then begins and the cross-section becomes elliptical.

11.19.3 Touchdown Plume

Touchdown occurs when the plume envelope first touches the ground. The plume cross-section is then modeled as a circular segment, the centroid of which falls until the cross-section becomes a semicircle, at which point the touchdown regime ends.

11.19.4 Slumped Plume

The model for the slumped plume consists of the same set of six differential equations as for the touchdown plume.

$$A = \frac{e\pi}{8} D^2 \tag{11.74}$$

$$e = \frac{e\pi(z/D)}{2|\cos \phi|} \tag{11.75}$$

where D is the length of the major axis of the ellipse and e is the eccentricity of the ellipse.

11.20. CONCENTRATION AND CONCENTRATION FLUCTUATIONS

Concentration fluctuations are relevant to both flammable and toxic gas releases. For flammable gas clouds, fluctuations may cause the gas concentration equivalent to the lower flammability limit to reach further than the boundary given by the mean value. For a toxic gas cloud, fluctuations may result in toxic effects associated with concentrations higher than the mean value.

Accounts of concentration fluctuations have been given by a number of authors, including Gosline (1952), Gifford (1960), Long (1963), Slade (1968), Barry (1971), Ramsdell and Hinds (1971), Csanady (1973), Chatwin (1982), C.D. Jones (1983), S.R. Hanna (1984), Ride (1984a,b), and J.K.W. Davies (1987, 1989a,b).

11.20.1 Experimental Studies and Empirical Features

It has been shown in experiments on plumes from a continuous point source by a number of workers that the instantaneous concentration measured at a fixed point fluctuates about the mean value and that the ratio of the maximum instantaneous value to the time mean value, or peak-to-mean ratio, can be quite large.

11.20.2 Causes of Concentration Fluctuation

Fluctuations of concentration have a number of causes. One of these is meandering of the plume due to the variability of wind direction. But even without meander, concentration fluctuations occur due to the normal effects of turbulence.

11.20.3 Measures of Concentration Fluctuation

Concentration fluctuation may be characterized in a number of ways. These include:

1. peak-to-mean ratio;
2. intermittency;
3. intensity;
4. distributional properties.

11.20.4 Time Mean Concentration

As discussed in Section 11.16, the mean concentration obtained by averaging over a particular sampling interval is a function of that interval. The correction used by D.B. Turner (1970) for this effect is

$$\frac{C}{C_r} = \left(\frac{t_r}{t}\right)^p \tag{11.76}$$

where C is the concentration, t is the sampling time interval, p is an index, and subscript r denotes reference. The value of p is between 0.17 and 0.2 where the sampling reference time t_r is of the order of 10 min and the other sampling time t is up to 2 h.

Another approach is that described by the CCPS (1987/2).

11.20.5 Models for Concentration Fluctuation

Most of the approaches used to model dispersion may also be applied to the modeling of concentration fluctuations. A review by S.R. Hanna (1984) describes Gaussian, gradient transfer, similarity, meandering plume, and distribution function models.

11.21. TOXIC GAS CLOUDS

The effect of exposure to a cloud of toxic gas is a function of the concentration time profile and of the toxic characteristics of the gas.

11.21.1 Toxic Load and Probit

The inhalation toxicity of a gas is characterized by the toxic load L which produces some defined toxic effect, which is typically of the form

$$L = \int_0^{\mathcal{T}} C^n \, dt \qquad (11.77a)$$

$$= C^n T \qquad (11.77b)$$

$$= \sum_{i=1}^{p} C_i^n t_i \qquad (11.77c)$$

where C is the concentration, L is the toxic load, t is the time, T is the time of exposure, and n is an index.

The proportion of the exposed population which will suffer, or the probability that a single individual will suffer, this defined degree of injury is generally expressed in terms of a probit equation of the form

$$Y = k_1 + k_2 \ln L \qquad (11.78)$$

where Y is the probit and k_1 and k_2 are constants. The probability P of injury is a function of the probit Y.

There is usually little information on its applicability to combinations of values of (C,t) where C is high and t is short.

11.21.2 Intermittency and its Effects

Griffiths and co-workers (R.F. Griffiths and Megson, 1984; R.F. Griffiths and Harper, 1985) have drawn attention to the implications of concentration intermittency when combined with the toxic load relation. The intermittency γ is defined as the fraction of time for which the concentration is zero:

$$\gamma = \frac{t_s - t_{pe}}{t_s} \qquad (11.79)$$

where γ is intermittency and subscripts pe and s denote exposure and sampling, respectively. These authors define intermittency as the proportion of time and the concentration is zero. The average concentration $\overline{C}$ is

$$\overline{C} t_s = C_p t_{pe} \qquad (11.80)$$

where $\overline{C}$ is the average concentration and C_p is the concentration over time t_{pe}. Then from Equations 11.81,

$$Y = k_1 + k_2 \ln \left[\left(\frac{1}{1-\gamma} \right)^{n-1} \overline{C}^n t_s \right] \qquad (11.81)$$

For the usual case of $n > 1$, the probit Y, and hence the probability P of injury, increases as the intermittency increases. The effect can be pronounced.

11.22. DISPERSION OVER SHORT DISTANCES

It is necessary in some applications, particularly in relation to plant layout, to estimate dispersion over short distances.

11.22.1 Dispersion of Passive Plume

The relationships for passive dispersion are commonly used over relatively large distances, but the question of the range of his model was specifically considered by O. G. Sutton (1950). He describes experiments on the dispersion of a vertical jet of hot gas for which he obtained a value of the diffusion coefficient C. This value is similar to that obtained in the dispersion of smoke from a smoke generator over a distance of 100 m, provided that for comparability use is made in the latter case of the instantaneous rather than the time width of the plume.

11.22.2 Dispersion of Dense Gas Plume

Some guidance on the dispersion of a dense gas plume over short ranges has been given by McQuaid (1980) in the context of the design of water spray barriers.

11.22.3 Dispersion from Vents and Reliefs

It is common practice in the process industries to discharge material to the atmosphere. Intended discharges occur through chimneys, vents, and relief devices.

Accounts of safe discharge include those given in the various editions of API RP 520 and RP 521 and by Long (1963), Loudon (1963), Bodurtha et al. (1973), and Gerardu (1981). The topic is also frequently discussed by authors of models developed to assist with this problem.

11.22.4 ASME Equation

A formula for plume rise from a momentum source widely used in design for safe discharge is that given by the American society of Mechanical Engineers (ASME, 1969/1). This is

$$b = D \left(\frac{v_s}{u} \right)^{1.4} \qquad (11.82)$$

where D is the internal diameter of the stack (m), Δh is the plume rise (m), v_s is the velocity of the stack gas (m/s), and u is the wind speed at stack height (m/s).

The applicability of the ASME equation is to a stack gas which is essentially of neutral density.

11.23. HAZARD RANGES FOR DISPERSION

It is convenient for certain purposes, such as comparison of models or preliminary hazard assessment, to have simple expressions for the downwind extent of particular concentrations, such as lower flammability limits or toxic concentrations; in other words, for the hazard range.

The variables which enter into the simpler correlations are generally the concentration, the downwind distance, the mass rate of release, and the wind speed, and the treatment here is confined to these.

11.23.1 Passive Dispersion

For passive dispersion of an instantaneous release from a point source the relevant expression is the Pasquill–Gifford model. Considering concentration at the center of the cloud at ground level so that $x = ut$, $y = 0$, and $z = 0$, this equation reduces to

$$\chi \propto \frac{Q^*}{\sigma_x \sigma_y \sigma_z} \qquad (11.83)$$

From the correlations of Slade noting that it is usual to take $\sigma_x = \sigma_y$,

$$\sigma_x = \sigma_y \propto x^{0.92} \qquad (11.84a)$$

$$\sigma_z \propto \chi^{0.7} \qquad (11.84b)$$

Then relations 11.83, 11.84a and 11.84b give for the hazard range x.

$$\chi \propto Q^{*0.39} \qquad (11.85)$$

In this case, the hazard range is a function of the mass released, but not of the wind speed.

11.23.2 Dense Gas Dispersion

It is less easy to derive analytical expressions for the hazard range of a dense gas. One approach is to correlate results from a number of runs of a dense gas dispersion model. The *Second Canvey Report* gives for an instantaneous release the following correlation derived from DENZ for propane and butane:

$$R = kM^{0.4} \qquad (11.86)$$

where M is mass of release (te), R is the downwind range (km), and k is a constant. It also gives the following additional relations for distance (km): maximum width = k_1R; distance to maximum width, $X = k_2R$, and upwind range = k_3R.

11.24. SOURCE AND DISPERSION MODELING: CCPS GUIDELINES

The CCPS has issued two publications giving practical guidance on emission and dispersion of hazardous materials.

11.24.1 Guidelines for Use of Vapor Cloud Dispersion Models

The CCPS *Guidelines for the Use of Vapor Cloud Dispersion Models* (*the CCPS Vapor Cloud Dispersion Model Guidelines*) (1987/2) covers both emission and dispersion. The Guidelines give an overview of release scenarios; particularly pipe ruptures, vessel rupture, and reactor venting; review the dominant phenomena of initial acceleration and dilution, buoyancy, and atmospheric turbulence; and give a decision tree for handling the scenarios.

11.24.2 Workbook of Test Cases for Vapor Cloud Source Dispersion Models

The CCPS Vapor Cloud Dispersion Model Guidelines are supplemented by the CCPS Workbook on Test Cases for Vapor Cloud Source Dispersion Models (the CCPS Vapor Cloud Source Dispersion Model Workbook) (1989/8). The Workbook gives a set of emission, pool vaporization, and jet models, and describes methods of matching the output of these models with the input required by selected dispersion models.

11.25. VAPOR RELEASE MITIGATION: CONTAINMENT AND BARRIERS

There are a number of methods of preventing or mitigating the dispersion of gases. They include the use of

1. bunds;
2. foam;
3. solid barriers;
4. fluid barriers
   a. water spray barriers,
   b. steam curtains.

Guidelines for vapor release mitigation have been given by Prugh (1985, 1987a,b).

11.25.1 Bunds

A bund around a storage tank is generally designed to contain the contents of the tank. A bund with a smaller floor area but higher walls presents a smaller surface to heat up and vaporize the liquid. The higher wall also acts as a barrier to the flow of the vapor.

Further, the provision of a bund facilitates the use in suitable cases of foam, which can effect a further large reduction in evaporation.

11.25.2 Foam

The use of foam is an effective means of reducing the rate of evaporation, where a suitable foam is available. Foam acts by insulating the surface of the spill and preventing vaporization. Other modes of action include absorption of the vapor and scrubbing out of aerosol and particulate matter. Any foam used must be suitable for the application; the wrong foam can do more harm than good. A guide to the use of foam on hazardous materials has been published by Norman (1987). Likewise, water should only be used advisedly; incorrect use can make things worse.

Foam should be applied gently, possibly with continuous application or frequent reapplication. Personnel should be trained and should have suitable protective equipment, including self-contained breathing apparatus.

at an angle of approximately 45° to the horizontal. Water sprays and curtains are used mainly against flammable gases, but may be applied against toxic gases also.

11.25.5 Steam Curtains

A steam curtain is used as a permanent installation to contain and disperse leaks of flammable gas heavier than air. A steam curtain system has been described by Cairney and Cude (1971) and Simpson (1974).

11.25.6 Water Spray Barriers vs Steam Curtains

Several authors have compared the use of water spray barriers and steam curtains. A detailed comparison of the two devices has been made by McQuaid (1980), as illustrated in Table below:

Predictions of CRUNCH of Effect of Water Spray Barrier on Plume from Continuous Release of Carbon Dioxide

Barrier Downwind Distance(m)	Pasquill Stability Category	Plume Dimensions (m)			Airflow (kg/s) due to	
		Width	Height Before Barrier	Height After Barrier	Plume Entrainment	Spray Barrier
11	D	14.4	1.2	2.3	27	53
30	D	25.6	2.5	3.4	172	93
11	F	38.2	0.75	2.94	15	143
30	F	88.3	0.66	2.87	30	319

Air entrainment rate = 3.6 kg/s m.
Source: After McQuaid and Fitzpatrick (1983). Courtesy of the Institution of Chemical Engineers.

11.25.3 Solid Barriers

A solid barrier such as a fence or wall can serve either to contain a gas cloud entirely or to effect an appreciable dilution. An impermeable barrier designed for this purpose and erected within the works is generally known as a vapor barrier. A barrier does not need to be impermeable to mitigate a gas release. A plantation of trees may provide a worthwhile degree of dilution of a gas cloud.

11.25.4 Water Spray Barriers

The systems used include fixed water spray installations and mobile water spray monitors. Fixed installations are typically a set of spray nozzles several meters off the ground with the spray directed downward. Such systems are used both in the open air and in buildings. A typical mobile monitor system is a set of spray nozzles inclined

11.26. VAPOR CLOUD MITIGATION: CCPS GUIDELINES

The CCPS has published guidance on the mitigation of vapor clouds and this is now described.

11.26.1 Guidelines for Vapor Release Mitigation

The CCPS *Guidelines for Vapor Release Mitigation* (the CCPS *Vapor Release Mitigation Guidelines*) (1988/4) cover a wide range of approaches to mitigation. Prugh (1985, 1987a,b) has given a number of accounts of vapor release mitigation which foreshadow the Guidelines.

The main headings of the Guidelines may be summarized as

1. overview;
2. inherently safer design;

3. engineering design;
4. process safety management;
5. early vapor detection and warning;
6. countermeasures;
7. on-site emergency response;
8. off-site emergency response;
9. selection of measures.

11.27. FUGITIVE EMISSIONS

The fugitive emissions consist of leaks from flanges, valves, pump and compressor seals, pressure relief valves, and so on.

An account of fugitive emissions is given in *Fugitive Emissions of Vapors from Process Equipment* by the British Occupational Hygiene Society (BOHS) (the *BOHS Guide*) (1984 TG3), *Health Hazard Control in the Chemical Process Industry* (Lipton and Lynch, 1987), *Fugitive Emissions and Controls* (Hesketh and Cross, 1983), and *Handbook of Health Hazard Control in the Chemical Process Industry* (Lipton and Lynch, 1994).

11.28. CLASSIFICATION OF MODELS

There is a wide selection of models to be used when dispersion is going to be modeled, but it is important to mention that most of them belong to four different groups:

1. Workbooks/correlations
2. Integral models
3. Shallow layer models
4. Computational fluid dynamics (CFD) models.

The workbooks/correlations models are used mainly for rapid calculations. Using empirical relations, the model assumes that the relationships that describe a single set of conditions will have the same behavior for all the different conditions.

The integral models use simple differential equations in order to describe the dispersion phenomena as a whole. These models allow idealized atmospheric states, and some of them permit the inclusion of variations in the soil and other ambient characteristics. The shallow layer models have properties from both integral and CFD models. Therefore, they tend to model dense gas dispersions as low wide clouds, where the main parameters are the lateral dimensions. Some allow to incorporate the complexity of the terrain into the model. For the solution algorithm, these models make use of Navier−Stokes equations and one-dimensional integral models (Luketa-Hanlin, 2006). Computational fluid dynamics (CFD) models make use of the Navier−Stokes equations. These models give a three-dimensional time-dependent solution, incorporating features that not many models did not

before, such as complex geometry and dispersion effects (e.g., dispersion including dykes and bunds). The addition of such characteristics means the models are more time consuming and use more resources than the other groups in order to find a numerical and graphical solution.

11.29. TOXIC EFFECTS

11.29.1 Modes of Exposure

Toxic chemicals enter the body in three ways: (1) inhalation, (2) ingestion, and (3) external contact. Generally, gases, vapors, fumes, and dusts are inhaled, and liquids and solids are ingested. Entry may also occur through the intact skin or the mucous linings of the eyes, mouth, throat, and urinary tract.

11.29.2 Effects of Exposure

The effects of exposure to toxic chemicals may be acute or chronic. Acute effects result from a single exposure to a high concentration of the chemical; chronic effects result from exposure to low concentrations, perhaps over a large part of a working lifetime. It is also possible for the effects of a single exposure to a high concentration to be latent.

A toxic chemical may induce a graded or a quantum response. A graded response refers to the symptoms shown by an individual, which become progressively more severe as the dose is increased. A quantum response, on the other hand, refers to the effect of a toxic chemical on a population, in which some individuals suffer the defined injury and others do not.

Different toxic chemicals affect different sites in the body. The effect of such chemicals depends on the target organ.

11.30. TOXIC SUBSTANCES

Some factors relevant to toxic substances include generation of substance, toxic concentrations, effects of exposure, detectability by odor, precautions in handling, leak detection, and first aid.

Most toxic substances that present a hazard in the chemical industry are chemicals that are deliberately produced, but some are generated as by-products by accident. The number of chemicals used in industry is very large and grows each year.

The problems posed by toxic chemicals have generally been perceived primarily in terms of noxious effects resulting from chronic exposure to chemicals that possess a degree of toxicity which has not been appreciated. More recently, there has been increased concern over the threat of large-scale acute poisoning from the accidental release of toxic chemicals.

11.31. TOXICITY ASSESSMENT

Comprehensive accounts of occupational health risks involving toxic hazards are given in a number of texts. There is a good deal of guidance available on the handling of individual toxic substances.

However, information on toxicity is often incomplete or non-existent. Therefore, in some cases, it is necessary to conduct tests in order to obtain information on toxicity. Toxicity testing is now a well-established activity which is conducted to ensure safety not only in the manufacture of chemicals but also in the use of food, drugs, and cosmetics. Methods that are used to assess toxicity include (1) microorganism tests and (2) animal experiments. Studies of the effects of chemicals on microorganisms are used for screening chemicals, particularly for possible carcinogenic, mutagenic, or teratogenic effects. Animal experiments allow the use of the normal techniques of controlled experimentation. The object of toxicity testing is to obtain quantitative information on toxic effects. The simple classification of substances as 'toxic' or 'non-toxic' is of little value.

Another approach to the assessment of toxicity is epidemiology. Epidemiological studies are based on comparisons of disease or abnormality between the group under study and a control group. This approach is applicable to situations where a number of people have been exposed. Epidemiological studies have the great drawback that they yield information on the existence of a toxic effect only after people have fallen victim to it. This drawback is most serious where the effects are latent rather than acute. Despite this, epidemiology is an important tool for toxicity assessment.

The correlation of toxicity data requires the definition of the toxic load. This is the independent variable in terms of which toxic injury is expressed. Toxic load is thus a form of injury factor. A correlation may be sought between the toxic load and the proportion of the population suffering a defined degree of injury. This correlation is the toxic load−response relation. There are relatively few toxic load−response relations established for toxic substances.

Statistical methods have an important part to play in the design and interpretation of animal experiments, in the interpretation of toxic load−response data and in estimating the parameters of correlations. For obvious reasons, the number of animals, which can be used in gas toxicity experiments, has to be kept as low as possible and the statistical interpretation of the results is therefore crucial. Thus for a given confidence level, it is necessary to use more animals to determine an LC_{10} or LC_{90} than an LC_{50}.

The application to man of results from experiments on animals is an area of considerable difficulty and uncertainty. With many substances, different species exhibit similar reactions and extrapolation to man may be made. But there are also many examples of different reactions to the same chemical in different species. One general principle is that the toxic effect is likely to be similar only if the target organ is the same in the two species. Another general principle is that what matters is the quantity of the toxin which reaches the target organ, rather than simply the quantity which enters the body. These two principles are applicable in the interpretation of the chronic effects of chemicals such as carcinogens. A widely used rule-of-thumb in toxicology is that if consistent results are obtained for three animal species, they may be treated, with caution, as applicable to humans.

The information obtained from toxicity assessments of the kind just described may be used to make a toxic risk assessment for a particular chemical or plant. There are two rather different kinds of toxic risk assessment. One is the risk assessment undertaken by the regulatory agency in order to determine the precautions to be taken and to set the hygiene limits. The other is that carried out by a manufacturer in order to define the requirements for plant design and operation.

Clearly the existence or otherwise of a threshold value below which any noxious effect is negligible is a key issue. The definition of the level of exposure to be expected is also difficult but important. Once the risks have been assessed, they may be evaluated using suitable risk criteria.

11.32. CONTROL OF TOXIC HAZARD: REGULATORY CONTROLS

There is a worldwide trend toward much stricter regulatory control of toxic chemicals. Elements of such control include determination of the properties of the substances, limitation of emissions to the atmosphere, setting of limits for airborne concentrations, monitoring and control of airborne concentrations, monitoring of health of workers, and assessment of risk to workers.

The control of the chronic toxic hazard needs to be based on a coherent strategy. The outline of such a strategy is given by Lowrance (1984), in a review which is concerned specifically with carcinogens, but which is of wider applicability. He argues that there is need for a framework that allows different risks to be compared and that the approach taken should be more explicit.

11.33. HYGIENE STANDARDS

There are two main types of toxic limit. For exposure over a working lifetime, there are hygiene standards in the form of OELs, while for emergency exposure there are emergency exposure limits. Occupational hygiene standards are considered in this section, and emergency exposure limits are considered in Section 18.9.

Three principal sets of occupational hygiene standards are the TLVs used in the United States, the OELs used in the United Kingdom, and the MAK-Werte used in Germany.

A set of threshold limit values (TLVs) is published in the United States by the ACGIH. The TLV system is widely used, not only in the United States but also in many other countries.

As far as concerns the limits for workplace exposure to toxic chemicals in the United Kingdom, for many years use was made of a modified version of the US TLV system. In 1980, the United Kingdom moved to a system of OELs. The latter system is described in the following section.

The German system of hygiene standards is the MAK-Werte system.

11.34. HYGIENE STANDARDS: OCCUPATIONAL EXPOSURE LIMITS

The system of OELs that are used in the United Kingdom is described in EH 40/94 *Occupational Exposure Limits* 1994. Two sets of limits are used. These are (1) the maximum exposure limits (MELs) and (2) the occupational exposure standards (OESs). The difference between these two types of limit is that, whereas an OES is set at a level at which there is no indication of risk to health, for an MEL a residual risk may exist.

11.34.1 Maximum Exposure Limits

An MEL is the maximum concentration, averaged over a reference period, to which an employee may be exposed by inhalation in any circumstances. There are two reference periods: a long-term period and a short-term period. The long-term period is an 8 h TWA period. The short-term period given in EH 40/94 is 15 min.

11.34.2 Occupational Exposure Standards

As already stated, an OES is a concentration at which there is no indication of risk to health. OESs are listed in EH 40/94, which is referred to in the COSHH Regulations as the list of Approved OESs. The reference periods are the same as for the MELs, the long-term one being an 8 h TWA and the short-term one being 15 min.

11.34.3 Occupational Exposure Limit System

Maximum exposure limits and OESs are set by the HSC on the recommendations of the ACTS, following assessment by the Working Group on the Assessment of Toxic Chemicals (WATCH).

A substance is assigned an OES if it meets all three of the following criteria: (1) The available scientific evidence allows for the identification, with reasonable certainty, of a concentration averaged over a reference period, at which there is no indication that the substance is likely to be injurious to employees if they are exposed by inhalation day after day to that concentration. (2) Exposure to concentrations higher than that derived under criterion 1 and which could reasonably occur in practice are unlikely to produce serious shorter long-term effects on health over the period of time it might reasonably take to identify and remedy the cause of excessive exposure. (3) The available evidence indicates that compliance with the OES, as derived under criterion 1, is reasonably practicable.

A substance is assigned an MEL if it meets either of the following criteria: (4) The available evidence on the substance does not satisfy criterion 1 and/or 2 for an OES and exposure to the substance has, or is liable to have, serious health implications for workers. (5) Socioeconomic factors indicate that although the substance meets criteria 1 and 2 for an OES, a numerically higher value is necessary if the controls associated with certain uses are to be regarded as reasonably practicable.

11.34.4 Application of Occupational Exposure Limits

An MEL is a maximum limit. The long-term, 8 h limit attracts requirements for monitoring under the COSHH Regulations, unless an assessment shows such monitoring to be unnecessary. The Regulations also require that the exposure of personnel be kept as far below this level as reasonably practicable. The short-term, 15 min, MEL should never be exceeded.

Control to or below an OES can always be regarded as adequate under the COSHH Regulations. However, it is still incumbent on the occupier to follow good occupational hygiene practice and it is prudent to reduce exposure below the OES to allow for concentration fluctuations in the workplace.

11.34.5 Limitations of Occupational Exposure Limits

Occupational exposure limits are intended to be used for normal working conditions in factories and other workplaces. Their application is not to be extended to other situations, and specifically they should not be used as limits for either emergencies or pollution.

11.35. DUSTS

There are many industrial chemicals that are used in powder form and can give rise to dust in the atmosphere. Dust may also be generated by the attrition of materials.

Dusts are covered by the OEL system, and OELs for dusts have been described in Section 18.6.

Some dusts are not classified as having any specific toxic effect but can cause irritation. Excessive quantities of such dust are unpleasant to inhale, can deposit in the eyes and ears, and can injure the skin or mucous membranes by chemical or mechanical action, or through the cleaning procedures. Other dusts can cause injurious effects. One such dust is silica in the crystalline, or quartz, form.

Fibrosis can also be caused by asbestos dust. In addition, this dust is a carcinogen. Metal dusts that are sufficiently fine can enter the lungs and give rise to the toxic symptoms associated with the parent metal. Other dusts which are themselves non-fibrogenic may produce an allergic response in the lungs that eventually leads to fibrosis. Such conditions are 'farmer's lung' and byssinosis, an asthmatic condition prevalent in the cotton industry. Process workers involved with enzyme washing powders have experienced similar asthmatic conditions.

11.36. METALS

The toxic effects of metals and their compounds vary according to whether they are in inorganic or organic form, whether they are in the solid, liquid, or vapor phase, whether the valency of the radical is low or high, and whether they enter the body via the skin, lungs, or alimentary tract.

Some metals that are harmless in the pure state form highly toxic compounds. Nickel carbonyl is highly toxic, although nickel itself is fairly innocuous. The degree of toxicity can vary greatly between inorganic and organic forms.

Hazard arises from the use of metal compounds as industrial chemicals. Another frequent cause of hazard is the presence of such compounds in effluents, both gaseous and liquid, and in solid wastes. Fumes evolved from the cutting, brazing, and welding of metals are a further hazard. Such fumes can arise in the electrode arc welding of steel. Fumes that are more toxic may be generated in work on other metals such as lead and cadmium.

11.37. EMERGENCY EXPOSURE LIMITS

Limits for occupational exposure are complemented by limits for emergency exposure. A number of types of limit for exposure in an emergency have been defined by various bodies, but it is fair to say that there is no system with a status comparable to that of the TLV or OEL systems.

Acute Exposure Guideline Levels (AEGLs) are intended to depict the risk to humans resulting from once-in-a-lifetime, or rare, exposure to airborne chemicals. AEGLs represent threshold exposure limits for the general public and are applicable to emergency exposure periods ranging from 10 min to 8 h.

The National Research Council (NRC) Committee on Toxicology has, since the 1940s, submitted to the Department of Defense EEGLs for toxic chemicals. An EEGL is a concentration that is judged acceptable and which will permit exposed individuals to perform specified tasks during emergency conditions lasting from 1 to 24 h.

There are also indices for acute toxic exposures that are not a simple concentration value but take account of other parameters also. One such is the Chemical Exposure Index (CEI) developed by the Dow Chemical Company (1994).

11.38. GAS TOXICITY

11.38.1 Lethal Concentration and Load

In general, the injurious effect of the inhalation of a toxic gas is a function of concentration and of time that may be expressed by the relation

$$c^n t^m = \text{Constant} \qquad (11.87)$$

where c is the concentration, t is the time, and m and n are indices.

If the exposure time is constant, a lethal concentration LC_i may be defined such that for this exposure time C_i is the concentration that is lethal at the $i\%$ level. Widespread use is made of the LC_{50} value and also of other values such as LC_{10}, LC_{05}, and LC_{01}.

If the exposure time is not constant, but the injurious effect is proportional to the product ct of the concentration and time ($m = n = 1$), and hence to the dosage D, a lethal dosage LD_i may be defined with

$$D = ct \qquad (11.88)$$

If the injurious effect is proportional to some other function ($m \neq n$), it is necessary to use the concept of toxic load L and to define a lethal load LL_i with

$$L = ct^m \qquad (11.89)$$

An alternative toxic load L^* may also be defined

$$L^* = c^n t \qquad (11.90)$$

with $m = 1/n$. This second form of the lethal load is that most often used in hazard assessment studies.

11.38.2 Experimental Determination

The main source of information on the lethal toxicity of gases is experimentation on animals, particularly mice. In a typical study, groups of mice are exposed to different concentrations of gas for a single exposure period and the mortality is determined over a given period of observation after the exposure is over.

For a particular gas, assuming there are any data, there will typically be between one and half a dozen studies quoted in the literature that appear applicable. There may be one or two in which the exposure period has been varied. There may also be one or two studies with other species such as rats, guinea pigs, rabbits, and, in older work mainly, cats and dogs.

The determination of the lethality of a toxic gas by inhalation experiments with animals is a difficult undertaking and is subject to various sources of error. In addition to the concentration of the gas and the exposure time, other important variables are the caging conditions, the breed, sex, age and health of the animals, and their behavior, including their breathing rate. The animals may not die immediately and it is necessary to observe delayed deaths over a period of time, usually 10 days, and to record both immediate and delayed deaths. A sufficient number of animals need to be used to obtain results with a high level of confidence, and pathological examinations should be conducted. The toxicity data sought are usually for a given exposure time and comprise the LC_{50} together with suitable values nearer the extremes of mortality such as the LC_{10} and LC_{90}.

The lethal toxicity estimates required for hazard assessment are essentially the LC_{50}, the slope of the concentration mortality line, which may be expressed in terms of the ratio LC_{90}/LC_{10}, and the toxic load function, which defines the equivalence between concentration and time.

Usually, if there are any data at all, there will be enough to permit some estimate to be made not only of the LC_{50} but also of the ratio LC_{90}/LC_{10}, but the latter estimate will generally be such as to yield much less confidence in the LC_{10} than in the LC_{50}.

Extrapolation of results obtained on one particular species to another species is beset with many difficulties, but it is an unavoidable step in the estimation of toxicity. The crucial question is whether or not the toxic effects are the same, or at least sufficiently similar, in the two species, thus providing a basis for extrapolation. Other important features are the relative modes and rates of inhalation and the mechanisms and rate of elimination, or metabolism.

In many cases, extrapolation from animals to humans is supported by an argument based on the ratio of physiological quantities such as the minute volume or body weight. Extrapolation from animals to man is usually done in the first instance for healthy young adults. It may be necessary, however, to allow for vulnerable members of the population.

It may be preferable to derive separate estimates of the lethal toxicity for the regular and vulnerable populations. This makes it possible to allow for differences in the numbers and composition of the exposed population at different times of day.

11.38.3 Physiological Factors

Any attempt to model the toxic effects of inhaled gases requires some understanding of the respiratory process.

Accounts of the respiratory system, the effects of gases on this system, and the absorption of gases into the blood include those given in standard texts on general physiology, respiratory system, and the inhalation of gases. The account given here describes in outline the respiratory process and some of the quantitative data and relations in respiratory physiology. This information is the starting point for modeling the effect of toxic gases, both irritant gases that attack the lungs and other gases which enter the blood. The respiratory system is illustrated in Figure 11.10. Air entering the lung passes through the trachea, then down the bronchioles and through the alveolar duct into the alveolar sac. Interchange between the air and the blood occurs at the surface of the alveoli.

The walls of the capillaries lining the alveoli are extremely thin. The equilibrium partial pressure of a solute gas in the alveoli is effectively the same as that of the blood, neglecting the membrane. Mass transfer between the air and the blood is very rapid.

The volume of air moved in one respiratory cycle is less than the total capacity of the lungs. Moreover, not all the air moved enters the alveolar space, since there is dead space.

The flow of blood through the capillaries of the lung depends on the cardiac output, which is a function of the heart rate and the stroke volume. It is the function of the lung to absorb oxygen from the air into the blood and to desorb carbon dioxide from the blood into the air. The measure of the ability of the lung to do this is pulmonary diffusion capacity D_L. The diffusion capacity of a gas is proportional to its solubility and inversely proportional to the square root of its molecular weight.

The respiratory system provides the front-end part for a toxicokinetic model describing the absorption and distribution of the toxin in the body. Such toxicokinetic models for an inhaled gas may be developed by modeling the absorption of gas in the lung into the bloodstream. The difference between the mass inhaled and that exhaled equals the mass transferred across the membrane of the lung and this in turn equals the mass deposed in the body.

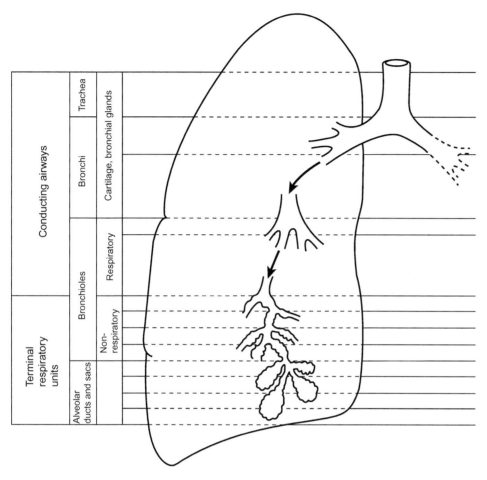

FIGURE 11.10 The structure of the respiratory tree in the lung.

Then if the chemical enters the main bloodstream, its concentration in the blood will be a function of the rates of absorption and of elimination. In simple terms, this situation may be modeled as a single exponential stage with a time constant that is a function of the apparent volume of distribution. The equilibrium back pressure of the chemical at the lung surface will depend on the concentration in the blood.

A toxicokinetic model of this general type is applicable where the toxic gas does not attack the respiratory system but enters into the bloodstream and attacks some other organ in the body. There is, however, a lack of models relevant to hazard assessment.

In most cases, the inhalation rate is a dominant factor in the entry of toxic gas into the system. It is therefore of great importance both in modeling the effects of and in extrapolating from experiments on toxic gases. In applying to human beings the results of animal experiments on gas toxicity, it is necessary to make allowance for the effect of inhalation rate. If the base case for comparison between animals and man is that each has the inhalation rate which is normal at rest, there are two separate allowances, or factors, which need to be applied. The first is

between the inhalation rate of the animal at rest and in the experiment, the second between that of man at rest and in the accident conditions envisaged in the hazard assessment.

11.38.4 Toxicity Data

Before considering the toxicity of some of the principal industrial gases, attention is drawn to some of the general sources of data on toxicity, particularly toxicity to humans. Many of the main industrial gases have also been used in warfare, particularly in the First World War. Accounts of the military use of these gases do not generally provide usable quantitative data, but some give pointers to the effects of the gases on humans that constitute a collateral check on data from animal experiments, which is the main source.

11.38.5 Vulnerability Model

As just stated, some of the first correlations for the toxicity of industrial gases relevant to major releases were those given in the vulnerability model by Eisenberg et al.

(1975). These workers analyzed data for the toxicity of chlorine and ammonia and obtained probit equations for the lethality of these two gases.

Eisenberg, Lynch, and Breeding took the causative factor V in the probit equation for chlorine as

$$V = \sum C_i^n T_i \qquad (11.91)$$

where C is the concentration (ppm), T is the time interval (min), and n is an index. From analysis of the data, they obtained a value of $n = 2.75$. The probit equation for chlorine lethality derived by these authors is then

$$Y = -17.1 + 1.69 \ln\left(\sum C_{Ci}^{2.75} T_i\right) \qquad (11.92)$$

where Y is the probit.

Derive the following probit equation for ammonia lethality:

$$Y = -30.57 + 1.3851n\left(\sum C_i^{2.75} T_i\right) \qquad (11.93)$$

The relationships described apply to healthy adults. The population also includes more susceptible individuals such as infants, old people, and people with advanced pulmonary/cardiovascular disease.

Eisenberg, Lynch, and Breeding also consider non-lethal injury from chlorine. In this case, the causative factor is taken simply as $V = C$. The probit equation for non-lethal injury from chlorine is then

$$Y = -2.40 + 2.90 \ln C \qquad (11.94)$$

Non-lethal injury is here taken to mean hospitalization without or with lasting impairment of health.

A further set of probit equations for lethality for use in the vulnerability model was derived by Perry and Articola (1980). The support for the probit equations in the vulnerability model is in some cases rather slight and the level of physical activity to which they are applicable is not well defined.

11.38.6 Major Industrial Gases

In general, toxic hazard from industrial gases arises from their manufacture, storage, and use in the manufacture of other chemicals. In some cases, hazard arises from another cause. For irritant gases, an important parameter is the solubility in water, which largely determines the part of the respiratory system attacked. The toxicity of two major industrial gases is now considered in more detail.

There are two industrial toxic gases which in liquefied form are handled on a large scale and are readily dispersed and which therefore present a major toxic hazard. Chlorine is one, and ammonia the other.

Chlorine is a highly toxic gas. The long- and short-term OESs are 0.5 and 1 ppm, respectively. The values represent a small reduction from the TLV (TWA) of 1 ppm that held for many years. Physiologically, chlorine is an irritant gas. The effects of a single exposure include irritation of the mucous membranes, attack of the respiratory tract and pulmonary edema.

The other principal industrial toxic gas is ammonia. Like chlorine, it is a severe toxic hazard because it is handled on a large scale, it is a liquefied gas and is therefore readily dispersed, and it is highly toxic. Ammonia has long- and short-term OESs of 25 and 35 ppm, respectively.

Physiologically, ammonia is an irritant gas. The effects of a single exposure include irritation of the mucous membranes, attack of the respiratory tract and pulmonary edema. Ammonia is very much more soluble in water than chlorine and therefore tends to attack particularly the upper respiratory tract, stripping the lining, and inducing laryngeal edema.

There is little evidence of serious long-term effects of non-lethal single exposure, except where the exposure has been so serious that death has been prevented only by medical attention, when respiratory deficiencies may occur.

11.38.7 MHAP Studies

The toxicity of three gases—chlorine, ammonia, and phosgene—is the subject of a set of monographs by the MHAP, convened under the auspices of the Institution of Chemical Engineers (IChemE).

11.38.7.1 Chlorine

The Chlorine Toxicity Monograph (MHAP, 1987) is in two parts: the First Report of the Working Party and an account by Davies and Hymes of the HSE of the approach by that body to the development of criteria for hazard assessment.

The working party reviews the data on chlorine toxicity. Many of the data quoted for humans are derivative and appear poorly founded. It is therefore necessary to resort to the use of animal data. The authors present data for the LC_{50} for different exposures and for a number of species.

In respect of the concentration lethal to man, the authors find that the effects of acute inhalation of chlorine are similar in animals and humans, the prime lethal effect being pulmonary edema; the factors bearing on differences between species may be divided into those which govern delivery of the chemical to the target organ and those which govern the effect on this and other organs; and that the effect of chlorine is to cause damage to the respiratory system itself, so that metabolic factors do not come into play. They conclude that there is little basis for supposing that the sensitivity of humans to chlorine is significantly different from that of small animals.

The contribution by Davies and Hymes states the HSE view that for the purpose of land use planning a toxic load criterion corresponding to a much less severe level of injury than 50% lethality is appropriate and quotes the Major Hazards Assessment Unit (MHAU) as using for this purpose the criterion

$$C^{2.75}t = 3.2 \times 16^6 \qquad (11.95)$$

where C is the concentration (ppm) and t is the time (min).

11.38.7.2 Ammonia

The *Ammonia Toxicity Monograph* (MHAP, 1988) follows a broadly similar format. The data on ammonia toxicity are reviewed and animal data are presented for the LC_{50} for different exposures and for a number of species. With regard to the applicability of these data to man, the authors conclude that it is reasonable to apply the animal data to humans.

The monograph also gives an account of medical measures for handling of cases of gassing by ammonia, brief descriptions of case histories involving the gas, and summaries of some 24 papers on its toxicity.

11.38.7.3 Phosgene

The *Phosgene Toxicity Monograph* (MHAP, 1993) again rejects many of the data quoted for humans as being second-hand and unreliable. The same applies to many of the animal data, but a small number of data sets are considered to be of high quality.

In respect of the application of the animal data to humans, the authors argue that the main acute effect of phosgene for both animals and humans is on the lung, that large differences in response between species would not be expected and that this is borne out by the similarity of the results obtained for the various species, including human being. They conclude that the animal data may be applied to humans.

The monograph also gives an account of medical measures for handling of cases of gassing by phosgene and brief descriptions of case histories involving the gas.

11.38.8 Probit Equations

As described above, probit equations are available for a number of the toxic gases of industrial interest. In addition to the inherent limitations of probit equations, it is also necessary to consider in each particular case both the

situation to which it is intended to apply and the degree of conservatism incorporated. The basic data are generally obtained from experiments on animals that tend to remain passive, but the use of the correlation is presumably for humans in an emergency situation. In some cases, the probit equation is explicitly conservative, while in others it is not, reflecting a difference of philosophy as to the point at which any safety factor should be applied.

11.38.9 HSE Dangerous Dose

The HSE uses for the purpose of land use planning the concept of a 'dangerous dose.' For toxic substances, the interpretation of the dangerous dose concept and the methodology by which a value is derived is described in *Assessment of the Toxicity of Major Hazard Substances* (R.M. Turner and Fairhurst, 1989).

The authors discuss the paucity of data on the effect of toxic gases on man and the problems of probit equations. They propose for land use planning the use not of probit equations but of a toxic load value that they call the SLOT. This is expressed in the form of an equation for the dangerous toxic load (DTL) and of a set of values for the SLOT computed from this relation.

The SLOT is defined as a value associated with a situation in which (1) almost all persons suffer severe distress; (2) a substantial fraction require medical attention; (3) some persons are seriously injured, requiring prolonged treatment; or (4) any highly susceptible person might be killed. Typically, the SLOT used corresponds to a mortality of 1−5%.

The methodology described by Turner and Fairhurst for the derivation of a SLOT is broadly as follows. Since there is a lack of toxicity data for humans, it is necessary to resort to the use of data from experiments on animals. The most reliable data generally relate to the LC_{50} values and the associated exposure times, and it is this that is taken as the starting point. Data are gathered for a number of species and usually the data adopted are those for the most sensitive species. Next, estimates are made of the LC_{50} or LC_{01} values. In some cases, the data available permit an extrapolation based on probits. If this is not possible, an alternative approach is to use an empirical value of the ratio LC_{50}/LC_{05}, or LC_{50}/LC_{01}, as the case may be.

At this stage, therefore, the method yields a single value of the SLOT LC_{50} or LC_{01} with an associated exposure time. It is then necessary to determine the trade-off between the concentration c and the time t, or, in other words, the value of the index in the toxic load cnt. Essentially this requires data on concentration vs time for a given level of effect, such as the LC_{50}.

The SLOTs obtained are then correlated as a DTL equation. These SLOTs and the DTL are applicable to an animal species. Collateral evidence may then be sought to confirm that it is reasonable to apply the relation to humans.

For a uniformly distributed population, the number of fatalities caused by a toxic release may be approximated by estimating the number of people exposed to an LD_{50} dose (i.e., SLOD DTL) inside the concentration contour. The approximation results according to the assumption that those people inside the SLOD contour who do not die (due to factors such as physiology and fitness levels) will be balanced by an approximately equal number outside the SLOD contour who do die.

Moreover, the number of people injured (serious and minor) by the release will also closely equal the number of people estimated to be between the SLOD and SLOT DTL contours.

When assessing the numbers of people affected, the first factor to take into account is the proportion of the population that is indoors. When compared to the proportion outdoors, there will be a degree of protection against the effects of the release. Usually the level of protection is determined by the rate measured by air changes per hour (ACH) when air and toxic materials enter a building.

11.38.10 Combustion Gases

A somewhat separate problem concerns the toxicity of combustion products. Much of the research interest in such gases centers on the combustion of furnishings, but work in this area is also relevant to toxic clouds from combustion of materials on plant and in storage, particularly warehouses.

11.39. PLANT DESIGN FOR TOXIC SUBSTANCES

Plants handling toxic substances need to be designed to minimize both large accidental releases and fugitive emissions.

As far as concerns large toxic releases, this is not solely a matter of the mechanical design of the plant, important though that is. Inherently safer design also has an important part to play, in minimizing the effects of any failure through the choice of the substances used in the process and the operating conditions.

There are a number of accounts, both by plant designers and by occupational hygienists, of the design and operation of plant handling toxic substances to counter fugitive emissions and the resultant chronic toxic hazard. Emissions of toxic substances to the atmosphere are not confined to continuous leaks from the plant. They also occur as a result of operations carried out on the plant, particularly those involving purging or breaking into equipment for operations or maintenance purposes.

The design of plants to minimize exposure to toxic substances is essentially a two-pronged one, based on the reduction of fugitive emissions and the provision of ventilation.

There has been a growing concern over the emission of hydrocarbons and volatile organic compounds (VOCs). Two points made there bear reiteration. One is that there is considerable variability between plants in the levels of emission, and the other that it is possible for plants to achieve very low levels of emissions, though as always this is easier in the design of a new plant.

Most of the data on fugitive emissions relate to hydrocarbons and VOCs. They provide a useful initial indication, but in view of the wide differences in the results obtained and the rather different nature of the problem in the case of toxic materials, care should be exercised in applying them to toxic emissions.

11.40. TOXIC GAS DETECTION

A release of toxic gas may be detected by the visual appearance of the cloud, by its odor, or by instrumentation.

Some gases have a characteristic color. Even if there is no such color, the gas may well form a visible fog. Some gases tend to form a fog by taking up moisture from the air to form an aerosol.

Many gases give a characteristic odor. The concentrations at which such an odor is detectable vary by orders of magnitude. In some cases, a person can become desensitized to an odor. The classic case is hydrogen sulfide.

In some cases, it is appropriate to install toxic gas detectors. Details of detectors and their response times are given in the CCPS Guidelines. The purpose of a toxic gas detection system should be clearly defined. One purpose may be to give a rapid warning of a major release. Another may be to detect fugitive emissions of a toxic gas for occupational hygiene purposes. In locating the detectors, there is a choice between monitoring specific potential leak sources and giving good area coverage.

The CCPS Guidelines emphasize that a toxic gas detection system with an insufficient number of detectors or inadequate maintenance may be worse than no system at all, particularly if it leads to less human surveillance or to confusion arising because a field report of a leak is not confirmed by the detection system.

11.41. TOXIC RELEASE RESPONSE

Essentially, the control of the toxic release hazard means, on the one hand, the prevention of serious loss of containment and, on the other, the elimination of hazardous concentrations in the environment. There are also the intermediate problems of dealing with small quantities of toxic materials arising mainly from leaks and spillages and from maintenance operations.

Elimination of hazardous concentrations in the working environment requires assessment of leaks and other sources of toxic substances. Engineering measures can then be taken to improve the leak-tightness of plant.

It is also necessary to have ventilation and monitoring of the atmosphere and medical checks on personnel. In addition, there should be procedures for handling of abnormal leaks and spillages.

11.41.1 Leaks and Spillages

If the plant is in the open, small leakages from the plant may be dispersed by the wind. If the plant is in a building, mechanical ventilation is necessary. The toxic hazard should be a principal consideration in deciding whether to put a plant in a building or in the open.

Provision should be made for handling larger emissions. It may be necessary to have emergency isolation valves and relief, blowdown and gas absorption facilities in order to reduce the amount likely to escape. In dealing with spillages, actions should be avoided which actually increase the rate of evaporation.

11.41.2 Emergency Action

The effect of a large toxic release can be greatly mitigated if the people exposed take the right action. In plant handling toxic materials, workers have protective buildings and equipment and are trained in emergency procedures. It is commonly considered that, provided these precautions are taken, workers in the factory itself are at no more risk than members of the public.

In order to prepare instructions to be issued to the public in the event of a large release, it is necessary to decide whether it is safer for a person to flee from the gas cloud or to stay indoors taking measures to prevent ingress of the chemical. Methods are available for the calculation of the rate at which a toxic gas diffuses into buildings.

The usual practice in assessing a toxic release hazard is to consider a range of scenarios. For some scenarios, the best course of action may be to evacuate, while for others it may be to stay indoors. It should be borne in mind, however, that any instruction to be issued to the public should be simple and clear. In most cases, the preferred advice to the public is to stay indoors and shut doors and windows.

11.42. TOXIC RELEASE RISK

In general, for all three major hazards—fire, explosion, and toxic release—the large number of fatalities given by some theoretical estimates, assuming the most

unfavorable and improbable circumstances and using models which may prove to be based on pessimistic assumptions, has been in contrast with the small number of fatalities shown by the historical record.

The alternative approach to the determination of the risk from a large toxic release is the use of hazard assessment involving assumed scenarios of release and with appropriate estimates of emission, dispersion, and toxic effects.

11.43. HAZARD ASSESSMENT METHODOLOGY

Developments in the methodology for the hazard assessment of toxic releases have occurred in a number of areas, including the following: (1) source term, (2) heavy gas dispersion, (3) concentration fluctuations, (4) mitigation by barriers and sprays, (5) mitigation by shelter and evacuation, (6) toxicity relations, (7) degree of injury, (8) specific gases, (9) population exposure, (10) plant layout, (11) warehouse fires, and (12) computer aids.

11.43.1 Source Term

One area in which the hazard assessment of toxic releases has become much more realistic is in the handling of the source term. Much early work was based on the outright failure of a pressure vessel and, often, the ejection of its entire contents.

It is recognized that the total failure of a pressure vessel is very rare. Better estimates are available of lesser failures such of those of pipework. Allowance is made for emergency action such as the closure of emergency isolation valves.

Methods have been developed to guide the choice between continuous and quasi-instantaneous release models, and thus reduce the proportion of cases where the more pessimistic quasi-instantaneous case is assumed.

Progress has been made in the modeling of the release of pressurized or refrigerated fluids. Models are available for the behavior of the inventory on rupture of a pressure vessel and for two-phase flow from leaks on pipework. The models for evaporation from the pool formed following emission of a liquid are also much improved.

11.43.2 Gas Dispersion

In most cases, the vapors of the principal toxic liquids exhibit heavy gas behavior and great strides have been made in heavy gas dispersion modeling. The methods now available, which include models for manual use such as the Workbook models, computer codes such as HEGADAS, and the three-dimensional models as well as physical modeling using wind tunnels, provide a set of tools capable of handling not only dispersion over flat, unobstructed terrain, but also the effects of slopes, buildings, barriers, and water sprays. The concentration estimates yielded by these heavy gas models tend to be quite different from those of passive gas dispersion models.

11.43.3 Concentration Fluctuations

It has long been appreciated that there are considerable fluctuations in the concentrations at a fixed point in a gas cloud. Methods are now available which allow estimates to be made of these concentration fluctuations. It is also appreciated that these fluctuations must have some influence on the effective toxic load.

11.43.4 Mitigation: Terrain, Barriers, and Sprays

As just stated, the heavy gas dispersion models available can now treat situations where mitigating features exist. Since the typical release occurs in a works, the gas dispersion will be strongly affected by the presence of buildings, which tend to enhance the dispersion. An illustration of a hazard assessment of a toxic release in the presence of buildings is given by Deaves (1987).

11.43.5 Mitigation: Shelter and Evacuation

Another major form of mitigation occurs where the exposed population remains indoors and thus benefits from shelter. Methods have been developed to estimate the indoor concentration−time profile, and hence toxic load, given the profile of the outdoor concentration.

11.43.6 Toxicity Relations

The form of injury relation generally used in the hazard assessment of toxic releases has been the probit equation. An account has been given above of the development of probit equations for this purpose.

The use of probit equations involves a number of problems. In large part, these reflect the basic difficulty of determining the human response to a toxic load. One problem is that a probit equation is available only for a relatively small proportion of toxic gases. Even where a probit equation exists, it may be subject to considerable uncertainty. The accuracy of the relation can be expected to be highest close to the LC_{50}, but relatively low when extended to an LC_{05} or LC_{01} level. Yet in many cases, the number of persons exposed to these lower loads is much higher than that for those exposed to the higher load; where there is a cordon sanitaire around the site, as there usually is to some degree, the difference is increased.

Particularly where the purpose of the hazard assessment is to identify the contour for a particular level of toxic effect, as in land use planning, use may be made not of a probit equation but of some fixed toxic load, such as the HSE SLOT value.

11.43.7 Degree of Injury

Whereas the effect commonly estimated in hazard assessments of toxic releases has tended in the past to be the number of fatalities, there is some trend to estimate lesser degrees of injury. Thus the HSE SLOT value is a toxic load that is estimated not to be fatal to the vast majority of persons exposed. However, this type of criterion tends to be used not to determine the total number of persons exposed to the toxic load but the location of the contour around the site where the load will occur.

11.43.8 Specific Gases

There have also been significant advances in the characterization of the toxicity of a number of common industrial gases. In particular, reference may be made to the CPD Green Book probit equations, the HSE SLOT values and the toxicity estimates, and probit equations of the MHAP.

11.43.9 Population Exposure

Another area that has become more sophisticated is the characterization of the exposed population. There now exist what are, in effect, population exposure models.

11.43.10 Plant Layout

Increasingly, the effect of flammable and toxic releases is taken into account in plant layout, although the extent to which this is practical is greater for flammable and explosive materials than for toxic ones.

11.43.11 Warehouse Fires

Another type of toxic hazard is that arising from warehouse fires. Most work on the quantitative treatment of this hazard is relatively recent.

11.43.12 Computer Aids

The computation either of the consequences of a toxic release or of the risk to the population around an industrial site or along a transport route is a natural application for computer codes, and a considerable number of codes are available.

REFERENCES

AIChE (American Institute of Chemical Engineers), 1992/149. Emergency Relief Systems for Runaway Reactions and Storage Vessels: A Summary of Multiphase Flow Methods (New York).

American Society of Mechanical Engineers, 1969. Recommended Guide for the Prediction of the Dispersion of Airborne Effluents. American Society for Mechanical Engineers, New York.

Appleton, P.R., 1984. A study of two-phase flashing jets. SRD report R303, UK AEA, Culcheth, Warrington, Cheshire, UK.

Baker, O., 1954. Designing for simultaneous flow of oil and gas. Oil Gas J. July, 185.

Baker, O., 1958. Multiphase flow in pipelines. Oil Gas J. November, 156.

Baroczy, C.J., 1966. A systematic correlation for two-phase pressure drop. In: Heat Transfer Los Angeles. American Institute of Chemical Engineers, New York, p. 232.

Barry, P.J., 1971. A note on peak-to-mean concentration ratios. Bound. Layer Meteorol. 2, 122.

Bell, R.P., 1978. Isopleth calculations for ruptures of sour gas pipelines. Energy Process/Canada. JulyAugust, 36.

Benjaminsen, M.W., Miller, J.G., 1941. The flow of saturated water through throttling orifices. Trans. ASME. 63 (5), 419.

Bettis, R.J., Nolan, P.F., Moodie, K., 1987. Two-phase flashing releases following rapid depressurisation due to vessel failure. Hazards from Pressure. Inst. Chem. Engrs.-IChemE, Rugby, p. 247.

Blackmore, D.R., Herman, M.N., Woodward, J.L., 1982. Heavy gas dispersion models. J. Hazard. Mater. 6 (1/2), 107 (See also in Britter, R.E., Griffiths, R.F., 1982, op. cit., p. 107).

Bodurtha, F.T., Palmer, P.A., Walsh, W.H., 1973. Discharge of heavy gas from relief valves. Loss Prev. 7, 61.

Burgess, D.S., Zabetakis, M.G., 1973. Detonation of a flammable cloud following a propane pipeline break: the December 9, 1970, explosion in Port Hudson, Mo. (Technical Report BM-RI-7752), Bureau of Mines, Pittsburgh, Pa. (USA). Pittsburgh Mining and Safety Research Center.

Burgess, D.S., Murphy, J.N., Zabetakis, M.G., 1970. Hazards of LNG spillage in marine transportation. SRC Rep. MIPR Z-70099992317 S-4105.

Burgess, D., Biordi, J., Murphy, J., 1972. Hazards of spillage of LNG into water. Bur. Mines. (PMSRC Rep. 4177).

Burnell, J.G., 1947. Flow of boiling water through nozzles, orifices and pipes. Eng. London. 164 (4272), 572.

Bushnell, D.M., Gooderum, P.B., 1968. Atomisation of superheated water jets at low ambient pressures. J. Spacecraft Rockets. 5, 231.

Butterworth, D., Hewitt, G.F., 1977. Two-phase Flow and Heat Transfer. Oxford University Press, Oxford.

Cairney, E.M., Cude, A.L., 1971. The safe dispersal of large clouds of flammable heavy vapours. Major Loss Prev.163.

Chan, S.T., Ermak, D.L., 1984. Recent results in simulating LNG vapor dispersion over variable terrain. In: Ooms, G., Tennekes, H., op. cit., p. 105.

Chan, S.T., Rodean, H.C., Ermak, D.L., 1984. Numerical simulations of atmospheric releases of heavy gases over variable terrain. In: de Wispelaere, C. (Ed.), op. cit., p. 295.

Chatwin, P.C., 1982. The use of statistics in describing and predicting the effects of dispersing gas clouds. J. Hazard. Mater. 6 (1/2), 213. (See also in Britter, R.E., Griffiths, R.F. (1982), op. cit., 213).

Chen, J.R., Richardson, S.M., Saville, G., 1992. Numerical simulation of full-bore ruptures of pipelines containing perfect gases. Process. Saf. Environ. 70B, 59.

Chisholm, D., 1983. Two-Phase Flow in Pipelines and Heat Exchangers. Godwin, London.

Chisholm, D., Watson, G.G., 1966. The Flow of Steam/Water Mixtures through Sharp Edged Orifices. National Engineering Lab, East Kilbride (Rep. NEL 213).

Chisholm, D., Sutherland, L.A., 1969. Prediction of pressure gradients in pipeline systems during two-phase flow. Proc. Inst. Mech. Eng. 184 (3c), 24.

Colenbrander, G.W., Puttock, J.S., 1984. Maplin Sands experiments 1980: interpretation and modelling of liquified gas spills onto the sea. In: Ooms, G., Tennekes, H. (Eds.), op. cit.

Coulson, J.M., Richardson, J.F., 1955. Chemical Engineering. Pergamon Press, Oxford (vol. 1, 1955; vol. 2, 1955; vol. 3, 1971).

Coulson, J.M., Richardson, J.F., 1977. Chemical Engineering, SI ed. Pergamon Press, Oxford.

Csanady, G.T., 1973. Turbulent Diffusion in the Environment. Reidel, Dordrecht.

Davies, J.K.W., 1987. A comparison between the variability exhibited in small scale experiments and in the Thorney Island Phase I trials. J. Hazard. Mater. 16, 339.

Davies, J.K.W., 1989a. The application of box models in the analysis of toxic hazards by using the probit doseresponse relationship. J. Hazard. Mater. 22 (3), 319.

Davies, J.K.W., 1989b. Probabilistic methods in the development of mathematical models of system failure. In: Cox, R.A. (Ed.) (1989), op. cit., p. 47.

Deaves, D.M., 1987. Incorporation of the effects of buildings and obstructions on gas cloud consequence analysis. Vapor Cloud Model.844.

Dow Chemical Company, 1994. Dow's Chemical Exposure Index. American Institute of Chemical Engineers, New York. NY.

Dukler, A.E., Wicks, M., Cleveland, R.G., 1964. Frictional pressure drop in two phase flow. AIChE J. 10, 38, and 44.

Eisenberg, N.A., Lynch, C.J., Breeding, R.J., 1975. Vulnerability Model: A Simulation System for Assessing Damage Resulting from Marine Spills. Rep. CG-D-136−75. Enviro Control Inc., Rockville, MD.

Ermak, D.L., Chan, S.T., 1986. A study of heavy gas effects on the atmospheric dispersion of dense gases. In: de Wispelaere, C., Schiermeier, F.A., Gillani, N.V. (Eds.), op. cit., p. 723.

Ermak, D.L., Chan, S.T., 1988. Recent developments on the FEM3 and SLAB atmospheric dispersion models. In: Puttock, J.S. (Ed.), op. cit., p. 261.

Ermak, D.L., Chan, S.T., Morgan, D.J., Morris, L.K., 1982. A comparison of dense gas dispersion model simulations with Burro series LNG spill test results. J. Hazard. Mater. 6 (1/2), 129.

Ermak, D.L., Goldwire, H.C., Hogan, W.J., Koopman, R., Mcrae, T.G. 1983. Results of 40-m^3 LNG spills on water. In: Hartwig, S. (1983), op. cit., p. 163.

Fauske, H.K., 1961. Critical two-phase, steam-water flows. In: Proceedings of the Heat Transfer and Fluid Mechanics Institute, p. 79.

Fauske, H.K., 1962. Contribution to the Theory of Two-Phase, One Component Critical Flow. Rep. ANL 6633. Argonne National Laboratory, Argonne, IL.

Fauske, H.K., 1963. A theory for predicting pressure gradients for two-phase critical flow. Nucl. Sci. Eng. 17, 1.

Fauske, H.K., 1964. The discharge of saturated water through tubes. Seventh National Heat Transfer Conference. American Institute of Chemical Engineers, New York, NY, p. 210.

Fauske, H.K., 1986. Disposal of two-phase emergency releases. In: Proceedings of the Fourth Miami International Symposium on Multi-Phase Transport and Phenomena. Miami Beach, FL, December 15–17.

Fauske, H.K., Min, T.C., 1963. A Study of the Flow of Saturated Freon 11 through Apertures and Short Tubes. Argonne National Laboratory, Argonne, IL. (Rep. ANL 6667).

Fauske, H.K., Grolmes, M.A., Henry, R.E., 1983. Emergency relief systems sizing and scaleup. Plant/Operations Prog. 2 (1), 27.

Feldbauer, G.F., Heigl, J.J., Mcqueen, W., Whipp, R.H., May, W.G., 1972. Spills of LNG on Water Vaporization and Downwind Drift of Combustible Mixtures. Esso Research and Engineering Co., Rep. EE61E-72.

Fletcher, B., 1982. Sudden discharge of superheated fluid to atmosphere. In: The Assessment of Major Hazards. Institution Chemical Engineers, Rugby, p. 25.

Fletcher, B., 1984a. Discharge of saturated liquids through pipes. J. Hazard. Mater. 8 (4), 377.

Fletcher, B., 1984b. Flashing flow through orifices and pipes. Chem. Eng. Prog. 80 (3), 76.

Fletcher, B., Johnson, A.E., 1984. The discharge of superheated liquids from pipes. In: The Protection of Exothermic Reactors and Pressurised Storage Vessels. Institution of Chemical Engineers, Rugby, p. 149.

Gerardu, N.H., 1981. Safe blowoff conditions for storage vessels. Loss Prev. 14, 134 (See also Chem. Eng. Prog. 1981, 77 (1), 49).

Gifford, F.A., 1960. Atmospheric diffusion calculations using the generalized Gaussian plume model. Nucl. Saf. 2 (2), 56.

Gifford, F.A., 1961. Use of routine meteorological observations for estimating atmospheric dispersion. Nucl. Saf. 2 (4), 47.

Gosline, C.A., 1952. Dispersion from short stacks. Chem. Eng. Prog. 48 (4), 165.

Griffiths, R.F., Megson, L.C., 1984. The effect of uncertainties in human toxic response on hazard range estimation of ammonia and chlorine. Atmos. Environ. 18, 1195.

Griffiths, R.F., Harper, A.S., 1985. A speculation on the importance of concentration fluctuations in the estimation of toxic response to irritant gases. J. Hazard. Mater. 11 (1/3), 369. See also in McQuaid, J., op. cit., p. 369.

Grolmes, M.A., Epstein, M., 1985. Vapor-liquid disengagement in atmospheric liquid storage vessels subjected to external heat source. Plant/Operations Prog. 4, 200.

Haddock, S.R., Williams, R.J., 1978. The density of an ammonia cloud in the early stages of atmospheric dispersion. United Kingdom Atomic Energy Authority Safety and Reliability Directorate Report SRD R103. Culcheth, Warrington, Cheshire, UK. (1978 SRD R103).

Haddock, S.R., Williams, R.J., 1979. The density of an ammonia cloud in the early stages of atmospheric dispersion. J. Chem. Technol. Biotechnol. 29, 655.

Hanna, S.R., 1984. Concentration fluctuations in a smoke plume. Atmos. Environ. 18, 1091.

Hanna, S.R., Drivas, P.J., 1987. Guidelines for use of vapour cloud dispersion models, Center for Chemical Process Safety CCPS (1987/2). AIChE, New York.

Hardee, H.C., Lee, D.O., 1974. Expansions of Clouds from Pressurised Liquids. Sandia Lab, Albuquerque, NM (Rep. SLA-73-7012).

Hardee, H.C., Lee, D.O., 1975. Expansion of clouds from pressurised liquids. Accidents Anal. Prev. 7, 91.

Havens, J.A., 1978. An assessment of the predictability of LNG vapor dispersion from catastrophic spills onto water. Fifth International Symposium on Transport of Dangerous Goods by Sea and Inland Waterways, paper C2.

Havens, J.A., 1985. The atmospheric dispersion of heavy gases: an update. In: The Assessment and Control of Major Hazards. Institution of Chemical Engineers, Rugby, p. 143.

Havens, J.A., 1986. Mathematical modeling of heavy gas dispersion an overview. Loss Prev. Saf. Promotion. 5, 32.2.

Havens, J., 1992. Review of dense gas dispersion field experiments. J. Loss Prev. Process Ind. 5 (1), 2841.

Henry, R.E., Fauske, H.K., 1971. The two-phase critical flow of one-component mixtures in nozzles, orifices and short tubes. J. Heat Transfer. 93 (May), 179.

Hesketh, H.E., Cross, F.L., 1983. Fugitive Emissions and Controls. Ann Arbor Science, Ann Arbor, MI.

Hess, K., Hoffmann, W., Stoeckel, A., 1974. Propagation processes after the bursting of tanks filled with liquid propane experiments and mathematical model. Loss Prev. Saf. Promotion. 1, 227.

High, W.G., 1976. Explosions: Occurrence, Containment and Effects. Course on Loss Prevention in the Process Industries. Department of Chemical Engineering, Loughborough University of Technology, Loughborough.

Humbert-Basset, R., Montet, A., 1972. Flammable mixtures penetration in the atmosphere from spillage of LNG. LNG Conf. 3.

Inkofer, W.A., 1969. Ammonia transport via pipeline. Ammonia Plant Safety. 11, 40.

Isbin, H.S., Moy, J.E., Da Cruz, A.J.R., 1957. Two-phase steam water critical flow. AIChE J. 3, 361.

Isbin, H.S., Fauske, H.K., Grace, T., Garcia, I., 1962. Two-phase steam-water pressure drops for critical flows. In: Symposium on Two-Phase Fluid Flow. Institution of Mechanical Engineers, London.

Jones, C.D., 1983. On the structure of instantaneous plumes in the atmosphere. J. Hazard. Mater. 7 (2), 87.

Jones, M.R.O., Underwood, M.C., 1983. An appraisal of expressions used to calculate the release rate of pressurised liquefied gases. Chem. Eng. J. 26, 251.

Jones, M.R.O., Underwood, M.C., 1984. The small-scale release rate of pressurised liquefied propane to the atmosphere. The Protection of Exothermic Reactors and Pressurised Storage Vessels. Institution of Chemical Engineers, Rugby, p. 157.

Kletz, T.A., 1977. Unconfined vapour cloud explosions. Loss Prev. 11, 50.

Kneebone, A., Prew, L.R., 1974. Shipboard jettison tests of LNG onto the sea. LNG Conf. 4., Algiers, Nov. 13-13.

Koopman, R.P., Bowman, B.R., Ermak, D.R., 1980. Data and calculations on 5 m3 LNG spill tests. In: Liquefied Gaseous Fuels Safety and Environmental Control Assessment Program. Second Status Report, Report P, Department of Energy.

Koopman, R.P., Mcrae, T.G., Goldwire, H.C., Ermak, D.L., Kansa, E.J., 1986. Results of recent large-scale NH3 and N2O4 dispersion experiments. In: Hartwig, S. (Ed.), op. cit., p. 137.

Koopman, R.P., Ermak, D.L., Chan, S.T., 1989. A review of recent field tests and mathematical modelling of atmospheric dispersion of large spills of denser-than air gases. Atmos. Environ. 23, 731.

Lemonnier, H., Selmer-Olsen, S., Aarvik, A., Fjeld, M., 1991. The effect of relaxed mechanical non-equilibrium on gas-liquid critical flow modeling. In: Proceedings of the Twenty-Seventh National Heat Transfer Conference. American Nuclear Society, Minneapolis, MN (July 28–31).

Levy, S., 1965. Prediction of two-phase critical flow rate. J. Heat Transfer. 87, 53.

Leung, J.C., 1986. A generalised correlation for one-component homogeneous equilibrium flashing choked flow. AIChE J. 32, 1743.

Leung, J.C., 1990a. Similarity between flashing and non-flashing two-phase flows. AIChE J. 36, 797.

Leung, J.C., 1990b. Two-phase flow discharge in nozzles and pipes a unified approach. J. Loss Prev. Process Ind. 3 (1), 27.

Leung, J.C., 1992. Size safety relief valves for flashing liquids. Chem. Eng. Prog. 88 (2), 70.

Lewitt, E.H., 1952. Hydraulics. ninth ed. Pitman, London.

Lipton, S., Lynch, J., 1987. Health Hazard Control in the Chemical Process Industry. Wiley, New York.

Lipton, S., Lynch, J., 1994. Handbook of Health Hazard Control in the Chemical Process Industry. Wiley, New York.

Lockhart, R.W., Martinelli, R.C., 1949. Proposed correlation of data for isothermal two-phase, two-component flow in pipes. Chem. Eng. Prog. 45, 39.

Long, V.D., 1963. Estimation of the extent of hazard areas around a vent. Chem. Process Hazard. II, 6.

Loudon, D.E., 1963. Requirements for safe discharge of hydrocarbons to atmosphere. API Proceedings, Division of Refining, 43, 418.

Lowrance, W.W., 1984. Reducing industrial cancer: the strategic agenda. In: Deisler, P.F. (Ed.),op. cit., p. 21

Luketa-Hanlin, A., 2006. A review of large-scale LNG spills: Experiments and modeling. J. Hazard. Mater. 132 (2–3), 119–140.

Macfarlane, K., 1991. Development of plume and jet release models (poster item). Modeling and Mitigating the Consequences of Accidental Releases, 657.

Major Hazards Assessment Panel, 1987. Chlorine Toxicity Monograph. Institution of Chemical Engineers, Rugby.

Major Hazards Assessment Panel, 1988. Ammonia Toxicity Monograph. Institution of Chemical Engineers, Rugby.

Major Hazards Assessment Panel, 1993. Phosgene Toxicity Monograph. Institution of Chemical Engineers, Rugby.

Massey, B.S., 1968. Mechanics of Fluids; (1970, second ed, Van Nostrand, London; 1975, third ed. 1979, fourth ed. Van Nostrand, New York; 1983, fifth ed. van Nostrand, Wokingham; 1989, sixth ed. Chapman & Hall, London).

Maurer, B., et al., 1977. Modeling of vapour cloud dispersion and deflagration after bursting of tanks filled with liquefied gas. Loss Prev. Saf. Promotion Process Ind. 2, 305.

Mcquaid, J., 1980. Dispersion of heavy gases by the wind and by steam or water-spray barriers. In: Course on Process Safety Theory and Practice. Teesside Polytechnic, Middlesbrough.

Mcquaid, J., 1984. Large scale experiments on the dispersion of heavy gas clouds. In: Ooms, G., Tennekes, H. (Eds.), op. cit., p. 129.

Moody, F.J., 1965. Maximum flow rate of a single component two-phase mixture. J. Heat Transfer. 87, 134.

Morgan, D.L., Kansa, E.J., Morris, L.K., 1984. Simulations and parameter variation studies of heavy gas dispersion using the slab model. In: Ooms, G., Tennekes, H. (Eds.), op. cit., p. 83.

Morris, S.D., 1988a. Choking conditions for flashing one-component flows in nozzles and valves a simple estimation method. Preventing Major Chemical and Related Process Accidents, p. 281.

Morris, S.D., 1988b. Offshore Relief Systems: Some Guidelines in Designing for Two-phase Flow. Energy Technology Unit, Heriot-Watt University, Edinburgh (Rep. ETU-027).

Morris, S.D., 1990a. Discharge coefficients for choked gas-liquid flow through nozzles and orifices and applications to safety devices. J. Loss Prev. Process Ind. 3 (3), 303.

Morris, S.D., 1990b. Flashing flow through relief lines, pipe breaks and cracks. J. Loss Prev. Process Ind. 3 (1), 17.

Morrow, T.B., Bass, R.L., Lock, J.A., 1982. An LPG pipeline break flow model. Fifth Annual Energy Sources Technology Conference, ASME paper I00149. American Society of Mechanical Engineers, New York.

Norman, C., 1987. A guide to the use of foam on hazardous material spills. Hazard. Mater. Waste Manage. September/October.

NTSB (National Transportation Safety Board), 1972. Phillips Pipeline Company, Propane Gas Explosion, Franklin County, Missouri, December 9, 1970. NTSB (1972 PAR-72-01).

Nyren, K., Winter, S., 1983. Two phase discharge of liquefied gases through pipes. Field experiments with ammonia and theoretical model. Loss Prev. Saf. Promotion. 4 (1), E1.

Ooms, G., Duijm, N.J., 1984. Dispersion of a stack plume heavier than air. In: Ooms, G., Tennekes, H. (Eds.), op. cit., p. 1.

Opschoor, G., 1979. Methods for the Calculation of the Physical Effects of the Escape of Dangerous Materials (Report). TNO, Voorburg, The Netherlands.

Pasqua, P.F., 1953. Metastable flow of Freon 12. Refrig. Eng. 61, 1084A.

Pasquill, F., 1961. The estimation of the dispersion of windborne materials. Met. Mag. 90 (1063), 33.

Pasquill, F., 1962. Atmospheric Diffusion. van Nostrand, London.

Pasquill, F., 1965. Meteorological Aspects of the Spread of Windborne Contaminants. In: Proceedings of the conference on Protection of the Public in the Event of Radiation Accidents. World Health Organization, Geneva, Switzerland, p. 43.

Pasquill, F., 1970. Prediction of Diffusion over an urban area—current practice and future prospects. In: Proceedings of Symposium on Multiple-Source Urban Diffusion Models. Environmental Protection Agency, Research Triangle Park, NC, p. 3.1.

Pasquill, F., 1976a. Atmospheric Dispersion Parameters in Gaussian Plume Modeling, Pt II: Possible Requirements for Change in Turner Workbook Values. Environmental Protection Agency, Research Triangle Park, NC (Rep. EPA-600/4-76-030b).

Perry, R.H., Green, D.W., 1984. Perry's Chemical Engineers Handbook. sixth ed. McGraw-Hill, New York.

Perry, W.W., Articola, W.P., 1980. Study to Modify the Vulnerability Model of the Risk Management System. Rep. CG-D-22-80. Environmental Control Inc., Rockville, MD.

Picard, D.J., Bishnoi, P.R., 1988. The importance of real-fluid behaviour and non-isentropic effects in modeling decompression characteristics of pipeline fluids for application in ductile fracture propagation analysis. Can. J. Chem. Eng. 66, 3.

Picard, D.J., Bishnoi, P.R., 1989. The importance of real-fluid behaviour in predicting release rates from high pressure sour gas pipeline ruptures. Can. J. Chem. Eng. 67, 3.

Pilz, V., van Herck, W., 1976. Chemical engineering investigations with respect to the safety of large chemical plants. European Federation of Chemical Engineering, Third Symposium on Large Chemical Plants, Antwerp.

Prugh, R.W., 1985. Mitigation of vapor cloud hazards. Plant/Operations Prog. 4, 95, 1986, 5, 169.

Prugh, R.W., 1987a. Guidelines for vapor release mitigation. Plant/Operations Prog. 6, 171.

Prugh, R.W., 1987b. Evaluation of unconfined vapor cloud explosion hazards. Vapor Cloud Model. 712.

Puttock, J.S., 1987c. The development and use of HEGABOX/HEGADAS dispersion models for hazard analysis. Vap. Cloud Model.317.

Puttock, J.S., et al., 1982. Field experiments on dense gas dispersion. J. Hazard. Mater. 6 (1−2), 13−41.

Puttock, J.S., Colenbrander, G.W., Blackmore, D.R., 1983. Maplin Sands experiments 1980: dispersion results from continuous releases of refrigerated liquid propane. In: Hartwig, S. (Ed.), op. cit., p. 147.

Puttock, J.S., Colenbrander, G.W., Blackmore, D.R., 1984. Maplin Sands experiments 1980: dispersion results from continuous releases of refrigerated liquid propane and LNG. In: Wispelaere, C. (Ed.), op. cit., p. 353.

Ramsdell, J.V., Hinds, W.T., 1971. Concentration fluctuations and peak-to-mean concentration ratios in plumes from a ground-level continuous point source. Atmos. Environ. 5, 483.

Reed, J.D., 1974. Containment of leaks from vessels containing liquefied gases with particular reference to ammonia. Loss Prev. Saf. Promotion. 1, 191.

Ride, D.J., 1984a. An assessment of the effects of fluctuations on the severity of poisoning by toxic vapours. J. Hazard. Mater. 9 (2), 235.

Ride, D.J., 1984b. A probabilistic model for dosage. In: Ooms, G., Tennekes, H. (Eds.), op. cit., p. 267.

Roberts, O.F.T., 1923. The theoretical scattering of smoke in a turbulent atmosphere. Proc. R. Soc. Ser. A. 104, 640.

Roberts, A.F., 1981. Thermal radiation hazards from releases of LPG from pressurised storage. In: Health and Safety Executive Seminar on Thermal Radiation and Overpressure Calculations. Health and Safety Executive, London.

Sallet D.W., 1979a. Pressure relief valve sizing for vessels containing compressed liquefied gases. In: Institute of Mechanical Engineers, Conference on Pressure Relief Devices, London, pp. 85−96.

Schmidli, J., Bannerjee, S., Yadigaroglu, G., 1990. Effects of vapour/aerosol and pool formation on rupture of vessels containing superheated liquid. J. Loss Prev. Process Ind. 3 (1), 104.

Schweppe, J.L., Foust, A.S., 1953. Effect of forced circulation rate on boiling heat transfer and pressure drop in a short vertical tube. In: Heat Transfer Atlantic City. American Institute of Chemical Engineers, New York, p. 77.

Shapiro, A., 1953. The Dynamics and Thermodynamics of Compressible Fluid Flow. Ronald Press, New York (vol. 1, 1953; vol. 2, 1954).

Simpson, H.G., 1974. The ICI vapour barrier. Power Works Eng. May (8).

Slade, D.H. (Ed.), 1968. Meteorology and Atomic Energy (report). Office of Information Services, Atomic Energy Commission, Washington, DC.

Spicer, T.O., Havens, J.A., 1986. Development of a heavier-than-air dispersion model for the US Coast Guard Hazard Assessment Computer System. In: Hartwig, S. (Ed.), op. cit., p. 73.

Spicer, T.O., Havens, J.A., 1987. Field test validation of the DEGADIS model. J. Hazard. Mater. 16, 231.

Sutton, O.G., 1950. The dispersion of hot gases in the atmosphere. J. Met. Prog. 7, 307, 7, 122.

Sutton, O.G., 1953. Micrometeorology. McGraw-Hill, New York, NY.

Swift, I., 1984. Developments in emergency relief system design. Chem. Eng. London. 406, 30.

Taylor, G.I., 1915. Eddy motion in the atmosphere. Phil. Trans. R. Soc. Ser. A. 215, 1.

Te Riele, P.H.M., 1977. The atmospheric dispersion of heavy gases emitted at or near ground level. Loss Prev. Saf. Promotion. 2, 347.

Turner, D.B., 1970. Workbook of Atmospheric Dispersion Estimates. Office of Air Programs. Environmental Protection Agency, Research Triangle Park, NC.

Turner, R.M., Fairhurst, S., 1989. Assessment of the Toxicity of Major Hazard Substances. Health and Safety Executive, Bootle (Specialist Inspectors Rep. 21).

Van Den Akker, H.E.A., Snoey, H., Spoelstra, H., 1983. Discharges of pressurised liquefied gases through apertures and pipes. Loss Prev. Saf. Promotion. 4 (1), E23.

Wallis, G.B., 1969. One Dimensional Two-Phase Flow. McGraw-Hill, New York.

Wallis, G.B., 1980. Critical two-phase flow. Int. J. Multiphase Flow. 6, 97.

Wallis, G.B., Richter, H.J., 1978. An isentropic streamline model for flashing two-phase vaporliquid flow. J. Heat Transfer. 100, 595.

Westbrook, G.W., 1974. The bulk distribution of toxic substances: a safety assessment of the carriage of liquid chlorine. Loss Prev Safety Promotion. 1, 197.

Wheatley, C.J., 1986. Factors affecting cloud formation from releases of liquefied gases. In: Refinement of Estimates of the Consequences of Heavy Toxic Vapour Release. Institution of Chemical Engineers, Rugby, p. 5.1.

Wilson, D.J., 1979. The Release and Dispersion of Gas from Pipeline Ruptures. Rep. Alberta Environment Contract 790686. Department of Mechanical Engineering, University of Alberta, Edmonton, Alberta.

Zaloudek, F.R., 1961. The Low Pressure Critical Discharge of Steam/Water Mixtures from Pipes. Rep. HW-68934.

Zeman, O., 1982a. The dynamics and modeling of heavier than-air cold gas releases. Atmos. Environ. 16, 741.

Zeman, O., 1982b. The Dynamics and Modeling of Heavier than air Cold Gas Releases. Lawrence Livermore National Laboratory, Livermore, CA (Rep. UCRL-15224).

Fire

Chapter Outline

12.1. FIRE

The first of the major hazards in a process plant is fire. Fire in the process industries causes more serious accidents than explosion or toxic release, although the accidents in which the greatest loss of life and damage occur are generally caused by explosion. Fire is normally regarded as having a disaster potential less than explosion or toxic release. One of the worst explosion hazards, however, is usually considered to be that of an explosion of a vapor cloud that has drifted over a populated area, and in this case the difference in the number of casualties caused by a flash fire rather than an explosion in the cloud may be relatively small. Fire is, therefore, a serious hazard.

Vapor clouds, or flash fires and fireballs, radiate intense heat that can be lethal. Another lethal effect is the depletion of oxygen in the atmosphere caused by a flash fire. Although the Flixborough disaster was primarily a vapor cloud explosion, a large flash fire in part of the cloud also accompanied the latter and this fire was responsible for some of the deaths that occurred. Flash fires can also do considerable damage to the plant. In buildings, fire is the main threat and can cause great damage as well as loss of life.

In the United Kingdom, there are a number of bodies concerned with fire and fire protection. These include fire services and organizations such as the British Fire Services Association, professional institutions such as the

Institution of Fire Engineers (IFE), insurance organizations such as the Fire Offices Committee, research establishments such as the Fire Research Station (FRS), and educational institutions such as the Fire Protection Association (FPA), and the Department of Fire Engineering at Edinburgh University. In the United States, the relevant bodies include in insurance, the National Board of Fire Underwriters; in research, the Underwriters Laboratories (UL), the Bureau of Mines (BM), and the Combustion Institute; and in the formulation of codes and research, the National Fire Protection Association (NFPA).

12.1.1 The Combustion Process

Fire (or combustion) is a chemical reaction in which a substance combines with oxygen and heat is released. Usually, fire occurs when a source of heat comes into contact with a combustible material. If a combustible liquid or solid is heated, it evolves vapor and if the concentration of vapor is high enough, it forms a flammable mixture with oxygen in the air. If this flammable mixture is then heated further to its ignition point, combustion starts. Similarly, a combustible gas or vapor mixture burns if it is heated to a sufficiently high temperature.

A useful starting point for consideration of combustion is the propagation of the flame as an adiabatic planar combustion wave, as shown in Figure 12.1. At the front of the combustion zone, the temperature is that of the burned gas T_b. From this maximum value, the temperature then falls off. Within the wave, there is a plane at temperature T_1 where transition occurs from a heat source to a heat sink in that the heat supplied by the reaction now falls below the heat lost to the unburned gas behind. The temperature

then falls further until it reaches the temperature T_u of the unburned gas behind the wave. The first part of the combustion zone, in front of T_1, is the reaction zone and the second part, behind T_1, the preheat zone.

There are three conditions essential for a fire: (1) fuel, (2) oxygen, and (3) heat. These three conditions are often represented as the fire triangle as shown in Figure 12.2. If one of the conditions is missing, fire does not occur; if one of them is removed, fire is extinguished.

Normally, the heat required is initially supplied by an external source and then provided by the combustion process itself. The amount of heat needed to cause ignition depends on the form of the substance. A gas or vapor may be ignited by a spark or small flame, while a solid may require a more intense heat source.

Ignition of a combustible gas or vapor mixture may occur in two ways. In the first, the energy for ignition is supplied by a local source such as a spark or small flame at a point within the mixture, as shown in Figure 12.3(a). In the second, the bulk gas mixture is heated up to its ignition temperature, as shown in Figure 12.3(b).

The three conditions of the fire triangle also provide clues on how fires may be fought. The first method is to cut off the fuel. This is particularly relevant for fires caused by leaks at a process plant. The second method is to remove heat. Putting water on the fire usually does this. The third method is to stop the supply of oxygen. This may be affected in various ways, including the use of foam or inert gas.

Fire normally grows and spreads by direct burning, which results from impingement of the flame on combustible materials, by heat transfer or by travel of the burning material. The three main modes of heat transfer are (1) conduction, (2) convection, and (3) radiation. All these modes are significant in heat transfer from fires. Conduction is important particularly in allowing heat to

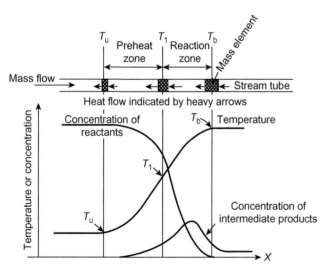

FIGURE 12.1 Scheme of a planar combustion wave (Lewis and von Elbe, 1961). *Source: Courtesy of Academic Press.*

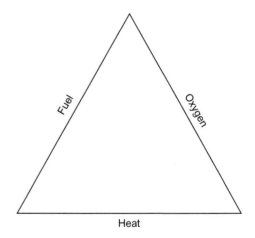

FIGURE 12.2 The fire triangle.

pass through a solid barrier and ignite material on the other side.

Most of the heat transfer from fires, however, is by convection and radiation. It is estimated that in most fires, some 75% of the heat emanates by convection. The hot products of combustion rising from a fire typically have a temperature in the range of 800–1200°C and a density a quarter that of air. On an open plant, much of the heat is dissipated into the atmosphere, but in buildings it is transferred to the ceiling.

Radiation is the other main mode of heat transfer. Although it usually accounts for a smaller proportion of the heat issuing from the fire, radiated heat is transferred directly to nearby objects, does not go preferentially upward, and crosses open spaces. For these reasons, it is generally the most significant mode of heat transfer on an open plant.

The other way in which fire spreads is by travel of bulk materials, such as burning liquids or solid brands. With fires in buildings particularly, a stage is generally reached when the materials have been heated up to the point where they produce flammable vapors. The rapid spread of fire which occurs at this point is called 'flashover.'

12.1.2 Classification of Fires

There are several classification systems for fires. In the United Kingdom, the British Standard (BS) classification is given in BS EN2: 1992 *Classification of Fires*. This classification is:

Class A	Fires involving solid materials, usually of an organic nature, in which combustion takes place with the formation of glowing embers.
Class B	Fires involving liquids or liquefiable solids.
Class C	Fires involving gases.
Class D	Fires involving metals.

BS EN2: 1992 replaces BS 4547: 1972.

In the United States, the National Fire Protection Association (NFPA) classification is given in the NFPA Codes. NFPA 10: 2002 contains the following classification:

Class A	Fires in ordinary combustible materials, such as wood, cloth, paper, rubber, and many plastics.
Class B	Fires in flammable or combustible liquids, oils, greases, tars, oil-based paints, solvents, lacquers, alcohols, and flammable gases.
Class C	Fires that involve energized electrical equipment.
Class D	Fires in combustible metals, such as magnesium, titanium, zirconium, sodium, lithium, and potassium.
Class K	Fires in cooking appliances that involve combustible cooking media (vegetable or animal oils and fats).

This classification has obvious relevance to the extinguishing medium to be used. For a Class C fire, when electrical equipment is de-energized, extinguishers for Class A and B fires may be used.

With regard to fires in the process industries specifically, fires may be classified broadly into the following categories:

1. Vapor cloud fires
   a. fires with no explosion,
   b. fires resulting from explosion,
   c. fires resulting in explosion;
2. Fireballs
3. Jet flames.
4. Liquid fires
   a. pool fires,
   b. running liquid fires;
5. Solids fires
   a. fires of solid materials,
   b. dust fires;
   c. warehouse fires;
   d. fires associated with oxygen.

The following sections provide an overview of different topics of fire and fire protection methods. For a detailed discussion about the fundamentals of the combustion phenomena, readers are encouraged to refer to the

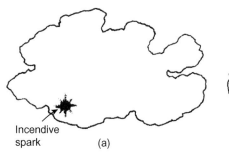

Incendive spark   (a)

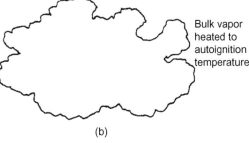

Bulk vapor heated to autoignition temperature

(b)

FIGURE 12.3 Ignition of a flammable mixture: (a) local ignition and (b) bulk gas ignition.

full version of Lees' *Loss Prevention in the Process Industries*, fourth edition.

12.2. FLAMMABILITY OF GASES AND VAPORS

12.2.1 Flammability Limits

A flammable gas burns in air only over a limited range of composition. Below a certain concentration of the flammable gas, the lower flammability limit (LFL), the mixture is too lean; while above a certain concentration, the upper flammability limit (UFL), it is too rich. The concentrations between these limits constitute the flammable range. The LFLs and UFLs are also sometimes called, respectively, the lower and upper explosive limits (LELs and UELs).

In general, the most flammable mixture corresponds approximately, but not exactly, to the stoichiometric mixture for combustion. It is frequently found that the concentrations at the LFLs and UFLs are roughly one-half and twice that of the stoichiometric mixture, respectively, for hydrocarbons.

Unfortunately, the word 'flammable' is ambiguous. On one hand, it denotes a combustible in air and on the other it refers to an explosive without any further addition of air or oxidizers. The database CHEMSAFE® (Sass and Eckermann, 1992) uses the wording 'explosion limit' of gas mixtures instead of 'flammability limit.' The term 'explosion' is used for describing the explosive properties of gas mixtures in order to avoid misunderstanding in Europe. In the narrower sense, an explosion limit refers to the limiting flammable gas fraction that is mixed with air where a combustion reaction (flame) fails to propagate. The range between the lower explosion limit (LEL) and the upper explosion limit (UEL) is the so-called explosion range. They are distinct from the detonability limits.

Flammability limits are affected by pressures, temperatures, direction of flame propagations, gravitational fields, surroundings, and other conditions. The limits are determined experimentally and the precise values obtained depend, therefore, on the particular test method.

Although there is no universally used equipment, one which has been utilized for the measurement of many flammability limit data is the Bureau of Mines (BM) apparatus, which is described by Coward and Jones (1952) and which consists of a cylindrical tube 5 cm diameter and 1.5 m high. The tube is closed at the top but open at the bottom. Gas−air mixtures of different compositions are placed in the tube and a small ignition source is applied at the bottom. The flame initiated by the source of ignition travels one-half of the full length of the tube

and then the lower and upper limit concentrations are determined.

Normal variations of atmospheric pressure do not have any appreciable effect on flammability limits. The effect of larger pressure changes is not simple or uniform, but is specific to each mixture. For most gases, a decrease in pressure below atmospheric can narrow the flammable range by raising the LFL and reducing the UFL until the two limits coincide and the mixture becomes nonflammable. Conversely, an increase in pressure above atmospheric can widen the flammable range by reducing the LFL and raising the UFL. This effect is shown in Figure 12.4 for natural gas. It may be noted that the effect is more marked on the upper than on the LFL.

An increase in temperature tends to widen the flammable range. This effect for methane is shown in Figure 12.5.

Similar results are obtained for the effect of temperature on the LFLs of other paraffinic hydrocarbons. For these, the approximate flame temperature can be taken as

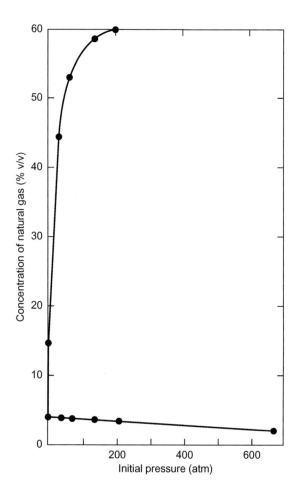

FIGURE 12.4 Effect of pressure on inflammability limits of natural gas in air (Zabetakis, 1965). *Source: Courtesy of the Bureau of Mines.*

1300°C. Then for the first 10 members of the series from methane to decane

$$\frac{L_t}{L_{25}} = 1 - \frac{t - 25}{1300 - 25} \qquad (12.1)$$

where L_t is the LFL at $t°C$ (%v/v), L_{25} is LFL at 25°C (%v/v), and t is the temperature (°C).

The data may also be fairly well correlated by the modified Burgess–Wheeler law, suggested by Zabetakis et al. (1959):

$$\frac{L_t}{L_{25}} = 1 - \frac{0.75(t - 25)}{\Delta H_c} \qquad (12.2)$$

where ΔH_c is the net heat of combustion (kcal/mol). Likewise the modified Burgess–Wheeler law may correlate the effect of temperature on the UFL of these hydrocarbons, in the absence of cool flames:

$$\frac{U_t}{U_{25}} = 1 + \frac{0.75(t - 25)}{\Delta H_c} \qquad (12.3)$$

where U_t is the UFL at $t°C$ (%v/v) and U_{25} is the UFL at 25°C (%v/v).

A more recent study of the effect of pressure and temperature on flammability limits is that of Gibbon et al. (1994), who investigated a limited number of solvents.

Their work confirmed that an increase in temperature causes a decrease in the LFL and an increase in the UFL. They also found that the UFL increased with increase in pressure. Unexpectedly, the LFL also increased with increase in pressure. In sum, the effect of pressure on the flammability limits is much less predictable than that of temperature. In particular, increase in pressure causes in some cases a decrease in the LFL and in others an increase.

Flammability limits are also affected by the addition of an inert gas such as nitrogen, carbon dioxide, or steam. This effect for methane is shown in Figure 12.6. Minimum inert gas contents for suppression of flammability of selected substances in air are given in Table 12.1. In general, carbon dioxide causes a greater narrowing of the flammable range than does nitrogen. For many flammable gas–air systems, the mixtures can be rendered nonflammable by the addition of about 30% of carbon dioxide or about 40% of nitrogen.

The flammability limits described are those for mixtures of a flammable gas in air. There are also flammability limits for combustion in pure oxygen. In general, the

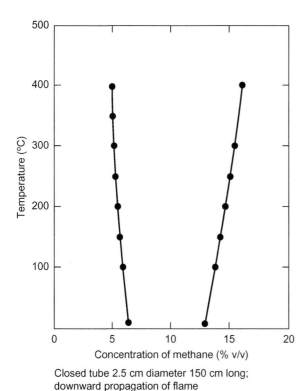

FIGURE 12.5 Effect of temperature on flammability limits of methane in air (Coward and Jones, 1952). *Source: Courtesy of the Bureau of Mines.*

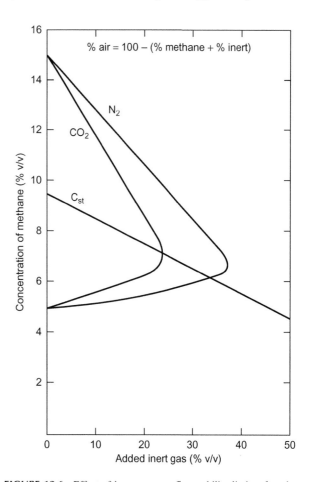

FIGURE 12.6 Effect of inert gases on flammability limits of methane in air (Zabetakis, 1965).

TABLE 12.1 Minimum Inert Gas Content for Suppression of Flammability of Selected Substances in Air

	Nitrogen (%v/v)	Carbon Dioxide (%v/v)
Methane	38	24
Ethane	46	33
Propane	43	30
n-Butane	41	28
n-Pentane	43	29
n-Hexane	42	29
Ethylene	50	41
Propylene	43	30
Benzene	45	32

After Burgoyne (1965). Courtesy of the Institution of Marine Engineers.

TABLE 12.2 AIT for the Selected Paraffinic Hydrocarbons

	AIT (°C)		AIT (°C)
Methane	537	n-Hexane	223
Ethane	515	n-Heptane	223
Propane	466	n-Octane	220
n-Butane	405	n-Nonane	206
n-Pentane	258	n-Decane	208

After Zabetakis (1965) BM Bull. 627.

LFL of a gas is almost the same in oxygen as in air, but the UFL is much greater in oxygen than in air. The flammable range in oxygen is thus wider than it is in air. The applicability of flammability limits is not confined, however, to air and oxygen. There are flammability limits for substances that burn in chlorine. The database CHEMSAFE® contains measured explosion/flammability limits for approximately 3000 pure substances and many mixtures under not only atmospheric condition, but also elevated initial pressures (up to 100 bar) and temperatures (up to 400°C), and with other oxidizing substances, e.g., pure oxygen and chlorine. Flammability limits for a fuel mixture that contains several flammable gases may be calculated from the equation of Le Chatelier.

For the lower limit:

$$L = \frac{1}{\sum_{i=0}^{n}(y_i/L_i)} \quad (12.4a)$$

where L is the LFL of air-free fuel (%v/v), L_i the LFL of fuel component i (%v/v), and y_i is the concentration of fuel component i (mole fraction). For the upper limit:

$$U = \frac{1}{\sum_{i=0}^{n}(y_i/U_i)} \quad (12.4b)$$

where U is the UFL of air-free fuel (%v/v) and U_i, the UFL of fuel component i (%v/v).

Le Chatelier's equation is an empirical one and it is not universally applicable. Its limitations are discussed by Coward and Jones (1952). Further information on and examples of the calculation of the flammability limits of fuel mixtures are given in BS 5345: Part 1: 1976. It is necessary to exercise care in using flammability limit data, since the limit may vary with pressure, temperature, and other conditions. The LFL required to determine the amount of flammable gas−air mixture resulting from a plant leak can probably be obtained quite readily, because usually the data are available for ambient conditions and the accuracy called for is not high, but that required to maintain the concentration of a flammable gas−oxygen mixture entering a reactor just below the flammable range under other pressure and temperature conditions needs greater consideration, because the data may not be available so readily for these other conditions, but they must be accurate.

12.2.2 Ignition Temperature

If the temperature of a flammable gas−air mixture is raised in a uniformly heated apparatus, it eventually reaches a value at which combustion occurs in the bulk gas. For the range of flammable mixtures, there is a mixture composition that has the lowest ignition temperature. This is the minimum spontaneous ignition temperature (SIT) or autoignition temperature (AIT).

The AITs for the first 10 paraffinic hydrocarbons are given in Table 12.2 (Zabetakis, 1965).

It may be noted that there is a break between the AIT for n-butane and that for n-pentane and that the rate of decrease of these temperatures after n-pentane is low. Some substances have quite low AITs. The AIT of carbon disulfide, for example, is 90°C.

The AIT temperature is a property that is particularly liable to variations caused by the nature of hot surfaces. The values normally quoted are obtained in laboratory apparatus with clean surfaces. The AIT may be reduced by as much as 100−200°C for surfaces that are lagged or are contaminated by dust. Other factors, such as test vessel volume, detection criteria, and pressure, also have significant effects on AIT measurement. Autoignition temperatures of selected substances are given in Table 12.3.

TABLE 12.3 Flammability Limits, AITs, and Flashpoints of Selected Substances in Air at Atmospheric Pressure

	Flammability Limit (%v/v)		Autoignition Temperature (°C)	Flashpoint (°C)	
	Lower	Upper		Closed Cup	Open Cup
Acetone	2.6	13	465	−18	−9
Acetylene	2.5	100	305	–	–
Ammonia	15	28	651[a]	–	–
Benzene	1.4[a]	8.0[a]	562[a]	−11	–
n-Butane	1.8	8.4	405	−60	–
Carbon disulfide	1.3	50	90	−30	–
Carbon monoxide	12.5	74	–		
Cyclohexane	1.3	7.8	245	−20	–
Ethane	3.0	12.4	515	−135	–
Ethylene	2.7	36	490	−121	–
Ethylene dichloride	6.2[a]	15.9[a]	413[a]	13	18
Ethylene oxide	3[a]	100[a]	429[a]	–	−20
Hydrogen	4.0	75	400	–	–
Methane	5.0	15.0	540	–	–
Propane	2.1	9.5	450	<−104	–
Propylene	2.4	11	460	−108	–
Styrene	1.1[a]	6.1[a]	490[a]	32	38
Toluene	1.3[a]	7.0[a]	536[a]	4	7
Vinyl chloride	4[a]	22[a]	472[a]	–	−78

[a]FMEC (1967).
Flammability limits: Zabetakis (1965), except as given in footnote. AIT: Zabetakis (1965), except as given in footnote.
Flashpoints: FMEC (1967).

12.2.3 Ignition Time Delay

Ignition of a flammable mixture raised to or above the temperature at which spontaneous ignition occurs is not instantaneous; there is a finite time delay before ignition takes place. This time delay decreases as the ignition temperature increases, and the reduction has been correlated by Semenov (1959) using the equation:

$$\ln \tau = \frac{k_1 E}{T} + k_2 \qquad (12.5)$$

where E is the apparent activation energy, T is the absolute temperature, τ is the time delay before ignition, and k_1 and k_2 are constants. The time delay may be as little as

a fraction of a second at higher temperatures or several minutes close to the AIT.

This effect may be of significance in flow systems where the fluid is a flammable mixture that comes in contact with a hot surface above the AIT, but for a short time only.

12.2.4 Flashpoint

The flashpoint of a flammable liquid is the temperature at which the vapor pressure of the substance is such as to give a concentration of vapor in the air that corresponds to the LFL. The relationships between vapor pressure, flammability limits, flashpoint, and AIT are illustrated in Figure 12.7.

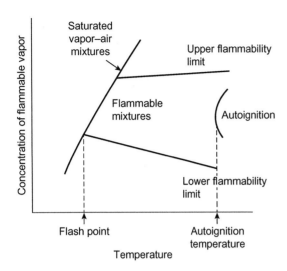

FIGURE 12.7 Relationships between vapor pressure, flammability limits, flashpoint and autoignition temperature. *Source: After Zabetakis (1965). Courtesy of the Bureau of Mines.*

There are two methods of measuring flashpoint, the closed cup test and the open cup test. Some standard test methods are:

Closed cup	Pensky Martens	ASTM D 93−61
		BS 2839: 1969
	Tagliabue	ASTM D 56−61
Open cup	Cleveland	ASTM D 92−57

The open cup flashpoint is usually a few degrees higher than the closed cup flashpoint, and the closed cup flashpoint method is the only one which is used for the classification of flammable substance in the Globally Harmonized System of Classification and Labelling of Chemicals (GHS).

A liquid that has a flashpoint below ambient temperature and can thus give rise to flammable mixtures under ambient conditions is generally considered more hazardous than one with a higher flashpoint. The flashpoint is a main parameter, therefore, in hazard classification of liquids and in government regulations based on these. Thus, in aviation, there has been a movement to replace fuel JP4 with JP1. The former is akin to petroleum and has a low flashpoint, while the latter is more like kerosene and has a higher flashpoint, which is therefore safer. Obviously, however, a higher flashpoint liquid also becomes more hazardous if it is heated up to a temperature above its flashpoint. The flashpoints of selected substances are given in Table 12.3.

12.2.5 Fire Point

The fire point of a flammable liquid is the lowest temperature at which the liquid, when placed in an open container, will give off sufficient vapor to continue to burn

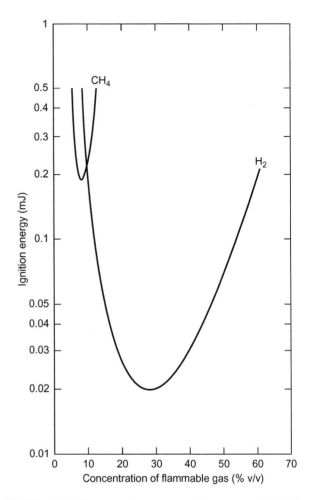

FIGURE 12.8 Effect of mixture composition on electrical ignition energy of methane and of hydrogen in air. *Source: Factory Mutual Engineering Corporation (1967).*

once ignited. The fire point is usually a few degrees above the open cup flashpoint.

12.2.6 Ignition Energy

If a flammable gas−air mixture is to be ignited by a local source of ignition, however, it is not sufficient (as it is with autoignition) to raise a volume of the mixture to a certain temperature for a certain time. There is also a minimum volume of mixture so treated that is required in order to give rise to a continuing flame through the rest of the mixture. There is in effect an ignition energy requirement.

Similarly, ignition of a flammable gas−air mixture by a local source of ignition by electrical discharge occurs only if the latter possesses a certain energy. The variation of ignition energy with composition is shown for hydrogen and for methane in Figure 12.8.

There is a minimum ignition energy (MIE) which usually occurs close to the stoichiometric mixture. Minimum

TABLE 12.4 Minimum Ignition Energies for Selected Substances in Air

	Minimum Ignition Energy (mJ)
Carbon disulfide	0.01–0.02
Hydrogen	0.019
Acetylene	0.02
Methane	0.29
Ethane	0.24
Propane	0.25
n-Butane	0.25
n-Hexane	0.25
Ethylene	0.12
Benzene	0.22
Ammonia	up to > 100

The recommended value of minimum ignition energy for ammonia is 14 mJ in CHEMSAFE®. This is measured at the most ignitable atmosphere; concentration of ammonia gas is 20% (v/v) in air. The measurement was carried out in the Physikalisch–Technische Bundesanstalt in Braunschweig, Germany.
Burgoyne (1965); FMEC (1967).

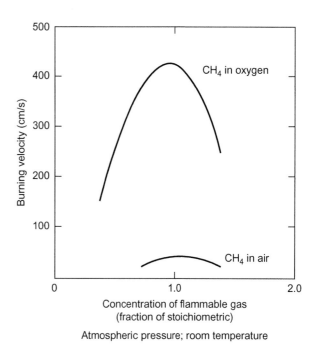

FIGURE 12.9 Effect of mixture composition on burning velocity of methane in air and in oxygen (Zabetakis, 1965). *Source: Courtesy of the Bureau of Mines.*

ignition energies (MIEs) for selected substances are given in Table 12.4.

12.2.7 Burning Velocity

The burning velocity affects the rate of combustion of a flammable mixture. The burning velocity is a property of the mixture. It is in effect the velocity at which a gaseous fuel–air mixture issuing from a burner burns back on to the burner. The burning velocity of a fuel–air mixture is usually determined by measurement of a premixed bunsen burner flame in the laminar region. The burning velocity S_u is obtained by dividing the volumetric gas flow V by the area A of the flame front cone:

$$S_u = \frac{V}{A} \qquad (12.6)$$

Alternatively, the burning velocity may be obtained from the velocity u of the gas and the half-angle α of the apex of the flame front cone:

$$S_u = u \sin \alpha \qquad (12.7)$$

If the flame front cone has radius r, height h, and slant height l, the area A of the cone surface and the area A' of the cone triangular cross section are, respectively:

$$A = \pi r l \qquad (12.8a)$$

$$= \pi r (r^2 + h^2)^{1/2} \qquad (12.8b)$$

$$A' = rh \qquad (12.9)$$

Also

$$A = \frac{\pi A' l}{h} \qquad (12.10)$$

The gas velocity u is:

$$u = \frac{V}{\pi r^2} \qquad (12.11)$$

The flame front is usually not a perfect cone and, in consequence, the equations just quoted are subject to errors of varying size. The choice of equations for the calculation of burning velocity is discussed by Lewis and von Elbe (1961) and by Fells and Rutherford (1969). The maximum burning velocity obtained at atmospheric pressure and temperature is referred to as the maximum fundamental burning velocity.

Burning velocities for methane in air and in oxygen are shown in Figure 12.9. In general, burning velocities for paraffinic hydrocarbons range from a few centimeters per second near the flammability limits up to about 45 cm/s near the stoichiometric mixture. For these hydrocarbons in oxygen the corresponding values are about 125 and 425 cm/s, respectively.

Burning velocities are affected by pressure and temperature. The effect of pressure changes is variable. A decrease in the burning velocity of stoichiometric methane—air mixtures has been found with increasing pressure in the range 0.5—20 atm, but with stoichiometric methane—oxygen mixtures an increase in pressure in the range 0.2—2 atm gave an increase in burning velocity.

The effect of temperature on burning velocity is more consistent and is sometimes described by the equation:

$$S_u = A + BT^n \qquad (12.12)$$

where T is the absolute temperature, A and B are constants, and n is an index. A value of 2.0 for the index n has been found for some paraffinic hydrocarbons (Zabetakis, 1965). Maximum fundamental burning velocities of selected substances in air and in oxygen are given in Table 12.5.

The actual flame speed in combustion of a flammable mixture is greater, and sometimes very much greater, than the burning velocity. Whereas the burning velocity is a property of the mixture only, the flame speed depends on other factors such as turbulence and pressure waves. If detonation occurs, the flame speed is greater by orders of magnitude than the burning velocity.

12.2.8 Adiabatic Flame Temperature

The heat radiated by combustion of a flammable mixture depends on the flame temperature, the theoretical maximum value of which is the adiabatic flame temperature. The adiabatic flame temperature is the temperature attained by combustion under adiabatic conditions. It is normally determined for the stoichiometric mixture and values quoted generally refer to this unless otherwise stated.

The adiabatic flame temperature can be calculated, but for an accurate calculation this is not entirely straightforward. The adiabatic flame temperature may be determined for constant pressure or constant volume conditions. The value normally quoted is for the former. For this case the enthalpy balance is:

$$\Delta H_2 + \Delta H_1 + \Delta H_c = 0 \qquad (12.13)$$

or

$$\Delta H_2 + \Delta H_1 = -\Delta H_c \qquad (12.14)$$

with

$$\Delta H_1 = \sum_{i-1}^{r} n_i \int_{T_1}^{T_0} c_{pi}(T)dT \qquad (12.15)$$

$$= C_{pmr}(T_0 - T_1) \qquad (12.16)$$

TABLE 12.5 Maximum Fundamental Burning Velocities of Selected Substances in Air and in Oxygen

	Maximum Fundamental Burning Velocity,[a] S_u	
	(cm/s)	(ft/s)
In air		
Methane	36.4[b]	1.2
Ethane	40.1	–
Propane	45[b]	1.5
n-Butane	40.5	1.3
n-Hexane	38.5	1.3
Ethylene	68.8[b]	2.3
Town gas[c]	–	3.7
Acetylene	173	5.8
Hydrogen	320	11.0
Benzene	40.7[b]	–
In oxygen		
Methane	393[d]	–
Propane	390	–
Ethylene	550	–
Acetylene	1140	–
Hydrogen	1175	–

[a]The values given in the first column (cm/s) are those of Fiock (1955) and in the second column (ft/s) those of the HSE (1965 HSW Bklt 34). The values quoted from Fiock are for initial pressure atmospheric and initial temperature room temperature and, in most cases, dry gas, and generally are selected from several values listed.
[b]Some higher values are also listed by Fiock.
[c]For town gas containing 63% H_2.
[d]For stoichiometric mixture.

$$\Delta H_2 = \sum_{i=1}^{P} n_i \int_{T_0}^{T_2} c_{pi}(T)dT \qquad (12.17)$$

$$= C_{pmp}(T_2 - T_0) \qquad (12.18)$$

where c_p is the specific heat at constant pressure of component i (kJ/kmol°C), c_{pm} is the mean specific heat over the range of interest (kJ/kmol°C), ΔH_c is the heat of combustion at stage 0 (kJ/mol fuel), ΔH_1 is the enthalpy change between states 1 and 0 (kJ/kmol of fuel), ΔH_2 is the enthalpy change between states 0 and 2 (kJ/kmol fuel), n is the number of moles per mole of fuel, p is the number of products, r is the number of reactants, subscripts i, p, and r denote component i, products and reactants, respectively, and 0, 1, and 2 denote the standard state, initial state, and final state, respectively.

TABLE 12.6 Constants for Specific Heat at Constant Pressure (Barnard and Bradley, 1985)

Species	Constant[a]				Valid Over Temperature Range (K)
	a	$10^2 b$	$10^5 c$	$10^9 d$	
H_2	29.09	-0.1916	0.4000	-0.870	273–1800
O_2	25.46	1.519	-0.7150	1.311	273–1800
N_2	27.32	0.6226	-0.0950	–	273–3800
CO	28.14	0.1674	0.5368	-2.220	273–1800
CO_2	22.24	5.977	-3.499	7.464	273–1800
H_2O	32.22	0.1920	1.054	-3.594	273–1800
CH_4	19.87	5.021	1.286	-11.00	273–1500
C_2H_2	21.80	9.208	-6.523	18.20	273–1500
C_2H_4	3.95	15.63	-8.339	17.66	273–1500
C_2H_6	6.895	17.25	6.402	7.280	273–1500
C_3H_8	-4.042	30.46	-15.71	31.71	273–1500
C_6H_6	-39.19	48.44	-31.55	77.57	273–1500
CH_3OH	19.04	9.146	-1.218	-8.033	273–1000
NH_3	27.55	2.563	0.9900	-6.686	273–1500
NO	27.03	0.9866	0.3223	0.3652	273–3800
SO_2	25.76	5.791	-3.809	8.606	273–1800

[a]Constants in the equation $c_p = a + bT + cT^2 + dT^3$, where c_p is the specific heat (kJ/kmol°C) and T is the absolute temperature (K).
Source: Courtesy of Chapman and Hall.

The specific heat c_{pi} of a reactant or product is a function of the temperature T. It is commonly represented in the form

$$c_{pi} = a_i + b_i T + c_i T^2 + d_i T^3 \qquad (12.19)$$

where a_i (kJ/kmol°C), b_i (kJ/kmol°C K), c_i (kJ/kmol°C K^2), and d_i (kJ/kmol°C K^3) are constants.

Then from Equations 12.17–12.19, the mean specific heat of product i is

$$c_{pmi} = \frac{(a_i T + (b_i T^2/2) + (c_i T^3/3) + (d_i T^4/4))\big|_{T_0}^{T_2}}{T_2 - T_0} \qquad (12.20)$$

and the mean specific heat of the product mixture is

$$c_{pmp} = \sum_{i=1}^{P} n_i c_{pmi} \qquad (12.21)$$

If necessary, equations similar to Equations 12.20 and 12.21, derived from Equations 12.15 and 12.19, may be used to obtain c_{pmr}, but generally the initial temperature T_1 is close to the standard temperature T_0 so that a point

value of c_{pmr} may be used. If in fact $T_1 = T_0$, then $\Delta H_1 = 0$.

Table 12.6 gives values of the constants in Equation 12.19 for the determination of point values of the specific heats c_p. Table 12.7 gives values of the mean specific heat c_{pm} over the interval $0 - t°C$.

In the foregoing the implicit assumption is that the products of combustion are those given by the simple stoichiometric equation, which for hydrocarbons yields only CO_2 and H_2O. In fact, for flame temperatures above about 1370°C, dissociation of the products of combustion occurs. At adiabatic flame temperatures, dissociation and ionization absorb a significant amount of energy and result in an appreciable lowering of the temperature of the flame. This is illustrated in Table 12.8, which shows that for propane the values obtained for the adiabatic flame temperature are 2219 K using a comprehensive set of equilibriums, 2232 K using a simplified set, and 2324 K considering only the formation of CO_2 and H_2O.

Calculations of the adiabatic flame temperature allowing for dissociation are complex. A computer program

TABLE 12.7 Mean Specific Heats (kJ/kmol°C) at a Constant Pressure

Final Temperature, t (°C)	Mean Specific Heat (kJ/kmol°C)			
	O_2	N_2	CO_2	H_2O
2000	35.27	33.47	54.85	43.67
2100	35.42	33.61	55.14	44.05
2200	35.55	33.72	55.43	44.42

[a]Mean specific heat over the temperature interval 0–t°C. Values based on data for zero pressure given by the authors and based on those of Wagman (1953).
Source: After Hougen et al. (1954).

that performs such calculations has been described by Gordon and McBride (1971, 1976). An account of its use is given by Kuo (1986).

Table 12.9 gives some theoretical and experimental values of flame temperature quoted by Lewis and von Elbe (1987) and Siegel and Howell (1991). The theoretical values given in the table are adiabatic flame temperatures at constant pressure. Details of the experimental temperatures are given in the footnote to the table. Lewis and von Elbe state that the experimental flame temperatures were probably influenced somewhat by heat loss and other factors, but that the error is small; Siegel and Howell describe the experimental values as 'maximum temperatures.'

TABLE 12.8 Complete Thermodynamic Treatment of Adiabatic Flame Temperature of Propane in Air (Barnard and Bradley, 1985)

Species	Concentration (Mole Fraction)			
	Constant Pressure			Constant Volume
	Case 1	Case 2	Case 3	
CO_2	0.1004	0.1003	0.1111	0.0914
H_2O	0.1423	0.1439	0.1481	0.1374
N_2	0.7341	0.7347	0.7407	0.7276
CO	0.0099	0.0100		0.0182
H_2	0.0032	0.0033		0.0053
O_2	0.0048	0.0055		0.0075
NO	0.0020	0.0022		0.0052
CH_2O	$<10^{-5}$			$<10^{-5}$
C_2H_4	$<10^{-5}$			$<10^{-5}$
C_3H_6	$<10^{-5}$			$<10^{-5}$
N_2O	$<10^{-5}$			$<10^{-5}$
CHO	$<10^{-5}$			$<10^{-5}$
CH_3	$<10^{-5}$			$<10^{-5}$
C_2H_5	$<10^{-5}$			$<10^{-5}$
$i\text{-}C_3H_7$	$<10^{-5}$			$<10^{-5}$
H	0.0035			0.0009
O	0.0020			0.0008
N	$<10^{-5}$			$<10^{-5}$
OH	0.0027			0.0058
HO_2	$<10^{-5}$			$<10^{-5}$
Final temperature (K)	2219	2232	2324	2587

[a]Calculations are for a stoichiometric mixture. Case 1 refers to calculations based on a comprehensive set of equilibriums, Case 2 to calculations using a smaller set of equilibriums, and Case 3 to calculations made considering only the formation of CO_2 and H_2O.
Source: Courtesy of Chapman and Hall.

TABLE 12.9 Adiabatic Flame Temperatures of Selected Substances in Air at Atmospheric Pressure

	Theoretical[a] (K)		Experimental (K)	
	Complete Combustion	With Dissociation and Ionization	1[a]	2[b]
Methane	2285	2191	2148	2158
Ethane	2338	2222	2168	2173
Propane	2629	2240	2198	2203
n-Butane	2357	2246	2168	2178
Ethylene	2523	2345	2248	2253
Propylene	2453	2323	2208	2213
Acetylene	2859			2598
Benzene	2484			
Carbon monoxide	2615			
Hydrogen	2490			2318

[a]Given by Siegel and Howell (1991), quoting Barnett and Hibbard (1957) and Gaydon and Wolfhard (1960). Values for combustion in dry air at 298 K.
[b]Given by Lewis and von Elbe (1987), quoting Loomis and Perrott (1928), Jones et al. (1931).

12.2.9 Degree of Flammability

Comparisons are often made between the flammability either of different substances or of different mixtures of the same substance in air. There is no single parameter that defines flammability, but some that are relevant are (1) the flashpoint, (2) the flammability limits, (3) the AIT, (4) the ignition energy, and (5) the burning velocity.

The flashpoint of a substance is often treated as the principal index of flammability, a substance being regarded as highly flammable if it has a low flashpoint. The other flammability characteristics are also important, however. The flammability of the substance is increased by wide flammability limits and by a low minimum AIT, low MIE, and a high maximum burning velocity. The mixture that has the lowest AIT and ignition energy and the highest burning velocity tends to occur near to, but normally not exactly at, the stoichiometric composition.

12.2.10 Quenching Effects

Flame propagation is suppressed if the flammable mixture is held in a narrow space. Thus, there is a minimum diameter for apparatus used for determination of flammability limits such that below this diameter the flammable range measured is narrower and inaccurate. Ultimately, if the space is sufficiently narrow, flame propagation is suppressed completely. The largest diameter at which flame propagation is suppressed is known as the quenching diameter. For an aperture of slot-like cross section, there is a critical slot width.

The term 'quenching distance' is used sometimes as a general term covering both quenching diameter and critical slot width and sometimes meaning the latter only. An account of quenching distances is given by Potter (1960). He gives the following empirical relation between quenching diameter D_o and critical slot width $D_{\parallel}$:

$$D_{\parallel} = 0.65 \, D_o \qquad (12.22)$$

He also discusses the effect of pressure and temperature on these variables.

There is a maximum experimental safe gap (MESG) that avoids the transmission of an explosion occurring within a container to a flammable mixture outside the container. The MESG is measured in a standard apparatus. One apparatus consists of a spherical vessel with the gap between a pair of flat equatorial flanges, the breadth of the gap being 1 in. A value of less than 1 in causes a decrease in the MESG, but a greater value has no effect.

Critical slot widths and MESGs for selected substances in air are given in Table 12.10. It is emphasized that these values relate to a stationary flame. If the gas flow is in the direction of flame propagation, a smaller gap is needed to quench the flame and, conversely, if the gas flow is in the opposite direction, a larger gap will effect quenching. If the gas velocity is high enough, a condition can occur in which a flame propagating against the flow is stabilized at a constriction and causes local overheating.

TABLE 12.10 Critical Slot Widths and Maximum Experimental Safe Gaps for Selected Substances in Air

	Critical Slot Width[a,b] (mm)	Maximum Experimental Safe Gap[b,c] (mm)
Acetone	–	1.01
Acetylene	0.52	0.37[d]
Ammonia	–	3.18
Benzene	1.87	0.99
n-Butane	–	1.07[e]
Carbon disulfide	0.55	0.20
Carbon monoxide	–	0.91[d]
Cyclohexane	3.0	0.94
Ethane	–	0.91
Ethylene	1.25	0.65
Ethylene dichloride	–	1.82
Ethylene oxide	1.18	0.59
Hydrogen	0.50	0.20
Methane	2.16	1.14
n-Pentane	2.07	0.89[e]
Propane	1.75	0.92
Propylene	–	0.91
Vinyl chloride	–	0.96

[a]Critical slot widths are for stoichiometric mixtures and are corrected where necessary to atmospheric pressure and 25°C.
[b]Where several values of critical slot width or maximum experimental safe gap are given in original reference, the value quoted here is the smallest.
[c]Maximum experimental safe gaps are corrected where necessary to atmospheric pressure and 20°C.
[d]Data for acetylene and carbon monoxide should be interpreted with care. For acetylene under certain ill-defined circumstances, external ignitions have been reported at flange gaps too small to measure. For carbon monoxide, data are for moist (not saturated) gas. Addition of moisture greatly increases the burning velocity of carbon monoxide.
[e]Burgoyne (1965). Maximum experimental safe gap at atmospheric pressure.
Critical slot width: Potter (1960). Maximum experimental safe gap: Lunn and Phillips (1973), except as given in footnotes.

For some substances, notably acetylene, carbon disulfide, and hydrogen, the quenching diameters and distances are very small. These quenching effects are important in the design of flameproof equipment and of flame arresters.

12.2.11 Flammability in Oxygen

The flammability of a substance depends strongly on the partial pressure of oxygen in the atmosphere. The oxygen

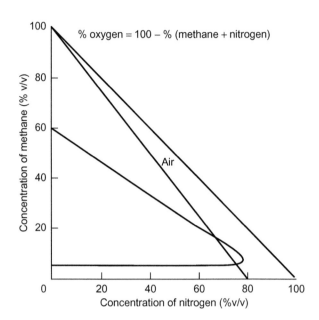

FIGURE 12.10 Effect of mixture composition on burning velocity of methane in air and in oxygen (Zabetakis, 1965). *Source: Courtesy of the Bureau of Mines.*

concentration affects both the flammability limits and the other flammability parameters. In general, increasing oxygen content alters the LFL only slightly, but its effect on the UFL is marked. Figure 12.10 illustrates the effect of oxygen concentration on the flammability limits of methane–oxygen–nitrogen mixtures. Conversely, there is a minimum oxygen content to support combustion. For the system shown in Figure 12.10, this is approximately 12%. Increasing the oxygen content reduces the ignition energy and the quenching distance of methane–oxygen–nitrogen mixtures as shown in Figures 12.11 and 12.12, respectively. Again the effect is a marked one.

The increase in burning velocity of methane when air is replaced by oxygen was illustrated in Figure 12.9.

12.3. FLAMMABILITY OF AEROSOLS

It is also relevant to consider the combustion of aerosols such as fogs, mists, and sprays. Such aerosols may be produced by condensation of a saturated vapor or by atomization of liquid by mechanical forces. The former may be referred to as a condensed fog or mist and the latter as a mechanical spray. Normally in a condensed mist, the diameter of most of the drops is less than 10 μm, while in a mechanical spray it is greater than 100 μm.

A large proportion of work in this field is directed toward the combustion of atomized fuels at burners in furnaces and boilers. Some of the work deals with the combustion of single droplets of fuel, and some deals with the combustion of an aerosol cloud.

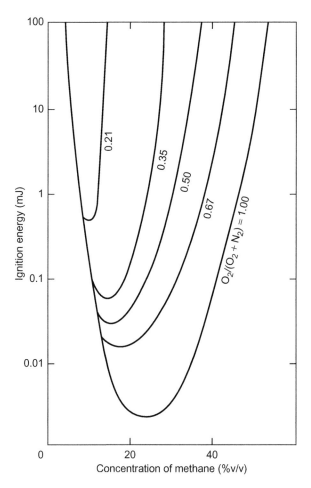

FIGURE 12.11 Minimum spark ignition energies in methane—oxygen—nitrogen mixtures at 1 atm pressure (Lewis and von Elbe, 1961). *Source: Courtesy of Academic Press.*

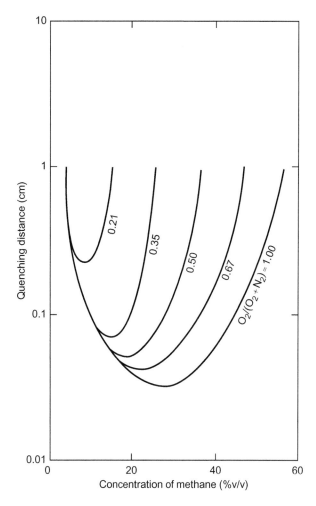

FIGURE 12.12 Quenching distances in methane—oxygen—nitrogen mixtures at 1 atm pressure (Lewis and von Elbe, 1961).

12.3.1 Combustion of Aerosols

A suspension of finely divided droplets of flammable liquid in air can give a flammable mixture, which has many of the characteristics of a flammable gas—air mixture and which can burn or explode.

In classic experiments using aerosols of tetrahydronaphthalene, or tetralin, with droplets of closely controlled diameter in the range 7—55 μm, Burgoyne and Cohen (1954) found that below a droplet diameter of 10 μm, the aerosol behaved like a vapor in respect of the LFL and the burning velocity. With a droplet diameter above 40 μm, the behavior of the aerosol was different. The droplets were observed to burn individually in their own air envelopes, one burning droplet igniting the next.

If the droplet size, and hence the distance between droplets, exceeds a critical value, the flame does not propagate. The critical distance is of the same order of magnitude as the radius of the sphere of air required for the combustion of a droplet.

Coarser aerosols are capable of sustaining a flame at substantially lower fuel—air ratios than fine aerosols and vapors. The difference lies in the ability of the droplets to move in relation to the ambient air. Coarse particles are responsive to acceleration and move randomly, and thus communicate flame more readily. Burgoyne's work has also shown that the burning velocity and the quantity of inert gas required for the suppression of flammability in aerosols with small droplet diameters are those of the equivalent vapor—air mixture.

12.3.2 Burning of Single Droplets

The combustion of the individual liquid droplets of an aerosol has been the subject of a good deal of work, particularly in relation to spray combustion in gas turbines and rockets.

Research in this area includes work on the formation of droplets by atomization and other mechanisms, and the size distribution and velocities of the droplets produced; the evaporation of and the drag on the droplets; and the mass burning rate of the droplets.

The mass burning rate $\dot{m}_F$ of a liquid droplet is

$$\dot{m}_F = -\frac{d}{dt}\left(\frac{\pi}{6}d_L^3 \rho_L\right) \qquad (12.23)$$

where d_L is the diameter of the droplet and ρ_L is the density of the liquid. This equation can be rearranged to give

$$-\frac{d}{dt}(d_L)^2 = \frac{4\dot{m}_F}{\pi d_L \rho_L} \qquad (12.24)$$

It is found experimentally that

$$-\frac{d}{dt}(d_L)^2 = K \qquad (12.25)$$

where the constant K is known as the burning constant. This constant is therefore

$$K = \frac{4\dot{m}_F}{\pi d_L \rho_L} \qquad (12.26)$$

There are a number of correlations of the burning constant, predominantly for conditions of forced convection, as in spray combustion. One of the most widely used is that of Agoston (1957).

12.3.3 Flammability Limits

Burgoyne and coworkers, as described above, have studied the flammability limits of the aerosols of flammable liquids. Burgyone (1963) quotes early work by Haber and Wolff (1923), who obtained for the LFL the results given in Table 12.11.

As already described, Burgoyne and Cohen (1954) found in their studies on tetralin a difference in behavior between aerosols with a droplet diameter d_L less than 10 μm and those with a droplet diameter more than 40 μm. The LFLs were found to be 46 mg fuel/l of aerosol for $d_L < 10$ μm and 18 mg fuel/l of aerosol for $d_L > 40$ μm; for intermediate diameters, the LFLs change linearly between these two concentrations.

For aerosols in which the droplet diameter is less than 10 μm, the LFL of the liquid and vapor suspension is virtually the same as that of the substance wholly in vapor form at the somewhat higher temperature necessary for vaporization. Above a droplet diameter of 10 μm, the lower limit of flammability decreases as the drop diameter increases.

Above a droplet diameter of 20 μm, another phenomenon was identified as significant, namely the rate of sedimentation of the droplet. For this effect, Burgoyne and Fauls (1963) give the following treatment. They define a flame front concentration C_f that is related to the volumetric concentration C_v as follows:

$$\frac{C_f}{C_v} = \frac{V_f + V_a + V_s}{V_f + V_a} \qquad (12.27)$$

where C_f is the volumetric flame front concentration, C_v is the volumetric concentration, V_a is the downward velocity of the air, V_f is the upward velocity of the flame through the suspension, and V_s is the sedimentation velocity of the drops relative to the air.

The LFL concentrations obtained by experiment were close to the flame front concentrations as defined by Equation 12.27. In some experiments with larger drop diameters, the LFL measured was less than one-tenth of that for the equivalent vapor−air mixture. The burning velocity and the quantity of inert gas required for the suppression of flammability are also affected by large drop sizes.

In general, for flammable aerosols in air, it is found that for small droplet diameters, the LFLs, measured as mass of fuel per unit volume of aerosol, are similar to those for vapor−air mixtures. Some values of the LFLs of vapors are quoted by Burgoyne and Fauls (1963). For most of the substances which they list the LFLs of the vapor lies in the range 0.04−0.10 g/l (or oz/ft^3). Further details are given in Section 12.2.

TABLE 12.11 Lower Flammability Limits of Some Vapors and Mists in Air

	Temperature (°F)	Mist Concentration (oz/ft³)	Temperature (°F)	Vapor Concentration (oz/ft³)
Petroleum 356−428°F	75	0.044	140	0.043
Tetralin	84	0.0409	212	0.0416
Quinoline	97	0.0662	230	0.064

Source: After Haber and Wolff (1923).

12.3.4 Minimum Ignition Energy

A method of obtaining the MIE of a dust cloud, which is also applicable to a vapor and an aerosol, has been given by Ballal (1983b). Ballal has correlated MIEs for a number of mixtures of gases, vapors, liquid droplets, and dusts with air at atmospheric pressure, the size of the liquid and solid particles being of the order of 50 μm. The correlation is in terms of the Spalding mass transfer number B:

$$B = \frac{q_{st}H + c_{pa}(T_g - T_b)}{L + c_p(T_b - T_s)} \qquad (12.28)$$

where c_p is the specific heat of the fuel, c_{pa} is the specific heat of air, H is the heat of combustion, L is the latent heat of vaporization, q is the mass ratio of fuel to air, and subscripts b, g, s, and st denote the boiling point of the fuel, gas, the surface of the fuel, and stoichiometric, respectively.

Figure 12.13(a) gives an approximate relation between the MIE E_{min} and B for homogeneous and two-phase mixtures of the type described. For a more accurate estimate, Ballal gives the MIE E_{min} as a function of the quenching diameter d_q:

$$E_{min} = \pi C_p \rho_a \Delta T d_q^3 \qquad (12.29)$$

with

$$d_q = (8\alpha)^{1/2} \left\{ \left[\frac{C_3^3 \rho_p D_{32}^2}{8C_1 f^2 (k/C_p)\phi \ln(1+B)} + \frac{12.5\alpha}{S_u^2} \right]^{-1} \right.$$

$$\left. - \frac{9q C_1^2 \varepsilon \sigma T_p^4}{c_p \rho_p C_3^3 f D_{32} \Delta T_{st}} \right\} \qquad (12.30)$$

where c_{pg} is the specific heat of the gas, C_1 is the ratio of the surface mean area to the Sauter mean diameter, C_3 is the ratio of the volume mean diameter to the Sauter mean diameter, D_{32} is the Sauter mean diameter, f is the swelling factor of the fuel, k is the thermal conductivity of the fuel, S_u is the laminar burning velocity, α is the thermal diffusivity of the particle, E is the emissivity of the particle, ρ is the density, ΔT is the temperature difference, ϕ is the Stefan−Boltzman constant, ϕ is the equivalence ratio, and subscript p denotes fuel.

12.3.5 Burning Velocity

Ballal (1983a) has also given a method of obtaining the laminar burning velocity of a dust cloud, which is also

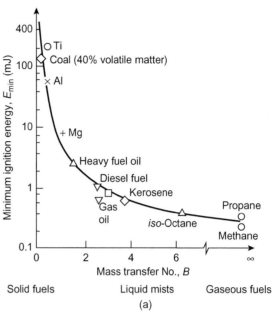

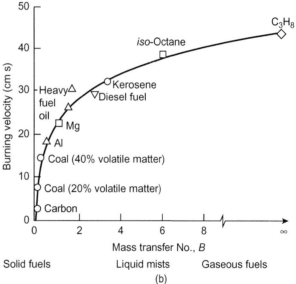

FIGURE 12.13 Some properties of mixtures of air with flammable vapors, aerosols, or dusts: (a) minimum ignition energy and (b) fundamental burning velocity. *Source: After Ballal (1983b). Courtesy of Combustion and Flame. After Ballal (1983a). Courtesy of the Combustion Institute.*

applicable to a vapor and an aerosol. Figure 12.13(b) gives an approximate relation between laminar burning velocity S_u and B for homogeneous and two-phase mixtures. For a more accurate estimate, Ballal gives the laminar burning velocity as a function of the thickness δ_r of the reaction zone:

$$S_u = \frac{\alpha \Delta T_r}{\alpha_r \Delta T_{pr}} \qquad (12.31)$$

with

$$\delta_r = \alpha_g^{0.5} \left\{ \left[\frac{C_3^3 \rho_f D_{32}^2}{8f^2 C_1 (c_p/k)_g \phi \ln(1+B)} + \frac{\alpha \Delta T_r}{S_g^2 \Delta T_{pr}} \right]^{-1} \right.$$

$$\left. - \frac{9q C_1^2 \varepsilon \sigma T_p^4}{c_{pg} \rho_f C_3^3 f D_{32} \Delta T_r} \right\}^{-0.5} \qquad (12.32)$$

where c_{pg} is the specific heat of the gas, S_g is the laminar burning velocity of the gases liberated from the particles, α_g is the thermal diffusivity of the gas, and subscripts f, g, pr, and r denote the fuel, gas, pre-reaction zone, and reaction zone, respectively.

These correlations of Ballal are subject to a number of qualifications, as discussed by the author and by Nettleton (1987), who nevertheless concludes that the approach is applicable to many practical situations. Nettleton quotes some typical ranges of the laminar burning velocity S_u in air. These are $0.05 < S_u < 0.2$ m/s for particulate suspensions; $0.2 < S_u < 0.35$ m/s for fogs; $0.3 < S_u < 0.4$ m/s for hybrid fog and vapor; and $S_u < 0.5$ m/s for mixtures with gas or vapor. For stoichiometric mixtures of acetylene or hydrogen in air, S_u is greater than 1.0 m/s.

12.3.6 Chemical Decomposition

If a condensed mist is formed from the saturated vapor of a liquid hydrocarbon of high boiling point, prolonged contact with a source of heat can result in the formation of cracking products such as hydrogen or acetylene, which reduce the LFL and increase the burning velocity and the quantity of inert gas required for the suppression of flammability.

12.4. IGNITION

12.4.1 Ignition Sources

Potential ignition sources in process plants include flames, hot work, hot surfaces, hot particles, friction and impact, chemical energy, hot materials and gases, reactive/unstable/pyrophoric materials, engines, vehicles, lightning, radio frequency emissions, smoking, arson and sabotage, self-heating, static electricity, and electrical equipment. Each of the above sources can cause ignition of a flammable gas–air mixture outside the plant. Insofar as they can occur inside the plant, these sources can also cause ignition of a flammable gas–air mixture inside it. In addition, ignition of a flammable gas–air mixture inside the plant may occur due to:

1. autoignition;
2. compression effects.

This section provides an overview of common ignition sources. Details about the physics and chemistry of the ignition process and the modeling of ignition can be found in the full version of Lee's *Loss Prevention in the Process Industries*, 4th edition.

12.4.1.1 Flames

The flames of burners in fired heaters and furnaces including boiler houses and flame in a flare stack may be sources of ignition in process plants. Such flames cannot be eliminated. It is necessary, therefore, to take suitable measures such as care in location and use of trip systems.

Burning operations such as solid waste disposal, rubbish bonfires, and smoldering materials may also act as sources of ignition. The risk from these activities should be reduced by suitable location and operational control.

12.4.1.2 Hot Work

Hot work, such as welding, cutting, and grinding activities, is a potential source of ignition. In welding, for example, this applies not only to the welding flame or arc, but also to material ignited by the welding. It is necessary to exercise close control of hot work by training, supervision, and use of a permit system.

12.4.1.3 Hot Surfaces

Surfaces of plant equipment are frequently hot and some may be potential ignition sources. Hot surfaces include hot process equipment and distressed machinery. It states that as a rule of thumb, and based on open air tests, ignition by a hot surface should not be assumed unless the surface temperature is at least 200°C above the normally accepted minimum ignition temperature.

A hot surface is a potential ignition source inside as well as outside the plant. The rule of thumb just quoted is not intended to apply in this case, although induction time and gas velocity are again relevant factors.

12.4.1.4 Friction and Impact

Essentially, incendive sparks caused by impact and friction are due to the generation of a surface hot spot. A falling object can give an incendive spark. The temperature rise may correlate only weakly with the kinetic energy but can be sufficient to cause ignition. It is normal to adopt procedures to prevent damage to the plant by falling objects. However, dropping of objects can occur, which may not be controlled by these measures such as drums and hand tools.

Ignition by a falling object is likely to be most serious in a space that already contains a flammable atmosphere.

Objects such as tank washing machines or cathodic protection anodes in ships' cargo tanks illustrate this. Cigarette lighter flints can also cause a mechanical spark if the lighter is dropped.

One situation that is liable to cause ignition in frictional impact is where rusty steel with an aluminum smear is hit by a hard object. Under these conditions the thermite reaction occurs. This reaction, which can reach a temperature of 3000°C, is an effective ignition source. The source of the aluminum may be aluminum paint or the impacting object itself.

An aspect that has received a good deal of attention over the years is the use of 'non-sparking tools.' Materials used in such tools characteristically do not readily oxidize and have high thermal conductivity.

Reviews of the use of non-sparking tools have been given by Powell (1969) and Cross and Farrer (1982). Essentially, some investigators have concluded that regular steel tools are no more hazardous than non-sparking tools, while others believe there is a benefit to safety in the use of non-sparking tools. Anfenger and Johnson (1941) performed experiments on the incendivity of sparks produced by steel on a grinding wheel, which ignited carbon disulfide but not petrol vapor.

On the basis of this work and practical experience, the American Petroleum Institute (API) in *Sparks from Hand Tools* (1956) concluded that there is no significant increase in safety from the use of non-sparking tools. API Publ. 2214: 1989 *Spark Ignition Properties of Hand Tools* states that nothing essentially new has been learned since the 1956 publication and that the fire records of more and more companies which have never used, or have ceased to use, non-sparking tools amply confirm the position then taken. Other research such as those at the Department of Scientific and Industrial Research (1963) and Riddlestone and Bartels (1965) draw attention to ignitions caused by the striking of metals on concrete or stone and point out that it is in this situation that the ignition risk is significant and that metals used in non-sparking tools are also liable to give ignition in this case. For this rather different reason, they conclude that non-sparking tools offer little benefit.

Another point that detracts from the potential benefit of non-sparking tools is that they tend to be soft, so that hard particles can become embedded in the tool. Other workers have seen a benefit in the use of non-sparking tools. On the basis of impact and grinding experiments, Fischer (1965) recommends the use of steel tools for work where there may be Class I gases, but not where there may be Class II gases. Certain gases such as hydrogen, acetylene, and carbon disulfide are particularly sensitive to ignition by impact and friction.

12.4.1.5 Chemical Energy

There are several forms of chemical energy that may give ignition. They include (1) the thermite reaction, (2) reactive, unstable, and pyrophoric materials, and (3) catalytic instruments. The thermite reaction is obtained from aluminum smears on rusty steel, as described above. Reactive, unstable, and pyrophoric materials are considered below.

Some instruments contain catalytic elements, which measure the temperature rise resulting from combustion of a flammable gas. Generally, these instruments (such as flammable gas detectors) are intended to operate in hazardous areas and are therefore designed so that they should not act as an ignition source.

12.4.1.6 Hot Materials and Gases

Hot materials such as hot ash, hot used catalyst or hot process material, and hot gases may, in principle, act as ignition sources. Generally, however, operations involving such hot materials are excluded from hazardous areas and they do not figure as a significant ignition source.

12.4.1.7 Reactive, Unstable, and Pyrophoric Materials

Reactive, unstable, or pyrophoric materials may act as an ignition source by undergoing an exothermic reaction so that they become hot. In some cases, the material requires air for this reaction to take place, in others it does not.

The most commonly mentioned pyrophoric material is pyrophoric iron sulfide. This is formed from reaction of hydrogen sulfide in crude oil in steel equipment. If conditions are dry and warm, the scale may glow red and act as a source of ignition. Pyrophoric iron sulfide should be damped down and removed from the equipment. No attempt should be made to scrape it away before it has been dampened. A reactive, unstable, or pyrophoric material is a potential ignition source inside as well as outside the plant.

12.4.1.8 Engines

Engines in process plants are another ignition source and tend to figure significantly in fires and explosions. One type of engine in use for a variety of purposes is the diesel engine. One potential ignition source in a diesel engine is the hot gas. Exhaust gas temperatures can be as high as 500°C. Ignition by hot exhaust gas involves turbulent stream ignition, which can occur at a temperature less than the average temperature of the gas mixture. A safety factor is therefore required. Codes typically recommend that the temperature of the exhaust gas should not exceed 0.6–0.8 of the AIT of any flammable gas that may be present.

Another important part of the engine is the induction and exhaust systems. These systems are designed to avoid detonation but to handle deflagration. There are minimum design pressures, which tend to lie in the range 8–10 bar, that for the United Kingdom being 10 bar and that for the United States 8.6 bar. Induction and exhaust systems are provided with flame arresters; exhausts are provided with spark arresters. Crank cases are another feature of a diesel engine that requires protection. The preferred start-up arrangements for the engine in a hazardous area are pneumatic, hydraulic, or manual rather than electrical.

Diesel engines are generally equipped with shut-down devices and alarms activated by overspeed and by failures of the various subsystems. Depending on the application, the device may act to cut off the fuel or to shut off the air intake and effect of gradual gas path deflection.

12.4.1.9 Vehicles

A chemical plant may contain at any given time considerable numbers of vehicles. These vehicles are potential sources of ignition. Instances have occurred in which vehicles have had their fuel supply switched off, but have continued to run by drawing in, as fuel, flammable gas from an enveloping gas cloud. It is necessary, therefore, to exclude ordinary vehicles from hazardous areas and to ensure that those that are allowed in cannot constitute an ignition source.

Vehicles that are required for use on process plant include cranes and forklift trucks. Various methods have been devised to render vehicles safe for use in hazardous areas and these are covered in the relevant codes.

12.4.1.10 Lightning

Lightning is another potential ignition source in process plants. Information on lightning and on lightning protection can be found in *Installation of Lightning Protection Systems* (NFPA 780).

Lightning has traditionally been a significant ignition source for storage tank fires. A major incident of this kind occurred in Beaumont, Texas, in 1970. Guidance on lightning protection is given in BS 6651: 1985 Code of Practice for Protection of Structures against Lightning.

12.4.1.11 Radio Frequency Transmissions

The possibility exists that RF transmissions from strong sources such as large military transmitters may act as an ignition source in process plants. This has been recognized for some time as instanced by the existence of a British Standard BS 4992: 1974 *Guide to Protection Against Ignition and Detonation Initiated by Radio Frequency*.

A situation where concern was expressed about such possible interaction arose in Britain in the late 1970s in respect of the transmitter at Crimond operated by the Royal Navy and located about 4 miles from the natural gas terminal at St Fergus. The naval transmitter responded by reducing its power output. The British Standard guidance was revised and issued as BS 6656: 1986 Prevention of Inadvertent Ignition of Flammable Atmospheres by Radio-frequency Ignition, and revised again in 1991.

The conditions for RF ignition of a flammable mixture to occur are (1) electromagnetic radiation of sufficient intensity, (2) a structure capable of acting as a receiving aerial, and (3) a mechanism for creating an incendive spark.

The transmitter may be fixed or mobile. Mobile transmitters include vehicles, ships, and aircraft. The standard gives details of typical transmitter frequency ranges and power outputs. Transmission may be continuous wave (CW) or modulated. The latter includes pulsed radar.

The degree of hazard depends on the frequency of the transmission. The hazardous range of interest is 15 kHz to 35 GHz. There is little hazard at frequencies below 15 kHz. For frequencies below 30 MHz, the most efficient receiver is the loop configuration. At higher frequencies, all structures are large compared to the wavelength. A part of a structure may behave as an efficient aerial and is then treated as a long dipole. An aerial can concentrate the power in a particular direction and is said to have gain in that direction.

The structures primarily considered in BS 6656:1991 are loop-type structures and vertical structures. Some typical loop-type structures are illustrated in Figure 12.14. A loop structure has maximum efficiency and is self-resonant when its internal perimeter is about half one wavelength, but structures with a smaller perimeter can be brought to resonance if there is a discontinuity with stray capacitance across it. Cranes are particularly efficient receivers in this context and require particular attention.

For any potential discontinuity in the structure, it is possible to determine the maximum extractable power given: (1) the structure perimeter, (2) the transmission frequency, and (3) the incident field strength.

Vertical structures include vents, flares, and columns. Freestanding structures of this kind are not classed as among the more efficient receivers, since even a concrete base has a low impedance path to ground. They can generally be disregarded, except where the vertical structure is part of a loop.

An RF discharge occurs most readily when two surfaces are drawn apart, thus giving a break-spark. Discharges across a fixed gap are not considered to be a significant problem. It is activities such as maintenance and handling and phenomena such as flexing and

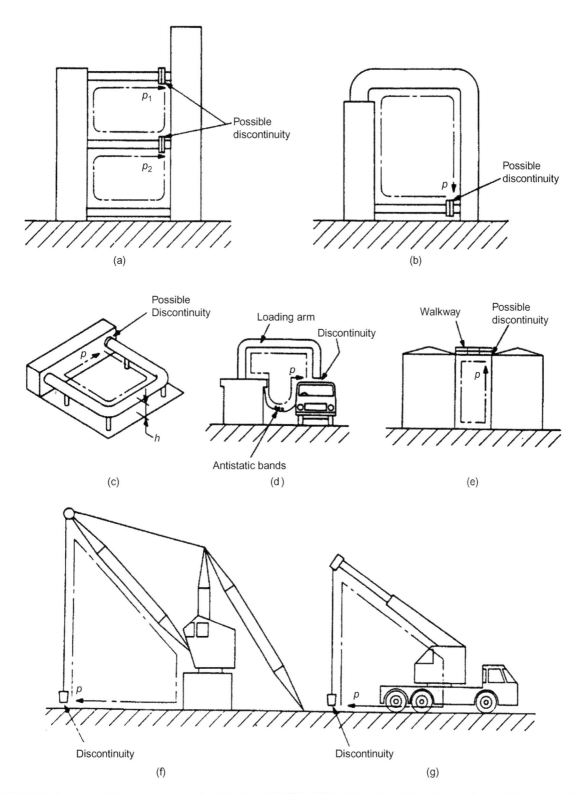

FIGURE 12.14   Some typical loop-type structures for RF ignition (BS 6656: 1991): (a) loop formed by columns and pipes; (b) loop formed by columns and pipes; (c) horizontal loop; (d) tanker loading facility; (e) storage tanks; (f) fixed crane; and (g) mobile crane, h, height of loops; p, internal perimeter of loop. *Source: Courtesy of the British Standards Institution.*

vibration of pipe work or thermal expansion of structures, which are liable to give rise to an incendive discharge.

The factor determining RF ignition is the thermal initiation time. For times less than this ignition is governed by the discharge energy, and for times greater than this it is governed by its power. The thermal initiation times are approximately 100 μs for methane and ethylene and 20 μs for hydrogen, which are representative of the Class IIA, IIB, and IIC gases, respectively. The standard gives ignition criteria for continuous transmissions and for pulsed radar transmissions.

The potential RF ignition hazard exists only in a few locations. There are few reported incidents of such ignition. BS 6656: 1991 adopts a graded approach to assessment of the hazard. An initial assessment is used for screening and a full assessment is undertaken only if a potential hazard is found to exist.

The initial assessment is based on the search areas given in Table 12.12 and proceeds as follows. The search area around the plant is determined from the table. If there is no transmitter within the search area, there is no hazard. If there is a transmitter within the search area, a check is made as to whether the plant is within the vulnerable zone of the transmitter. The standard gives information on the vulnerable zones for various types of transmitter. If the plant is within the vulnerable zone of a transmitter, a full assessment should be undertaken.

BS 6656:1991 gives the procedures for a full assessment, including a flow chart for a theoretical assessment and one for an assessment based on plant measurements, methods of taking measurements, methods of performing the calculations, and worked examples.

If a hazard exists, the principal countermeasures are (1) bonding, (2) insulating, (3) reducing the RF efficiency of structures, and (4) de-tuning of structures. Typical applications where bonding might be used are where thermal expansion may result in intermittent contacts and across a pair of flanges prior to their parting in maintenance work. Bonding to ground is not suitable. Insulation may sometimes be used where bonding is impractical; it is a possible solution for intermittent contact by thermal expansion. The RF efficiency of a structure may be reduced at the design stage by altering the internal perimeter. Other measures for reducing efficiency include breaking the loop into smaller sections by the use of conductors and covering the entire area of the loop with a sheet of metallic mesh bonded to the structure at points around the perimeter. If there is one major transmission frequency causing the problem and other measures are not practical, it may be possible to de-tune the structure by connecting to it reactive components.

Special cases considered in the standard include cranes, mobile and portable transmitters, ships and offshore platforms. The Orford Ness transmitter is accorded special treatment.

12.4.1.12 Smoking

Smoking and smoking materials are potential sources of ignition. Ignition may be caused by a cigarette, cigar, or pipe or by the matches or lighter used to light it. A cigarette itself may not be hot enough to ignite a flammable gas—air mixture, but a match is a more effective ignition source.

It is normal to prohibit smoking in a hazardous area and to require that matches or lighters be given up on entry to that area. The 'no smoking' rule may well be disregarded; however, if no alternative arrangements for smoking are provided. It is regarded as desirable, therefore, to provide a room where it is safe to smoke, though whether this is done is likely to depend increasingly on general company policy with regard to smoking.

12.4.1.13 Autoignition

Strictly, ignition of a bulk flammable gas—air mixture by heating the mixture to its AIT is the alternative to ignition by a local ignition source. It is convenient, however, to deal with it at this point. Heating of the bulk gas—air mixture outside the plant will occur by contact with a hot surface, as described above. Autoignition as such may therefore be regarded as a form of ignition, which occurs inside the plant. The autoignition may occur inside the plant at essentially atmospheric pressure or at some higher pressure. If, however, the heating occurs due to

TABLE 12.12 Maximum Radius of Search Area (m) for Initial Assessment of an RF Ignition Hazard (BS 6656: 1991)

Gas Group	Maximum Radius (m)	
	All Loop Structures of Inside Perimeter ≤ 40 m and Horizontal Loops of Height ≤ 5 m	All Other Loop Structures
I or IIA	4100	11,500
IIB	5200	14,200
IIC	6500	17,500

*a*Table does not apply to locations to the seaward side of Orford Ness, Suffolk.
Source: Courtesy of the British Standards Institution.

compression, the phenomenon is one of compression ignition, which is described below.

12.4.1.14 Compression Effects

A mixture of flammable gas and air may be heated to its AIT by compression. This effect is often referred to as the diesel effect. There are a number of conditions in process plant that can give rise to such compression.

Compression of a mixture of flammable gas and air may be caused by pumping against a closed valve, by water hammer in a pipeline or by a liquid slug traveling down a line.

Another form of compression ignition can occur due to compression of air bubbles in a flammable liquid. The temperature rise inside such bubbles can be high.

12.4.1.15 Self-Heating

Materials in process, storage, or transport may undergo self-heating. The self-heating is due to the exothermic reaction of slow oxidation of the material. If conditions are critical, this self-heating results in ignition. Examples are materials handled in process equipment such as driers, materials stored in piles in warehouses or in the open, or materials transported in large containers as in ships. A well-known example is the spontaneous combustion of coal stored in piles on the ground.

In some cases, the hazard is intensified by the fact that the material enters the storage relatively hot. This can occur, for example, with material that has just been passed through a drier.

Unstable materials may also undergo an exothermic reaction so that self-heating occurs.

Self-heating is liable to occur in oil-soaked lagging. In this case it is the oil which undergoes oxidation. Self-heating may also occur in dust layers. In both cases the smoldering material may then act as an ignition source. Self-heating can also occur in oil rags left on steam pipes, in dirty cotton waste put in a boiler suit pocket, or in damp clothing stowed away in a locker. The substance that undergoes reaction must be reactive and may be bulk solid such as coal or a reactive substance on a substrate such as oil-soaked lagging.

Prevention of self-heating depends on recognition of the hazard, design features, and good housekeeping and adherence to procedures. Materials leaving the process hot can be cooled before being sent to storage. Self-heating of materials in storage and transport can be prevented by using smaller containers and/or by remixing the material periodically. Where oxidative self-heating is involved, measures can be taken to reduce the oxygen content of the ambient gas.

12.4.1.16 Static Electricity

Static electricity is an important source of ignition in process plants. There have been many apparently mysterious explosions the cause of which was eventually traced to static electricity. The industrial situations in which undesired static charges are generated are largely those in which two surfaces move relative to each other, with initial contact followed by subsequent separation. When the surfaces are separated, one body tends to be left with a positive charge and the other with a negative charge. If the bodies are good conductors of electricity, the charge moves quite freely and both bodies are effectively restored to their original uncharged states through the last points of contact at separation. But if one or both of the bodies are poor conductors, the charge does not flow freely and both bodies retain charge after separation.

There are many industrial processes that involve surface contact, movement, and separation of poorly conducting materials. The hazard of static electricity occurs in the process industry in (1) fluid handling operations such as pipeline flow, settling of drops, agitation, filling of storage tanks, filling of tankers; (2) powder and dust, and powder handling operations such as grinding, sieving, and pneumatic conveying; (3) in sprays and mists such as in steam cleaning and steam leaks; (4) moving equipment such as conveyor belts and bucket elevators; and (5) the human body.

Static electricity can act as a source of ignition giving rise to a fire or explosion only under the following conditions: a flammable atmosphere exists; an electrostatic charge is generated, accumulates, and produces an electric field strength which exceeds the critical value for breakdown; the resultant incendive discharge has an energy greater than the minimum ignition energy of the flammable atmosphere. Thus precautions against static electricity aim to eliminate one or more of the above factors. These precautions tend to be based, therefore, on the following approaches: elimination of a flammable atmosphere, control of charge generation, control of charge accumulation, and minimization of incendive discharge.

It is a general principle in handling flammable materials to make avoidance of a flammable atmosphere the first line of defense where this is practical. It is necessary also to try to eliminate sources of ignition, but it is much more difficult to do this reliably.

Grounding and bonding comprise one of the principal methods of providing protection against static electricity. Bonding involves making an electrical connection between two conducting objects and the ground. The effect of bonding is to maintain the two objects connected at the same potential and that of grounding is to drain away to the ground the charge on the object connected. Grounding and bonding systems are also required to give protection against electrical systems and against lightning, but the systems considered here relate only to static electricity.

The purpose of grounding and bonding is to prevent an incendive discharge. The conditions for an incendive discharge to occur are that the field strength reaches the breakdown value and that the energy in the discharge equals the relevant ignition energy.

Grounding is usually achieved by means of copper strips or wire attached to a special point on the object to be grounded. Bonding across flanged joints to ensure better electrical continuity than that through the bolts has been widely practiced, but many companies have satisfied themselves by tests that this is not necessary and have discontinued the practice. A good grounding system also requires that it be recognizable as such and that it be checked periodically and after maintenance operations.

Further details about precautions Against Static Electricity can be found in the British Standard: BS 5958: Parts 1 and 2: 1991 or in the United States standard NFPA 77: 2000 Static Electricity.

12.4.1.17 Electrical Equipment

Electrical equipment is widely used in process plant and may be a source of ignition unless close control is exercised. The basis for the control of electrical equipment to prevent its acting as a source of ignition is the classification of plant according to the degree of hazard. Originally this procedure was commonly termed the 'electrical area classification,' but it is now known by the more comprehensive term HAC, or classification of hazardous locations (CHL). Hazardous area classification is the subject of a number of codes and standards. These include BS 5345 *Code of Practice for the Selection, Installation and Maintenance of Apparatus for Use in Potentially Explosive Atmospheres (Other than Mining Applications or Explosive Processing and Manufacture)*, including Part 1: 1989 General Recommendations and, in particular, Part 2: 1983 (1990) *Classification of Hazardous Areas*. Influential codes have been those of the successive IP codes and the ICI *Electrical Installations in Flammable Atmospheres Code* (the ICI *Electrical Installations Code*). The current IP code is *Area Classification Code for Petroleum Installations* (the IP *Area Classification Code*) (1990 MCSP 15). In the United States, NFPA codes include NFPA 70: 2002 *National Electrical Code*, NFPA 497: 1997 *Classification of Flammable Liquids, Gases, or Vapors and of Hazardous (Classified) Locations for Electrical Installations in Chemical Process Areas*. API RP 500: 1991 *Recommended Practice for Classification of Locations for Electrical Installations at Petroleum Facilities* consolidates previous separate codes (RP 500A, 500B, and 500C). The following classifications of hazardous locations are defined in NFPA 70: 2002 Article 500.

Class I	in which the combustible material is a gas or vapor.
Class II	in which the combustible material is a dust.
Class III	in which the combustible material is easily ignitible fibers or flyings such as rayon and cotton.

Each class is divided into Division 1 and Division 2, the distinction between the two being based on the frequency with which the substance may be present in the atmosphere.

There are a number of methods available for the safeguarding of electrical equipment so that it is suitable for use in a hazardous area, but it is a fundamental principle that electrical equipment should not be located in a hazardous area if it is practical to site it elsewhere. The methods of safeguarding traditionally used in the United Kingdom are listed in the ICI *Electrical Installations Code*, such as (1) segregation, (2) flameproof enclosures, (3) intrinsically safe systems, (4) approved apparatus and apparatus with type of protection 's', (5) pressurizing and purging, (6) apparatus with type of protection 'N', (7) Division 2 approved apparatus, apparatus with type of protection 'e', (8) non-sparking apparatus and totally enclosed apparatus, (9) apparatus for use in dust risks. The overall procedure for selecting electrical apparatus for flammable atmospheres can be found in BS 5345: Part 1: 1989 (UK).

Guidance on Electrical Surface Heating (ESH) devices (e.g., flexible heaters, heating cable, panel heaters, parallel circuitry heaters), including their use in hazardous areas, is given in BS 6351 *Electrical Surface Heating*, of which Part 1: 1983 deals with specification, Part 2: 1983 deals with design, and Part 3: 1983 deals with installation, testing, and maintenance.

Further recommendations for ESH are given in the ICI *Electrical Installations Code*.

12.4.2 Probability of Ignition

The information available on the probability of ignition is mostly in the form of expert estimates. In the context of vapor cloud explosions, Kletz (1977) states that in polyethylene plants the leaks are mostly very small and that about one leak in 10,000 ignites, due probably to good jet mixing with air. He also states that in a series of plants handling a hot mixture of hydrogen and hydrocarbons about one leak in 30 ignites. He argues that the probability of ignition increases with the size of leak and suggests that for large leaks (>10 ton), the probability of ignition is greater than 1 in 10 and perhaps as high as 1 in 2.

Browning (1969) has given a set of estimates of the relative probabilities of ignition. For ignition under conditions of no obvious source of ignition and with explosion-proof electrical equipment, he gives the following probabilities of ignition in Table 12.13.

Elsewhere Browning (1980) gives a table of probabilities, which are evidently absolute probabilities. The table includes an estimate of the probability of ignition of flammable gas−liquid spills of 10^{-2} to 10^{-1}.

The First Canvey Report (HSE, 1978) gives probability of ignition for LNG vapor clouds in Table 12.14.

The Second Canvey Report (HSE, 1981) gives the following 'judgment' values for ignition on site in Table 12.15.

For offshore locations, Dahl et al. (1983) have analyzed ignition data for gas and oil blowouts, which may be regarded as massive releases. The data are shown in Table 12.16 Section A. They cover both drilling rigs and production platforms. The overall probability of ignition of blowouts is similar for the two cases.

Cox, Lees, and Ang draw on the foregoing to make estimates of the probability of ignition. They define a minor leak as one <1 kg/s and take for this an average leak flow of 0.5 kg/s. They define a massive leak as one >50 kg/s and take for this an average leak flow of 100 kg/s. They estimate the probability of ignition of a minor leak of either gas or liquid as 0.01, that of a massive leak of gas as 0.03, and that of a massive leak of liquid as 0.08. From these estimates, they derive those given in Table 12.17 and Figure 12.15.

12.4.3 Probability of Explosion

The information available on the probability of explosion is also mostly in the form of expert estimates. In his account of vapor cloud explosions, Kletz (1977) has also given estimates of the probability of explosion (Table 12.18). He quotes the following figures:

Frequency of serious vapor cloud fires = 5/year
Frequency of serious vapor cloud explosions = 0.5/year
and derives from these for a large vapor cloud
Probability of explosion given ignition = 0.1

He also gives estimates which are evidently for the probability of explosion given a leak. These are a probability >0.1 for a large vapor cloud (>10 ton) and 0.0001−0.01 for a medium vapor cloud (1 ton or less).

For offshore locations, Dahl et al. (1983) have analyzed blowouts as shown in Table 12.16, Section B.

Sofyanos (1981) has given for fires and explosions in the Gulf of Mexico the data shown in Table 12.19.

For the estimation of the probability of explosion, Cox, Lees, and Ang use for massive leaks the value of 0.3 given by Dahl et al. for blowouts and for minor leaks the value of 0.04 given by Sofyanos for Category V leaks. From these estimates, they derive those given in Table 12.20.

12.4.4 Distribution of Leaks

To model the occurrence of fire and explosion on a plant, it is also necessary to have information on the frequency of leaks. Table 12.21 gives the estimates of leak frequency used by Cox et al.

12.5. FIRE IN PROCESS PLANT

Fires in process plant are a serious hazard to both life and property. It is essential, therefore, to understand the ways in which fire can occur and develop. Normally, fire

---

TABLE 12.13 Relative Probability of Ignition to Different Flammables (Browning, 1969)

	Relative Probability of Ignition
Massive LPG release	10^{-1}
Flammable liquid with flashpoint below 110°F or with temperature above flashpoint	10^{-2}
Flammable liquid with flashpoint 110−200°F	10^{-3}

TABLE 12.14 Relative Probability of Ignition of LNG Vapor Clouds (HSE, 1978)

	Probability of Ignition
Limited releases	10^{-1}
Large releases	1

TABLE 12.15 Relative Probability of Ignition on Site (HSE, 1981)

Ignition Sources	Probability of Ignition
None	0.1
Very few	0.2
Few	0.5
Many	0.9

TABLE 12.16 Probability of Ignition and of Explosion for Offshore Blowouts (Cox et al., 1990)

A. Probability of Ignition

Blowout Fluid	No. of Blowouts	No. of Ignitions	Probability of Ignition
Gas	123	35	0.3
Oil	12	1	0.08

B. Probability of Explosion, Given Ignition

Blowout Fluid	No. of Blowouts	No. of Ignitions	No. of Explosions	Probability of Explosion Given Ignition
Gas	123	35	12	0.34
Oil	12	1	0	0

Source: After Dahl et al. (1983). Courtesy of the Institution of Chemical Engineers.

TABLE 12.17 Estimates of Probability of Ignition for Leaks of Flammable Fluids (Cox et al., 1990)

Leak (kg/s)	Probability of Ignition	
	Gas	Liquid
Minor (<1)	0.01	0.01
Major (1−50)	0.07	0.03
Massive (>50)	0.3	0.08

Source: Courtesy of the Institution of Chemical Engineers.

TABLE 12.18 Estimates of Probability of Explosion Given Ignition (Kletz, 1977)

	Probability of Explosion Given Ignition
Large vapor clouds	1
Smaller clouds of gases other than methane	0.1
Smaller clouds of methane	0.01

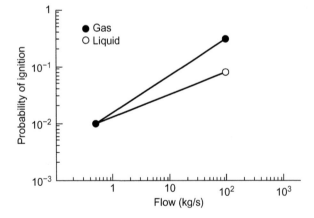

FIGURE 12.15 Estimated probability of ignition for leaks of gas or liquid (Cox et al., 1990). *Source: Courtesy of the Institution of Chemical Engineers.*

TABLE 12.19 Probability of Explosion Given Ignition for Blowouts in the Gulf of Mexico (Cox et al., 1990)

Damage Category[a]	No. of Incidents with Fire and Explosion	No. of Incidents with Explosion Only	Probability of Explosion Given Ignition
I	9	5	0.55
II	13	3	0.23
III	33	6	0.18
IV	128	22	0.18
V	143	6	0.042
Total	326	42	0.13[b]

[a]The damage categories are in decreasing order of severity, Category I being loss of platform and Category V being an incident of no consequence.
[b]Weighted average.
Source: After Sofyanis (1981). Reproduced by permission of the Institution of Chemical Engineers.

occurs as a result of a leakage or spillage of fluid from the plant. Larger leaks may occur due to the failure of a vessel, pipe, or pump, and smaller ones from flanges, sample, and drain points and other small bore connections.

Combustion of material, which has leaked from a plant, may take a number of forms. A leak of gas or liquid may be ignited at the point of issue so that it behaves like a flame on a burner. In some circumstances, this flame may be directed like a blowtorch at another part of the plant.

If the leak gives rise to a gas or vapor cloud which grows for a period before it is ignited, the resultant effect may be either a vapor cloud, a flash fire, or a vapor cloud explosion. In a flash fire the gas cloud burns, but does not explode. A typical flash fire may cause quite extensive damage, particularly to vulnerable items such as electric cabling, but may leave the main plant equipment relatively unharmed. However, a flash fire does cause a sudden depletion of oxygen, and this effect can be lethal to personnel.

If the leak forms a liquid pool on the ground, this may ignite and burn. The flame may be substantial and may do damage by direct impingement or by radiation. If the release results from fire engulfment of a vessel, a fireball may be formed. Prevention of fire in process plant is primarily a matter of preventing leaks and avoiding sources of ignition.

TABLE 12.20 Estimates of Probability of Explosion Given Ignition for Leaks of Flammable Gas (Cox et al., 1990)

Leak (kg/s)	Probability of Given Ignition
Minor (<1)	0.04
Major (1–50)	0.12
Massive (>50)	0.3

Source: Courtesy of the Institution of Chemical Engineers.

TABLE 12.21 Estimates of Leak Frequency Used in Ignition Model (Cox et al., 1990)

A. Pipework: Leak Frequency (leaks/m year)

	Pipe Diameter (m)			
	0.025	0.050	0.100	0.300
Rupture leak	0.005 (0.01)	0.005 (0.01)	0.0015 (0.003)	0.0005 (0.001)
Major leak	0.05 (0.1)	0.05 (0.1)	0.015 (0.06)	0.005 (0.03)
Minor leak	0.5 (1)	0.5 (1)	0.15 (0.3)	0.05 (0.1)

B. Flanges (All Pipe Diameters): Leak Frequency (leaks/year)

Major leak	0.3 (1)
Minor leak	3 (10)

C. Valves: Leak Frequency (leaks/year)

	Pipe Diameter (m)			
	0.025	0.050	0.100	0.300
Rupture leak	0.01 (0.1)	0.01 (0.1)	0.01 (0.1)	0.005 (0.1)
Major leak	0.1 (1)	0.1 (1)	0.1 (1)	0.05 (1)
Minor leak	1 (10)	1 (10)	1 (10)	0.5 (10)

D. Pumps (All Pipe Diameters): Leak Frequency (leaks/year)

Rupture leak	0.3 (0.3)
Major leak	3 (3)
Minor leak	30 (30)

E. Small Bore Connections: Leak Frequency (leaks/year)

Pipe diameter = 0.01 m
Rupture leak 1 (5)
Major leak 10 (50)

[a]All values to be multiplied by 10^{-4}.
Source: Courtesy of the Institution of Chemical Engineers.

In addition to fires arising from leakage in general, there are certain characteristic types of fire in process plants. These include (1) pump fires, (2) flange fires, (3) lagging fires, (4) duct fires, (5) cable tray fires, and (6) storage tank fires.

Guidance on the modeling of fire events is given in the CCPS Guidelines for Evaluating the Characteristics of Vapor Cloud Explosions, Flash Fires, and BLEVEs (1994/15) (the CCPS Fire and Explosion Model Guidelines). The *Guidelines* deal with (1) vapor cloud fires, (2) vapor cloud explosions, (3) BLEVEs, and (4) fireballs. For each of the three phenomena, the *Guidelines* outline basic physical concepts, experimental work and available models and correlations for the phenomena themselves and for damage and injury, and give sample problems. They also contain a set of case histories.

Thermal effect is the dominant consequence from fires. In general, the radiant heat flux emitted by a flame is given by the relation

$$E = \varepsilon \sigma T^4 \tag{12.33}$$

$$= \frac{Q_r}{A_f} \tag{12.34}$$

with

$$Q_r = F_r Q_c \tag{12.35}$$

where T is the absolute flame temperature, σ is the Stefan–Boltzmann constant, ε is the emissivity, A_f is the surface area of the flame, E is its surface emissive power, F_r is the fraction of heat released which is radiated, Q_c is the total heat released by combustion, and Q_r is the total heat radiated.

The heat received by the target is

$$I = \alpha \tau F E \tag{12.36}$$

where F is the view factor, I is the heat radiation intensity received by the target, α is absorptivity, and τ is the atmospheric transmissivity. These equations are used in conjunction with relations for the geometry of the flame and for the heat radiated per unit area of the flame surface.

There are alternative approaches to the estimation of the heat radiated Q_r. One is to work in terms of the flame temperature and emissivity. Another is to use empirical values of the surface emissivity of the flame envelope. A third is to calculate the total heat generated and to apply a factor F_r for the fraction of heat radiated. For further details on thermal radiation modeling from fires, readers are encouraged to refer to the full version of this book: Lee's *Loss Prevention in the Process Industries*, 4th edition. An overview of the various types of fire is discussed below.

12.5.1 Pump Fires

Pumps tend to leak at the gland or the seal and the leakage frequently ignites causing a fire.

A fire at a pump can do considerable damage. It is important, therefore, to assess the effect of a fire on equipment above the pump. In some fires, the damage has been much reduced by a concrete floor above the pump alley. Electrical and instrument cabling is particularly liable to be damaged by a pump fire.

The provision of protection against pump fires is a good illustration of the loss prevention approach. The need for measures, such as improved pump reliability, an emergency isolation valve, or protected cabling, should be assessed by considering for each case the frequency and consequences of a pump fire.

12.5.2 Flange Fires

Pipe flanges tend to leak and sometimes the leakage is ignited so that there is a fire. Leakage at flanges is induced mainly by temperature transients, which put elements of the flange assembly under stress and may cause them to yield.

12.5.3 Lagging Fires

Lagging on plant equipment frequently becomes impregnated with oils and other liquids. If the lagging is hot, self-heating and ignition may occur, leading to a lagging fire. A lagging fire is essentially a self-heating phenomenon. The conditions for a lagging fire to occur are that there should be sufficient fuel, oxygen, and heat. Factors relevant to the occurrence of a fire are (1) oil, (2) leak, (3) insulation material, (4) insulation sealing, and (5) insulation geometry and temperature.

The most important factor in a lagging fire is the oil itself. For significant self-heating to occur, the oil needs to be involatile (a volatile oil vaporizes too easily). An intrinsically reactive unsaturated oil is more prone to self-heating than is a saturated mineral oil, but any combustible involatile oil may self-heat. Typical leakage points which may cause a lagging fire are pumps, flanged joints, and sample and drain points.

The material of insulation is another factor in determining self-heating, although it is generally less important than the nature of the oil. A good insulating material has a low thermal conductivity based on a porous structure of low density. It is precisely these features which favor self-heating. The important aspects for self-heating are the extent to which the material provides surface area on which the oil is exposed, allows air to diffuse in, and prevents heat from being conducted away.

Frequently the main insulating material is covered with an impervious cement finish or sealing material. This greatly reduces the extent to which oxygen can diffuse into the insulation.

The temperature which can be attained in the lagging depends on its geometry and on the pipe temperature. In some cases it is possible to estimate the lagging temperature resulting from self-heating by using theoretical methods in combination with suitable experimental tests.

There are a number of precautions which can be taken against lagging fires. One is to prevent the lagging becoming soaked with oil. This means primarily a high standard of operation and maintenance to avoid leaks. But, in addition, measures may be adopted such as not lagging flanged joints and protecting lagging at sample points with sheet metal collars.

Methods of making the lagging surface impervious to air include the use of a cement finish, bituminous coating material, or aluminum foil. If such sealing is used, it is necessary to ensure that the sealing is maintained, particularly at the ends of the lagging.

Another approach is to use not the usual type of insulating material but other insulation such as foam glass or crimped aluminum sheeting. Foam glass is relatively brittle and requires additional care. It is more expensive, although the price differential appears to be narrowing. Aluminum sheeting is another alternative, although it is not a particularly good insulation. It has been suggested that the use of aluminum introduces the hazard of aluminum−iron smear ignition.

In some plants with high risk and difficult self-heating problems, the approach adopted is to do without insulation altogether. In such cases, expanded metal 'stand-offs' may be used to protect personnel from the hot pipework.

The use of temperature sensitive paints has been suggested as a means of detection, but this is not likely to be particularly effective since the lagging surface temperature is not a very reliable indicator of self-heating.

The hazard from a lagging fire may be that of flames issuing from the lagging. But frequently the worst hazard occurs when the lagging is opened up to remove the smoldering material. The fire may then grow, sometimes very strongly, particularly if it has been deprived of air previously.

Flames from a lagging fire should normally be extinguished with small quantities of water. Water should also be available to extinguish fire when smoldering lagging is removed. Fire extinguishing agents, which do not contain water, are generally less effective because of the high risk of re-ignition. Regard should be paid, however, to any hazard which the use of water may involve, such as that associated with electrical equipment. Any lagging which might re-ignite should be removed to a safe place. Further

precautions may be required if the lagging contains asbestos material or evolves toxic fumes.

12.5.4 Duct and Cable Fires

Ducts of various kinds are common in process plants and buildings. They include ducts used for (1) the conveyance of fluids, (2) extraction and other ventilation, (3) pipes, and (4) electrical cables.

Such ducts may have walls, which are combustible and/or may contain combustible materials. In ordinary ventilation systems, the duct walls may be plastic. In fume extraction systems, combustible deposits may build up on the duct walls. Cable and pipe coverings may be combustible. Other sources of combustible material include leaks from pipes and debris left in the duct.

If combustible material in a duct is ignited, fire growth can be rapid, because the heat does not escape as readily as in an open fire and much of it serves to preheat surfaces further down the duct, making the fire much more severe and causing combustion of materials, which would not normally make much contribution to a fire.

A fuel rich fire is the more serious hazard. There is a greater rate of flame advance and heat release, the combustion gases still contain a large proportion of flammable material and they are also toxic. In a fuel rich fire the flame advances some 10 times as rapidly as in an oxygen-rich fire. The combustion gases contain a large proportion of carbon monoxide. Evidence indicates that fuel-rich combustion is favored by narrow ducts, high air velocities, high fuel loading, the presence of obstructions, and large ignition sources.

In an established duct fire, the transition from oxygen-rich to fuel-rich conditions tends to increase the hazard. Reduction of the air flow may thus actually make things worse. Since the combustion gases leaving the duct may still be highly flammable, they present the hazard of further fire and explosion. These gases also contain large amounts of smoke and toxic gas, which are not only hazardous, but also tend to severely hamper attempts to fight the fire.

There are a number of precautions which may be taken against duct fires. Good housekeeping can reduce the amount of combustible material in the duct. Fire detectors may be provided to give early warning. Fire stops may be installed to prevent fire spread. The duct may be designed to allow access for firefighting. Fire protection systems of various kinds are available. Proper working practices may be enforced by the use of a permit-to-work system.

A type of duct fire, which is of particular concern in process plants, is cable tray fires. This is an important matter, because damage to cable systems often results in a long outage.

A single cable may not burn very readily, but a number of cables will often burn vigorously, particularly in a vertical duct which is favorable to fire spread. Once heated up, cables may not need a flame to ignite them. Rubber-covered cables can be ignited by hot air or radiant heat.

PVC-insulated cables present a particular problem. They burn well and readily spread fire in vertical ducts, with the plastic melting and releasing burning droplets. The PVC also decomposes to give large quantities of hydrogen chloride gas, which is toxic and renders corrosive the water used in firefighting. In one power station fire, about a third of the damage was attributed to this cause.

Where power and control cables are mixed, short circuits can occur from high current lines to lines which are not rated for such currents.

Cable ducts form a means by which fire may spread. This is especially serious where the cable ducts lead to a vulnerable point such as a control room. Damage to cables may disable the emergency system, including the telephones required to summon assistance. In one incident, notification of the fire brigade was delayed because the telephone circuits had failed.

There are a number of precautions which may be taken against cable fires. Cables may be segregated from other services and the different types of cable, particularly power and instrument cables, may be segregated. The criticality of all cable runs should be assessed and an appropriate degree of protection provided for each. The hazard of fire spread through cable ducts to vulnerable points should be minimized.

Fire protection for petrochemical plants and oil platforms has been discussed by Corona (1984) who states that a fire protection system for cabling should be based on a hydrocarbon pool fire at not less than 1800°F (980°C), should maintain the cable temperature within operating limits, which on ordinary PVC cables is usually below 300°F (149°C), and should do this for a time period long enough for emergency operation of equipment, which might be of the order of 20 min. Another time period often quoted for cable protection in petrochemical plants is 15 min.

Use may be made of fire resistant cabling. If reliance is placed on such fire resistance, it is important to select the right cable. This is partly a matter of specifying the degree of fire resistance required and partly one of assessing whether a given cable meets this specification. Several workers have emphasized that any tests conducted should be on a sufficiently large scale as to be realistic.

Spillage of oil into the cable duct may be minimized by sealing the cable holes. Protection against the spread of fire through the duct may be provided by fire stops.

Smoke detection devices may be used to give early warning. The cable duct may be designed to give access for firefighting with a maximum distance between access points. Points may be provided to receive high expansion foam from mobile generators. Use is also made of sprinklers and other fixed fire protection systems. The application of such protection to electrical cables is a specialist matter. Protection against cable fires is required during construction as well as during operation. There have been several serious cable fires during the construction phase.

12.5.5 Storage Tank Fires

Storage tank fires are not infrequent in process plants and other sites. The quantities of material involved, and consequently the losses, tend to be large. A single 'jumbo' storage tank in the petroleum industry may have a capacity of 500,000 barrels or more. In many tank fires, more than one tank is involved. The frequency of fires/explosions in fixed roof tanks containing volatile hydrocarbons has been estimated by Kletz (1971) as once in 833 tank-years. The estimated frequency for tanks holding non-hydrocarbons is one-tenth of this value. Comparable data for floating roof tanks are not given.

In some cases, there is a fire/explosion of a flammable mixture in the vapor space of the tank. In others, a vapor cloud forms outside the tank and then ignites. Other cases are fire following liquid slop-over or liquid spillage due to tank rupture.

One of the most frequent causes of tank fires/explosions is overfilling of the tank. This is usually due to defects in operating procedures, failure of instrumentation, and/or operator error. Failure of ancillary equipment, such as pumps, and strikes of lightning are other common causes.

If there is an initial explosion which blows the tank roof off, a fire may be established in the tank and may burn there without spreading. If there is a spillage of liquid into the bund around the tank, due to overfilling, a vapor cloud may form, find a source of ignition, and flash back. This may ignite the liquid leaving the tank and/or that in the bund. If there is a spillage of liquid in the bund from ancillary equipment, such as a pump, and the spillage again ignites, a general fire may occur. Initially, at least, there may be no fire in the tank.

Once a fire is established, it frequently causes failures which feed the fire. Thus a fire on equipment such as pumps or pipework within the bund can cause a pipe failure, which then results in spillage of the tank contents into the bund. Experience shows that pipework exposed to a strong fire in a bund usually fails within about 10–15 min and heat radiation from a tank fire may cause other nearby tanks to fail.

There are a number of measures which can be taken to reduce the risk of tank fires. Much can be done by good plant layout. There should be generous spacing between tanks to reduce the risk from radiant heat. Tanks should be provided with water spray systems, which drench them in water and keep them cool. To some extent, there is a trade-off which can be made between these two measures.

Frequently, the fire engulfs all tanks in the bund. It is highly desirable, therefore, to have a separate bund for each tank, particularly for large tanks. Pipework inside bunds should be kept to a minimum with as few flanges, valves, and other fittings as possible. Pipework can be buried but then tends to corrode. Pumps should be installed outside bunds, both because they are sources of leakage and of ignition, and because they are often needed to fight the fire by pumping out the tank.

Measures should be taken to prevent overfilling of the tank. A high-level alarm is normally a minimum requirement and often a high-level trip is desirable. This instrumentation should be backed up by appropriate operating procedures.

If a fire occurs in a tank, the water sprays should be activated to protect the other nearby tanks. It is often appropriate to pump down to a suitable receiver the tank which is on fire, but it should be borne in mind that in the later stages of this operation the liquid may be very hot and may create a hazard at the receiver.

Tank fires are fought with water and/or foam. The quantities required are very large. It is essential, therefore, that the fire water mains be adequately sized, both for the fire pumps and for the drench water sprays. The tank farm should not be a backwater in this respect. Similarly, there should be a substantial storage of foam.

A tank fire can be difficult to fight for several reasons. The conditions favor the formation of vapor clouds. Metal surfaces become and remain very hot. In consequence, there is frequently flashback of fire to an area where it had appeared to have been extinguished.

Some atmospheric storage tanks contain refrigerated liquefied flammable gas. There is much less experience, however, with fires on such tanks. Liquefied flammable gases are also stored in pressure vessels. Again, at a pressure storage vessel, overfilling is one of the most frequent causes of fire. A fire that develops around a storage vessel containing a liquid under pressure can cause the pressure in the vessel to build up so that there is an explosion. It is normal practice to provide a pressure relief valve on such vessels and it is essential for it to be properly designed and maintained, although this does not fully protect against the hazard. If the vessel is overheated, it may rupture even though the relief valve has operated. Measures should be taken to minimize fire and explosion on pressure storage vessels. The ground underneath the vessel should be sloped away to prevent accumulation of flammable liquid. There should be water sprays and/or fireproof thermal insulation to give protection against fire exposure and there should be safeguards against overfilling.

12.5.6 Vapor Cloud Fires

A vapor cloud fire, or flash fire, occurs when a vapor cloud forms from a leak and is ignited, but without creation of significant overpressure. If such overpressure occurs, the event is a vapor cloud explosion (VCE) rather than a vapor cloud fire (VCF).

Release of flammable vapor from a process plant followed by ignition is not an uncommon occurrence. If the ignition is prompt, the cloud may be modest in size, but if the cloud has time to spread over an appreciable part of the site and is then ignited, a major vapor cloud fire may result. The conditions favoring a vapor cloud fire are a prolonged release in conditions of poor dispersion.

The combustion of the vapor cloud involved first burning of the pre-mixed part and then diffusive burning of the fuel-rich part. The flame in the pre-mixed burning did not propagate quickly across the top of the cloud, but remained as a 'wall of fire.'

Expansion of the combustion products was principally in the vertical direction. The unburned gas was not pushed ahead of the flame front to any significant extent. Where a pool fire occurred, the height of the flame was appreciably greater.

The vapor clouds were made visible by the associated water fog. In the case of the LNG clouds, the contour of the lower flammability limit (LFL) lay within that of the fog, and combustion was entirely within the visible fog; some of the visible cloud remained unburned. For the propane clouds, the LFL contour lay outside the fog and combustion took place in part outside it.

The flame speeds measured during the combustion were relatively low, and far removed from the figure of 150 m/s often quoted as necessary for the generation of appreciable overpressure.

Strictly, a vapor cloud combustion, which generates overpressure, has to be classed as a vapor cloud explosion. In practice, the latter term tends to be reserved for cases where the explosion causes significant destruction. For vapor cloud fires, the overpressures vary from the imperceptible up to those which may cause some window damage.

A vapor cloud fire may cover a wide area, perhaps some thousands of square meters. It results in scorching and depletion of oxygen, with potential for injury and damage. It may initiate BLEVEs and other releases of flammable material which then feed the fire. It also deposits soot.

The hazard from a vapor cloud fire is usually assessed by considering dispersion of the vapor cloud and ignition of this cloud and making some relatively simple assumption concerning the effects inside and outside the cloud.

In many assessments, no explicit model of a vapor cloud fire has been utilized. Instead it has been assumed that the contours of the burning cloud are those of the lower flammability limit concentration; persons inside the cloud suffer a defined degree of injury, generally a fatal injury; and those outside the cloud are subject to a level of thermal radiation based on an assumed surface emissive power at the edge of the burning cloud.

The modeling and calculations of parameters of vapor cloud fires can be found in the full version of Lee's *Loss Prevention in the Process Industries*, 4th edition.

12.5.7 Fireballs

The ignition of a release on a liquefied gas pipeline or an eruption in hot oil giving rise to a release of burning vapor may give rise to a 'fireball.' A fireball from the bursting of a pressure vessel containing liquefied gas is observed to pass through a number of fairly well-defined phases. Crawley (1982) has described the development of the fireball through three phases: (1) growth, (2) steady burning, and (3) burnout. In the first interval, the flame boundary is bright with yellowish-white flames and the fireball grows to about half its final diameter. In the second interval of the growth phase, the fireball grows to its final volume. The fireball begins to lift off and changes to the familiar mushroom-shape. In the third phase, the fireball remains the same size, but the flame become less sooty and more translucent. Figure 12.16 illustrates the typical development of a fireball as a function of time.

The modeling of fireballs covers the fireball regime, the mass of fuel in the fireball, the fireball development and timescales, the fireball diameter and duration, the heat radiated, and the view factor. The case of prime interest of the modeling is a fireball resulting from the bursting of a pressure vessel. The other case is a fireball from the burning of a stationary vapor cloud at atmospheric pressure. These two situations constitute quite different regimes. For the situation where initially there is high momentum, a change of regime occurs as the momentum declines and gravity slumping begins.

The mass of fuel in the fireball depends on the fraction of fuel which flashes off and on the further fraction which forms liquid spray. For propane, the relation between the theoretical adiabatic flash fraction and the liquid temperature and vapor pressure is discussed by Roberts, (1981).

The correlation derived by Roberts (1982) provides a general correlation that could be used for fireballs that the fraction of the fuel released which participates in the fireball is three times the flash fraction. This is the method used by the CCPS (1994) to determine the mass of fuel in the fireball. An essentially similar approach is taken in the treatment given by Marshall (1987) with a distinction between summer and winter conditions.

The several distinct stages of the life cycle of a fireball are discussed by Roberts (1981):

Stage 1: involves the rapid mixing of the fuel with air and rapid combustion of the fuel.
Stage 2: the residual fuel is mixed with air already in the cloud or entrained into it and is burned.
Stage 3: combustion essentially complete the fireball rises due to buoyancy, entraining further air and cooling.

The heat radiated by a fireball can be estimated based on the heat evolved and radiated, the surface emissive power, and the flame temperature and emissivity. The heat evolved may be obtained from the heat of combustion of the fuel. A rule of thumb for the fraction of heat radiated is that given by Hymes (1983) which is that the fraction of heated radiated for a vessel bursting below the set pressure of the pressure relief valve may be taken as 0.3 and that for one bursting above this pressure as 0.4. A commonly used value for the fraction of heat radiated from a fireball is 0.3.

Roberts estimated the heat flux at the surface of the fireballs in the range $141-196 \text{ kW/m}^2$ with individual values up to 450 kW/m^2. The fraction of heat radiated is used in conjunction with the point source model of a fireball and the surface emissive power is used in conjunction with the solid flame model. With regard to fireball temperature, Roberts stated that the values of the surface emissive power are in the range consistent with flame temperatures of $1000-1400°C$, a flame emissivity of unity, and $50-100\%$ excess air.

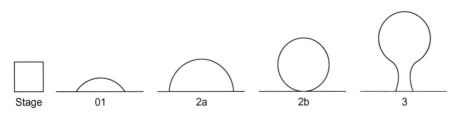

FIGURE 12.16 Development of a typical fireball from a source at ground level.

Stage   01   2a   2b   3

When modeling the thermal radiation from a fireball, it is commonly assumed that the fireball is a sphere with its base just touching the ground and that its diameter and duration are given by one of the sets of correlations just described. A target close to the fireball may be engulfed by it and it is of interest to know the furthest distance at which engulfment occurs. The fireball can be assumed essentially spherical but slightly settled on the ground or a hemispherical fireball. Crocker and Napier (1988) have given relations for the view factor for a spherical fireball just touching the ground for the three cases of a large vertical target, a differential vertical target, and a differential horizontal target. A set of view factors for fireballs covering other situations has been given by the CCPS (1994).

One of the simplest practical models for a fireball is the point source model. Hymes (1983) has given a version of this model, which effectively incorporates a combustion rate and an allowance for the effect on this of the mass of fuel. The other main alternative for practical use is the solid flame model.

The fireball model, which is most widely used, is probably that of Roberts (1981). The model comprises both fundamental and correlation models and covers the whole range of features of practical interest. Roberts considers fireball development, fireball volume and diameter, duration time and heat radiated.

A model for a fireball was given in the CCPS QRA Guidelines (1989). A summary of this model has been published by Prugh (1994). A more recent model is that included in the CCPS Fire and Explosion Model Guidelines (1994), where it is included as part of the treatment of BLEVEs. The method covers the mass of fuel participating in the fireball, the diameter and duration of the fireball, point source and solid flame models of the fireball, the surface emissive power of the fireball, the view factor, and the atmospheric transmissivity.

A type of fireball which somewhat resembles one arising from a BLEVE is that which can occur following sudden rupture of a vessel, such as a reactor, with release of flammable contents and immediate ignition. An instance of this is given by Cates (1992), who terms the incident which he describes a 'congested fireball.'

Fireballs from high explosives tend to be of shorter duration than those of hydrocarbons for a given mass. Generally, the duration time of the fireball is short and the diameter also appears to be less than for the same mass of hydrocarbons. On the other hand, the fireball temperatures are appreciably higher, particularly in the early stages.

The high explosive experiments of Gayle and Bransford (1965) consisted of some 14 explosions involving TNT, composition C-4, and pentolite. Gayle and Bransford did not give a correlation for the duration time of a high explosive fireball. Stull (1977) quotes for TNT

the following model and provided correlations for two separate sources: for fireball diameter, the work of van Dolah and Burgess (1968); and for the fireball duration, that of Strehlow and Baker (1975).

A model for the fireball from a high explosive has been given by Gilbert et al. (1995d). The model is based partly on theoretical and partly on empirical considerations. The combustion processes in the fireball pass through several stages: (1) expansion following detonation; (2a) growth to the maximum diameter as a hemisphere; (2b) formation of a sphere at ground level, lift-off and rise to maximum height as a sphere, and (3) persistence at this height as a sphere until cooled.

12.5.8 Pool Fires

A pool fire occurs when a flammable liquid spills onto the ground and is ignited. A fire in a liquid storage tank is also a form of pool fire, as is a trench fire. A pool fire may also occur on the surface of flammable liquid spilled onto water. There is considerable experimental literature related to pool fires. Apart from the large number of publications on pool fires themselves, there are many relevant studies on flames such as those on combustion processes in, and heat radiation from, flames. Figure 12.17 illustrates a typical pool fire.

A pool fire is a complicated phenomenon and the theoretical treatment is correspondingly complex. A pool fire burns with a flame which is often taken to be a cylinder with a height twice the pool diameter. In still air the

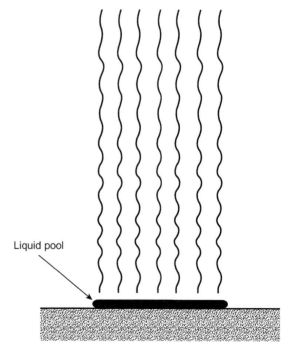

FIGURE 12.17 A typical pool fire.

flame is vertical, but in wind it tilts. Wind also causes the base of the flame to extend beyond the downwind edge of the pool, thus exhibiting flame drag. With some pool fires, blowout can occur at a wind speed of about 5 m/s. The characteristics of a pool fire depend on the pool diameter. The liquid burning rate increases with diameter until for large diameters it reaches a fixed value. The heat radiated from the flame behaves similarly. Some fuels, such as LNG, burn with a relatively clear flame, while others, such as kerosene and LPG, give a smoky flame.

The modeling of pool fires covers the aspects of flame geometry, liquid burning rate, flame characteristics, heat radiated, and view factor. There are three different ways of determining the heat radiated: (1) the use of a value of the fraction of the heat radiated; (2) the use of a value for the surface emissivity; and (3) the estimation from the flame properties such as flame temperature and emissivity. Hottel (1958) analyzed the work of Blinov and Khudiakov and showed that as the pan diameter increases, the fire regime changes from laminar to turbulent. The flame height decreases with increasing pan diameter in the laminar and transition regimes and then remains constant with pan diameter in the turbulent regime.

The flame on a pool fire has often been assumed to be an upright cylinder with a length L twice the diameter D. The correlations or equations for flame length had been discussed by Thomas (1963), Steward (1970), Heskestad (1983).

The review by Hottel (1959) of the work of Blinov and Khudiakov includes a simple analysis of the heat q transferred from the flame to the pool. An equation for the regression rate of a burning liquid surface under windless conditions has been discussed by Burgess and Zabetakis (1962). As shown in Figure 12.18, the contribution of convection falls and that of radiation rises as the pool diameter increases, until both contributions reach a constant value.

An extensive review of liquid burning rates is given by Hall (1973), who considers among other things the effects of the fuel, fuel mixtures, fuel containing

dispersed water, the liquid surface temperature, the liquid temperature distribution, the pool diameter, the heat transfer from the flame to the liquid, and the wind speed.

Work on the burning rates of fuels of solid plastics has been reviewed by de Ris (1979) in the context of pool fires. Babrauskas (1986) distinguished various burning modes of pool fires as given in Table 12.22.

Babrauskas (1986) also addresses some of the factors affecting the liquid burning rate, including pool diameter, lip effects, and wind speed. Values of various liquid burning rates are given in the ICI LFG Code (ICI/RoSPA 1970 IS/74) and by Hearfield (1970) and Robertson (1976). A liquid burning rate of 0.75−1.0 cm/min is quoted in the fourth edition of the Dow *Guide* (Dow Chemical Company, 1976).

Important characteristics of the flame on a pool fire are the nature of the flame, its temperature, and its emissivity. The difference between LNG and LPG pool fires was illustrated by the work of Mizner and Eyre (1982). The LPG flames were smoky, the LNG flames much less so. Figure 12.19 illustrates the difference.

Hottel (1959) had referred to a flame temperature of 1100 K and Craven (1976) had discussed that this value is not necessarily the maximum value which may be attained. If the flame is smoky, the unobscured parts of the flame will emit radiation to the target, while the obscured parts will not. A method of taking account of obscuration of the flame has been given by Considine (1985).

There are three methods of estimating the heat radiated by a pool fire. These are based on heat evolved and radiated, surface emissive power, and flame temperature and emissivity. The heat evolved in the flame may be determined from the liquid burning rate. The heat radiated from the flame is then usually calculated as a fraction of the heat evolved by using an empirical factor.

The intensity of heat radiation may then be calculated by dividing the heat radiated by the area of the flame envelope. The second method is to use directly information on the heat radiation intensity, or emissive power, of the flame

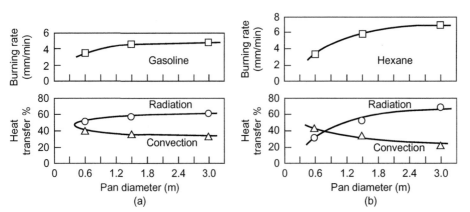

FIGURE 12.18 Liquid burning rate and heat transfer to the liquid surface in large pool fires: (a) gasoline fire and (b) hexane fire. *Source: After Yumoto (1971). Courtesy of Combustion and Flame.*

surface. The third method is to determine the heat radiated from the temperature and emissivity of the flame.

Values of surface emissive power of hydrocarbon fires have been reviewed by Nash (1974). Moorhouse and Pritchard (1982) conclude that for large pool fires of hydrocarbons, the surface emissive power is unlikely to exceed $60 \, kW/m^2$. Moorhouse and Pritchard quote for pool fires of LNG the surface emissivities of the order of $200 \, kW/m^2$ obtained by Raj et al. (1979).

Mudan (1984) suggests that for hazard assessment, the surface emissive power be taken as a weighted value for the unobscured and obscured portions of the flame. A further discussion of the intensity of heat radiation is given in the ICI *LFG Code*. In order to determine the heat incident on a target, it is necessary to take into account the target absorptivity, the atmospheric transmissivity, and the view factor.

The targets for heat radiated from a pool fire include both the liquid surface and objects outside the flame. For heat radiation from a flame treated as a vertical or tilted cylinder to a target, the three principal models are point source model, solid flame model, and equivalent radiator model. These models and their application have been described by Crocker and Napier (1986). The point source model has been used by a number of workers (Hearfield, 1970) for pool and tank fires. A collection of view factors is given by Howell (1982). For a vertical cylinder, an expression based on the more approximate subtended solid angle method has been given by Stannard (1977). Stannard has given an expression based on the subtended solid angle method for a tilted cylinder. For a vertical cylinder, the view factor has been given by Hamilton and Morgan (1952). A set of graphs for the view factors for pool fires of LPG are given in the IP *LPG Code*.

A review of pool fire models has been given by Mudan (1984) who describes the principal elements of pool fire models but does not present a preferred selection. Modak (1977) has described a relatively complex model consisting of a fairly large set of equations. The model is capable of determining the heat radiation to the liquid surface and to objects outside the flame. A type of pool fire which is of special interest is a storage tank fire. The differences are that the liquid surface, and hence the base of the flame, is elevated and that it is surrounded by a metal wall. Another important feature is that the targets of interest usually include other storage tanks. Burgoyne and Katan (1947) conducted experiments on the heat wave moving through the liquid in a tank fire. Crocker and Napier (1986) have reviewed the main elements of pool fire models as they apply to tank fires and have considered the implications for storage tank layout.

The slot, trench, or channel fire is a fire, which may occur on a spill of flammable liquid which has entered a trench or channel of any kind. A characteristic feature is that this type of fire may occur where the burning of a liquid spill is directed to a high aspect ratio catchment

TABLE 12.22 Burning Modes at Different Pool Sizes (Babrauskas, 1983)

Pool Diameter (m)	Burning Mode
<0.05	Convective, laminar
<0.2	Convective, turbulent
0.2–1.0	Radiative, optically thin
>1.0	Radiative, optically thick

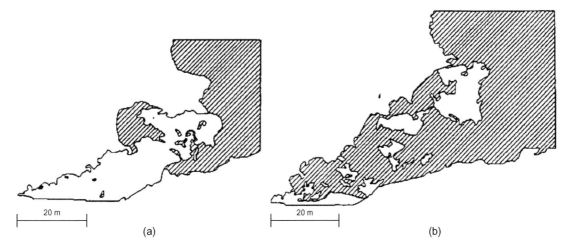

FIGURE 12.19 Smokiness of flames in large pool fires: (a) LNG fire and (b) LPG fire. *Source: After Mizner and Eyre (1982). Courtesy of the Institution of Chemical Engineers.*

area. Slot fires are discussed briefly by Moorhouse and Pritchard (1982). Moorhouse and Pritchard (1982) state that the work on LNG pool fires which he describes included some limited tests on LNG fires with an aspect ratio of 1:2.5. Work on trench fires of LNG with aspect ratios up to 30 has been conducted by Mudan and Corce (1984) and is summarized by Mudan (1989).

12.5.9 Flares

A jet flame occurs when flammable gas issuing from a pipe or other orifice is ignited and burns on the orifice. Conventionally, flares have been designed to give a flame which is vertical in still air. The use of flares on offshore oil production platforms, where helicopter access is necessary, has prompted the development of flares with an inclined flame.

A series of experimental studies of flares has been reported by Brzustowski and coworkers. The correlations obtained by Brzustowski and Sommer (1973) are the basis of the more elaborate flare model given in API RP 521. Experiments on flares have been conducted by Shell and are described by Chamberlain (1987).

A thermal radiation model of a flame on a flare stack had been developed by Hajek and Ludwig (1960). For the determination of radiation exposure, the current version of API RP 521 gives detailed equations. Another flare model is that by de Faveri et al. (1985), which had been discussed by Crocker and Napier (1988) in relation to multiple point-source modeling of flare radiation. The flare model given by Chamberlain (1987) of Shell is based on the flame geometry determined by the parameters as functions of the velocity ratio.

12.5.10 Jet Flames

Ejection of flammable fluid from a vessel, pipe, or pipe flange can give rise to a jet flame if the material ignites. Scenarios involving jet flames are not easy to handle, since a large jet flame may have a substantial 'reach,' sometimes up to 50 m or more.

Generally, the length of a jet flame over the range of practical interest is approximately proportional to the square root of the mass flow. This is the implication of the API correlation for the flame length on a flare and of the work on propane jet flames just described. For diffusion flames in general, the flame temperatures have been discussed by Gugan (1976).

The cases mainly treated for modeling the jet flame are that of a vertical flame on an upward pointing jet in calm conditions or in wind, jet points upward not vertically but at an angle and horizontal jet, for which again the wind direction may be confluent, opposed, or across. Different workers have assumed different geometries for the jet; a cone (e.g., Craven, 1972), a cylinder (e.g.,

Croce and Mudan, 1986), and a frustrum (e.g., Chamberlain, 1987). A model of a jet flame was formulated by Burke and Schumann (1928a,b) and was later taken up by Savage (1962). Another early model is that of Hawthorne et al. (1949). They envisage the flame as an inverted cone with the apex on the orifice and the equations are provided for the length of the flame and for the diameter at the top of the cone.

In some applications, it may be necessary to use a more accurate view factor and it had been discussed in detail by Crocker and Napier (1988). For the model by Hawthorne et al. (1949) of a vertical flame, Crocker and Napier refer to the work of Tunc and Venart (1984), who treat the flame as conical. In the model of Chamberlain (1987) for a flare, the jet flame is treated as a frustrum with the widest end furthest from the jet exit. Crocker and Napier also give the results of a numerical investigation of the values of the radiation intensity on a target for a propane jet flame burning at a vent, as obtained for vertical and tilted flames using for the vertical flame the model of Hawthorne, Weddell, and Hottel and the Becker relations, and for the tilted flame a point source treatment based on the model of Brzustowksi and Sommer, a multiple *point* source treatment based on the model of de Faveri et al. and the solid flame model of Chamberlain with the Croce and Mudan view factor.

A jet flame may occur within an enclosure, which may be relevant to ignited releases in offshore modules. Work on this aspect has been described by Chamberlain (1994). The situation investigated is strongly similar to the compartment fires which have been much studied in work on fires in buildings. A jet flame in an enclosure differs from the pool fire case in at least two important respects. One is the much greater degree of mixing and the other the absence of 'flashover'. In addition to the heat effects on the wall and the contents of the enclosure, other hazards are the thermal radiation from the external flame and the generation of smoke and carbon monoxide.

12.5.11 Engulfing Fires

The engulfment of vessels in fire may occur in various ways and the problem has been studied from several different angles. There have been a number of experiments on rail tank cars and on fixed storage vessels engulfed in fire, and various models have been developed for heat transfer to such vessels, for the response of the temperatures and pressures in the vessel, for pressure relief of the vessel and for bursting of the vessel.

A theoretical study of the effect of an engulfing fire on a vessel filled with methane has been made by Nylund (1984). Some of the results obtained by Nylund show the time response of the gas temperature, the gas pressure, and the rupture pressure for a jet fire, for a jet with an effective diameter of 1 m. Nylund also discusses pressure relief and depressurization. In the case of the jet fire, the rupture occurs at gas pressures well below the relief valve set pressure. In the pool fire case, the gas pressure does reach the set pressure, but the relief valve capacity is insufficient to vent the pressure quickly enough.

There have been several experimental investigations of the response of vessels containing liquefied flammable gas to an engulfing fire. A test on fire engulfment of a 64 te LPG rail tank car has been described by Anderson and Norris (1974) and Charles (1974). Measurements were made of the heat flux from the fire, of the vessel wall temperatures in the liquid and vapor spaces and the relief valve flare. A series of tests on vessels for rail tank cars have been carried out to investigate the effectiveness of the internal thermal radiation protection by Appleyard (1980). One objective of these tests was to investigate the effectiveness of the internal thermal radiation protection explosive.

Models for radiant heat transfer to a horizontal cylindrical vessel from (1) an engulfing fire and (2) a relief valve flare have been given by Tunc and Venart (1984).

The behavior of the wall temperature in the liquid space in these tests points to nucleate rather than film boiling, and estimates of the heat flux indicate that the high heat fluxes necessary for film boiling did not occur. Metallurgical examination revealed no indication of failure due to cracks and was consistent with failure by hoop stress in the wall. In association with these tests, a theoretical model for the behavior of a vessel engulfed in fire has been developed by Hunt and Ramskill (1985).

There is relatively little available data on the effect on equipment of jet flame attack, as opposed to pool fire engulfment, but the EC currently has in hand a major program of work on this, an account of which is given by Duijm (1994). The objectives of the program are (1) the modeling of unobstructed jet fires, (2) the modeling of obstructed jet fires, (3) the modeling of the thermal response of pressure vessels, (4) the assessment of mitigation techniques, and (5) the modeling of the failure modes of the pressure vessels.

For unobstructed single-phase jet flames, the work involves the development of the models incorporated in the UMPFIRE and TORCIA codes, as described by Crespo (1991) and Bennett (1991), with extension of the treatment to two-phase jet flames. For obstructed jet flames, a distinction is made between small targets engulfed by the flame and large targets. Techniques such as CFD (Hernandez and Crespo, 1992) and wind tunnel tests (Verheij and Duijm, 1991) are used to model the heat transfer.

12.6. EFFECTS OF FIRE: DAMAGES AND INJURIES

Fire causes damage to property and injury to people. The modeling of fire damage and injury from fire is treated in this section. Broadly, fire may be treated in terms of ignition, steady- and unsteady-state combustion, an extinction, and of flame spread, both over surfaces and within buildings.

The susceptibility of a flammable liquid to ignition is related to the flashpoint. For high flashpoint liquids, the fire point is also relevant. The fire point is the temperature at which the liquid not just ignites but also supports a flame and it is therefore a higher temperature than the flashpoint. The fire point is determined experimentally and the information on the fire point of liquids is relatively sparse.

It was found in the work of Roberts and Quince (1973) that for a flame to be sustained on the surface of the liquid, it is necessary for the vapor concentration to be above the stoichiometric value. The maintenance of the flame on a liquid surface has been interpreted by Rasbash (1975) in terms of the flame heat balance. Traditionally, a liquid has been classed as highly flammable if it has a low flashpoint. For a low flashpoint liquid, the hazard is that a flammable mixture will form above the liquid surface and will be ignited. If the temperature of the liquid is high enough, a vapor cloud may form. For a high flashpoint liquid to be ignited, it must be heated above its fire point. It has been shown by Sirignano and Glassman (1970) that if such flame impingement occurs, a surface tension effect occurs and convection currents are set up which have the effect of cooling the volume of liquid which is being heated. The liquid pool will eventually ignite but, as shown by Burgoyne and Roberts (1968), only after the transfer of a substantial amount of heat. Ignition of a high flashpoint liquid occurs much more readily, however, if there exists at the surface of the liquid a wick. This effect has been studied by Burgoyne and Roberts (1968).

In the ignition of combustible solids, it is usual to distinguish between spontaneous, or unpiloted, ignition and piloted ignition. With spontaneous ignition, also termed unpiloted ignition or autoignition, ignition occurs when the material has been heated up to the point where not only does it evolve vapors of volatile material, but it is not enough to ignite these vapors. With piloted ignition, by contrast, ignition of these vapors is by the pilot flame. A further distinction may be made between the case where the pilot flame actually impinges on the surface of the solid and that where it does not. Simms (1962) refers to these situations as piloted ignition (in the more restricted sense) and surface ignition.

A classic study of the ignition of wood by thermal radiation is that described by Lawson and Simms (1952a,b), who subjected samples of dry wood to thermal radiation from a heated panel, with and without a pilot flame playing half an inch from the surface of the sample. The effect of water content has been studied by Simms and Law (1967). Water content influences ignition in several ways. It affects the thermal inertia of the wood, the overall heat transfer, the minimum ignition energy, the minimum thermal radiation intensity, and the ignition time for both spontaneous ignition and piloted ignition without contact. The properties of the wood itself affect thermal radiation ignition. The absorptivity of the wood increases as charring occurs. With regard to the thermal radiation incident on the wood, it has been shown by Kashiwagi (1979) that there can be a significant attenuation due to the volatile matter issuing from it. A comparison of the minimum ignition temperature for heat transfer by radiation and convection has been made by Kanury (1972).

The growth of fire is frequently governed, particularly in the early stages, by flame spread over the surface of solids or liquids. The two situations considered here are flame spread across a liquid surface and a solid surface. In both cases the spreading flame has two effects: it serves as a source of heat and it causes pilot ignition.

The work by Sirignano and Glassman (1970) has demonstrated the role of surface tension effects in the circulation flows set up in the bulk liquid. It has been found by Burgoyne and Roberts (1968) that for a stoichiometric mixture the apparent rate of spread reaches a limiting value of some four to five times the fundamental burning velocity.

Work on flame spread over ground soaked with fuel has been described by Ishida (1986). Flame spread over solid surfaces is a function of the orientation and geometry of the surface, the direction of propagation, the thickness of the combustible material, the properties of the material, and the environment factors.

The behavior of a flame on a vertical surface differs from that of one on a horizontal surface. The direction of propagation also makes a radical difference. The rate of upward flame spread on a vertical surface tends to increase exponentially. It can be characterized by the time taken for the rate of spread to double, or the doubling time, as described by Alpert and Ward (1984). Flame spread is also influenced by environmental factors, including the initial temperature of the fuel, the air velocity over the surface, the imposed thermal radiation, and an oxygen enriched atmosphere. Thermal radiation can cause a marked enhancement of the rate of flame spread, as evidenced by the work of Alvares (1975), Kashiwagi (1976), Fernandez-Pello (1977a,b), Hirano and Tazawa (1978), Quintiere (1981), and others.

The effect of air velocity on flame spread depends on the direction of air flow. The spread of flame across surfaces has been modeled by Williams (1977). The rather different problem of flame spread through open fuel beds has been studied by Thomas et al. (1965), who conducted a classic set of experiments on wooden cribs. In practical cases some factors which are of particular importance in determining flame spread over a solid surface are the thickness and composition of the surface, the imposed thermal radiation, and the air movement over the surface.

There is a requirement in hazard assessment for relationships which permit heat radiation intensity to be translated into the equivalent burn injury. Correlations for both fatal and non-fatal burn injury have been derived in the vulnerability model by Eisenberg et al. (1975). Further correlations are given in the *Green Book* (CPD, 1992). The effect of thermal radiation on human depends very much on the source of radiation. Some source of thermal radiation is flare, fireball, typically as a part of a BLEVE.

Experimental work on injury from thermal radiation has been done on humans and on animals. Henriques (1947) carried out experiments in reddening and blistering in human exposed to thermal radiation.

A first degree burn involves the epidermis. There is reddening but no blistering. A second degree burn may be superficial or deep. A third degree burn destroys both the epidermis and the dermis. The terms second and third degree burn have largely given way to characterization in terms of the depth of burn, reference being made to full depth, and other burns.

Burn injury correlates closely with skin temperature. Pain is experienced at a temperature of 44°C and above this temperature injury occurs. Relations for the unsteady-state temperature profile of the skin suddenly exposed to a source of heat radiation have been given by Buettner (1951a) and Stoll and Chianta (1971). Stoll and Chianta describe work on burn in rats. Inquiry due to exposure to a short but strong pulse of thermal radiation may be correlated in several ways. The most convenient is to take the injury factor as the thermal dose which is the product of the thermal radiation intensity and time. It is found, however, that this somewhat understates the effect of very high intensities of thermal radiation and that better correlation is obtained if these are more highly weighted.

Stoll and Greene (1959), Buettner (1951b), and Mudan (1984) have correlated thresholds of pain and blistering. The results were obtained for the time to the threshold of pain in human. In work by Hinshaw (1957) at the University of Rochester, pigs were subjected to thermal radiation exposures. Histological examination in these experiments showed that the thermal radiation was sufficiently intense in some tests to cause steam 'bleb' formation and mitigation of the tissue damage. The data have been analyzed by Hymes and correlation has been provided in terms of the heat injury factor. The depth of skin burn in Hinshaw's experiments at thermal radiation intensities below those which cause steam bleb has been correlated by Hymes.

Human response to a fire event depends on the nature of the event and the awareness of the person. The event primarily considered in relation to burn injury is a fireball, a flare, and other types of event such as a pool fire. Accounts of burn injury such as those by Hymes and in the *Green Book* quote a human response time of 5 s. The application of this reaction time to a fireball scenario is that the person faces the fireball for 5 s and then turns his/her back and seeks to escape. Hymes gives an account of the BLEVE at the Lowell Gas Company at Tewkesbury, Massachusetts, in 1972.

The degree of protection offered by clothing depends on the fraction of the body which is so protected. Data on the fraction of the body area represented by the different parts of the body are given in the *Green Book*. The *Green Book* takes for the Dutch population a maximum value of the body surface exposed of 20% in normal conditions.

If the thermal radiation is so intense as to ignite the clothing, a different situation pertains. A detailed discussion of the protection afforded by clothing is given by Hymes. The degree of protection afforded by a building to a person within it depends on the location of that person. In many cases, this will be such that the building affords complete protection.

In estimating burn injury, it is important to utilize an appropriate value of the thermal radiation. The effective intensity is determined by the atmospheric transmissivity and the geometry of the body as presented to the radiation source. The importance of the geometry of the body has been emphasized by Marshall (1987). There is a tendency in hazard assessments to treat the whole of the exposed area of the body as subject to the radiation nominally incident at the particular point. The thermal radiation incident on the exposed area is appreciably less than that calculated as falling on the projected area.

The protection afforded by clothing is limited by the fact that at thermal radiation intensities within the range of interest here, it is prone to ignite. For clothing subject to thermal radiation, there is also a distinction to be made between spontaneous and piloted ignition. Hymes suggests that in certain scenarios, notably a BLEVE, the probability of piloted ignition is quite high. As already stated, work on ignition of fabrics has been carried out by Wulff and co-workers. From this work, Hymes has given relations for the determination of the times to unpiloted and to piloted ignition together with tables of parameters for common fabrics.

Information on the mortality in persons admitted to hospital with burn injuries has been given in a series of papers spanning some 40 years by Bull, Lawrence and co-workers, the first being that of Bull and Squire (1949) and the most recent that of Lawrence (1991), the

intermediate publications being those of Bull and Fisher (1954) and Bull (1971). Burn injury is an area where there have been great improvements in medical treatment over the years. Early set of relations for thermal injury were those given in the vulnerability model by Eisenberg et al. (1975). For fatal injury, these authors utilize an adaptation of data given by White (1971) for the thermal injuries caused by nuclear weapons dropped on Japan. Using a formula given by Glasstone (1962) for the relationship between weapon yield and pulse duration.

For fatal injury, Hymes utilizes the correlations from Eisenberg, Lynch, and Breeding. He also gives a number of thermal injury effects. The work of Hymes covers most of the relevant aspects of burn injury, including clothing ignition, thresholds of pain and blistering, and burn fatalities.

A set of relations for thermal injury, including the non-fatal injuries, are given in the *Green Book*. For fatal injury, the correlation from Eisenberg, Lynch, and Breeding has been adopted and modification has been made based on the fact that thermal radiation from nuclear weapons is in the UV range, whereas that from hydrocarbons is in the IR range.

Another model for injury by thermal radiation has been developed by Lees (1994). The model is essentially a set of relationships which may be combined in various ways depending on the scenario modeled and which have been utilized by the author to produce a relation for fatal injury from a sudden heat release such as a fireball, subject to defined assumptions. Clothing ignition is taken in the model as resulting in the doubling of the effective exposed area.

A model for burn injury from exposure to a fireball has been given by Prugh (1994). He presents a summary of data on burn injury, which for fatality draws on the work of Eisenberg et al. (1975), Lees (1980), Roberts (1982), and the Society of Fire Protection Engineers (1988 NFPA/22). Combining the fireball model and the injury relationships, Prugh derives for a propane fireball a graphical correlation for mortality, which is a function of the mass of fuel and the distance of the person exposed.

12.7. FIRE PROTECTION OF PROCESS PLANT

In general, fire prevention and protection measures constitute either passive prevention and protection or active protection.

12.7.1 Passive Fire Protection

Passive prevention and protection comprises measures which are taken in order to prevent a fire occurring and to limit its spread. As a broad generalization, passive fire protection has the great advantage that it is very much less dependent on the intervention of protective devices or of humans, both of which are liable to fail, and is therefore that much less vulnerable to management failings. One of the principal passive fire protection measures is fireproofing. Often passive fire protection can limit fire spread and can 'buy time' in which the firefighting resources can be mobilized. Measures of fire prevention and passive fire protection are built into the basic plant layout and design.

There are certain plant configurations which assist fire spread, by avoiding or modifying these configurations, fire spread may be limited. In a fire, a chimney effect can develop in the space between a table top and a tall equipment or structure, or between a table top and the bottom of a large column. Such an effect tends to increase damage extent to a higher level. Where a potential chimney cannot be avoided, it may be appropriate to install a water spray system to control fire in the area. To avoid fire spreading by the flow of a burning liquid, bunds are put around storage tanks. But there may be other situations which can give rise to liquid flow; drains, sewers, and pipe trenches are features which are particularly liable to aggravate this hazard.

Many items of equipment are fitted with thermal insulation. This is to be distinguished from fire insulation proper. Much process plant is thermal insulated, as are low temperature storage tanks.

Some form of structural protection against fire, in the form of fire insulation or fireproofing, is normally provided for supporting members and also for vessels and pipe work. A general account of fireproofing has been given by Waldman (1967). Fireproofing is sometimes limited to support of major items such as large columns or vessels, but it is often extended to support other items, such as heat exchangers and pipework. The main object of fireproofing is to prevent the failure in a fire of items containing flammable material which can feed the fire. Fireproofing may not be needed, therefore, for items with a small inventory, unless failure could lead to large loss of containment from some other source. A fireproofing system normally consists of a bulk insulation material which gives protection against fire. Materials which are widely used as the bulk fire insulation include concrete and magnesium oxychloride cement.

Protection against fire may also be obtained by the use of a mastic, which may be asphaltic or vinyl acrylic. Fireproofing should provide sufficiently effective insulation to keep the temperature of the structural member to be protected below a specified temperature under defined fire conditions. The mechanical strength of the fireproofing should be sufficient to offer reasonable resistance to damage both from normal plant activities and from water from fire hoses. The bulk fire insulation should not be corrosive to the substrate or to the reinforcing material.

There is frequently a requirement to test a proposed fireproofing system. There is no standard test appropriate to structural elements in a flammable liquid fire, which is the typical situation in process plant. Many experimenters have resorted, therefore, to the ASTM E-119 standard time−temperature test, which was devised for building fires. It has long been recognized that the ASTM curve tends to underestimate the temperatures which occur in fires of hydrocarbons. An early proposal for a higher temperature relation was made by Waldman (1967), who gave two alternative curves. The unsatisfactory nature of the ASTM E-119 test in this application has been demonstrated by Castle (1974). Much equipment and pipework is provided with insulation, which serves primarily to conserve heat or cold in the process. Experiments on insulations commonly used for pipework have been described by McMillan (1974). Detailed experimental work on the behavior of polyurethane foams in fires is described by Boult et al. (1972) and by Boult and Napier (1972). Fireproofing is discussed in the ICI *LFG Code* (ICI/RoSPA 1970 IS/74). For pressurized storage vessels, the code suggests that one suitable form of fireproofing is vermiculite cement.

A quite different method of fireproofing is the use of some form of a reactive coating. The reactive coating includes magnesium oxychloride, intumescent coatings, and subliming coating. The reactive coatings are lightweight and durable. The applications where they are most likely to be economic are those where alternative softer materials are less suitable. Some practicalities of the use of reactive materials are discussed by Kawaller (1980).

Turning to the passive fire protection of particular plant features, protection of structural steelwork is of some importance. Design guidance for fire protection of steelwork is given in *Designing Fire Protection for Steel Columns* (American Iron and Steel Institute, 1980), *Fire Protection for Structural Steel in Buildings* (CONSTRADO, 1983) (the *Yellow Book*), and *Design Manual on the European Recommendations for the Fire Safety of Steel Structures* (ECCS, 1985) and by Law (1972, 1991), and Barnfield (1986).

Valves are another type of equipment for which good fire resistance is highly desirable. Fire resistant, or fire safe, valves are available. Relevant standards are API Std 589: 1993 Fire Test for Evaluation of Valve Stem Packing, API Spec. 6FA: 1994 Specification for Fire Test for Valves and BS 6755: Testing of Valves, Part 2: 1987 Specification for Fire-Type Testing Requirements.

12.7.2 Active Fire Protection

Active fire protection measures are also provided in the basic plant design but are effective only when activated in response to a fire.

12.7.2.1 Fire Detection and Alarm

The initiation of active fire protection measures, whether fixed systems or mobile firefighting systems, depends on the fire detection and alarm arrangements. Relevant codes and standards are the NFPA 72 series, particularly NFPA 72E: 1990 *Automatic Fire Detectors*, BS 5445: 1977– *Components of Automatic Fire Detection Systems*, BS 5839: 1988– *Fire Detection and Alarm Systems for Buildings* and, until recently, BS 6020: 1981– *Instruments for the Detection of Combustible Gases*, now superseded by the series BS EN 50054–50058 which has the general title *Electrical Apparatus for the Detection and Measurement of Combustible Gases*. When a fire occurs, two things which are of prime importance are to ensure the safety of personnel and the prompt initiation of action to deal with the fire. The most effective way of ensuring rapid action is generally the use of an automatic system. But in situations where personnel may be present, the arrangements need to ensure that they are not put at risk by the action of the automatic system. There are several different types of system for the detection of leakage and/or fire. These systems and the devices used are listed in Table 12.23.

The most widely used type of system is the combustible gas detection method. Descriptions are given by several authors (Bossart, 1974; Johnson, 1976).

A single instrument system is more suitable for detecting toxic gases, where some time delay is acceptable, that may pose a long-term hazard. The instrument scale is usually calibrated as 0–100% of the lower explosive limit (LEL).

There tends to be some trade-off between the speed of response and the number of spurious signals obtained. It is suggested by Johnson that the specification for response time is generally too tight and that an allowance of a few seconds delay reduces the number of false alarms.

There are various approaches to laying out the detection points. Generally, it is desirable to have detectors around the periphery. Other detectors are then sited according to principles such as uniform coverage, location at leak sources, location between leak, and ignition sources.

The optimum height of the sample point depends on the density of the gas. A combustible gas leakage detection system may be used to provide alarm signals in the control room or to initiate protective action by water sprays or by a steam curtain.

Detection of a flame is done by infrared or ultraviolet detecting instruments. There are a number of devices which detect heat, which include temperature measuring instruments, quartz bulbs, and devices which explode, melt or change characteristics when exposed to fire. A quartz bulb detector figured prominently in the Flixborough inquiry (Parker, 1975 and Figure 12.8). Another method of detecting fire is by smoke detectors. These are widely used in buildings, but appear likely to be less effective for fires on open plant.

12.7.2.2 Emergency Material Transfer

It is necessary to have arrangements so that in the event of fire it is possible to transfer flammable material away from the parts of the plant affected. A relief header leading to a vent stack or flare stack allows vapor to be vented safely from pressure vessels, removing liquid from a pressure vessel may be catered for by facilities for blowdown, and pumps may be considered on atmospheric storage tanks to allow liquid to be transferred out.

12.7.2.3 Firefighting Agents

The main types of firefighting agents are water, foam, dry chemicals, vaporizing liquids, inert gases, and other agents. The supply of fuel to the flame may be reduced by cooling the liquid, by diluting or emulsifying it and so reducing the partial pressure of the vapor, or by blanketing the liquid with some inert material. There are a number of hazards in the use of firefighting agents. They are associated primarily with discharge of the agent, incompatibility between the agent and the material on fire, an electric shock from high voltage equipment, and toxic and asphyxiant effects from the agent or its breakdown products. The compatibility of firefighting agents should

TABLE 12.23 Some Detection Systems for Fire Protection

Basis of Detection	Detector
All modes	Humans
Leakage	Flammable gas detector
Flame	Infrared detector
	Ultraviolet detector
Heat	Temperature measuring instrument
	Quartz bulb detector
	Gun cotton bridge and wire
	Fusible link and wire
	Temperature sensitive resistor
	Air line
Smoke	Smoke detector

be considered in two distinct aspects, the compatibility with the material and/or a fire of the material and compatibility with other agents. The most widely used firefighting agent is water, but it is not suitable in all cases. It is important that full information is available on the proper agents for fighting fires with the chemicals used in the process, that the right agents are selected and are available on the plant, and that personnel understand the nature of the agents and their use.

It should be appreciated that where use is made of agents which are available in limited amounts only, it is essential to kill the whole fire first time. It is necessary to establish that the firefighting agent selected is capable of extinguishing a fire on the particular chemical concerned and this may require that tests be conducted.

12.7.2.3.1 Fire Protection Using Water: Extinguishment, Control, and Exposure Protection

Water is used in extinguishment of fire, control of fire, fire exposure protection, and fire and explosion prevention. Fire protection using water is treated in the NFPA *Handbook* by Hodnett (1986e). Two principal codes are NFPA 13: 1991 *Installation of Sprinkler Systems* and NFPA 15: 1990 *Water Fixed Systems*. Water is most effective against fires of high flashpoint liquids. The principal mechanism is the cooling of the liquid so that its vaporization is reduced. The cooling effect is strong, because water has a high latent heat. Another important mechanism by which water extinguishes liquid fires is vaporization to form steam, which blankets the fire and cuts off the oxygen. If the liquid is miscible with water (e.g., ethylene oxide), addition of water reduces the vapor concentration. The water used to fight a fire may be large in volume and is liable to become contaminated. Its removal can pose a serious problem. Work has been done by Rasbash and coworkers on the extinction of liquid fires (Rasbash and Rogowski, 1957; Rasbash and Stark, 1962; Rasbash, 1960; Rasbash, 1962).

The water drops must enter the liquid in sufficient quantity despite the updraught of the flames and the evaporation of the spray in the flames. Factors which assist in achieving this are positioning of the spray near the liquid surface, high spray impetus, and large liquid drop size. When a deep pool of liquid in this class is burning, it may form after a period a 'hot zone' and water below the surface may then boil and cause a 'slopover,' thus spreading the fire. It is necessary to try to detect and extinguish fires before a hot zone occurs.

The second class of liquid may be extinguished by diluting the surface layer with water until the fire point of the mixture is raised sufficiently to cause extinction. The third class of liquid may be extinguished by cooling the flame by heat transfer to the water droplets falling through

it. Such liquids include petrol and, again as a borderline case, kerosene. Although water is the most common firefighting agent, the situation becomes hazardous when the material or fire is incompatible with water. The use of water on live electrical equipment involves the hazard of electrical shock. The use of water in large quantities and contamination of this water can pose a problem of disposal.

Water is delivered to the fire mainly either by fixed systems such as water sprinkler and water spray systems and fixed water monitors or by mobile systems such as fire hoses. Fixed water sprinkler systems are used to protect indoor areas and vulnerable items, loading and unloading facilities, including jetties, and storage vessels in the outdoor environments. Exposure protection is effected using fixed water spray systems. Proper cooling requires that water be applied such that the whole surface protected is wetted by the water rate specified. The spray nozzles should be located so that the extremities of the spray pattern at least meet, and for wetting below a vessel equator reliance should not be placed on rundown.

Fire protection by fixed water sprinkler systems is treated in the NFPA *Handbook* by several authors, notably Hodnett who deals with automatic sprinklers (1986a), automatic sprinkler systems (1986b), and water supplies for sprinkler systems (1986d). Relevant NFPA codes are NFPA 13: 1991 Installation of Sprinkler Systems, NFPA 13A: 1987 Inspection, Testing and Maintenance of Sprinkler Systems, NFPA 16: 1991 Deluge Foam−Water Sprinkler and Spray Systems, and NFPA 16A: 1988 Installation of Closed-Head Foam−Water Sprinkler Systems. Other NFPA publications on sprinklers include Automatic Sprinkler and Standpipe Systems (1990/25), Automatic Sprinkler Systems Handbook (1994/37), and Installation of Sprinkler Systems (1994/39). The relevant British Standard is BS 5306 Fire Extinguishing Installations and Equipment on Premises, Part 2: 1990 Specification for Sprinkler Systems. The elements of a fixed water sprinkler system include the sprinkler alarms, the sprinkler heads, the water distribution pipework, the valves and other fittings, and the water supply system and disposal arrangements.

As just indicated, fixed water spray systems are distinguished by the fact that the spray is characteristic of the type of nozzle used. Fire protection by fixed water spray systems is treated in the NFPA *Handbook* by Hodnett (1986c). Relevant codes and standards are NFPA 15: 1990 Water Spray Fixed Systems and NFPA 16: 1991 Deluge Foam−Water Sprinkler and Spray Systems. Recommended water rates for the extinguishment or control of fire are given in the NFPA 15 in detail for fire extinguishment, control of burning, exposure protection, and fire and explosion prevention. A critique of water rates traditionally recommended has been given by Fritz and Jack (1983). Fritz and Jack discuss the work of

Rasbash and co-workers on fire, which draw attention to the influence of the water nozzle and spray characteristics on the water rates needed for extinguishment and to the fact that these characteristics are not specified in the NFPA codes. The amount of water needed for firefighting and cooling presents a considerable problem. There should be a secure water supply to the water spray system. There are a number of ways in which a water spray system can be rendered ineffective. One of these is freezing up, as just mentioned. Another is leaving valves closed, particularly after maintenance. In addition, it is not uncommon for a water spray system to be knocked out by an explosion. The disposal of fire water, particularly contaminated water, can be a serious problem, when sufficient water disposal systems are not present. Essentially, exposure protection is provided for large flammable gas or liquid inventories which are liable to be exposed to a fire. The purpose of providing such protection is to cool the metal and so prevent loss of strength. For exposure protection by cooling, NFPA 15 treats the following cases: vessels, structures and miscellaneous equipment, transformation, and belt conveyors. Fritz and Jack (1983) discussed that the NFPA recommended water rates for fire extinguishment are too low, and exposure protection cases are too high. The background works on the recommended rates on vessels engulfed in fire are listed in API RP 520: 1990. Other recommended water rates are provided by the Fourth Edition of the Dow *Guide* (Dow Chemical Company, 1976) and ICI *LFG Code* (ICI/RoSPA 1970 IS/74).

For vessels, the code states that the rules given assume that there is a relieving capacity based on a maximum allowable heat input of 6000 BTU/ft^2 h (18.9 kW/m^2). For a vertical or inclined vessel surface, the water rate should not be less than 0.25 USgal/ft^2 min (10.2 l/m^2 min) of exposed uninsulated surface. The code also gives various other provisions. Fritz and Jack (1983) have reviewed the water rates recommended in NFPA 15 for fire extinguishment or control. These authors also review the NFPA recommendations on water rates for exposure protection.

Fritz and Jack have determined that the NFPA water rates recommendation is too low for fire extinguishment, but that for exposure protection they are too high. Other recommended water rates are discussed by the Fourth Edition of the Dow *Guide* (Dow Chemical Company, 1976) and ICI *LFG Code* (ICI/RoSPA 1970 IS/74).

A tank exposed to fire, including an engulfing fire, may be cooled by applying water. The requirements for exposure protection generally greatly exceed those for extinction of fire. The water needed for exposure protection is greatly reduced if application is restricted to the tank roof and ullage sections. A stringent fire protection requirement for a jet flame impinging directly on a target

and model for a water spray system to protect against jet flames are discussed by Lev and Strachan (1989) and Lev (1991). A continuous film of water over the metal surface is capable of intercepting radiant heat, but as the heat flux is increased the protection so afforded is gradually lost as the water film thins and then disintegrates.

Information on the flame temperature and surface emissive power in jet flames has been reviewed by Lev (1991). In order to deliver this cooling water under conditions where there is likely to be in-flight evaporation and crossflow losses, the droplets need to be large and to be projected at high velocity. Lev presents results from the model showing the fraction of water droplets reaching the target as a function of crossflow velocity and droplet diameter.

12.7.2.3.2 Fire Protection Using Foam

Foams are widely used against liquid fires. Initially the foam acts as a blanketing agent and then, as the water drains from the foam, as a cooling agent. A foam for firefighting should possess certain general properties, which include expansion, cohesion, stability, fluidity, fuel resistance, and heat resistance. Three quantitative criteria for foam are expansion, fluidity, and drainage time. A study by Dimaio and Lange (1984) detected deleterious effects from contaminants such as corrosion inhibitors and antifouling agents. Another important application of foam is the suppression of vaporization from toxic liquid spills. This use of foam is treated in ASTM-F-1129-88 *Standard Guide for Using Aqueous Foams to Control the Vapor Hazard from Immiscible Volatile Liquids*.

Fire protection using foam is treated in the NFPA *Handbook* by Lockwood (1986). Relevant codes and standards are NFPA 11: 1988 Low Expansion Foam and Combined Agent Systems, NFPA 11A: 1988 Medium and High Expansion Foam Systems, NFPA 16: 1991 Deluge Foam−Water Sprinkler and Spray System, NFPA 16A: 1988 Installations of Closed Head Foam−Water Sprinkler Systems, NFPA 11C: 1990 Mobile Foam Apparatus and BS 5306: 1976 Fire Extinguishing Installations and Equipment on Premises, Part 6: Foam, Section 6.1: 1988 Specification for Low Expansion Foams and Section 6.2: 1989 Specification for Medium and High Expansion Foams.

Chemical foam is produced by reacting aqueous solutions of sodium bicarbonate and aluminum sulfate in the presence of a foam stabilizer. The reaction generates carbon dioxide, which both forms foam and ejects the mixture from the apparatus. Mechanical foam is generated by mechanical aeration of aqueous solutions of certain chemicals, which usually have a protein base. This type of foam is the most widely used for both fixed and mobile apparatus. High expansion foam is generally similar to standard foam but has an approximately 1000:1 expansion.

It is very light and is easily blown away, and is thus more suitable for fires in containers such as tanks or ships' holds than for those in open situations such as in a bund. Medium expansion foam is again generally similar to standard foam but has an approximately 100–150:1 expansion. This type of foam is also light, but is not so easily blown away as high expansion foam. The fluorochemical foam includes 'light water' foam, which contains a straight-chain fluorocarbon surface active agent. This has the effect that as the water drains from the foam it spreads in a thin film over the liquid and seals it. The fluoroprotein foam contains a branched chain fluorocarbon. It is less expensive and appears in many cases to be more effective than light water foam. Synthetic detergent foam is generated by mechanical aeration of aqueous solutions containing 2–3% of detergent. The most effective method of usage appears to be massive application is a knockout attack. Aqueous File Forming Foam (AFFF) has low viscosity and spreads easily over liquid fires. Another useful property of AFFF is that it does not need elaborate foaming devices and can be utilized in many water sprinkler and spray systems. File Forming Fluoroprotein Foam (FFFP) is another type of foam which has low viscosity and good spreading properties and can be used in many water spray systems. FFFP foam tends to drain rapidly and is therefore less reliable in maintaining a foam blanket. Regular air foams do not perform well on liquids which are of the polar solvent type, notably alcohols, for these AR foams have been developed. A foam for firefighting should possess certain general properties, which include expansion, cohesion, stability, fluidity, fuel resistance, and heat resistance.

Three quantitative criteria for foam are the expansion, fluidity, and drainage time. Standard tests using the foam had been conducted by the Underwriters Laboratories and Det norske Veritas (DnV) tests and the MIL Specification MIL-F.24385, the draft ISO/DP 7203, and the United Kingdom, Defence Standards 42–21, 42–22, 42–24, and BS 6535: Part 6: 1988. Such tests are discussed by Evans (1988) who compares the performance of different foams in different tests.

Foam is water based and, to this extent, hazards associated with water apply to foam also. They include increased vaporization of low boiling liquids, reaction with incompatible materials and electric shock from live with incompatible materials, and electric shock from live electrical equipment. The rupture of the foam blanket and burnback, which may put firefighters at risk, is another hazards. The ignition of hydrocarbons in a storage tank roof by static electricity from foam injection is described by Howells (1993).

Foam is delivered to the fire by means essentially similar to those used for water, involving three stages: proportioning of the foam concentrate, foam generation, and foam distribution. A detailed account of methods of proportioning and distribution is given by Lockwood (1986).

Fire extinction by blanketing may be achieved using foam. It is just as unsuitable as water for fighting fires involving electrical equipment or substances which have undesirable reactions with water. Other conditions for the use of foam are that the liquid surface be horizontal and that the temperature of the liquid be not too high. Foam application rates are discussed by Lockwood (1986). Low expansion foam is mainly used to prevent, extinguish, or control fires in storage tank tops and bunds and on spills. Medium and high expansion foams are used to prevent, extinguish, or control fire in spaces such as ships' holds.

Fixed foam systems are used for fire prevention, extinguishment, and control in bunds or on spills. The expansion foams also had been used around the LNG facility for two different purposes. The expansion foams can be applied to cover the surface of LNG and suppress the vapor generation and in case of LNG pool fire, the expansion foam can be applied to reduce the radiation heat. Yun et al. (2011) have conducted a small- and medium-scale experiments, and have verified that the expansion foam allows the control of vapor generation, and allows further emergency procedures to take place. For pool fire application, it was evaluated that the expansion foam application reduced the flame heights by 61% and with reduced mass burning rate. The thermal hazard distance reduced up to 52%.

A particularly important application of foam is the protection of storage tanks. A relevant code is NFPA 11. For fixed roof tanks, some principal arrangements are foam chambers, internal tank distributors, and subsurface foam injection. Foam chambers are installed at intervals on the outside near the top of the tank wall, providing an over-the-top foam generation (OFG). An alternative is internal distributors fitted inside the tank. Foam trucks are the principal mobile mode of delivery for foam. Relevant codes are NFPA 1901 which covers basic water systems with a foam option, and NFPA 11C which covers foam trucks. Foam trucks carry a supply of foam concentrate and delivery hoses and can be equipped with telescopic booms of articulated towers. There are a variety of mobile devices that can be used to apply foam to the top of a storage tank which is on fire.

Another important application is the suppression of vaporization from toxic liquid spills. This use of foam is treated in ASTM-F-1129-88 *Standard Guide for Using Aqueous Foams to Control the Vapor Hazard from Immiscible Volatile Liquids.*

12.7.2.3.3 Fire Protection Using Dry Chemicals

An alternative to water or water-based foam is dry chemicals or dry powders. Fire protection using dry chemicals is treated in the NFPA *Handbook* by Haessler (1986). Relevant codes and standards are NFPA 17: 1990 *Dry Chemical Extinguishing Systems* and BS 6535: 1989– *Fire Extinguishing Media*, Part 3: 1989 *Specification for Powder*.

The principal mechanism by which dry chemicals act against fire is to cause chain termination of the combustion reactions. Standard dry chemical consists principally (over 90%) of sodium bicarbonate with additives to improve fluidity, non-caking, and water repellent characteristics. General purpose powder is a mixture of ammonium dihydrogen phosphate, diammonium hydrogen phosphate, and ammonium sulfate. Monnex is a potassium alkali-based material. Its effectiveness is about five to six times that of standard dry chemical.

Dry chemical formulations may be ranked with respect to their effectiveness in extinguishing fires according to their performance in test. This performance is a function both of the chemical composition and the particle size. In storage, dry chemicals are stable at normal and low temperatures but deteriorate at high temperature. Different dry chemicals should be segregated.

Some of the hazards from the use of dry chemicals are the sudden release of the agent, unexpected reignition, and toxicity from the combustion process. Dry chemicals are delivered to the fire mainly by fixed systems, from cylinders or by mobile systems, and portable extinguishers. Dry chemicals are utilized on class A fires (combustible materials), Class B fires (flammable liquids), and Class C fires (electrical equipment) in the NFPA classification. They find application in extinguishment of large outdoor fires, protection of spaces indoors, and in portable extinguishers. Vapors from a flammable liquid may be reignited by a hot surface, a combustible solid may be reignited by smoldering material, and an electrical fire may reignite due to continued arcing.

12.7.2.3.4 Fire Protection Using Vaporizing Liquids

Another type of firefighting agent is the vaporizing liquid. Fire protection using vaporizing liquids is treated in the NFPA *Handbook* by Moore (1986). Relevant codes and standards are NFPA 12A: 1992 Halon 1301 Fire Extinguishing Systems, NFPA 12B: 1990 Halon 1211 Fire Extinguishing Systems, NFPA 2001: Clean Agent Fire Extinguishing Systems, BS 5306: 1976– Fire Extinguishing Installations and Equipment on Premises, Part 5: 1982– Halon Systems and BS 6535: 1989– Fire Extinguishing Media, Part 2: 1989– and Halons, Section 2.1: 1989 Specification for Halon 1211 and Halon 1301, and Section 2.2: 1989 Code of Practice of Safe Handling and Transfer Procedures. The substances used as vaporizing liquids are halogenated hydrocarbons which classified as halon (H), hydrofluorocarbon (HFC),

hydrochlorofluorocarbon (HCFC), and perfluorocarbon (PFC). The halons contain bromine and the perfluorocarbons contain fluorine as the only halogen. Vaporizing liquids are available which are much more effective extinguishants than carbon dioxide. The relative effectiveness of various vaporizing liquids in extinguishing a hexane test fire is discussed in Hearfield (1970).

One of the most efficient extinguishing agents is Halon 1301. Halon 1211, or bromochlorodifluoromethane (BCF), is now the principal vaporizing liquid. Halon 1211 is a very effective extinguishing agent for flammable liquid and electrical fires. The concept is that the halon is injected into the space and develops an approximately uniform concentration capable of extinguishing a fire anywhere in the enclosure. The alternative method of deployment of halon systems is local application in which the jet of halon is directed at the seat of the fire.

Vaporizing liquids may be ranked in respect of their effectiveness in extinguishing fires according to their performance in tests. Properties of a vaporizing liquid which bear on its effectiveness as a firefighting agent include chemical action, vapor pressure, and density.

The ability to arrest combustion by interfering with the chain reactions is obviously a prime property. The vapor pressure determines the ability of the agent to discharge under its own pressure. The density of the agent affects the mixing in the space into which is it discharged.

There are certain toxic hazard effects associated with the use of vaporizing liquids which are the agent, breakdown products of the agent, and combustion products of the fire. Moore (1986) gives the approximate lethal concentrations of the natural vapor and of the decomposed vapor for a 15 min exposure. The toxicity of the combustion products of the fire is another significant factor in firefighting using vaporizing liquids.

Vaporizing liquids are delivered to the fire mainly by fixed systems. They are utilized for this purpose in enclosed spaces and in portable extinguishers. They are also used to suppress incipient combustion inside plant. Fixed vaporizing liquid systems may be total flooding systems or local application systems. Computer programs for halon hydraulics have been developed by suppliers. The corrosion issue of halon system is discussed by Peissard (1982). A feature of interest for a space protected by a halon system is the retention time of the halon.

If the gaseous extinguishing system is an automatic one activated by rapid response fire detectors, two separate detection signals should be provided in order to reduce the frequency of false activations. The space protected should be provided with warning signs at all entrances. If the system is a manually activated one, the release devices should be so located such that the person operating the device is not at risk from the fire or the discharge of the agent.

The installation should be covered by a safe system of work. The system should be subject to regular inspection by a competent person. GS 16 gives guidance on the selection of the agent and the system. Account should be taken of the possibility that discharge of the agent may raise a dust cloud, with the attendant risk of a dust explosion.

The discovery by Molina and Rowland (1974) that chlorine can act as a catalyst for the conversion of ozone to molecular oxygen in the upper atmosphere initiated a growing concern for the state of the ozone layer. This has led to the Montreal Protocol on Substances that Deplete the Ozone Layer, 1987, which lays down a program for the phasing out of halons. The original Montreal Protocol provides that the production of halons be reduced to 50% of the 1986 level by 1994 and to 0% by 1999.

With regard to the continued use of existing halon systems and to the recycling of halons for use in such systems, national policies vary. Amendments to the Montreal Protocol also cover HCFCs, the production of which is to be frozen at 1989 levels and phased out by 2030, or possibly earlier. Alternatives to halons have been reviewed by the Halons Technical Options Committee (Taylor, 1991) of the UN Environment Program (UNEP). NFPA 2001: 1994 deals with clean agent alternatives to halons. A review of vaporizing liquids used or proposed for use as fire extinguishing agents is given by Senecal (1992), who covers HCFCs, HFCs, and PFCs as well as CO_2, with a comparative tabulation of relevant properties and a detailed discussion of each agent.

12.7.2.3.5 Fire Protection Using Inert Gas

The last of the principal firefighting agents is inert gas. Relevant codes and standards are NFPA 12: 1993 Carbon Dioxide Extinguishing Systems and BS 5306: 1976— Fire Extinguishing Installations and Equipment on Premises, Part 4: 1986 Specification for Carbon Dioxide Systems and BS 7273: Code of Practice for the Operation of Fire Protection Measures, Part 1: 1990 Electrical Actuation of Gaseous Total Flooding Extinguishing Systems.

Extinguishment of fire by an inert gas is effected by reducing the concentrations of the fuel and oxygen. Design of an inert gas system is based on reducing the concentration of oxygen below that which will support combustion and the flame temperature below that necessary for combustion. The principal individual inert gases used in fire protection applications are carbon dioxide and nitrogen. Use is also made of inert gas mixture formulations.

Properties of an inert gas relevant to its use for fire protection include density, vapor pressure, minimum concentration required for extinguishment, and gas specific heat.

The two principal inert gases used are carbon dioxide and nitrogen and the main hazards for both gases is that of asphyxiation. The density of the gas is relevant to the asphyxiation hazard; the density of nitrogen is close to that of air, while that of carbon dioxide is appreciably greater, so that it may accumulate in low-lying spaces. Inert gases are delivered to the fire by fixed systems but not normally by mobile systems. The inert gas carbon dioxide is one of the agents used in portable extinguishers. In a total flooding system, the inert gas is injected into an enclosure in sufficient quantity to ensure the minimum concentration for extinguishment. In a local application system, a jet of inert gas is directed at the surface protected, which is typically a tank surface but may also include the adjacent floor area. For carbon dioxide, hand-held hose lines from cylinders are also used.

The essential requirement for applying the inert gas is to ensure that operation of the system does not expose personnel to hazardous concentrations.

12.7.2.3.6 Fire Protection Using Portable Extinguishers

It is a fundamental principle of firefighting that a fire should be hit hard and quickly, preferably before it has a chance to take hold. Conditions for success in the use of portable fire extinguishers are that the fire be small, that extinguishers be accessible, that they be of the right type, that they operate when activated, and that personnel capable of using them are present. Fire protection using portable fire extinguishers is treated in the NFPA *Handbook* by Petersen (1986). Relevant codes and standards are NFPA 10: 1990 *Portable Fire Extinguishers* and BS 5423: 1987 *Specification for Portable Fire Extinguishers* with its associated standard BS 6643: 1985— *Recharging Fire Extinguishers*. There are also a number of Underwriters Laboratories (UL) standards.

Fires of combustible materials (NFPA Class A), flammable liquids (Class B), electrical equipment (Class C), and metals (Class D) can all be dealt with by the use of suitably chosen portable fire extinguishers. A detailed tabulation of agents and their applications is given by Petersen (1986). For water and foam extinguishers, the three traditional types have been stored: pressure, pump tank, and inverting. Other portable extinguishers utilize stored pressure, cartridge, or self-expulsion methods.

12.7.2.3.7 Other Methods

Other types of firefighting methodology include solids and powders, wet water or wet water foam, and combined agent systems. Fire protection using such special methods is treated in the NFPA *Handbook* by Johnson (1986).

In addition to dry chemicals, use is also made of various kinds of dry powder, sand, and other solid materials. One group of solid agents comprises those used for metal fires, for example, talc, soda ash, and graphite. Glass granules are used for the control of LNG bund fires (Lev, 1981). The carbon microspheres can be used for extinguishment of metal fires (McCormick and Schmitt, 1974). The wetting agents can be applied to modify the properties of water, thus producing 'wet water.' A relevant code is NFPA 18: 1990 *Wetting Agents*. In this context a wetting agent is one which, when added to water, effects a substantial change in its surface tension, thus increasing its ability to penetrate and spread. If the wetting agent has suitable foaming qualities, a wet water foam is produced. Fixed water sprinkler and water spray systems, as described in NFPA 13 and NPFA 15, may be designed to operate on wet water. Different agents can be combined to provide fire protection. Some agent combinations used are water and foam, carbon dioxide and foam, and certain dry chemicals and foam. In some cases, the agents are applied simultaneously, in others they are used in rapid succession.

12.7.3 Dow Fire and Explosion Index

The Dow Fire and Explosion Index assists in the selection of preventive and protective features.

The detailed set of basic and recommended minimum preventive and protective features are listed in the Fourth Edition of the *Guide* (Dow Chemical Company, 1976). The Dow Fire and Explosion Index is widely used as an aid to the selection of fire preventive and protective features, and is therefore a relatively well-proven method. Nevertheless, it is a generalized method and, if used, it should be supplemented by specific hazard identification and assessment studies which may reveal the need for other preventive and protective features.

12.8. FIRE PROTECTION APPLICATIONS

A type of fire common in all occupancies is a combustible solids fire (NFPA Class A). The NFPA codes covering particular solid materials are NFPA 43A (oxidizers), NFPA 43B (organic peroxides), NFPA 490 (ammonium nitrate), NFPA 654 (powders and dusts), and NFPA 655 (sulfur). Where solids are handled indoors, regard should be had to the potential difficulties of fighting the fire once it has taken hold and is producing large volumes of smoke. This implies that any fixed fire protection system should respond rapidly and strongly.

Another type of fire common in the process industries particularly is a flammable liquids fire (NFPA Class B). Fixed fire protection systems are provided for process operations and areas vulnerable to releases and spills, for loading and unloading facilities and, above all, for storage tanks and vessels. Storage tanks are commonly provided with a bund. It is therefore necessary to protect not only against a fire in the tank itself but also against one in the bund. There are also certain special types of fire that require treatment, one of which is the electrical fire (NFPA Class C). If there is a fire at a piece of live electrical equipment, it is usually appropriate to de-energize the equipment, unless there are good reasons for not doing so.

Among the agents used on electrical fires are water, dry chemicals, vaporizing liquids, and inert gases, but not all agents are suitable for every type of electrical fire.

Fires of electrical equipment are mainly fought using portable carbon dioxide or dry chemical extinguishers. Another type of fire which requires special treatment is metal fires (NFPA Class D). The NFPA codes on the storage of certain metals are NFPA 48 (magnesium), NFPA 481 (titanium), and NFPA 482 (zirconium). The hazards associated with the metal fires are extremely high temperatures, steam explosion, hydrogen explosion, toxic combustion products, and oxygen depletion. The agent used should be matched to the metal: an agent suitable for dealing with a fire of one metal may be quite unsuitable for that of another.

Prominent proprietary agents are Pyrex G1 powder, which is a coke powder with an organic phosphate additive. Another agent is Met-L-X powder, which has a sodium chloride base, again with additives. Turning to the fire protection of particular plant and storage, protection of storage tanks is a prominent theme. Fire protection of storage tanks is treated in the NFPA *Handbook* for flammable liquids by Henry (1986) and for chemicals by Bradford (1986). Relevant codes and standards for storage tanks, including low temperature storages, are NFPA 30: 1990, API Std 620: 1990 and API Std 650: 1993, and BS 799: 1972–, BS 2654: 1989, Bs 4741: 1971, and BS 5387: 1976. There are also codes for the Storage of particular chemicals such as NFPA 43A: 1990, which covers liquid oxidizers.

Among the facilities, which may be provided for the fire protection of a storage tank, are fire resistant thermal insulation or fire insulation, a fixed water system for cooling, a fixed foam system for the tank top, a fixed foam system for the bund, fixed water/foam monitors, and mobile water/foam monitors. The fire protection of storage vessels, particularly those containing liquefied flammable gas, is equally important. Fire protection of storage vessels for LFG is treated in the NFPA *Handbook* by Walls (1986). Relevant codes and standards for storage vessels for LFG are NFPA 54: 1992, NFPA 58: 1992, NFPA 50: 1990, NFPA 59A: 1990, the ASME *Boiler and Pressure Vessel Code*, Section VIII and BS 5500: 1991. NFPA 50: 1990 covers liquid oxygen and NFPA 50A:

1989 and NFPA 50B cover gaseous and liquid hydrogen, respectively.

Warehouses are another type of storage which contain large quantities of combustible material. A relevant code is NFPA 231C: 1991 *Rack Storage of Materials.*

The earlier FRS program also showed that conventional sprinklers do not respond quickly enough to counter the fire before it gains hold. An alternative system was developed in which the bays are zoned with sets of sprinklers dedicated to a zone and actuated by a fire detection wire 'laced' through the racking. The later program, described by Field (1985), concentrated on the response of the sprinkler head.

12.8.1 Firefighting in Process Plant

In most firefighting operations in process plants, it is necessary not only to fight the fire but also to protect the vessels exposed to it. A fundamental principle of firefighting is to attack the fire at as early a stage as possible. Firefighting in the process industries and with chemicals is discussed in Fire Protection Manual for Hydrocarbon Processing Plants (Vervalin, 1964) and by several other authors, particularly Risinger (1964).

A liquid fire is contained if the liquid is held in a well-defined container and in some cases, it may even be flowing. A liquid with high flashpoint can generally be extinguished by cooling with water, whereas one with a low flashpoint usually needs to be blanketed by foam or dry chemicals. Contained fires of either high or low flashpoint liquids can normally be extinguished by foam. With uncontained fires, the first step is generally not to extinguish the fire but to cut off the flow of liquid from the containment.

An uncontained fire of a high flashpoint liquid can generally be extinguished by the use of water. For a fire on a low flashpoint, liquid, foam, or dry chemicals should normally be used. Often where dry chemicals are used, it is necessary to use water also. Typically, it is necessary to apply water to cool this metal, then use dry chemicals to extinguish the fire, and then apply water to cool the liquid.

In some cases, there are simultaneous contained and uncontained fires. In such a case the procedure is to put out the uncontained fire first. It is not always either practical or desirable to extinguish a liquid fire. The procedure in such cases is to cut off the flow of liquid to the fire, to cool the container and other exposed surfaces, and to seek not to put out the fire itself but to contain it. Further information is given by Risinger (1964d).

Fires of liquefied flammable gas (LFG) are often fought using dry chemicals or foam, or both in combination. The objective is to control the fire and so reduce the heat radiation from it. Large-scale tests on the

effectiveness of foam and of dry powder on fires have been carried out as part of the experiments by MITI (1976). In another test, the fire extinguishing agent used was dry powder of ammonium phosphate discharged from two vehicles near the bund.

A fire on a large LNG spillage radiates intense heat and can do extensive damage. Firefighting will often be conducted with the aim of containing and controlling rather than extinguishing the fire. Early work by Burgoyne and Richardson (1948) showed that foam forms a frozen blanket on burning liquid methane, leaving only small tears which can be readily extinguished with dry chemicals. Walls (1973) stated that foams and water will not extinguish LNG fires and that extinction can be completed in the short application period, which is difficult to achieve. The limitation of the effects of LNG fires is described by Wesson et al. (1973a). Work at Lake Charles, Louisiana, in 1961 on the extinction of LNG fires has been reported by Burgess and Zabetakis (1962). An extensive series of tests on the extinction of large LNG fires conducted by the University Engineers was sponsored by the American Gas Association (AGA) and has been reported by Wesson and co-workers. The control of LNG fires using foam is described by Wesson et al. (1973a). Wesson et al. (1973b) also report work on the extinction of LNG fires using dry chemicals. The water spray system can be applied in enhancing the vapor dispersion in case of an LNG spill. The outdoor LNG spill experiment from Mary Kay O'Connor Process Safety Center at Texas A&M University have applied the water spray directly in the pathway of vapor dispersion. Rana et al. (2010) and Rana and Mannan (2010) have identified various dilution mechanisms involved in the commercial water spray applications. The effectiveness of the firefighting agents depends on the fire conditions, including the weather, the ground, and the process plant.

A fire in a storage tank, which has lost its roof, may be dealt with by trying to extinguish it or let it burn itself out. Such a fire is normally extinguished by the use of water or foam. There is a risk, however, that the addition of water or foam may cause a slopover of oil—water mixture. This can occur if there is a deep layer of hot oil at the burning liquid surface. Methods of reducing the hazard include mixing the hot oil in the tank with cold oil, by pumping the liquid out and in again, or by using only small quantities of water or foam until frothing subsides. Alternatively, the fire may simply be allowed to burn itself out. In some cases, a tank fire can be controlled by agitating the liquid with air, which is described by Johnson (1986). In general, the fighting of storage tank fires involved not only extinction of the fire itself but also protection of other exposed tanks. Storage tank fires are discussed by Risinger (1964a).

In certain situations, the requirement is not so much to project water onto the fire as to permit the firefighters to advance toward the fire. In such cases, use may be made of a fog nozzle, which is more suited for this purpose than a regular hose nozzle.

A technique, which finds application in certain situations involving a leak or fire of a flammable material, is the use of a water curtain. One application of a water curtain is to disperse a leak of flammable vapor, before it can ignite, which has been described by Beresford (1981). Another use of water curtains is to attenuate the thermal radiation from a fire.

The issues to be determined were the stability of the curtain and its capability to attenuate thermal radiation. Models were developed for the behavior of the water curtain with respect to both stability and attenuation; the results for the latter show close agreement with the experimental values.

Firefighting is always a hazardous task, but this is particularly so in a process plant. Many process plant fires involve liquids. These are difficult to extinguish and tend to reignite, particularly if there are areas of hot metal. Fires often cause explosions in vessels, particularly those containing liquids. It is standard practice to cool vessels with water to lessen the chance of their rupturing or exploding, but nevertheless explosions do occur. Large quantities of toxic fumes are frequently involved in fires. Some materials produce more insidious fumes. Outright asphyxiation due to lack of oxygen is another hazard in such fires. Fire causes loss of containment of other chemicals, such as corrosive substances, which constitute further hazards to personnel.

12.8.2 Fire and Fire Protection in Buildings

The most serious fires, with respect both to loss of life and to damage, are those which occur in buildings. The problem of fire in buildings is hazard to life, damage to the building, and exposure of nearby buildings. The maximum size of fire which can develop in a building depends on the amount of material available for combustion. In a classic study, described below, Ingberg (1928) put forward the concept of 'fire load' as the determinant of the severity of a fire. The fire load is the total heat, which can be generated by the combustible material within the building, which can be used to define the severity of a potential fire. The fire load density is the heat per unit area of floor which can be generated. A discussion of fire load is given by Langdon-Thomas (1972).

It has been shown that under standard conditions, there is a close relationship between the time for which a building fire burns and the temperature which it attains. This was first demonstrated in 1928 by Ingberg (1928),

who carried out a series of test burns in an experimental building at the National Bureau of Standards (NBS). A standard time−temperature curve, which has been widely used is that of ASTM E-119 and a British Standard curve, given formerly in BS 476: Part 8: 1972 and now in Part 20: 1987. The area under the time−temperature curve corresponds effectively to the fire load density. The standard time−temperature curves are widely used for the testing of structural components for fire grading.

The proposition put forward by Ingberg is that two fires should be regarded as of equal severity if the area under the time−temperature curve, above a certain datum temperature, is equal, a concept which has become known as the 'equal area hypothesis.'

A scheme for the fire grading of buildings in the United Kingdom, which is based essentially on the concepts of fire load and the time−temperature curve, was proposed in *Fire Grading of Buildings*, Part I, *General Principles and Structural Precautions* (Ministry of Works, 1946). A different approach to fire grading is that developed by Law (1971), who investigated the relation between fire severity and fire resistance.

Ignition is important not only in the initiation of fire but also for its spread. After the fire has started, its growth depends on the ease with which further combustible material is ignited.

Once ignition has occurred, the immediate growth of the fire is affected by the form of the nearby materials. Flame spread across a combustible surface differs somewhat according to whether it is vertical or horizontal. Fire spread through a continuous fuel bed requires little explanation except that this description includes spread between any types of object which can act as a fuel, including whole buildings. With a discontinuous bed, by contrast, it is necessary for the fire to spread by one of the modes of heat transfer and to cause ignition by heating other objects to their ignition temperature without the aid of direct flame contact.

The sudden propagation of flame through the unburned gas below the ceiling and all the combustible surfaces involved in the fire is known as 'flashover.' If the air supply is suddenly increased, as by the cracking of windows, the opening of a door or other means, there will be a sudden increase in the rate of burning.

The spread of fire through a building depends on a number of factors. Staircases, lift shafts, and open corridors encourage the rapid spread of flames. Fire spread from a building across an open space is generally almost entirely by radiation, although there may sometimes be a contribution from convection and there may also be flying brands. The spread of fire between buildings depends on the distance between them, the sizes and angles of the radiating surfaces, the contents of the buildings, and the construction of the buildings. The hazards of fire in a building are toxic gas, smoke, and heat. Toxic gases produced by building fires are principally the normal products of combustion, that is, carbon monoxide and carbon dioxide. Other more toxic gases may be produced by the combustion of materials such as plastics and rubber. Smoke has the effect of reducing visibility and hindering both escape and rescue. The quantity of smoke generated is greater if the air supply is restricted and the burning material is wet.

Most modeling has been directed to the characterization of the fire in a compartment. A compartment fire may burn out, self-extinguish due to lack of air, or progress to flashover. 'Flashover' has been used with a number of meanings, as reviewed by Thomas (1982).

In the pre-flashover stage, the mass loss rate is at first relatively low. The fire regime is one of fuel-controlled burning. As the fire grows, the quantity of air required increases markedly, and the regime becomes one of ventilation-controlled burning. The works on mass loss rate are provided by Thomas et al. (1964) and Kawagoe (1958).

There are a number of models of the post-flashover compartment fire. If the ventilation is increased, there comes a point at which the regime becomes fuel controlled. For this regime, there is no single criterion for transition and no single expression for the mass loss rate. The transition from ventilation to fuel control has been discussed by Harmathy (1972). Bullen and Thomas (1979) have obtained a correlation which includes some of the important variables for mass loss rate.

Flame projected from an opening may cause fire to spread, impinging on an object outside and is in any event a source of radiant heat. Flame projection has been correlated by Thomas and Law (1972).

The spread of fire from a compartment into other parts of a building has been received by Quintiere (1979). A fire spread from a room into a corridor has been described by Quintiere et al. (1978). Another route for the spread of flames is through an opening and up to the next storey. For narrow windows, flame spread may occur with only a relatively small projection, whereas for wide windows a greater projection is required.

A quite different type of fire model used to investigate the response of temperature measuring devices used for fire detection is given by Budnick and Evans (1986).

Smoke is produced by smoldering fires and by flaming combustion, but the types of smoke are very different. The carbon particles in smoke are the result of incomplete combustion, as is carbon monoxide. Other gases, such as hydrogen chloride or phosgene, may also be present, depending on the materials burned. From the practical viewpoint, the significant feature is the rate of production of smoke, which is governed by the burning rate. One method of estimating the rate of smoke production is to infer it from the rate of air entrainment, described by Thomas et al. (1963) and Butcher and Parnell (1979).

A number of computer codes have been developed incorporating models of fires in buildings. An account is given by Budnick and Walton (1986). Enclosure fire models include not only deterministic models but also probabilistic ones. There are also a number of special purpose models which address topics such as detector actuation, smoke control, structural response, and evacuation.

The Building Regulations 1976, particularly Part E, contain requirements that cover both materials and construction. Other requirements are designed to limit fire spread between buildings and deal with distances between other buildings and boundaries and with the construction and combustibility of external walls and roofs. These regulations were revoked by the Building Regulations 1985, which are much less detailed (31 pages as opposed to 295). The material and component properties which are important are indicated by the tests specified in BS 476: 1970− *Fire Tests on Building Materials and Structures.* BS 476: Part 8: 972 is obsolete, being replaced by BS 476: Parts 20−23: 1987, which cover general principles, load and non-loading bearing elements and certain special components.

Compartments are spaces in buildings which are enclosed by fire-resistant walls and floors, and have all openings protected by fire-resistant construction. The fire resistance of the floors and walls should be capable of containing the most severe and prolonged fire which is likely to occur. The nature of such a fire depends on the likely contents of the compartment. The openings between compartments should be minimal. In some explosives factories, openings are virtually eliminated. But, in general, openings are acceptable, provided appropriate precautions are taken. Linings for walls and ceilings should be of non-combustible material or should have low flame spread characteristics. Roofs should be constructed so that fire does not spread either internally beneath the roof or externally over it. Vents should be provided for the release of hot gases and smoke. These control the spread of fire and smoke and assist firefighting. The design of vents depends on the type of building. The Building Regulations 1976, Regulation E5, specify minimum periods of fire resistance, which are a function of the type and size of building and the part of the building. The required minimum periods do not exceed 2 h, except for basements in large buildings and for large storage buildings.

Sprinkler systems for buildings are usually automatic, with a detection system which actuates the sprinklers and sounds the fire alarm. Heat sensitive detectors are dealt with in BS 5445: Part 5: 1977. In a large proportion of cases where a sprinkler installation has not been effective in protecting a building, this has been due to the fact that it was disabled in some way.

Smoke control may be applied to a space where smoke from a fire may occur and also to escape routes. Methods of smoke control include removal by venting and exclusion by maintaining a positive pressure. An escape route may be protected from smoke by the use of closed doors, by applying positive pressure, and/or by venting the smoke from the space where it is being produced. Where reliance is placed on closed doors, it must be ensured by means of suitable measures, including training, that they remain closed in fire conditions. Other fire precautions which should be taken in buildings include fire alarm system, escape routes, firefighting equipment, fire notices, and fire drills.

12.8.3 Fire Protection in Transport

In large part, the fire hazard in road, rail, and marine transport is addressed by measures to prevent accidents and to ensure appropriate design and maintenance. The principles of fighting fires arising from transport accidents are the wide variety of chemicals and the relative lack of familiarity of the local authority fire services. A considerable effort has been made, therefore, to furnish information to the fire services in advance about the hazards of chemicals which they may encounter. *Dangerous Substances. Guidance on Dealing with Fires and Spillages* (Home Office, 1972) and *Hazardous Loads* (IFE, 1972) provide some information on this aspect.

Firefighting in road transport requires prompt action in using portable extinguishers, alerting of the fire services, alerting of the public, and discouragement of spectators. Fire protection in road transport is treated in the NFPA *Handbook* by McGinley (1986). Relevant codes for road transport are NFPA 385: 1990 *Tank Vehicles for Flammable and Combustible Liquids*, NFPA 1901: 1985 *Automotive Fire Apparatus*, NFPA 512: 1990 *Truck Fire Protection* and NFPA 513: 1990 *Motor Freight Terminals*, and, for explosive materials NFPA 495: 1992 *Explosive Materials Code* and NFPA 498: 1992 *Explosives Motor Vehicles Terminals*.

Fire arrangements for rail transport are not well served with national codes, the matter tending to be left to the carrier. An account of fire protection and firefighting in rail transport is given in *Manual of Firemanship*, Book 5, *Incidents Involving Aircraft, Shipping and Railways* (Home Office, 1985). Fire protection in rail transport is treated in the NFPA *Handbook* by Fitch (1986).

Fire protection in marine transport is treated in the NFPA *Handbook* by Keller et al. (1986). Fire control is an important feature of the fighting capability of a warship. An account is given by Dimmer (1986). A relevant code is NFPA 306: 1993 *Control of Gas Hazards on Vessels*. The much larger storages of flammable materials in marine transport provide more scope for, and necessitate, fire protection arrangements more similar to those used in fixed installations.

The fire protection arrangements on chemical tankers and gas carriers are discussed by Keller et al. (1986).

12.9. FIRE HAZARD

The hazard of a large industrial fire may be assessed by consideration of the historical record of such fires and their effects and/or of assumed scenarios using appropriate frequency estimates and hazard and effects models.

Hazard assessment of fire in process plants generally involves the development of a set of scenarios for fire, an estimation of the frequency/probability of the initial events and of the events in the branches in the event trees and modeling of the outcome events including both physical events and damage/injury. Overviews of generic process plant fire scenarios have been given for LPG by Rasbash (1979) and for LNG by Napier and Roopchand (1986).

A large proportion of such releases will give rise to a vapor cloud. A jet, or torch, fire, and an engulfing fire tend to be of interest primarily as events which may cause a vessel to suffer a BLEVE. The First Canvey Report (HSE, 1978) included assessments of the hazard at the specific installations at Canvey from oil spillage over the bund at storages, a vapor cloud fire from the British Gas LNG storage, and vapor cloud fires from several LPG storages. A general account of hazard assessment for warehouses is described by Atkinson et al. (1992).

The hazard range of a fire depends on the type of fire concerned. Fires other than a vapor cloud or flash fire occur at a fixed point, and therefore the range of the fire is essentially determined by radiant heat transfer. The hazard range of a fire depends on the type of fire concerned. Considine and Grint (1985) who discuss the relations for the hazard range of a flash fire are based on their relations for dense gas dispersion. The authors derive their results for a flash fire from a flash fire code in which the flame is assumed to travel radially away from the ignition source. The thermal load at a given location is then evaluated numerically.

For fireballs, Considine and Grint use the model of Roberts (1981) to describe the lethality relations. The CCPS QRA Guidelines (1989) give the hazard distance for a fireball. Another set of hazard distances and lethal distance are those given for a propane fireball by Prugh (1994).

For jet flames of LPG, the model used by Considine and Grint (1985) provides two sets of hazard range relations, one for persons end on and one for those sides on to the flame.

REFERENCES

Agoston, G.A., Wise, H., Rosser, W.A., 1957. Dynamic factors affecting the combustion of liquid spheres. Combustion. 6, 708.

Alpert, R.L., Ward, E.J., 1984. Evaluation of unsprinklered fire hazards. Fire Saf. J. 7, 127.

Alvares, N.J., 1975. Some experiments to delineate the conditions of flashover in enclosure fires. In: International Symposium on Fire Safety of Combustible Materials. University of Edinburgh, Edinburgh.

American Petroleum Institute, 1956. Sparks from Hand Tools. American Petroleum Institute Pamphlet, March 1956. New York, NY.

American Iron and Steel Institute, 1980. Designing Fire Protection for Steel Columns, third ed.

Anderson, C., Norris, E.B., 1974. "Fragmentation and Metallurgical Analysis of Tank Car RAX 20", Report No. FRA-OR&D 75-30, US DoT, FRA, April

Anfenger, M.B., Johnson, C.W., 1941. Proc. Am. Petrol. Inst. 22, 54.

Appleyard, R.D., 1980. Testing and Evaluation of EXPLOSAFE System as a Method of Controlling the Boiling Liquid Expanding Vapour Explosion. Rep. TP 2740. Transport Canada.

Atkinson, G.T., Jagger, S.F., Kirk, P.G., 1992. Assessment of individual risks from fires in warehouses containing chemicals. Hazard Identification and Risk Analysis, Human Factors and Human Reliability in Process Safety, 141.

BS 6656, 1991. Guide to prevention of inadvertent ignition of flammable atmospheres by radio-frequency radiation. ISBN 0580324664, published by BSI, Nov. 1991

Babrauskas, V., 1983. Estimating large pool fire burning rates. Fire Technol. 19, 251.

Babrauskas, V., 1986. Pool fires: burning rates and heat fluxes. In: Cote, A.E., Linville, J.M. (Eds.), Fire Protection Handbook, pp. 21–36.

Ballal, D.R., 1983a. Flame propagation through dust clouds of carbon, coal, aluminium and magnesium in an environment of zero gravity. Proc. R. Soc. Ser. A. 385, 21.

Ballal, D.R., 1983b. Further studies on the ignition and flame quenching of dust clouds. Proc. R. Soc. Ser. A. 385, 1.

Barnard, J.A., Bradley, J.N., 1985. Flame and Combustion. Chapman & Hall, London.

Barnett, H.C., Hibbard, R.R., 1957. Basic Considerations in the Combustion of Hydrocarbon Fuels with Air. Rep. NACA 1300. National Association of Corrosion Engineers, Houston, TX.

Barnfield, J., 1986. The UK procedure for the appraisal of fire protection of structural steelwork. CTICM, Proceedings of the EGOLF Symposium, Brussels.

Bennett, P.A., 1991a. Safety. In: McDermid, J. (Ed.), op. cit., p. 60/1.

Bennett, P.A., 1991b. Specification, design and testing of safety critical software. The Fellowship of Engineering, op. cit., p. 21.

Beresford, T.C., 1981. The use of water spray monitors and fan sprays for dispersing gas leakages. In: Greenwood, D.V. (Ed.), op. cit., paper 6.

Bossart, C.J., 1974. Monitoring and control of combustible gas detectors below the lower explosive limit. In: Twentieth ISA Analysis Instrumentation Symposium. Instrument Society of America, Pittsburgh, PA.

Boult, M.A., Napier, D.H., 1972. Behaviour of polyurethane foams under controlled heating. Fire Prev. Sci. Technol. 3, 13.

Boult, M.A., Gamadia, R.K., Napier, D.H., 1972. Thermal degradation of polyurethane foams. Chem. Process Hazards. 4, 56.

Bradford, W.J., 1986. Storage and handling of chemicals. In: Cote, A.E., Linville, J.L. (Eds.), Fire Protection Handbook. National Fire Protection Association, Quincy, MA, pp. 11–49.

Browning, R.L., 1969. Estimating loss probabilities. Chem. Eng. 76 (December 15), 135.

Browning, R.L., 1980. The Loss Rate Concept in Safety Engineering. Dekker, Basel.

Brzustowski, T.A., Sommer, E.C., 1973. Predicting radiant heating from flares. Proc. API Div. Refining. 53, 865.

Budnick, E.K., Evans, D.D., 1986. In: Cote, A.E., Linville, J.M. (Ed.), Hand calculations for enclosure fires. Fire Protection Handbook, 16th ed, pp. 19–21.

Budnick, E.K., Walton, W.D., 1986. Computer fire models. In: Cote, A.E., Linville, J.M. (Ed.), Fire Protection Handbook, 21/25.

Buettner, K., 1951a. Effects of extreme heat and cold on human skin. I, Surface temperature, pain and heat conductivity in experiments with radiant heat. J. Appl. Physiol. 3, 703.

Buettner, K., 1951b. Effects of extreme heat and cold on human skin. II, Surface temperature, pain and heat conductivity in experiments with radiant heat. J. Appl. Physiol. 3, 703.

Bull, J.P., 1971. Revised analysis of mortality due to burns. Lancet. November (20), 1133.

Bull, J.P., Fisher, A.J., 1954. A study of mortality in a burns unit: a revised estimate. Ann. Surg. 139, 269.

Bull, J.P., Squire, J.R., 1949. A study of mortality in a burns unit. Ann. Surg. 130, 160.

Bullen, M.L., Thomas, P.H., 1979. Compartment fires with non-cellulosic fuels. Combustion. 17, 1139.

Burgess, D., Zabetakis, M.G., 1962. Fire and explosion hazards of LNG. US Bureau of Mines Investigation Report 6099. U.S. Department of Commerce, Washington, DC.

Burgoyne, J.H., 1965. Accidental ignitions and explosions of gases in ships. Instn. Mar. Engrs Trans. 77 (5), 129.

Burgoyne, J.H., Cohen, L., 1954. The effect of droplet size on flame propagation in liquid aerosols. Proc. R. Soc. Ser. A. 225, 375.

Burgoyne, J.H., Fauls, J., 1963. Dust explosion prevention in the United Kingdom with special reference to starch and allied industries. Die Starke. 15 (7), 260.

Burgoyne, J.H., Katan, L.L., 1947. Fires in open tanks of petroleum products: some fundamental aspects. J. Inst. Petrol. 33, 158.

Burgoyne, J.H., Richardson, J.F., 1948. Fire and explosion risks associated with liquid methane. Fuel. 27 (2), 37.

Burgoyne, J.H., Roberts, A.F., 1968. The spread of flame across a liquid surface: III, A theoretical model. Proc. R. Soc. Ser. A. 308, 69.

Burke, S.P., Schumann, T.E.W., 1928a. Diffusion flames. Combustion. 1, 2.

Burke, S.P., Schumann, T.E.W., 1928b. Diffusion flames. Ind. Eng. Chem. 20, 998.

Butcher, E.G., Parnell, A.C., 1979. Smoke Control in Fire Safety Design. Spon, London.

CCPS, 1989. Guidelines for Chemical Process Quantitative Risk Analysis. American Institute of Chemical Engineers, New York, NY, ISBN 0-8169-0402-2, p. 585.

CCPS, 1994. Guidelines for Evaluating the Characteristics of Vapor Cloud Explosions, Flash Fires, and Bleves. American Institute of Chemical Engineers, New York, NY, ISBN 0-8169-0474-X, p. 387.

Castle, G.K., 1974. Fire protection of structural steel. Loss Prev. 8, 57.

Cates, A.T., 1992. Shell Stanlow fluoroaromatics explosion—20 March 1990: assessment of the explosion and of blast damage. J. Hazard. Mater. 32, 1.

Chamberlain, G.A., 1987. Developments in design methods for predicting thermal radiation from flares. Chem. Eng. Res. Des. 65, 299.

Chamberlain, G.A., 1994. An experimental study of large-scale compartment fires. Hazards. XII, 155.

Charles, A., 1974. The Effects of a Fire Environment on a Rail Tank Car Filled with LPG. Rep. PB-241358. Department of Transportation, London.

Considine, M., Grint, G.C., 1985. Rapid assessment of the consequences of LPG releases. Gastech. 84, 187.

Constrado, 1983. Fire Protection for Structural Steel in Buildings. Constrado, Croydon.

Corona, A., 1984. *Fireproofing in refineries and petrochemical plants and offshore platforms.* Am. Petrol. Inst. Symp..

Coward, H.F., Jones, G.W., 1952. Limits of flammability of gases and vapors. Bureau Mines Bull. 503.

Cox, A.W., Lees, F.P., Ang, M.L., 1990. Classification of Hazardous Locations. Institution of Chemical Engineers, Rugby.

CPD, 1992. Methods for the Determination of Possible Damage to People and Objects Resulting from Release of Hazardous Materials – Green Book. Director General of Labour, The Hague.

Craven, A.D., 1972. Thermal radiation hazards from the ignition of emergency vents. Chem. Process Hazards. IV, 7.

Craven, A.D., 1976. Fire and explosion hazards associated with the ignition of small-scale spillages. Process Ind. Hazards.39.

Crawley, F.K., 1982. The effects of the ignition of a major fuel spillage. In: The Assessment of Major Hazards, Institute of Chemical Engineers, Rugby.

Crespo, A., 1991. One-Dimensional Model of Torch Fires. Rep. UPM-FIRE. University of Politecnica de Madrid, Madrid, Spain.

Croce, P.A., Mudan, K.S., 1986. Calculating impacts for large hydrocarbon fires. Fire Saf. J. 11 (1), 99.

Crocker, W.P., Napier, D.H., 1986. Thermal radiation hazards of liquid pool fires and tank fires. Hazards. IX, 159.

Crocker, W.P., Napier, D.H., 1988. Assessment of mathematical models for fire and explosion hazards of liquefied petroleum gases. J. Hazard Mater. 20, 109.

Cross, J., Farrer, D., 1982. Dust Explosions. Plenum Press, New York, NY.

Dahl, E., Bern, T.I., Golan, M., Engen, G., 1983. Risk of Oil and Gas Blow-out on the Norwegian Continental Shelf. Rep. STF 88A82062. SINTRF, Trondheim.

De Faveri, D.M., Fumarola, G., Zonato, C., Ferraiolo, G., 1985. Estimate flare radiation intensity. Hydrocarbon Process. 64 (5), 89.

De Ris, J., 1979. Fire radiation—a review. Combustion. 17, 1003.

Department of Scientific and Industrial Research, 1963. Report on 'non-sparking' hand tools. J. Inst. Petrol. 49 (474), 180.

Design Manual on the European Recommendations for the Fire Safety of Steel Structures, Brussels.

Dimaio, L.R., Lange, R.F., 1984. Effect of water quality on fire fighting foams. Plant/Operations Prog. 3, 42.

Dimmer, P.R., 1986. Fire fighting and damage control in warships. Fire Int. 97, 33.

Dow Chemical Company, 1976. Dow Fire and Explosion Index Hazard Classification Guide, 4th ed., Dow Chemical Company, Midland, MI.

Duijm, N.J., 1994. Jet flame attack on vessels. J. Loss Prev. Process Ind. 7, 160.

Eisenberg, N.A., Lynch, C.J., Breeding, R.J., 1975. Vulnerability Model: A Simulation System for Assessing Damage Resulting from Marine Spills. Rep. CG-D-136−75. Enviro Control Inc., Rockville, MD.

ECCS, European Convention for Constructional Steelwork, 1985. Design Manual on the European Recommendations for the Fire Safety of Steel Structures, Brussels.

Evans, J.L., 1988. Foams for flammable liquid fires: choice and evaluation. Fire Surveyor. 17 (5), 5.

Factory Mutual Engineering Corporation, 1967. Handbook of Industrial Loss Prevention. McGraw-Hill, New York, NY.

Fells, I., Rutherford, A.G., 1969. Burning velocity of methane−air flames. Combust. Flame. 13, 130.

Fernandez-Pello, A.C., 1977a. Downward flame spread under the influence of externally applied thermal radiation. Combust. Sci. Technol. 17, 1.

Fernandez-Pello, A.C., 1977b. Upward laminar flame spread under the influence of externally applied thermal radiation. Combust. Sci. Technol. 17, 87.

Field, P., 1985. Effective sprinkler protection for high racked storage. Fire Surveyor. 14 (5), 9.

Fiock, E.F., 1955. Measurement of burning velocity. In: Lewis, B., Pease, R.N., Taylor, H.S. (Eds.), *High Speed Aerodynamic and Jet Propulsion*, vol. IX, pt 2. Geoffrey Cumberledge/Oxford University Press, London, p. 409.

Fischer, W., 1965. Chem. Tech. (Berlin). 17 (5), 298.

Fitch, R.K., 1986. Rail transportation systems. Fire Protection Handbook, 16th ed. (ed. by Cote A.E. and Linville J.M.). p13−31.

Fritz, R.H., Jack, G.G., 1983. Water in loss prevention: where do we go from here? Hydrocarbon Process. 62 (8), 77.

Gayle, J.B., Bransford, J.W., 1965. Size and Duration of Fireballs from Propellant Explosions. Rep. NASA TM X-53314. George C. Marshall Space Flight Center, Huntsville, AL.

Gibbon, H.J., Wainwright, J., Rogers, R.L., 1994. Experimental determination of flammability limits of solvents at elevated temperatures and pressures. Hazards. XII, 1.

Glasstone, S., 1962. The Effects of Nuclear Weapons, revised ed. Atomic Energy Commission, Washington, DC.

Gordon, S., Mcbride, B.J., 1971. Computer Program for Calculation of Complex Chemical Equilibrium Compositions, Rocket Performance, Incident and Reflected Shocks, and Chapman−Jouguet Detonations. Rep. SP-273. National Aeronautics and Space Admininistration.

Gordon, S., Mcbride, B.J., 1976. Computer Program for Calculation of Complex Chemical Equilibrium Compositions, Rocket Performance, Incident and Reflected Shocks, and Chapman−Jouguet Detonations SP-273. Interim Revision, Rep. N78-17724. National Aeronautics and Space Administration.

Gugan, K., 1976. Flixborough—a combustion specialist's viewpoint. Chem. Eng. (London). 309, 341.

Haber, F., Wolff, H., 1923. Mist explosions. Z. Angew. Chem. 36, 373.

Haessler, W., 1986. Dry chemical agents and application systems. In: Cote, A.E., Linville, J.L. (Eds.), op. cit., p. 19−24.

Hajek, J.D., Ludwig, E.E., 1960. How to design safe flare stacks. Petro/ Chem. Eng. XXXII (6), C31−C44.

Hall, A.R., 1973. Pool burning. Oxidation Combust. Rev. 6, 169.

Hamilton, D.C., Morgan, W.R., 1952. Radiant-Inter-change Configuration Factors. Tech. Note 2836. National Aeronautics and Space Administration.

Harmathy, T.Z., 1972. A new look at compartment fires. Fire Technol. 8, 196.

Hawthorne, W.R., Weddell, D.S., Hottel, H.C., 1949. Mixing and combustion in turbulent gas jets. Combustion. 3, 266.

Health and Safety Executive, 1978. Canvey: An Investigation of Potential Hazards from Operations in the Canvey Island/Thurrock Area. HM Stationery Office, London.

Health and Safety Executive, 1981. Canvey: A Second Report. A Review of the Potential Hazards from Operations in the Canvey Island/Thurrock Area Three Years After Publication of the Canvey Report. HM Stationery Office, London.

Hearfield, F., 1970. Design for Safety—Fire and the Modern Large-scale Chemical Plant (Course). Department of Chemical Engineering, Teesside Polytechnic, Middlesbrough.

Henriques, F.C., 1947. Studies of thermal injury. V, The predictability and the significance of thermally induced rate processes leading to irreversible epidermal injury. Arch. Path. (Lab Med.). 43, 489.

Henry, M.F., 1986. Flammable and combustible liquids. In: Cote, A.E., Linville, J.M. (Eds.), Fire Protection Handbook, 5. National Fire Protection Association, Quincy, MA, p. 28.

Hernandez, J., Crespo, A., 1992. Phoenics. J. Comput. Fluid Dynamics Appl. 2, 205.

Heskestad, G., 1983. Luminous heights of turbulent diffusion flames. Fire Saf. J. 5, 103.

Hinshaw, J.R., 1957. Histological Studies of some Reactions of Skin to Radiant Thermal Energy. ASME Paper 57-SA-21. American Society of Mechanical Engineers, New York, NY.

Hirano, T., Tazawa, K., 1978. A further study of the effects of external thermal radiation on flame spread over paper. Combust. Flame. 32, 95.

Hodnett, R.M., 1986a. Automatic sprinklers. In: Cote, A.E., Linville, J.L. (Eds.), Fire Protection Handbook, 18. National Fire Protection Association, Quincy, MA, p. 36.

Hodnett, R.M., 1986b. Automatic sprinkler systems. In: Cote, A.E., Linville, J.L. (Eds.), Fire Protection Handbook, 18. National Fire Protection Association, Quincy, MA, p. 2.

Hodnett, R.M., 1986c. Water spray protection. In: Cote, A.E., Linville, J.L. (Eds.), Fire Protection Handbook, 18. National Fire Protection Association, Quincy, MA, p. 77.

Hodnett, R.M., 1986d. Water supplies for sprinkler systems. In: Cote, A. E., Linville, J.L. (Eds.), Fire Protection Handbook, 18. National Fire Protection Association, Quincy, MA, p. 27.

Hodnett, R.M., 1986e. Water and water additives for fire fighting. In: Cote, A.E., Linville, J.L. (Eds.), Fire Protection Handbook, 17. National Fire Protection Association, Quincy, MA, p. 2.

Home Office, 1972. Emergency Services. Circular 1. Home Office, London.

Home Office, 1985. Manual of Firemanship: Incidents Involving Aircraft, Shipping and Railways Bk. 4: Survey of the Science of Fire-fighting. Home Office, London.

Hottel, H.C., 1958. Certain laws governing diffusive burning of liquids by V.I. Blinov and G.N. Khudiakov (book review). Fire Res. Abs. Rev. 1, 41.

Hougen, O.A., Watson, K.M., Ragatz, R.A., 1954. Chemical Process Principles, vol. 1, 1954; vol. 2, 1959. Wiley, New York, NY.

Howell, J.R., 1982. A Catalog of Radiation Configuration Factors. McGraw-Hill, New York, NY.

Howells, P., 1993. Electrostatic hazards of foam blanketing operations. Health, Saf. Loss Prev.312.

Hunt, D.L.M., Ramskill, P.K., 1985. The behaviour of tanks engulfed in fire—the development of a computer program. The Assessment and Control of Major Hazards. Institution of Chemical Engineers, Rugby.

Hymes, I., 1983. The Physiological and Pathological Effects of Thermal Radiation, Report N. SRD R 275. UK Atomic Energy Authority, Warrington, UK.

IFE, Institution of Fire Engineers, 1972. Hazardous Loads, second ed. Leicester.

Ingberg, S.H., 1928. Tests of the severity of building fires. NFPA Quart. 22, 43.

Ishida, H., 1986. Flame spread over fuel soaked ground. Fire Saf. J. 10 (3), 163.

Ishida, H., 1988. Flame spread over ground soaked with highly volatile liquid fuel. Fire Saf. J. 13, 115.

Johnson, P.F., 1986. Special systems and extinguishing techniques. In: Cote, A.E., Linville, J.L. (Eds.), op. cit., pp. 19–55.

Johnson, S.C., 1976. US EO/EG—past, present and future. Hydrocarbon Process. 55 (6), 109.

Jones, G.W., Lewis, B., Friauf, J.B., Perrott St, G.J., 1931. Flame temperatures of hydrocarbon gases. J. Am. Chem. Soc. 53, 869.

Kanury, A.M., 1972. Ignition of cellulosic materials: a review. Fire Res. Abstracts Rev. 14, 24.

Kashiwagi, T., 1976. A study of flame spread over a porous material under external radiation fluxes. Combustion. 15, 255.

Kashiwagi, T., 1979. Effects of attenuation of radiation on surface temperature for radiative ignition. Combust. Sci. Technol. 20, 225.

Kawagoe, K., 1958. Fire Behaviour in Rooms. Rep. 27. Building Research Institute, Tokyo.

Kawaller, S.I., 1980. Chemically reactive coatings for the protection of structural steel. Fire Surveyor. 9 (4), 36.

Keller, C.L., Kerlin, D.J., Loeser, R., 1986. Motor vehicles. In: Cote, A. E., Linville, J.L. (Eds.), Fire Protection Handbook, sixteenth ed. National Fire Protection Association, Quincy, MA, pp. 18–27.

Kletz, T.A., 1971. Hazard analysis—a quantitative approach to safety. Major Loss Prev. 111.

Kletz, T.A., 1977. Unconfined vapour cloud explosions. Loss Prev. 11, 50.

Kuo, K.K., 1986. Principles of Combustion. Wiley, New York, NY.

Langdon-Thomas, G.J., 1972. Fire Safety in Buildings, Principles and Practice. Black, London.

Law, M., 1971. A Relationship Between Fire Grading and Building Design and Contents. Fire Res. Note. 877, Fire Res. Station, Borehamwood.

Law, M., 1991. Fire grading and fire behaviour. Fire Saf. J. 17, 147.

Lawrence, J.C., 1991. The mortality of burns. Fire Saf. J. 17, 205.

Lawson, D.I., Simms, D.L., 1952a. The ignition of wood by radiation. Br. J. Appl. Phys. 3 (9), 288.

Lawson, D.I., Simms, D.L., 1952b. The ignition of wood by radiation. Br. J. Appl. Phys. 3 (12), 394.

Lees, F.P., 1980. Accident fatality number—a supplementary risk criterion. Loss Prev. Saf. Promotion 3. 2, 426.

Lees, F.P., 1994. The assessment of major hazards: a model for fatal injury from burns. Process Saf. Environ. 72B, 127.

Lev, Y., 1981. A novel method for controlling LNG pool fires. Fire Technol. 17, 275–284.

Lev, Y., 1991. Water protection of surfaces exposed to impinging LPG jet fires. J. Loss Prev. Process Ind. 4, 252.

Lev, Y., Strachan, D.C., 1989. A study of cooling water requirements for the protection of metal surfaces against thermal radiation. Fire Technol. 25 (3), 213.

Lewis, B., Von Elbe, G., 1961. Combustion, Flames and Explosions of Gases. second ed. Academic Press, New York, NY.

Lewis, B., Von Elbe, G., 1987. Combustion, Flames and Explosions of Gases, third ed. Academic Press, Waltham, MA.

Lockwood, N.R., 1986. Foam extinguishing agents and systems. In: Cote, A.E., Linville, J.L. (Eds.), op. cit., pp. 19–32.

Loomis, A.G., Perrott St, G.J., 1928. Measurement of the temperatures of stationary flames. Ind. Eng. Chem. 20, 1004.

Lunn, G.A., Phillips, H., 1973. A Summary of Experimental Data on the Maximum Safe Gap. Safety in Mine Res. Estab., London.

Marshall, V.C., 1987. Major Chemical Hazards. Ellis Horwood, Chichester.

McCormick, J.W., Schmitt, C.R., 1974. Extinguishment of selected metal fires using carbon microspheres. Fire Technol. 10, 197.

McGinley, H., 1986. Motor vehicles. Fire Protection Handbook, 16th ed. (ed. by Cote A.E. and Linville J.M.). p13–20.

McMillan, W.J., 1974. Testing pipe insulation systems. Loss Prev. 8, 48.

Ministry of Works, 1946. Fire grading of buildings, General Principles and Structural Precautions. Post-War Building Studies 20, pt. 1. HM Stationery Office, London.

MITI, Ministry of International Trade and Industry, 1976. A Report on the Experimental Results of Explosion and Fires of Liquid Ethylene Facilities (Tokyo). Insurance Technical Bureau (English translation).

Mizner, G.A., Eyre, J.A., 1982. Large-scale LNG and LPG pool fires. The Assessment of Major Hazards. Institution of Chemical Engineers, Rugby.

Modak, A.T., 1977. Thermal radiation from pool fires. Combust. Flame. 29, 177.

Molina, M., Rowland, S., 1974. Stratospheric sink for chlorofluoromethanes: chlorine catalysed destruction of ozone. Nature. 249 (June), 810.

Moore, D.W., 1986. Halogenated agents and systems. In: Cote, A.E., Linville, J.L. (Eds.), op. cit., pp. 19–11.

Moorhouse, J., Pritchard, M.J., 1982. Thermal radiation hazards from large pool fires and fireballs—a literature review. The Assessment of Major Hazards. Institution of Chemical Engineers, Rugby.

Mudan, K.S., 1984. Thermal radiation hazards from hydrocarbon pool fires. Prog. Energy Combust. Sci. 10, 59.

Mudan, K.S., Croce, P.A., 1984. Thermal radiation model for LNG trench fires. Winter Annual Meeting. American Society of Mechanical Engineers, New York, NY.

Mudan, K.S., 1989. Evaluation of fire and flammability hazards. Encyclopaedia of Environmental Control Technology. Gulf, Houston, TX.

Napier, D.H., Roopchand, D.R., 1986. An approach to hazard analysis of LNG spills. J. Occup. Accid. 7 (4), 251.

Nash, P., 1974. The fire protection of flammable liquid storages with water sprays. Chemical Process Hazards. 51.

Nettleton, M.A., 1987. Gaseous Detonations. Chapman & Hall, London.

Nylund, J., 1984. Fire survival of process vessels containing gas. The Protection of Exothermic Reactors and Pressurised Storage Vessels. Institution of Chemical Engineers, Rugby.

Parker, R.J., 1975. The Flixborough Disaster. Report of the Court of Inquiry. HM Stationery Office, London.

Peissard, W.G., 1982. Corrosion from halon: the true extent of the problem. Fire Int. 74, 88.

Petersen, M.E. 1986a, 1986b. The role of extinguishers in fire protection. Fire Protection Handbook, 16th ed. (ed. by Cote A.E. and Linville J.M.). p20–22.

Potter, A.E., 1960. Flame quenching. Prog. Combust. Sci. Technol. 1, 145.

Powell, F., 1969. Ignition of gases and vapours. Ind. Eng. Chem. 61 (12), 29.

Prugh, R.W., 1994. Quantitative evaluation of fireball hazards. Process Saf. Prog. 13, 83.

Quintiere, J., Mccaffrey, B.J., Kashiwagi, T., 1978. A scaling study of a corridor subject to a room fire. Combust. Sci. Technol. 18, 1.

Quintiere, J.G., 1979. The spread of fire from a compartment: a review. In: Smith, E.E., Harmathy, T.Z. (Eds.), Design of Buildings for Fire Safety. ASTM STP 685. American Society of Testing Materials, Philadelphia, PA, p. 139.

Quintiere, J.G., 1981. A simplified theory for generalising results for a radiant panel rate of flame spread apparatus. Fire Mater. 5, 52.

Raj, P. K., Moussa, A.N., Aravamudan, K., 1979. Experiments Involving Pool and Vapor Fires from Spills of LNG on Water, NTIS # AD-A077073, USCG Report. Washington, DC 20590.

Rana, M., Mannan, M.S., 2010. Forced dispersion of LNG vapor with water curtain. J. Loss Prev. Process Ind. 23 (6), 768–772.

Rana, M., Guo, Y., Mannan, M.S., 2010. Use of water spray curtain to disperse LNG vapor clouds. J. Loss Prev. Process Ind. 23 (1), 77–78.

Rasbash, D.J., 1960. Mechanisms of extinction of liquid fires with water spray. Combust. Flame. 4 (3), 223.

Rasbash, D.J., 1962. The extinction of fires by water sprays. Fire Res. Abstr. Rev. 4 (122), 28.

Rasbash, D.J., 1975. Interpretation of statistics on the performance of sprinkler systems. Control. May, 63.

Rasbash, D.J., 1979. Review of explosion and fire hazard for liquefied petroleum gas. Fire Saf. J. 2 (4), 223.

Rasbash, D.J., Rogowski, Z.W., 1957. Extinction of fires in liquids by cooling with water sprays. Combust. Flame. 1, 453.

Rasbash, D.J., Stark, G.W.V., 1960. Extinction of fires in burning liquids with water sprays from hand lines. Inst. Fire Eng. Quart. 20 (3), 55.

Rasbash, D.J., Stark, G.W.V., 1962. Some aerodynamic properties of sprays. Chem. Eng. (London). December, A83.

Riddlestone, H.G., Bartels, A., 1965. The relative hazards of ferrous and non-sparking tools in the petroleum industry. J. Inst. Petrol. 51 (495), 106.

Risinger, J.L., 1964a. Controlling floating roof tank fires. In: Vervalin, C.H. (Ed.), op. cit., p. 309.

Risinger, J.L., 1964b. Controlling high vapor-pressure fires. In: Vervalin, C.H. (Ed.), op. cit., p. 320.

Risinger, J.L., 1964c. How to control and prevent crude oil tank fires. In: Vervalin, C.H. (Ed.), op. cit., p. 313.

Risinger, J.L., 1964d. How to fight refinery fires. In: Vervalin, C.H. (Ed.), op. cit., p. 300.

Roberts, A.F., 1981. Thermal radiation hazards from releases of LPG from pressurised storage. Fire Saf. J. 4 (3), 197.

Roberts, A.F., 1982. The effect of conditions prior to loss of containment on fireball behaviour. The Assessment of Major Hazards. Institution of Chemical Engineers, Rugby.

Roberts, A.F., Quince, B.W., 1973. A limiting condition for the burning of flammable liquids. Combust. Flame. 20, 245.

Robertson, R.B., 1976. Spacing in chemical plant design against loss by fire. Process Ind. Hazards.157.

Sass, R., Eckermann, R., 1992. CHEMSAFE—a user friendly information system of safety values. Loss Prev. Saf. Promotion. 3 (7), 145–1.

Savage, L.D., 1962. The enclosed laminar diffusion flame. Combust. Flame. 6, 77.

Semenov, N.N., 1959. Some Problems of Chemical Kinetics and Reactivity. Pergamon Press, London.

Senecal, J.A., 1992. Halon replacement: the law and the options. Plant/Operations Prog. 11, 182.

Siegel, R., Howell, J.R., 1991. Thermal Radiation Heat Transfer. third ed. Hemisphere, New York, NY.

Simms, D.L., 1962. Damage to cellulosic materials by thermal radiation. Combust. Flame. 6, 303.

Simms, D.L., Law, M., 1967. Ignition of wet and dry wood by radiation. Combust. Flame. 11, 377.

Sirignano, W.A., Glassman, I., 1970. Flame spreading across liquid fuels: surface tension-driven flows. Combust. Sci. Technol. 1, 307.

Sofyanos, T., 1981. Causes and Consequences of Fires and Explosions on Offshore Platforms: Statistical Survey of Gulf of Mexico Data. Rep. 81−0057. Det norske Veritas, Oslo.

Stannard, J.H., 1977. Thermal radiation hazards associated with marine LNG spills. Fire Technol. 13, 35.

Steward, F.R., 1970. Prediction of the height of turbulent diffusion buoyant flames. Combust. Sci. Technol. 2, 203.

Stoll, A.M., Greene, L.C., 1959. Relationship between pain and tissue damage due to thermal radiation. J. Appl. Physiol. 14, 373.

Stoll, A.M., Chianta, M.A., 1971. Heat transfer through fabrics as related to thermal injury. Trans N.Y. Acad. Sci. Ser. 2. 33, 649.

Strehlow, R.A., Baker, W.E., 1975. The Characterisation and Evaluation of Accidental Explosions. Rep. CR 134779. Aerospace Safety Research and Data Institute, Lewis Research Center, NASA, Cleveland, OH.

Stull, D.R., 1977. Fundamentals of Fire and Explosion. American Instituse of Chemical Engineers, New York, NY.

Taylor, G., 1991. Report of the Halons Technical Options Committee. Technology and Economics Review Panel, UN Environment Programme. Nairobi, Kenya.

Thomas, P.H., 1963. The size of flames from natural fires. Combustion. 9, 844.

Thomas, P.H., 1982. Modelling of compartment fires. Sixth International Fire Protection Seminar, Verein. zur Forderung des Deutschen Brandschutzes, p. 29.

Thomas, P.H., Law, M., 1972. The Projection of Flames from Burning Buildings. Fire Research Note 921. Fire Research Station, Borehamwood.

Thomas, P.H., Hinckley, P.L., Theobald, C.R., Simms, D.L., 1963. Investigations into the Flow of Hot Gases in Roof Venting. Fire Research Technical Paper 70. HM Stationery Office, London.

Thomas, P.H., Simms, D.L., Wraight, H., 1964. Fire Spread in Wooden Cribs. Part 1. Fire Research Note 537. Fire Research Station, Borehamwood.

Thomas, P.H., Simms, D.L., Wraight, H., 1965. Fire Spread in Wooden Cribs. Part 2. Fire Research Note 799. Fire Research Station, Borehamwood.

Tunc, M., Venart, J.E.S., 1984. Incident radiation from an engulfing pool fire to a horizontal cylinder—Part I. Fire Saf. J. 8, 81.

Van Dolah, R.W., Burgess, D.S., 1968. Explosion Problems in the Chemical Industry. American Chemical Society, Course Notes.

Verheij, F.J., Duijm, N.J., 1991. Wind Tunnel Modelling of Torch Fire. Rep. 91-422. TNO-IMET, Apeldoorn, The Netherlands.

Vervalin, C.H., 1964. Fire Protection Manual for Hydrocarbon Processing Plants. Gulf, Houston, TX.

Vervalin, C.H., 1973. Fire Protection Manual for Hydrocarbon Processing Plants, second ed. Gulf, Houston, TX, Risinger.

Wagman, D.D., 1953. Selected values of chemical thermodynamic properties. National Bureau Standards. US Govt Printing Office, Washington, DC.

Waldman, S., 1967. Fireproofing in chemical plants. Loss Prev. 1, 90.

Walls, W.L., 1973 Coping with LNG plant fires. In: Vervalin, C.H. (Ed.), op. cit., p. 377.

Walls, W.L., 1986. In: Cote, A.E., Linville, J.M. (Eds.), Gases. Fire Protection Handbook, sixteenth, ed. pp. 5−39.

Wesson, H.R., Welker, J.R., Brown, L.E., 1972. Control LNG-spill fires. Hydrocarbon Process. 51 (12), 61.

Wesson, H.R., Welker, J.R., Brown, L.E., Sliepcevich, C.M., 1973a. Fight LNG fires with foam. Hydrocarbon Process. 52 (10), 165.

Wesson, H.R., Welker, J.R., Brown, L.E., Sliepcevich, C.M., 1973b. Fight LNG spill fires with dry chemicals. Hydrocarbon Process. 52 (11), 234.

White, C.S., Jones, R.K., Damon, E.G., Fletcher, E.R., Richmond, D.R., 1971. The Biodynamics of Air Blast. Rep. DNA-2738-T. Defense Nuclear Agency, Department of Defense, Washington, DC.

Williams, R.L., 1977. Systems reliability analysis using the GO methodology. In: Gangadharan, A.C., Brown, S.J. (Eds.), op. cit., p. 103.

Yumoto, T., 1971. Heat transfer from flame to fuel surface in large pool fires (letter). Combust. Flame. 17, 108.

Yun, G.W., Ng, D., Mannan, M.S., 2011a. Key findings of liquefied natural gas pool fire outdoor tests with expansion foam application. Ind. Eng. Chem. Res. 50 (4), 2359−2372.

Yun, G.W., Ng, D., Mannan, M.S., 2011b. Key observations of liquefied natural gas vapor dispersion field test with expansion foam application. Ind. Eng. Chem. Res. 50, 1504−1514.

Zabetakis, M.G., Lambiris, S., Scott, G.S., 1959. Flame temperatures of limit mixtures. Seventh Symposium. Combustion. Butterworths, London, p484.

Zabetakis, M.G., 1965. Flammability Characteristics of Combustible Gases and Vapors Bulletin 627. (Bureau of Mines, Pittsburgh).

Explosion

13.1. EXPLOSIONS

The second of the major hazards is explosion. Explosion in the process industries causes fewer serious accidents than fire but more than toxic release. When it does occur, however, it often inflicts greater loss of life and damage than fire. Explosion is usually regarded as having a disaster potential greater than that of fire but less than that of toxic release.

An explosion is a sudden and violent release of energy. The violence of the explosion depends on the rate at which energy is released. There are several kinds of energy which may be released in an explosion. Three basic types are (1) physical energy, (2) chemical energy, and (3) nuclear energy. Physical energy may take forms such as pressure energy in gases, strain energy in metals, or electrical energy and thermal energy. Chemical energy derives from a chemical reaction. Nuclear energy is not considered here. In the present context, it is chemical explosions, and in particular explosions resulting from combustion of flammable gas, that are of prime interest.

There are two kinds of explosions from combustion of flammable gas: (1) deflagration and (2) detonation. In a deflagration, the flammable mixture burns at subsonic speeds. In a detonation, the flame front travels as a shock wave followed closely by a combustion wave which releases the energy to sustain the shock wave. At steady state, the detonation front reaches a velocity equal to the velocity of sound in the hot products of combustion; this is much greater than the velocity of sound in the unburnt mixture. A detonation generates greater pressures and is more destructive than a deflagration. Whereas the peak pressure caused by the deflagration of a hydrocarbon—air mixture in a closed vessel is of the order of 8 bar, a detonation may give a peak pressure of the order of 20 bar.

A deflagration may turn into a detonation, particularly when traveling down a long pipe. The sequence of events in the DDT is summarized by Kuo (1986) as follows: (1) an accelerating laminar flame, with flame

wrinkling and compression waves; (2) formation of a shock front; (3) creation of a turbulent flame brush; (4) onset of an explosion within an explosion, with transverse waves; (5) development of a spherical shock behind the shock front; (6) interactions of the transverse waves with the shock front, detonation wave, and reaction zone and (7) establishment of a steady wave, leading to a shock-deflagration ensemble, or detonation.

13.2. DETONATION

A particularly severe form of explosion occurs when an explosive substance detonates. Detonation can occur in liquid and solid explosives, in explosive gas mixtures, and in vapor clouds.

In a detonation, a detonation wave passes through the explosive substance. The detonation wave may develop by a process of transition. This transition may be illustrated by considering the combustion in a tube of a flammable gas—air mixture which is initially at constant pressure. If ignition occurs and energy is released at one end of the tube, the burnt gases expand. The deflagration front moves at a flame speed which is the sum of the burning velocity and the velocity of the burnt gases. If the flame speed is low enough, the combustion continues at essentially constant pressure, but if the flame speed is sufficiently high for momentum changes to exercise a significant effect, pressure disturbances are created. In this latter case, the flame front accelerates and travels as a combustion wave proceeded by a shock wave. Further acceleration of the flame front may cause the deflagration to turn into a detonation. The detonation wave then travels with a velocity greater than that of sound in the unburnt gas.

Different models for the analysis of the behavior of shock waves are included on Table 13.1.

The detonability of fuel—air mixtures is characterized in terms of the detonation limits and ignition sources. Based on Burgess et al. (1968), almost any flammable gas can be detonable using a sufficiently

TABLE 13.1 Models for the Analysis of the Behavior of Shock Waves

Model	Description
Unidimensional models	Analysis of the behavior of shock waves with and without reaction was for many years conducted mainly in terms of models in one dimension.
Non-reactive shock wave Rankin–Hugoniot conditions	It is convenient to consider first the unidimensional model of a planar shock wave in a non-reactive medium. Conventionally, the model is derived for a shock wave acting like a piston compressing the gases before the wave front. It is expressed first with a coordinate system moving with the wavefront as stated by Lewis and von Elbe (1987, p. 535).
Rankin–Hugoniot curves for ideal gases	The R–H equations show the relation between the initial and final states for a given change of enthalpy across the shock front. It is formulated based on an ideal gas equation with a constant ratio of specific heats. A detailed discussion of the properties of the R–H curves is given by Kuo (1986).
Reactive shock wave	Includes the modeling of a planar shock wave in a reactive medium or detonation wave. The model includes the heat input term q, which is the energy addition per unit mass in the flow behind the shock front. Then, a family of R–H curves for $q = 0$ and a family of R–H curves for various fractions of completion of the combustion reaction can be obtained.
Chapman–Jouguet model	The CJ model explains why detonations for a given fuel travel at a constant velocity. The shock wave profile yielded by the CJ model includes three main points: the shock wave travels as a sharp front. At the instant when the front arrives there occurs the von Neumann spike (point A). Very close to this in time is a lower pressure peak at the CJ plane (point B). Following this the pressure decays to a plateau value (point C). More detailed accounts of the full CJ model equations are given by Lewis and von Elbe (1987, p. 538) and Kuo (1986). The latter also gives a calculation scheme for the determination of the CJ parameters. Some correlations such as Zeldovich approximation (Zeldovich and Kompaneets, 1960) may be used to obtain approximate estimates of the CJ parameters. The CJ model has proved relatively successful in predicting the detonation pressure, density, and velocity in readily detonable mixtures in straight pipes.
Zeldovich–von Neumann– Doring model	A further advance came with the development of a model which takes account of the finite rate of reaction and of heat release. The model was formulated independently by Zeldovich (1940), von Neumann (1942), and Döring (1943). In this model, there is no reaction immediately behind the shock wave and there is an incubation period before the reaction begins. The ZND model is more realistic and provides a firmer basis for the development of unidimensional models. One particular use is in predicting the pressure–time profile of the detonation, including the von Neuman spike.
Taylor expansion wave	The product gases behind the CJ plane expand isentropically and accelerate, so that there is a distribution of particle velocities. This distribution was investigated by Taylor (1950). The Taylor expansion wave theory provides predictions for the velocity decay behind the CJ wave front.

strong ignition source. However, there is some evidence that detonation may be a stochastic process (Terao, 1977) and there are certain features which cast a degree of uncertainty over detonation limits. One is the phenomenon of cool flames. Another is the existence of compounds which are capable of detonation in the absence of an oxidant (e.g., gaseous acetylene, ethylene, hydrogen peroxide, and ozone (Bretherick, 1985). Further, a distinction is made in respect of detonation between confined and unconfined situations. The later situations are more difficult to measure.

Separate detonation limits are quoted for these two situations (Nettleton, 1980).

13.3. EXPLOSION ENERGY

As already stated, an explosion is a sudden and violent release of energy. The energy released in an explosion on a process plant is normally one of the following: (1) chemical energy, (2) fluid expansion energy, and/or (3) vessel strain energy.

13.3.1 Chemical Energy

The energy release in a chemical explosion is a function of the nature and state of the reactants and of the products. In considering the energy release in a chemical explosion, it is convenient to consider first condensed phase explosives, or high explosives, and then flammable gases and liquids. A condensed phase explosive contains its own oxygen so that it can explode even in the absence of air. A flammable organic gas or vapor which explodes in excess of air is normally assumed to yield incombustible gases such as CO_2 and H_2O. The same applies to a condensed phase explosive which undergoes combustion in excess air.

The quantities which characterize a high explosive are discussed by Kinney and Graham (1985). Those of interest here are (1) the energy of explosion, (2) the heat of explosion, and (3) the heat of combustion. The energy in an explosion is of two different kinds: (1) thermal energy and (2) work energy. That part of the energy transferred due to temperature difference is termed the heat of explosion. The heat of combustion is determined by causing the substance to burn in air in a bomb calorimeter and measuring the heat evolved. The heat of explosion is determined in a similar manner, but using inert gas instead of air. The substance is placed in a bomb calorimeter, air is replaced with inert gas, the substance is initiated, and the heat evolved is measured. Measurement of the energy of explosion is carried out by encasing the substance in a heavy sheath of metallic gold. The explosion flings the gold casing against the bomb wall, where it gives up its kinetic energy as heat, so that in this case the heat measured now includes this energy as well as that transferred by temperature difference.

The relative magnitude of these quantities may be illustrated by considering the values for TNT as given by Kinney and Graham (1985): heat of combustion = 15,132 J/g, energy of explosion = 4850 J/g, heat of explosion = 2710 J/g.

The heat of explosion corresponds to the internal energy change for the explosion ΔE. In order to estimate the effects of an explosion, it is necessary to know the energy of explosion. This is the work energy, or work. This work is done by the expansion of the gas and is given by

$$W = -\int_1^2 P \, dV \qquad (13.1)$$

where P is the absolute pressure, V is the volume, and W is the work of expansion. This is the energy transferred in the explosion as work done on the blast wave and missiles. The integral of Equation 13.1 is difficult to evaluate, and it is more convenient to work directly in terms of the initial and final values of the thermodynamic properties. The quantity generally used is the Helmholtz free energy change ΔA:

$$-\int_1^2 P \, dV \approx \Delta A \qquad (13.2)$$

Treatments of the relationship between the work and the Helmholtz free energy change are given by Kiefer et al. (1954) and Moore (1962).

The actual energy released in an explosion tends to be somewhat less than the Helmholtz free energy change. One reason is that the process is not reversible. Another is that the products of combustion of the explosion are usually at a temperature higher than ambient.

Often, data are not available on the Helmholtz free energy change for a compound but are available for the Gibbs free energy change. For many substances, including hydrocarbons, the difference is not great and the error involved in using the Gibbs instead of the Helmholtz free energy is small.

The following thermodynamic relations, which apply to a reversible process at constant temperature and pressure, are useful in explosion calculations. For absolute quantities

$$H = E + PV \qquad (13.3)$$

$$A = E - TS \qquad (13.4)$$

For isothermal change in a system

$$\Delta H = \Delta E + \Delta(PV) \qquad (13.5)$$

$$\Delta A = \Delta E - T \, \Delta S \qquad (13.6)$$

$$\Delta F = \Delta H - T \, \Delta S \qquad (13.7)$$

where ΔA is the Helmholtz free energy change, ΔE is the internal energy change, ΔF is the Gibbs free energy change, ΔH is the enthalpy change, ΔS is the entropy change, and T is the absolute temperature.

The energy of explosion is usually calculated at standard conditions of 25°C and atmospheric pressure.

The internal energy change for the explosion reaction is

$$\Delta E = (\Delta E_f^o)_p - (\Delta E_f^o)_r \qquad (13.8)$$

where the superscript o denotes the standard state and the subscript f the formation value, p the products, and r the reactants. The enthalpy change for the reaction is

$$\Delta H = (\Delta H_f^o)_p - (\Delta H_f^o)_r \qquad (13.9)$$

The entropy change for the explosion reaction, or entropy of explosion, is

$$\Delta S = (S)_p - (S)_r \qquad (13.10)$$

The Helmholtz free energy change ΔA for the explosion reaction is given by Equation 13.6, or alternatively by

$$\Delta A = (\Delta A_f)_p - (\Delta A_f)_r \qquad (13.11)$$

For the entropy and Helmholtz free energy per mole of the reactants and products, assuming ideal gases,

$$S = S^\circ - R \ln p \qquad (13.12)$$

$$A = A^\circ + RT \ln p \qquad (13.13)$$

where R is the universal gas constant. For liquids and solids, the second term on the right-hand side of Equations 13.12 and 13.13 is negligible.

The Helmholtz free energy change may be calculated from thermodynamic data in several ways. One is to calculate ΔE from Equation 13.8 and ΔS from Equation 13.10 using Equation 13.12 to obtain S, and then to calculate ΔA from Equation 13.6. Alternatively, ΔA may be calculated from Equation 13.11, using Equation 13.13 to obtain ΔA_f.

The entropy term $T \Delta S$ is significant where there is a large difference between the number of moles of the reactants and of the products. This is the case for condensed phase explosives, but usually not for flammable gases and vapors. Where the change in the number of moles is not significant

$$\Delta A \approx \Delta E \qquad (13.14)$$

the enthalpy change may also be written as

$$\Delta H = \Delta E + \Delta n RT \qquad (13.15)$$

where Δn is the change in the number of moles. Where the change in the number of moles is not significant

$$\Delta H \approx \Delta E \qquad (13.16)$$

Thus, for explosions of flammable gases and vapors, it is often possible to make certain approximations which simplify the task of obtaining the necessary data on thermodynamic properties.

13.3.2 Gas Expansion Energy

Explosions can also be caused by gas or liquid under high pressure. The energy released in an explosive expansion of a compressed gas is again given by Equation 13.1.

Considering an ideal gas, for 1 mol of the gas, the limiting value of the energy of explosion, assuming an ideal gas and isothermal expansion, is

$$W = \int_1^2 P \, dV \approx T \, \Delta S = RT \ln(P_1/P_2) \qquad (13.17)$$

where the subscript 1 denotes the initial state and 2 the final state.

Assuming isentropic expansion, the limiting value of the energy of explosion is

$$W = \frac{P_1 V_1 - P_2 V_2}{\gamma - 1} \qquad (13.18)$$

or eliminating V_2 using $PV^\gamma = $ constant

$$W = \frac{P_1 V_1}{\gamma - 1} \left[1 - \left(\frac{P_2}{P_1} \right)^{(1-\gamma)/\gamma} \right] \qquad (13.19)$$

where γ is the ratio of the gas specific heats.

13.4. DEFLAGRATION INSIDE PLANT

The conditions for a deflagration to occur are that the gas mixture is within the flammable range and that there is a source of ignition or that the mixture is heated to its auto-ignition temperature.

The sources of ignition usually considered are those outside the process plant, but ignition sources can occur inside vessels and pipework also. These include flames and hot surfaces; sparks; chemicals (unstable compounds, reactive compounds, and catalysts, pyrophoric iron sulphide); static electricity; compression.

If a flammable mixture may be present, precautions should be taken to eliminate all ignition sources. But it is prudent to assume that, despite these efforts, a source of ignition will at some time occur. Ignition can also occur if the flammable mixture is heated to its autoignition temperature.

13.4.1 Deflagration in Vessels and Pipes

For a deflagration at constant volume in a sphere, the maximum explosion pressure is

$$\frac{P_2}{P_1} = E \qquad (13.20)$$

with

$$E = \frac{n_2 T_2}{n_1 T_1} \qquad (13.21a)$$

$$= \frac{M_1 T_2}{M_2 T_1} \qquad (13.21b)$$

where E is the expansion ratio, M is the molecular weight of the gas mixture, n is the number of moles in the gas mixture, P is the absolute pressure, T is the absolute temperature, and the subscripts 1 and 2 denote the initial and final states, respectively.

The values quoted for the final pressure attained in deflagration in a closed vessel are often underestimated due to the use of constant pressure rather than constant volume assumptions. On the correct basis of constant volume combustion, for hydrocarbon−air mixtures, the ratio of the maximum explosion pressure to the initial pressure is approximately as follows:

$$\frac{P_2}{P_1} \approx 10 \qquad (13.22)$$

For conventionally designed pressure vessels, the bursting pressure P_b is of the order of

$$\frac{P_b}{P_1} \approx 4 - 5 \qquad (13.23)$$

Hence in the absence of explosion relief, the deflagration of a hydrocarbon−air mixture is easily capable of bursting a pressure vessel. If the explosion relief is to be effective, it must act within a period which is generally much less than 1 s.

If an explosion occurs in a compartmented system in which there are separate but interconnected spaces, a situation can arise in which the pressure developed by the explosion in one space causes a pressure rise in the unburnt gas in an interconnected space, so that the enhanced pressure in the latter becomes the starting pressure for a further explosion. This effect is known as pressure piling or cascading. The peak pressure enhancement in compartmented tube systems is less predictable, and such systems should be avoided. Pressure piling may be significant not only in the main process plant but in other equipment also.

Whereas in a vessel, deflagration results in a pressure rise which is uniform throughout the vessel, this is not so in a pipe, where the explosion velocity and explosion pressure can change along the length of the pipe.

An account of deflagration in pipes has been given by Bartknecht (1981). He defines three cases as given in Table 13.2.

In order to understand the phenomena involved, it is necessary to consider all three cases. The explosion velocity is given by the relation

$$u_{ex} = \phi u_n + u_d \qquad (13.24)$$

TABLE 13.2 Types of Deflagrations in Pipes (Bartknecht, 1981)

Case 1	Pipe open at one end, ignition at open end
Case 2	Pipe open at one end, ignition at closed end
Case 3	Pipe closed at both ends

where u_d is the displacement velocity, u_{ex} the explosion velocity, u_n the normal burning velocity, and ø the ratio of the area of the flame front to the cross-sectional area of the pipe. The displacement velocity u_d is the velocity of the unburned mixture, and the explosion velocity is the visible flame velocity. The equation applies to all three cases, but for that of an open pipe with ignition at the open end, the displacement velocity is zero. For ignition at the closed end (case 2) it is high, being some 80−90% of the explosion velocity. In this latter case only part of the gas mixture, theoretically 1/7, is burned within the pipe, the rest being ejected from the open end and burned outside. The high displacement velocity also results in much higher turbulence, a higher flame front area and a higher value of ϕ, in this latter case.

For an open pipe with ignition at the closed end with large diameter pipes ($\geq$400 mm), the explosion velocity exhibits an approximately linear increase along the length of the pipe. With smaller diameter pipes, this continuous increase is no longer observed. As the pipe diameter is decreased, a critical diameter is reached at which the flame propagation no longer occurs.

If the flow is highly turbulent and if the pipe is long enough, the explosion velocity can reach a value such that detonation, or quasi-detonation, occurs.

For a pipe closed at both ends, the explosion pressure again increases linearly with the explosion velocity, but since all the gas is burned the explosion pressure is several bar higher than for the open pipe case.

13.4.2 Plant Design

The hazard of an explosion should in general be minimized by avoiding flammable gas−air mixtures inside a plant. It is bad practice to rely solely on elimination of sources of ignition.

If the hazard of a deflagrative explosion nevertheless exists, the possible design policies include (1) design for full explosion pressure, (2) use of explosion suppression or relief, and (3) the use of blast cubicles. It is sometimes appropriate to design the plant to withstand the maximum pressure generated by the explosion. Often, however, this is not an attractive solution. Except for single vessels, the pressure piling effect creates the risk of rather higher maximum pressures. This approach is liable, therefore, to be expensive.

An alternative and more widely used method is to prevent overpressure of the containment by the use of explosion suppression or relief. In some cases, the plant may be enclosed within a blast resistant cubicle. Total enclosure is normally practical for energy releases up to about 5 kg TNT equivalent. It is more difficult to design for a detonative explosion. A detonation generates much higher explosion pressures. Explosion suppression and relief methods are not normally effective against a detonation. Usually, the only safe policy is to seek to avoid this type of explosion.

13.4.3 Effect of Congestion on Explosions

The effect of turbulence and turbulence-generating obstacles and structure can be a dominant factor in determining the flame speed and peak overpressure in a deflagration. Rasbash (1976) suggest that for indoor explosions, the turbulent flame speed is typically 1.5−5 times the laminar flame speed. For outdoor or partially confined deflagrations, empirical estimation methods developed by Baker et al. (1994, 1997) and the multienergy method by van den Berg (1985) and van Wingerden et al. (1989a,b) apply a similar factor to account for congestion. These factors range over three orders of magnitude to predict the peak overpressure. Risk-planning studies usually incorporate a detailed determination of the congestion and turbulence-inducing structure in a plant setting.

13.5. DETONATION INSIDE VESSELS AND PIPES

Detonation of a flammable gas−air mixture may occur by direct initiation of detonation by a powerful ignition source or by transition from deflagration.

The transition from deflagration to detonation requires a strong acceleration of the flame front. It occurs in pipelines but is very improbable in vessels. In practical terms, the peak pressure ratio P_2/P_1 resulting from a detonation of a hydrocarbon−air mixture in a containment such as a vessel is about 20:

$$\frac{P_2}{P_1} \approx 20 \qquad (13.25)$$

This is the order of pressure rise.

The effects of detonation in a pipe system vary. In some cases, damage is confined to blank ends and sudden changes of direction. In other cases, the pipe may be ripped open from end to end.

13.5.1 Plant Design

Essentially, the preferred approach in plant design to the hazard of detonation in vessels and pipes is prevention rather than protection. The pressures predicted for detonations inside plant tend to be so high that containment is impractical except for certain limited cases such as straight pipes which are of small diameter and/or are open-ended.

In designing against explosions, an approach usually considered is to accept a degree of plastic deformation. A discussion of the planned deformation approach and of

the practical problems in implementing it is given by Nettleton (1987). Essentially, it is limited to the early stages of flame acceleration. Nettleton also gives guidance on the local strengthening of plant at vulnerable points.

To the extent practical, it is desirable to keep pipelines small in diameter and short, to minimize bends and junctions, and to avoid abrupt changes of cross-section and turbulence promoters.

For protection, the following strategies are described by Nettleton (1987): (1) inhibition of flames of normal burning velocity, (2) venting in the early stages of an explosion, (3) quenching of flame−shock complexes, (4) suppression of a detonation, and (5) mitigation of the effects of a detonation.

13.5.2 Some Limitations

There are two principal mechanisms which create pressures which exceed those predicted by the unidimensional models and against which it is difficult to design. One is the pressure piling and wave interaction effects described above which occur during the development of the detonation wave. The other, which applies to the detonation condition itself, arises from instabilities in the combustion behind the leading front. The effects are enhanced in marginally detonable mixtures and by interactions with the confinement.

13.6. EXPLOSIONS IN CLOSED VESSELS

The deflagration of a flammable gas mixture in a closed vessel is important in itself and in relation to venting of the vessel. It is closely related to the combustion of a flammable dust mixture in a vessel. Quantification requires the measurement of flame speed and burning velocity. The two parameters of main interest in explosion in a closed vessel are the maximum pressure P_m and the rate of pressure rise, particularly the maximum rate $(\mathrm{d}P/\mathrm{d}t)_{max}$.

13.6.1 Energy Release and Final Conditions

Combustion in a closed vessel is a constant volume process. The heat of reaction and the gas specific heats applicable to it are those at constant volume. The relevant thermodynamic quantity is the internal energy. Thus, the final temperature should be obtained by equating the internal energy of the products to the internal energy of the reactants and the heat of reaction at constant volume.

The computation of the final temperature is frequently done making the incorrect assumption of constant pressure conditions. The problem is discussed by Richardson et al. (1990). It is a common, but incorrect practice to take the final temperature as the adiabatic flame temperature. This is incorrect because the adiabatic flame temperature is a constant pressure quantity. The constant volume flame temperature is appreciably higher.

13.6.2 Factors Influencing Closed Vessel Explosions

The maximum pressure and the rate of pressure rise in a closed vessel explosion are affected by a number of factors, including vessel size and shape, fuel, fuel–air ratio, vessel fractional fill, initial pressure, initial temperature, initial turbulence, ignition source, Bartknecht (1981), Harris (1983), Nagy and Verakis (1983), and Lunn (1984, 1992).

A factor which may affect both explosion parameters is heat transfer from the flame and the hot gas to the vessel wall. This can be significant if the vessel has a high aspect, or length/diameter, ratio or if the combustion is relatively slow. Both the maximum pressure and rate of pressure rise have maximum values at a fuel concentration close to the stoichiometric value, and decrease as the concentration moves away from this toward the explosibility limits. The effect is shown in Figure 13.1.

If the vessel is only partially filled by the fuel–air mixture, the maximum pressure varies in an approximately linear manner with the fractional fill. The initial temperature affects the maximum pressure, the rate

of pressure rise, and the final temperature. The maximum pressure decreases as the initial temperature increases is due to the decrease in density of the fuel–air mixture.

In a spherical vessel, the location of the ignition source has a slight effect on the maximum pressure, but a marked effect on the rate of pressure rise. The rate of pressure rise is greatest for central ignition. The rate of pressure rise increases with the strength of the ignition source.

13.7. EXPLOSIONS IN BUILDINGS AND LARGE ENCLOSURES

Many process plants are in buildings. Therefore, it is also necessary to consider explosions occurring inside buildings. A leak of flammable gas or liquid may create a flammable atmosphere inside a building and give rise to an explosion. Such leaks may occur from plant processing flammable fluids, from activities involving such fluids or from fuel gas supplies. In enclosed conditions, dispersion of the leaked gas is poor and the hazard is therefore much enhanced.

Additionally, the space in a building tends to contain obstructions. There are usually multiple compartments which generate turbulence.

The building itself is generally not very strong. An explosion pressure of 7 N/m^2 (1 psi) is often quoted as that at which a typical brick building may be destroyed. On the other hand, a normal building will have walls which contain weaker members which will fail and in so doing provide vents so that the explosion pressure does not rise as high as it otherwise would.

13.7.1 Gas Accumulation and Mixing in Buildings

If a leak occurs, the part of the enclosure which is affected depends on the density of the gas, the height of the leak source, and the ventilation pattern.

Ventilation is the movement of air in a building may be due to natural or forced draught ventilation. Natural draught ventilation may be either thermal driven or wind driven. In the former case, the movement of air is caused by the temperature difference between the inside and the outside of the building, while in the latter it is caused by the pressure of the wind on the side of the building. Human comfort sets a limit to the ventilation rate which can be used under normal conditions. A typical rule of thumb is that the upper limit of the air velocity for comfort is about 0.5 m/s. However, in case of an emergency situation, additional ventilation may be brought into play.

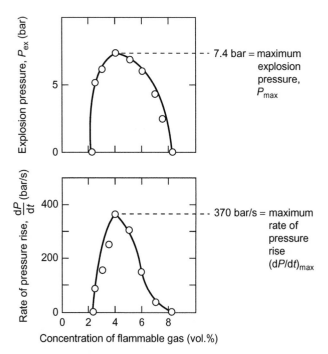

FIGURE 13.1 Explosions in closed vessels: effect of gas concentration (Bartknecht, 1981). *Source: Courtesy of Springer-Verlag.*

For practical purposes, the space liable to have a high concentration may be assumed to be perfectly mixed, and the gas concentration in the space may be determined from the equation for a single, perfectly mixed stage, which may be written for this case as

$$\tau \frac{dC}{dt} C_i - C \tag{13.26}$$

with

$$C_i = Q_g/Q \tag{13.27}$$

$$Q = Q_g + Q_a \tag{13.28}$$

$$\tau = V/Q \tag{13.29}$$

where C is the volumetric concentration in the space, C_i is an effective inlet volumetric concentration defined by Equation 13.27, Q is the total volumetric flow rate, Q_a is the volumetric flow of air, Q_g is the volumetric flow of contaminant gas, t is the time, V is the volume of the space, and τ is the volume/throughput ratio, or time constant.

Then for an increase in concentration starting from the time of release

$$\frac{C}{C_i} = 1 - \exp\left(-\frac{t}{\tau}\right) \tag{13.30}$$

and for a decrease in concentration starting from cessation of the release with a concentration C_o at that time

$$\frac{C}{C_i} = \exp\left(-\frac{t}{\tau}\right) \tag{13.31}$$

In this latter case, the value of Q to be used in Equation 13.28 is $Q = Q_a$, since $Q_g = 0$.

13.7.2 Explosions in Large Enclosures

The type of large enclosure which is envisaged here is exemplified by an offshore module. It is with such enclosures that the account given in this section is primarily concerned.

An offshore module tends to have a high aspect, or length/diameter ratio, and to contain a number of obstructions in the form of equipment and pipework.

The overpressure generated in the combustion of a vapor cloud is due to two effects. One contribution to the overpressure comes from the production of a large quantity of hot burned gas, the volume production. The other is due to the effect of the flame speed. The overpressure which might occur due to the volume production would be up to about 8 bar, whereas the overpressure due to the flame speed effect might have any value up to that associated with a detonation.

13.8. EXPLOSION PREVENTION AND PROTECTION

Explosion prevention methods are widely used to avoid that the explosions occur. However, if the explosion takes place, explosion protection methods are used to reduce the explosion consequences.

13.8.1 Explosion Prevention

Prevention of gas and vapor explosions in general depends on (1) avoidance of flammable mixtures and (2) elimination of sources of ignition. These general requirements for explosion prevention have already been discussed. For plant, or closed, systems, an important method of eliminating flammable mixtures which merits further consideration is (3) atmosphere control. This is discussed below.

Atmosphere control: The hazard of explosion can be much reduced by control of the atmosphere to render it non-flammable. This control is often affected by the use of an inert gas. Although, in some cases, it is more appropriate to operate in the fuel-rich region. In general, the use of atmosphere control is a much more reliable method than the elimination of sources of ignition, which is very difficult to achieve.

Some operations and equipment in which atmosphere control is practiced are (1) reactors, (2) storage tanks, (3) head tanks, (4) centrifuges, (5) driers, and (6) pneumatic conveyors.

In general, the quantity of inert gas used must not be excessive if inerting is to be economic. It is important to seek to minimize inert gas consumption. For a perfectly mixed system, the concentration changes effected by a ventilating air or purge gas flow are given by the equation

$$\frac{c}{c_o} = \exp\left(-\frac{QT}{V}\right) \tag{13.32}$$

where c is the volumetric concentration of the component in the system atmosphere, Q is the volumetric flow of gas through the vessel, t is the time, V is the volume of the vessel, and the subscript o denotes the initial value. Equation 13.32 may be rewritten as

$$\log_{10}\left(\frac{c}{c_o}\right) = -\frac{E}{2.3} \tag{13.33}$$

with

$$E = Qt/V \tag{13.34}$$

where E is the number of changes of atmosphere. Thus, the concentration of a component in the system atmosphere can be reduced by a factor of 10 by 2.3 changes and by a factor of 100 by 4.6 changes.

13.8.2 Explosion Protection

Explosion protection and relief include containment, separation, flame arresters, automatic isolation, automatic explosion suppression, explosion venting of vessels, explosion venting of pipes and ducts, explosion relief of buildings, explosion relief of large enclosures, and venting of chemical reactors among others. Some of these explosion protection alternatives are briefly described below.

Containment: Protection by containment of the explosion is a potential design option, but it is usually not practicable except for small-scale plants. Another method of containment is the use of blast walls and barricades, and of blast cubicles.

Flame arresters: A flame arrester, or flame trap, is a device used to prevent the passage of a flame along a pipe or duct. A flame arrester is generally an assembly of narrow passages through which gas or vapor can flow, but which are too small to allow the passage of flame.

Flame arresters are generally distinguished as end-of-line or in-line arresters. An end-of-line arrester is designed to prevent the passage of a deflagration from the downstream to the upstream side. An in-line arrester should be able to stop either a deflagration or a detonation passing in either direction. However, even if a flame arrester prevents the passage of the detonation flame, it does not stop the detonation shock wave.

Automatic isolation: In some systems, the passage of an explosion from one section to another may be prevented by automatic high speed isolation. An account of this method is given in FPA FS 6012: 1989 *Flammable Liquids and Gases: Explosion Control* and NFPA I2A (1997).

A typical installation is illustrated in Figure 13.2. Very rapid detection and valve closure are necessary.

The design of such a system requires information on the rate of pressure rise caused by the expected explosion. From the data on the rate of pressure rise, a suitable high speed detector can be chosen. The shut-off valve must

also operate very quickly. Isolation protects only the section isolated and not the section in which the explosion occurs.

Automatic explosion suppression: A developing explosion may be detected and suppressed using an automatic high speed suppression system. Accounts of automatic explosion suppression are given in FPA FS 6012: 1989 and FS 6015: 1974 *Explosible Dusts, Flammable Liquids and Gases: Explosion Suppression* and by Grabowski (1965) and Palmer (1973).

The basic data for the design of an explosion suppression system are given by the explosion pressure curve for the gas. A typical curve is shown in Figure 13.3 as curve 1.

This information is used to estimate the required response time of the detector and of the suppressant in order to limit the explosion pressure to a specified value. The pressure response typically obtained with explosion suppression is shown in curve 2 of Figure 13.3.

The size and shape of vessel to which explosion suppression can be effectively applied is not unlimited. FS 6015: 1974 stated that explosion suppression is applicable to vessels with a volume of up to 115 m^3 and for flammable gas mixtures for which the maximum explosion pressure is generated in not less than 40 ms.

With an automatic explosion suppression system operating, the maximum pressure of the suppressed explosion is typically reached in about 10 ms.

Explosion relief and venting: A quite different approach is to relieve the explosion by venting. Explosion venting is applied to (1) vessels, (2) pipes and ducts, (3) buildings, (4) large enclosures, (5) reactors, and (6)

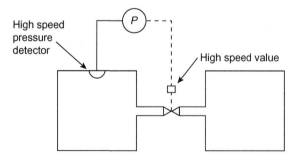

FIGURE 13.2　System for automatic isolation installed in a pipe.

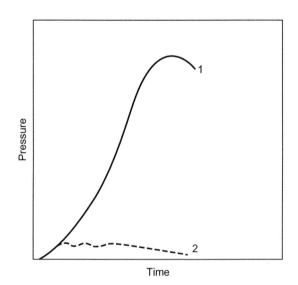

FIGURE 13.3　Typical explosion pressure curves for an automatic suppression system: curve 1, explosion pressure in absence of suppression; cure 2, explosion pressure with suppression operating.

storage vessels. If this option is adopted, it is necessary to consider not only the venting itself, but also the disposal of the material vented, which may not be a trivial problem. Additional information about explosion venting is included in the following session.

13.9. EXPLOSION VENTING OF VESSELS

One of the options for the protection of a vessel against a gas explosion is the use of explosion venting, also referred to as explosion relief.

Accounts of explosion venting are given in Explosions (Bartknecht, 1981), Development and Control of Dust Explosions (Nagy and Verakis, 1983), Gas Explosions in Buildings and Heating Plants (Harris, 1983), Venting Gas and Dust Explosions—A Review (Lunn, 1984, 1992), and the NFPA 68: 2007 *Deflagration Venting*.

The notation used in explosion venting is particularly liable to cause confusion. The term 'maximum explosion pressure' refers generally to a closed, or unvented, explosion, but occasionally it refers to a vented explosion. In the account given here, the maximum pressure in a vented system is referred to as the reduced pressure, P_{red}.

The various factors which influence the maximum pressure and the rate of rise of pressure in an explosion in a closed vessel are relevant to vented explosions also. One factor which assumes particular significance for a vented explosion is the location of the ignition source. Different models have been developed for vented explosions with different locations of the ignition source, as described below.

13.9.1 Empirical and Semi-Empirical Methods

The venting of an explosion is a complex process. It is therefore difficult to model theoretically. This has led to the development of a number of empirical and semi-empirical methods and scaling laws. Some semi-empirical methods used to calculate reduced pressures in vented scenarios (P_{red}) are listed below:

(a) Rasbash method (1969): This method is based on a collation of information on experimental work on venting using particularly propane and referred to work on explosion relief of ducts (Rasbash and Rogowski, 1960b). P_{red} is obtained using the venting ratio (ratio between the area of the smallest cross-section of the enclosure (A_v) to the vent area (A_v)), and pressure within the building space at which the vent opens (P_v). In subsequent work, Rasbash (1976) modified their empirical correlations to include a

term for the vent inertia and the mass of the vent cover (W_v). This method has been widely used for the explosion relief of buildings.

(b) Cubbage and Simmonds method: In contrast to Rasbash's method, Cubbage and Simmonds's method is based on separate equations for the two pressure peaks. The first of their equations depends only on the inertia w of the vent panel, the second only on the vent coefficient K, while the vent opening pressure P_v does not appear.

(c) Runes method: In contrast to the two empirical correlations just described, the equation given by Runes (1972) for the vent area of a large container or building has some theoretical basis. It is based on equating the volume production rate and the volumetric vent outflow.

(d) Decker method: (1971): This method is based on somewhat similar considerations to that of Runes. The approach is again to equate the volume production rate and the volumetric vent outflow.

(e) Dragosavic method (1973): This method provides equations for the first and second peak pressures in explosions in rooms in dwellings.

(f) Cubbage and Marshall method (1973): This method expresses the reduced pressure (P_{red}) as a function of the venting coefficient (K), the vent opening pressure (P_v), the maximum fundamental burning velocity (S_u), the volume of the vessel (V), and the inertia of the vent (w).

13.9.2 Venting Phenomena and Ducting Effects

If the vent is an open one, venting will result in the emergence of flame. The flame may be substantial. van Wingerden (1989a,b) has reported a jet flame 18 m long.

If the vent has a duct, there is potential for unburned gas to undergo further combustion in the duct, possibly resulting in overpressure in the duct. The ducting needs to be designed to avoid this.

It has been found that the presence of a vent duct can have a marked effect on the maximum pressure in a vented explosion. Figure 13.4 is a graph from the work of Pineau et al. (1978), showing the pressure in the vessel and in the duct as a function of duct length for a 10 m^3 vented vessel.

Typically, the maximum pressure rises rapidly with the initial increase in the length of the tube and then reaches a plateau or increases or decreases at a much slower rate. This effect was described by Sagalova and Resnick (1962) for dust explosions and has been observed by Kordylewski and Wach (1986, 1988) for gas.

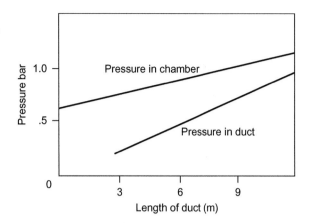

FIGURE 13.4 Explosion venting of vessels: effect of vent duct on pressure in a 10 m³ vessel. *Source: Cross and Farrer (1982); after Pineau et al. (1978).*

13.9.3 Vent Design

Design of a vent system based on a vent panel includes consideration of the following features: (1) reduced explosion pressure, (2) vent opening pressure, (3) vent area, (4) vent distribution, (5) vent opening, and (6) vent panel.

The reduced explosion pressure for a vented explosion will be determined by the mechanical strength of the containment. Explosion venting is used for low-strength as well as high-strength equipment.

The time of opening of the vent can affect the pressure developed. A sudden opening, such as occurs with a bursting disk, tends to increase turbulence. A smooth opening is desirable to minimize turbulence.

The design of the vent panel itself is important. Some features which need to be taken into account in the design of the vent panel itself are (1) panel material, (2) panel inertia, (3) relieving fasteners, (4) panel restraint system, (5) guarding, and (6) maintainability.

Detailed guidance on vent panel design is given in the *HSE Guide* and in NFPA 68: 1994.

13.10. EXPLOSION VENTING OF DUCTS AND PIPES

In general, where combustion of a flammable gas mixture occurs in a duct or pipe, there is a possibility that if the duct is long enough the flame front will accelerate to the point where the deflagration turns into a detonation.

Accounts of explosion venting of ducts are given in the *Guide to the Use of Flame Arresters and Explosion Reliefs* (HSE, 1965 HSW Bklt 34), *Explosions* (Bartknecht, 1981), *Venting Gas and Dust Explosions* (Lunn, 1984), and in NFPA 68: 1994 *Deflagration Venting*.

13.10.1 *HSE Guide* Method

The methods given in the *HSE Guide* are based on the work of Rasbash and Rogowski (Rasbash and Rogowski, 1960a,b; Rogowski and Rasbash, 1963).

Unless otherwise stated, *HSE Guide* methods are applicable to gas mixtures at atmospheric pressure. They are applicable to ducts with an *L/D* ratio of 6 or more. Other methods are available for vessels with an *L/D* ratio of 3 or less. For containers with an *L/D* ratio greater than 3 but less than 6, the use of methods for vessels with an *L/D* ratio of 3 or less is increasingly conservative as the *L/D* ratio approaches 6.

In designing explosion vents for ducts, the following cases are distinguished:

1. Gas velocity <10 ft/s
   a. straight unobstructed ducts with *L/D* < 30,
   b. straight unobstructed ducts with *L/D* > 30,
   c. ducts containing obstacles;
2. Gas velocity 10−60 ft/s
   a. unobstructed ducts,
   b. ducts containing obstacles.

The design of vent areas is essentially based on the relations between the maximum pressure *P* and the vent coefficient *K*.

If the gas is stationary or has a velocity less than 10 ft/s, the duct is straight and unobstructed and the *L/D* ratio is less than 30, only one relief opening is generally sufficient. If a single vent is used, it should be placed as near as possible to the most probable location of the ignition source, but if this latter is uncertain, the vent should be placed near the center of the duct.

For this vent system, a vent weighing not more than 2 lb/ft² and held by springs or magnets is suitable. An alternative relief is a bursting disk designed to burst at a pressure one half of the maximum pressure given in Equation 13.35.

If the gas has a velocity less than 10 ft/s and the duct is straight and unobstructed, but the *L/D* ratio is greater than 30, it is necessary to have more than one relief opening.

The open end of a long duct may be regarded as an explosion relief opening. In this context, an open end is an end leading without restriction to atmosphere or to a vessel which is itself provided with explosion reliefs or to a room of volume 200 times greater than that of the duct.

If the end of a duct is not open or if it may be regarded as restricted, an explosion relief should be located as near as possible to the end. For this vent system also, vents weighing not more than 2 lb/ft² of vent area and held by springs or magnets are suitable.

If at a gas velocity of less than 10 ft/s, the duct contains obstacles or features such as sharp right angle elbows or tees, the maximum pressure resulting from an explosion is greatly increased. An obstacle blocking only 5% of the cross-sectional area of the duct can increase the maximum pressure by a factor of 2–3, while an orifice blocking 30% of the duct area can increase the pressure by a factor of 10. Any bend sharper than a long, sweeping, smooth bend and any obstruction blocking more than 5% of the cross-sectional area of the duct should be regarded as an obstacle.

For this vent system, the weight of vents depends on the velocity of the gas. For gas velocities of 25 and of 25–60 ft/s, the vents should weigh not more than 10 and 5 lb/ft^2, respectively. The vents may be held by springs or magnets.

If at a gas velocity of 10–60 ft/s the duct contains obstacles, additional explosion relief openings are required. If there is a long straight duct connected to an obstacle, explosion reliefs should be located at 3 diameters on either side of the obstacle and again at 6 diameters on either side of it. Thereafter, the straight unobstructed section of the duct may be treated in the usual way. For this case, the maximum duct diameter for which information was available is 1.5 ft.

For gases other than propane, the maximum pressure may be calculated from the maximum pressure for propane using the relation

$$P_2 = \frac{S_u^2}{22} P_1 \qquad (13.35)$$

where P_1 is the maximum pressure resulting from an explosion for propane (lb$_f$/in^2), P_2 is the maximum pressure resulting from an explosion for a gas other than propane (lb$_f$/in^2), and S_u is the maximum fundamental burning velocity of the gas other than propane (ft/s). The value of 2.2 derives from the square of the maximum fundamental burning velocity of propane, which is 1.5 ft/s.

Alternatively, for a given maximum pressure, the distance between neighboring vents L_1 may be calculated using the relation

$$L_2 = \frac{22}{S_u^2} L_1 \qquad (13.36)$$

where L_1 is the distance between relief openings for propane (ft) and L_2 is the distance between relief openings for a gas other than propane (ft).

Vent closures should be designed so that a degree of deterioration or lack of maintenance does not cause the design maximum pressure resulting from the explosion to be exceeded. Closures should be robust and leak-tight. Methods of holding vent closures include springs, magnets, and hinges. Further details of vent closure are given in the *HSE Guide*.

It is pointed out by Lunn (1984) that the experimental data on which these recommendations are based are sparse. He suggests that the maximum pressure correlates as well with vent ratio A_v/V as with vent coefficient K. Figure 13.5 shows the data given by Lunn for vented ducts in support of this argument.

He is also critical of the use of Equation 13.35 to extrapolate from propane to other gases. Figure 13.5 shows a curve for the maximum pressure in hydrogen–air mixtures calculated from Equation 13.35 which differs appreciably from the experimental curve also shown in the figure.

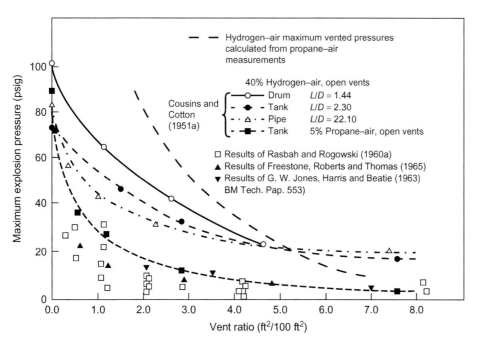

FIGURE 13.5 Explosion venting of ducts and pipes: maximum explosion pressure in vessels of different aspect ratios (Lunn, 1984). *Source: Courtesy of the Institution of Chemical Engineers.*

On the basis of the work of Cousins and Cotton (1951), he suggests as an alternative that if explosion pressures for a particular gas have been measured in a compact enclosure, then for a duct with $L/D < 30$ and with the same vent ratio these pressures represent upper bounds on the pressures likely to occur in the duct.

Relief devices which can be used include bursting disks and vent doors. The latter have the advantage that they seal the vent opening after effecting relief. However, relief of a pipe is a more demanding duty for a vent door and experience shows that it is not easy to get a suitable design. Many of those tested have either blown off or failed to reseat to give a tight closure.

13.10.2 NFPA 68 Method

The method given in NFPA 68: 1994 which is described here is that for the explosion venting of pipes, ducts, and elongated vessels operating at or near atmospheric pressure. The pressure is given in the code in terms of the reduced pressure P_{red}, which is expressed as a gauge pressure. The treatment applies to pipes in which the reduced pressure P_{red} is limited to 0.2 barg.

For ducts of non-circular cross-section, the relevant diameter is taken as the hydraulic mean diameter.

Two basic situations are distinguished: (1) a pipe with a single vent consisting of an open end and (2) a pipe with multiple vents. NFPA 68 gives several graphs to assist in the design of vents for the relief of deflagrations in pipes. The figures apply to flow in smooth, straight pipes. Some are applicable only to propane, but a formula is given which permits them to be applied to certain other gases also, as described below. They are also applicable to dusts, but with restrictions on the St class.

Figure 13.6 is used to determine the maximum allowable length of smooth, straight pipe or vessel which is closed at one end and vented at the other. If the L/D ratio is greater than that shown, there is a risk of detonation. The graph is applicable to propane and to dusts, a distinction being made between dusts with $K_{st} \leq 200$ and $K_{st} > 200$.

Figure 13.7 gives the reduced pressure for deflagration of gases and dusts in a pipe with multiple vents for an initial velocity <2 m/s. The limits for gas are a burning velocity ≤ 1.3 times that of propane and for dusts $K_{st} \leq 300$.

For gases other than propane which have a burning velocity not exceeding 1.3 times that of propane, the following formula are given:

$$L_x = \left(\frac{S_p}{S_x}\right)^2 L_p \qquad (13.37)$$

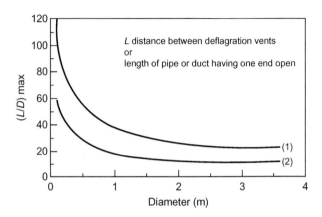

FIGURE 13.6 Explosion venting of ducts and pipes: maximum allowable length of a pipe vented at one end, or maximum allowable distance between vents, for gases and dusts (NFPA 68: 1994): (1) dusts with $K_{st} \leq 200$ bar m/s and (2) propane, dust with $K_{st} > 200$ bar m/s. See text for details. *Source: Reproduced with permission from NFPA 68 Deflagration Venting,* © *1994, National Fire Protection Association, Quincy, MA 02269.*

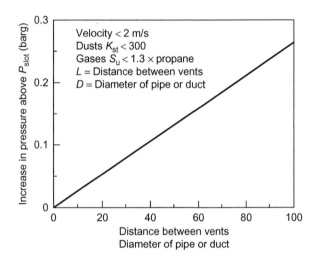

FIGURE 13.7 Explosion venting of ducts and pipes: reduced explosion pressure in pipe with multiple vents for deflagration of gases and dusts (NFPA 68: 1994). Initial flow velocity <2 m/s, gas burning velocity $\leq 1.3 \times$ burning velocity of propane, dust $K_{st} \leq 300$ bar m/s. See text for details. *Source: Reproduced with permission from NFPA 68 Deflagration Venting, 1994, National Fire Protection Association, Quincy, MA 02269.*

$$P_{red, x} = \left(\frac{S_x}{S_p}\right)^2 P_{red, p} \qquad (13.38)$$

where L is the maximum distance between vents, P_{red} is the reduced pressure, S is the maximum fundamental burning velocity and the subscripts p and x denote propane and the gas in question, respectively.

When the initial velocity exceeds 20 m/s, the gas burning velocity exceeds 1.3 times that of propane or the dust K_{st} value exceeds 300, the distance between vents should not exceed 1−2 m.

With regard to the vents, the total vent area at each location should be at least equal to the cross-sectional area of the pipe. This area may be provided by one or more vents. For an individual vent, an area exceeding the cross-sectional area of the pipe is not effective in further reducing the pressure of a deflagration.

The release pressure of the vents should be set as far below the reduced pressure P_{red} as practical, but in any event should not exceed half the reduced pressure. The mass of the vent closures should not exceed 2.5 lb/ft^2.

NFPA 68 gives a number of worked examples illustrating the application of the above guidance. These include an example involving the specification of the vents on a drier system consisting of the drier itself and a dust collector.

13.11. EXPLOSION RELIEF OF BUILDINGS

Accounts of explosion relief of buildings are given in *Gas Explosions in Buildings and Heating Plants* (Harris, 1983) and Venting Gas and Dust Explosions (Lunn, 1984) and by Rasbash (1969), Runes (1972), and Howard (1972). A relevant code is NFPA 68: 1994 *Deflagration Venting*.

A building is usually too weak to withstand a high explosion pressure and may be blown apart by a sustained pressure of 1 psig (0.07 bar) or even less. Nevertheless, for buildings containing flammable materials, it is often appropriate to provide explosion venting. This gives the building a degree of protection against explosions resulting from causes such as leaks and spillages.

Explosion relief of buildings may be regarded as intermediate between explosion relief of vessels and explosion relief of large enclosures such as offshore modules.

13.11.1 Empirical and Semi-Empirical Methods

The methods available for the explosion relief of buildings are essentially a subset of those already given for explosion venting of vessels.

Two methods which have been widely used are those of Rasbash (1969) and Runes (1972). Both methods were in fact developed in the first instance for the relief of buildings.

The methods preferred by Harris (1983) are those of Cubbage and Simmonds (Cubbage and Simmonds 1955a−c; Simmonds and Cubbage 1960; Cubbage and Marshall, 1973). Harris gives a theoretical model for the initial rate of pressure rise and initial pressure in an explosion in a building, but not for a vented explosion in

a building; for the latter he refers to the model of Fairweather and Vasey (1982). These methods have all been developed by British Gas and thus may be considered to constitute a coherent set of methods for this problem.

Rasbash method (Rasbash (1969) and Runes Method (Runes (1972)) have been widely used for the explosion relief of buildings. Runes considered that his equation gives conservative but reasonable vent areas. He presented several worked examples showing that by comparison with the then current version of NFPA 68 his equation gave reasonable values and avoided the more extreme vent areas which the latter sometimes gave. He also gave a comparison with the results of the Bromma explosion venting tests.

Following an explosion incident, Howard (1972) made proposals for modification of the Runes equation. The explosion occurred in a building which contained process plant equipment and had an area of 5335 ft^2 of open windows and of weak components which would act as vents. The end wall was estimated to be able to withstand a static pressure of 7 kPa. This wall was blown off, some portions falling as integral sections, but not forming missiles. Part of the side walls were also blown out and the other end wall was cracked.

Howard first took one of Runes's examples and substituted numerical values. Then, he proposed that the value of the Runes constant be increased to 2.6, in effect calibrating the Runes constant from this incident as shown in the following equation

In SI units

$$A_v = C \frac{L_1 L_2}{P^{1/2}} \qquad (13.39)$$

where A_v is the vent area (m^2), L_1 is the smallest dimension of the rectangular building enclosure to be vented (m), L_2 is the second smallest dimension of that building (m), P is the maximum internal building pressure (kPa), and C is a constant. The maximum internal building pressure P is that which can be withstood by the weakest building member which it is desired should not vent or break.

Equation 13.39 is believed to apply to buildings with a nominal length/width, or L/D, ratio up to 3. In a rectangular building where L_1 and L_2 are unequal, D should be taken as $(L_1 L_2)^{1/2}$. For a building in which the L/D ratio is greater than 3, the space should be divided into units each of which has an L/D ratio of not more than 3.

NFPA 68: 1978 gives for C a value in metric units of 6.8 for gases such as propane and others with a similar burning velocity, but a higher value for gases with a higher burning velocity. The suggested value for ethylene is 10.5 and that for hydrogen is 17.

13.11.2 NFPA68 Method

Explosion relief of buildings is treated in NFPA 68: 2002. The approach adopted has evolved over the years. That used in the 1974 edition was to calculate the vent area A_v using both the Rasbash and the Runes equations and to take the larger of the two areas. The 1978 edition gave only the Runes equation. For the venting of low strength enclosures, including buildings, NFPA 68: 2002 gives the following equation:

$$A_v = C \frac{A_s}{P_{red}^{1/2}} \quad P_{red} \leq 0.1 \quad (13.40)$$

where A_s is the internal surface area of the enclosure (m^2), A_v is the vent area (m^2), P_{red} is the maximum internal overpresssure which can be withstood by the weakest structural element not intended to fail (bar), and C is a venting constant (bar$^{1/2}$).

For the venting constant, the code gives in metric units a value of 0.045 bar$^{1/2}$ for gases which have a fundamental burning velocity less than 1.3 times that of propane. The value of the constant for methane is 0.037 bar$^{1/2}$.

NFPA 68: 1994 gives an illustrative example of the use of Equation 13.40 involving a building of fairly complex shape.

13.11.3 Vent Design

It is preferable that the vent areas should be located on different walls of the building rather than concentrated all on one wall.

Vent panels for buildings should have the lowest practicable inertia. The release pressure for a vent panel should be as low as possible relative to the expected wind pressures. An appropriate value is usually in the range 1−1.5 kPa.

A vent panel should be restrained by a device such as a hinge or chain to prevent it flying off the building if it does open. Consideration should be given to the space into which the panel is to open.

Materials which shatter into pointed fragments should not be used for vent panels. In particular, asbestos cement-type board is unsuitable.

It may be necessary to provide railings along the edge of the floor near a vent panel to prevent people accidentally knocking the panel open and falling out.

13.12. VENTING OF REACTORS

The reactors which are considered here are liquid-phase reactors, essentially those capable of an exothermic reaction runaway. It has long been the practice to protect such reactors against overpressure by the use of some form of relief such as a bursting disk or pressure relief valve.

In this section, an account is given of overview about methods of sizing the vent for a reactor.

Table 13.3 gives the methods of sizing the vent for a reactor. Methods have progressed from early empirical correlations to methods based on various types of model, both simplified analytical models and full unsteady-state models embodied in computer codes.

One point is worth emphasizing at the outset. This is that in reactor venting it is the volumetric flow which is important.

Vent sizing of reactors is a complex and specialist matter. The account given here is no more than an overview.

An alternative approach to the vent sizing of liquid phase reactors is scale-up from experimental tests. Such extrapolation is valid, however, only if similarity exists between the test and full-scale systems.

Some of the conditions which must be met before similarity can be assumed are listed by Duxbury (1976). Thus, for polymerization reactions, he suggests that the basis for scaling up, though not for scaling down, might be a constant ratio of vent area to reactor volume. This accords quite well with the FIA method in which vent area increases with reactor volume to the power 0.92.

The flow entering the vent may be a homogeneous vapor−liquid equilibrium mixture or, if vapor disengagement occurs, a vapor. The size of vent required is different for the two cases.

When the reactor pressure is reduced due to venting, the liquid becomes superheated and vapor bubbles are formed, causing liquid swell. Unless vapor disengagement occurs, a two-phase mixture will enter the vent.

In certain cases, as discussed by Duxbury (1976), containment of a possible runaway reaction may be an attractive alternative to venting. This may be so, for example, if the peak pressure attainable is close to the working pressure or if it is less than the minimum design pressure for a vessel of that size or if there are special venting problems such as blockage of the vent lines.

13.13. VENTING OF REACTORS AND VESSELS: DIERS

It was recognized toward the end of the 1970s that understanding of reactor relief was deficient and that methods of vent sizing were inadequate and a major international cooperative project, Design Institute for Emergency Relief Systems (DIERS), was initiated to remedy this. Overviews of the project have been given by Fisher (1985, 1991).

The work is described in the *DIERS Project Manual*, which gives an overview and contains chapters on the following aspects: vapor disengagement; relief system flow; large scale tests; high viscosity flashing flow; mechanical

TABLE 13.3 Methods of Sizing the Vent for a Reactor

Method	Definition	Correlation
Factory Insurance Association (FIA)	The method is an empirical one in which vent area is plotted as a function of reactor volume and reaction class.	$$A = KV^{0.92}, \qquad (13.41)$$ where A is the vent area (in^2), V is the reactor capacity (US gal), and K is a costant.
Monsanto correlation (Howard, 1973)	This method is one of those included in a review given by Fauske (1984a). The correlation is used for sizing vents for runaway phenol–formaldehyde reactions.	$$D = 0.3V^{1/2}, \qquad (13.42)$$ where D is the diameter of the vent (in) and V is the volume of the reaction mass (US gal).
All vapor venting	This involves calculating the maximum allowable liquid temperature, which is related by the vapor pressure to the maximum allowable pressure, and then determining the rate of reaction and of heat generation at this temperature, the rate of vapor evolution and hence the flow of vapor to be generated.	$$A = W/Y \qquad (13.43)$$ with $$W = \frac{r(-\Delta H_\tau)}{\Delta H_V}, \qquad (13.44)$$ where A is the vent area (m^2), G is the mass flux (kg/m^{2}s), ΔH_r is the heat of reaction (kJ/kg), ΔH_v is the latent heat of vaporization (kJ/kg), r is the rate of reaction (kg/m^{3}s), V is the volume of the reaction mass (m^3), and W is the mass flow (kg/s).
All liquid venting	Some cases the fluid is vented as a liquid or a vapor–liquid mixture. An alternative approach, therefore, is to size the vent on the basis of discharging the whole reaction mass as a liquid. The required flow is calculated from the rate of reaction and of heat generation at the maximum liquid temperature and the time required to heat the reaction mass to this temperature.	$$W = m_0/\Delta t, \qquad (13.45)$$ with $$\Delta t = \frac{m_0 C \, \Delta T}{r(\Delta H_r)V}, \qquad (13.46)$$ where C is the specific heat of the reaction mass (kJ/kg K), m_0 is the initial mass in the vessel, or the reaction mass (kg), Δt is the time interval to the maximum allowable pressure (s), and ΔT is the temperature difference at equilibrium between the initial pressure and the maximum allowable pressure (K).
Boyle method (1967)	Is based on the pressure–time curve for the reactor. This may be obtained from the reaction rate or by experiment. Effectively, the method separates the calculation of the vent area from that of the vent flow. A modified Boyle method which takes account of two-phase flow has been described by Duxbury (1976). The two-phase flow calculations are performed using a fluid flow code.	A statement of the Boyle model has been given by Leung (1986b) as follows: $$W = \frac{m_0}{\Delta t_p}, \qquad (13.47)$$ where Δt_p is the emptying time (s). Leung also refers to the relation given by Fauske (1984b) for the emptying time $$\Delta t_p = \frac{\Delta T}{(\mathrm{d}T/\mathrm{d}t)_s}, \qquad (13.48)$$ where ΔT is the temperature rise corresponding to the overpressure ΔP (°C) and $(\mathrm{d}T/\mathrm{d}t)_s$ is the self-heat rate at the set pressure (°C/s).
Huff method	A method of vent sizing based on a much more comprehensive reactor model, including reaction kinetics and two-phase flow relations and embodied in a computer code, has been described by Huff (1973). Huff also gives a comparison of the different methods of calculating vent sizes for liquid-phase reactors.	A statement of the Huff model has been given by Leung (1986b) as follows: $$W = \frac{m_0}{\tau_t} - \frac{\beta}{2\tau_t^2}\left[\left(1 + \frac{4m_0\tau_t}{\beta}\right)^{1/2} - 1\right], \qquad (13.49)$$ where m is the reaction mass in the vessel (kg), β is a parameter, τ_t is the turnaround time for temperature (s), and the subscripts m and o denote peak and initial pressure (or temperature).

design, containment, and disposal; bench-scale apparatus and the SAFIRE computer code.

The DIERS project involved an extensive investigation of the behavior of the fluid venting through the vent line. It included a study of the vapor disengagement in the vessel and of the fluid flow through the vent. The work showed that there are different types of reaction systems and that the behavior of these systems is different. A major finding was that the flow at the vent tends to be a two-phase vapor–liquid flow. Methods of ERS design were developed for the different systems. These included simplified methods, a computer model, and a scale-up method. A bench-scale apparatus was designed capable of determining the principal design parameters.

The overall venting behavior of reaction system depends on a number of factors. They include (1) reaction regime, (2) vapor disengagement, (3) fluid viscosity, and (4) vent line length. It is necessary to allow for these factors in design, and in particular to take them into account in scale-up from bench scale to full scale.

The different types of reaction system, or reaction regimes, distinguished in the work are vapor-generating system, gas-generating system, and hybrid system.

A *vapor-generating system* is one in which the total pressure is the system vapor pressure. In other words, the system contains one or more high vapor pressure components. When the vent opens, tempering of the reaction occurs due to the cooling associated with the latent heat of vaporization. Such a system is also referred to as a high vapor pressure, or vapor pressure, system or a tempered system.

In a *gas-generating system*, by contrast, there is no high vapor pressure component, and the total pressure is that of the gaseous components. The reaction is not tempered and its course is essentially independent of the venting process. Such a system is also referred to as a gassy system or a non-tempered system.

The intermediate case is that of *a hybrid system*, and its behavior is intermediate between that of a pure vapor pressure system and a pure gassy system. A degree of tempering occurs.

The nature of the fluid which enters the vent depends on the flow regime in the vessel when the vent opens, and the vessel is depressurized. There are two flow regimes, the bubbly, or foaming, regime and the churn turbulent regime. In the former, there is little vapor disengagement and the vapor–liquid mixture entering the vent line is essentially the same as that in the vessel itself. In the latter regime, there is significant vapor disengagement.

13.14. VENTING OF REACTORS AND VESSELS: VENT FLOW

The account given in this section is confined to those models used for the venting of reactors and storage vessels.

Leung (1986b) has reviewed flow models for venting of a vapor pressure, or tempered, system. He considers frictionless flow and a short vent line ($L/D < 50$). The models include the non-flashing liquid flow, the homogeneous equilibrium model, the equilibrium rate model, the homogeneous non-equilibrium model, and the slip equilibrium model.

For a simple orifice in the side of a vessel, the outflow may be determined by treating the fluid as a non-flashing liquid.

For other cases, the homogeneous equilibrium model (HEM) is generally the most appropriate. It gives lower flows over the entire two-phase region and is to this extent conservative for ERS design. However, it is realistic rather than overly conservative and agrees well with experimental data. The original HEM model is somewhat complex and requires extensive thermodynamic data, but a more convenient version requiring only stagnation properties has been developed by Leung (1986a).

An alternative model is the equilibrium rate model (ERM) of Fauske (1985b), who has developed the model to deal with a number of different situations.

The equilibrium condition is closely approached for pipes longer than 0.1 m. Where the vent line is shorter than this, the homogeneous non-equilibrium model of Henry and Fauske (1971), the HFM model, may be used to estimate the flow.

A family of flow models for venting of vapor pressure systems has been described by Fauske (1984a,b; 1985a,b) in the context of DIERS, in particular the ERM. The base case is choked non-equilibrium flashing frictionless flow. For this case, Fauske gives

$$G \approx \frac{h_{fg}}{V_{fg}} \left(\frac{1}{NTC} \right)^{1/2} \qquad (13.50)$$

with the dimensionless quantity N:

$$N \approx \frac{h_{fg}^2}{2\,\Delta P \rho_1 K^2 v_{fg}^2 TC} + 10L \qquad (13.51)$$

where K is the discharge coefficient, L is the length of the vent pipe (m), ΔP is the total available pressure drop (Pa), T is the absolute temperature (K), ρ_1 is the density of the liquid (kg/m³), and N is a non-equilibrium parameter. For a sharp-edged orifice, K is 0.61 and C is the system heat capacity.

The family of flow relations given above is that of Fauske and the principal model is the ERM. An alternative model for equilibrium flow is the HEM.

This model has been formulated in convenient form by Leung (1986b), who gives the following relations. For the high and low quality regions, respectively, the equations are

$$\frac{G}{(P/v)^{1/2}} = \frac{0.66}{\omega^{0.39}}, \quad \omega < 4.0 \qquad (13.52a)$$

$$\frac{G}{(P/v)^{1/2}} = \frac{0.6055 + 0.1356 \ln \omega - 0.0131(\ln \omega)^2}{\omega^{0.3}}, \quad \omega \geq 4.0$$

$$(13.52b)$$

with

$$\omega = \frac{x v_{fg}}{v} + \frac{CTP}{v}\left(\frac{v_{fg}}{h_{fg}}\right)^2 \qquad (13.53)$$

$$v = v_f + x v_{fg} \qquad (13.54)$$

where P is the pressure (Pa), v is the specific volume of the mixture (m^3/kg), and ω is the critical flow scaling parameter.

For an all liquid inlet flow, Equation 13.52a and 13.52b may be approximated by

$$G \approx 0.9 \frac{h_{fg}}{v_{fg}}\left(\frac{1}{TC}\right)^{1/2} \qquad (13.55)$$

13.15. VENTING OF REACTORS AND VESSELS: VENT SIZING

The bench-scale apparatus used in some of the methods are described in detail in Lees Book full version, in this section an account is given an overview of methods developed in DIERS.

For a vapor pressure system, it is necessary to know the self-heat rate. One option for obtaining this is the use of the bench-scale apparatus. But it may be possible to determine it without resort to experiment.

Given the self-heat rate, the following simplified method has been described by Fauske (1984b) for the vent area:

$$A = \frac{V_\rho}{G \Delta t_v} \qquad (13.56)$$

with

$$\Delta t_v = \frac{\Delta TC}{q_s} \qquad (13.57)$$

where q_s is the energy release rate at the set pressure (kW/kg), Δt_v is the venting time (s), and ρ is the density of the reaction mass (kg/m^3).

Combining Equations 13.56 and 13.57 with Equation 13.58

$$G_{ERM} = \frac{\Delta P}{\Delta T}\left(\frac{T}{C}\right)^{1/2} \qquad (13.58)$$

gives

$$A = V_\rho (TC)^{-1/2} \frac{q_s}{\Delta P} \qquad (13.59)$$

Equations 13.58 and 13.59 are based on the assumption that the flow in the vent line is homogeneous equilibrium flow and that it is turbulent and intended for use where there is a modest overpressure in the range 10−30%, say 20%. For a frictionless vent line, Equation 13.59 predicts a vent area larger by a factor of less than 2 than that given by the integral model, assuming homogeneous vessel behavior and homogeneous equilibrium flow.

There are different vapor pressures systems, for instance high viscosity systems, Fauske monograph, revised Fauske monograph, extended method for vapor disengagement, Gassy systems, Gassy systems, Fauske nomograph, and hybrid systems. Some of these systems are briefly discussed below.

Fauske monograph, a generalized vent sizing nomograph based on Equation 13.59 was published in 1984 by Fauske (1984a) and revised by Fauske, Grolmes, and Clare in 1989. This initial nomograph was based on the assumptions of turbulent flow and a modest overpressure.

Extended method for vapor disengagement, as described by Grolmes et al. (1983), interpretation of the DIERS data has shown that during two-phase venting, relatively large deviations in the equilibrium between the vapor and the liquid may exist. The implications are discussed by Fauske (1989). It is virtually impossible to predict the vapor/liquid ratio entering the vent line.

For *gassy systems*, it is necessary to know the peak gas generation rate. It is usually necessary to use the bench-scale apparatus to obtain this.

A method for gassy systems, and the associated nomograph, has been given by Fauske et al. (1989). The relation given for the vent area is

$$A = \left(\frac{1}{2}\right)^{1/2}\frac{Q_g}{F}\left(\frac{\rho_1}{P - P_a}\right)^{1/2} \qquad (13.60)$$

where P is the absolute pressure (Pa) and P_a is the absolute atmospheric pressure (Pa). (In the authors' paper, the term $P - P_a$ is written as P where the latter is a gauge pressure.)

For *hybrid systems*, the situation is more complex. The behavior of the system depends on whether it is tempered or non-tempered. Untempered hybrid systems have generally to be treated as gassy systems.

A relation for the vent area at initiation of homogeneous two-phase venting, reflecting the worst case, is given in the Technology Summary (Equation 2.5) as follows:

$$A = \frac{(Q_g + Q_v)\rho_1(1 - \alpha)}{G} \qquad (13.61)$$

where Q_g is the peak volumetric gas evolution rate (m^3/s) and Q_v is the peak volumetric vapor evolution rate (m^3/s).

Tempered hybrid systems may be treated as tempered systems, provided it is assured that the system is really tempered and will continue to be so until the reaction is complete.

As already described, the bench-scale equipment may be used in three different ways: (1) to determine the required flow, (2) to determine the parameters for the calculation of flow, and (3) to effect direct scale-up.

A method for direct scale-up from the bench-scale apparatus, applicable only to vapor pressure systems, has been described by Fauske (1984c).

13.16. VENTING OF REACTORS AND VESSELS: LEUNG MODEL

Another model for vent sizing both of reactors and of storage vessels which has found wide acceptance has been given by Leung (1986b). The model gives relations for vent sizing. The equations for vent area contain the vent flow mass velocity, but the model may be regarded as independent of the vent flow correlation used. The author does, however, state his own homogeneous equilibrium model (Leung, 1986a).

The basic relations of the model are defined by the mass and energy balances in unsteady-state for the vessel reactor or storage vessel. These balances are show below:

$$\frac{d(\rho V)}{dt} = -W \qquad (13.62)$$

$$\frac{d(\rho V u)}{dt} = Q - Q\left(u_1 \frac{P}{\rho_1}\right) \qquad (13.63)$$

where P is the absolute pressure, Q is the heat generation or input rate, T is the absolute temperature, u is the specific internal energy, V is the volume of the vessel, W is the mass flow from the vessel, ρ is the density, and the subscript 1 denotes vent inlet. After mathematical treatment, the model obtained corresponds to Equation 13.64 given in Table 13.4.

13.17. VENTING OF REACTORS AND VESSELS: ICI SCHEME

Accounts of the approach to vent sizing for reactors used in ICI have been given by Duxbury and Wilday (1987, 1989, 1990). These provide a useful insight into the way in which one company has integrated the DIERS methods with other methods.

Set pressure and overpressure, an effective, or redefined, set pressure is defined as the pressure at which the relief device is known to be fully open. For a bursting disk, this will correspond to the nominal set pressure plus any tolerances or to the maximum specified bursting pressure. For a safety valve, the redefined set pressure will often be 10% above the nominal set pressure.

Vent flow, the methods used for estimation of vent flow described by Duxbury and Wilday (1990) are those applicable to vapor pressure systems. If venting is through a safety valve, it is usually sized using the ERM, but where the pipework is significant, a fluid flow code is used to check pressure drops upstream and downstream. If venting is through a bursting disk, the vent line flow correction factor provided for use with the ERM model is used for preliminary estimates, but since it is applicable only to a vent line with constant diameter and no static head changes, the final calculation is done using the fluid flow code.

Vapor pressure systems, for vapor pressure systems, use is made of Leung's method,

$$A = \frac{m_0 q}{G\left[\left(\frac{V}{m_0}\frac{h_{fg}}{v_{fg}}\right)^{1/2} + (C\,\Delta T)^{1/2}\right]^2} \qquad (13.72)$$

The assumption of homogeneous equilibrium flow is safe and generally realistic and the method gives a safe and usually acceptable vent size. An alternative and sometimes more convenient formulation of Equation 13.62,

$$A = \frac{m_0 q}{G\left[\left(\frac{V}{m_0}T_m\left(\frac{dP}{dT}\right)_m\right)^{1/2} + (C\,\Delta T)^{1/2}\right]^2} \qquad (13.73)$$

where subscript m denotes the mean value between the set pressure (as redefined) and the maximum allowable pressure.

The authors term this the Leung long-form equation (method J).

Where it is suitable, use is also made of Fauske's method for vapor pressure systems. This is the enhanced method which takes account of vapor disengagement. The Fauske method is used in conjunction with the Leung method. In the first instance it provides a check. If the two methods are significantly different, the calculations are reviewed. A significantly smaller answer from the Fauske method would not be accepted without rechecking the conditions for applicability and checking that the vent is large enough for all vapor venting.

The vent sizes from the two methods so obtained are then compared. The final vent size is taken as the smaller of the two.

Before applying either method, it is necessary to review its applicability.

The Fauske method is based on the ERM with a correction factor for vent line length. It is limited to overpressures in the range 10−30% and to turbulent flow.

TABLE 13.4 Reactor and Storage Vessels Models (13.43)

$$mC\frac{dT}{dt} = Q - Wh_{fg}\left(x_1\frac{v_f}{v_{fg}}\right), \tag{13.64}$$

where C is the specific heat at constant volume of the liquid, and the subscript f denotes liquid and fg liquid–vapor transition. Equation 13.64 is used to derive expressions for the vent area for the specific cases. The vent area is given by the relation

$$W = GA, \tag{13.65}$$

where A is the area of the vent and G is the mass velocity of the vent flow.

1. Reactor venting

Homogeneous venting: With $x_1 = x$ and $v_1 = v = V/m$, Equation 13.64 becomes

$$mC\frac{dt}{dt} = mq - GA\frac{V}{m}\frac{h_{fg}}{v_{fg}}, \tag{13.66}$$

On certain assumptions and utilizing Equations 13.62 and 13.65, Equation 13.66 can be integrated and with zero overpressure to yield $A_o = \frac{m_o q_s v_{fg}}{Gvh_{fg}}$, where A_o is the area of the vent for zero overpressure and m_o is the initial mass in the vessel.

All vapor or all liquid venting: With $v_i = v_g$ all vapour venting and $v_i = v_f$ all liquidventing, Equation 13.64 becomes

$$mC\frac{dT}{dt} = mq - GAv_i\frac{h_{fg}}{v_{fg}} \tag{13.67}$$

This equation can be integrated to give:

$$T_m - T_s = \frac{m_o q}{GAC}\left(1 - \frac{A}{A_o}\right) + \frac{v_i h_{fg}}{v_{fg}C}\ln\left(\frac{A}{A_o}\right), \tag{13.68}$$

with $A_o = \frac{m_o q_s v_{fg}}{Gv_i h_{fg}}$. For zero overpressure, Equation 13.68 reduces limit with $A = A_o$.

2. Storage vessel venting

Homogeneous venting: With $Q = Q_T$, where Q_T is the total heat input. Hence, Equation 13.64 becomes

$$mC\frac{dt}{dt} = Q_T - GA\frac{V}{m}\frac{h_{fg}}{v_{fg}}, \tag{13.69}$$

this equation can be integrated to give

$$T_m - T_s = \frac{Q_t}{GAC}\left[\left(\ln-\frac{m_o}{V}\frac{Q_T}{GA}\frac{V_{fg}}{h_{fg}}\right) - 1\right] + \frac{vh_{fg}}{m_o Cv_{fg}} \tag{13.70}$$

For zero overpressure $A_o = \frac{m_o q_s v_{fg}}{Gv_i h_{fg}}$

All vapor or all liquid venting: With $v_i = v_g$ all vapour venting and $v_i = v_f$ all liquid venting, Equation 13.64 becomes

$$mC\frac{dT}{dt} = Q_T - GAv_i\frac{h_{fg}}{v_{fg}}. \tag{13.71}$$

For all vapor venting, it is assumed that the design is for no overpressure to yield $A_o = \frac{Q_T v_{fg}}{Gv_g h_{fg}}$, and for liquid venting it is assumed that temperature turnaround occurs at a point where the vessel contains 10% of its initial inventory. Equation 13.71 can then be integrated to yield

$$A = \frac{Q_T}{G\left(v_f\frac{h_{fg}}{v_{fg}} + \frac{C\Delta T}{e}\right)}$$

In most cases, the Fauske method is used without taking vapor disengagement into account and thus with α_d set equal to unity.

For those cases where credit is to be taken for vapor disengagement, it is necessary to perform the following check. The Fauske method is regarded as potentially unsafe if early vapor disengagement occurs, which in this context means disengagement occurring before the pressure would otherwise have turned around during two-phase venting. The method is therefore used only if disengagement would have occurred after the turnaround. Duxbury and Wilday (1989) have derived the following criterion which must be satisfied for safe use of the method:

$$q < \frac{GAh_{fg}v_f^2}{Vv_{fg}(1-\alpha_d)^2} \qquad (13.74)$$

where α_d is the void fraction at disengagement.

Gassy systems, for gassy system, the principal method of vent sizing are based on maintaining the pressure constant. The vent is sized so that the two-phase volumetric flow at the maximum allowable pressure exceeds the peak volumetric gas generation rate at that pressure. For gassy systems, use may also be made of direct scale-up, which may well give a smaller vent area due to early loss of reactant. Alternatively, the vent may be sized by simulation using a numerical model. Again, for the reason just given, this may give a smaller vent.

A note of caution is sounded in that for gassy systems it may be unsafe to make the assumption of homogeneous two-phase venting. This assumption gives the maximum rate of emptying the reactor. If some disengagement occurs, liquid will remain in the reactor and the peak gas generation rate may be reached. A larger vent may then be required than for the relatively low gas generation rate occurring earlier in the relief process.

Hybrid systems, for hybrid systems, the appropriate method depends on whether the reaction is tempered. For tempered hybrid systems, it is first necessary to check that the reaction will remain tempered throughout. Reference is made to the treatment of such systems by Leung and Fauske (1987). Untempered hybrid systems usually have to be treated as gassy systems.

13.18. VENTING OF REACTORS: RELIEF DISPOSAL

The problem of reactor relief does not stop with the venting. There remains the question of the disposal of the vented materials, which is often not trivial. The engineering of relief disposal is therefore a topic in its own right. An account is given by Kneale (1984).

The legal requirement is that any venting should be to 'a safe place.' Discharge to atmosphere may meet this requirement, but such discharge is increasingly constrained.

The options for disposal are (1) discharge and dispersion, (2) destruction, and (3) containment.

A full definition of the relief requirements is as important for the design of the disposal system as for that of the relief device itself. Cases which should be considered include cooling failure during normal reaction, runaway of reaction for other reasons, explosion due to runaway or thermal instability, fire, overpressure by gas, and overfilling by liquid.

The relieving process may be violent, and all equipment connected to the relief device, however remotely, should be designed for the hydraulic and mechanical forces released. This includes not only the inlet and outlet lines to the device but vessel and line supports, foundations and foundation bolts, and building structures.

Some common causes of deficiencies in relief disposal systems include failure to allow for (1) the presence of another phase, (2) the presence of fine particles, and (3) variation in emission with time.

It should be appreciated, however, that emergency discharge from a reactor is a transient process and that toward the end of this process the discharge will lose momentum and will tend to slump. This stage may well be the most hazardous, but also the most difficult to model.

Gas relieved is often sent to flare. The efficiency of destruction of combustible gas in a flare can be high; studies have shown combustion efficiencies well in excess of 99% (Davis, 1983). An alternative means of dealing with gas is scrubbing. A major factor in the choice of system is whether the scrubber has the function of handling a normal process gas stream or of treating the emergency vent gas only. In the latter case, the options are to run with liquid constantly circulating or to start up automatically when venting starts. The cost of continuous circulation can be high, but it is not usual to rely on automatic start-up. Problems which can occur with scrubbers are boiling due to the high heat load imposed by the vent gas and carryover of spray.

Another option for handling the vented material is containment in a separate containment vessel or catch-tank. If the vent stream contains liquid or solid, this is the appropriate design. Systems vary widely in complexity depending on the duty.

Apart from the cost, some problems which arise with containment are (1) reaction forces, (2) undetected leakage, (3) testing, and (4) plugging of lines.

The reaction forces affect the containment vessel as they do the whole system. A leak occurring undetected across the relief system may fill the containment. Testing of the system requires careful thought.

Studies have also been made on particular configurations. The forces on a system consisting of a reactor and a containment vessel separated by a bursting disk have been analyzed by Porter (1982), and the interactive effects of bursting disks in a relief manifold by Beveridge and Jones (1984).

In particular, more guidance has become available on knockout and catchment facilities, for instance, treatments of the design of knockout drums and catchtanks are given in API RP 521: 1990, by the BPF (1979) and by Grossel (1986, 1990). API RP 521 gives a method for the sizing of a knockout drum which has been widely used. The basis is the provision of a cross-sectional area sufficient to ensure that the liquid droplets are not entrained in the vapor.

13.19. VENTING OF STORAGE VESSELS

It is necessary to protect storage vessels against overpressure and this is done by fitting vents. Principal sources of overpressure are operational deviations and fire.

Fire is not the sole potential source of heat input to a storage vessel. Many chemicals held in storage have the potential to undergo exothermic reaction. Any external heat source, including fire, may act as an initiator. Addition of water may be another. This effect should be taken account of in considering the venting requirements for storage.

13.19.1 Fire Behavior of Atmospheric Storage Tanks

Work on the fire behavior and on vent sizing for storage vessels has been described by Fauske et al. (1986) and Fauske (1987).

Vent sizing for atmospheric storage tanks has traditionally been based on all vapor flow. The aim of the study was to check whether this approach is adequate. The implication of the need to consider two-phase flow would be a large increase in vent size. As a first approximation, the increase in vent area required would be of the order of the square root of the ratio of the vapor and liquid densities $(\rho_g/\rho_l)^{1/2}$.

This concern was addressed in an extension of the DIERS project. It was concluded that liquid swell in an externally heated storage tank is essentially limited to the boiling two-phase boundary layer and that for a non-foaming liquid all vapor venting would appear adequate, provided the tank is not completely filled with liquid.

A criterion for the onset of liquid entrainment in a storage vessel exposed to fire has been given by Epstein et al. (1989). They consider a hemispherical surface centered on a vent at the center of the top of a vertical cylindrical vessel.

Further relationships for the vent area for storage vessels with non-foamy liquid have been given by Fauske et al. (1989). For an atmospheric storage tank,

$$A = \frac{Q_v}{3Fh_{fg}[\sigma g(\rho_1 - \rho_g)]^{1/4}\rho_g^{1/2}} \qquad (13.75)$$

where F is the flow reduction factor, Q_v is the energy release rate, g is the acceleration due to gravity, ρ_1 is the density of the liquid, σ is the surface tension of the liquid, and h_{fg} is the latent heat of vaporization. This equation is based on all vapor flow with no entrained liquid.

For a storage vessel capable of withstanding substantially higher pressure,

$$A = \frac{Q_v(R_m T)^{1/2}}{0.62Fh_{fg}P_s} \qquad (13.76)$$

where P_s is the absolute set pressure, R_m is the mass basis gas constant, and T is the absolute temperature. This equation is based on all vapor critical flow.

13.20. EXPLOSIVE SHOCK IN AIR

One of the main effects of an explosion is the creation of a shock wave or blast wave. This blast wave generates overpressures which may injure people and damage equipment and buildings.

The situation of prime interest here is that of a chemical explosion at the ground surface and it is this which is mainly considered. The description given of the explosion is based, unless otherwise stated, on an explosion of high explosive such as TNT.

The blast wave, an explosion in air is accompanied by a very rapid rise in pressure and by the formation of a shock wave. An account of the phenomena involved is given by Glasstone (1962).

The shape of the pressure profile near the center depends on the type of explosion involved. The initial shape differs for explosions of high explosives, nuclear weapons, and flammable vapor clouds.

The initial pressure profile for nuclear explosions is probably the most readily defined. The pressure at the edge of the fireball is approximately twice that at the center.

With detonations, the shock wave travels outward with the higher pressure parts moving at higher velocities. After it has traveled some distance, the shock wave reaches a constant limiting velocity which is greater than the velocity of sound in the air, or in the unburnt gas in the case of a vapor cloud. The shock wave has a profile in which the pressure rises sharply to a peak value and then gradually tails off. As the shock wave travels outward, the peak pressure at the shock front falls.

At some distance from the explosion center, the region of positive pressure, or overpressure, in the shock wave is followed by a region of negative pressure or underpressure. The underpressure is quite weak and does not exceed about 4 psi.

An idealized representation of the blast wave is shown in Figure 13.8. This shows the pressure pulse as a function of distance from the explosion center with time as the parameter. The shock wave reaches points A, B, C, and D at times 1, 2, 3, and 4, respectively, and at these times its pressure profile is as illustrated. The shape of the curve at point A is not shown, since it depends on the type of explosion. As the wave moves outward, however, the influence of the nature of the explosion declines and the wave establishes a profile which is common to all types of explosion. The curves at points B−D and times 2−4 show the decrease in peak overpressure. The curve at point D and time 4 shows both positive and negative pressures.

The variation of overpressure with time at such a point is illustrated in Figure 13.9(a) and (b). Important parameters are the peak overpressure $p°$, the arrival time t_a, the duration time t_d, which is the duration of the positive phase, and the decay parameter a, which defines the shape of the decay curve in the positive phase. For a deflagration, the rising pressure is gradual and for a detonation it is abrupt.

The peak overpressure $p°$ is more correctly described as the peak side-on overpressure or peak incident overpressure, in order to distinguish it from other peak overpressures such as the peak reflected overpressure described below. The peak side-on overpressure is the peak overpressure occurring at the side of a structure being passed by the blast wave.

There are several equations which are used to describe the positive phase of the overpressure decay curve of

Figure 13.9. A widely used one is the modified Friedlander equation (Kinney, 1962)

$$p = p°(1 - t/t_d\exp(-\alpha t/t_d)) \tag{13.77}$$

where p is overpressure, $p°$ is the peak overpressure, t is the time, t_d is the duration time, and α is the decay parameter. Another common equation applicable at or below about 10 psi is Equation 13.77 with the decay parameter set equal to unity. Information on the decay of overpressure is given by Glasstone (1962, p. 124).

Other important properties of the blast wave, described by Glasstone (1962), are the shock velocity, the particle velocity, or peak wind velocity, behind the shock wave, the peak dynamic pressure and the peak reflected overpressure. The shock velocity is given by

$$U = c_o\left(1 + \frac{\gamma+1}{2\gamma}\frac{p°}{p_a}\right)^{1/2} \tag{13.78a}$$

where c_o is the velocity of sound in air, p_a is the absolute ambient pressure (ahead of the shock front), U is the shock velocity, and γ is the ratio of specific heats of air. For air $\gamma = 1.4$ and hence

$$U = c_o\left(1 + \frac{6p°}{7p_a}\right)^{1/2} \tag{13.78b}$$

The particle velocity is given by

$$u = \frac{c_o p°}{\gamma p_a}\left(1 + \frac{\gamma+1}{2\gamma}\frac{p°}{p_a}\right)^{-1/2} \tag{13.79a}$$

where u is the particle velocity. For $\gamma = 1.4$

$$u = \frac{5c_o p°}{7p_a} + \left(1 + \frac{6p°}{7p_a}\right)^{-1/2} \tag{13.79b}$$

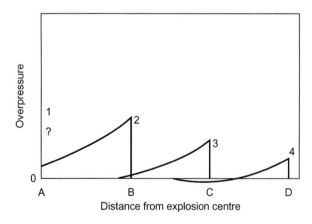

FIGURE 13.8 Idealized representation of development of detonation blast waves.

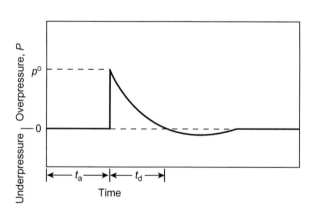

FIGURE 13.9 Variation of overpressure from a blastwave with time at a fixed point.

The peak dynamic pressure q° is defined as

$$q^\circ = \frac{1}{2}\rho u^2 \tag{13.80}$$

where q° is the peak dynamic pressure and ρ is the density of air (behind the shock front). This can be shown to be

$$q^\circ = \frac{p^{\circ 2}}{2\gamma p_a + (\gamma - 1)p^\circ} \tag{13.81a}$$

For $\gamma = 1.4$

$$q^\circ = \frac{5}{2}\frac{p^{\circ 2}}{7p_a + p^\circ} \tag{13.81b}$$

The peak reflected overpressure p_r°, which occurs if the blast wave strikes a flat surface at normal incidence, is

$$p_r^\circ = 2p^\circ + (\gamma + 1)q^\circ \tag{13.82a}$$

For $\gamma = 1.4$

$$p_r^\circ = 2p^\circ + 2.4q^\circ \tag{13.82b}$$

Then, substituting for q° from Equation 13.81b

$$p_r^\circ = 2p^\circ\left(\frac{7p_a + 4p^\circ}{7p_a + p^\circ}\right) \tag{13.83}$$

As stated earlier, the distinction between the peak incident overpressure, or simply the peak overpressure, p° and the peak reflected overpressure p_r° is important, and it should always be made clear which overpressure is being referred to.

The peak dynamic pressure is less than the peak overpressure at low values of these two pressures, but greater at higher values. Equation 13.81b shows that the cross-over point is about 70 psi (4.83 bar).

The peak reflected overpressure is shown by Equation 13.82b to approach a value of twice the peak overpressure for weak shocks in which the peak dynamic pressure is negligible, but to approach a value of 8 times the peak overpressure for strong shocks in which the peak dynamic pressure is dominant.

This maximum factor of 8 by which the peak reflected pressure exceeds the peak overpressure is frequently quoted. It depends, however, on the validity of the assumptions underlying Equation 13.82b.

13.21. CONDENSED PHASE EXPLOSIONS

An explosion of a high explosive is referred to as a condensed phase explosion. The explosive which serves as the reference standard is TNT.

In this section, consideration is given primarily to overview of the characteristics of the explosion itself. First is given one incident involving high explosive.

Incidents involving munitions and commercial explosives include the explosion of the munition ship Mont Blanc in the harbor at Halifax, Nova Scotia, in 1917, in which 1963 people were killed (Case History A3). Within a radius of about three quarters of a mile, destruction was almost complete. It was estimated that 95% of the glass in the city was broken.

13.21.1 TNT Explosion Model

The blast characteristics of a TNT explosion are important in their own right and are also sometimes used in modeling other types of explosion. There is much more information available on explosions of explosives, particularly TNT, than for other cases.

In principle, the blast characteristics of a TNT explosion depend on the mass and shape of the charge and also on the point(s) of ignition.

For a TNT explosion, the two standard types of reference data are those for explosions from a point source and from a spherical charge. The blast parameters for the two explosions are slightly different. For overpressure, for example, the initial effect of a finite charge size is to reduce the overpressure very close to the charge relative to that from a point source. This effect holds for a distance of some five charge diameters. At greater distances, however, the overpressure from a spherical charge is greater, being approximately equivalent to that from a point source with one-third greater energy release.

Other features which affect the blast characteristics include the location relative to sea level and the meteorological conditions.

The two reference standards for data on a TNT explosion are a free air burst (where the blast wave is not affected by interaction with the ground), which has spherical symmetry, and a surface burst, which has hemispherical symmetry.

Kinney and Graham have given their values of the peak overpressure, impulse and duration time from an explosion of 1 kg of TNT in the form of network equations. In SI units for the scaled peak overpressure p_s

$$p_s = \frac{808[1 + (z/4.50)^2]}{[1+(z/0.048)^2]^{1/2}[1+(z/0.32)^2]^{1/2}[1+(z/1.35)^2]^{1/2}} \tag{13.84}$$

for the impulse

$$i_p = \frac{0.067[1+(z/0.23)4]^{1/2}}{z^2[1+(z/1.55)^3]^{1/2}} \tag{13.85}$$

and for the scaled duration time

$$\frac{t_d}{W^{1/3}} = \frac{980[1+(z/0.54)^{10}]}{[1+(z/0.02)^3][1+(z+0.74)^6][1+(z/6.9)^2]^{1/2}} \tag{13.86}$$

where p_s is the scaled peak overpressure, i_p is the impulse (bar ms), t_d is the duration time (ms), W is the mass of explosive (kg), and z is the scaled distance (m/kg$^{1/3}$).

It has already been mentioned that the blast parameters for a surface burst (hemispherical symmetry) can be obtained from those for a free-air burst (spherical symmetry) by using in the relations for the latter a charge weight suitably modified to account for the concentration of energy into the hemisphere rather than the sphere. The presumption is that this yield ratio, or symmetry factor, should be 2, and it is this value which is commonly used.

There are different correlation basis in TNT explosion model, for instance methods of Baker et al. and TNT equivalent model. These works are explained in detail in the Lees Book full version.

13.22. VESSEL BURST EXPLOSIONS

Turning now to types of explosion more typical of process plant, consideration is given first to vessel bursts in which the energy of explosion derives from the pressure in the vessel and the explosion is essentially physical rather than chemical. Boyer et al. (1958) performed tests on the bursting of small gas-filled glass spheres.

This work brings out several relevant empirical features.

The work of Pittman (1972a,b), which involved high pressures, demonstrates the need to take into account in estimating the energy of explosion for such cases the non-ideal behavior of the gas. The differences between the behavior of the blast wave from a condensed phase explosion and that from a vessel burst are illustrated in the work of Esparza and Baker (1977), who found that for the latter the blast wave tended to have a lower overpressure, a longer positive phase duration, a larger negative impulse, and a stronger second shock.

The explosion energy in a vessel burst is a function of the initial pressure, which in turn depends on the failure scenario. The higher the pressure, the larger the explosion.

Three cases may be distinguished: (1) overpressure, (2) mechanical failure, and (3) fire engulfment.

If the vessel fails because it is exposed to a high operating pressure and because the pressure relief has failed, the vessel will burst at a pressure which is some factor of the design pressure, typically for mild steel a factor of 4. This is the worst case.

If a mechanical failure of the vessel occurs, which may be due to a metallurgical defect, corrosion or impact, the burst pressure will be the operating pressure.

If the vessel fails due to fire engulfment, but with the pressure relief operating, the pressure at burst will be the accumulation pressure. This is the typical scenario of a BLEVE.

The estimation of the energy of the explosion in a vessel burst is discussed below, but it is appropriate first to consider briefly the distribution of this energy.

In principle, the energy available will be partitioned as energy in (1) vessel expansion, (2) vessel rupture, (3) blast, and (4) fragments.

There have also been studies involving the numerical and analytical modeling of vessel bursts. In some cases, the main interest of the work has been with other types of explosion for which the bursting of a gas-filled sphere serves as a limiting case.

The study most relevant here is that of Baker et al. (1975), who conducted a parametric exploration of vessel bursts and developed a method of estimating the blast parameters analogous to that for a condensed phase explosion.

A method for the estimation of the blast parameters for the bursting of a gas-filled pressure vessel is now described. The method is that of Baker et al. (1975) and relates to a free-air burst of a massless spherical vessel containing an ideal gas.

This method is oriented to the estimation of the blast parameters in the far field, and its accuracy in the near field is limited. There are two checks which should be made with regard to the latter.

The first check is on the value of the scaled distance $\overline{R}$. If this is less than 2, use should be made of a modification to the method appropriate for the near field, which is described below. The other check is on the value of the peak side-on overpressure p_s, discussion of which is deferred.

The near field modification is as follows. First an effective radius for the equivalent hemisphere formed by the initial gas volume is calculated:

$$r_o = \left(\frac{3V_1}{2\pi} \right)^{3/4} \tag{13.87}$$

where r_o is the effective radius of the equivalent hemisphere (m) and V_1 is the volume of the gasified vessel (m^3).

$$\frac{P_1}{P_0} = (\overline{P}_{so} + 1) \left(1 - \frac{(\gamma_1 - 1)(a_o/a_1)\overline{P}_{so}}{\{2\gamma_o[2\gamma_o + (\gamma_o + 1)]\overline{P}_{so}\}^{1/2}} \right)^{-2\gamma_1/(\gamma_1 - 1)} \tag{13.88}$$

where a is the velocity of sound (m/s), p_o is the absolute pressure of the ambient air (Pa), p_1 is the absolute initial pressure of the gas (Pa), $\overline{P}_{so}$ is the scaled peak side-on overpressure at $\overline{R}$, γ is the ratio of the specific heats, and subscripts o and 1 denote ambient air and compressed gas in the vessel, respectively.

13.23. VAPOR CLOUD EXPLOSIONS

When a cloud of flammable vapor burns, the combustion may give rise to an overpressure or it may not. If there is no overpressure, the event is a vapor cloud fire, or flash fire, and if there is overpressure, it is a vapor cloud explosion (VCE).

A feature of a vapor cloud is that it may drift some distance from the point where the leak has occurred and may thus threaten a considerable area.

The relative importance of the VCE hazard has grown in recent years. Whereas in the early days of the chemical industry, the largest disasters tended to be those caused by explosives, including ammonium nitrate, many of the largest disasters are now due to VCEs.

There are many gaps in the state of knowledge on VCEs and there is no completely satisfactory theoretical model, although much progress has been made in modeling. The practical approaches tend to combine theoretical and empirical inputs.

In this section, consideration is confined to the characteristics of the explosion itself. The effects of the explosion are considered below.

There are a number of compilations of major vapor cloud explosion incidents (VCEs). Marshall (1987) gives details of some 11 such compilations. The principal lists are those of Strehlow (1973), Davenport (1977, 1983), Gugan (1979), the ACMH (Harvey, 1979), Wiekema (1983a,b, 1984), and Lenoir and Davenport (1992). The account by Davenport (1977) also lists incidents in which a vapor cloud formed but did not explode.

13.23.1 Empirical Features

Some observations and generalizations can be made by analyzing explosion incidents.

A compilation of VCE incidents and corresponding statistical data has been presented by Davenport (1977). These data relate to 43 incidents where overpressures were created, of which 32 were in industrial plant, 8 in transport and 3 elsewhere. Of those in industrial plant, 8 were in refineries and 24 in the petrochemical industries.

Some questions which are capable of resolution by an empirical approach are frequency of release, quantity of material released, fraction of material vaporized, probability of ignition of distance traveled by cloud before ignition, time delay before ignition of cloud, probability of explosion rather than fire, existence of a threshold quantity for explosion, and efficiency of explosion.

An early but fairly comprehensive treatment of the empirical features of vapor cloud explosions was given by Kletz (1977). He gives estimates of the frequency of such explosions as given in Table 13.5. A special pipeline is one on which special care can be taken, such as a short

TABLE 13.5 Estimated Frequencies of VCE

		Frequency (explosions/plant-year)[a]
(1)	Caused by failure of pressure vessel	10^{-5b}
	Pipeline	$10^{-5}-10^{-4}$
	Special pipeline	
	Normal pipeline	$10^{-4}-10^{-3}$
	Pump	10^{-2}
	Normal pump	
	Severe duty pump	10^{-1}
	Reciprocating compressor	10^{-1}
(2)	Caused by leak from Batch reactor	$10^{-2}-10^{-1}$
	Tanker-filling house	$10^{-2}-10^{-1}$

[a]Plant is defined as a major unit such as ethylene or aromatics plant or collection of smaller units—something smaller than a typical UK works or US plant.
[b]The estimated failure rate of a pressure vessel is 10^5/vessel-year.
Source: After Kletz (1977).

pipe without pumps or other intervening equipment between two vessels. A severe duty pump is one on a very hot or very cold liquid.

In open situations with few sources of ignition, the vapor cloud may drift much further. In the Port Hudson explosion (NTSB 1972 PAR−72−01), the cloud traveled 1500 ft in a long plume before igniting. In the flash fire at Austin (NTSB 1973 PAR−73−04), ignition did not occur until the cloud had traveled 2400 ft. Data on the distance traveled by the vapor cloud in 81 rail tank car spills have been given by G.B. James (1947−48), who states that 58% found a source of ignition within 50 ft and 76% within 100 ft and all which ignited did so within 300 ft.

A separate but related question is the time delay before ignition. Strehlow (1973) states that a delay as long as 15 min has been reported. The quantity of material which can accumulate in this time may be very large. The explosion can be a large one, however, even if the time delay is short. The delay before ignition at Flixborough was probably in the range 30−90 s. An estimate of 45 s has been widely quoted.

The question of whether there is a minimum size of cloud below which an explosion will not occur has been much discussed, and a threshold quantity of 10−151 has been suggested. It is argued by Kletz that there is no theoretical basis and no detailed survey to support the assumption that there is a threshold quantity for an explosion to occur.

Kletz' account of the empirical features of VCEs is intended primarily as a guide to plant design. It may be regarded as an integrated treatment such that using the various empirical values given the result obtained is reasonable. The application of the values may be less valid if used as part of an eclectic approach.

13.23.2 Combustion in Vapor Clouds

The combustion of vapor clouds in the open air was already a subject of research before the four major explosions just mentioned, but their occurrence provided a further stimulus to work on what was then generally referred to as unconfined VCEs.

'Open-air' explosions usually occur with at least part of the vapor cloud in a congested region of a processing plant. Except in rare instances, these explosions are deflagrations. Deflagrations can be highly damaging.

In the early work, much attention centered on the question of detonation of the vapor cloud. Detonations are more damaging than deflagrations, but the distinction can often be expressed as detonations produce smaller fragments.

It is known that flame speeds obtained in vapor clouds are greater than the normal burning velocity. Strehlow quotes a burning velocity of 9 m/s for stoichiometric hydrogen—oxygen mixtures as measured on a bunsen burner and velocities of 68—120 m/s obtained using balloons. These velocities are typical of a flash fire rather than an explosion. The acceleration of the flame in a vapor cloud and the influence of factors such as apertures and turbulence have been the subject of many experiments. A major focus is to determine what degree of turbulation induction will accelerate flame speed to a deflagration let alone a detonation.

Some of the problems associated with detonation are which fuels have a high reactivity leading to detonations, existence of detonation limits, occurrence of unconfined detonation, direct initiation of detonation, characteristics of flame propagation, and transition from deflagration to detonation.

Only a few fuels are considered highly reactive and can detonate lacking confinement. These include ethylene, acetylene, ethylene oxide, and methyl nitrate.

It has been shown, however, that detonation limits are in fact distinct from flammability limits, but they are more complex and more difficult to measure. Information on detonation limits for spherical detonation is especially sparse and the limits are a function of the strength of the ignition source.

The violence of some VCEs suggests that they might well be detonations, but the occurrence of detonation as opposed to deflagration in an unconfined vapor cloud was for a long time considered to be highly unlikely, even impossible. However, in an account of the Port Hudson explosion Strehlow (1973) stated, 'The Port Hudson explosion is a proven example of an accidental vapor—cloud detonation.' And again, 'There is no question but that detonation has occurred in accidental explosions, or that

detonation can be initiated without delay in an unconfined cloud using a sufficiently strong ignition source.' He also stated, 'More commonly, the vapor cloud simply deflagrates. However, deflagration velocities are commonly observed to be quite high and extensive blast damage can occur even for this type of vapor—cloud combustion, particularly if the cloud contains a sufficiently large volume of combustible mixture at the time of ignition.'

It is possible that a deflagration may turn into a detonation. The dominant view has been that this is difficult to envisage in a truly unconfined situation, but the shock waves to promote the transition might arise from interaction with structures.

Although much work has been done, and continues to be done, on the detonation of vapor clouds, the focus of attention has gradually shifted to the effect of confinement on the flame and, in particular, on the conditions which promote its acceleration to very high flame speeds.

Another feature which is better appreciated is the importance of the initial combustion zone. This may be envisaged as a spherical zone in which the flame front expands until it breaks through the envelope of the cloud. Once this happens, relief occurs and the pressure falls. It follows from this that the overpressure developed depends on features such as cloud height and vertical confinement.

Traditionally, vapor clouds have been treated as hemispherical, but in practice many vapor clouds are pancake shaped. Increasingly, this feature is being taken into account in studies of vapor cloud combustion.

It will be apparent from the foregoing that VCEs are not yet well understood. Experimental studies have therefore been undertaken to improve understanding, while design approaches have been developed based on correlation of empirical features.

13.23.3 Experimental Studies

There is now a considerable body of experimental work on VCEs. Reviews include those of Lee (1983), Pikaar (1985), Bull (1992), and the CCPS (1994/15).

The review by Pikaar (1985) covers (1) release phenomena, (2) vapor cloud generation and dispersion, (3) vapor cloud combustion, and (4) interactions. Vapor cloud combustion is divided into (1) combustion of premixed clouds and (2) combustion of fuel-rich clouds. The combustion of premixed clouds covers (1) combustion in open terrain, (2) influence of partial confinement and obstacles, and (3) pressure waves. The interactions considered are (1) the effect of deliberate dispersion by water sprays and (2) the effective lower flammability limit following dispersion.

In his review, Bull (1992) covers large scale work on both vapor clouds in the open and vapor mixtures in large

TABLE 13.6 Clasification of Large Scale Work on Both Vapor Clouds in the Open and Vapor Mixtures (Bull, 1992)

State		Phase	
		Single	Aerosol +hybrid
'Unconfined'	Intially quiescent	A	B
	Initially turbulent	C	D
With congestion	Intially quiescent	E	(F)
	Initially turbulent	G	(H)
With congestion	Intially quiescent	I	J
	Initially turbulent	K	(L)
With congestion and confinement	Intially quiescent	M	(N)
	Initially turbulent	O	(P)

enclosures such as modules and gives the classification given in Table 13.6.

No large scale work was found in the classes shown in parentheses.

The review by the CCPS (1994/15) is under the following headings: (1) unconfined deflagration under controlled conditions, (2) unconfined deflagration under uncontrolled conditions, and (3) partially confined deflagration have been found a long list of the experimental studies for the category A (combustion of dispersed vapor clouds in unconfined state) for instance: Work on fuel−air explosion, early field trials of combustion in unconfined vapor clouds, early work on detonation in unconfined vapor clouds, combustion in unconfined vapor clouds: with or without field trials, combustion in vapor clouds with obstructions: closely controlled tests or field trials, combustion in partially confined vapor clouds without or with obstructions: work using tubes, parallel plates or using channels.

13.23.4 Vapor Cloud Explosion modeling

Although there are many gaps in the understanding of VCEs, considerable progress has been made in modeling such explosions.

Work on fundamental models has been described by a number of authors, including Taylor (1946), Williams (1965), Kuhl et al. (1973), Strehlow (1975, 1981), Munday (1975a,b, 1976a,b), Gugan (1979), Wiekema (1980), Baker et al. (1983), and van den Berg (1985).

Other work has been directed to the correlation of experimental results and to the development of semi-empirical models. Some of the approaches to modeling have been described by Munday (1976a). These are point source models (TNT equivalent model and self-similarity model); fuel-air cloud model; bursting vessel model; piston model (constant velocity piston model and accelerating piston model).

In the TNT equivalent model, the explosion is taken to be equivalent to that of a TNT explosion with the same energy of explosion. This model is therefore an empirical one, but it was for some time virtually the only practical model available. The TNT equivalent model has a single parameter, the mass of TNT. It can be made more flexible by the introduction of a second parameter, the height above ground zero at which the explosion occurs. The effect of increased height is to reduce the overpressures near the center.

In the self-similarity model, the blast parameters such as peak overpressure are correlated in terms of the ratio of radial distance to time. In its simplest form, the model gives a power law relation for the variation of peak overpressure with distance.

In the fuel−air cloud model, it is assumed that a detonation propagates through the fuel−air mixture without any expansion of the cloud. However, there is some expansion of a burning cloud during a deflagration. A shock wave with a high peak overpressure is produced at the cloud boundary. With combustion thus complete, subsequent decay of the shock wave is similar to that for the point source models.

The constant velocity piston approach has also been used by Wiekema (1980) at TNO. His model is a practical one and is quite widely used.

Van den Berg (1985) at TNO has developed this work to produce a multienergy (ME) model, which reflects the fact that the main energy release is from the confined parts of the cloud.

Finally, TNO have presented an empirical model for the damage circles caused by a VCE.

The estimation of the mass of fuel within the flammable range in the vapor cloud is not entirely straightforward. For a defined release scenario, the mass of fuel within the flammable range may be estimated using gas dispersion models. This has been reviewed by a CCPS book (Woodward, 1998).

Generally, dense gas dispersion models will be more appropriate. The typical dense model is a box or slab model in which the gas cloud is defined in terms of its radius and height. For an instantaneous release box model, the gas concentration within the cloud is assumed to be uniform. In such a model, the available mass of fuel remains constant, until at some time the concentration falls below the lower flammability limit.

Or for a continuous release, a slab model takes the variation of concentration with distance into account so that the available mass of fuel changes with distance.

The energy of explosion in a VCE is usually only a small fraction of the energy available as calculated from the heat of combustion. As stated there, the explosion efficiency, or yield factor, may be quoted on the basis of the total mass released or of the flammable mass. Although the latter is the more fundamental basis, the former is the more convenient and the more common.

13.23.5 TNT Equivalent Model

The TNT equivalent model has been widely used to model VCEs. The TNT equivalent model has been widely applied to VCEs, as described by van den Berg (1985). It is one of the types of model used in the CCPS methodology. It also has its critics, who regard it as obsolete. An example of this viewpoint is the account by Pasman and Wagner (1986).

There are, however, important features of a VCE which differentiate it from one of TNT. These include (1) the large volume of the cloud, (2) the lower overpressure at the explosion center, (3) the different initial shape of the blast wave, and (4) the longer duration time of the blast wave.

The overpressure at the explosion center of a vapor cloud is much less than at that of a TNT explosion. The report suggests that a theoretical upper limit might be calculated related to the pressure which would be achieved if the equivalent mass of TNT after detonation were confined within particular boundaries under adiabatic conditions. For a TNT explosion, the boundaries would be those of the TNT charge and the pressure developed would be about 0.5 million bar. For a VCE, the boundaries would be those of the cloud and the pressure developed would be about 8 bar. In both cases, the practical upper limit of overpressure would be some fraction of the theoretical maximum value, particularly for a vapor cloud. The report suggests that on the basis of the rather meagre data available, the practical upper limit is probably about 1 bar at the center and about 0.7 bar at the boundary of the cloud.

In a VCE, the shape of the initial blast wave is different from that in a TNT explosion. But it is frequently assumed that after the blast wave has traveled a certain distance, it becomes indistinguishable in form from the wave of a TNT explosion.

The apparent TNT equivalence of a VCE changes with distance. In the far field, the quantity of TNT required to obtain a fit for the overpressure tends to be appreciably higher than in the near field.

The blast wave from a VCE also differs from that of a TNT explosion in duration time. The duration time of a VCE is generally considered to be appreciably longer than that of the equivalent TNT explosion.

Thus, TNT gives a 'hard' explosion with high overpressure and short duration time, while a vapor cloud gives a 'soft' explosion with low overpressure and long duration time. The lower overpressure reduces the relative destructiveness, but the longer duration time increases it.

There is, however, little information on the duration time for VCEs. This is a serious deficiency, because a knowledge of duration time is needed for the design of works buildings.

Some of the various TNT equivalent models developed tend to emphasize a particular influence on or feature of VCEs. These include (1) the reactivity of the fuel, (2) the effect of cloud size, (3) the effect of obstructions, (4) the variation of TNT equivalence with distance, and (5) the peak overpressure at the edge of the cloud. The treatments of these factors are of value quite apart from their use in the particular models.

There are different approaches in instance: Two different approaches to the effect of obstructions are used in the methods of Exxon (n.d.) and Harris and Wickens (1989), the variation of TNT equivalence with distance is taken into account by Prugh (1987) by the use of a virtual distance method in which the virtual distance is a function of the mass of fuel involved in the explosion. The concept of virtual distance is also utilized by Harris and Wickens (1989). Detailed information is in the Lees Book full version.

All these approaches have something to offer, but there does not appear to be any single method which incorporates them all.

In addition to those just described, several other TNT equivalent models for VCEs have been developed, published or unpublished. They include those of Exxon (n.d.), Eichler and Napadensky (1977), Prugh (1987), the FMRC (1990), and IRI (1990). These are reviewed by the CCPS (1994/15).

Table 13.7 concludes three different kinds of TNT equivalent models (Brasie and Simpson model (1968), ACMH (Harvey, 1979) and Harris and Wickens model (1989)), which tend to differ in the methods used for (1) the mass of vapor participating in the explosion and (2) the yield factor of the explosion. Other differences relate to (3) the value used for the energy of explosion of TNT, and (4) the correlation used for the peak overpressure of TNT.

There are other different models; some of them have piston model approach, for instance an early model of a VCE was given in a classic paper by Taylor (1946). He considers small pressure disturbances, as in acoustics, so that the system equations can be linearized. The treatment which he describes is in terms of an expanding piston.

Another is VCE model given by Strehlow (1981) also follows the acoustic approach, but utilizes a constant velocity piston. It was shown by Stokes (1849) that a source of mass with mass flow $\overline{m}(t)$ generates a sound

TABLE 13.7 TNT Equivalent Models

TNT Equivalent Model	Definition	Correlation
Brasie and Simpson model (1968)	Brasie and Simpson noted that the values quoted for the energy of explosion of TNT varied between 1800 BTU/lb (4190 kJ/kg) and 2000 BTU/lb (4650 kJ/kg), and adopted the latter.	$$W_{TNT} = \alpha \frac{W \, \Delta H}{E_{TNT}}, \qquad (13.92)$$ where E_{TNT} is the energy of explosion of TNT (kJ/kg), ΔH_c is the heat of combustion of the hydrocarbon (kJ/kg), W is the mass of hydrocarbon (kg), W_{TNT} is the equivalent mass of TNT (kg), and α is the yield factor
ACMH (Harvey, 1979)	The *Second Report* of the ACMH (Harvey, 1979) discusses the problem of VCEs. In particular, it gives an equation for the cloud size, a suggested value for the explosion efficiency and an overpressure curve based on the TNT equivalent model.	The curve peak side-on overpressure for a vapor cloud explosion, explains the ACMH correlation. This curve utilizes a damage classification based on bomb damage to dwelling houses in the Second World War.
Harris and Wickens model (1989)	In the Harris and Wickens model, the mass of gas participating in the VCE is restricted to the confined and congested part of the plant, and that in the unconfined part of the cloud is neglected.	From the relation $$M_{TNT} = \alpha \frac{M_{gas} E_{gas}}{E_{TNT}}, \qquad (13.93)$$ the authors obtain $$M_{TNT} = 2M_{gas}, \qquad (13.94)$$ where E_{gas} is the heat of combustion of the hydrocarbon (MJ/kg), E_{TNT} is the energy of explosion of TNT (MJ/kg), M_{gas} is the mass of gas (in the confined/congested region) (te), M_{TNT} is the mass of TNT (te), and α is the yield factor.

wave in three dimensions which has an overpressure that is proportional to the mass addition acceleration $\bar{m}(t)$.

A number of workers have obtained approximate analytical solutions of the constant velocity piston case based on self-similarity. The model of Williams (1976a,b), also discussed by Anthony (1977), is illustrative of this approach.

In the model it is assumed that ignition occurs at a point source, that the flame front which develops travels out from the 'core' at a flame speed S and that the pressure waves produced by the flame generate a weak shock which travels ahead of the flame at a velocity $V(t)$ which varies with time. It is also assumed that the pressure and density in the 'shell' between the flame and shock fronts are constant.

13.23.6 TNO Model Principles

Three models of VCE have been developed by TNO. The first of these is the shock wave model described in the Yellow Book and by Wiekema (1980). The model is also known as the expanding piston or piston blast model. It allows the peak overpressure and the duration time of the explosion to be estimated.

The model is a composite one derived from separate models for deflagration and detonation in the cloud. Considering first the model for deflagration, it has been shown by Strehlow et al. (1973) that given the pressure–time profile it is possible to derive the energy of an explosion. Wiekema uses this result, inverting it to obtain the pressure–time profile from an arbitrary energy release.

The second TNO VCE model is the correlation model; this too is described in the Yellow Book. The model allows an estimate to be made of the radius of defined damage circles. It does not give explosion parameters such peak overpressure or duration time.

The model is based on a correlation for VCE incidents of damage effects outside the gas cloud. The correlation is based on the energy content E of the part of the cloud within the explosive range.

The third TNO VCE model is the ME model described by van den Berg (1985). This model allows the peak overpressure, peak dynamic pressure, and duration time to be estimated.

The starting point for this model is recognition of the role of partial confinement in VCEs. It is assumed in the model that the explosion in those parts of the cloud which are confined is of much higher strength than in those where it is unconfined. The method involves estimating the combustion energy available in the various parts of the cloud and assigning to each part an initial strength.

13.23.7 Hydrogen Explosions

Vapor clouds of hydrogen are somewhat unusual and need separate treatment. An account of hydrogen VCEs has been given by Bulkley and Jacobs (1966).

These authors describe a number of incidents in which hydrogen−air explosions have occurred. Of particular importance is an explosion which took place during the intentional release of hydrogen from Los Alamos Scientific Laboratory at Jackass Flats, Nevada, in 1964 and which has been described by Reider et al. (1965). The incident was recorded by high speed photography. The discharge rate had been at a peak of 430,000 lb/h but had been cut back to 125,000 lb/h when spontaneous ignition occurred. The resulting explosion was estimated to have caused a pressure wave of 0.5 psi at buildings less than 200 ft way. From the photographs it was estimated that the cloud of hydrogen−air mixture which took part in the explosion was 30 ft in diameter and 150 ft high and contained 200 lb of hydrogen and that the flame speed reached 100 ft/s. The flame speed was thus considerably above the normal burning velocity but had not reached detonation velocity.

13.23.8 Explosion Yield Limit

As already stated, in VCE work, the fraction of the energy of combustion which is converted to blast energy is frequently expressed as a TNT equivalent.

The generally quoted limit for this fraction is about 40%. This is the value given, for example, by the CCPS (1994/15), which also refers to numerical simulation work in which the highest fraction obtained was 38%, as described above.

In his work on VCEs, Gugan (1979) gave equations and listed incidents for which the TNT equivalent was cited as above 50%. Ale and Bruning (1980) took issue with this on the basis that it violated the law of thermodynamics governing the conversion of energy into mechanical work. The matter was further considered by Phillips (1981).

The difficulty appears to lie in the use of the TNT equivalent. In an investigation of an explosion incident, the equivalent mass of TNT is obtained by examining the damage effects. It does not follow, however, that for a given level of damage, the explosion energies in a TNT explosion and a VCE are the same. A TNT explosion involves a very high drop in overpressure in the near field and a shorter impulse. In other words, the energy released

in a TNT explosion may be less effective in causing damage than that in a VCE.

Further, there are uncertainties in the peak overpressure for a TNT explosion and in the correlation of damage effect with peak overpressure. Phillips states that differences in blast correlations can introduce a factor of as much as an order of magnitude.

13.24. BOILING LIQUID EXPANDING VAPOR EXPLOSIONS

Another of the most serious hazards in the process industries is the BLEVE. Generally, this occurs when a pressure vessel containing a flammable liquid is exposed to fire so that the metal loses strength and ruptures. In this section, an account is given of overview about BLEVE definition, empirical features, and finally modeling of BLEVEs.

Accounts of BLEVEs have been given by Kletz (1977), Reid (1979), A.F. Roberts (1981/82, 1982), Manas (1984), Pietersen (1985), Skandia International (1985), Blything and Reeves (1988 SRD R488), Selway (1988 SRD R492), Johnson and Pritchard (1991), and the CCPS (1994/15).

When a vessel containing liquid under pressure is exposed to fire, the liquid heats up and the vapor pressure rises, increasing the pressure in the vessel. When this pressure reaches the set pressure of the pressure relief valve, the valve operates. The liquid level in the vessel falls as the vapor is released to the atmosphere. The liquid is effective in cooling that part of the vessel wall which is in contact with it, but the vapor is not. The proportion of the vessel wall which has the benefit of liquid cooling falls as the liquid vaporizes. After a time, metal which is not cooled by liquid becomes exposed to the fire; the metal becomes hot and weakens and may then rupture. This can occur even though the pressure relief valve is operating correctly.

A BLEVE of a vessel containing a flammable liquid gives rise to the following effects: (1) blast wave, (2) fragments, and (3) fireball.

Following Flixborough, much attention was focused on VCEs. It was pointed out by Kletz (1977) that BLEVEs can cause as many casualties as VCEs and that by comparison they were being relatively neglected. This view has been amply justified by the subsequent record.

The essential features of a BLEVE are that (1) the vessel fails, (2) the failure results in flash-off of vapor from the superheated liquid, and, if the liquid is flammable, (3) the vapor ignites and forms a fireball.

The accompanying effects are (1) blast, (2) fragments, and, for flammable liquids, (3) a fireball.

The BLEVE creates an overpressure. The phenomena associated with this are (1) the expansion of the vapor,

(2) the flash vaporization of the liquid, and, for flammable liquids, and (3) the combustion of the vapor. These events are not completely simultaneous but have been measured as separate effects.

13.24.1 Empirical Features

It is also helpful to consider some features of BLEVEs from an empirical viewpoint. Those discussed here are (1) the time to BLEVE, (2) the mode of rupture, (3) the blast effects, (4) the fireball, (5) the missiles, and (6) the release of flammable fluids. Time to BLEVE is treated briefly below.

With regard to the mode of rupture, cylindrical tanks usually rupture longitudinally, though some rupture circumferentially. The latter particularly tend to take off like rockets and may travel long distances. This was true of some of the cylindrical tanks at Mexico City. Spheres often explode, but in some cases may simply split at the top as occurred both at Feyzin and Mexico City.

The pressure at the instant of rupture must approximate to that in the vessel. Generally, therefore, it will be appreciably greater than that at the center of a VCE, which is commonly estimated to have a maximum overpressure of about 1 bar. Thus, in a BLEVE, the overpressure at the vessel at the instant of rupture may be an order of magnitude higher than that at the center of a VCE. The blast wave from a BLEVE can cause damage.

The liquid in the vessel does not necessarily vaporize completely. At Tewkesbury, Massachusetts, in 1972, some 35% of the propane flashed off, the remaining liquid being scattered in all directions.

BLEVEs do not generally give rise to VCEs. But they may scatter flammable liquid spray which gives a spray fire and/or falls on people and property, rendering them more flammable and liable to ignition. Hot missiles may also cause fires.

The time between the occurrence of an engulfing or torch fire and BLEVE is of significance not just for the general characterization of BLEVEs but for the design of fire systems to protect against them and for fire fighting. For the sphere at Feyzin in 1966, the time between ignition of the leak and vessel rupture was about an hour and a half. This time to BLEVE is much longer than in most other incidents.

More commonly, for storage vessels, the time to BLEVE has been of the order of 5−30 min. A period of 3−10 min was observed for some vessels at Montreal, McKittrick, and Mexico City. A time of some 30 min was observed for other vessels at these three incidents.

The energy of the explosion which is part of the BLEVE event depends on the conditions in the vessel. In particular, under certain conditions, explosive flashing of the superheated liquid can occur, giving a large release of energy.

13.24.2 Modeling of BLEVEs

There are several different approaches to the modeling of BLEVEs. They include (1) the superheated liquid explosion model, (2) the cloud formation model, and (3) the bursting vessel model. Here one of these models is treated briefly.

Models for *vapor cloud formation* following depressurization have been given by Hardee and Lee (1975), Maurer et al. (1977), and Giesbrecht et al. (1980) and have been compared by Roberts (1982). The first model is based on conservation of momentum with the momentum created by the liquid release appearing explicitly as a function of the initial conditions; the second model is based on turbulent diffusion and the initial conditions do not appear explicitly. Both models give similar predictions of cloud growth for releases of the order of 100 kg and have been verified experimentally at this level, but diverge for larger releases. Roberts bases his treatment on the model of Hardee and Lee.

The CCPS *Fire and Explosion Model Guidelines* (1994/15) give a method for the treatment of the various effects of a BLEVE. The treatment covers (1) blast, (2) fragments, and (3) fireball.

The foregoing models mentioned the background for the estimation of the energy release and blast overpressure in a BLEVE. The two principal sources of energy are the compressed vapor in the vapor space and the superheated liquid as it undergoes flashing. The superheat limit temperature model provides guidance on the extent to which liquid flashing is a significant contributor. A further discussion of the energy release in a BLEVE is given by Prugh (1991).

13.25. EXPLOSIONS IN PROCESS PLANT

Some of the principal types of explosion which occur on process plant have already been described. These are (1) explosions in chemical reactors, (2) explosions of high pressure gases inside plant, (3) explosions of flammable gas−air mixtures inside plant and in buildings, (4) VCEs, and (5) BLEVEs.

In addition, there are certain types of explosion which may occur in particular processes, operations or equipment. Some of these have already been mentioned, such as explosions in driers, centrifuges and vaporizers, and in oxygen and chlorine plants.

Other explosion hazards include (1) aerosol explosions, (2) crankcase explosions, (3) superheated liquid explosions, (4) air system explosions, and (5) molten metal—water explosions. Some of these types of explosion are briefly discussed.

For aerosol explosions, and under the conditions of an accident, a cloud of finely divided drops of flammable liquid in air, in the form of a mist or spray, may be produced which has many of the characteristics of a flammable gas—air mixture and which can burn or explode. If the droplet size of the cloud is sufficiently small, the lower flammability limit, minimum ignition energy, and burning velocity are essentially the same as those of a vapor—air mixture of the same concentration, measured in mass per unit volume.

For superheated liquid explosions, there are two different cases, the first one is called *explosions in plant*, and suggests that in some cases the release of vapor is rather more violent. A layer of water at the bottom of the tank of hot oil, for example, may became superheated so that there is a sudden evolution of vapor. It has been suggested by King (1975a—c, 1976a,b, 1977, 1990) that such an occurrence may have contributed to the Flixborough disaster.

The second one is that which can occur when a refrigerated liquefied hydrocarbon is spilled onto water and is called *explosion on water*. Explosions of this kind are variously known as physical vapor explosions, flameless vapor explosions, or simply vapor explosions, and as rapid phase transition (RPT) explosions.

13.26. EFFECTS OF EXPLOSIONS

In this section, an account is given of overview about the possible effects caused by explosion.

An explosion may give rise to the following effects: (1) blast damage, (2) thermal effects, (3) missile damage, (4) ground shock, (5) crater, and (6) injury. Below are briefly described.

Not all these effects are given by every explosion. An aerial blast, for example, tends not to form a crater.

Many of the data on the effects of explosions come, not surprisingly, from studies of military and industrial explosives, but an increasing amount of information is available from the investigation of process plant explosions.

Information on the effects of explosions has been given in *Explosions, Their Anatomy and Destructiveness* (Robinson, 1944), *Structural Defence* (Christopherson, 1946), the *Textbook of Air Armaments* (Ministry of Supply, 1952), and *The Effects of Nuclear Weapons* (Glasstone, 1962; Glasstone and Dolan, 1980). Full references are described in the Lees Book full version.

The blast damage is one of the principal effects of an explosion in the creation of a blast wave, and much of the energy of the explosion is expended on this. A description of blast effects must take into account both the nature of the blast wave and the damage caused by the blast to structures.

Missile damage is another principal effect of an explosion in the generation of missiles, and this takes up most of the energy not transmitted to the blast wave. A description of missile damage involves both the generation and flight of missiles and the damage caused by missiles to structures.

The combustion process involved in a chemical explosion can give rise to intense local heat radiation called *thermal effects*, which may cause damage or injury.

In industrial explosions, *ground shock* effects are usually small and less than those due to blast and assessment of such effects is generally not undertaken.

The *ground shock* produced by an explosion may be regarded as a sinusoidal disturbance and may be characterized by its frequency and amplitude and thus also by its maximum velocity and maximum acceleration.

A condensed phase explosion tends to give rise to a crater. For industrial explosions, this effect is not usually assessed beforehand but may provide information useful in accident investigation. Accounts of cratering are given by Robinson (1944) and Clancey (1972). The factors which affect the crater produced by an explosion are the position of the charge relative to the ground surface, the nature of the ground, and the type and quantity of explosive. A charge exploded at the ground surface gives a wider and shallower crater than one exploded just beneath the surface.

Cratering can occur even without an explosion. The V-2 rockets, which had very high kinetic energy, were capable of giving a large crater even when not armed.

The effects of an explosion on people include injury caused by (1) blast, (2) missiles, (3) thermal effects, and (4) toxic effects.

13.27. EXPLOSION DAMAGE TO STRUCTURES

13.27.1 Air Blast Loading

The loading of structures by air blast is described by Glasstone (1962) and Glasstone and Dolan (1980). In general, the effective pressure p_e on a given face of a structure is the sum of the overpressure p and the drag pressure p_d:

$$p_e(t) = p(t) + p_d(t) \qquad (13.89)$$

The drag pressure p_d is a function of the dynamic pressure q:

$$p_d(t) = C_D q(t) \qquad (13.90)$$

where C_D is the drag coefficient.

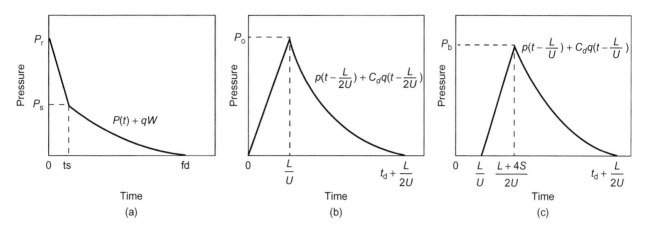

FIGURE 13.10 Blast loading of structures—average pressure loading on a closed cuboid structure (Glasstone and Dolan, 1980): (a) average front face loading; (b) average side and top loading; and (c) average back face loading. *Source: Courtesy of Castle House Publications.*

If a blast wave reaches the structure and its incidence on the front face is normal, the overpressure p_f at that face rises very rapidly to the peak reflected overpressure p_r^o. As the wave moves around the structure, the front face pressure falls rapidly to the stagnation pressure p_s at the stagnation time t_s. Thereafter, the front face pressure falls:

$$p_f = p(t) + q(t) \qquad (13.91)$$

The front face pressure is shown in Figure 13.10(a). The average side face and top pressure is obtained from Equations 13.89 and 13.90 and showed in Figure 13.10(b) and (c). L is the structure length, U is the shock velocity, and S is the distance through which pressure relief is obtained.

The net horizontal loading is the front face loading less the back face loading. This is shown in Figure 13.11.

The response of the walls depends primarily on the loadings on the individual faces, while the response of the frame depends on the net loading.

13.27.2 Diffraction and Drag Loading

In the air blast loading of structures, a distinction is made between diffraction loading and drag loading. Diffraction loading is determined primarily by the peak overpressure, drag loading by the dynamic, and hence the drag, pressure.

The loading is a function both of the air blast and the structure. A given structure may display diffraction- or drag-type behavior, depending on the duration. A short duration tends to load by diffraction, a long one by drag.

Certain types of structure are more prone to damage by diffraction, others by drag. Diffraction-type targets tend to be closed or semi-closed structures such as buildings with small window areas and storage tanks. Drag-

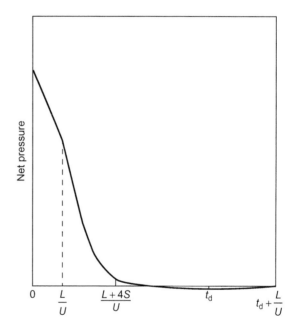

FIGURE 13.11 Blast loading of structures—net pressure loading on a closed cuboid structure (Glasstone and Dolan, 1980): net horizontal loading. *Source: Courtesy of Castle House Publications.*

type targets tend to be tall, thin objects such as telegraph poles and lamp posts. Buildings with large window areas may also constitute drag-type targets.

Drag loading is relatively more important for explosions which give long duration times, such as nuclear explosions, than for those giving short duration times, such as TNT.

13.27.3 Structure Response

These relationships may be represented graphically in the form of a force−impulse, or F−I, diagram, as illustrated

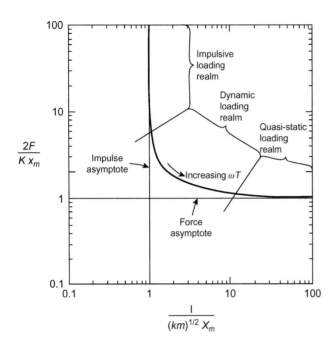

FIGURE 13.12 Blast loading of structures: force—impulse (F—I) diagram of a structure (Strehlow and Baker, 1976). *Source: Reprinted from R.A. Strehlow and W.E. Baker, Progress in Energy and Combustion Science, 2, 27, © 1976, with permission from Pergamon Press.*

in Figure 13.12, which shows again the three domains of impulsive loading and quasi-static loading separated by dynamic loading.

The response of buildings to blast is generally analyzed using a static load of long duration. The authors state that the metal frame buildings common in chemical plant tend to have greater impulsive and dynamic resistance than the static analysis indicates. It is quite common for plant buildings to experience a dynamic response in which the duration of the load is close to the natural period of the structure. The conventional plant buildings which perform best are those which allow plastic response without failure. The thrust of the design recommendations is to enhance this plastic behavior.

Some parameters which govern the structure response are (1) the natural period, (2) the elastic limit, (3) the ductility ratio, and (4) the resistance function.

A typical load—displacement diagram is illustrated in Figure 13.13(a) and other related diagrams are shown in Figure 13.13(b)—(d).

13.27.4 Overpressure Damage Data

Structural damage caused by blast waves from explosions has traditionally been correlated in terms of the peak overpressure of the explosion. There are available a large number of tables and graphs giving damage levels in terms of the peak overpressure.

The applicability of such data is greater for quasi-static loading than it is for impulsive loading. In the latter regime, this approach is liable to result in considerable overestimate of the damage.

Thus, Baker et al. (1983) recommend against the use of simple overpressure damage data. Preferred methods are the use of pressure—impulse, or P—I, diagrams and distance—charge, or R—W, correlations.

However, useable data in these alternative forms are often not available. In addition, some refinement of the overpressure approach has been effected. It still appears worthwhile, therefore, to consider first this approach.

13.28. EXPLOSION DAMAGE TO HOUSING

Explosion damage to housing is usually expressed in the form of distance—charge, or R—W, relations or of pressure—impulse, or P—I, relations. In the description just given of these two methods, the application considered is housing damage.

During the Second World War, it became the practice in the United Kingdom to assign air raid housing damage to the categories A, B, Cb, Ca, and D.

There are several definitions of these categories. Table 13.8 gives the categories given by Jarrett.

13.28.1 Jarrett Equation

The paper by Jarrett (1968), already mentioned, was the first open publication and gave wider currency to Jarrett equation. He also gave explicit constants for the A, Cb, Ca, and D damage categories. In British units, Jarrett's equation is

$$R = \frac{kW^{1/3}}{[1+(7000/W)^2]^{1/6}} \tag{13.95a}$$

where R is the distance (ft) and W is the mass of explosive (lb). In SI units, this equation is

$$R = \frac{kW^{1/3}}{[1+(3175/W)^2]^{1/6}} \tag{13.95b}$$

where R is the distance (m) and W is the mass of explosive (kg). Equation 13.95a is frequently referred to as the Jarrett equation. The values of the constant k given in British units by Jarrett and the SI equivalents are given in Table 13.9(A) and (B), respectively. The RB ratios are also given.

13.28.2 P—I Diagrams

The housing damage correlations just given are R—W correlations. As described above, it is also possible to correlate housing damage in terms of a P—I diagram.

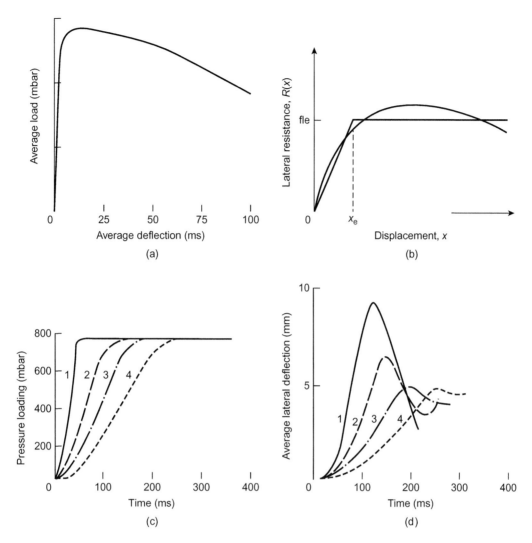

FIGURE 13.13 Blast loading of structures: load−displacement and related diagrams: (a) load−displacement diagram; (b) resistance−displacement diagram; (c) load−time diagram; and (d) displacement−time diagram. 1−4, rise times.

The P−I diagram for housing damage based on the Jarrett equation given by Baker et al. (1983) has been given above as Figure 13.14. The figure shows iso-damage curves for B, Cb, and Ca category damage.

For many charge weights of interest, both peak overpressure and impulse are relevant. For this reason, housing damage is correlated by Gilbert, Lees, and Scilly directly in terms of charge weight rather than in terms of overpressure.

For the same reason, they take the scaled distance appropriate to housing damage as

$$z' = \frac{R}{W^{1/3}[1 + (3175/W)^2]^{1/6}} \quad (13.96)$$

where z' is a modified scaled distance (m/kg$^{1/3}$).

13.29. EXPLOSION DAMAGE BY MISSILES

If the explosion occurs in a closed system, fragments of the containment may form missiles. In addition, objects may also be turned into missiles by the blast.

A significant source of large missiles in process plant incidents is BLEVE. Studies of the missiles from such sources have been described by Holden and Reeves (1985) and Pietersen (1985).

The generation, size, and range of missiles, and the damage which they can cause, have already been discussed to some extent above, mainly in relation to BLEVEs. Missiles are generally classified as primary and secondary. Primary missiles are those resulting from the bursting of a containment so that energy is imparted to the fragments which become missiles.

Secondary missiles occur due to the passage of a blast wave which imparts energy to objects in its path, turning them into missiles. Missiles from the bursting of a wall due to an internal explosion are treated here as primary missiles.

As far as concerns primary missiles from plant, three cases are commonly distinguished: case 1, bursting of a vessel into a large number of relatively small fragments; case 2, separation and rocketing of a vessel or vessel end; case 3, ejection of a single item.

13.29.1 Parameters of Missiles

There are a number of different approaches to the estimation of the initial velocity of a missile. They include consideration of (1) the force or pressure on the missile, (2) the transfer of momentum to the missile, and (3) the transfer of energy to the missile.

The first case considered (case 1) is that of the bursting of a gas filled vessel into a large number of frag-

---

TABLE 13.8 Housing Damage Categories

Jarrett Damage Categories (Jarrett, 1968)	
A	Almost complete demolition
B	50–75% external brickwork destroyed or rendered unsafe and requiring demolition
Cb	Houe uninhabitable—partial or total collapse of roof, partial demolition of one to two external walls, severe damage to load-bearing partitions requiring replacement
Ca	Not exceeding minor structural damage, and partitions and joining wrenched from fittings
D	Remaining inhabitable after repair—some damage to ceilings and tiling, more than 10% window panes broken

---

TABLE 13.9 Constants and RB Distances for the Jarrett Equation

Damage Category	Constant, k	RB Ratio
A. Eauation constants (British units) and RB ratios (Jarrett, 1968)		
A	9.5	0.675
B	14	1
Cb	24	1.74
Ca	70	5
D	140	10
B. Equation constants (SI units) and RB ratios (after Jarrett, 1968)		
A	3.8	0.675
B	5.6	1
Cb	9.6	1.74
Ca	28	5
D	56	10

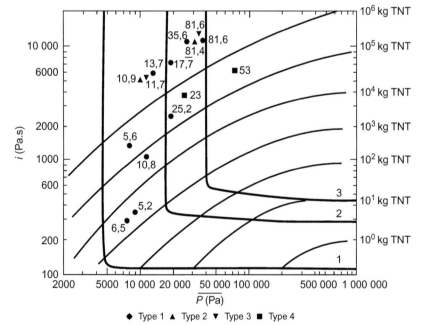

◆ Type 1 ▲ Type 2 ▼ Type 3 ■ Type 4

FIGURE 13.14 Damage effects of explosions: P–I diagram for housing damage—1. Curves are those given by the first set of authors and are as follows: curve 1, threshold for minor structural damage; curve 2, threshold for major structural damage; and curve 3, threshold for partial demolition. Experimental points are those given by the second set of authors and refer to the percentage damage (expressed as building costs) for house types 1–4: type 1, two-floor wooden house; type 2, two-floor masonry house; type 3, single-floor wooden house; and type 4, two-floor masonry house. *Source: After Baker et al., 1983; Mercx et al., 1993.*

ments. An account of the initial velocity of fragments from vessels has been given by Ardron et al. (1977). They distinguish two main fundamental treatments, one based on the stored energy and the other on fluid mechanics. The models of Moore (1967) and Munday (1980), respectively, exemplify these two approaches.

The second case (case 2) is that of the separation of a gas filled vessel into two parts and the rocketing of one or both of these.

The third case considered here (case 3) is the ejection of a fitting such as a valve stem by a high pressure jet of gas. This is a relatively well-defined situation, and modeling based on fluid mechanics is appropriate. Models have been developed by a number of workers, including Cottrell and Savolainen (1965), Gwaltney (1968), and Ardron et al. (1977).

To determine the flight of a missile, it is necessary also to have information on the angle of departure. The extent to which this is known depends on the particular case. For ejection of a fitting, the direction of the gas jet and, hence, the angle of departure of the fitting is usually defined. Likewise, this will generally be so for the two parts in the separation and rocketing of a gas filled vessel.

The angles of departure in the case of the bursting of a gas filled vessel into a large number of fragments are less well defined. The common assumption is that the fragments are projected uniformly in all directions. Alternatively, the more conservative approach may be adopted in which the spatial density of the fragment is assumed to be greater in the direction of the vulnerable targets.

For some methods, the results for an individual fragment can be sensitive to the angle of departure. The CCPS (1994/15) report this to be the case for the model of rocketing fragments.

The behavior of the fragment in flight depends on its shape. Some fragments are chunky, with dimensions similar along the three main axes, while others are less symmetrical, with dissimilar lengths along these axes.

A fragment in flight is acted on by a lift force normal to the trajectory and a drag force along the trajectory. These forces are defined in Equations 13.97 and 13.98 below in terms of a lift coefficient C_L and a drag coefficient C_D.

In ballistic terms, a fragment may be characterized as a drag-type fragment if it is chunky so that for any orientation $C_D >> C_L$ and as a lifting type fragment if it is such that for some orientation $C_L \geq C_D$.

The force resisting the flight of the missile is a function of its velocity, but the relationship is not a simple one. The velocity range of interest for projectiles is termed the ballistic range. At very low velocities, below

the ballistic range, the resistance is proportional to the velocity. At ballistic but subsonic velocities, the resistance is proportional to the square of the velocity. At supersonic velocities, the resistance is a complex function.

The flight of a projectile is a standard problem in mechanics and is treated in texts on this topic (Synge and Griffiths, 1960; Smith and Smith, 1968; Brown, 1969).

Three cases are considered here: case 1, no air resistance, case 2, air resistance proportional to the velocity, and case 3, air resistance proportional to the square of the velocity. The first case is not realistic and results in gross overestimates of velocity and range but provides a useful illustration of the general approach. The third case is realistic for ballistic velocities in the subsonic range.

A fragment in flight is acted on by a lift force, which is normal to the trajectory and opposes gravity, and a drag force, which is along the trajectory. These forces are

$$F_L = C_L A_L \frac{\rho u^2}{2} \qquad (13.97)$$

$$F_D = C_D A_D \frac{\rho u^2}{2} \qquad (13.98)$$

where A_D is the drag area, A_L is the lift area, C_D is the drag coefficient, C_L is the lift coefficient, F_D is the drag force, F_L is the lift force, u is the velocity of the fragment, and ρ is the density of air.

A fundamental approach to the estimation of fragment range has been developed by Baker et al. (1975, 1978) and is described by Baker et al. (1983). The range is obtained by solving the equations of motion for acceleration of the fragment in the horizontal and vertical directions, utilizing for the drag and lift forces Equations 13.97 and 13.98. The results are correlated in terms of the dimensionless, $\bar{u}_i$, velocity defined as

$$\bar{u}_i = \frac{C_D A_D \rho_o u_o^2}{Mg} \qquad (13.99)$$

and the dimensionless range $\bar{R}$

$$\bar{R} = \frac{C_D A_D \rho_o R}{M} \qquad (13.100)$$

and the lift/drag ratio

$$= \frac{C_L A_L}{C_D A_D} \qquad (13.101)$$

where g is the acceleration due to gravity, M is the mass of the fragment, R is its range, u_o is its initial velocity, and ρ_o is the density of air. The treatment is based on the assumption that the fragment is spinning and applies to fragment velocities up to Mach 1 or about 340 m/s. The equations were solved using the computer code FRISB.

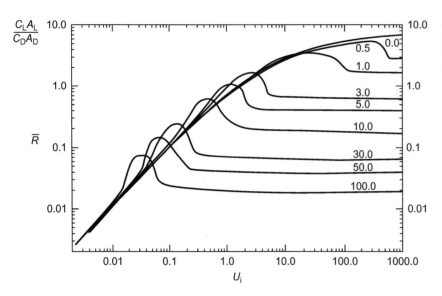

FIGURE 13.15 Missile effects of explosions: distance traveled by missiles (Baker et al., 1983). *Source: Courtesy of Elsevier Science Publishers.*

The correlation obtained is shown in Figure 13.15. For values of the lift/drag ratio not given, interpolation can be used, but it becomes inaccurate on the steep portions of the curve.

Many fragments from explosions are chunky and have a lift coefficient C_L, and hence a lift/drag ratio, of 0. Other fragments may be pieces of plate for which the lift/drag ratio is more complex.

13.29.2 Missile Damage Method of Scilly and Crowther

A method for assessment of the risk of missile impact on a process plant site has been described by Scilly and Crowther (1992). The method which they describe involves for a given target estimation of the following features: (1) the number of fragments, (2) the range distribution, (3) the mass distribution, (4) the orientation factor, (5) the effective range interval, and (6) the probability of a strike.

They illustrate the method by reference to the case of a distillation column 50 m high, 2.4 m in diameter, and 12.5 mm thick subject to a sudden pressure increase due to a rapid decomposition.

In general, the number N of fragments depends on a variety of factors, including the vessel size and shape, the material of construction, the operating temperature, and the rate of pressure rise. The authors obtain the estimate of the number of fragments N by inspection of Table 13.10. For their illustrative example, they take N as 35.

As already described, the distribution of fragments may exhibit directionality. The authors suggest there is some justification for allowing for this by the use of a correction factor of 2, applied to N.

For the distribution of the ranges of the fragments, they utilize not the fragment with maximum range R_{max} but that with the next farthest range, or penultimate range, R_{pen}, arguing that the latter is more meaningful. The range distribution is found to be log-normal and is fitted by the authors using a probit equation. They give the following tentative relations:

$$R_{med} = 2.8P_v \qquad (13.102)$$

$$R_{pen} = 4.1R_{med} \qquad (13.103)$$

where P_v is the vessel pressure (barg), R_{med} is the median range (m), and R_{pen} is the range of the penultimate fragment (m). A probit equation for the probability P of a fragment falling at a range R may then be constructed from the pair of points ($P = 0.5$, R_{med}) and ($P = N_{pen}/N$, R_{pen}), where $N_{pen} = N - 1$.

Given that for vessels of size exceeding 20 m³, the great majority of fragments will have energy sufficient to penetrate steel containments, which are unlikely to have a wall thickness greater than 15 mm, the distribution of mass is much less important. However, for the case where it is desired to estimate the mass distribution for vessels of such size, the authors propose that the median mass of the fragments be taken as that of the vessel divided by 73 and that the slope of the associated probit equation be taken as 1.5.

The orientation factor F_{or} is

$$F_{or} \approx \frac{W}{2\pi R} \qquad (13.104)$$

where R is the range of the fragment (m) and W is the combined width of the target and the fragment (m). The width of a fragment is taken in the illustrative example as 4 m.

TABLE 13.10 Behavior of Fragments In Some Vessel Explosions—2

Incident[a]	Vessel Details			Vessel Pressure		No. of Fragments	Missile Range			Missile Mass		
	Diameter (m)	Volume (m³)	Thickness (mm)	Operating (barg)	Burst (barg)		Median	Penultimate/ Median (m)	Probit Slope	Median	Vessel/ Median (kg)	Probit Slope
1a	3.1	90		16	75–85	29	94	4.4	1.60			
1b	7.2	1560	63	16	75–85	14	95	3.7	1.03	16,900	27	1.95
2	1.7	210	38	9	155	3						
3	2.7	57	13	0	63	10	156	3.4	1.87	170	52	1.92
4	2.3	200/53[b]	16	3	60	≥50						
5	3.0	73	19	8	50	21	117	4.8	1.20	173	100	1.10
6	2.5	162	12, 16, 19	3	32[c]	35	140	4.0	2.46	539	67	1.63
7	0.94	1.3	6.5	−1	34	24	14	11.4	3.60	4	77	1.21

[a]Incidents are as follows: 1a and 1b, separator and reactor respectively, at Whiting, Indiana, 1957—fuel/air combustion probably leading to detonation; 2, Ponca City, 1959—probable fuel/air combustion; 3, Doe Run, 1962—probable liquid phase polymerization and gas phase decomposition of ethylene oxide; 4, Texas City, 1969—liquid phase decomposition of vinyl acetylene; 5, Grangemouth, 1987—breakthrough of high pressure (155 bar) gas; 6, Antwerp, 1987—decomposition of ethylene oxide during distillation; 7, Wendstone Chemicals, 1988—liquid phase decomposition of o-nitrobenzaldehyde containing nitrate esters.
[b]Distillation column had a volume of 200 m³, but only the central section of volume 53 m³ disintegrated.
[c]Authors suggest that this value is understood and that it should be at least 43 barg.
After Scilly and Crowther (1992). Courtesy of American Institute of Chemical Engineers

The target is vulnerable to any fragment falling between a minimum range R and a maximum range $R + L$, where L may be termed the effective range interval. For a target consisting of a sphere supported on legs, the authors derive by geometry for fragments directed, respectively, at the midpoint and at the edge of the sphere

$$L = (H - r + r \sec \theta)\tan \theta + r \quad \text{sphere midpoint} \tag{13.105}$$

$$= (H - r)\tan \theta + r \quad \text{sphere edge} \tag{13.106}$$

where H is the height of the top of the sphere (m), L is the effective range interval (m), r is the radius of the sphere (m), and θ is the angle of descent (°). For the angle of descent, the authors suggest 45° as a conservative value. They use 60° in their illustrative example.

The probability P_{ri} that a fragment falls within the effective range interval L is then determined from the probit equation for the range distribution by taking the difference between the probabilities of a fragment reaching $R + L$ and of one reaching R.

The probability of a strike P_{st} is then

$$P_{st} = F_{or}NP_{ri} \tag{13.107}$$

As already stated, the authors effectively take as unity the probability of puncture of the target given a strike.

For an explosion in the vessel used to illustrate the method, described above, and for a target consisting of a sphere 12 m in diameter supported on 3 m legs at 150 m distance from the exploding vessel, the authors obtain a calculated burst pressure $P_v = 45$ barg, a number of fragments $N = 35$, a range of fall between 150 and 199 m, the range probit $Y = -1.44 + 3.066 \log_{10} R$, a probability that a fragment falls within the effective range interval of 0.14, an orientation factor $F_{or} = 0.016$, and a probability of strike $P_{st} = 0.08$.

13.29.3 Impact Effects

It is convenient to quote first some of the empirical equations for determining fragment penetration given in the *High Pressure Safety Code* by Cox and Saville (1975),[1] which may be regarded as a coherent set selected by the authors for plant design purposes. A treatment of the impact of missiles on pipework is given in Section 17.35.

The equations quoted in the *High Pressure Safety Code* assume penetration normal to the surface. They are valid for projectile velocities not exceeding 1000 m/s, above which there is often a quite different mechanism in which target and projectile are melted by the impact energy.

TABLE 13.11 Parameters K and the Indices n_1 and n_2 for Different Target Materials

	K	n_1	n_2
Concrete (crushing strength 35 MN/m^2)	18×10^{-6}	0.4	1.5
Brickwork	23×10^{-6}	0.4	1.5
Mild steel	6×10^{-5}	0.33	1.0

For penetration by small fragments, the depth of penetration is given by the relation

$$t = Km^{n_1}V^{n_2} \tag{13.108}$$

where m is the mass of the fragment (kg), t is the thickness of the barricade needed just to stop the fragment (m), V is the velocity of the fragment (m/s), K is a constant, and n_1 and n_2 are indices. The values of the constant K and the indices n_1 and n_2 for different target materials are given in Table 13.11.

Equation 13.108 is valid for compact blunt steel fragments such as solid cylinders with the length equal to the diameter with a mass of not more than 1 kg.

13.29.4 Barricade Design

In some cases, use is made of barricades around a potential source of explosion. Barricades are utilized particularly in work with explosives, but may also be used around other potential explosion sources such as high pressure equipment. There is also considerable interest in barricades in the nuclear industry.

Accounts of the design of barricades include those given in the *High Pressure Safety Code* by Cox and Saville (1975) and by Loving (1957), High (1967), and Moore (1967).

The essential first step in design of a barricade is to determine the failures against which protection is required. In other words, it is necessary to start with hazard identification.

In barricade design, provision of protection against an explosion of flammable gas is generally treated as a special case. Excluding this, the design resolves into provision of a barricade which offers resistance to (1) blast and (2) fragments. Broadly, design against blast is often based on an equivalent static pressure approach and design against fragments on fragment penetration correlations.

Design methods for barricades to protect against vessel rupture are discussed in the *High Pressure Safety Code*. It is relatively easy to provide a barricade for vessels with energy contents in the range $10^3 - 10^5$ J, but it becomes

---

1. Reproduced with permission of the High Pressure Technology Association.

progressively more difficult as the energy content rises, and for energy contents capable of giving a shock wave of 50×10^6 to 100×10^6 J it is usually impractical.

The preferred method is a closed cubicle. For protection against blast using the equivalent static pressure method, the *Code* gives the relevant pressure as

$$P = 7.6[(E \times 10^{-6})/V]^{0.72}, \quad P < 70 \qquad (13.109)$$

where E is the shock wave energy (J), P is the equivalent static pressure (bar), and V is the volume of the enclosure (m^3). Equation 13.109 is applicable where the ratio of the maximum to the minimum dimension of the enclosure does not exceed 2.

Cylindrical cubicles can be designed as thin-walled pressure vessels. Small cubicles can be made of angle iron and steel plate. Large cubicles should be of reinforced concrete. Since the shock wave has positive and negative phases, reinforcement is required on both inner and outer faces.

Some transmission of the shock wave occurs through the walls of the cubicle. This creates the hazard that persons outside may be injured by eardrum rupture. The Code gives the following method of estimating this. The impulse incident on an isolated wall is given as

$$I = 0.008E^{1/3}(R/E^{1/3})^{-1/2}, \quad 0.004 < (R/E^{1/3}) < 0.2 \qquad (13.110)$$

where I is the impulse per unit area (N s/m^2) and R is the distance (m). Also for an isolated wall, the initial velocity of the wall, assumed unconstrained, is

$$u = I/m \qquad (13.111)$$

where m is the mass of the wall per unit area and u is the velocity (m/s). For the wall of a cubicle, the impulse is taken as 3 times the value given by Equation 13.110 and the initial velocity of the wall is then that given by Equation 13.111.

If the value of the initial velocity u so calculated is less than 10 m/s, the shock wave generated outside should be too small to cause eardrum rupture.

If the plant in the cubicle contains flammables or toxics, an opening may be necessary to allow dispersion of small leaks by ventilation. The *Code* states that the design of a vented cubicle is essentially the same as that for a closed cubicle.

With a vented cubicle, however, the escaping shock wave will tend to cause overpressure similar to that which would exist if there was no cubicle and the explosion was in the open. The *Code* gives the following equation for the estimation of the peak incident overpressure:

$$P = 8.45 \times 10^{-5}(E^{1/3}R)^{2.3} - 0.16(R/E^{1/3}) + 0.06, \quad 0.004 < (R/E^{1/3}) < 0.2 \qquad (13.112)$$

where P is the peak incident overpressure (bar).

The cubicle should also withstand the fragments generated. Since the distances are usually short, such a fragment will tend to be at its initial velocity. Much of the work on fragment penetration, described above, has been done to assist with barricade design.

Safety walls may be used instead of a cubicle. For a safety wall, the impulse from the blast wave is as given by Equation 13.110 without application of the factor of 3 and the initial velocity as given by Equation 13.111. There will, however, be diffraction of the shock wave around the wall which is difficult to estimate.

The thickness of the barricade should be such as to enable it to withstand both shock wave and fragment penetration.

The barricade may be required to survive the simultaneous assault of the shock wave and of fragments. This problem is discussed in the *Code* and also by Moore.

Information on viewing ports for cubicles is given by Moore (1967).

13.29.5 Plant Design

The hazard from missiles has not generally been a major factor in plant design, but rather more attention is now paid to it.

Design of plant for the hazard of missiles may focus on the source or on the target. On the whole, it makes more sense to concentrate on the source, since the probability that a given plant item will be the target of a missile, even if a missile-generating explosion occurs, is very low.

Attention should be paid in design to features such as doors, vents, and other fixtures on plant which may become missiles, and measures should be taken to minimize this.

Where a serious missile hazard exists, use may be made of barricades and safety walls as described above.

13.30. EXPLOSION

13.30.1 Explosion of a Cased Explosive

Another event which can give rise to missiles is the explosion of a cased explosive. This is considered in this section. While the discussion has most relevance to an explosion involving munitions, there are features which have wider application.

The exploding object primarily considered is a cased high explosive, typically a shell or bomb. The explosive could be some combination of TNT and RDX. The following treatment, therefore, is in terms of weapons.

The fragment mass distribution of this latter type of casing is obtained experimentally by conducting an explosion and collecting the fragments formed. In some cases, information is also obtained on the fall of the fragments.

An early method of representing the FMD was that of Mott (1947). That described here is the method developed by Held (1968, 1979, 1990). The correlation is

$$M(n) = M_0[1 - \exp(-Bn^\lambda)] \qquad (13.113)$$

where $M(n)$ is the cumulative fragment mass or overall mass of the fragments of number n, starting with the largest fragment, M_0 is the total mass of fragments, n is the number of the nth largest fragment, or cumulative fragment number, and B and λ are constants.

The mass m of the nth fragment is obtained from Equation 13.113 by differentiation:

$$m = \frac{dM(n)}{dn} = M_0 B \lambda n^{\lambda-1} \exp(-Bn^\lambda) \qquad (13.114)$$

Held (1990) gives as an example the fragment mass distribution of the Hispanic Suiza 30 mm $\times$ 170 mm incendiary high explosive projectile. The parameters which he obtains for this weapon are $M_0 = 288.7$ g, $B = 0.511$, and $\lambda = 0.7318$.

The risk from explosion of a cased explosive depends on the number and size of the fragments. As the number of fragments increases, the risk initially increases but eventually decreases as the fragments become too small to cause injury.

Methods exist for the estimation of the angle at which the fragments from a single weapon are projected. Treatments include those of Christopherson (1946), Taylor (1963a,b), and Karpp and Predebon (1974).

A study of the fragmentation of tubular bombs was conducted by Taylor (1963a,b). He gives the following equation for the angle θ formed by the deforming case relative to the direction of the detonation wave:

$$2 \sin \phi = u_0/D \qquad (13.115)$$

where D is the velocity of detonation (VOD), u_0 is the initial velocity of the fragment (VOF), and ϕ is the angle to the axis of the tube at which fragments are projected. ϕ is known as the Taylor angle.

In practice, these models of projection angle appear of limited use in accident modeling, where the appropriate approach depends on the particular case.

Methods of determining the initial velocity of the fragments from a weapon include those given in the *Textbook*, in Structural Defence, and by Gurney (1943), Clancey (1972), Kamlet and Finger (1979), and Hirsch (1986).

One approach to the estimation of the initial velocity of fragments is the use of empirical rules. The *Textbook* suggests values for the initial velocities of 4000 ft/s for bomb fragments and 3000 ft/s for shell fragments.

One equation for the initial velocity of the fragments is that given in Structural Defence by Christopherson (1946):

$$V_0 = \{8.22 \times 10^7[1 - \exp(-0.69E/C)]\}^{1/2} \qquad (13.116)$$

where C is the mass of the uniform cylindrical case per unit length (lb), E is the mass of charge per unit length (lb), and V_0 is the initial velocity (ft/s). Equation 13.116 is a generalization of an empirical equation obtained by Payman for small tetryl filled bombs.

Following the *Textbook*, the number n of incapacitating fragments is

$$n(x) = \int_0^\infty q(v, x) r(x) dv \qquad (13.117)$$

where n is the number of incapacitating fragments, q is the number of fragments of equivalent velocity v at a distance x, r is the proportion of fragments of equivalent velocity v at a distance x capable of causing incapacitation, v is the velocity of the fragment (ft/s), and x is the distance (ft).

In the foregoing treatment, the exploding object is taken as a single weapon. In many situations of practical interest, it is likely to be a stack of weapons. There is a potential stack effect on each of the following features: (1) the number of fragments, (2) fragment mass distribution, (3) the projection angle of the fragments, and (4) the initial velocity of the fragments.

There is little published information which can be used to quantify these effects, but the following qualitative comments can be made. In a stack explosion, a significant proportion of the fragments hit other objects within the stack and do not leave the stack. This potentially affects not only the number but also the mass distribution of fragments leaving the stack. The angle of projection of the fragments depends on the nature of the stack, which may be an ordered stack or a disordered pile. The former has to be dealt with on a case by case basis, while for the latter the common assumption is that the fragment density is the same in all directions. There may also be stack effects on the initial velocity in so far as fragments may be further accelerated by the hot gases from the explosion.

13.30.2 Explosion of an Explosive Load

Another situation which it may be necessary to model is the explosion of a load of condensed phase explosives, either uncased or cased. The former might typically be blasting explosives, the latter munitions.

The exploding object is envisaged as a load of explosives, typically uncased civil explosives or cased explosives in the form of munitions such as shells and bombs.

The elements of the scenario are the events prior to the situation of imminent risk, the explosion itself, the persons at risk, the physical characteristics of the location, and the human behavior prior to the explosion.

The approach to scenario development takes as its starting point the persons at risk, in other words the targets of the explosion. Categories of persons exposed are defined which are intended to be exhaustive over the whole course of the scenario and estimates made of the numbers in each category at the start of the scenario and of the change in these numbers with time until the explosion occurs.

If the explosion is caused by a sudden event such as impact, the start is the explosion itself. If it is caused by a more gradual event such as fire, the start is the occurrence of an event observable either by the vehicle crew or the public.

The principal effects of a condensed phase explosion such as the explosion of a load of cased explosives on a lorry are (1) blast, (2) fireball, (3) missiles (a, primary fragments; b, secondary missiles), (4) crater, (5) building damage.

These explosion effects act on an exposed population. For this particular type of incident, it is necessary to model this population in some detail. The exposure model used covers the population density, the population's disposition indoors and outdoors, the categories of exposure, the incident scenarios, and the vulnerabilities of persons in each exposure category in given scenarios.

For exposure, a basic distinction is between persons indoors and those outdoors, since this affects vulnerability to the individual injury mechanisms. For example, persons outdoors are more vulnerable to fragments from the casing and those indoors to building collapse.

13.31. EXPLOSION INJURY

13.31.1 Explosion Injury to Persons Outdoors

For persons in the open, there is a large amount of information available on specific explosion injury causes and modes.

The causes of explosion injury to a person in the open include the following: (1) blast, (2) whole body displacement, (3) missiles, (4) thermal effects, and (5) toxic effects.

The modes of explosion injury due to these causes include (1) eardrum rupture, (2) lung hemorrhage, (3) whole body displacement injury, (4) missile injury, (5) burns, and (6) toxic injury.

In the case of outdoor exposure, it is generally possible to apply the injury relations for the individual physical causes, provided that where applicable due allowance is made for the effects of more than one cause.

There are several ways in which injury may be correlated. One is in terms of single values of the injurious physical effect, of which the threshold value and the value for 50% probability of injury are particularly important. Another is a probit equation, which also generally correlates the probability of injury with a single physical effect, typically overpressure. A third is in the form of a P–I diagram.

13.31.2 Explosion Injury to Persons Indoors

The causes of explosion injury to a person indoors include those causing injury to a person in the open, namely (1) blast, (2) whole body displacement, (3) missiles, (4) thermal effects, and (5) toxic effects but, in addition, (6) falling masonry, and (7) asphyxiating dust.

There are also certain differences between the outdoor and indoor situations which modify the effect of some of these causes. Indoors the distances which the body can travel before impact are short and there are many more objects which can become missiles, including walls and windows.

The modes of explosion injury due to these causes include, for the causes common to outdoors and indoors, (1) eardrum rupture, (2) lung hemorrhage, (3) whole body displacement injury, (4) missiles injury, (5) burns, and (6) toxic injury and, for those specific to indoors, (7) crushing, and (8) asphyxiation.

A large proportion of indoor casualties are due to the two causes specific to the indoor situation, falling masonry and asphyxiating dust, and a large proportion of the casualties are therefore in the two injury modes of crushing and asphyxiation. Injury due to these two causes, and in these two modes, is difficult to model by any means other than by correlation with structural damage.

13.31.3 Explosion Injury from Flying Glass

The shattering of window glass is an important blast damage effect, since flying glass can cause severe injury.

A model for the behavior of flying glass should cover the following features: (1) breaking pressure, (2) fragment characteristics, (3) velocity of fragments, (4) spatial density, and (5) distance traveled by fragments. These features are now considered in turn.

The pressure at which windows break is generally quoted in terms of the peak side-on overpressure.

The strength of window glass is very variable. It is not uncommon for an explosion to cause isolated window breakages at very considerable distances, and correspondingly low overpressures, while bangs from supersonic aircraft sometimes break windows at overpressures much less than those usually regarded as necessary to cause damage. Such instances, however, rarely cause injury.

Fletcher et al. (1974) characterize glass fragments in terms of the mean mass m_{50} and mean frontal area A_{50}.

Their expression for the latter, in the form given by Fletcher et al. (1980), is

$$\ln A_{50}^* = 4.2643 - (12.5 + 0.00343 p_e^{o^*})^{1/2} \quad (13.118)$$

where A_{50}^* is the mean frontal area (cm^2) and $p_e^{o^*}$ the peak effective overpressure (kPa).

The mean mass m_{50} may be then obtained as

$$m_{50} = A_{50} \rho_{g1} t \quad (13.119)$$

where A_{50} is the mean frontal area of the fragment (m^2), m_{50} is the mean mass of the fragment (kg), t is the thickness of the fragment (m), and ρ_{gl} is the density of glass (kg/m^3).

A treatment of the velocity of glass fragments is given by Glasstone (1962). He gives a correlation in terms of the geometric mean velocity V_{50}' or antilogarithm of the mean of the logarithms of the velocities. He uses a scaled geometric mean velocity V_{50}', which allows for the thickness of the glass and which is defined as

$$\overline{V}_{50}' = \frac{V_{50}'}{0.83 + 0.019(t' - 0.03)^{-0.93}} \quad (13.120)$$

where t' is the pane thickness (in), V_{50} is the geometric mean velocity (ft/s), and $\overline{V}_{50}$ is the scaled geometric mean velocity (ft/s). The correlation relates the scaled geometric mean velocity to the peak effective overpressure applicable to conventional explosions of 15–500 t. Glasstone states that the correlation is based on experiments with various types of glass ranging from 0.25 in thick plate glass, through various standard thicknesses of single- and double-strength glass, to non-standard glass panes 0.064 in thick. This correlation may be represented as

$$\ln \overline{V}_{50}' = 3.746 + 0.546 \ln p_e^{'o} \quad (13.121)$$

where $p^{'o}$ is the peak effective overpressure (kPa).

A correlation for the spatial density of glass fragments from window breakage based on the Eskimo trials is given by Fletcher et al. (1974). Their expression for the latter, in the form given by Fletcher et al. (1980), is as follows. A scaled spatial density $\overline{\rho}_{sd}'$ is defined as

$$\overline{\rho}_{sd}' = \frac{\rho_{sa}}{4.91 \exp(-5.0121) + 22.28} \quad (13.122)$$

and gives the correlation

$$\ln \overline{\rho}_{sd}' = 3.1037 + 0.05857 p_e^{o^*} \quad (13.123)$$

where $p_e^{o^*}$ is the peak effective overpressure (kPa), ρ_{sd} is the spatial density (fragments/m^2), and t is the glass thickness (cm).

Hadjipavlou and Carr-Hill (1986) have analyzed data from animal experiments and have proposed the following probit equations for laceration and for penetration:

$$Y = -12.23 + 0.83 \ln mu' \quad \text{laceration} \quad (13.124)$$

$$Y = -8.35 + 0.6 \ln mu' \quad \text{penetration} \quad (13.125)$$

where m is the mass of the fragment (g) and u' is its velocity (ft/s).

The question of injury from flying glass is considered in the Second Report of the ACMH (Harvey, 1979). The report is particularly concerned with the risks from a VCE.

In general, much of the concern about possible injury from flying glass relates to injury to people indoors. The shattering of glass as a result of an explosion has occurred at distances up to 20 mi. In such cases, however, the energy of the fragments is very low. The evidence appears to indicate that there are surprisingly few injuries to people from glass fragments even in buildings where most of the windows have been shattered by blast.

It is concluded in the report that there is ample justification for regarding as negligible the risk of injury from flying fragments of window glass for an explosion which gives a peak overpressure outside the building of 0.6 psi (0.04 bar) or less.

13.31.4 Explosion Injury from Penetrating Fragments

One of the principal modes of injury from an explosion is wounding by penetrating fragments.

This problem is of particular interest to the military, and much of the data are from military sources. These include the *Textbook* and the work of Gurney (1944), Dunn and Sterne (1952), and Beyer (1962). The work of the Stockholm International Peace Research Institute (SIPRI) is also relevant.

Table 13.12 gives the injury classification used.

A serious injury is defined more specifically as one involving the following:

Head and neck:	Perforation of the skull
Thorax:	Penetration of the [vulnerable area]
Abdomen:	Penetration of the [vulnerable area]
Limbs:	Penetration of the [vulnerable area].

The abdomen is taken to mean the abdomen proper and that part of the thorax which is not protected by bone, while the thorax is taken as that part of the chest which is so protected.

Traditionally the criterion for military incapacitation has been expressed in terms of a kinetic energy. Later it became recognized that the presented area of the fragment was also relevant.

It has become usual to work, therefore, in terms of the following causative factor X:

$$X = mu^2/A \quad (13.126)$$

TABLE 13.12 Classification of Injury

K	Injury which is fatal, either immediately or in hospital.
S	Injury which is serious, involving perforation of the skin, and which necessitates medical attention, implying hospitalization.
M	Injury which is generally not serious, involving perforation of skin, but which deserves medical attention.
T	Injury which is trivial, possibly penetrating the skin but not perforating it, or no injury at all.

TABLE 13.13 Penetration of Vulnerable Areas by Fragments

Skull[a]

Normal Impact Velocity (ft/s)	Adjusted Impact Velocity,[b] (ft/s)	Causative Factor, mu^2/A (J/m^2)	Probability of Perforation, $P(P_{skull})$ (%)
312	343	1.40×10^6	0
408	449	2.40×10^6	12.7
507	558	3.70×10^6	75.2
591	650	5.03×10^6	82.1
769	846	8.52×10^6	100

where A is the projected area of the fragment (m^2), m is its mass (kg), u is its velocity (m/s), and X is the causative factor (J/m^2).

A distinction is made between skin penetration, or epidermis perforation, and skin perforation, or full-depth perforation.

The problem of skin perforation is relatively complex. In a review of work on this topic, Di Maio (1981) makes a distinction between experiments carried out on samples of skin with, and without, subcutaneous tissue attached. Skin seems to be more resistant to fragment penetration when it is *in situ*, with underlying muscle and fat still attached.

Information on penetration of the skull by fragments is given in the *Textbook*. A series of 54 shots were made using eight fresh skull caps obtained at autopsy at velocities in the range 280−960 m/s. The data given are for penetration by a 53 mg fragment and are given in Table 13.13. The probit equation is

$$Y(P_{skull}) = -40.5 + 3.03 \ln(mu^2/A) \quad (13.127)$$

The *Textbook* gives information on British military casualties in the First World War as given in Table 13.14. These data have been used in relation to fatal injuries of the trunk.

Surveys of battle casualties generally conclude that rapid first aid, medical treatment, and modern surgery can greatly reduce mortality from wounds to the extremities, or limbs, although this is of little help in the case of wounds to the head. This is supported by statistics which show that US battle deaths from extremity wounds were reduced from about 13% of all deaths in the Second World War to 7.4% in Vietnam.

The probability of fatality given penetration to a vulnerable part of a limb $P(K|P_{vul, limb})$ during the Second World War is therefore reduced in the model to half its crude estimated value by taking credit for medical advances. No other credit for medical advances is taken in the model.

Clothing may in principle affect the probability of skin perforation. The work of French and Callender (1962) indicates that although exhibiting a finite threshold velocity for perforation, light clothing is much less resistant than skin. It is assumed in the model that the type of clothing worn by civilians offers very little protection against missiles and hence bare skin criteria are used.

13.32. DUST EXPLOSIONS

Another explosion hazard in the process industries is the explosion of flammable dusts. Explosions of dust suspensions have characteristics in common with gas explosions, but there are also some important differences. Fire and explosions can also occur in dust layers.

Many industrial materials are at some stage handled as dusts or powders and many final products are in dust/powder form. Some typical industrial dusts/powders are wood, coal, food (e.g., starch, flour, sugar, cocoa, feedstuffs), chemicals (e.g., drugs, dyestuffs), plastics (e.g., urea formaldehyde resin, polyethylene, polystyrene) and metals (e.g., aluminum, magnesium).

The hazard of a dust explosion or fire exists wherever flammable dusts are handled. Generally, a dust explosion occurs only if the dust is dispersed in air, but transition from a fire to an explosion can occur, and vice versa. If a burning dust is disturbed, a dust suspension may be formed and ignited. This initial explosion may generate further dust clouds, which in turn explode. On the other hand, burning particles from an explosion may act as the source of ignition for a fire of other flammable materials.

The explosive effect of a dust explosion is caused by the rapid release of heat and the accompanying rapid pressure rise or expansion of the hot gases.

In dust explosions, the combustion process is very rapid. The flame speed is high, comparable with that in gas deflagrations. Maximum explosion pressures are often

TABLE 13.14 Fatal Injury by a Fragment Penetrating to a Vulnerable Area

British Military Casualties in the First World War[a]

	Proportion of Effective Wounds in Each Part of Body (%)			Proportion of Fatal or Serious Injury (%)	Conditional Probability of Fatal Injury, $P(K_i \mid P_{vul, i})$
	Dead	Seriously Wounded	Slightly Wounded		
Head and neck	25.8	27.8	46.4	53.6	0.481
Thorax	59.6	20.7	19.7	80.3	0.782
Abdomen	61.1	23.1	15.7	84.2	0.726
Upper limb	3.0	36.3	60.7	39.3	0.076
Lower limb	3.9	44.4	51.7	48.3	0.081

[a]Source: Textbook of Air Armament (Ministry of Supply, 1952). Section A: table 3; section B: table 2.

close to the theoretical values calculated assuming no heat loss during the explosion.

The sequence of events in a serious industrial dust explosion is often as follows. A primary explosion occurs in an item of plant. The explosion protection is not adequate to prevent the flame issuing from the plant, due either to rupture of the plant or to poor explosion venting. The air disturbance disperses the dust in the work room and causes a secondary explosion. The quantity of dust in the secondary explosion often exceeds that in the primary one. Moreover, the building in which the secondary explosion occurs may be weaker than the plant itself. The secondary explosion is thus often more destructive than the primary one.

In some cases, the primary explosion also occurs in the open and disturbs dust deposits, and this causes a secondary explosion. In other cases, the primary explosion occurs in one unit of the plant and the explosion propagates within the plant to other units.

13.32.1 Dust Explosibility Characteristics

Important explosibility characteristics of dust suspensions are:

1. explosibility classification;
2. minimum explosive concentration;
3. minimum ignition temperature;
4. minimum ignition energy;
5. maximum permissible oxygen concentration to prevent ignition;
6. explosion pressure characteristics
   a. maximum explosion pressure,
   b. maximum rate of pressure rise,
   c. average rate of pressure rise.

An account is given first of the factors which influence dust explosibility, then of the tests used to determine the explosibility parameters and then of the explosibility parameters themselves.

13.32.2 Factors Influencing Dust Explosibility

The explosibility of dust suspensions in air is characterized by parameters similar to those which define the flammability of gas—air mixtures, but there are some significant differences.

Three elements are required to obtain fire: fuel (combustible dust), oxygen, and an ignition source. Two more elements are required to obtain a dust explosion: dispersion of dust particles and confinement. The dust concentrations must be above the minimum explosive concentration and in general, each of these five elements should be present in a specific range.

Based on Eckhoff (2003, 2009) for organic dust, the values of mass of powder/dust per unit volume for a maximum acceptable hygienic exposure are typically between 10^{-3} g/m^3 to 10^{-2} g/m^3, for dust explosions the range of dust concentration in air are around $100-1000$ g/m^3 and for dust deposit combustions the concentration values are higher than 10^6 g/m^3 at normal temperature and pressure.

Some factors which influence dust explosibility are:

Chemical Composition: There are certain chemical groups such as COOH, OH, NH$_2$, NO$_2$, C≡N, C = N, and N = N which tend to be associated with higher dust explosibility and certain others such as Cl, Br, and F with lower explosibility. Dusts of pure metals generally react with air to form metallic oxides. In this case, the explosive increase in pressure is due to expansion of the nitrogen of the air caused by the heat release. In some cases, metals actually react violently with the nitrogen itself to form a metallic nitride. Volatile matter in the dust tends to enhance the explosibility.

Particle Size: Dust explosibility is strongly affected by particle size. Particle size is usually defined in terms of an equivalent particle diameter. Generally, a dust with a particle diameter greater than 500 μm is unlikely to be responsible for initiation of an explosion, though it may undergo combustion in one already occurring. A dust usually contains a range of particle sizes. A relatively small proportion of fine particles enhance the explosibility of a dust. Moreover, attrition caused by handling the dust tends to generate fine particles. Figure 13.16 illustrates the effect of particle size on some of the more important dust explosibility parameters. These results are for atomized aluminum, which is a rather extreme case, but nevertheless they illustrate general trends.

Moisture Content: Moisture content has a strong effect on dust explosibility, although the effect is generally weak for moisture contents below 10%. At the other end of the range, dust with a moisture content greater than 30% is unlikely to be responsible for initiation of an explosion.

Oxygen Concentration: The oxygen concentration in the surrounding atmosphere has a strong effect on dust explosibility, which increases as the oxygen concentration increases. Conversely, the explosibility decreases as the oxygen concentration decreases and the inerts concentration increases.

Inert gas and inert dust concentration: Although the effect is generally weak for inert dust concentrations, below 10−20%.

13.32.3 Dust Explosibility Tests

The tests which are carried out on dust explosibility vary between countries and a large variety of tests are in use. It is convenient to start with those developed at the Bureau of Mines (BM) in the United States, since these have also been widely used in other countries, including the United Kingdom.

Accounts of the various tests in the Bureau of Mines scheme have been given in a series of reports, in particular that by Dorsett et al. (1960 BM RI 5624). A summary has been given by Nagy and Verakis (Nagy and Veraki, 1983). Some of the apparatus used in these tests is illustrated in Figure 13.17.

Dust explosibility classification is performed using the vertical tube, shown in Figure 13.17(a), the horizontal tube, and the inflammator apparatus, shown in Figure 13.18(b) and (c), respectively. The dust is classed as explosible if a positive result is obtained in any one of the three types of test. The minimum explosive concentration is measured in the vertical tube apparatus, the minimum ignition temperature in a modified form of the Godbert−Greenwald furnace, shown in Figure 13.17(d), and the minimum ignition energy in a modified form of the vertical tube apparatus, called MIKE 3 equipment.

The maximum permissible oxygen concentration to avoid ignition used to be measured in the Godbert−Greenwald furnace but is now measured in the vertical tube apparatus at ambient temperature. The dust is dispersed by a blast of the reduced oxygen mixture.

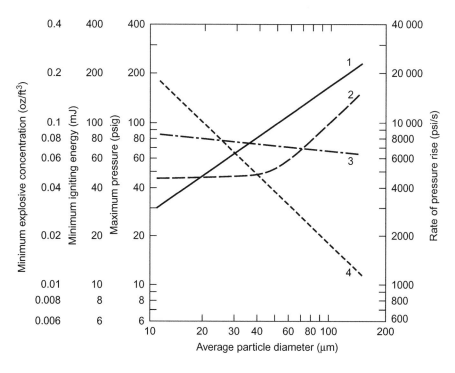

FIGURE 13.16 Dust explosibility characteristics: effect of particle size on some principal parameters for atomized aluminum (Nagy and Verakis, 1983). 1, minimum ignition energy; 2, minimum explosive concentration; 3, maximum explosion pressure; and 4, maximum rate of pressure rise.

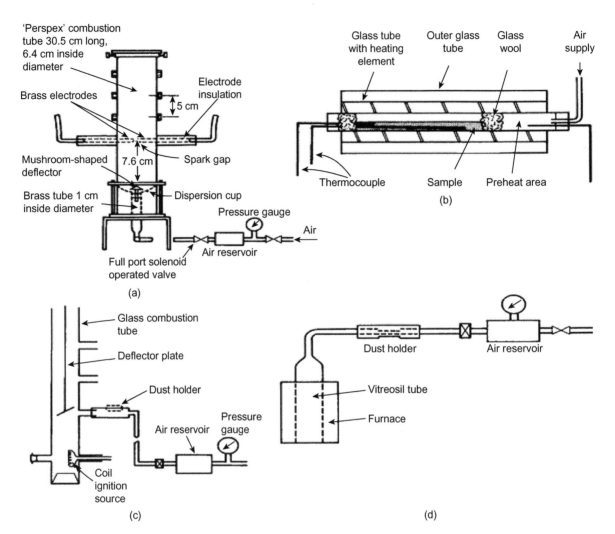

FIGURE 13.17 Dust explosibility characteristics—Bureau of Mines test apparatus (Cross and Farrer, 1982): (a) Hartmann vertical tube apparatus; (b) horizontal tube apparatus; (c) inflammator apparatus; and (d) Godbert–Greenwald furnace. *Source: Courtesy of Plenum Publishing Corporation.*

The ignition source is a hot coil or spark igniter. Any propagation of flame is observed.

The explosion pressure parameters are obtained by measuring the pressure–time profile in a stronger form of the vertical tube apparatus. However, various workers have shown that the Hartmann vertical tube test involves wall effects and gives less than ideal dust dispersion. In this scheme, therefore, basic test for dust explosibility is performed in a 20 l spherical vessel, shown in Figure 13.19. This test is an alternative to the Hartmann vertical tube test in the BM scheme.

In the 20 l sphere test, the dust is injected into the sphere from a separate container. The ignition source is located in the center of the sphere and is usually a chemical igniter with an ignition energy of 10 kJ. A standard time delay is used between injection and ignition.

The 20 l sphere test is used to determine whether a dust is explosible and to measure the maximum explosion pressure and the maximum rate of pressure rise.

The general approach taken in this scheme is to use these spheres to determine also the other explosibility characteristics. Thus, the spheres are used to determine the minimum explosible concentration and the minimum ignition energy.

The analysis of the results is similar to that used for gas explosions. The cubic root law is written in the form

$$\left(\frac{dP}{dt}\right)_{max} V^{1/3} = K_{st} \qquad (13.128)$$

where K_{st} is the dust explosibility constant (or K_{st} value), P is the absolute pressure, t is the time, and V is the volume of the vessel. The K_{st} value is the basis of the dust explosibility classification and is a measure of the maximum rate of pressure rise. It corresponds to the K_G value for gases.

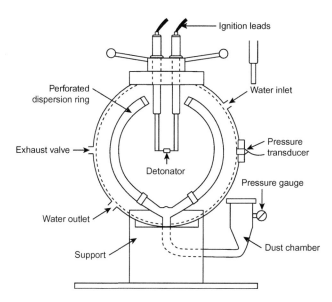

FIGURE 13.18 Dust explosion venting: reduced explosion pressure in deflagration of dust/air mixture in a smooth, straight pipe or duct closed at one end (NFPA 68: 1994).

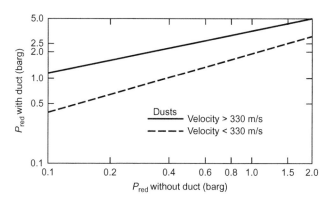

FIGURE 13.19 Dust explosibility characteristics: 20 l sphere test apparatus. *Source: Field (1982); reproduced by permission of Elsevier Science Publishers.*

TABLE 13.15 Dust Explosibility Characteristics: the K_{st} Classification

Dust Explosion Classes

Dust Explosion Class	K_{st}^a(bar m/s)	Explosion Features
St 0	0	No explosion
St 1	>0−200	Weak
St 2	201−300	Strong
St 3	>300	Very strong

[a]These figures are for a strong ignition source (10,000 J). For a weak ignition source (10 J), the corresponding figures are given by NFPA 68: 1978 as 0, >0−100, 101−200, and >200.

The actual K_{st} classification is given in Table 13.15.

Another parameter of interest in relation to dust explosions is the burning velocity. Knowledge of this parameter would allow the use of the large number of methods and models developed for gas explosions and gas explosion venting. The burning velocity is not the subject of standard tests, but some work has been done to estimate burning velocities of dusts.

Maisey (1965) has used an 'equivalent burning velocity' for dust mixtures. His basic approach is to compare the maximum rate of pressure rise for dust mixtures with that obtained for gases with known burning velocity and thus to calibrate the dusts.

13.32.4 Hybrid Dust−Gas Mixtures

If flammable gas is present in addition to the dust, the explosibility of the dust is enhanced. Increase in the concentration of flammable gas results in decrease in the minimum explosive concentration, minimum ignition temperature, minimum ignition energy, and increase in the maximum rate of pressure rise.

The presence of flammable gas can therefore render explosive a dust−gas mixture at a dust concentration which is below the normal lower explosive limit for the dust and at a gas concentration below the normal lower explosive limit for the gas. Another important effect is to make explosive a dust with a particle size large enough to keep it normally non-explosive.

The presence of flammable gas also affects the other explosibility characteristics. In particular, it affects the maximum rate of pressure rise and hence the K_{st} value. Bartknecht gives data showing the effect of different concentrations of methane and propane on the dust explosion class of polyvinyl chloride dust. Dust of class St 0 changes to classes St 1, 1/2, 2, and 3 at methane concentrations of 1%, 3%, 5%, and 7%, respectively, and to classes 1, 2/3 and 3 at propane concentrations of 0.9%, 2.7%, and 4.5%, respectively.

For hybrid mixtures, the minimum ignition energy is also lower than that of the dust alone.

13.32.5 Dust Ignition Sources

The ignition sources for dusts are broadly similar, although there are certain points specific to dusts and dust-handling plant. Some principal ignition sources for dusts include flames and direct heat, hot work, incandescent material, hot surfaces, electrostatic sparks, electrical sparks, friction sparks, impact sparks, self-heating, static electricity, and lightning.

As with gases, it is not easy to compare ignition effects. Different ignition sources are naturally characterized by different quantities such as temperature, energy, and power, as reflected in the laboratory tests performed, so that comparison of different ignition sources is difficult. However, in general terms, dusts can be ignited by low energy as well as high energy ignition sources.

Ignition sources which can occur inside the plant are of particular importance. They include incandescent material, hot surfaces, sparks, self-heating, and static electricity.

Some information of ignition sources for primary explosion is given by Spiegelman (1981), information about ignition sources for dust explosions can be found in Palmer (1981), Jacobsen (1981), and NFPA (1957) for all dusts; Verkade and Chiotti (1978) for grain elevator and mills; and information about probable ignition sources in elevator incidents can be found in Cross and Farrer (1982) and Department of Agriculture (1980).

Further statistical data on ignition sources for dust explosions are given by Lunn (1992). The HSE survey for 1979–1988 gives the principal ignition sources as friction/mechanical failure, overheating/spontaneous heating, flames/flaming material, tramp material, and welding/cutting, with 56 (18%), 51 (17%), 44 (15%), 21 (7%), and 20 (7%) events, respectively; 83 (27%) events were classified in the category 'unknown.' In the BIA survey, the main ignition sources were mechanical sparks, smouldering clumps, mechanical heating, and electrostatic discharge, with 26.1%, 11.3%, 8.9%, and 8.7% events, respectively; 16% of events were classified in the category 'unknown.'

13.32.6 Dust Explosion Prevention and Protection

As for flammable gases, control of dust explosions may be approached by way of prevention or protection.

There are several approaches to the prevention of dust explosions:

1. Use of a dust-free process: A fundamental solution to the dust explosion problem is to use a dust-free process. In particular, it may be possible to process the materials wet rather than dry, so that dust suspensions do not occur at all. This approach is in effect an application of the principle of inherently safer design. Where it is applicable, it is one of the most satisfactory methods. In general, inherent safety principles can be useful in preventing dust explosions by minimization, substitution, moderation, and simplification of dust explosion hazards, as it is given by Amyotte (2009).

2. Avoidance of flammable dust suspensions: If flammable dust has to be handled dry, it is generally not possible to prevent the occurrence of dust concentrations above the lower explosive concentration in some parts of the plant. Nevertheless, much can be done to minimize the volume of any dust clouds formed and to reduce the probability of formation. Dust should be removed regularly to prevent accumulation of dust deposits.

3. Elimination of sources of ignition: Control of ignition sources needs to be addressed both in the design and operation of the plant. Measures which can be taken in design include the location or elimination of direct firing and the avoidance of situations where static electricity can give rise to incendive sparks.

4. Maintenance of equipment to minimize the fault conditions which could constitute ignition sources is another significant aspect of control of ignition sources not only outside but inside equipment.

5. Inerting: The suspension of a flammable dust in air may be rendered non-explosive by the addition of inert gas. The main gases used for inerting of dust-handling plant are nitrogen, carbon dioxide, flue gas, and inert gas from a generator. One factor is any hazards associated with the use of the gas. One such hazard is reaction with the dust: carbon dioxide can react violently with aluminum dust, and nitrogen can react at high temperature with magnesium dust. Carbon dioxide can also generate static electricity. Other relevant factors are the availability and cost of supply. The inerting solution is particularly useful in handling dusts of very high explosibility ($K_{st} > 600$ bar/s). Inerting is not necessarily effective in eliminating dust fires. Inerting may be used in combination with dust explosion suppression or venting.

Methods of protection against dust explosions include:

1. Explosion containment: In some ways containment is an attractive option, since it is an essentially passive method and avoids the problem of relief disposal. It is not usually practicable, however, to design the whole of a dust-handling plant so that it can withstand the pressures generated by dust explosions. Containment may be practicable, on small scale units and on particular equipments. One basic principle is to use rotational symmetry and to avoid large flat surfaces and angular parts. An alternative to full containment is partial containment. This involves the use of a stronger vessel combined with explosion relief.

2. Explosion isolation: The three basic methods of isolation. The first type of isolation is the automatic isolation is applied to a pipe and involves the use of a quick acting shut-off valve. Detection of the explosion is by means of pressure and/or optical sensors. The former are usually preferred, since an optical detector can be blinded. On the other hand, a pressure sensor may not detect a weak pressure wave. A common threshold pressure setting is 0.1 bar. The second type of isolation is the automatic explosion suppression applied to a pipe; the explosion is detected by instrumentation similar to that just described for automatic

isolation. The suppressant barrier is located some 5−10 m from the detectors. Quick acting valves operate most effectively on pipes up to about 0.5 m in diameter. Suppressant barriers have been found effective in pipes up to 2.5 m in diameter. The third type of isolation is the use of a material choke. This is applicable where it is necessary to have a flow of dust between units. A treatment of this method is given in the HSE Dust Explosion Guide. Two commonly used types are rotary valves and worm conveyors.

3. Explosion suppression: The general principle is similar to that for suppression of explosions of flammable gases. Design of an explosion suppression system is based on the maximum rate of pressure rise in the explosion. It may be characterized in terms of the dust St class. In applying the basic data, account should be taken of the features of the particular application, in the light of knowledge of the factors which influence the violence of such an explosion. These include the initial pressure and turbulence and the vessel aspect ratio. Explosion suppression requires the use of a control system, which has several functions. These are (1) to detect the explosion and inject suppressant, (2) to shut down the plant, and (3) to prevent restart of the plant unless it is safe to do so. Detection of the explosion is generally by means of a pressure sensor. Detectors are available which are robust to most materials, to condensation and corrosion, and to shock. In some cases, use is made of two detectors oriented in different planes. Activation occurs when the pressure reaches its threshold value, typically of the order of 0.05 bar. In some systems, use is also made of the rate of pressure rise. The mechanisms of suppression of the explosion are (1) quenching, (2) free radical scavenging, (3) wetting, and (4) inerting. Of these, the principal mechanism is quenching or abstraction of heat. The contribution of free radical scavenging is specific to the particular explosion reaction. Wetting of unburned particles is applicable to liquid suppressants. There is also some inerting effect. The effectiveness of suppression depends in large measure on the injection system. The requirement on this is that it be capable of injecting a large quantity of suppressant in a very short time and with adequate reach to all parts of the space protected. An injection system should be capable of a high mass discharge rate, a high discharge velocity, and hence good 'throw' and good angular coverage. An effective suppression system requires rapid detection, rapid injection, and an adequate quantity of suppressant. The IChemE *Guide* gives a number of graphs of the course of failed suppressions which illustrate these points.

4. Explosion Venting: This part will be illustrated in the following section.

13.32.7 Dust Explosion Venting

The use of explosion venting is generally an effective and economic method of providing protection against dust explosions and is the method normally considered.

Venting is suitable only if there is a safe discharge for the material vented. Preferably the plant should be in the open. If it is in a building, it should be possible to effect a discharge through a short duct. Any such duct will have an effect on the maximum vented pressure and the whole problem should therefore be considered at an early stage of the design.

The venting solution is not appropriate if the plant contains toxic dusts, or other associated toxic substances, which cannot be vented to atmosphere, or is awkwardly sited so that safe discharge is not possible.

An account of the factors which affect dust explosibility was given in Section 13.32.8. There are in addition other factors which influence the strength of an explosion of a dust of given explosibility, including

1. Vessel size and shape
2. Dust concentration
3. Initial pressure
4. Initial temperature: The initial temperature may have several effects. These include reduction of the mass of air available for combustion and reduction of the moisture content of the dust. The net effect of an increase in initial temperature may therefore be to reduce rather than increase the maximum explosion pressure. An increase in initial temperature does, however, increase the maximum rate of pressure rise through its effect on the combustion rate and also through any reduction in moisture content.
5. Initial turbulence: As already indicated, it is difficult to envisage a dust explosion without a degree of initial turbulence. Turbulence tends, however, to be nonuniform and difficult to measure or quantify. In general terms, an increase in initial turbulence will have only a weak effect on the maximum explosion pressure, but a strong effect on the maximum rate of pressure rise. Situations associated with high turbulence include grinding operations.
6. Ignition source: The effect of the ignition source on the strength of the explosion is complex and depends essentially on the nature as well as the strength of the ignition source. Some experiments have shown that similar K_{st} values are obtained for ignition by condenser discharge with an ignition energy in the range 0.005−8 J and by chemical detonator with an ignition energy of 10,000 J.
7. The presence of flammable gas, inert gas, or dust. The effects of the presence of flammable gas and/or inert

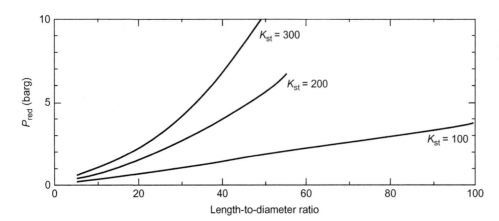

FIGURE 13.20 Dust explosion venting: effect of velocity in vent duct on reduced pressure (NFPA 68: 1988).

gas or inert dust have already been described in discussing dust explosibility.

There are different methods for the estimation of the vent area, including Heinrich method (1966a,b 1974), Palmer method (1974), Rust method (1979), and Nagy and Verakis method (1983).

The theoretical methods all rely more or less heavily on experimental data. Lunn finds that the Heinrich method gives upper limits of the reduced pressure for St 1 and 2 dusts but should not be used for St 3 dusts. The Palmer method gives good predictions for St 1 and 2 dusts but tends to underestimate the reduced pressure for St 3 dusts in larger vessels. The Rust method is better for St 1 dusts than for St 2 and 3 dusts and can be highly inaccurate in some cases. The Pittsburgh method requires information on the burning velocity and turbulence factor which are generally not available and is of limited application.

For silos, Lunn states that the only available method is that given in VDI 3673, which tends to overestimate the vent area. He suggests that the Rust method may also be used for quiescent conditions.

In his later treatment, Lunn (1992) concentrates on the Bartknecht nomograph, the vent ratio, and venting coefficient, or K factor, methods.

If the vent is an open one, venting will result in the emergence of a flame. With a dust, the flame tends to be considerably larger than with a gas. This is illustrated by the series of photographs given by Bartknecht (1981) for gas and dust explosion venting.

If the vent has a duct, there is the potential of unburned dust to undergo combustion, possibly resulting in overpressure in the duct.

The vent duct can have a marked effect on the maximum pressure in a vented explosion.

The general recommendations given by Bartknecht for vent ducts apply to gases and dusts. The vent duct should be straight, as short as practicable and preferably cylindrical. It should have the same pressure rating as the vessel.

Since detonations may develop in pipes of 10−30 m in length, the vent duct should be limited to 10 m. Reaction forces should be allowed for.

There are available several correlations to assist in allowing for the effect of the vent duct. Figure 13.20 shows the relationship between the reduced pressure with the duct and that without the duct. There are two lines, the parameter being the gas velocity.

The correction given in the NFPA 69: 2002 edition for ducts of lengths 2−6 m (10−20 ft) is

$$P'_{red} = 0.172(p_{red})^{19.36}$$

where P'_{red} is the reduced pressure resulting with a vent duct (bar or psi).

Figure 13.18, also given in NFPA 68: 1994, shows the maximum explosion pressure developed in deflagration of dust in a smooth, straight pipe open at one end with an initial flow velocity of less than 2 m/s.

13.32.8 Dust-Handling Plants

Operations in which dusts are generated or handled include size reduction, conveying (manual, mechanical, pneumatic), separation (settling chambers, cyclones, filters, scrubbers, electrostatic precipitators), driers, screening and classifying, mixing and blending, storage, packing, and fired heaters.

All plants handling flammable dusts should have a combination of sufficient strength and explosion protection to withstand a dust explosion safely.

Some other basic principles in the design and operation of dust-handling plants are (1) minimization of space filled by dust suspensions, (2) maintenance of design operating conditions, (3) minimization of mechanical failure and overheating, (4) precautions against static electricity, and (5) precautions against passage of burning dust.

Operations involving screening and classifying or mixing and blending tend to create dust suspensions.

Plants for such operations should be enclosed and preferably run under a slight vacuum. Precautions should be taken against static electricity and explosion protection should be provided.

Where dusts such as pulverized coal are used as fuels for fired heaters or furnaces, an explosive dust suspension is burned under controlled conditions. As with other fuels, the principal hazard is loss of ignition followed by reignition and explosion.

Accounts of the safe design of dust-handling plants are given by Palmer (1973), Bartknecht (1981), and Field (1982).

13.32.9 Dust Fires

It is convenient to deal at this point with dust fires. An account of dust fires is given by Palmer (1973).

Dust fires occur in dust deposits and are of two types—laming and smouldering fires.

Detection of dust fires is difficult. There are no special types of detector for dust fires. The effects given by smouldering are usually weak and difficult to detect with the sensors normally used in automatic fire protection systems. It may be appropriate, however, to monitor the temperature in the interior of large dust piles with thermocouples.

This delay between ignition and outbreak of flaming can create hazards. Fire may break out unexpectedly in a factory shut-down overnight or at the weekend, or the cargo of a ship may be discovered to be on fire when it is being unloaded. Hazards of dust fires include those of a dust explosion resulting from the formation and ignition of a dust suspension, of the ignition of other flammables, and of the evolution of toxic combustion products.

In general, dust deposits around the factory should be minimized by good housekeeping. As already indicated, the dust layer does not need to be very thick to sustain smouldering. Extinction of the dust fire may be affected by letting the fire burn itself out, by applying extinguishing agents or by starving the fire of oxygen. Whichever approach is used, however, it is essential to avoid disturbing the dust in such a way as to allow a suspension to form and ignite.

Water is the usual extinguishing agent and is suitable unless it reacts with the dust or electrical equipment is involved. The water should not be applied, however, as a high pressure jet, which could raise a dust cloud, but as a low pressure spray which simply dampens the dust deposit. The penetration of this water into the dust layer can be assisted by the addition of about 2% of a wetting agent such as a detergent.

The use of fire fighting foam generally offers little advantage over water. Foam has the additional capability of cutting off air to the fire, but this is unlikely to have much effect in a slow smouldering combustion which requires very little air anyway. Since the foam is largely water, it should not be used in applications where water is unsuitable.

The other non-gaseous fire fighting agents are of limited application. Dry powder is appropriate if the dust is one which reacts with water, as do some metal dusts. Vaporizing liquids are appropriate if electrical equipment is involved but should not be used on reactive metal dusts.

Inert gases may be used as extinguishing agents and can be effective if the dust is held in a relatively gas tight container such as a hopper or a ship's hold. It is necessary, however, not only to cut off the supply of oxygen, but also to effect sufficient cooling to prevent reignition when the air supply is restored. Thus, the inerting may need to be maintained for a long period.

In some situations, however, the fire cannot be extinguished sufficiently rapidly by fire extinguishing agents alone. A smouldering fire in a ship's hold, for example, could immobilize the whole ship. In such cases, it may be necessary to dig the material out. This involves a number of hazards and suitable precautions should be taken against them. The operations should be conducted in such a way as not to raise a dust cloud. The atmosphere should be monitored and any necessary breathing equipment worn. The danger of subsidence of the dust due to the creation of burnt-out hollows beneath the surface should be allowed for.

In other instances, a more gradual extinction of the fire is acceptable. This might be the case, for example, with a fire smouldering inside a tip. The methods used in such situations are generally based on excluding air and include covering the heap with a layer of noncombustible such as earth or pumping a limestone slurry into fissures in the heap. A hazard in the first method is subsidence due to hollows, and in the second, movement in the tip due to the slurry. Alternatively, but generally as a last resort, the heap may be dug out.

13.33. EXPLOSION HAZARD

The types of explosion typical of the process industries are those described above. The hazard of a large industrial explosion may be assessed by consideration of assumed scenarios using appropriate hazard and effects models or of the historical record of explosions and their effects.

13.33.1 Historical Experience

The main classification of process industries explosions are (1) physical explosions, (2) condensed phase explosions,

(3) VCEs, (4) BLEVEs, (5) confined explosions with reaction, (6) vapor escape into, and explosions in, buildings, and (7) dust explosions.

The process industries have suffered major explosions in virtually all the categories of the classification in Section 13.1. Considering these in turn, and starting with physical explosions, these may occur as (1) a mechanical failure, (2) overpressure, (3) underpressure, (4) overtemperature, or (5) undertemperature of the system. The essential distinction is that in the first case the failure occurs while the process conditions are within the design envelope, so that the failure is due to a mechanical defect, and in the other cases it occurs because the process conditions have been taken outside the design envelope.

Condensed phase explosions in the process industries include explosions of (1) high explosives, (2) ammonium nitrate, (3) organic peroxides, and (4) sodium chlorate.

In recent years, the most destructive explosions in the process industries have tended to be VCEs or BLEVEs.

Some of the principal VCEs include those at the Ludwigshafen works in 1953 and 1958, the refinery at Lake Charles in 1967, the refinery at Pernis in 1968, a pipeline at Port Hudson in 1970, a rail tank car at East St Louis in 1972, the works at Flixborough in 1974, the rail tank car at Decatur in 1974, a petrochemical plant at Beek in 1975, and a petrochemical plant at Pasadena in 1989. Flixborough and Pasadena are described in Appendices 2 and 6, respectively.

Major incidents in which the BLEVE occurred due to release from the vessel involved or a similar, adjacent vessel include those at a storage in Montreal in 1957, the refinery at Feyzin in 1966, a rail tank car at Crescent City in 1970, a refinery in Rio de Janeiro in 1972, and a storage at Texas City in 1978.

13.33.2 Mortality Index for Explosions

The relation between the size of an explosion and the number of people killed has been investigated by Marshall (1977), who has developed a mortality index (deaths/t) for explosions.

Data on the relation between the size of the explosion and the number of fatalities, and on the mortality index, are plotted in Figure 13.21. The solid squares which do not have a specific reference are accidental explosions. The full line is the best fit for these accidental explosions.

Mortality data on the number of casualties by shell explosions are available from the First World War. According to one account, the belligerents used 2.23×10^6 t of high explosive in 1.39×10^9 shells and caused 10^7 casualties. The latter figure includes wounded as well as dead, and the deaths are estimated at 2.4×10^6. Thus, on average a shell contained 1.6 kg of explosive. Allowing for 10% dud

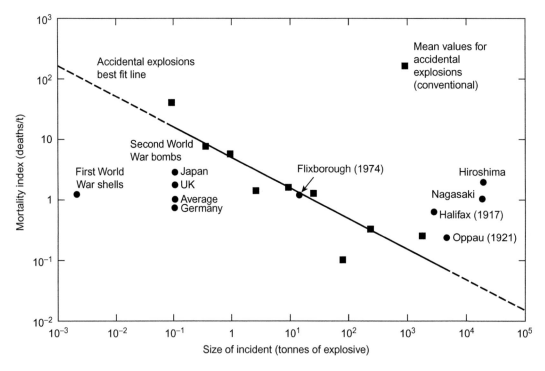

FIGURE 13.21   Mortality index for explosions. *Source: After V.C. Marshall (1977); Courtesy of the Institution of Chemical Engineers.*

shells, the fatalities per shell are approximately 0.0019, which gives a mortality index of 1.2.

As Figure 13.21 shows, the mortality index decreases as the size of the explosion increases. The line has a slope of about −0.5. This differs somewhat from the slope suggested by theoretical considerations. From the blast scaling laws, the radius r at which a given blast intensity occurs is related to the mass of explosive W by the relation

$$r \propto W^{1/3} \tag{13.129}$$

The number of fatalities f is proportional to the area:

$$f \propto r^2 \propto W^{2/3} \tag{13.130}$$

The mortality index MI is the ratio of the number of fatalities to the mass of explosive:

$$\mathrm{MI} \, \alpha \, \frac{f}{W} \, \alpha \, W^{-1/3} \tag{13.131}$$

Marshall suggests that a possible explanation is that small explosions which do not kill people tend not to be well reported

13.33.3 Hazard Assessment

Hazard assessment of explosion on process plants tends to be rather more varied than that of fire. In principle, a hazard assessment involves formulation of a scenario and a full assessment of the frequency and consequences of the event. An assessment which is confined either to the estimation of the frequency of the event or to the modeling of the consequences is a partial one. Hazard assessments may be generic or may relate to a specific situation.

The scenario which has received most attention is the release and ignition of a vapor cloud, which may result either in a vapor cloud fire or a VCE. Thus, the hazard assessment of VCEs involves the development of a set of scenarios for explosion, by release and ignition, and of the associated event trees; the estimation of the frequency/probability of the initial events and of the events in the branches of the event trees from a mixture of historical data and synthesis of values, often using fault trees, and the modeling of the outcome events in the event tree, including both physical events and damage/injury.

Up to the point of damage and injury, the assessment of a VCE involves essentially the same set of models as those used for the modeling of vapor cloud fires. If the event gives an overpressure, it is a VCE, otherwise it is a vapor cloud fire.

A hazard assessment of the kind just described constitutes a full hazard assessment. The types of explosion which are most fully represented are high explosives and ammonium nitrate explosions, VCEs, and BLEVEs.

An early, essentially generic, hazard assessment of explosion is that given in the vulnerability model by Eisenberg et al. (1975), who studied the area affected by a flammable vapor cloud and the lethality of a VCE in such a cloud. The events of principal interest which they consider are marine spillage of 20,750, 83, and 0.8 te of LNG on water with drifting of the vapor cloud toward populated areas.

For condensed phase explosions, the hazard assessments of explosions of high explosives are represented by those in the ACDS study and of ammonium nitrate explosions by those in the Canvey Reports and the ACDS study. These are for specific situations, although in the ACDS study the results are then generalized.

13.34. HAZARD RANGE OF EXPLOSIONS

The estimation of the hazard range of an explosion depends on the type of explosion considered.

The two main types of explosion of interest here are condensed phase explosions such as those of TNT, together with explosions which can be characterized in terms of TNT equivalent, and VCEs.

13.34.1 Condensed Phase Explosions

The blast wave from a condensed phase explosion is characterized by the peak side-on overpressure and the impulse.

The variation of peak overpressure of a condensed phase explosion with distance is given by the scaling law and values of the overpressure may be obtained from graphs.

$$p^o = f(z) \tag{13.132}$$

with

$$z = \frac{r}{W^{1/3}} \tag{13.133}$$

where p^o is the peak overpressure, r is the distance, W is the mass of explosive, and z is the scaled distance. Over a limited range, the curve of p^o vs z may be approximated by a straight line so that

$$\frac{(p^o)r}{p^o} = \left(\frac{r_r}{r}\right)^n \tag{13.134}$$

where n is an index and subscript r is reference value. From the Baker curve for the overpressure range 1.0−0.1 bar, the index n is approximately 1.7. A similar approach may be used for the determination of the impulse.

13.34.2 Vapor Cloud Explosions

For a VCE, it is necessary to take into account the movement of the cloud, which is likely then to be the dominant factor.

The variation of cloud size with distance may be estimated from gas dispersion relations. An account is given there of the method of Considine and Grint (1985) which allows an estimate to be made of the size of the cloud for an LPG release.

Considine and Grint have also given for LPG methods for estimating the hazard range for fires and explosions. For a VCE, they use the ACMH method based on the TNT equivalent. They give for the overpressure the relations

$$p^o = 138 \left(\frac{I^{1/3}}{r_c} \right)^{1.558} \quad r < r_c; \quad 0.05 < p^o < 1 \quad (13.135a)$$

$$p^o = 138 \left(\frac{I^{1/3}}{r_c} \right)^{1.558} \quad r < r_c; \quad 0.05 < p^o < 1 \quad (13.135b)$$

where I is the mass of vapor in the cloud (te), p^o is the peak overpressure (bar), and r_c is the radius of the cloud (m).

These authors quote from the *First Canvey Report* the estimates of 50% and 1% fatalities at overpressures of 0.36 and 0.18 bar, respectively. Within the cloud, 100% fatalities are assumed.

For a quasi-instantaneous release, the cloud range R and the cloud radius r_c vary with time. The mass m of vapor in the cloud may be taken as the total mass released. The cloud is flammable, however, only while its concentration is within the flammable range.

For a quasi-continuous release, the approach taken is to treat the plume as if it were fully established. In this case, the cloud is non-circular and the following relation is given for the equivalent value of cloud radius r_c:

$$r_c = \left(\frac{1.2 k^* R_{LFL}^{5/3}}{\pi} \right)^{1/2} \quad (13.136)$$

where R_{LFL} is the distance to the LFL (m) and k^* is a constant defining crosswind range.

REFERENCES

Ale, B.J.M., Bruning, F., 1980. Unconfined vapour cloud explosions (letter). Chem. Eng. Lond. 352, 47.

Amyotte, P.R., Eckhoff, R.K., 2009. Dust explosion causation, prevention and mitigation: an overview. J. Chem. Health Saf. 17 (1), 15–28.

Anthony, E.J., 1977. Some aspects of unconfined gas and vapour explosions. J. Hazard Mater. 1 (4), 289.

Ardron, K.H., Baum, M.R., Lee, M.H., 1977. On the evaluation of dynamic forces and missile energies arising in a water reactor loss-of-coolant accident. J. Br. Nucl. Energy Soc. 16, 81.

Baker, Q.A., Tang, M.J., Scheier, E., Silva, G.J., 1994. Vapor cloud explosion analysis. AIChE 28th Annual Loss Prevention Symposium, 17–21 April, Houston, TX.

Baker, Q.A., Doolittle, C.M., Fitzgerald, G.A., Tang, M.J., 1997. Recent developments in the Baker-Strehlow VCE analysis methodology. AIChE 31st Loss Prevention Symposium, 9–13 March, Houston, TX, paper 42f.

Baker, W., Kulesz, J., Ricker, R., Westine, P., Parr, V., Vargas, L., Mosely, P. 1978. Workbook for estimating the effects of accidental explosion in propellant handling systems. NASA Contractor report (3023).

Baker, W.E., Kulesz, J.J., Ricker, R.E., Bessey, R.L., Westline, P.S., 1975. Workbook for Predicting Pressure Wave and Fragment Effect of Exploding Propellent Tanks and Gas Storage Vessels (No. 02-4130). Southwest Research Inst, San Antonio, TX.

Baker, W.E., Cox, P.A., Westine, P.S., Kulesz, J.J., Strehlow, R.A., 1983. Explosion Hazards and Evaluation. Elsevier, Amsterdam.

Bartknecht, W., 1981. Explosions: Course, Prevention, Protection. Springer, Berlin.

Beveridge, H.J.R., Jones, C.G., 1984. Shock effects on a bursting disc in a relief manifold. The Protection of Exothermic Reactors and Pressurised Storage Vessels. IChemE Symposium. Series, 85, 207–214 (IChemE, Rugby, UK).

Beyer, J.C., 1962. Wound Ballistics. Office of the Surgeon General, Department of the Army, Washington, DC.

Blything, K., Reeves, A., Britain, G., 1988. An initial prediction of the BLEVE frequency of a 100 te butane storage vessel: UKAEA Safety and Reliability Directorate.

Boyer, D.W., Brode, H.L., Glass, I.I., Hall, J.G., 1958. Blast from a Pressurized Sphere. Rep. UTIA 48. Institute of Aerophysics, University of Toronto.

BPF, 1979. Guidelines for the Safer Production of Phenolic Resins, British Plastics Federation, London.

Brasie, W.C., Simpson, D.W., 1968. Guidelines for estimating damage explosion. Loss Prev. 2, 91.

Bretherick, L., 1985. Handbook of Reactive Chemical Hazards, third ed. Butterworths, London.

Brown, B., 1969. General Properties of Matter. Butterworths, London.

Bulkley, W.L., Jacobs, R.B., 1966. Hazard of atmospheric releases of large volumes of hydrogen. Facilities Subcommittee, American Petroleum Institute. October (4).

Bull, D.C., 1992. Review of large-scale explosion experiments. Plant/Operations Prog. 11, 33.

Burgess, D.S., Murphy, J.N., Hanna, N.E., van Dolah, R.W., 1968. Large scale studies of gas detonation. Bureau of Mines, Investigations Report 7196.

CCPS, 1994. Guidelines for Evaluating the Characteristics of Vapor Cloud Explosions, Flash Fires, and Bleves. American Institute of Chemical Engineers, New York, NY, ISBN 0-8169-0474-X, p. 387.

Christopherson, D.G., 1946. Structural Defence. Rep. RC 450. Ministry of Home Security, London.

Clancey, V.J., 1972. Diagnostic features of explosion damage. Sixth International Meeting on Forensic Science, Edinburgh.

Considine, M., Grint, G.C., 1985. Rapid assessment of the consequences of LPG releases. Gastech. 84, 187.

Cottrell, W.B., and Savolainen, A.W., 1965. US Reactor Containment Technology. A Compilation of Current Practice in Analysis, Design,

Construction, Test, and Operation. No. ORNL-NSIC-5, vol. I and II. Oak Ridge National Lab., Tenn. and Bechtel Corp., San Francisco, CA.

Cousins, E.W., Cotton, P.E., 1951. Design closed vessels to withstand internal explosions. Chem. Eng. 58 (8), 133.

Cox, B., Saville, G., 1977. High pressure safety code: High Pressure Technology Association.

Cross, J., Farrer, D., 1982. Dust Explosions. Plenum Press, New York, NY.

Cubbage, P.A., Marshall, M.R., 1973. Pressures Generated by Gas–Air Mixtures in Vented Enclosures. Institution of Gas Engineers, London.

Cubbage, P.A., Simmonds, W.A., 1955a. An Investigation of Explosion Reliefs for Industrial Drying Ovens. Res. Comm. GC23. Gas Council, London.

Cubbage, P.A., Simmonds, W.A., 1955b. Investigation of explosion reliefs for industrial drying ovens. Gas World. 142 (3719), 1380.

Cubbage, P.A., Simmonds, W.A., 1955c. An investigation of explosion reliefs for industrial drying ovens—I, Top reliefs in box ovens. Trans. Inst. Gas Eng. 105, 470.

Davenport, J.A., 1977. A survey of vapor cloud incidents. Loss Prev. 11, 39.

Davenport, J.A., 1983. A study of vapor cloud incidents—an update. Loss Prev. Saf. Promotion. 1 (4), C1.

Davis, B.C., 1983. Flare efficiency studies. Plant/Operations Prog. 2, 191.

Decker, D.A., 1971. Explosion venting guide. Fire Technol. 7 (3), 219.

Department of Agriculture, ca. 1980. In: Cross, J., Farrer, D., op. cit., p. 7.

Di Maio, V.J.M., 1981. Penetration and perforation of skin by bullets and missiles. Am. J. Forensic Med. Pathol. 2 (2), 107.

Doring, W., 1943. On detonation processes in gases. Ann. Phys. 43, 421.

Dorsett, H.G., 1960. Laboratory Equipment and Test Procedures for Evaluating Explosibility of Dusts. Department of the Interior, Bureau of Mines, Washington, D.C.

Dragosavic, M., 1973. Structural measures against gas explosion in high rise blocks of flats. Heron. 19, 5.

Dunn, D.J., Stern, T.E., 1952. Hand Grenades for Rapid Incapacitation. Rep. BRL R-806. Ballistic Research Laboratory, Aberdeen Proving Ground, MD.

Duxbury, H.A., 1976. Gas vent sizing methods. Loss Prev. 10, 147.

Duxbury, H.A., Wilday, A.J., 1987. Calculation methods for reactor relief: a perspective based on ICI experience. Hazards Press.175.

Duxbury, H.A., Wilday, A.J., 1989. Efficient design of reactor relief systems. Runaway React.372.

Duxbury, H.A., Wilday, A.J., 1990. The design of reactor relief systems. Process Saf. Environ. 68B, 24.

Eckhoff, R.K., 2003. Dust Explosions—Origin, Propagation, Prevention, and Mitigation: An Overview, third ed. Gulf Professional Publishing, Burlingtonp, pp. 1–156.

Eckhoff, R.K., 2009. Understanding dust explosions. The role of powder science and technology. J. Loss Prev. Process Ind. 22 (1), 105–116.

Eichler, T.V., Napadensky, H.S., 1977. Accidental Vapor Phase Explosions on Transportation Routes Near Nuclear Plants. Rep. J6405 ANL-K77-3776-1. Argonne National Laboratory, Argonne, IL.

Eisenberg, N.A., Lynch, C.J., Breeding, R.J., 1975. Vulnerability Model: A Simulation System for Assessing Damage Resulting from Marine Spills. Rep. CG-D-136−75. Enviro Control Inc., Rockville, MD.

Epstein, M., Fauske, H.K., Hauser, G.M., 1989. The onset of two-phase venting via entrainment in liquid-filled storage vessels exposed to fire. J. Loss Prev. Process Ind. 2 (1), 45.

Esparza, E.D., Baker, W.E., 1977. Measurement of Blast Waves from Bursting Pressurized Frangible Spheres. Rep. NASA CR-2843. NASA Science and Technology Information Office, Washington, DC.

Exxon, Exxon, n.d. Damage estimates from BLEVEs. Unpublished work, quoted by CCPS.

FS 6012, 1989. Flammable Liquids and Gases: Explosion Control. Fire Protection Agency, London.

FS 6015, 1974. Explosible Dusts, Flammable Liquids and Gases: Explosion Suppression. Fire Protection Agency, London.

Fairweather, M., Vasey, M.W., 1982. A mathematical model for the prediction of overpressures generated in totally confined and vented explosions. Combustion 19, 546.

Fauske, H.K., 1984a. Generalized vent sizing monogram for runaway chemical reactions. Plant/Operations Prog. 3, 213.

Fauske, H.K., 1984b. A quick approach to reactor vent sizing. Plant/ Operations Prog. 3, 145.

Fauske, H.K., 1984c. Scale-up for safety relief runaway reactions. Plant/ Operations Prog. 3, 7.

Fauske, H.K., 1985a. Emergency relief system (ERS) design. Chem. Eng. Prog. 81 (8), 53.

Fauske, H.K., 1985b. Flashing flows—some practical guidelines for emergency releases. Plant/Operations Prog. 4, 132.

Fauske, H.K., 1987. Pressure relief and venting: some practical considerations related to hazard control. Hazards from Pressure. Institution of Chemical Engineers, Rugby.

Fauske, H.K., 1989. Emergency relief system design for runaway chemical reaction: extension of the DIERS methodology. Chem. Eng. Res. Des. 67, 199.

Fauske, H.K., Epstein, M., Grolmes, M.A., Leung, J.C., 1986. Emergency relief vent sizing for fire emergencies involving liquid filled atmospheric storage vessels. Plant/Operations Prog. 5, 205.

Fauske, H.K., Grolmes, M.A., Clare, G.H., 1989. Process safety evaluation applying DIERS methodology to existing plant operations. Plant/Operations Prog. 8, 19.

Field, P., 1982. Dust Explosions. Elsevier, Amsterdam.

Fisher, H.G., 1985. DIERS research program on emergency relief systems. Chem. Eng. Prog. 81 (8), 33.

Fisher, H.G., 1991. An overview of emergency system design practice. Plant/Operations Prog. 10, 1.

Fletcher, E.R., Richmond, D.R., Richmond, D.W., 1974. Air blast effects on windows in buildings and automobiles on the Eskimo II event. Explosives Safety Board, Department of Defense. Proceedings of the Sixteenth Explosives Safety Seminar, vol. 1, p. 185.

Fletcher, E.R., Richmond, D.R., Yelverton, J.T., 1980. Glass Fragment Hazard from Windows Broken by Air Blast. Rep. DNA 5593T. Defense Nuclear Agency, Department of Defence, Washington, DC.

Factory Mutual Research Corporation, 1990. Private Communication to CCPS. Factory Mutual Research Corporation.

French, R.W., Callender, G.R., 1962. Ballistic characteristics of wounding agents. In: Beyer, J.C. (ed.) (1962), op. cit., chapter 2.

Giesbrecht, H., Hess, K., Leuckel, W., Maurer, B., Stoeckel, A., 1980. Analysis of the potential explosion effects of flammable gases during short time release into the atmosphere. In: Hartwig, S., op. cit., p. 207.

Glasstone, S., 1962. The Effects of Nuclear Weapons, revised ed. Atomic Energy Commission, Washington, DC.

Glasstone, S., Dolan, P.J., 1980. The Effects of Nuclear Weapons, third ed. Castle House Publications, Tonbridge Wells.

Grabowski, G.J., 1965. Suppression by design. Chem. Eng. Prog. 61 (9), 38.

Grolmes, M.A., Leung, J.C., Fauske, H.K., 1983. Transient two-phase flow discharge of flashing liquids following a break in a long transmission pipe line. In: Veziroglu, T.N., op. cit., p. 299.

Grossel, S.S., 1986. Design and sizing of knock-out drums/catchtanks for reactor emergency relief systems. Plant/Operations Prog. 5, 129.

Grossel, S.S., 1990. An overview of equipment for containment and disposal of emergency relief system effluents. J. Loss Prev. Process Ind. 3 (1), 112.

Gugan, K., 1979. Unconfined Vapour Cloud Explosions. Institution of Chemical Engineers, Rugby.

Gurney, R.W., 1943. The Initial Velocities of Fragments from Bombs, Shells and Grenades. Rep BRL 405. US Army, Ballistics Research Laboratory, Aberdeen Proving Ground, MD.

Gurney, R.W., 1944. A New Casualty Criterion. Rep BRL 498. US Army, Ballistics Research Laboratory, Aberdeen Proving Ground, MD.

Gwaltney, R.C., 1968. Missile Generation and Protection in Light Water Cooled Power Reactor Plants. Rep. ORNL-NSIC-22. Oak Ridge National Laboratory, Oak Ridge, MD.

Hadjipavlou, S., Carr-Hill, G., 1986. A Review of the Blast Casualties Rules Applicable to UK Houses. Publication 34/86. Science Research Development Branch, Home Office, London.

Hardee, H.C., Lee, D.O., 1975. Expansion of clouds from pressurised liquids. Accid. Anal. Prev. 7, 91.

Harris, R.J., 1983. Investigation and Control of Gas Explosions in Buildings and Heating Plant. Spon, London.

Harris, R.J., Wickens, M.J., 1989. Understanding Vapour Cloud Explosions—An Experimental Study. Comm. 1408. Institution of Gas Engineers, London.

Harvey, B.H., 1979. Second Report of the Advisory Committee on Major Hazards. HM Stationery Office, London.

Heinrich, H.J., 1966a. Determination of relief openings for protection of plants subject to explosion in the chemical industry. Chem. Ing. Tech. 38, 1125.

Heinrich, H.J., 1966b. Sizing of pressure relief openings for the explosion protection of plants in the chemical industry. Chem. Ing. Tech. 38 (11), 1125.

Heinrich, H.J., 1974. Pressure relief of dust explosions. Arbeitsschutz. 11, 314.

Held, M., 1990. Fragment mass distribution of HE projectiles. Propell. Explos. Pyrotech. 15, 254.

Henry, R.E., Fauske, H.K., 1971. The two-phase critical flow of one-component mixtures in nozzles, orifices and short tubes. J. Heat Transfer. 93 (May), 179.

High, W.G., 1967. The design and scale model testing of a cubicle to house oxidation or high pressure equipment. Chem. Ind. June, 899.

Hirsch, E., 1986. Improved Gurney formulas for exploding cylinders and spheres using 'hard core' approximation. Propell. Explos. Pyrotech. 11, 81.

Holden, P.L., Reeves, A.B., 1985. Fragment hazards from failures of pressurised liquefied gas vessels. The Assessment and Control of Major Hazards. Institution of Chemical Engineers, Rugby.

Howard, W.B., 1972. Interpretation of a building accident. Loss Prev. 6, 68.

Howard, W.B., 1973. Reactor Relief Systems for Phenolic Resins. Monsanto Company.

HSE (Health and Safety Executive), 1965. Guide to the Use of Flame Arresters and Explosion Reliefs. A Guide to the Health and Safety at Work Act, Vol. 34. HM Stationery Office, London.

Huff, J.E., 1973. Computer simulation of polymerizer pressure relief. Loss Prev. 7, 45.

IRI, 1990, IR Information Manual. Industrial Risk Insurers, Hartford, Connecticut.

Jacobsen, M., 1981. Historical background of dust explosion research in the United States 1910–1970. The Hazards of Industrial Explosion from Dusts. Oyez/IBC, London.

Jarrett, D.E., 1968. Derivation of the British explosives safety distances. Ann. N.Y. Acad. Sci. 152 (Art. 1), 18.

Johnson, D.M., Pritchard, M.J., 1991. Large scale experimental study of boiling liquid expanding vapour explosions (BLEVEs). Gastech. 90, 3.3.

Kamlet, M.J., Finger, M., 1979. An alternative method of calculating Gurney velocities. Combust. Flame. 34, 213.

Karpp, R.R., Predebon, W.W., 1974. Calculation of fragment velocities from fragmentation munitions. First International Symposium on Ballistics, IV–145.

Kiefer, P.J., Kinney, G.F., Stuart, M.C., 1954. Principles of Engineering Thermodynamics. Wiley, New York, NY.

King, R., 1975a. Flixborough—the role of water reexamined. Process Eng. September, 69.

King, R., 1975b. Major fire and explosion hazards in hydrocarbon processing plants. Protection. 12, 9.

King, R., 1975c. A mechanism for transient pressure increase. Symposium on Technical Lessons of Flixborough, Institution of Chemical Engineers, London.

King, R., 1976a. Accidents caused by water in process plants. Ind. Saf. January, 6.

King, R., 1976b. Danger—Beware of Water. Fire. February, 464.

King, R., 1977. Latent superheat—a hazard of two phase liquid systems. Chem. Process Hazards. VI, 139.

King, R., 1990. Safety in the Process Industries. Butterworth-Heinemann, London.

Kinney, G.F., 1962. Explosive Shocks in Air. Macmillan, New York, NY.

Kinney, G.F., Graham, K.J., 1985. Explosive Shocks in Air, second ed. Springer, Berlin.

Kletz, T.A., 1977. What are the causes of change and innovation in safety? Loss Prev. Saf. Promotion. 2, 1.

Kneale, M., 1984. The engineering of relief disposal. The Protection of Exothermic Reactors and Pressurised Storage Vessels. Institution of Chemical Engineers, Rugby.

Kordylewski, W., Wach, J., 1986. Influence of ducting on the explosion pressure. Combust. Flame. 66, 77.

Kordylewski, W., Wach, J., 1988. Influence of ducting on explosion pressure: small scale experiments. Combust. Flame. 71, 51.

Kuhl, A.L., Kamel, M.M., Oppenheim, A.K., 1973. Pressure waves generated by steady flames. Combustion. 14, 1201.

Kuo, K.K., 1986. Principles of Combustion. Wiley, New York, NY.

Lee, J.H.S., 1983. Gas cloud explosion—current status. Fire Saf. J. 5, 251.

Lenoir, E.M., Davenport, J.A., 1992. A survey of vapor cloud explosions-second update. American Institute of Chemical Engineers. Twenty-Sixth Symposium on Loss Prevention (New York), paper 74d.

Leung, J.C., 1986a. A generalised correlation for one-component homogeneous equilibrium flashing choked flow. AIChE J. 32, 1743.

Leung, J.C., 1986b. Simplified vent sizing equations for emergency relief requirements in reactors and storage vessels. AIChE J. 32, 1622.

Leung, J.C., Fauske, H.K., 1987. Runaway system characterization and vent sizing based on DIERS methodology. Plant/Operations Prog. 6, 77.

Lewis, B., von Elbe, G., 1987. Combustion, Flames and Explosions of Gases, third ed. Academic Press, Orlando, FC.

Loving, F.A., 1957. Barricading hazardous reactions. Ind. Eng. Chem. 49, 1744.

Lunn, G.A., 1984. Venting Gas and Dust Explosions—A Review. Institution of Chemical Engineers, Rugby.

Lunn, G.A., 1992. Guide to Dust Explosion Prevention and Protection—Part 1, Venting, second ed. Institution of Chemical Engineers, Rugby.

Maisey, H.R., 1965. Gaseous and dust explosion venting. Chem. Process. Eng. 46 (10), 527.

Manas, J.L., 1984. BLEVEs—their nature and prevention. Fire Int. 87, 27.

Marshall, V.C., 1977. How lethal are explosions and toxic escapes? Chem. Eng. Lond. 323, 573.

Marshall, V.C., 1987. Major Chemical Hazards. Ellis Horwood, Chichester.

Maurer, B., 1977. Modeling of vapour cloud dispersion and deflagration after bursting of tanks filled with liquefied gas. Loss Prev. Saf. Promotion Process Ind. 2, 305.

Mercx, W.P.M., Van Wingerden, C.J.M., Pasman, H.J., 1993. Venting of gaseous explosions. Process Saf. Prog. 12, 40.

Ministry of Supply, 1952. Textbook of Air Armaments, Chapter 12: Vulnerability of human targets to fragmenting and blast weapons. London (now declassified).

Moore, C.V., 1967. The design of barricades for hazardous pressure systems. Nucl. Eng. Des. 5, 81.

Moore, W.J., 1962. Physical Chemistry, fourth ed. Longmans, London.

Mott, N.F., 1947. Fragmentation of shell cases. Proc. R. Soc. Ser. A. 189, 300.

Munday, G., 1975a. The Sizes of Possible Flammable Vapour Cloud Discharged From Section 25A. Rep. to Flixborough Court of Inquiry, Cremer and Warner Rep. 9. Imperial College, London.

Munday, G., 1975b. Unconfined Vapour Cloud Explosions. Symposium on Technical Lessons of Flixborough. Institution of Chemical Engineers, London.

Munday, G., 1976a. Unconfined vapour cloud explosions. Chem. Eng. Lond. 308, 278.

Munday, G., 1976b. Unconfined vapour cloud explosions: a reappraisal of TNT equivalence. Process Ind. Hazards.19.

Munday, G., 1980. Initial velocities attained by plant-generated missiles. Nucl. Eng. 19, 165.

NFPA 68, 1994. Standard on Explosion Protection by Deflagration Venting, National Fire Protection Association.

NFPA I2A, 1997. Standard on Halon 1301 Fire Extinguishing Systems. National Fire Protection Association.

NFPA 68, 1998 Standard on Explosion Protection by Deflagration Venting, National Fire Protection Association.

NFPA 68, 2007 Standard on Explosion Protection by Deflagration Venting.

NFPA, 1957. Report of Important Dust Explosions, a Record of Dust Explosions in the United States and Canada Since 1860, National Fire Protection Association.

Nagy, J., Verakis, H.C., 1983. Development and Control of Dust Explosions. Dekker, New York, NY.

Nettleton, M.A., 1980. Detonation and flammability limits of gases in confined and unconfined situations. Fire Prev. Sci. Technol. 23, 29.

Nettleton, M.A., 1987. Gaseous Detonations. Chapman & Hall, London.

Palmer, K.N., 1973. Dust Explosions and Fires. Chapman & Hall, London.

Palmer, K.N., 1974. Relief venting of dust explosions. Loss Prev. Saf. Promotion. 1, 175.

Palmer, K.N., 1981. Dust explosions: basic characteristics, ignition sources and methods of protection against explosions. The Hazards of Industrial Explosion from Dusts. Oyez/IBC, London.

Pasman, H.J., Wagner, H.G., 1986. Explosion. A review of the state of understanding and predictive capabilities to safeguard against industrial explosion incidents. Loss Prev. Saf. Promotion. 5 (I), 1–8.

Phillips, H., 1981. Unconfined vapour cloud explosions: a new look at Gugan's book. Chem. Eng. London. 369, 286.

Pietersen, C.M., 1985. Analysis of the LPG Incident in San Juan Ixhuatepec, Mexico City, 19 November 1984. Rep. 85-0222. TNO, Apeldoorn, The Netherlands.

Pikaar, M.J., 1985. Unconfined vapour cloud dispersion and combustion—an overview of theory and experiments. Chem. Eng. Res. Dev. 63, 75.

Pineau, J., Giltaire, M., Dangreaux, J., 1978. Effectiveness of Venting in the Case of Dust Explosions. Cahiers de Notes Documentaires, no. 90, 1er trimester, Institute of National de Recherche et Securite, p. 37 (Note 1095-90-79).

Pittman, J.F., 1972a. Blast and Fragment Hazards from Bursting High Pressure Tanks. Rep. NOLTR 72-102. Naval Ordnance Lab, White Oak, Silver Spring, MD.

Pittman, J.F., 1972b. Pressures, fragments and damage from bursting pressure tanks. Department of Defense, Explosives Safety Board, Minutes of Fourteenth Explosives Safety Seminar, p. 1117.

Porter, W, 1982. The thrust on the supports of a two chamber vessel when the bursting disc in the duct connecting the chambers is ruptured. The Assessment of Major Hazards. Institution of Chemical Engineers, Rugby.

Prugh, R.W., 1987. Evaluation of unconfined vapor cloud explosion hazards. Vapor Cloud Model.712.

Prugh, R.W., 1991. Quantify BLEVE hazards. Chem. Eng. Prog. 87 (2), 66.

Rasbash, D.J., 1969. The relief of gas and vapour explosions in domestic structures. Struct. Eng. 47 (10), 404.

Rasbash, D.J., 1976. Gas and vapour explosions in enclosed spaces. In: Course on Fire and Explosion Safety in the Chemical Industry. Department of Fire Safety Engineering, University of Edinburgh.

Rasbash, D.J., Rogowski, Z.W., 1960a. Gaseous explosion in vented ducts. Combust. Flame. 4, 301.

Rasbash, D.J., Rogowski, Z.W., 1960b. Relief of explosions in duct systems. Chem. Process Hazards. I, 58.

Rasbash, D.J., Drysdale, D.D., Kemp, N., 1976. Design of an explosion relief system for a building handling liquefied fuel gases. Process Ind Hazards.145.

Reid, J.D., 1979. Markov chain simulations of vertical dispersion in the neutral surface layer for surface and elevated releases. Bound. Layer Meteorol. 16, 3.

Reider, R., Otway, H.J., Knight, H.T., 1965. An unconfined large volume hydrogen/air explosion. Pyrodynamics. 2, 249.

Richardson, S.M., Saville, G., Griffiths, J.F., 1990. Autoignition—occurrence and effects. Piper Alpha: Lessons for Life Cycle Management. Institution of Chemical Engineers, Rugby.

Roberts, A.F., 1981. Thermal radiation hazards from releases of LPG from pressurised storage. Fire Saf. J. 4 (3), 197.

Roberts, A.F., 1982. The effect of conditions prior to loss of containment on fireball behaviour. The Assessment of Major Hazards. Institution of Chemical Engineers, Rugby.

Robinson, C.S., 1944. Explosions. Their Anatomy and Destructiveness. McGraw-Hill, New York, NY.

Rogowski, Z.W., Rasbash, D.J., 1963. Relief of explosions in propane—air mixtures moving in a straight unobstructed duct. Chem. Process Hazards. II, 21.

Runes, E., 1972. Explosion venting. Loss Prev. 6, 63.

Rust, E.A., 1979. Explosion venting of low pressure equipment. Chem. Eng. 86 (November), 102.

Sagalova, S.L., Resnick, V.A., 1962. Teploenergetika. 12, 19.

Scilly, N., Crowther, J., 1992. Methodology for predicting domino effects from pressure vessel fragmentation. In Proc. CCPS Int. Conf. on Hazard identification and risk analysis, human factors and human reliability in process safety, New Orleans, LA.

Selway, M., 1988. Safety and reliability directorate, SRD/HSK/R 492 United Kingdom Atomic Energy Authority, Culcheth, Warrington WA3 4NE, UK.

Simmonds, W.A., Cubbage, P.A., 1960. The design of explosion reliefs for industrial drying ovens. Chem. Process Hazards. I, 69.

Skandia International, 1985. BLEVE! The Tragedy of San Juanico. First. 2 (November).

Smith, R.C., Smith, P., 1968. Mechanics. Macmillan, London.

Spiegelman, A., 1981. Risk management from an insurer's standpoint. The Hazards of Industrial Explosion from Dusts. Oyez/IBC, London.

Stokes, 1849. On some points on the received theory of sound. Phil. Mag. Ser. 3 (XXXIV), 52.

Strehlow, R.A., 1973. Unconfined vapour cloud explosions—an overview. Combustion. 14, 1189.

Strehlow, R.A., 1975. Blast waves generated by constant velocity flames: a simplified approach. Combust. Flame. 24, 257.

Strehlow, R.A., 1981. Blast wave from deflagrative explosions: an acoustic approach. Loss Prev. 14, 145.

Strehlow, R.A., Baker, W.E., 1976. The characterisation and evaluation of accidental explosions. Prog. Energy Combust. Sci. 2, 27.

Strehlow, R.A., Savage, L.D., Vance, G.M., 1973. On the measurement of energy release rates in vapour cloud explosions. Combust. Sci. Technol. 6, 307.

Synge, J.L., Griffiths, B.A., 1960. Principles of Mechanics, third ed. McGraw-Hill, New York, NY.

Taylor, G.I., 1946. The air wave surrounding an expanding sphere. Proc. R. Soc. Ser. A. 186, 273.

Taylor, G.I., 1950. The dynamics of the combustion products behind plane and spherical detonation fronts in explosives. Proc. R. Soc. Ser. A. 200, 235.

Taylor SIR, G.I., 1963a. In: Batchelor, G.K. (Ed.), Analysis of the Explosion of a Long Cylindrical Bomb Detonated at One End. Cambridge University Press, Cambridge.

Taylor SIR, G.I., 1963b. In: Batchelor, G.K. (Ed.), The Fragmentation of Tubular Bombs. Cambridge University Press, Cambridge.

Terao, K., 1977. Explosion limits of hydrogen-oxygen mixture as a stochastic phenomenon. Jpn. J. Appl. Phys. 16 (1), 29—38.

Van Den Berg, A.C., 1985. The multienergy method. A framework for vapour cloud explosion blast prediction. J. Hazard. Mater. 12 (1), 1.

Van Wingerden, C.J.M., 1989a. On the venting of large-scale methane-air explosions. Loss Prev. Saf. Promotion. 6, 25.1.

Van Wingerden, C.J.M., Van Den Berg, A.C., Opschoor, G., 1989b. Vapour cloud explosion blast prediction. Plant/Operations Prog. 8, 234.

Verkade, M., Chiotti, P., 1978. Literature Survey of Dust Explosions in Grain Handling Facilities: Causes and Prevention. Report. Energy and Mineral Resources Research Institute, Iowa State University.

Von Neumann, J., 1942. Progress report on the theory of detonation waves. OSRD Rep.549.

Wiekema, B.J., 1980. Vapour cloud explosion model. J. Haz. Materials. 3 (3), 221.

Wiekema, B.J., 1983a. An analysis of vapour cloud explosions based on accidents. In: Hartwig, S., op. cit., p. 237.

Wiekema, B.J., 1983b. Analysis of vapour cloud incidents. Euredata.4.

Wiekema, B.J., 1984. Vapour cloud incidents—an analysis based on accidents. J. Haz. Materials. 8 (295), 313.

Williams, F.A., 1965. Combustion Theory. Addison-Wesley, Reading, MA.

Williams, F.A., 1976a. Qualitative theory of non-ideal explosions. Combust. Sci. Technol. 12, 199.

Williams, F.A., 1976b. Quenching thickness for detonations. Combust. Flame. 26, 403.

Woodward, J.L., 1998. Appendix I: Evaluating Flammable Mass for Gaussian Dispersion Models: Instantaneous, Point Source. Estimating the Flammable Mass of a Vapor Cloud: A CCPS Concept Book, 241—243.

Zeldovich, Y.B., 1940. On the theory of the propagation of detonation in gaseous systems. J. Exp. Theor. Phys. (USSR). 10, 542.

Zeldovich, Y.B., Kompaneets, S.A., 1960. Theory of Detonation. Academic Press, New York, NY.

Plant Commissioning and Inspection

14.1. PLANT COMMISSIONING

The commissioning, or initial start-up, is a period when the plant is particularly at risk. It is also a time when equipment may be maltreated or damaged so that its subsequent operation is affected. Moreover, delays in bringing the plant up to full output can have a marked effect on the economics of the plant. For all these reasons it is essential to organize the commissioning of the plant efficiently and to allocate sufficient resources to it.

An account of plant commissioning is given in *Process Plant Commissioning* (Horsley and Parkinson, 1990), *Process Plant Design and Operation* (Scott and Crawley, 1992) and by Troyan (1960), Finlayson and Gans (1967), Kingsley et al. (1968−1969), Gans and Benge (1974), Unwin et al. (1974), Fulks (1982), Gans et al. (1983), Fraylink (1984), Ruziska et al. (1985), and Smet (1986).

In the following sections an account is given of plant commissioning and of some of its problems and hazards. It should be said at the outset, however, that in many cases these have been very successful overcome by good organization and engineering so that, for example, large ethylene plants have been brought up to full output within about 3 days (Chementator, 1977a,b).

It is not uncommon for the commissioning period to involve a series of crises. In fact, as it has been addressed by Horsley (1974), an efficient start-up without the emergence of numerous challenging technical problems would probably be slightly dull and disappointing to technologists and experts.

14.1.1 Commissioning Phases

The overall process of plant commissioning passes through a number of phases. In broad terms, the activities are (1) safety precautions, preparation of utilities, gland packing, preparation for line flushing; (2) commissioning utilities, machine rotation, pressure and leak testing, vessel cleaning, line flushing, relief value testing, and re-installation; (3) chemical cleaning, machine alignment, lubrication systems, machine short running tests, instrumentation checks; (4) final preparations for performance testing; (5) charging of feedstock; and (6) performance testing.

14.1.2 Contract

The responsibility for the construction and commissioning of the plant is normally shared between the eventual operator and its contractors, but the division of responsibility may vary considerably. The different types of contracts are discussed by Horsley and Parkinson (1990). The three areas of responsibility are supervision, trades and labor, and inspection. The project may be predominantly in-house, utilizing a small number of subcontractors, or it may be undertaken as a 'turnkey' by a contractor. Model conditions for different types of contracts available and those of the Institution of Chemical Engineers (IChemE) are widely used for process plants.

14.1.3 Organization

It is important to organize the commissioning thoroughly and to provide the appropriate personnel. There is no simple model for the organization and personnel for commissioning. These depend on the nature of the process, the size of the plant, the involvement of other parties, such as licensors and contractors, and so on. What is essential is that the management allocates sufficient resources to do the job efficiently and effectively. The commissioning of a large plant is a major enterprise and failure to recognize this is likely to increase costs and hazards.

The main conditions for success in commissioning, as listed by Kingsley (1990), are an ably led, well balanced, well trained, and committed commissioning team; a well-structured relationship at senior level with site construction management and future operational management; thorough planning and implementation of preparations; adequate involvement in design stages and safety studies; meticulous attention to hazards and safety; ready availability of assistance from supporting disciplines; and an expeditious approval system for plant modifications.

14.1.4 Personnel

Responsibility for the commissioning should rest with a single individual. It is normal to appoint a commissioning manager who has this responsibility.

Where the project is done in-house, a common practice is to appoint a plant manager for the new plant some time in advance to let them familiarize themselves with it and to make them the commissioning manager. The job of the project manager is normally a separate one. Experience indicates that it is usually not good practice to attempt to combine the roles of project manager and commissioning manager. The commissioning team for an in-house project should be a strong one, with good technical capability and relevant experience. The team may consist of the personnel assigned to the plant

supplemented by others who are brought in for the limited period of the commissioning only. An example team organization can be found in the *Guidelines for Safe Process Operations and Maintenance* (CCPS, 1995).

It is necessary that the commissioning manager should be able to command the resources necessary to carry through the commissioning. Unforeseen problems can often result in considerable demands on resources and impact the schedule. It may be desirable, particularly on large projects, to create an additional troubleshooting team. The organization of an extra fieldwork team for this purpose has been described by Horsley (1974). Some of the many variations in commissioning team composition are described by Troyan (1960), Finlayson and Gans (1967), and Garcia (1967).

14.1.5 Planning and Scheduling

Plant commissioning requires the coordination of a large number of activities carried out by many different people. Some aspects, for which planning is particularly important, include commissioning activities (pre-start-up and initial start-up), budgets, documentation, recruitment, and training.

The design and construction of a large process plant is normally scheduled using such methods as the Project Evaluation and Review Technique (PERT) and Critical Path Scheduling (CPS), but it is more common with commissioning than it is with design to do the scheduling by hand, using aids such as bar charts and arrow diagrams.

14.1.6 Preparation and Training

Management: The effectiveness of the commissioning depends to a considerable extent on the thoroughness of preparation by management. The commissioning manager normally prepares by studying the process and the plant, by visiting or working on similar plants, by involvement in the project through work on design committees, by preparing plant documentation and operator training schemes, and so on. It should not be assumed that it is only the process operators who need preparation and training for commissioning. These are required by management also.

Operations: The operation of process plant requires the creation of a large number of systems and procedures. Many of these have to be formulated by the plant manager prior to and during commissioning. These should be drafted early, typically using operating personnel so that they can be used for operator training, but they are likely to need some modification in the light of operating experience. It is also necessary to create a system of process records.

Further accounts of process operator training during commissioning are given by Finlayson and Gans (1967), de Regules (1967), Kingsley et al. (1968–1969), and Parsons (1971).

Maintenance: Prior to and during commissioning the maintenance engineers should set up the appropriate maintenance systems and documentation. Each major item of equipment should be given an identification code, which should be marked on the equipment itself.

The maintenance manuals for the equipment should be obtained from the manufacturer and reviewed. The maintenance of the equipment, both preventive and breakdown maintenance, should be planned and recorded. The quality of the data available for analyzing reliability, availability, and maintenance of the plant depends critically on the equipment records. Schedules for regular preventive maintenance should be developed, typically using one of several commercially available software systems. A system of maintenance records should be instituted, containing details of the equipment in the plant and of the maintenance and modification work done. Maintenance personnel should check the plant during construction for accessibility and ease of maintenance. It may be necessary to make some alterations to the layout, to the equipment, or to the lifting arrangements. The maintenance function is also involved in many of the pre-start-up activities such as checking and testing, and suitable preparations need to be made for this.

14.1.7 Pre-Commissioning Documentation

It is difficult to overemphasize the importance of having a comprehensive and up-to-date set of documentation, with process in place of maintaining it. It is appropriate to emphasize certain aspects that are of particular importance in commissioning. The documentation available at the pre-commissioning stage should include operating manual, operating instructions, safety instructions, permit system documents, pressure vessel register, pressure piping systems register, protective device register, maintenance systems documents, and checklists.

The plant should be covered by a comprehensive system of permits-to-work. These are needed during the commissioning as much as during operation and should be developed in good time. A major part of commissioning is the checking of systems, both of hardware and of software. This is greatly assisted if suitable checklists have been prepared.

14.1.8 Modification Control

There should be a system for the control of modifications during commissioning, so as to ensure that a safe design is not rendered unsafe by a modification. The system should ensure that proposed changes are identified, routed

to the design authority, any changes approved by appropriate management, and that there is a prompt response.

14.1.9 Mechanical Completion and Pre-Commissioning

A particular aspect of pre-commissioning is mechanical completion. The term is applied to individual items of equipment and is used to cover the activities between installation and process commissioning.

Mechanical completion ensures that the installed equipment is ready for commissioning and involves checking that it is installed correctly, that the component parts operate as specified and that any ancillary equipment is installed and working. The plant should be given a thorough visual inspection. A check should be made on all plant equipment and piping to ensure that it is installed in accordance with the engineering drawings. Any discrepancies should be marked on drawings to show the 'as built' condition. These discrepancies should be reviewed and corrected if necessary. The inspection should check for items such as loose bolts or missing valve wheels and for construction aids or debris.

The process of checking involves the marking up of drawings and the completion of test certificates and various forms of contractual certificate.

There are also a number of pre-commissioning activities to be carried out on the mechanical systems. These activities need to be coordinated with both the site construction manager and the plant operations and maintenance management.

A list of some of the checks and tests carried out on process machinery, such as pumps, compressors, and centrifuges, is given by Gans and Benge (1974).

14.1.10 Control Systems

The installation of the instrument and control system cannot be completed until most of the mechanical equipment is installed. Its installation tends to overlap with the commissioning of the rest of the plant.

The pre-commissioning of the control system should be governed by formal procedures. Test forms should be specified for each system to be tested. Guidance on the installation and testing of instrument and control systems with model forms is given in BS 6739: 1986 *Code of Practice for Instrumentation in Process Control Systems: Installation Design and Practice*.

Installation of the instrumentation is followed by checking. The responsibility for this normally lies with the contractor, but involvement of the user's instrument personnel assists familiarization.

The checks on the instrumentation should be extended where practical to cover operation with process fluids and conditions that are realistic but nevertheless safe. Checklists for the functional testing of a control loop and of a sequence are given by Horsley and Parkinson (1990).

Documentation for the instrument and control system should include: the system manuals, piping and instrument diagram, loop diagrams, wiring/circuit diagrams, termination rack layouts, tag number lists, database tables, sequence flow diagrams, etc.

14.1.11 Process Commissioning

Process commissioning is begun only when the pre-commissioning is complete and the defects identified have been corrected to the extent judged necessary. This commissioning may be undertaken on equipment or sections of plant as they become available, but only under close control so as to prevent hazardous interaction between the two activities of plant construction and process operation. One method of control is to require for the equipment a formal handover certificate verifying mechanical completion, with minor deficiencies listed. Before the process commissioning starts, it is usual to carry out a final check. Checklists applicable at this stage are one of those most commonly given.

The order in which the process commissioning is done may vary. The typical sequence cited by Horsley and Parkinson (1990) is (1) utilities, (2) laboratory, (3) raw material storage, (4) ancillary equipment, (5) reaction system, (6) work-up system, and (7) product storage. Where there is no reaction system, the overall process system may be substituted for stages (5) and (6).

14.1.12 Handover: Contractor Project

The conditions for acceptance of the plant from the constructor by the operator are normally specified in the contract. A handover certificate is usually issued which contains a list of reservations of items on which further work is required. The handover needs to be carried out formally and with particular regard to safety. The plant should be in a safe condition when it is handed over.

With a large plant it is common to have selective handover of plant systems as they are completed. This applies particularly to such systems as steam-raising plant, steam pipework, and cooling water systems.

If parts of the plant are to be operated while other units are still under construction, it is essential to take steps to ensure that these two activities do not interact in such a way as to create a hazard. A typical reservations checklist is given by Pearson (1977).

14.1.13 Start-Up and Performance Testing

With the end of the process commissioning, the next stage is to charge the feedstocks and start-up and operate the whole plant and subject it to performance testing in order to determine whether it meets its specification. This initial start-up of the plant should be thoroughly prepared. It should be recognized that the start-up may be prolonged and arrangements made to relieve personnel so that they do not have to work excessively long hours and consecutive workdays. It is valuable to keep fairly comprehensive records of the start-up, and personnel should be briefed on the recording requirements.

All the formal systems for control of hazards should be operational. These include in particular the permit-to-work systems, as well as the systems for the control of slip plates and of vents and drains. Checklists examples are given by Pearson (1977).

14.1.14 Handover: In-House Project

For a turnkey contract, the system of formal handover sets a term to the commissioning period. For an in-house project, the possibility exists that the commissioning period will become unduly extended.

Some criteria for termination are given by Kingsley et al. (1968–1969) as follows: (1) competence of operating staff, (2) reasonable level of plant reliability, (3) attainment of acceptable quality standards, (4) satisfactory use of resources (staff, materials, utilities, throughput), and (5) acceptable level of maintenance.

14.1.15 Commissioning Problems

Many of the problems associated with commissioning have been mentioned explicitly or are implicit in the comments made, but it is appropriate at this point to list some of the problem areas in commissioning. These include: (1) lack of process information, (2) lack of expert advice, (3) design changes, (4) unsuitable equipment, (5) lack of spares and supplies, (6) construction and maintenance errors, (7) operating errors, (8) safe fluid testing, and (9) water traces. Some specific problems arising during commissioning have been reviewed by Kingsley et al. (1968–1969).

14.1.16 Safety Audit and Commissioning Hazards

The process and plant designs should already have been subjected to the various checks. In particular, the HAZOP study conducted should have covered operations that are carried out only in commissioning. There are certain specific hazards associated with commissioning. The plant should be given a comprehensive safety audit during the commissioning period. The audit should cover both hardware and software aspects, as well as procedures.

14.1.17 Post-Commissioning Documentation

During commissioning a large amount of information is generated which needs to be properly documented. Modifications are made to the plant; a wide variety of tests and examinations are performed on individual items of equipment and on the plant as a whole; the computer software is modified and parameters entered; modifications are made to the systems and procedures and the associated documentation; reviews are made of safety and environmental features and there are matters to be carried forward.

14.2. PLANT INSPECTION

Plant inspection is an essential aspect of the fabrication, construction, commissioning leading to successful operations. Accounts of plant inspection are given in *Pressure Vessel Systems* (Kohan, 1987) and *Inspection of Industrial Plant* (Pilborough, 1989) and by Erskine (1980).

Relevant codes are the *Pressure Vessel Examination Code* (IP, 1993 MCSP Pt 12) and the *Pressure Piping Systems Examination Code* (IP, 1993 MCSP Pt 13). These are codes with international application but are aligned with the Pressure Systems and Transportable Gas Containers Regulations, 1989.

Another code is the *Registration and Periodic Inspection of Pressure Vessels Code* (ICI/RoSPSA, 1975 IS/107) (the ICI *Pressure Vessel Inspection Code*). This is now out of print but remains a good illustration of the fundamental principles of such codes.

In the United States, the *Guide for Inspection of Refinery Equipment* by the API (1962) has long been a principal inspection code but is now out of print. It has been replaced by other publications. One of these is API RP 510: 1989 *Pressure Vessel Inspection Code: Inspection, Rating, Repair and Alteration*.

The following description is concerned mainly with the inspection of pressure vessels, pipework, and protective devices, but it illustrates general principles for inspection which are relevant also to many other types of equipment.

14.2.1 Regulatory Requirements

It is normal industrial practice to exercise close control of all parts of a pressurized system, including both pressure vessels and other components, and to do this throughout the life of the system, starting with design and continuing through fabrication, installation, commissioning, operation, inspection, maintenance, and modification, by means of external and in-house standards and codes. The inspection system is the main means of exercising this control after the design stage.

The statutory controls are not the only external influence which industry has to consider. Adherence to a system similar to the statutory one is generally a condition of obtaining insurance. Much inspection of the process plant is in fact carried out by insurers.

14.2.2 Approval and Inspection Organizations

The inspection organization is responsible for the initial inspection of new equipment during its fabrication and construction, and for the periodic inspection of operating equipment throughout its working life. Its framework is determined by the design and operating authorities. The design authority should determine the parameters of the plant operation, specify the design codes and should carry out the actual design. It should also specify the standards for fabrication, construction, and testing and should prescribe the documentation required on these aspects. The operating authority should provide a code for the regular inspection of the plant. In addition, it should create a system to control both plant and process modifications.

There are a number of parties who may conduct inspection of a pressure vessel during its manufacture. They include (1) government, (2) the manufacturer, (3) the user, (4) an insurer, and (5) a consultant. The same set of parties, except for the manufacturer, may undertake inspections during operation.

14.2.3 Competent Persons

Inspection is a specialist matter and should be done only by a qualified inspector.

The question of the competent person for the inspection of boilers and pressure vessels was considered in the *Report of the Advisory Committee on the Examination of Steam Boilers in Industry* (Honeyman, 1960; the *Honeyman Report*). The report states in connection with the requirement for a competent person: What appears to be contemplated is that the person should have such practical and theoretical knowledge and actual experience of the type of machinery or plant which he has to examine as will enable him to detect defects or weaknesses which it is the purpose of the examination to discover and to assess their importance in relation to the strength and functions of the particular machinery or plant.

14.2.4 Inspections

Inspection activities cover the stages of (1) manufacture, (2) commissioning, and (3) operation. It is hardly necessary to emphasize that the existence of an inspection system and the presence of an inspector have an important influence on the quality of the work done.

The results of inspections should be recorded as detailed inspection/test reports, but in addition it is normal to issue release notes or inspection statements, which are less detailed, or, in the case of non-acceptance, a rejection note. A certain amount of information is available in the human factors literature on errors in inspection. The main finding is that the probability of detecting a defect decreases as the probability of the defect decreases.

An inspection system exercises control of weld quality by (1) approval of welders, (2) specification of welding materials, (3) specification of welding methods, and (4) examination/testing of welds.

Inspection of other features: Other features that are examined in pressure vessel inspection include: equipment dimensions; base metals; surface conditions; wear situations; high stress situations; dissimilar metals; stray electric currents; gaskets, seals, and joints; lagging; protective finishes; venting and draining; and access.

Inspection register and records: It is essential to keep a register of the equipment to be inspected and records of the results of the inspections. The precise contents of the register depend on the item concerned, but in general should include: identification number, order number, and drawing number; specification, design parameters, process fluids; inspection/test reports during manufacture; inspection category, interval, method; special features relevant to deterioration, failures; materials/parts list; design life/remnant life prediction; and date of entry into service.

The specific information requirements are given in the Institute of Petroleum (IP) *Pressure Vessel Examination Code* for the registration of pressure vessels.

14.3. PRESSURE VESSEL INSPECTION

The inspection of pressure vessels and their protective devices is crucial to the maintenance of the integrity of the pressure system and is a principal activity of the inspection authority.

Pressure vessels come within the Pressure Systems Regulations, 1989 and are covered in the guidance HS(R) 30 and in COP 37. A relevant code is the *Pressure Vessel Examination Code* by the IP (1993 MCSP Pt 12). The account of pressure vessel inspection given here is based primarily on this code.

14.3.1 Authority and Competent Person

The IP *Pressure Vessel Examination Code* defines the competent person as the person or body authorized by the user to draw up or approve the scheme of examination and to perform the examination.

Design authority: The design of the pressure vessel is the responsibility of the design authority. The IP *Pressure Vessel Examination Code* states that the design authority may be a vessel design group responsible to the user, an authorized design contractor, an independent design consultant, the engineering authority, or the competent person. In some countries, the design authority is a state agency.

Engineering authority: The engineering authority is responsible for the maintenance of the pressure vessel and is authorized to do this by the user.

14.3.2 Equipment

Definition: In considering an inspection scheme, it is necessary first to define the equipment which is to be brought within the scope of the pressure vessel inspection system. The definition of a pressure vessel and other devices can be found in the IP *Pressure Vessel Examination Code*. The ICI Code also lists equipment which, subject to approval, may be excluded from the requirements of the code itself, although it may still require some degree of inspection.

Registration: A pressure vessel identified as such should be registered before it is brought into service. The IP *Pressure Vessel Examination Code* can be consulted for registration systems.

Classification: A pressure vessel is generally classified according to whether or not it is subject to legal requirements for inspection. The IP *Pressure Vessel Examination Code* assigns a vessel to Classification A or B. Class A includes all vessels and their protective devices that are subject to periodic examination in accordance with national or regional legal requirements. The extent of the classification is therefore a function of the country or state where the vessel is to be used. Class B includes all vessels and their protective devices not assigned to Class A. It covers all vessels which are not subject to legal requirements.

Grading: The IP *Pressure Vessel Examination Code* requires that, where legal requirements permit, the pressure vessel and its protective device be allocated to a grade. The grade indicates the maximum interval that may elapse between major examinations.

The principles of grading are that the vessel receives a pre-commissioning examination before entering service; it is initially allocated to Grade 0 and is given a first thorough examination after a relatively short period of service; it is then either retained in Grade 0 or allocated to Grade 1 or 2; subsequently, after a second thorough examination, it is allocated to Grade 1, 2, or 3 and, as it approaches the end of its design life or predicted remaining life, it is reallocated to a lower grade if necessary.

14.3.3 Examination

Examination Intervals:

The maximum examination intervals are given by the IP *Pressure Vessel Examination Code*. The longest of these maximum intervals is therefore 144 months or 12 years. On the basis of the classification of equipment, it is possible to specify inspection categories that define the interval between inspections.

Examination principles:

The IP *Pressure Vessel Examination Code* states that the purpose of examination is:

To ensure that equipment remains in a satisfactory condition for continued operation consistent with the prime requirements of safety, compliance with statutory regulations and economic operation until the next examination.

The ICI *Pressure Vessel Inspection Code* defines the objectives of inspection as follows:

The objective of vessel inspection is to detect any deterioration such as corrosion, cracking or distortion indicating possible weaknesses that may affect the continued safe operation of the vessel.

The inspection should also include an examination of protective devices.

Examination practices: pressure vessels

The maximum interval between examinations is set by the grade to which the vessel is allocated. The IP *Pressure Vessel Examination Code* gives a number of additional factors that are to be taken into account in deciding on the actual interval. These include: any regulatory requirements, the works policy, the severity of the duty, the performance of other vessels on similar duty, the ability to carry out meaningful on-stream inspection, and the remaining design life and the predicted remaining life. Other factors in particular applications are catalyst life and regeneration intervals and performance of internal linings. It is also necessary to consider the consequences of failure.

The IP Code also refers by way of caution to three aspects that may require special attention: (1) lined vessels; (2) internal fittings; and (3) external lagging, cladding, and fireproofing. The matters to be recorded following the examination are given in the IP Code. The Code also gives an appendix containing examples of examination reports.

Examination practices: protective devices

The IP *Pressure Vessel Examination Code* divides protective systems into pressure-actuated devices and other devices. In the former category, it discusses regular safety valves, pilot-operated safety valves, and burst disks.

The *Code* also discusses the need to maintain adequate overpressure protection of the vessel at all times and the isolation practices necessary to achieve this.

14.3.4 Modification and Repair

The IP *Pressure Vessel Examination Code* requires that any modification necessary on a pressure vessel should be approved by the design authority and the competent person, and that the design and execution of any modification or repair be under the control of the latter. The effect on any protective devices on the piping system of any modifications or repairs on that system should be considered. Records should be kept of any modifications or repairs carried out and should include the original approval procedure documents together with details of materials and techniques, drawings and test certificates.

14.3.5 Vessel Testing

The IP *Pressure Vessel Examination Code* discusses five forms of testing: (1) strength testing, (2) leak testing, (3) NDT, (4) destructive testing, and (5) materials analysis. Any of these may be used at the various stages of the vessel's life, but the last two are used mainly at the construction stage or in connection with modification or repair.

An account of defects and failures in pressure vessels is given by Pilborough (1989). Some of the features that may be revealed by an inspection are internal corrosion, surface defects, weld defects, wear defects, deposits and debris, high stress situations, inadequate drainage, external corrosion.

14.4. PRESSURE PIPING SYSTEMS INSPECTION

Most releases occur not from pressure vessels but from the associated piping systems. Inspection of piping systems is therefore important. Pressure piping systems come within the Pressure Systems Regulations, 1989 and are covered in the guidance HS(R) 30 and in COP 37. A relevant code is *Pressure Piping Systems Examination Code* (IP, 1993 MCSP Pt 13). The account of pressure piping systems inspection given here is based primarily on this code. The provisions of this code largely mirror those of the *Pressure Vessel Examination Code* (IP, 1993 MCSP Pt 12), and to the extent that this is so they are not repeated here.

The criteria for selecting piping for registration are that there is a legal requirement or that the piping is known or suspected to deteriorate and its failure would give rise to an unacceptable situation.

REFERENCES

American Petroleum Institute (API), 1962.
American Petroleum Institute (API), RP 510, 1989.

Center for Chemical Process Safety (CCPS), 1995. Guidelines for Safe Process Operations and Maintenance. AIChE, New York, NY.

Chementator,1977a. Chemical Engineering. June 6, p. 67.

Chementator, 1977b. Chemical Engineering. October 24, p. 73.

De Regules, S.E., 1967. A production engineer's provision for start-up. Chem. Eng. Prog. 63 (12), 61.

Erskine, R., 1980. Vessel inspection procedure and philosophy. Ammonia Plant Saf. 22, 98.

Finlayson, K., Gans, M., 1967. Planning the successful start-up. Chem. Eng. Prog. 63 (12), 33.

Fraylink, J.R., 1984. How to start up major facilities. Oil Gas J. 82 (December), 112.

Fulks, B.D., 1982. Planning and organizing for less troublesome plant startups. Chem. Eng. 89 (September), 96.

Gans, M., Benge, J.N.., 1974. Scheduling the Commissioning Operation. Institution of Chemical Engineers, Symposium on Commissioning of Chemical and Allied Plant, London.

Gans, M., Kiorpes, S.A., Fitzgerald, F.A., 1983. Plant start-up—step by step by step. Chem. Eng. 90 (October), 74.

Garcia, E.A., 1967. Minimising start-up difficulties. Chem. Eng. Prog. 63 (12), 44.

Honeyman, G.G., 1960. Report of the Advisory Committee on the Examination of Steam Boilers, Cmnd 1173. HM Stationery Office, London.

Horsley, B., 1974. Commissioning a methanol plant. Institution of Chemical Engineers, Symposium on Commissioning of Chemical and Allied Plant, Swansea, London.

Horsley, D.C.M., Parkinson, J.S., 1990. Process Plant Commissioning. Institution of Chemical Engineers, Rugby.

Kingsley, R.J., 1990. In: Horsley, D.C.M., Parkinson, J.S., op. cit. (Foreword).

Kingsley, R.J., Kneale, M., Schwartz, E., 1968–1969. Commissioning of medium-scale process plants. Proc. Inst. Mech. Eng. 183 (pt 1), 205.

Kohan, A.L., 1987. Pressure Vessel Systems. McGraw-Hill, New York, NY.

Parsons, R., 1971. Guidelines for plant start-up. Chem. Eng. Prog. 67 (12), 29.

Pearson, L., 1977. When its time for start-up. Hydrocarbon Process. 58 (8), 116.

Pilborough, L., 1989. Inspection of Industrial Plant.

Pressure Piping Systems Examination Code, 1993. (IP, MCSP Pt 13).

Pressure Systems Regulations, 1989.

Pressure Systems and Transportable Gas Containers Regulations, 1989.

Pressure Vessel Examination Code, 1993. (IP, MCSP Pt 12).

Registration and Periodic Inspection of Pressure Vessels Code, 1975. (ICI/RoSPSA IS/107).

Ruziska, P.A., Song, C.C., Wilkinson, R.A., Unruh, W., 1985. Exxon chemical low energy ammonia process start-up experience. Ammonia Plant Saf. 25, 22.

Scott, D., Crawley, F., 1992. Process Plant Design and Operation. Institution of Chemical Engineers, Rugby.

Smet, E.J., 1986. Sequential completion of plant construction to allow early production. Plant/Operations Prog. 5, 108.

Troyan, J., 1960. How to prepare for plant startups in the chemical industries. Chem. Eng. 67 (September), 107.

Unwin, A.J.T., Robins, J.A., Page, S.J.R., 1974. Commissioning of a major materials handling offsite facility for a petrochemical complex. Institution of Chemical Engineers, Symposium on Commissioning of Chemical and Allied Plant, London.

Plant Operation

Accidents at process plants arise as often from deficiencies in operations as from those of design. It is difficult, therefore, to overstress the importance of plant operations in safety and loss prevention.

15.1. INHERENTLY SAFER DESIGN TO PREVENT OR MINIMIZE OPERATOR ERRORS

The concept of inherently safer design is critical in creating a safer process and work environment and should be applied throughout all stages of process development and design (conceptual, basic engineering, and detailed engineering design). An inherently safer design will take into account eliminating or minimizing the process hazard via different tools and methodologies including intensification, substitution, attenuation, limiting the effects and impact, and/or simplification (Wallace, 2007).

Amyotte et al. (2007) indicated that an inherently safer design approach should address the following factors:

- Minimize the use and storage of hazardous materials and use a less hazardous material when possible. Moreover, the storage of hazardous materials should be as far as possible from people and equipment to minimize impact in case of an accident.

- If possible, use the just in time delivery of hazardous materials.

- When a hazardous material is no longer needed, remove it or dispose of it properly so that it does not pose any hazard to the facility and personnel.

- Workers' schedules and shift rotations should be optimized to minimize fatigue.

- Consider implementing ergonomics for product and equipment movement.

- Explore the possibility of minimizing hazards resulting from operating at elevated operating conditions (e.g., pressures, temperatures, flow rates, and volumes). Can the process be operated at less hazardous conditions without compromising the economic feasibility of the process?

- Ensure that the documentation (e.g., manuals, guides, procedures) are clear and easy to understand.

- Explore the use of safeguards to prevent the incorrect or unintentional operation of equipment, units, and control systems.

- Ensure that the equipment and electrical installations are isolated and properly protected.

The human factor should be incorporated when considering the implementation of an inherently safer design approach. The design should take into consideration reasonable expectations from operators and should not cause unrealistic demands on the operators that will contribute to increasing the possibility of human error and fatigue. Task analyses that take into consideration the location, time frame, possible errors, and the consequences should be thoroughly performed (MacCollum, 2008; Widiputri et al., 2009).

15.2. OPERATING DISCIPLINE

The safe operation of a process plant requires adherence to a strict operating discipline. This discipline needs to be formulated, along with the safety precautions, during design and then enforced in operation. An account of operating discipline is given by Trask (1990).

The steps involved in developing the operating discipline are (1) to identify the important operating parameters, specify operating limits on these parameters, and arrange for them to be measured, controlled, and monitored, using both trend records and alarms; (2) to ensure mass and energy balances; and (3) to develop and document operating procedures, safety precautions, and remedial actions.

15.3. BEST OPERATING PRACTICES

Best operating practices (BOP) should be captured, documented, followed, and continuously improved upon and updated. Drennan indicates that a detailed description of the job, materials, equipment, methods, and measures against which the outcome is judged should be documented and included in the BOP. The vulnerability points should be identified so the operator is aware of them and how to avoid them. Moreover, safety measures should be put in place to assure the operator's safety and enable them to implement the BOP and increase efficiency and productivity.

15.4. OPERATING PROCEDURES AND INSTRUCTIONS

Fundamental to the safe operation of a plant is the development of suitable operating procedures. The procedures are normally formulated during the plant design and are modified as necessary during the plant commissioning and operation.

Accounts of the generation of operating procedures and instructions have been given by King (1990) and Sutton (1995). In designing a plant, the designer has in mind the way in which it is to be operated. It is desirable that this design intent be explicitly documented; otherwise it has to be inferred from the equipment provided. Practice in the formulation of operating procedures varies. The lead is normally taken by the plant manager. In the case of a new plant, the commissioning manager is often

the prospective plant manager and will therefore undertake the task. Operating procedures may be developed using a team of experienced people who examine successively the procedures necessary for testing and pre-start-up activities, normal start-up, normal shut-down, emergency shut-down, etc.

A systematic approach to the development of operating procedures is given in *Developing Best Operating Procedures* (Jenkins et al., 1991 SRD SRDA-R1) (the SRDA *Operating Procedures Guide*). The SRDA *Guide* distinguishes between two basic types of documentation to support operator performance: (1) manuals and (2) job aids. Essentially a manual is a resource, which is a store of information to which managers, engineers, and trainers may refer.

Operating instructions are commonly collected in an operating manual. The writing of the operating manual tends not to receive the attention and resources which it merits. Operating instructions should be written so that they are clear to the user rather than so as to absolve the writer of responsibility. The attempt to do the latter is a prime cause of unclear instructions. Sutton describes a method of writing the operating instructions and of creating the operating manual. Basic requirements for the operating instructions are that they should be complete, up-to-date, properly indexed, and easy to use. A distinction is made between information, rules, procedures, and checklists. Information is provided for reference, but selectively. Rules are stated, although these are likely to be of more interest to managers and supervisors than to operators. Checklists are given as appropriate.

Instructions are given for each procedure. The procedure may be based on a generic procedure, such as that for pump start-up, but with specific modifications for the operation in question. The level of the instructions should be such as to allow an experienced operator to run the plant with minimal reference to the supervisor. Each instruction should have an imperative verb in the active tense. Warnings, cautions, and notes are inserted in the text of the procedure, as appropriate; a warning should precede the procedure. Attention should be paid to the intended users of the operating instructions. These include: (1) novice operators, (2) operators whose experience has been in other plants but who are being trained for the plant in question, and (3) operators experienced on that plant. The organization of the operating manual into successively more detailed levels goes some way toward meeting their different needs. An experienced operator may require operating instructions mainly for safety critical operations or operations performed only rarely.

The writing of the operating instructions is a significant task. Some principal sources of information are the quantities flowsheet, the engineering line diagram, the design manual, and vendors' manuals. The contractors may provide very little and vendor information may become available only late in the day. The situation is eased if an operating manual is available from a previous plant.

The operators will require training in the operation of the plant based on the operating instructions. It is necessary, therefore, to provide a training manual. This may draw on the operations manual. Typically, it contains an operating instruction section together with other material such as a training plan, workbook exercises, an assessment plan, and performance criteria. The need for a training manual may be regarded as a principal justification for the creation of the operating manual.

15.5. EMERGENCY PROCEDURES

A set of operating procedures of particular importance are the emergency procedures. It is generally recognized that it is not possible to provide detailed procedures for all possible emergency scenarios. First, careful consideration should be given to what major emergencies can be reasonably foreseen, and the detailed procedure for tackling these (few) serious situations should be dealt with specifically both by training sessions and special instructions.

Second, since it is impossible to cover all eventualities by instructions, which will be effectively remembered or found for reference during an emergency, it is advisable to give a series of general instructions, amply backed by training, which will provide guidance for the correct action in those emergency or potential emergency situations, which have not been dealt with by specific cases. Such aspects as equipment isolation, the recognition of hazardous conditions, and communications are fundamental matters which can be treated in this way.

Emergency procedures are commonly treated in codes under the following headings: (1) leak detection and characterization, (2) raising the alarm and associated actions, (3) isolation of leaks, (4) handling of leaks and spillages, (5) action against fire, (6) emergency equipment, and (7) special features of particular chemicals. In dealing with emergencies, protective equipment is of utmost importance. Emergency equipment commonly listed in codes includes: (1) protective clothing, (2) breathing apparatus, (3) emergency tool kits, and (4) fire fighting equipment.

When considering emergency situations, two very important occurrences that must be considered are leaks and spillages. Leaks and spillages are a relatively common occurrence in process plants and it is necessary to have procedures for dealing with them. Plant management should formulate procedures for dealing with leaks and spillages. A small leak may often be stopped relatively easily. If the leak is more serious, other measures are required. The ICI LFG Code (ICI/RoSPA 1970 IS/74) lists these for liquefied flammable gases. If it is practical to do so, the leak should be isolated. Personnel involved in controlling the situation should avoid entering the vapor cloud. The leak should be approached only from upwind. Other personnel should not be allowed in the area. There should be some method of warning and evacuating them, and also of sealing off the area.

Measures should be taken as appropriate to disperse or contain the leakage. The area to which a flammable vapor cloud is likely to spread should be cleared of sources of ignition, which include not only activities such as hot work but also traffic. It may be necessary to close roads and railways that the vapor could reach. The emergency services should be alerted as appropriate.

For large releases, detection of a leak and characterization of the resultant cloud are important. These are discussed in the CCPS *Guidelines for Vapor Release Mitigation* (1988/3).

For detection the first line of defense is gas detectors. The *Guidelines* discuss the detectors available and their positioning. In most plants there are personnel who may detect the presence of the gas. If a leak occurs, which gives rise to a large vapor cloud, it assists in handling the emergency if personnel are able to estimate the size and movement of the cloud. Their ability to do so depends on its visibility. A small number of substances yield, at a sufficiently high concentration, a vapor cloud which is visible by virtue of its color. A more common form of visible cloud is that caused by fog formation. Liquefied flammable gases and other liquefied gases such as ammonia, hydrogen chloride, and hydrogen fluoride may give rise to fog, depending on the conditions of release and the atmospheric humidity.

One method of dealing with a liquid spillage is the application of foam. Regular foam is mainly water and its use will often actually increase the rate of vaporization of the spilled liquid, thus making the situation worse. On the other hand, there are legitimate uses of foam. If the chemical is completely miscible with water, water may be applied to dilute it. However, the number of industrial chemicals which pose a significant hazard and which are water miscible is limited. The response to a leak or spillage emergency depends also upon the specific characteristics of the chemical released. The CIA codes and guidelines give some guidance on handling leaks and spillages of other substances.

A large proportion of plant fires are due to hydrocarbons. If the fire consists of an ignited leak burning as a jet flame on the plant, the usual practice is to let it burn. It is generally not desirable to seek to extinguish the flame on such a leak, since this then turns the risk of creating a vapor cloud, which might then explode. There is a limited amount of guidance in the CIA codes and guidelines on action against fire of other substances. The guidance given for pool fires of vinyl chloride (CIA, 1978 PA15) illustrates some of the principles.

15.6. HANDOVER AND PERMIT SYSTEMS

Most large process plants operate round the clock on a shift system. It is essential that information on the state of the plant be communicated by the outgoing shift to the incoming shift. A formal handover system is necessary to ensure this. There should be a formal and detailed procedure for shift handover. It should cover both operating and maintenance personnel. An aspect of handover which is of particular importance is the status of permits-to-work. Much of the information which needs to be communicated at handover is likely to relate to such permits.

Another important form of communication is the permit-to-work system just mentioned. It is sufficient to note that the actions of handing over a section of plant to maintenance, and receiving it back, are important not only for maintenance but also for the operating function. Work permit system documentation should clearly include the purpose of the work, the appropriate authorization to conduct the work, qualifications of the person conducting the work and any certifications that need to be obtained, responsibilities, forms, and requirements to display the work permit, record keeping, and auditing (Guidelines for Process Safety Documentation, Center for Chemical Process Safety (CCPS)).

15.7. OPERATOR TRAINING AND FUNCTIONS

The process operator runs the plant and deals with the faults which arise in it. He has many training needs which must be met if he is to do the job properly.

As regards the content of operator training, consideration needs to be given not just to the operating procedures and safety training, but also to other topics where the operator has some responsibility. The training of process operators should not be a once-and-for-all exercise, but should involve updating as appropriate.

One of the principal functions of operators in process plants is the detection of faults. The detection of faults by the process operator in the control room has already been discussed. Other faults, particularly leaks, are detected by operators during routine patrols.

Routine patrolling is therefore important, and specific arrangements need to be taken to ensure that it is carried out. In particular, steps may be taken to ensure that certain vulnerable features are regularly passed by an operator.

15.8. OPERATION, MAINTENANCE, AND MODIFICATION

A process rarely operates for long without undergoing some modification. There are almost always changes in raw materials, in the operating conditions, such as flows, pressures and temperatures, concentrations, or in other aspects. A system is required, therefore, for the management of change in the process as well as in the plant.

Maintenance and modification work may present hazards both to the plant and to the workers involved. It is important, therefore, for the plant manager to give full consideration to the implications of such work being carried out on his plant and for him to take appropriate precautions.

15.9. START-UP AND SHUT-DOWN

As takeoff and landing are more hazardous operations for an aircraft, so in a process plant the hazard is greater during start-up and shut-down. General principles of plant start-up and shut-down are discussed by Scott and Crawley (1992). The procedures for the start-up, operation, and shut-down of particular types of plant are considered in the following sections.

Most accounts of the operation of process plants distinguish the following modes:

1. normal start-up;
2. normal operation;
3. normal shut-down;
4. emergency shut-down.

The actual procedures for start-up and shut-down depend on the process and vary somewhat.

15.9.1 Normal Start-Up

Start-up requires that the plant be taken through a predetermined sequence of stages. It is important that this sequence be planned so that it is safe and avoids damage to the plant and so that it is flexible enough to handle difficulties which may arise. The personnel involved in the start-up should understand the reasons for the sequence chosen and should adhere to it.

Start-up should be preceded by a series of pre-start-up checks. These should be governed by a formal system. The plant is first checked to ensure that it is mechanically complete; that there are no missing items such as valves, pipes, or instruments; that the status of vent, drain, and sample points is correct; and that slip plates have not been erroneously left in pipework. A check is then made to ensure that the instruments are working. The availability of the utilities is confirmed. The start-up period is one when serious damage can be inflicted on plant equipment. Thus, mention has already been made of the danger of using up a large proportion of the creep life during commissioning. Similar damage can be inflicted in regular start-ups.

Start-up is a time when there is a much higher than average risk of getting unwanted materials such as air or water into the plant. It is necessary to pay particular attention to these hazards.

Instruments are of particular importance during start-up, but unfortunately are likely to be less accurate or reliable than during normal operation. This should be borne in mind if an instrument failure could give rise to a hazardous condition.

It is an essential requirement, therefore, that there be a formal but practical system to control start-ups, that there be proper documentation, and that the personnel be fully trained in the procedures. If it is necessary to shut the plant down on account of a fault, the cause and implications of the fault should be established and any necessary measures taken before the plant is started-up again.

15.9.2 Normal Operation

The normal operation of the plant is not considered in detail in this section, but there are several points that may be restated. The plant management should ensure that the objectives of, and constraints on, the operation of the plant are clearly defined and well understood by the process operators. In any event, management must make it clear beyond doubt that any conflict between production and safety is to be resolved in favor of the latter.

15.9.3 Normal Shut-Down

Normal shut-down starts from the condition of normal operation and its preceding condition is therefore relatively well defined. This makes it practical to formulate shut-down procedures more fully than for some of the start-up situations mentioned.

15.9.4 Emergency Shut-Down

Emergency shut-down may be effected by an automatic protective system or by the operator. In the first case, the operator's function is to forestall the activation of the automatic system by averting the threatened parameter excursion, while in the second he has the additional function of performing the shut-down if it becomes necessary.

The emergency shut-down procedure is generally designed so that the conditions which should trigger it are unambiguously identified and the actions to be taken are clearly defined. These emergency shut-down procedures are to be distinguished from the procedures required to deal with escapes from the plant, which are generally referred to as emergency procedures.

15.9.5 Prolonged Shut-Down

If shut-down is prolonged, precautions should be taken to prevent deterioration of the plant. Typical examples are inspection of equipment to check for external corrosion and 'turning over' of pumps to avoid 'brinelling'. Brinelling is a surface failure caused by contact stress that exceeds the material limit.

15.10. OPERATION OF STORAGE

A large number of incidents occur associated with the bursting or collapse of atmospheric storage tanks due to mal-operation. Some ways in which overpressure can occur include: (1) pumping in liquid too fast, (2) an increase in the temperature of the liquid contents, (3) pumping hot liquid into water, and (4) blowing in air, steam, or gas. Similarly, ways in which vacuum can occur include: (1) pumping out liquid or emptying liquid under gravity too fast, (2) a decrease in the temperature of the liquid contents, and (3) condensation of the steam or vapor contents or a depletion of the gas content.

15.11. SAMPLING

Sampling is carried out to monitor product quality, the material balance, or equipment operation, or for the purposes of troubleshooting. Sampling is a common cause of accidents to personnel in the chemical industry. It needs to be addressed by the provision of suitable equipment and of formal systems of operation. Two principal problems are corrosive liquids and flammable gas and liquids.

15.12. TRIP SYSTEMS

Trip systems are provided to protect the plant against certain hazardous situations by shutting it down if particular parameters go outside the specified limits. Disarming of trips is a permissible practice provided that it is done in accordance with the design intent and with proper procedures. Disarming is not the only way in which a trip can be rendered ineffective. It is essential, therefore, to ensure that there is no interference with trip settings. A trip system should be designed so that it does not reset itself when the trip condition disappears.

15.13. IDENTIFICATION MEASURES

It is essential, if errors are to be avoided, that plant vessels and equipment be given appropriate marking so that they are readily identified and any hazards associated with them are understood. Equipment identification is of particular importance for maintenance work and should be considered carefully.

15.14. EXPOSURE OF PERSONNEL

Of the two main ways of reducing the size of a potential disaster, mention has already been made of one, that is, the limitation of inventory in the process design. The other is the limitation of exposure of personnel in plant operation.

Exposure is most simply limited by not having people there in the first place. For those whose presence is essential much may be done to provide protection, particularly through the siting and design of buildings. In the event of an emergency, the shift personnel have specific instructions on the action which they should take and on the alarm or other instruction at which they should evacuate. The order to leave the plant is given to the day maintenance personnel and to construction workers on the appropriate alarm signal or on instruction from their respective supervisors.

15.15. SECURITY

Another important aspect of plant operation is security. Security systems management is described by Spranza (1981, 1982, 1991, 1992) as being essentially a blend of traditional security concepts and modern management methods. The management system should include an explicit requirement for a review of the security system. Central to security is control of access. There then needs to be physical arrangements to ensure that access to the site is controlled at all points. Another basic principle of security is to create conditions that assist in the rapid detection of any unauthorized deviation.

Expectations for security have greatly changed over the past few years. Industry is faced with the challenge of dealing with elevated levels of perceived threat. Therefore, there is often the need to make decisions under much uncertainty. Moore (2004) discusses the new security threats that arise from potential deliberate releases. These threats have created a need for a new risk management paradigm for chemical security. The problem is we are not all prepared to deal with the threat. Moore describes the strategy to deal with the issue in three basic steps. The first step is to accept that the threat exists. The second step is to analyze the threats and vulnerabilities. The third step is to define a security management system that meets the criteria.

ACRONYMS

CCPS   Center for Chemical Process Safety
CIA   Chemical Industries Association
ICI   Imperial Chemical Industries
LFG   liquefied flammable gas
RoSPA   Royal Society for the Prevention of Accidents
SRD   Safety and Reliability Directorate

REFERENCES

Amyotte, R.R., Goraya, A.U., Hendershot, D.C., Khan, F.I., 2007. Incorporation of inherent safety principles in process safety management. Process Saf. Prog. 26, 24.

Drennan, D., 2009. Making Best Practice Standard in Your Business. Good People Management International. <http://www.gpminternational.com/PDF%20files/Making%20BOPs%20Standard.pdf>.

ICI LFG Code (ICI/RoSPA 1970 IS/74).

Jenkins, A.M., Bardsley, A.S., Staff, SRD., 1991. Developing Best Operating Procedures: AEA Technology. Atomic Energy Research Establishment, CIA. Chemical Industries Association. 1978. PA15.

King, R., 1990. Safety in the Process Industries. Butterworth-Heinemann, London.

MacCollum, D.V., 2008. Construction Safety Planning. Wiley.

Moore, D.A., 2004. The new risk paradigm for chemical process security and safety. J. Hazard. Mater. 115 (1–3), 175–180.

Prugh, R.W., Johnson, R.W., 1988. Guidelines for Vapor Release Mitigation. Center for Chemical Process Safety of the American Institute of Chemical Engineers (CCPS-AIChE).

Scott, D., Crawley, F., 1992. Process Plant Design and Operation. Institution of Chemical Engineers, Rugby.

Spranza, F.G., 1981. Improve your plant security system. Hydrocarbon Process. 60 (9), 331.

Spranza, F.G., 1982. Set high security standards. Hydrocarbon Process. 61 (4), 269.

Spranza, F.G., 1991. Will terrorists hit your plant? Hydrocarbon Process. 70 (7), 102.

Spranza, F.G., 1992. Plant security checklist. Hydrocarbon Process. 71 (9), 71.

Sutton, I.S., 1995. Writing Operating Procedures for Process Plants. Southwestern Books. <http://www.amazon.com/Writing-Operating-Procedures-Process-Plants/dp/1575020572>.

Trask, M.N., 1990. Operating discipline. Plant/Operations Prog. 9, 158.

Wallace, S.J., 2007. The Case for Inherent Safety: A Review of the Principles of Inherent Safety and Case Studies of Tragedies. American Society of Safety Engineers (ASSE) Professional Development Conference, Orlando, FL.

Widiputri, D., Löwe, K., Loher, H., 2009. Systematic approach to incorporate human factors into a process plant design. Process Saf. Progr. 28 (4), 347–355.

Storage and Transport

16.1. GENERAL CONSIDERATIONS FOR STORAGE

16.1.1 Purpose of Storage

The purpose of storage is to smooth fluctuations in the flows in and out. If the quantity in stock does not vary, there is no point in having storage. The exception is where the storage is held purely as insurance. Thus, there may be various ways of satisfying the design objective. The types of storage, which are economic in these alternative designs, may be different and may have different safety implications also. There may be considerable differences, for example, in the pressure and in the inventory. An account which illustrates the interaction between storage requirements and storage method is given by Hower (1961).

16.1.2 Storage Conditions

The main sets of conditions for gas or liquid storage are:

1. liquid at atmospheric pressure and temperature (atmospheric storage);
2. liquefied gas under pressure and at atmospheric temperature (pressure storage);
3. liquefied gas under pressure and at low temperature (refrigerated pressure storage, semi-refrigerated storage);
4. liquefied gas at atmospheric pressure and at low temperature (fully refrigerated storage);
5. gas under pressure.

The fluids so stored are referred to for convenience as (1) volatile liquids, (2) flashing liquefied gases, (3) semi-refrigerated liquefied gases, (4) refrigerated liquefied gases, and (5) gases under pressure, respectively.

16.1.3 Storage Hazards

The hazards presented by storage depend on the material and the type of storage.

On very rare occasions, a vessel or tank fails catastrophically. This may occur due to mechanical or metallurgical defects. The vessel or tank may be overpressured by overfilling. A tank may be overpressured by too rapid filling and underpressured by too rapid emptying. More commonly, release occurs from other equipment or from pipework or fittings. Equipment which may leak includes, in particular, pumps. Release from pipework may occur due to a crack or pinhole or by full bore rupture, or by a leak or failure at a flange, gasket, or valve.

Release may occur due to an explosion in the tank or vessel. There are various ways in which this can happen. One is physical overpressure which causes the vessel or tank to burst. Another is the ignition of a flammable mixture. Another is evolution of gas due to the reaction of an impurity, material of construction, etc. A fourth is a runaway reaction within the vessel or tank.

Fire at a vessel or tank can cause it to fail. The fire may be a fire beneath it or a jet flame playing on it.

An operational activity which may cause either direct release or vessel or tank rupture is overfilling. Other operational events which may give rise to a release include draining and sampling operations. Maintenance activities may result in a release, generally by admission of fluid to a section which is not fully isolated.

Impact events which may cause loss of containment from storage include impact from a carried item, a dropped load, a vehicle, or an aircraft. A missile from an explosion is another form of impact but is primarily an escalating rather than a true initiating event.

Natural events which may cause loss of containment include high winds, rainstorms, flooding, tsunamis, and earthquakes, while lightning may start a fire.

Arson or sabotage is another cause of hazardous events. Sabotage may take the form of interference with the plant or direct initiation by impact, fire, or explosion.

16.2. STORAGE TANKS AND VESSELS

The main types of storage tanks and vessels for liquids and liquefied gases are (1) atmospheric storage tanks, (2) low pressure storage tanks, (3) pressure or refrigerated pressure storage vessels, and (4) refrigerated storage tanks.

Some of the main types of storage tanks and vessels are shown schematically in Figure 16.1. Fuller descriptions and illustrations of storage tanks and vessels are available in various sources (Hughes, 1970; HSW, 1973, Bklt 30; Myers, 1997). Tanks and vessels for the storage

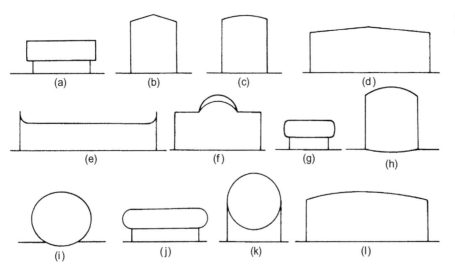

FIGURE 16.1 Types of storage tanks and vessels.

of LPG, liquefied natural gas (LNG), ammonia, and chlorine are considered further below.

16.2.1 Atmospheric Storage

Some typical atmospheric storage tanks are shown in Figure 16.1(a)–(f). Figure 16.1(a) is a horizontal cylindrical tank. These tanks usually have riveted flat or welded dished ends. Parts (b) and (c) in Figure 16.1 are vertical cylindrical fixed tanks with coned and domed roofs, respectively. Parts (d) and (e) in Figure 16.1 are, respectively, another larger fixed roof tank and a large floating roof tank. Figure 16.1(f) is a vapor dome tank with a flexible diaphragm in the dome. Atmospheric tanks are designed to withstand an internal pressure/vacuum of not more than 1 psig (70 mbar). Atmospheric tanks are used for the storage of liquids at ambient temperature and also for the storage of refrigerated liquids, as described below.

16.2.2 Low Pressure Storage

Some typical low pressure storage tanks are shown in Figure 16.1(g)–(i). Figure 16.1(g) shows a horizontal cylindrical tank with dished ends. Figure 16.1(h) shows a vertical cylindrical hemispheroidal tank. Figure 16.1(i) shows a spherical tank which has the shape of a squashed sphere. The two latter types are also made in noded as well as plain versions. Low pressure tanks are designed to withstand internal pressure in the range 0.5–15 psig. The design of low pressure tanks is governed by API Std 620. Low pressure tanks are suitable for the storage of liquids which are too volatile for atmospheric storage. Gasoline is a typical petroleum product to which this applies. Use is also made of low pressure tanks in refrigerated storage, as described below.

16.2.3 Pressure and Refrigerated Pressure Storage

Some typical pressure storage vessels are shown in Figure 16.1(j) and (k). Figure 16.1(j) shows a horizontal cylindrical pressure vessel and Figure 16.1(k) a spherical pressure vessel, or Horton sphere. Pressure storage vessels are regular pressure vessels and can be designed to high pressures as required. The lower end of the scale for pressure storage is 15 psig. Pressure storage vessels are suitable for the storage of liquefied gases such as LPG and ammonia. Pressure storage vessels are also used for refrigerated pressure, or semi-refrigerated, storage.

16.2.4 Refrigerated Storage

A typical refrigerated storage tank is shown in Figure 16.1(l). This is a domed roof, flat bottomed tank.

It is essentially an atmospheric tank, with a design pressure below 1 psig. Low pressure tanks may also be used for refrigerated storage.

16.2.5 Concrete Storage Tanks

The need for very large storage capacities for LNG has led to the development of pre-stressed concrete storage tanks which are protected by an earthen embankment, or berm, and are internally insulated.

16.2.6 Underground Cavity Storage

Underground cavities may also be used for storage. Natural gas has been stored for many years in underground reservoirs, which are frequently depleted oil or gas fields. Use may also be made of such man-made cavities as worked-out salt formations or mined cavities to store gases (Higgins and Byers, 1989). The well is usually operated with a brine system, but dry wells are also possible.

16.2.7 Earth Pit Storage

A method of storing LNG in pits, where a hole was excavated and the surrounding earth then frozen, was developed and used, but the operating problems were such that it has fallen into disuse.

16.2.8 Gas Storage

Gasholders are used to store gases such as hydrogen and acetylene close to atmospheric pressure. Storage of gas under pressure is usually done in gas cylinders or horizontal cylindrical pressure vessels. As already mentioned, gas may also be stored in underground cavities (Crowl and Louvar, 2002).

16.3. SELECTION OF MATERIALS FOR STORAGE TANKS

Carbon steel tanks are the most frequently used in industry. Other materials are also used for special applications, such as aluminum, stainless steel, nickel-steel, and fiberglass-reinforced plastic tanks.

Myers (1997) gives good account of storage material selection in *Aboveground Storage Tanks*. He defines four main principles of material selection: experience, code requirements, brittle fracture, and corrosion. In addition, other factors are also considered important in selecting the right material, such as material toughness and strength, availability, fabric ability, and cost.

Past experience with tanks operating at similar conditions is key in selecting materials, since there are materials known to be especially good for certain applications, that is, nickel-steel and aluminum for cryogenic applications.

It is very important to consider the vulnerability of the material to brittle fracture. Accidents due to brittle fracture are usually catastrophic due to the high speed of propagation of the flaws (about 7000 ft/s). Many accidents due to brittle fracture have occurred, but only a few have been well publicized. One of these well-reported accidents was the collapse of the Ashland storage tank in Floreffe, Pennsylvania, in 1988 (Gross and Smith, 1989). The tank, which was disassembled in Cleveland, Ohio, and reconstructed in Pennsylvania in 1986, collapsed while it was being filled for the first time. It was estimated that 1 million gallons of oil were spilled on the Monongahela River, contaminating water supplies of many towns (Gross and Smith, 1989). Other incidents include the collapse of a large grain-storage tank (Gurfinkel, 1989) and the failure of a polypropylene storage tank (Lewis and Weidmann, 1999). Also, Hayes (1996) summarizes some classic accidents due to brittle fracture, including tanks and other structures.

Myer states that three conditions have to be met simultaneously to have a brittle facture: notch-brittle steel at a given temperature, a notch that causes high local stresses, and stress at the notch. The probability of brittle fracture significantly decreases if one of the conditions is not met. Regardless of the quality of material, it is likely to find defects in the tank. Therefore, material should have *notch toughness*, the ability to absorb energy when a material with a notch is subject to stress (Myers, 1997). Several variables affect notch toughness, such as temperature and composition (carbon content), among others.

Regarding corrosion, different external and internal corrosion effects should be considered, such as pitting and stress corrosion cracking. Coatings are used in order to reduce external corrosion, mostly in carbon-steel tanks. Depending on the environment (e.g., seacoast), more corrosion-resistant materials such as stainless steel and aluminum might also need external coating (Myers, 1997).

According to Myers, in order to select an adequate material for given operating conditions, three steps should be followed. First of all, the Design Metal Temperature (DMT) has to be determined. Then, materials of appropriate toughness at the given conditions have to be selected and evaluated, and finally, the cost should be optimized in order to obtain a resistant and affordable material.

16.4. STORAGE LAYOUT

The siting and layout of storage in relation to the process are discussed in Chapter 7 in the original LEES book. Here the layout within the storage area is considered. As mentioned in Chapter 10 in the original LEES book, the storage, process, and terminals should be suitably arranged relative to one another, an appropriate layout being one in which the storage is located between the process and the terminals. The storage should be built on ground which is able to support the heavy load involved and with ground contours and wind characteristics which minimize the hazard of flammable liquid or vapors from storage collected in hollows or flowing across to the process and finding an ignition source.

16.4.1 Segregation

The segregation and separation of materials within the storage area is largely based on classification of the materials stored, secondary containment, hazardous area classification (HAC), and fire protection measures. The classification of liquids is described in Chapter 10 in the original LEES book. Principal classifications are those given in the IP *Refining Safety Code* and in National Fire Protection Association (NFPA) 30 (NFPA, 2000). Traditionally, the flashpoint classification has been used as a guide to segregation of liquids in storage. Using the earlier classification into classes A, B, and C, the main distinction was between Classes A/B and C. This was the basis of the system used in the 1965 version of the IP *Refining Safety Code*. The current code places less emphasis on this type of distinction as far as concerns segregation.

16.4.2 Separation Distances

Minimum recommended separation distances for storage are given in various codes and other publications. The separation distances for petroleum products given in the IP *Refining Safety Code* are given in Table 22.14 of the original LEES book. The code gives a number of layouts illustrating these separation distances.

Separation distances for flammable liquids are given in HS(G)176, 1998. Figure 16.2 shows a general layout given in HS(G)176. Recommended distances are given for single tanks, groups of small tanks, and large tanks of different capacities. These separation distances are for flammable liquids with flashpoints up to 55°C. There are, however, certain relaxations for liquids with a flashpoint in the range 32–55°C, which are described in the guide.

A group of small tanks (10 m in diameter or less) may be regarded as one tank in order to define separation distances from boundaries, buildings, process areas, and fixed sources of ignition. Such small tanks may be placed together in groups, no group having an aggregate capacity of more than 8000 m³. Recommended distances between individual tanks are given in Table 16.1 and for groups

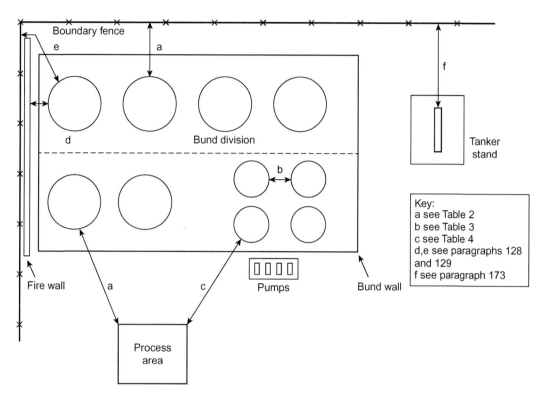

FIGURE 16.2   General layout given in HS(G)176.

of small tanks are given in Table 22.17 of original LEES book.

16.4.3 Separation Distances: Fire Models

An alternative approach is to base separation distances on engineering principles. Two main factors which should determine separation are (1) the heat from burning liquid and (2) the ignition of a vapor escape.

The application of such engineering calculations to the separation distances for storage tanks has been treated by Hearfield (1970) and Robertson (1976). Calculations on separation distances related to heat effects are usually based on direct flame impingement and heat radiation.

As an example, Raj (2005) used a radiation heat-transfer-based model to study the exposure of a liquefied gas container to an external fire. The purpose of the study was to give scientific foundations to the recommended distances in standards and codes such as NFPA 58.

Another recent study (Sengupta and Gupta, 2011) dealt with the location of tanks in a tank farm and safe separation distances. Different methods available in the literature for calculating the heat flux at various distances were used in order to find safe distances. The methods used were the point source model, ShokrieBeyler's method, and Mudan's method.

16.4.4 Secondary Containment

Some types of liquid storage tank are normally surrounded by a bund, or dike, and/or provided with a pit to retain any spillage of the liquid, or impounding basin. Bunds are made of earth or concrete. In general, bunds are provided for atmospheric storage tanks and for fully refrigerated storage tanks of liquefied gas, but not for pressure or semi-refrigerated storage of liquefied gas or for acid or alkali storage, although this generalization needs some qualification.

The object of bunding is to retain the liquid so that it can be dealt with in a controlled manner, by evaporation from a specially designed catchment/evaporation area, by foam blanketing or other measures. Thus, the relatively weak atmospheric storage tanks are generally provided with full bunds, while pressure storage vessels may not be. Bunds tend not to be used for pressure vessels because these rarely fail, the emission when it does occur is mainly in vapor/spray form and the dispersion of small leaks and spillages is hindered. Even where a full bund is not used, however, a low wall may be provided which gives the vessel some protection from damage by vehicles. Low walls may also be used to keep flammable liquids from some external source from reaching a storage vessel.

It may be noted that experience shows that in most cases of tank rupture, only a proportion of the liquid in the tank is lost and in many other cases most of the liquid is retained in the tank. Some practical aspects of the design of bunds for flammable liquids have been discussed by Hearfield (1970). He draws attention to the fact that a leak in the side of a tank may form a horizontal jet and may jump the bund if the latter is too close to the tank, an effect known as spigot flow. The bund wall should be far enough from the side of the tank to prevent a jet jumping over or, alternatively, the bund should be surrounded by an impervious surface sloped inward to the bund drain area. The corners of the bund should be rounded and not at a right angle. It is difficult to extinguish a fire in a 90° angle corner because of the air compression effect.

16.4.5 Hazardous Area Classification

Control of ignition sources in storage areas is exercised through the system of hazardous area classification (HAC). Guidance on HAC for refineries, including storage, is given in API RP 500: R2002 Recommended Practice for Classification of Locations for Electrical Installations at Petroleum Facilities Classified as Class I, Division I and Division 2. This supersedes the separates RPs 500A, 500B, and 500C, which covered refineries, offshore rigs and platforms, and pipelines, respectively. Guidance on HAC for storage is given in the Area Classification Code for Installations Handling Flammable Fluids of the IP (2005 EI IP MCSP P15) (the IP Area Classification Code). Some guidance of the Health and Safety Executive (HSE) is given in 1998 HS(G)176. More detailed guidance is given in the British Standard BS EN 60079-10-1:2009 Explosive atmospheres.

16.5. VENTING AND RELIEF

16.5.1 Atmospheric Vents

A fixed roof atmospheric storage tank is connected to the atmosphere by some form of a vent, generally either a simple free flow atmospheric vent or a pressure/vacuum (PV) valve. Vents for atmospheric storage tanks are dealt with in BS 14015: 2004 in annex L and in API STD 2000: 2009 Venting Atmospheric and Low Pressure Storage Tanks. Since atmospheric tanks can withstand pressure/vacuum of only a few inches WG, it is essential that atmospheric vents should remain free. Blockage can occur accidentally due to debris, icing up, solids formation, and polymerization.

A flame arrester on the end of a vent also serves to keep out debris. If there is no arrester, a coarse wire mesh guard is usually provided. But there is a danger of blockage of flame arresters and guards. The use of flame arresters is considered further below.

The tank may also suffer overpressure or underpressure if the capacity of the vent is not sufficient. Pressure changes occur mainly during filling and emptying, but also in other circumstances. Situations in which tank failures tend to occur due to accidental blockage or deliberate sealing off of a vent, or to lack of capacity in the vent, were described in Chapter 20 in the original LEES book.

16.5.2 Pressure/Vacuum Valves

A PV valve, also called a breather valve or conservation vent, actually has two vent valves—a pressure valve which opens to let vapor out and a vacuum valve which opens to let air in. A fixed roof tank tends to breathe fairly heavily and appreciable vapor loss can occur through an atmospheric vent, particularly for a volatile liquid. A PV valve is effective in reducing this loss. Avoidance of blockage and provision of adequate capacity are important for PV valves also. PV valves are subject to failure from icing up and deposition of material on the valve diaphragms or in the branch on the tank. The danger of ice formation is usually most severe just after the tank has been water tested, and the valve diaphragms are often removed for about 2 weeks until the water content of the liquid has reduced to normal, although the problem is less with newer diaphragm materials such as PTFE.

16.5.3 Flame Arresters

If the vapor space above the liquid in a fixed roof atmospheric storage tank contains a flammable mixture, there is a possibility that it will be ignited via the vent. A flame arrester may be used to prevent this. Guidance from HSE is given in 1996 HS(G)158 Flame arresters—Preventing the spread of fires and explosions in equipment that contains flammable gases and vapors. Guidance from API is given in API RP 2210: 2000 Flame Arrestors for Vents of Tanks Storing Petroleum Products and API RP 2028: 2002 Flame Arresters in Piping Systems. The discussion here is limited to the question of their use on storage tank vents.

The object of a flame arrester is to prevent a flame passing back through a vent to ignite a flammable mixture in the vapor space of the tank. Consideration needs to be given, therefore, to the conditions under which a flammable mixture can exist. The range of materials stored is such that for some the vapor mixture is below the lower flammability limit, while for others it is above the upper flammability limit, but for others again it is in the flammable range. The situation is also affected by the use of a floating deck. This reduces the concentration of vapor but does not necessarily eliminate flammable mixtures. The vapor space above the floating deck may in fact go from a mixture above the upper flammability limit near the deck to one below the lower flammability limit near the fixed roof.

16.5.4 Fire Relief

It is also necessary for fixed roof storage tanks to provide for the effect of fire which heats the liquid, increases its vaporization, and causes a rise in pressure. Tanks can normally withstand a pressure of only a few inches WG. It is normal not to size atmospheric vents or PV valves for fire relief but to provide for this by separate emergency venting.

Methods of providing emergency venting for fire are described in API 2210. They include larger or additional open vents, larger or additional PV valves, lifting manhole covers or hatches, and a weak roof-to-shell connection. This latter arrangement involves the use of a weak roof-to-shell attachment, otherwise known as a frangible joint or rupture seam. This involves making the seam between the shell and the roof deliberately weak so that it is the first to rupture, thus ensuring that the shell stays intact and contains the liquid. The incorporation of such a frangible joint as an emergency vent in accordance with API 2210 is allowed for in API Std 650. Not all companies consider it good practice to rely on a rupture seam for fire relief, although one may still be used to relieve explosions. The methods given in API Std 2000 for the determination of the heat input into storage tanks in a fire, and hence of the venting capacity required, are described in Chapter 16 in the original LEES book.

16.6. FIRE PREVENTION AND PROTECTION

Fire prevention in and protection of storage has several objectives. These are (1) to minimize the risk to personnel, (2) to minimize loss due to the initial fire, and (3) to prevent the spread of fire to other vessels and equipment. Personnel are at risk principally from an explosion or sudden spread of fire. Implementation of fire protection for storage takes the two forms: (1) fighting the fire and (2) protecting the storage vessels. The discussion there includes some treatment of storage, including the application rates of fire-fighting media.

16.6.1 Inerting of Storage Tanks

An effective method of fire prevention in storage is inerting. This is widely used to reduce the hazard of fire in atmospheric storage tanks. The value of inerting in reducing fire/explosion in storage tanks was discussed in Chapter 16 and methods of inerting were described in Chapter 17 in the original LEES book.

16.6.2 Fire Protection of Storage Tanks

There are two main effects of fire on an atmospheric storage tank. One is that those parts of the tank which are not cooled by the liquid inside may become hot and weaken. The other is that the liquid inside the tank is heated up and its vaporization is increased. In a fixed roof storage tank, these results in a pressure rise and, as explained earlier, relief of such pressure is now normally effected by the use of some form of emergency vent.

Fire protection for atmospheric storage tanks is provided by fixed water or foam sprays, which may be supplemented by fireproof thermal insulation, and by mobile water and foam sprays.

Fixed water sprays are effective in giving immediate cooling of exposed surfaces and are particularly useful where manpower is limited or access for mobile equipment is difficult. But the main water sprays are normally provided by mobile equipment and are used both to fight the fire and to cool exposed surfaces. There are also mobile water curtains which can be interposed to protect vessels if sufficient manpower is available (Manatt, 1985).

Foam is used to extinguish fires rather than to cool surfaces. The type of foam used is normally mechanical foam. Fixed foam pourers are used to direct foam to the inside of the tank shell so that the foam flows over the liquid surface. Another fixed device which puts foam on the liquid surface is the Swedish semi-subsurface foam system. This consists of a hose which is connected through the tank shell to a foam supply and is normally at the bottom of the tank. Injection of foam inflates the hose, the top end of which rises to the liquid surface and sprays the foam.

Mobile foam-spraying equipment consists of mobile monitors and portable foam towers. A monitor projects jets of foam in a manner similar to that of conventional fire water jets. A portable tower is a light tube which is put upon the side of the tank and pours foam on the top of the liquid. There are various types, some assembled manually and others of telescopic design extended mechanically or hydraulically.

16.6.3 Pipework and Fittings

Fire can also have a very damaging effect on pipework, which sometimes may withstand the fire for only about 10 min. Pipework is very vulnerable to fire and should be protected as far as possible. Fireproof insulation on pipework near storage vessels may be appropriate. A small fire at a leaking flange can have a torch effect and this possibility should be borne in mind in designing pipework. A fire of this type was one of the causes considered at the Flixborough Inquiry.

16.7. TRANSPORT HAZARDS

The transport of hazardous materials may present a hazard to (1) the transporter, (2) the crew, or (3) the public. The relative importance of these varies between the different modes of transport. The hazards presented by the transport of chemicals are:

1. fire,
2. explosion,
3. toxic release
4. conventional toxic substances,
5. ultratoxic substances.

As with fixed installations, so with transport, the most serious hazards arise from loss of containment. Modes of transport such as road, rail, waterway, and pipeline can give rise to release of flammable gas or vapor, which may result in a flash fire or a vapor cloud explosion, or of flammable liquid which may lead to a pool fire. Flammable liquid can also spread and then ignite, giving rise to a flowing fire. Jet flames may occur on the containers. The contained material may undergo a boiling liquid expanding vapor explosion (BLEVE) or other form of explosion. In general, a flammable fluid is more likely to give rise to fire than to explosion. An explosion hazard also exists with substances that are to some degree unstable. If the material is toxic, release may result in a toxic gas cloud. Toxic or corrosive materials may also spread as liquids. With waterway and sea transport, the release of vaporizable materials onto the water is liable to give rise to very rapid vaporization, resulting in a large vapor cloud. Transport accidents are particularly liable to cause pollution, since material spilled is often not recovered, but is dispersed into the environment.

The initiating factor in a transport accident may be (1) the cargo, (2) the transporter, or (3) the operations. The cargo may catch fire, explode, or corrode the tank; the transporter may be involved in a crash or derailment

or fire; the operations such as charging and discharging may be wrongly executed. Thus, the events which can give rise to hazards include particularly (1) container failure, (2) accident impact, and (3) loading and unloading operations. Although the attention in transport of hazardous materials by moving vehicles focuses on the injury caused by release from containment, fatalities also occur due to the movement of the vehicles themselves. This should also be taken into account, particularly in comparative studies of different modes of transport, since although the deaths from this cause in any one accident tend to be small in number, they may in total equal or exceed those from the releases.

16.8. SIZE OF UNITS FOR TRANSPORT

As with fixed installations, so with transport, the question arises as to whether it is safer to handle a given quantity of hazardous material in a few large units or in many smaller ones. In particular, there is a wide range of possibilities with sea transport. The choice made affects not only the size of the ship but also that of the associated storage on land. In general, a small number of large units appears safer in the sense that the product of scale and frequency tends to be less. This is because, with fewer units, it is easier to achieve a more uniformly high standard of design and operation, and the frequency of incidents is less. The scale of the most serious accident, however, is greater.

16.9. TRANSPORT CONTAINERS

16.9.1 Types of Container

The types of container used in the transport of dangerous goods include (1) tanks fixed to road vehicles and rail cars, (2) multi-modal tank containers, which can be transported by several different modes of transport, and (3) IBCs. There are also demountable tanks, which differ from tank containers in that they are not intended to be loaded and unloaded while fully charged.

The UN *Transport Code* defines a tank container as a tank having a capacity of not less than 0.45 m^3 whose shell is fitted with the items of service equipment and structural equipment necessary for the transport of dangerous liquids.

It defines an IBC as a rigid, semi-rigid, or flexible packaging having a capacity of not more than 3 m^3 which is designed for mechanical handling and is resistant to the stresses produced in handling and transport, as determined by tests. Tank containers are treated in the UN *Code* in terms of requirements for pressure vessels. IBCs are treated rather in terms of the requirements for packaging, with emphasis on tests for such packaging. There are pressure test requirements, however, for certain metal IBCs such as drums.

The ADR contains provisions for fixed tanks, or tank vehicles, and for tank containers.

The other main type of tank of interest here is fixed tanks on road tankers and rail tank cars. The general principles are similar for the two modes of transport and it is road tankers which are now considered. The requirements for road tankers, or tank vehicles, are treated in detail in the ADR. In the account that follows, the provisions quoted relate to the general requirements for tank vehicles. For some classes, there are also specific requirements.

16.9.2 Tank Shell Construction

The mechanical design of the tank shell on a tank vehicle is treated in detail in the ADR. The shell should be designed and constructed in accordance with a suitable code. The ADR gives certain minimum requirements covering, among other things, (1) the material of construction, (2) the minimum wall thickness, (3) the design pressure and test pressure, and (4) the transport shocks.

16.9.2.1 Material of Construction

The general requirement of the ADR is that the shell of a tank vehicle should be made of a suitable metallic material which is resistant to brittle fracture and stress corrosion cracking between $-20°C$ and $50°C$, except where another temperature range is prescribed for a particular class.

The usual material of construction for the shell is mild steel. Lined mild steel is also used. The usual linings are plastics and rubber. In this case, it is necessary to maintain the lining carefully, especially at the joints. Other materials are used for tanks for particular chemicals. For example, aluminum alloy is utilized for hydrogen peroxide.

16.9.2.2 Minimum Wall Thickness

ADR gives the following formula for the calculation of the thickness of the cylindrical wall of the shell of a tank vehicle and of the ends and cover plates:

$$e = \frac{PD}{20\sigma\lambda} \qquad (16.1)$$

where D is the internal diameter of the shell (mm), e is the thickness of the metal (mm), P is the calculation pressure (bar), σ is the permissible stress (N/mm^2), and λ is a coefficient, not exceeding unity, which allows for any weakening due to welds.

There is also a minimum wall thickness. For a shell not exceeding 1.8 m diameter, this minimum thickness is 5 mm for mild steel, and for one of larger diameter it is 6 mm. For other metals, an equivalent thickness is applicable, determined from a formula based on the general relation:

$$e_1 = e_0 \left(\frac{R_{m0}A_0}{R_{m1}A_1}\right)^{1/3} \qquad (16.2)$$

where A is the fractional minimum elongation of the metal on fracture under tensile stress, R_m is the minimum tensile strength, and subscripts 0 and 1 denote mild steel and the metal under consideration, respectively.

16.9.2.3 Design Pressure

For the design pressure of the shell of a tank vehicle, the general provisions of the ADR define four categories of duty. The fourth covers shells for use with substances having a vapor pressure of more than 1.75 bara at 50°C. For this case, the calculation pressure should be 1.3 times the filling or discharge pressure, but in any case not less than 4 barg.

16.9.2.4 Transport Shocks

The shell, and its fastenings, on a tank vehicle are required by the general provisions of the ADR to withstand the following forces:

1. In the direction of travel: twice the total mass.
2. At right angles to the direction of travel: the total mass.
3. Vertically upward: the total mass.
4. Vertically downward: twice the total mass.

16.9.3 Tank Equipment

The ADR general provisions detail the equipment which should be provided on the shell of a tank vehicle. This includes shut-off devices on bottom-discharge shells and pressure relief devices. A bottom discharge should be equipped with two mutually independent shut-off devices. The first should be an internal stop valve fixed directly to the shell and the second a sluice valve, or equivalent device, mounted in series, one at each end of the discharge pipe socket. In addition, the openings of shells should be capable of being closed by screw-threaded plugs, blank flanges, or equivalent devices. The internal shut-off valve should be operable from above or below, and its setting (whether open or closed), should if possible be verifiable from the ground. The valve should continue to be effective in the event of damage to the external control.

16.9.4 Pressure Relief

The general provisions of the ADR require that a shell for carriage of a liquid with a vapor pressure exceeding 1.1 bara at 50°C should either be fitted with a safety valve or be hermetically sealed. A hermetically sealed shell is one with hermetically closed openings and without a safety valve, bursting disk, or similar device, or one fitted with a safety valve preceded by a bursting disk.

The special provisions for Class 2, gases, state that a shell for gases in 1° to 6° and 9° may be fitted with safety valves. But they also state that for carriage of toxic gases in 1° to 9°, a shell should not have a safety valve unless it is preceded by a bursting disk. In this latter case, the arrangement should be satisfactory to the competent authority. These provisions also state that for shells intended for carriage by sea, the provisions of ADR do not prohibit the fitting of safety valves conforming to the regulations of the *IMDG Code*.

The fitting of a pressure relief valve to a road tanker is not a straightforward matter. It cannot be assumed that the fluid to be relieved will necessarily be vapor; in some cases, with the tanker lying on its side, the fluid may be liquid.

16.9.5 Filling Ratio

The general provisions of the ADR contain a number of formulae for the degree of filling (DF). These are of the general form

$$DF = \frac{\chi}{1 + \alpha(50 - t_F)} \tag{16.3}$$

with

$$\alpha = \frac{d_{15} - d_{50}}{35 \times d_{50}} \tag{16.4}$$

where d_{15} and d_{50} are the densities of the liquid at 15°C and 50°C, respectively, t_F is the mean temperature of the liquid at the time of filling (°C), α is the mean coefficient of cubical expansion of the liquid between 15°C and 50°C, and χ is a proportion (%) which has values dependent on (1) the nature of the fluid and (2) the pressure relief arrangements, these values ranging between 95% and 100%.

16.10. ROAD TRANSPORT

16.10.1 Regulatory Controls and Codes

At the international level, codes for the road transport of dangerous goods are the UN *Transport Code* and the ADR. In the United Kingdom, road transport is governed by three principal sets of regulations. The first set is the (CHIP) Regulations 2002, which cover these aspects as they apply to road transport. These regulations, and the associated ACOPs and guidance, have been described in Section 23.3 in the original LEES book. The second set is the Road Traffic (Carriage of Dangerous Substances in Packages) Regulations 1992 (PGR). The regulations deal with the design, construction, and maintenance of vehicles; the marking of vehicles; loading, stowage and unloading; precautions against fire and explosion; information to be provided to the operator; information for, instructions to, and training of drivers; supervision of parked vehicles; and information for the police. Regulation 11 places limitations on the carriage of certain substances. It contains requirements that an organic peroxide or flammable solid subject to self-accelerating decomposition be kept below its control temperature. The third set of regulations is the Road Traffic (Carriage of Dangerous Substances in Road Tankers and Tank Containers) Regulations 1992 (RTR). The regulations deal with the construction, maintenance, and testing of tankers and tank containers; the marking of vehicles with hazard warning panels; loading, filling, and securing of closures; information to be provided to the operator; information to be carried with the vehicle; precautions against fire and explosion; information for, instructions to, and training of drivers; parking of vehicles and their supervision when parked; and information for the police.

Both the PGR and RTR give definitions of dangerous substances, referring in particular to the *Approved List*. The PGR also contains a schedule which gives criteria by which to determine whether a substance not in the list is to be classed as a dangerous substance.

The regulations which preceded these two sets of road traffic regulations, PGR 1993 and RTR 1993, namely PGR 1986 and RTR 1981, were supported by a number of ACOPs and guidance documents. The ACOPs are COP 11 *Operational Provisions of the Dangerous Substances (Conveyance by Road in Road Tankers and Tank Containers) Regulations 1981* (HSE, 1983); COP 14 *Road Tanker Testing: Examination, Testing and Certification of the Carrying Tanks of Road Tankers and Tank Containers used for the Conveyance of Dangerous Substances by Road (in support of SI 1981 1059)* (HSE, 1985); and COP 17 *Notice of Approval of the Operational Provisions of the Road Traffic (Carriage of Dangerous Substances in Packages etc.) Regulations 1986 by the Health and Safety Commission* (HSE, 1987). The guidance is HS(R) 13 *Guide to the Dangerous Substances (Conveyance by Road in Road Tankers and Tank Containers) Regulations 1981* (HSE, 1981), and HS(R) 24 *Guide to the Road Traffic (Carriage of Dangerous Substances in Packages etc.) Regulations 1986* (HSE, 1987).

These documents are currently in the process of being updated, with a single L series document replacing both

ACOPs and the guidance. A single document will replace COP 11 and HS(R) 13, another will replace COP 14 and a third will replace COP 17 and HS(R) 24.

The design and construction of road tankers is covered in the following documents: L16 Design and Construction of Vented Non-pressure Road Tankers used for the Carriage of Flammable Liquids (HSE, 1993a); L17 Design and Construction of Road Tankers used for the Carriage of Carbon Bisulphide (HSE, 1993b); L18 Design and Construction of Vacuum Insulated Road Tankers used for the Carriage of Deeply Refrigerated Gases (HSE, 1993b); and L19 Design and Construction of Vacuum Operated Road Tankers used for the Carriage of Hazardous Wastes (HSE, 1993a).

The road transport of explosives is covered by the Road Traffic (Carriage of Explosives by Road) Regulations 1990. The associated ACOP is COP 36 *Carriage of Explosives by Road* (HSE, 1990).

16.10.2 Hazard Scenarios

The bulk of hazardous materials transported by road is carried as liquids or liquefied gases in road tankers. The main types of road accident associated with these are leaks due to tank puncture resulting from a collision or overturning or due to failure or maloperation of the tank equipment. Other causes of loss of containment are tank rupture due to overfilling, overheating, or a defect and due to fire.

If the tank is ruptured by overfilling or overheating, a physical explosion may occur, giving rise to a blast wave and to missiles. Material released from the tank by whatever cause may be flammable or toxic and in vapor or liquid form. A release of flammable vapor will form a vapor cloud which may ignite to give a flash fire or vapor cloud explosion. A flammable liquid spill may ignite to give a pool fire or flowing liquid fire. A toxic material will give a toxic gas cloud or toxic liquid spill. A fire at the tank itself may cause rupture of the tank and, if the tank contains flammable liquid, may lead to a BLEVE, with its associated fireball. Road tankers do not, however, suffer BLEVEs as commonly as rail tankers, because there is not the torch effect from relief valves on other tanks which occurs in rail crashes. Other hazardous materials are solids which are carried in goods vehicles. These include explosives and related substances. With these there is, therefore, the hazard of explosion.

There are several characteristics of road transport which bear on the nature of the hazard. The vehicle is moving through an environment over which the driver has relatively little control. If there is an accident, it may occur at a variety of points along the route with very different vulnerabilities. On the other hand, the quantity carried is limited, and it may be possible to move the vehicle

to a less vulnerable location. One important aspect of vulnerability is the number of people exposed at a particular location. The extent and composition of the population exposed varies widely with the location. Another aspect is the susceptibility to environmental pollution at that point. A road accident may pose a threat not only to life but also to the environment. A feature of road accidents is that access for the emergency services is generally relatively easy compared with rail. On the other hand, the accessibility of a road accident increases the likelihood that spectators will gather.

16.10.3 Road Tanker Design

Requirements for the design of road tankers are given in the Road Traffic (Carriage of Dangerous Substances in Road Tankers and Tank Containers) Regulations 1992 (RTR). The RTR are essentially a set of goal-setting regulations and do not specify matters such as wall thickness or overpressure protection. In particular, the requirements of the RTR for design and for testing and examination, given in Regulations 6 and 7(1), respectively, are very general, and much less detailed than those given in the ADR. The RTR states in Regulation 8(5) that the aforementioned regulations do not apply to the carrying tank of a road tanker or to a tank container where that tank and its fittings comply with the ADR, RID, or the IMDG *Code*. In other words, compliance with the ADR is in effect one way of meeting the requirements of the RTR. For the more hazardous substances, it is common practice to use top rather than bottom connections and to provide the vehicle with additional features such as side protection.

16.10.4 Carriage Tank Fittings

The provision of suitable carriage tank fittings can make a major contribution to minimizing any loss of containment, both in road accidents and in loading and unloading. One particular feature is the arrangements for protection against overpressure. For flammables, the practice is to fit pressure relief. HSE guidance is given in L16 and L18. L16 for flammable liquids requires each tank to be fitted with pressure relief devices, which it specifies. L18, for non-toxic, deeply refrigerated gases, likewise requires pressure relief devices on each tank and specifies them. For toxics, such provision is more controversial. In COP 38 on transportable gas containers, the HSE states that containers for toxic fluids should not normally be fitted with pressure relief devices.

16.10.5 Road Tanker Operation

The RTR also contains requirements for the operation of road tankers, which reflect some of the elements of good practice. The documentation supplied should contain the information necessary for all the parties concerned, which include the consignee, the driver, the police, and the fire service. The driver should be well trained in and provided with the information that he requires. The training should cover loading and unloading operations, including precautions against overfilling and measures to secure closures; arrangements at lorry stopover points; and handling emergencies, including both spillage and fire. The vehicle, the carriage tank, and its fittings need to be maintained to a high standard.

16.10.6 Filling Ratio

It is essential that a tank should not be overfilled, with the consequent hazard of rupture due to expansion of the liquid. The permissible extent of filling is specified in terms of the filling ratio. Guidance is given in the BS 5355: 1976 *Filling Ratios and Developed Pressures for Liquefiable Gases and Permanent Gases*. This standard, as amended in 1981, defines the filling ratio as:

The ratio of the mass of gas introduced into a container to the mass of water (water capacity) at 15°C that fills the container fitted as for use, i.e. complete with valve, dip tube, float, etc., as necessary.

BS 5355 gives in Appendix A formulae for the filling ratio (FR) in the original book. It distinguishes between a low pressure liquefiable gas, which has a critical temperature between −10°C and 70°C, inclusive, and a high pressure liquefiable gas, which has a critical temperature above 70°C. For single components, the FR formula is of the general form

$$FR = \frac{\chi \rho_1 (1 - C/100)}{\rho_w} \qquad (16.5)$$

where C is the confidence limits on the value of the liquid density (%), ρ_1 is the density of the liquid, ρ_w is the density of water, and χ is a parameter which has the value 0.97 for low pressure liquefiable gases and 0.98 for high pressure liquefiable gases. The liquid density is to be evaluated at the appropriate reference temperature. The standard also gives values of the filling ratio in tabular form for a large number of gases.

16.10.7 Reference Temperature

The reference temperature is the assumed maximum temperature which the tank contents could reach during carriage. It is used to determine the filling ratio and also the developed pressure. BS 5355: 1976 gives reference temperatures for the United Kingdom and also information which allows reference temperatures to be determined for other parts of the world. It classifies areas according to the maximum shade temperature. The United Kingdom is classified as a climatic area with a shade temperature <35°C. In addition to Class UK, there are Classes A−E, with maximum shade temperatures of 37.5°C, 42.5°C, 47.5°C, 52.5°C, and >52.5°C, respectively. Then for the United Kingdom, the reference temperatures for container volumes *(V)* over 1 m³ are given in Table 16.2.

For low pressure liquefiable gases in winter, the standard gives a relaxation in reference temperature for filling ratio for volumes exceeding 5 m³. The choice of reference temperature is a very debatable subject. If a single reference temperature is used, simplicity is achieved, but the temperature selected has to be much higher than is necessary and economic for a large proportion of cases. Discussions of this point are given in the Home Office *Containers Report* and by Clarke (1971). The latter states that reference temperatures as high as 65°C have been used in international codes. The reference temperatures provide overall guidelines, but it is also necessary to observe any requirements of international transport and of the country of delivery.

16.10.8 Hazardous Materials

For a given mode of transport, there are some hazardous materials which may not be conveyed. Road transport is a case in point. Writing in 1971, Clarke (1971) gave the following list of substances prohibited for road transport in West Germany:

- Class 2 Gases: chlorine, hydrogen bromide, hydrogen chloride, hydrogen fluoride, hydrogen sulfide, phosgene, sulfur dioxide.
- Class 3 Flammable liquids: carbon disulfide.

TABLE 16.2 Reference Temperatures for Specified Container Volumes in the United Kingdom

Type of Contents	Reference Temperature for Filling Ratio CO		Reference Temperature for Developed Pressure CO	
	$V = 1-5$ m³	$F > 5$ m³	$F = 1-5$ m³	$F > 5$ m³
Low pressure liquefiable gases	42.5	38	47.5	42.5
High pressure liquefiable gases			50	45

- Class 6 Poisons: acetone, cyanohydrin, acetonitrile, acrylonitrile, allyl alcohol, allyl chloride, aniline, epichlorohydrin, lead alkyls, organophosphorus compounds.
- Class 8 Corrosives: bromine, fluoboric acid, hydrazine, liquid acid halides, and liquid chlorides which give off acid fumes in contact with moist air (e.g., antimony pentachloride).

For certain chemicals which are transported by road special measures are required. These include:

1. substances liable to polymerization,
2. substances carried fused,
3. hydrogen peroxide,
4. organic peroxides,
5. sulfur trioxide,
6. bromine,
7. lead alkyls.

Some chemicals, such as butadiene or ethylene oxide, have a tendency to polymerize. It is essential to prevent polymerization from starting because if polymerization occurs heat is evolved and the reaction speeds up. One method of prevention is the use of an insulated tank, but this is increasingly being replaced by the alternative method, which is the use of polymerization inhibitors (Marshall et al., 1995).

Some substances which are solids at ambient temperature, such as sulfur, phosphorus, alkali metals, and naphthalene, are often carried in fused form in overdesigned and insulated tanks.

High strength hydrogen peroxide is transported in aluminum or special steel tanks. Materials which could react with the peroxide such as wood or valve lubricants are excluded from the vehicle. A water tank is carried to deal with any accident. Organic peroxides are a somewhat similar case. They are sometimes carried in refrigerated insulated tanks to minimize the hazard due to their instability.

Sulfur trioxide is a solid at ambient temperatures and is carried heated in overdesigned and insulated tanks. It is stabilized to present polymerization. Bromine is usually transported in lead-lined steel tanks. Lead alkyls are carried in overdesigned and externally protected tanks.

16.11. ROAD NETWORK AND VEHICLES

The assessment of major hazards arising in the road transport of hazardous materials requires the use of a wide range of data on the road transport environment.

The road transport environment varies somewhat from one country to another. Accounts relevant to the United Kingdom have been given by Appleton (1988) and Davies and Lees (1992), and accounts relevant to North America have been given by Glickman (1988), Harwood and Russell (1989), Harwood et al. (1989), Steward and van Aerde (1990), and Gorys (1990).

As in hazard assessment generally, two situations can arise with respect to the estimation of the frequency of particular accident scenarios. Either it is possible to estimate the frequencies of these scenarios from historical data or it is necessary to synthesize the frequencies, by methods such as modeling or fault tree analysis.

Thus, for scenarios such as release of materials which are transported in large quantities (e.g., gasoline and LPG), it may well be possible to obtain historical data. For other scenarios, such as a release of chlorine or an explosion of explosives during transport, it is much more difficult. Moreover, even where historical data exist, it may still be necessary to resort to modeling for reasons such as the need to adapt the data to the particular assessment or to explore the effect of possible mitigatory measures.

As will become apparent from the data given below, a large proportion of incidents involving hazardous materials are not due to traffic accidents, but to other causes. The prime concern here is with incidents which occur during transport rather than during loading and unloading or in temporary storage, but some of the data sets also cover the latter.

Hazardous goods are taken here to be goods defined as such under the United Nations classification and regulated by the Classification, Packaging and Labelling Regulations 1984 (CPL). These hazardous materials are mainly flammable and/or toxic liquids and liquefied gases, reactive chemicals, and explosives.

The principal source of information on road transport and on road accident statistics is the DTp. Other important sources are the Home Office and the Transport and Road Research Laboratory (TRRL), which is part of the DoE.

Unfortunately, as so often happens, there are difficulties in relating information from one source to that from another. For example, DTp statistics are for heavy goods vehicles (HGVs) with unladen weights of not less than 1.5 te, whereas the principal TRRL study of HGV fatal accidents deals with HGVs with unladen weights not less than 3 te.

16.11.1 Road Network

Roads in the United States are classified as paved or unpaved, with paved roads being subdivided into low, intermediate, and high classes or types. According to the Department of Transportation's Bureau of Transportation Statistics, the low type includes earth, gravel, or stone roadways that have a bituminous surface course less than 1" thick. The intermediate type includes mixed bituminous or bituminous penetration roadways on a flexible base with a combined surface and base thickness of less than 7". The high type is further divided as high-type flexible, high-type composite, and high type rigid. High type roads all have a combined surface and base layer thickness of at least 7".

16.11.2 Heavy Goods Vehicles

Vehicles may be classed as HGVs, light goods vehicles (LGVs), cars, and motor cycles (MCs). HGVs may be subdivided according to the number of axles. The vast majority of hazardous materials are carried in HGVs. The most common HGV is the rigid two-axle vehicle.

The mass unladen of a typical two-axle HGV is some 7.5 te and that of a typical three-axle HGV is some 11.5 te, while that of an LGV is some 1.5 te. The maximum load of a two-axle HGV is some 12 te. Approximately, 50% of HGVs carry full load and 25% is empty, with the remainder carrying part loads. The load of an LGV is some 1.5 te.

Truck data by weight class for the United State is provided. The average vehicle weight is the empty weight of the vehicle plus the average load of the vehicle. These numbers exclude vehicles owned by the federal, state, or local governments, ambulances, buses, motor homes, farm tractors, unpowered trailer units, and trucks reported to have been sold, junked, or wrecked prior to July 1 of the year preceding the 1992 and 1997 surveys and January 1, 2002, for the 2002 survey. Data obtained from the Bureau of Transportation Statistics, originally sourced from the US Census Bureau's 1997 and 2002 Economic Census: Vehicle Inventory and Use Survey: United States.

HGVs traveled a total of 221×10^8 km The average annual distance traveled per vehicle is thus 50,800 km ($221 \times 10^8/435,000$).

16.11.3 Speed in Built-up Areas

The severity of an accident depends on the speed of the vehicle(s). Of particular interest is the speed of vehicles in built-up areas (BUAs). Work on this has been done by the TRRL (Duncan, 1987). Their data for the speeds of good vehicles generally, LGVs as well as HGVs, are given in Table 16.3.

The figures refer only to the time when the vehicle is actually moving, the time spent stationary is discounted. With regard to the distribution of speeds, the assumptions made by TRRL are that the normal distribution applies and that the standard deviation is 1/5 of the mean. Using these assumptions, estimates may be made of the probability that a vehicle may be traveling at a particular speed.

TABLE 16.3 Average Speed of Vehicles, 1987

Time of Day	Average Speed (mph)	
	Small Towns	Large Towns
Peak period	13.1	20.5
Off-peak period	14.9	23.6

More specific data on vehicle speeds at the moment of impact may be obtained from tachograph records, as described below.

16.11.4 HGVs Conveying Hazardous Materials (HGV/HMs)

Some estimates are now made for HGVs conveying hazardous materials (HGV/HMs) in Great Britain.

It has been estimated by Kletz (1986) that in 1986 there were some 14,000 road tankers in operation. The authors have confirmed that this approximate figure is still valid. Kletz also gives the annual distance traveled per tanker as 60,000 miles (96,500 km).

There are, in addition, HGVs other than tankers carrying hazardous materials. The number of such vehicles which at some time transport hazardous materials may be quite large, but what matters in the present context is the number of equivalent 'full-time' vehicles. There appears to be no reliable source of information for this. An estimate has therefore been made. The data for hazardous cargoes in the United States indicate that the ratio of nontank truck to tank truck HGV/HMs is about 0.6. The former figure is regarded as a better guide. A roadside survey conducted by the authors is consistent with this estimate. The ratio of 0.6 is used for Great Britain also, which yields:

No. of tanker HGV/HMs = 14,000
No. of non-tanker HGV/HMs = 0.6 × 14,000 = 8400
No. of HGV/HMs = 14,000 + 8400 = 22,400.

For the distance traveled by HGV/HMs, loaded and unloaded:

Distance traveled by individual vehicle = 60,000 miles = 96,500 km
Distance traveled by fleet = 22,400 × 60,000 = 13.4×10^8 miles = 21.6×10^8 km.

This figure is probably an upper bound, since non-tankers may well not travel as high an annual distance as tankers.

16.12. WATERWAY TRANSPORT

The extent of waterway movement of hazardous materials is small in the United Kingdom, but it is considerable in Europe and the United States. The Rhine and the Mississippi are major arteries. In the United Kingdom, British Waterways bylaws apply to the transport of dangerous goods on canals. It issues a document on the terms and conditions of acceptance.

In the United States, the responsibilities of the US Coast Guard extend to inland waterways. The US Army Corps of Engineers is also involved through its responsibility for waterfront structures and embankments, canals, bridges, and dams.

Backhaus and Janssen quote the maximum size of an LNG barge which might be used on the Rhine as 108 m long × 11.4 m broad. Alternatively, use might be made of four barges with a push tug with total dimensions 185 m long × 22.8 m broad.

The construction of the barge tends to vary with the type of material which it carries. In the United States, for example, for petroleum, the barge shell is also the container wall, whereas for chlorine there are separate chlorine tanks within the barge structure.

A principal hazard of barge transport is collision. It is important to minimize the risk of this, not only when the barge is on the move but also while it is loading or unloading.

Another hazard of barge transport arises where barges are used to carry a variety of chemicals. While skilled people are generally available to assist at the loading point, the barge operator may well be on his own at the unloading point.

In the United States, the Coast Guard has attempted to improve the competence of barge crews by encouraging them to take specific training as chemical tanker men.

The NTSB annual reports describe measures to reduce collisions. In particular, efforts have been made to improve bridge-to-bridge communications.

16.13. PIPELINE TRANSPORT

For a pipeline carrying flammable gas, some principal hazard scenarios are a jet fire and a flammable vapor cloud, leading to a flash fire or vapor cloud explosion, while for one transporting liquid, the scenario is a liquid spillage leading to a pool fire or flowing liquid fire. The main hazard scenario for a pipeline carrying toxic gas or liquid is a toxic vapor cloud.

The majority of pipelines carry flammable gases such as natural gas, ethylene, or LPG or flammable liquids such as crude oil, oil products, or NGL. It is to hydrocarbon pipelines that the account given so far mainly relates.

There are some very extensive liquid anhydrous ammonia pipeline systems, especially in the United States. The ammonia is mainly used by farmers in the corn belt as fertilizer. The hazard of brittle fracture was given special consideration. Cases have occurred where brittle fracture has propagated along a pipeline at a velocity close to that of sound in the metal. Thus, a large length of line could be affected. The maximum transition temperature for the pipeline was set at 0°F (−17.8°C), which is well below the normal operating temperature range. Above the transition temperature, ductile shear failures may occur, but these were considered less serious. The composition of the anhydrous ammonia is closely controlled to prevent attack on the pipeline. Impurities in ammonia, such as air or carbon dioxide, can cause stress corrosion cracking. This is largely inhibited, however, if the ammonia contains 0.2% water and this water content is specified as a minimum.

The chlorine may be transported either as vapor or as liquid. A vapor line should be operated with vapor phase only and a liquid line with liquid phase only. The situations which might give rise to mixed phase conditions should be identified and measures taken to avoid them. The Chlorine Institute recommends that for any chlorine pipeline the maximum pressure should not exceed 300 psig (21 bar). However, this should not be construed as the design criterion. It also recommends that the maximum temperature at any section of the pipeline should not exceed 250°F (121°C).

16.14. MARINE TRANSPORT: SHIPPING

Marine transport is distinguished from the other modes of transport in a number of ways. These include its international nature, which has many aspects; the size of the cargoes carried; and the marine environment.

Marine safety needs to be viewed against the background of the shipping industry and its particular characteristics. Since 1945, international trade has experienced an explosive growth. One consequence has been a massive increase in vessel size. Contrary to what might be expected, the industry is a 'low entry' one. As Gilbert states, 'In boom times, banks and financial institutions have been only too willing to lend money to almost anyone who could submit a reasonable prospectus.'

A marine accident may pose a major threat not only to life but also to the environment. A loss of containment of a material may occur due to damage to the cargo tanks or during loading and unloading operations. Shipboard fire and explosion is another serious hazard and one which does occur periodically. The initiating event may be a fire in the engine room, the pump room, a cargo tank, or elsewhere. Or it may be an explosion in the engine room or cargo tank. A particular hazard exists if the ship carries explosive substances. Some of the most devastating explosions have occurred from the explosion of ammonium nitrate cargo.

Loss of cargo also presents a threat to the environment. Here oil cargoes present a particular threat, since oil is carried in large quantities and is liable to do damage to the marine environment which it is difficult to prevent.

16.15. TANK FARMS

A tank farm is a comparatively unsophisticated facility in that it normally lacks on site processing capabilities. The products that reach the depots are in their final form awaiting delivery to customers. In some cases, additives may be mixed with the stored products. While there is often a greater degree of automation on site, there have been few significant changes in the operational activities of oil depots over time.

There have been many accidents associated with tank farms throughout the history of the process industry (Chang and Lin, 2006). Accidents in tank farms can be especially dangerous considering the large quantities of material in one location and the tendency for a single tank to upset other tanks. Spacing issues between tanks, surrounding plants, and the community become central issues to the safety of a tank farm. Several of the most devastating industrial accidents have been exacerbated by communities encroaching on the buffer zone around a plant or tank farm.

16.15.1 Hazards in Tank Farms

For the tank storage area, there are hazards which must be assessed before construction, during operation, and after shutdown. For all parts included in this procedure, if the tank body and accessories are poorly designed, abused, or are not effectively inspected and maintained, incidents can occur.

We need to clearly understand the tank design, operation, and maintenance procedures. Also, for different kinds of tanks, spacing can affect the probability of incidents occurring and reduce the potential loss due to incidents.

16.15.1.1 Tank Design Failure

The design, body materials, and pressure/temperature resistance of a tank should be based on a chemical's storage properties and should follow all applicable codes.

Following are a few fundamental comments on low pressure tank layout and design.

Atmospheric tanks (internal design pressure >2.5 psig) should be vertical cylinders constructed above ground. A vertical, fixed-roof tank consists of a cylindrical metal shell with a permanently attached roof that can be flat, conical, or dome-shaped, among other styles. Fixed-roof tanks are used to store materials with a true vapor pressure (TVP) less than 1.5 psia. These tanks are less expensive to construct than those with floating roofs, and are generally considered the minimum acceptable for storing chemicals, organics, and other liquids.

Before designing or selecting a tank, the capacity needs to be determined first. The total capacity is the sum of the inactive capacity and the overfill protection capacity. The inactive working capacity is the volume below the bottom invert of the outlet nozzle, normally at least 10" above the bottom seam to avoid weld interference. The net working capacity is the volume between the low liquid level (LLL) and the high liquid level (HLL). For an in-process tank, the net working capacity is calculated by multiplying the required retention time of the liquid by its flow rate. In some cases, the required net working capacity may be divided up into multiple tanks if the size of a single tank is physically unrealistic, or if separate tanks are needed for other reasons, such as dedicated services or rundowns. The overfill protection capacity of a tank is that between the HLL and the designed liquid level. The designed liquid level is set higher than the normal operating liquid level to provide a safety margin. The overfill section is filled with vapor under normal operating conditions.

Mechanical design involves specifying the materials of construction, determining the dimensions of the tank and the plates used to build it, and sizing and positioning the nozzles and accessories. Mild-quality carbon steel (A-36, A-328) is the most widely used material for storage tanks (Amrouche, 2002). For corrosive surfaces, a suitable corrosion allowance is added to the thickness of the structure. Carbon steel tanks can be lined with corrosion-resistant materials such as rubber, plastic, or ceramic tile. Tanks can also be insulated for temperature control, personnel protection, energy conservation, or external condensation prevention. In these instances, fiberglass, mineral wool, expanded polystyrene, or polyurethane can be used. The chemical contained inside the tank should be analyzed for its corrosion hazards. In addition, a maintenance schedule should be created to assess and replace or repair eroded parts.

The venting pipes, breathing valves, drainage lines, and other external parts must be carefully designed and installed. Drilling systems, sprinkler systems, and communicating valves all need to be designed for the scale of the tank as well for emergency use. The wind and seismic loadings, available space, and soil-bearing strength determine the optimal height-to-diameter ratio. Shorter, wider tanks are preferred in windy or seismically active areas or when soil-bearing capacity is limited. As available plot space decreases and soil-bearing strength increases, tanks can be designed taller with smaller diameters (Lieb, 2006).

16.15.1.2 Overfilling

Tank hazards associated with operating procedures are important to consider when designing tank farms. Proper management of the storage area can prevent incidents.

Level measurement is vital to prevent overfilling. The loss of level control has contributed to three significant industrial incidents:

- In Australia, the Esso Longford explosion in September 1998 resulted in two fatalities, eight injuries, and more than $1 billion (AU $1.3 billion) in losses.
- In the United States, the BP Texas City explosion in March 2005 caused 15 fatalities and more than 170 injuries, profoundly affected the facility's production for months afterward, and created losses exceeding $1.6 billion.
- In the United Kingdom, the Buncefield explosion in December 2005 injured 43 people, devastated the Hertfordshire Oil Storage Terminal, and led to total losses of as much as $1.5 billion (£1 billion) (The Buncefield Major Incident Investigation Board, 2008; Mannan, 2009).

Tank filling requires proper procedures and protection systems. Overfilling usually leads to major accidents (Summers, 2010). The key causes are lack of hazard recognition, underestimation of overfilling frequency, insufficient training for operators, ill-defined safety fill limits, and lack of applicable mechanical integrity.

16.15.1.3 Static Electricity

Static electricity is dependent on two properties: generation and accumulation. Generation of static electricity occurs whenever two objects are rubbed against each other. The generation of static electricity cannot be completely prevented but should be reduced for tank safety. API 2003 mentions several means of static electric generation: flow generation, pumping, changes in vortex or pipeline diameter, flow through filters or fittings, splashing, spraying, and tank filling.

Under ideal conditions, though, the charge immediately dissipates. The generation of static electricity can be handled without hazard if accumulation does not occur. Accumulation occurs when the generation of electricity is greater than the dissipation. Electric accumulation can be prevented through rubber seal adaptations, shunts, or bonding cables that release the extra electrics. Other means include following good practices by draining free water from tanks regularly, taking adequate tank samplings, and avoiding overshot filling, and rapid free

falling droplets and flows. API RP 2003 identifies the proper flow rates and conditions that prevent static discharge in storage tanks.

16.15.1.4 Lightning

Flammable vapors mixed with oxygen in certain concentrations create the potential for explosions when presented with an ignition source; 0.16% of tanks have a rim fire every year, 95% of those are caused by lightning. The sparks are due to the accumulation of electricity and can lead to huge fires and losses.

16.15.1.5 Vent Fires

Vent fires occur as a result of the ignition of a plume created by hydrocarbon gases exiting tank vents, typically when the tank is being filled. These fires are usually caused by lightning; however, electrical arcing, static discharge, and human activities near the tank can all cause the ignition of a flammable mixture. Investigation of a 2003 tank fire in Glenpool, Oklahoma, found that a static charge was generated as a result of the operator using flow rates that were too high for the transfer operation. The subsequent static discharge ignited the vapors being vented from beneath the floating roof into the space between the internal floating roof and the fixed roof of the tank. Vent fires can occur in all tank types except external floating-roof tanks because they do not have vents.

16.15.1.6 Rim-Seal Fires

Rim-seal fires are the most common type of fire for floating-roof tanks, especially external floating-roof tanks. This shows that all regions of the world are subject to lightning strikes, although Europe and northern Asia have a lower probability of strikes.

To comply with NFPA 780, operators install roof shunts to dissipate the energy of the lightning to prevent fires. However, tests performed by the API RP 545 task group have shown that rather than reducing the risk of fire from lightning strikes, they may actually increase risks. Tests have shown that both above-roof and submerged shunts can produce arcing at the shunt−shell interface under all lightning conditions. Shunts above the roof produce a greater risk because the arcing occurs where there may be a flammable vapor−air mixture.

16.15.1.7 Full-Surface Fires

Full-surface fires occur when the entire liquid surface of the tank is on fire. They can be further divided into obstructed full-surface fires and unobstructed full-surface fires.

Obstructed full-surface fires occur when access to a portion of the burning surface is blocked by the roof or pan, and happen when the roof or pan sinks. Roof sinking occurs due to a variety of reasons, such as:

- Rain buildup on the roof where there is inadequate drainage due to plugged drains or rain that exceeds the design standards of the tank.
- Pontoon roofs where the pontoons have become filled with the tank liquid as a result of corrosion or other failures.
- Improper application of firefighting materials during a rim-seal fire that cause the roof to sink.

16.15.1.8 Unobstructed Full-Surface Fires

Unobstructed full-surface fires occur when the entire tank surface is easily accessible. For tanks 45 m or smaller in diameter, these are generally readily extinguished as long as there are sufficient resources (water, foam, etc.) and personnel available. In tanks greater than 45 m, the fires are generally very difficult to control due to the large amount of resources required to extinguish such a large tank. These fires usually occur in fixed-roof tanks without internal roofs, where, as a result of an incident, the weak roof-shell weld is broken and the roof is lifted off the tank.

16.15.1.9 Equipment/Instrument Failure

Storage tanks must be equipped with high temperature alarms, ventilation systems, high pressure alarms, and vapor concentration detectors. The failure of those devices will create larger issues when abnormal situations occur. When the material in a storage tank evaporates to the flammability/overpressure limits, the system should motivate the emergency ventilation instruments for relief. Level detectors should also be set to detect abnormal situations.

Linked valves (discharge valves, relief valves), pipelines, and other drilling conductions should be well maintained to prevent hazardous releases caused by equipment failures and abnormal heating/overpressure. All device parts should be regularly maintained.

16.15.1.10 Tank Crack and Rupture

There are several tank structure problems that result from chemical corrosion, overfilling, or heating process ruptures. The design and type of material for the tank should be carefully chosen. The building materials should be chosen based on the chemicals that will be stored in the tanks as well as the API codes. Monitors should be set to take continuous readings to avoid ruptures and cracks caused by deficiencies in the stability of the structure.

The pre-startup analysis should contain an evaluation of the tank structure's ability to bear the setting stream pressure, liquid level stress, temperature conditions, and possible component changes due to the chemical reactivity of the materials. It was found that in many cases tanks suffer during transportation, installation, and the loading/unloading of chemicals as well. These factors create major concerns regarding incidents where the tanks rupture or crack.

16.15.1.11 Facility Siting of Tank Farm Areas

When discussing the safety of tank farms, it is of upmost importance to take into account the siting of the entire area. The relative positions, distances, and levels control for each vessel are just some of the factors that should be taken into further considerations.

The impact of potential incidents may also be addressed by the following factors, among others:

- Adequately separating tanks
- Segregating different risks
- Minimizing the potential for an impact or explosion
- Minimizing the potential for and exposure to toxic releases
- Maintaining adequate spacing for emergency personnel, including firefighting
- Minimizing the exposure to fire radiation
- Considering the prevailing wind directions in site layout
- Considering potential future expansions during site layout.

The relative elevation of a site area is important to consider when designing site layout. Whenever practical, locate open flames (process units with heaters, direct fired utility equipment) at higher elevations than bulk quantities of flammables (tanks and storage); this minimizes the potential for the ignition of vapor releases or liquid spills as spills will migrate downhill. Where it is not feasible to locate storage tanks at elevations lower than process areas, increased protection measures may be required to offset the increased potential for ignition. These measures may include diking, high-capacity drainage systems, vapor detection, increased fire protection, shutdown systems, and other safety systems (Sawyer, 2010).

16.15.2 Prevention of Tank Farm Incidents

Based on the above discussions, the failure of liquid storage tanks originates from inadequate tank design, construction, inspection, and maintenance. Tank design and construction has already been discussed. The failure rate for tanks already in service can be reduced via tank maintenance and weld inspections. To minimize the severity of possible tank failures, there should be a secondary containment such as a dike surrounding the tank.

In terms of tank inspection, prior to start-up, a close external inspection of leaks and corrosion should be made. External inspections should be performed both regularly during normal operating rounds and periodically by individuals who assure the mechanical integrity of the tanks. Periodic internal inspections of the tanks are by far the most important. These inspections should include thickness readings of the walls and the tank floor.

The Chemical Safety Board (CSB) has identified hot works around tanks as significant ignition sources for tank fires (CSB, 2009). A proper hot work permit system must be in place to prevent possible ignition sources from encountering flammable atmospheres. These hot work permits should incorporate gas testing for flammables.

16.15.2.1 Overpressure and Vacuum Protection

Each fixed roof tank should be provided with proper over pressure and under pressure devices. These devices must be properly sized for the worse possible conditions and periodically inspected to assure that they are serviceable. Vents must be tamper proof. Vents must be checked for build up or choking. No alterations must be made to the vents or the relief devices without the use of Management of Change policies.

16.15.2.2 Fuel Tank Inerting

The elimination of flammable fuel vapors in commercial aircraft fuel tanks is a principal safety priority. Proper inerting of a fuel tank can significantly decrease the risk of explosions and fires. LNG carriers and ground level tank farms are required to have inerting systems to prevent such incidents. Inerting refers to the rendering of the ullage (the air above the fuel) unable to propagate a reaction given flammable conditions and an ignition source. In this case, it refers specifically to reducing the oxygen concentration in the tank. This effectively eliminates one side of the fire triangle. Usually, a pump is used to exhaust the oxygen rich waste to ambient pressure an inert gas generating system in order to improve the performance of the permeable membrane air separator component and to insure sufficient generation of inert gas when available source air pressure is low.

16.15.2.3 Nitrogen Inerting in Storage Tanks

Chemical plants that have fixed roof tanks and are concerned about flammable atmospheres or the inhalation of moisture may rely on nitrogen pads. Compressed nitrogen is piped into the vessels and maintains a few ounces of pressure. This avoids the potential for air to be sucked out

of the tank as the internal pressure drops due to cooling or liquid being pumped out.

16.15.2.4 Shunts

NFPA 780 requires that stainless steel shunts be spaced no more than 10 ft apart around the roof perimeter. These shunts are bolted to the edge of the floating roof and connect with the inside of the shell. Unfortunately, shunts do not bond well for several reasons. First, some components of heavy crude oil, such as wax, tar, and paraffin, tend to coat the inside of the tank wall, forming an isolating barrier between the shell and the shunts. Second, rust on the inside of the shell creates a high-resistance connection between the shells and the shunts. Third, 10−25% of tanks are painted on the inside, typically with an epoxy-based paint which insulates the shell from the shunts. Finally, large tanks may become elliptically distort by several inches, which can cause the shunts to pull away from the shell. Another method uses shunts submerged in the stored product. These submerged shunts may provide some benefits when arcing occurs, since no air is present; however, the submerged shunts still rely on pressure contact that is subject to all the conditions outlined above. In addition, submerged shunts are exceedingly difficult to inspect and maintain.

16.15.2.5 Walkways

Another attempt to create a bond between the roof and the shell relies on the tank's walkway. Nearly all tanks have a walkway or ladder where the upper end is attached to the rim of the tank, and the lower end is resting on the floating roof. The quality of this electrical connection is questionable. The upper connection is a bolted hinge subject to loosening, corrosion, and surface-covering paints. The lower end is a pressure connection with only two wheels resting on rails and is also subject to corrosion and surface-covering paints.

16.15.2.6 Roof-Shell Bonding Cable

Another method is to install a cable from the top of the shell to the middle of the roof, typically on the order of 250−500 MCM. The cable is connected to the top of the rim near the top of the internal ladder, suspended along the bottom of the ladder, and bonded to the center of the roof. The cable must be long enough to connect to the roof at its lowest position. Although at 60 Hz this cable has low impedance, at lightning frequencies it has very high impedance. For example, at 100 kHz, the impedance of 100 ft of 250 MCM cable is estimated at over 32 ohms. Therefore, when thousands of amps of electricity flow across the tank, the impedance of the roof-shell bonding cable is too high to prevent sustained arcing at the shunts.

REFERENCES

Amrouche, Y., 2002. General rules for above-ground storage tank design and operation. Chem. Process. December.

Appleton, A., 1988. The Consumer Protection Act 1987 and product liability. Chem. Eng. (London). 444, 11.

CSB (Chemical Safety Board), 2009. CSB Conducting Full Investigation of Massive Tank Fire at Caribbean Petroleum Refining. <http://www.csb.gov/newsroom/detail.aspx?nid = 295>.

Chang, J.I., Lin, C.C., 2006. A study of storage tank accidents. J. Loss Prev. Process Ind. 19, 51–59.

Clarke, A.W., 1971. Safe containment of dangerous goods during carriage. Major Loss Prev.175.

Crowl, D.A., Louvar, J.A., 2002. Chemical Process Safety Fundamentals with Applications, second ed. Prentice Hall International, Englewood Cliffs, NJ.

Davies, P.A., Lees, F.P., 1992. The assessment of major hazards: the road transport environment for conveyance of hazardous materials in Great Britain. J. Hazard. Mater. 32, 41.

Duncan, N., 1987. Private communication to P.A. Davies.

Glickman, T.S., 1988. Benchmark estimates of release accident rates in hazardous materials transportation by rail and truck. In: Transportation Research Record. Vol. 1193. Transportation Research Board, National Research Council, p. 22–28.

Gorys, J., 1990. Transportation of dangerous goods in the Province of Ontario. In: Transportation Research Record. Vol. 1264. Transportation Research Board, National Research Council, p. 57–68.

Gross, J.L., Smith, J.H., 1989. Ashland tank-collapse investigation. J. Perform. Construc. Fac. 3 (3), 144–162.

Gurfinkel, G., 1989. Brittle fracture and collapse of large grain-storage tank. J. Perform. Construc. Fac. 3 (3), 163–183.

HSE, 1981. Managing Safety. HM Stationery Office, London.

HSE, 1983. Health and Safety Statistics. HM Stationery Office, London.

HSE, 1987. Dangerous Maintenance. HM Stationery Office, London.

HSE, 1990. The Peterborough Explosion. A Report of the Investigation by the HSE into the Explosion of a Vehicle Carrying Explosives at Fengate Industrial Estate, Peterborough, on 22 March 1989. HM Stationery Office, London.

HSE, 1993a. The Costs of Accidents at Work. HM Stationery Office, London.

HSE, 1993b. The Fire at Allied Colloids Limited. HM Stationery Office, London.

HSE, 1998. HS(G) Guidelines 176. HM Stationery Office, London.

Harwood, D.W., Russell, E.R., 1989. Present Practices of Highway Transportation of Hazardous Materials. Rep. FHWA-RD-89-02. Research and Special Programs Administration, Department of Transportation.

Harwood, D.W., Russell, E.R., Viner, J.G., 1989. Characteristics of accidents and incidents in highway transportation of hazardous materials. Trans. Res. Record. 1245, 23.

Hayes, B., 1996. Classic brittle fractures in large welded structures. Eng. Fail. Anal. 3 (2), 115–127.

Health and Safety Commission, 1985. Plan of Work 1985/86 and Onwards. HM Stationery Office, London.

Health and Welfare Bklts, 1973. (HSW Bklt 30), New Series 21. HM Stationery Office, London.

Hearfield, F., 1970. Design for Safety—Fire and the Modern Large-Scale Chemical Plant (Course). Department of Chemical Engineering, Teesside Polytechnic, Middlesbrough.

Higgins, T.E., Byers, W., 1989. Leaking underground tanks: conventional and innovative cleanup techniques. Chem. Eng. Prog. 85 (5), 12.

Hower, N.R., 1961. Liquefied gases: cryogenic or pressure storage? Chem. Eng. 68 (May), 77.

Hughes, J.R., 1970. Storage and Handling of Petroleum Liquids—Practice and Law, second ed. Griffin, London.

Kletz, T.A., 1986. Will Cold Petrol Explode in the Open Air. The Chemical Engineer, IChemE.

Lewis, P.R., Weidmann, G.W., 1999. Catastrophic failure of a polypropylene tank Part i: primary investigation. Eng. Fail. Anal. 6 (4), 197–214.

Lieb, J., 2006. Updated on API 650 App. E: Seismic Zone Design, API Meeting Proceedings, September.

Manatt, S.A., 1985. Fuel tank inerting system, United States Patent 4556180.

Mannan S., 2009. A technical analysis of the buncefield explosion and fire. Proceedings of the HAZARDS XXI Conference, Manchester Conference Centre, Manchester, United Kingdom, November 10–12.

Marshall, J., Mundt, A., Hult, M., Mckealvy, T.C., Myers, P., Sawyer, J., 1995. The relative risk of pressurized and refrigerated storage for six chemicals. Process Saf. Prog. 14, 3.

Myers, P.E., 1997. Aboveground Storage Tanks. McGraw Hill, New York, NY.

NFPA 30 Flammable and Combustible Liquids Code, 2000. Edition, NFPA, New Jersey.

Raj, P.K., 2005. Exposure of a liquefied gas container to an external fire. J. Hazard. Mater. A122, 37–49.

Robertson, R.B., 1976. Spacing in chemical plant design against loss by fire. Process Ind. Hazards.157.

Sawyer, M., 2010. Tank Farm Safety. Presentation to the MKOPSC Steering Committee, August 3.

Sengupta, A., Gupta, A.K., 2011. Engineering layout of fuel tanks in a tank farm. J. Loss Prev. Process Ind. 24 (5), 568–574.

Steward, A.M., Van Aerde, M., 1990. Enhancements and updates to the RISKMOD risk analysis model. J. Hazard. Mater. 25 (1/2), 107.

Summers, A., 2010. Don't underestimate overfilling's risks. SIS-TECH solutions. Chem. Process.

The Buncefield Major Incident Investigation Board, 2008. The Buncefield Incident 11 December 2005, The final report of the Major Incident Investigation Board, Vol. 1.

Chapter 17

Emergency Planning

Chapter Outline

17.1. INTRODUCTION

Emergency response is the last layer of protection that is intended to control an event if possible or to reduce consequences in cases of loss of control. The three components of emergency planning are preparedness, response, and recovery. Emergency planning is an essential part of the safety and loss prevention strategy. Its objective is to mitigate the consequences of any incident that may occur.

The two main types of emergencies concerned to chemical industry are works emergency and transport emergency.

17.2. ON-SITE EMERGENCY PLANNING

17.2.1 An Overview

Emergency preparedness process begins with the identification of credible scenarios for which appropriate response strategies are developed. The analysis of resources and qualified facilities' personnel to respond to emergency scenarios is part of the preparedness stage. The development of resources is conducted according to the resources assessment and the potential of cooperation among site emergency responders, neighboring facilities, and neighboring communities.

The complex nature of emergency events requires a very clear hierarchy of command, and a procedure without any ambiguities.

The development of physical facilities infrastructure consists of: (1) development of shelters and safe havens, (2) establishment of emergency operations center (EOC), (3) development of emergency communication capabilities, and (4) development of appropriate medical support infrastructure.

As shown in Figure 17.1, emergency systems are developed in parallel with the development of physical facilities.

17.2.2 Identification of Credible Scenarios

17.2.2.1 The Problem

Preparedness for emergencies involving worst-case scenarios requires enormous resources and may overwhelm the business operability of the facility. Therefore, the outcome for each scenario should be evaluated based on the consequences and probabilities of the scenario. It is important to consider management controls. Incidents such as instantaneous loss of containment are a major concern in the process industries. According to this, there are measures such as control systems, overpressure reliefs, and alarms. Additionally, the mechanical measures such as non-destructive tests reduce the likelihood of development of such scenario.

It is impractical to plan for all emergencies, and therefore it is necessary to analyze and prioritize the scenarios. The process of scenario selection and prioritization is shown graphically in Figure 17.2.

17.2.2.2 Identification of Process Areas with High Hazards

A large number of techniques are available for the identification of areas of major hazards. Dow Fire and Explosion Index (F&EI) by AIChE (1994) and Dow Chemical Exposure Index (CEI) by AIChE (1998) are examples of techniques that can be used.

17.2.2.3 Credible Scenarios

The depth of analysis can vary from an informal review to a full Process Hazard Analysis session. The level of investigation is mainly dependent on the experience of the reviewers. Highly experienced reviewers are able to identify credible scenarios by an informal review only.

17.2.2.4 Consequence Assessments

A preliminary screening of incidents is a helpful phase in reducing the number of consequence assessments and in categorization of the incidents prior to the assessment stage. The following criteria could be helpful for the screening purpose:

1. Incidents with only minor effects and low probability for escalation should be eliminated.
2. Incidents with high similarity should be grouped.
3. Two representative incidents should be grouped from each of the groups: (1) incident with high probability of occurrence, which may result in severe illness/injury or major property damage and (2) a credible worst-case scenario with critical/catastrophic consequences.

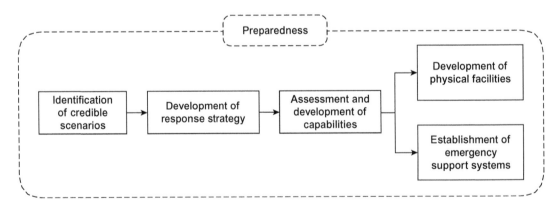

FIGURE 17.1   Emergency preparedness flow chart.

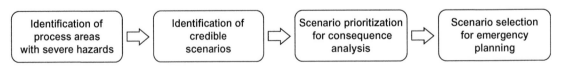

FIGURE 17.2   Process of scenario selection for emergency planning.

The screening procedure described above will establish a list of incidents for consequence assessment.

Consequence assessment tools help to estimate the impact of these incidents. Source term modeling is used to estimate the release rate and quantity released. Explosion modeling, fire and thermal modeling, and dispersion modeling are applied for the assessment of the impact of the incidents. The results of the consequence assessments lead to the determination of areas that will be affected and the level of toxicity/thermal load/overpressure in these areas.

17.3. RESOURCES AND CAPABILITIES

17.3.1 General

The development of strategy for emergency response requires consequence assessments from a list of credible incidents and data on resources and capabilities. Data should be gathered on: (1) emergency response equipment on-site, (2) emergency response equipment in neighboring facilities, (3) emergency management service in local communities, (4) on-site medical capabilities and medical facilities in local communities.

17.3.2 Facilities and Equipment in Emergency Planning

17.3.2.1 On-Site Shelters and Safe Havens

The process of emergency planning for shelters and safe havens should consider variables such as concentration of employees, distance from incident location, typical weather conditions, and alternate evacuation routes. When a building is selected as a shelter, the spaces areas should not have holes or cracks. Additionally, doors and windows must seal properly, and a ventilation system control should be available.

Both, shelters and safe havens should consist of support systems that are proportional to the estimated occupancy.

17.3.2.2 Emergency Power Supply

Planners should verify that vital systems and locations have a power supply for all situations. *NFPA 110— Emergency and Standby Power Systems* (NFPA, 2013a), *and NFPA 111—Stored Electrical Energy Emergency and Standby Power Systems* (NFPA, 2013b), provide guidelines and codes for these systems.

17.3.2.3 Adequate and Alternate Water Supply

Type of chemicals used in the plant, thermal load, demand of sprinkler systems, foam systems, and hose streams are factors that should be considered in calculation of water reserves.

Alternate water supply is necessary to assure water delivery in an emergency. In case of extreme cold weather, measures should be taken to avoid freezing.

17.3.2.4 Emergency Operations Center

It is common to designate a special room as an Emergency Operations Center (EOC). The distance of EOC from processing areas and storage is an important variable in its functionality. The distance of the EOC from the normal residency of the management should be considered as well. This will allow access to the alternate EOC in situations where access to the main EOC is not possible. The planner should designate an alternated EOC, which will be located opposite to the EOC in reference to the processing area. An optimal EOC is one designed as a safe haven. For more information, specifications and guidelines for designing an EOC, please refer to the *NFPA 101—Life safety code* (NFPA, 2012a).

17.3.2.5 Command Post Vehicle

Command Post Vehicle (CPV) allows the Incident Commander (IC) to control the efforts and activities in emergencies and be located at vantage points. The CPV should be selected to allow installation of all the equipment necessary and installation of power source that will be able to supply the consumption that this equipment demands.

17.3.2.6 Emergency Management Computation System

Emergency Management Computation Systems (EMCS) are used for organization of information, estimation of severity of the incident by using source and dispersion models, data collection, receiving, monitoring data, etc. This information is essential to determine the magnitude of the event, in order to make decisions as to announce an escalation and to determine the need of evacuation both on- and off-site.

17.3.2.7 Media Information Center

Sharing information and collaborating with the media could reduce the snowball effect that an emergency event may have. Moreover, media can assist by announcing the emergency, and increase the awareness of the general public in areas that might be affected. The Media information Center (MIC) should be located at a distance that will assure the safety of the reporters and should be designed as a shelter or safe haven.

17.3.2.8 Medical Facilities

On-site medical facility can have the capability of a first-aid room or a medical department. In general, the adequate capability is a function of parameters such as number of employees in the plant, capabilities of the local community medical centers, distance to these centers, and the anticipated consequences in an emergency. Local emergency centers should be made aware of the chemicals in the plant and the estimated number of injuries that are expected in an emergency.

17.3.2.9 Communication System

The design of a communication system should address the following: (1) maintain on-site communication, (2) establish communication with off-site agencies and neighboring facilities, and (3) allow communication among management, emergency team, and responders.

The planner must verify that equipment and supplies will meet the needs in an emergency. Response equipment and supplies can be divided to three main categories: (1) personal protection, (2) fire fighting, and (3) decontamination.

Personal protection equipment consists of two major types: (1) respiratory protection and (2) thermal and chemical protection.

The circumstance determines the level of respiration protection that is needed. NFPA standard 1981 *Open-Circuit Self-Contained Breathing Apparatus* determines the requirement of a self-contained breathing apparatus (SCBA) (NFPA, 2013c).

As for fire fighting gear, these systems supply reasonable thermal protection for most fires. An ice vest is used when high ambient temperature is imposed on the responders.

NFPA 1991—Standard on Vapor-Protective Ensembles for Hazardous Materials Emergencies (NFPA, 2005) and NFPA—1992 Standard on Liquid Splash-Protective Ensembles and Clothing for Hazardous Materials Emergencies (NFPA, 2012b) assist in determining the proper clothes according to the characteristics of the hazards.

A planner should determine the necessity of integrating heavy equipment.

17.4. DEVELOPING AN EMERGENCY PLAN

17.4.1 Overview

Emergency plan consists of the following major components: (1) establishment of a set of credible scenarios; (2) study emergency plans of neighboring facilities and local community; (3) study availability of resources and capabilities, and determining needs; (4) study emergency-related regulations and emergency authorities' requirement and responsibilities; (5) development of tactics to respond to fires and explosions, release of hazardous materials, natural hazards, terrorist threats, rescue, evacuation, and plan for medical response and emergency mitigation; (6) development of procedures to assess level of emergency; (7) training of emergency teams, other employees, and contractors; (8) development and execution of drills (include neighboring facilities and local communities); (9) development of drill-based improvement procedure; (10) communication of the emergency plan on- and off-site.

17.4.2 Emergency Organizations and Regulations

Emergency-related regulations, requirements, and responsibilities are different from country to country. The planner should study the requirements and regulations, in order to develop a plan to meet these requirements. An active Local Emergency Planning Committee (LEPC) in the United States can assist a great deal in coordinating between the plants, the local community authorities, and responding agencies.

17.4.3 Development of Response Tactic

The planner should establish procedures for assessment of severity of the incidents, and a set of criteria that will help in the decision to declare on escalation of the situation. The emergency team management must understand that the safety of the employees and responders is their first priority. Next is preventing the incident from escalating. Environmental concerns are next, and finally the prevention of property damage is the last priority among these.

The planner should develop procedures for the following indirect response components:

1. Warning and alerting;
2. Communication;
3. Distribution of managerial responsibilities;
4. Emergency shut-down;
5. Identification of missing employees;
6. Mutual assistance with the variety of external entities;
7. Reporting;
8. Declaration of escalation;
9. Security;
10. Dealing with the media;
11. Special notification.

Procedures for direct response efforts should include the following:

1. Evacuation;
2. Medical mitigation;

3. Rescue;
4. Fires and explosions;
5. Release of hazardous materials;
6. Terrorism threats;
7. High winds;
8. Flood;
9. Freeze.

Procedures for training, drills, effectiveness measurements, and post-drill change management should be considered part of the response stage as well as part of the preparedness.

17.5. TRAINING

Safety and Health Administration (OSHA), Environmental Protection Agency (EPA), and the Department of Transportation require training for emergencies. Other organizations established standards and good practices that are very useful.

17.5.1 Fundamental and Operators Training

The most fundamental emergency training is the use of fire extinguishing systems and equipment. Additional training that all employees should receive is familiarity with alert codes, evacuation routes, and the locations of shelters and safe havens.

The operators should be trained on the procedure, and to the criteria for proper shut-down. This type of training is extremely important, and more important is exercising it.

17.5.2 Response Teams

Response teams have to understand the inherent hazards that are characterizing the plant, and the list of credible incident that the emergency plan was developed for, hierarchy of command, reporting system, responsibilities, communication systems, and all other relevant information that the emergency plan consists of. The response teams are required to have knowledge and skills in several types of response categories: (1) fire fighting, (2) release of hazardous material, (3) terrorism threats, (4) rescue (natural hazards, man-made hazards, confined places), (5) medical, and (6) evacuation.

Several of the emergency response personnel should be trained to deal with the media and public relations.

17.6. ESSENTIAL FUNCTIONS AND NOMINATED PERSONNEL

17.6.1 Incident Commander

It is the Incident Commander (IC) responsibility to proceed to the scene of the incident and to coordinate the efforts. The IC's major responsibilities are (1) as soon as the IC becomes aware of an incident, he should assess its scale against predetermined criteria or emergency reference levels and decide whether a major emergency exists or is likely; (2) ensure that emergency services have been called; (3) direct the shutting down and evacuation of the other plant areas, likely to be affected; (4) ensure that key personnel have been notified; (5) direct all operation at the scene; (6) control the rescue and fire fighting operations activities until the arrival of the fire brigade; (7) search for casualties; (8) set up a communications point with radio, telephone, or messenger contact with the EOC; (9) give advice and information as requested to the emergency services; (11) brief the site emergency management and keep them informed of developments.

It is important that the IC be readily recognizable when he is at the scene of the incident.

17.6.2 Emergency Operation Manager

The function of the Emergency Operation Manager (EOM) is to take overall control of the emergency operation activities from the EOC. Among his responsibilities are (1) ensure that the emergency services have been called and the off-site plan activated; (2) ensure that key personnel are called in; (3) exercise direct operational control of those parts of the works outside the affected area; (4) continually review and assess possible developments to determine the probable course of events; (5) direct the shutting down of plants and their evacuation in consultation with the IC; (6) ensure that casualties are receiving adequate attention and that relatives are informed; (7) in case of emergencies that involve risk to outside areas from windblown materials, contact the local meteorological office to receive early notification of impending changes in weather; (8) liaise with chief officers of the fire and police services and provide advice on possible effects on areas outside the works, and ensure that personnel are accounted for; (9) control traffic movement within the works; (10) arrange for a log of the emergency to be maintained; (11) where the emergency is prolonged, arrange for the relief of site personnel and the provision of catering facilities; (12) issue authorized statements to the news media; (13) ensure that proper consideration is given to the preservation of evidence; (14) control rehabilitation of affected areas after the emergency.

The EOM should have the authority needed to take any necessary major decisions affecting the activities in the facilities, the neighboring facilities, and the outside services.

17.6.3 Other Functions and Personnel

There are numerous other functions that may have to be carried out in an emergency. Each of the key personnel should be familiar with all his responsibilities, all the relevant procedures, to whom he reports, and who those under his responsibilities are.

17.7. DECLARATION AND COMMUNICATION OF THE EMERGENCY

17.7.1 Raising the Alarm

When a serious incident occurs, it is very desirable for the alarm to be raised as quickly as possible. Crucial aspects of raising the alarm are the authority to raise the alarm, the training of personnel, and the alarm system.

17.7.2 Declaration of the Emergency

When the alarm has been raised, the emergency procedure is activated. The IC should visit the scene of the incident. He then makes the decision whether to declare an emergency. Other competent personnel may be vested with the authority to declare an emergency.

17.7.3 Communication of the Emergency

Once an emergency has been declared, it must be communicated to the following entities: (1) EOM, (2) personnel working within the areas affected, (3) personnel that are needed to be called in from outside, (4) off-site emergency services, (5) key personnel on call-in, (6) personnel in other threatened areas and facilities, (7) police, (8) medical services, (9) media, and (10) headquarters.

It is desirable to work out in advance fairly formal procedures for the notification of an emergency.

17.8. EVACUATION

It is critical that all organizations have evacuation planning as an integral part of their emergency preparedness. An effective emergency planning should include:

- Conditions under which an evacuation would be required. For more information on 'evacuation or not', and 'shelter-in-place', refer to OSHA website: http://www.osha. gov/SLTC/etools/evacuation/shelterinplace.html.
- A clear chain of command and designation of people who have the authorization to order an evacuation or shutdown. For more information about the roles of evacuation coordinators and wardens, refer to OSHA website: http://www.osha.gov/Publications/osha3088.pdf
- Clear and detailed evacuation procedures, including routes and exits.
- Specific procedures to assist people of special needs.
- Designation of authorized persons responsible for critical operations during an evacuation.

- A system for accounting for personnel following an evacuation.
- Shelters for employees, including 'shelter-in-place' and 'exterior shelters'.
- A plan for protection of vital documents, reports, and equipment.

The Federal Emergency Management Agency's website provides useful evacuation guidelines: http://www. fema.gov/plan/prepare/evacuation.shtm, http://www.fema. gov/pdf/areyouready/basic_preparedness.pdf.

Overall, an evacuation planning procedure typically includes the following steps:

- Vulnerability assessment;
- Analysis;
- Probability impact assessment;
- Review of plans, resources, codes, and regulations;
- Identification of internal resources;
- Identification of external resources.

17.9. COOPERATION AND DRILLS

17.9.1 Planning Cooperation with Off-Site Services

The first step is obviously establishment of agreements to cooperate in emergencies.

The planning should aim to clarify not only what is to be done but also who is to do it.

Cooperation benefits greatly if there is a full-time liaison officer, and this can be justified in an area where there is large potential for emergencies. The aim should be to plan broad areas of responsibility, chains of command, and systems of communication.

Off-site services will require their own communications. On-site, the practice has been for the police, fire, and medical services to be linked by radio to their own communications systems. The EOC should be available to them. Alternatively, they may wish to set up their own mobile control centers.

The off-site services will also normally need to tap into the on-site emergency internal communication. This should be allowed through the EOC. Each service should have its own emergency plans.

17.9.2 Emergency Drills

Emergency drills are effective in familiarizing personnel with their functions. While drills with on-site personnel can be dictated by a routine, real-time simulation that involves all forces, on- and off-site are much more complicated to perform. However, the effectiveness of the plan, as well as the performance of each of the entities in the response stage can be assessed only in full-scale drills.

17.10. PUBLIC RELATIONS

If a major incident occurs, it is important to provide the media with accurate information. It is mandatory to give a report in the press. A good emergency plan will consist of a procedure that provides guidelines to answering questions. Several questions might be anticipated. It will normally be appropriate that the long-term handling of major hazard policies be done by the company's head office with reference to national policies and standards. A periodic training of a spokesman can significantly improve the credibility and reliability of the firm's public relations.

17.11. OFF-SITE EMERGENCY PLANNING

The off-site emergency plan is similar in structure to the on-site emergency plan. However, coordination of off-site emergency plan requires much more energy and effort. For off-site plans, the EOM may have partial authority only.

Warning procedures should use common channels of communication to advise the public on an emergency. It is necessary to address special needs groups such as sight and hearing impaired and nursing home.

Provision of shelters and mass care is a major responsibility. There is a large number of parameters involved in the process of designation of shelters. The efforts in the United States should be coordinated with the American Red Cross.

The criteria for the decision of evacuation should be clear. Large number of procedures, modular plans, and templates are available for the disciplines listed above on the US Federal Emergency Management Agency and on states' Division of Emergency Management websites.

17.12. TRANSPORT EMERGENCY PLANNING

17.12.1 Background

It is essential for hazardous chemicals to be transported in tanks and containers designed and maintained to high standards, for transport personnel to have any necessary training, for comparative studies of the different means of transport to be done and routes carefully selected, and for hazard studies to be carried out (as described in Lees' book fourth edition, Chapter 23). Even so, it is still necessary to plan for possible transport emergencies.

Transportation emergency planning includes the following elements: (1) chemical data, (2) information and labeling, (3) incident control network, (4) emergency procedures, (5) emergency teams, (6) outside services, and (7) public relations.

The following sections discuss these elements.

17.12.1.1 Chemical Data

It is necessary to pay particular attention to any property that is especially important in a transport emergency. Material Safety Data Sheet (MSDS) should be part of the delivery.

17.12.1.2 Information and Labeling

The most accessible source of information is the labeling of the container. Then there is usually more detailed information carried somewhere on the vehicle, if an MSDS is not reachable. The shipper and outside services, such as police and fire services, should have manuals that contain similar data. Finally, there is the manufacturer who may be contacted through the Federal Emergency Management Agency, EPA, and others.

17.12.1.3 Emergency Procedures

The priority actions that should be taken by the person at the scene of the incident are (1) keep people away, (2) inform incident control, (3) contain the chemical, (4) avoid igniting the chemical, and (5) obtain chemical data.

The communication with the EOC is important. The center should be informed as follows: (1) place and time of the incident, (2) chemical involved, (3) container condition, (4) injuries or/and fatalities, (5) the surrounding area (urban, rural), (6) weather conditions, (7) the assistance available (police, fire services), and (8) the means of maintaining contact.

It is particularly important for the person on the spot to ensure that the control center can keep in contact with them. If possible, the chemical should not be allowed to get into sewers. A pool of liquid may often be rendered safer by covering it with a blanket of suitable foam.

Ignition of the material should be avoided if at all possible. Sources of ignition should be kept well away.

17.12.1.4 Emergency Teams

The personnel comprising the emergency team are usually thoroughly familiar with the chemical and are trained in handling incidents. The emergency team usually has expertise in dealing with leaks and fires and in emptying damaged containers and clearing up.

The emergency team is sent as soon as possible on receipt of notification.

17.12.2 External Services

A transport emergency differs from emergency activities on-site. There is therefore much less scope for advance cooperation between a manufacturer and the services in his local authority area. The variety of hazardous chemicals that the services may have to handle is much greater.

Another feature of a transport emergency is that it may well occur in an urban area. This means that measures to keep people away from the scene, to divert traffic, to maintain access for emergency vehicles, and possibly to evacuate the population assume particular importance.

Likewise, information on fighting fires of the chemical is important for the fire services. In particular, the latter need to know whether the chemical reacts violently with water.

17.12.3 Rail Transport

The foregoing account has been concerned primarily with road transport, but most of it is applicable to rail road transportation as well. In this case, the railway authority establishes its own procedures and trains its own personnel. Again the prime requirement from the manufacturer is full information about the chemicals.

Emergencies involving hazardous chemicals on railways tend to have some special features. One is the problem of access, particularly for fire engines. The track is often blocked by derailed rail vehicles. Another aspect is that there are usually a number of tank cars involved, with the danger of a spread of the fire/explosion.

17.13. EMERGENCY PLANNING FOR DISASTERS

17.13.1 The Importance of Emergency Planning for Disasters

Disasters, whether man-made or natural, cause great losses to human lives, property, and the environment. One way to alleviate the damages of disasters is to have an effective emergency planning. In addition to providing guidance during an emergency, having an emergency plan brings other benefits, such as: (1) discovering potential hazards which may occur in an emergency situation, (2) identifying deficiencies, for example, the lack of resources (equipment, trained personnel, supplies, etc.), (3) and promoting safety awareness and preparedness. In contrast, the lack of an emergency plan could lead to additional losses such as multiple casualties and possible financial collapse of the organization.

17.13.2 Effectiveness of Emergency Aid to Disaster Zone

When incidents happen, it is critical to respond effectively by identifying qualified personnel and deploying resources in order to save lives, protect property and (conserve/preserve) the environment. An effective response includes the following four steps: (1) obtain and sustain situational awareness, (2) activate and deploy resources and capabilities, (3) coordinate response actions, (4) and demobilize.

17.13.3 Counseling to the General Public After a Disaster

Providing proper counseling to disaster victims is a critical step of the recovery process.

17.14. SPECTATORS

The effects of spectators are two-fold. The number of people at risk may considerably increase, and routes of the emergency services can become congested, leading to delays. The control of spectators is therefore an essential part of an off-site emergency plan and a transport emergency plan.

17.15. RECOVERY

The recovery manager needs to establish a recovery team that represents all the disciplines in the plant. The team should include representatives from engineering, maintenance, production, purchasing, ES&H, legal department, and other, as the circumstances require.

Following the emergency phase, the site should be secured and preservation of evidence and data collection need to take place. Parallel to these activities, human resources should provide assistance to employees and victim families. Incidents that resulted in fatalities will require professional psychological support as well. A portion of the efforts of damage's assessment to environment, property, and incident investigation could be integrated to avoid repetition of some of the phases. Cleanup activities should be coordinated extremely carefully to avoid removal of evidence. It is recommended to consider involvement of the media in the process.

17.16. REGULATIONS AND STANDARDS

17.16.1 Regulations

In the United States, the OSHA established the Process Safety Management (PSM) requirements, following the issuance of the Clean Air Act section 112(r). The US EPA followed by issuance of the Risk Management Program (RMP), for Chemical Accidents Release Prevention.

The Health and Safety Executive in the United Kingdom established guidance for writing on- and off-site emergency plans 'HS (G) 191 Emergency planning for major accidents: Control of Major Accident Hazards (COMAH) regulations 1999 (HSE, 1999).' OSHA PSM standards consist of 14 elements (OSHA, 2004). CFR 1910.38 in the standard states the requirements for emergency planning. However, other OSHA requirements, such as CFR 1910.156 that establish requirements for training Fire Brigades and CFR 1910.146 that states the requirement for training emergencies in confined spaces, are related as well CFR 1910.156.

EPA RMP (EPA, 2004) rule is based on industrial codes and standards, and it requires companies to develop an RMP if they handle hazardous substances that exceed a certain threshold.

The EPA Emergency Planning and Community Right-to-Know Act (EPCRA) (EPA, 1984) establish requirements for the industry, federal, state and local governments, on reporting on hazardous and toxic chemicals.

Both the Department of Transportation and the Department of Energy address emergency planning. Information is available on their websites.

17.16.2 Standards

A large number of standards that addresses and are relevant to process safety are available. However, the NFPA published a variety of standards that are useful in emergency planning and training for emergencies. A list of NFPA publications can be found in the extended version of this book. Similarly, the National Institute of Safety and Health and the American National Standards Institute (ANSI) provide guidelines and standards that can be helpful in planning for emergencies.

REFERENCES

The references in the list below were considered during the review of this chapter:

AIChE, American Institute of Chemical Engineers, 1994. Dow's Fire & Explosion Index Hazard Classification Guide, 7th Edition. AIChE, NY.

AIChE, American Institute of Chemical Engineers, 1998. Dow's Chemical Exposure Index Guide, 1st Edition. AIChE, NY.

EPA, U.S. Environmental Protection Agency, 1984. Emergency Planning and Community Right-To-Know (EPCRA). 40 CFR 300-399 http://www.gpo.gov/fdsys/pkg/CFR-2002-title40-vol1/pdf/CFR-2002-title40-vol1.pdf.

EPA, U.S. Environmental Protection Agency, 2004. Risk Management Plan (RMP) Rule, 2004 Amended Rule. 40 CFR 68 (chemical accident prevention provision) subpart G http://www.gpo.gov/fdsys/pkg/CFR-2002-title40-vol1/pdf/CFR-2002-title40-vol1.pdf.

HSE, Health and Safety Executive, 1999. Emergency Planning for Major Accidents: Control of Major Accident Hazards (COMAH). Health and Safety Guidance. HSE Books.

NFPA, National Fire Protection Association, 2005. NFPA 1991 Standard on Vapor-Protective Ensembles for Hazardous Materials Emergencies, 2005 Edition. NFPA, NY.

NFPA, National Fire Protection Association, 2012a. NFPA 101: Life Safety Code, 2012 Edition. NFPA, NY.

NFPA, National Fire Protection Association, 2012b. NFPA 1992: Standard on Liquid Splash-Protective Ensembles and Clothing for Hazardous Materials Emergencies, 2012 Edition. NFPA, NY.

NFPA, National Fire Protection Association, 2013a. NFPA 110: Standard for Emergency and Standby Power System, 2013 Edition. NFPA, NY.

NFPA, National Fire Protection Association, 2013b. NFPA 111: Standard on Stored Electrical Energy Emergency and Standby Power Systems, 2013 Edition. NFPA, NY.

NFPA, National Fire Protection Association, 2013c. NFPA 1981: Standard on Open-Circuit Self-Contained Breathing Apparatus (SCBA) for Emergency Services, 2013 Edition. NFPA, NY.

Occupational Safety & Health Administration, 2004. U.S. Department of Labor OSHA. Principal Emergency Response and Preparedness. <https://www.osha.gov/Publications/osha3122.pdf>. Last access date: October 4, 2013.

Personal Safety

Loss prevention is particularly concerned with technical aspects, it is appropriate nevertheless to devote some consideration to hazards to the person and to their control. The same management discipline is required to deliver good performance in the personal safety area as in that of high technology. Three aspects appear particularly relevant: formal systems and procedures, hazard identification, and hazard assessment.

In addition to injuries from accidents, workers may also suffer impairment of health which sometimes becomes apparent only over a long period. It is necessary, therefore, to make some mention of this type of hazard and its control by occupational health measures.

The overall approach in questions of personal safety should always be, in order of descending preference, (1) hazard elimination, (2) hazard control, and (3) personal protection. The principle of hazard elimination and control before resorting to personal protection is a recurring theme.

Another recurring theme in personal safety is the need for thorough training and effective supervision. The objectives of training should include not only the ability to use equipment and perform tasks, but also the understanding and motivation to do the work properly and safely. Similarly, it is essential to have good supervision so that malpractices are stopped and procedures are enforced.

18.1. HUMAN FACTORS

Much can be done to reduce accidents by proper design of the work situation. This is the province of human factors. Human factors should be considered from the beginning of the project. Much of the contribution which it can make to accident prevention is in areas such as the allocation of function between man and machine. If it is brought in later for detailed equipment design or in a rescue role, many of the options for design are already foreclosed. Much work in human factors is concerned with human error. In fact, error rates are one of the main criteria used in scoring human factors experiments. Where an error leads to injury or other serious effect, it becomes an accident.

Some error is certainly attributable to the individual, and variations in error rates between individuals can be investigated. But management should concentrate primarily on the work situation, since this is not only the factor which is most important but also that which it can most readily alter. Principal influences on the frequency of errors, and hence of accidents, include: (1) equipment design, (2) working methods, (3) motivation, and (4) stress.

18.2. OCCUPATIONAL HEALTH

Occupational health is an established branch of medicine. It involves not only the investigation of industrial disease but also the provision of health services at the place of work. A chemical works usually has a medical center with facilities for medical tests and which is able to handle a small number of casualties. The center is typically staffed by a doctor working part time and a nurse working shifts. The medical personnel should be familiar with the chemicals and processes used in the particular factory. Information such as that given on material safety data sheets provides a useful starting point. Information on the medical history of individual workers is equally important.

In the United Kingdom, the Employment Medical Advisory service acts as the medical arm of the Health and Safety Commission (HSC), does work on medical matters for which there are statutory requirements, provides medical advice to the Factory Inspectorate and to employers, and furthers research on the epidemiology of industrial disease. It is able to advice on the effects of particular jobs on health, on the medical precautions to be taken in working with substances which are toxic or otherwise hazardous to health, and on medical examinations, investigations, and surveys.

Industrial chemicals tend to be noxious substances. It is essential, therefore, to take appropriate measures of occupational hygiene to ensure that those who work with them are not exposed to unacceptable risks. Occupational health and hygiene are the subjects of the Control of Substances Hazardous to Health Regulations 1988 (the COSHH Regulations).

18.3. GENERATION OF CONTAMINANTS

There is a variety of ways by which a contaminant may enter the workplace atmosphere. Four principal routes are (1) a contaminant-generating process or operation, (2) a leak, (3) opening up of plant, and (4) handling fabrics saturated with the material.

For a vapor contaminant, a principal determinant is the vapor pressure. If a liquid is being handled under conditions where there is a route to the atmosphere and it is at a temperature at which it exerts an appreciable vapor

pressure, a significant amount may escape. In relation to industrial hygiene, a significant measure is the ratio of the vapor pressure to the occupational exposure limit. A vapor hazard index (VHI) is defined as

$$VHI = \frac{Vapor\ pressure}{Occupational\ exposure\ limit}$$

where both the vapor pressure and the occupational exposure limit are measured in parts per million (ppm).

The conditions that must be met for a dust contaminant to enter the workplace atmosphere are that a route exists for it to do so and that it has imparted to it sufficient energy to follow that route. Some conditions in which a contaminant, vapor or dust, may enter the atmosphere include: (1) passage of air through the material, (2) free fall of the material, (3) pouring of the material, (4) agitation of the material, (5) transfer of the material, (6) opening of bags or drums of the material, (7) leak of the material, (8) opening up of plant containing the material, and (9) handling of fabrics saturated with the material.

A common source of contamination is leaks from the plant. A gross leak may be readily identified, but this may not be so where the leak is small, but still large enough to cause significant contamination. Most contaminants move with the air in the workplace. This air movement is therefore of great importance in developing controls. There are available a number of methods for the investigation of air movements. One is the use of a smoke tracer. Smoke is introduced from a smoke generator and its movement is observed or filmed.

There is a hierarchy of methods which may be used as the basis for the monitoring and control of noxious substances. These are the measurement and monitoring of: (1) the concentration of airborne contaminants in the workplace, (2) their concentration in workers' body tissues and body fluids, and (3) their effect on workers' health.

The minimization of exposure of workers to contaminants should be approached in a systematic manner and should be based on a hierarchy of measures of prevention and control. In accordance with the philosophy of inherently safer design, the first objective should be prevention. Resort should be made to control only after the potential for prevention has been fully explored.

The prevention stage should address the following possible approaches: (1) elimination or substitution of the substance, (2) modification of the process, (3) segregation of processes, (4) elimination of leaks, (5) modification of operations, and (6) arrangements for dealing with contaminated clothing.

When measures based on prevention have been exhausted, controls should be used to reduce exposure further. Also it is frequently necessary to use ventilation as one of the control measures taken to reduce exposure.

18.4. COSHH REGULATIONS 1988

A requirement for a comprehensive system for the protection of workers based on industrial hygiene principles is given in the COSHH Regulations 1988 (the COSHH Regulations). Some of the important factors of these Regulations are discussed below.

18.4.1 Substances Hazardous to Health

The same substance may be hazardous in one form but not in another. Thus, a substance in dust form may be hazardous but be safe in solid form. The fibrous form may also be hazardous. A substance may contain an impurity which is more hazardous than the substance which it contaminates. A substance may have been found by experience to be hazardous, even if the causative agent has not been identified. Some combinations of substances may have harmful additive, or synergistic, effects.

18.4.2 Suitable and Sufficient Assessment

The assessment of risks should identify the substances to which personnel are liable to be exposed, the effects of the substances on the body, the places where and the forms in which these substances are likely to be present, and the ways in which and the extent to which personnel could be exposed, allowing for foreseeable deterioration, or failure, of a control measure. It should include an estimate of the exposure given the engineering measures and systems of work adopted. This estimate should be compared with any available standards for adequate control.

If comparison indicates that control is likely to be inadequate, the assessment should go on to determine the measures which need to be taken to achieve adequate control. A record should be kept of the assessment, except in very simple and obvious cases which could be readily repeated and explained at any time.

Also *COSHH Assessment* identifies five possible outcomes of the assessment, which may be summarized as:

1. risks are insignificant now and for the foreseeable future;
2. risks are high now, and not adequately controlled;
3. risks are controlled now, but foreseeably could become higher;
4. risks are uncertain—uncertainty concerns extent and degree of exposure;
5. risks are uncertain—uncertainty due to lack of information.

Guidance is given on the action to be taken for each of these outcomes.

18.4.3 Competent Person

COSHH Assessment also gives guidance on the competence required of the person who is to undertake the assessment. The guide distinguishes between basic skills and additional skills. The basic skills are an understanding of the regulations and the abilities to make a systematic assessment of the risks and the exposures, to specify the measures which need to be taken in light of the assessment, and to communicate the findings in a report.

18.4.4 Occupational Exposure Limits

As already stated, the two types of occupational exposure limit mentioned in the regulations are the Maximum Exposure Limit (MEL) and the Occupational Exposure Standard (OES). The MELs are listed in Schedule 1 of the COSHH Regulations 1988 and the OESs are contained in the List of Approved OESs. This is given in EH 40 *Occupational Exposure Limits*, which is updated annually.

An MEL is the maximum concentration to which an employee may be exposed by inhalation under any circumstances. The concentration is averaged over a reference period which is specified in Schedule 1. Where an OES is assigned, exposure should be reduced to that standard. However, if exposure by inhalation exceeds the OES, the control may still be deemed adequate provided the reason for the excursion has been identified and appropriate steps are being taken to comply with the OES as soon as is reasonably practicable.

18.4.5 Prevention and Control

As far as is reasonably practicable, prevention or adequate control of exposure should be achieved by means other than PPE. Circumstances in which it may be necessary to resort to the latter include those: (1) where it is not technically feasible to achieve adequate control by process, operational, and engineering measures alone; (2) where a new or revised assessment necessitates the temporary use of personal protection until adequate control is achieved by other means; (3) where urgent action is required, such as that following a plant failure; and (4) where routine maintenance operations have to be done. In deciding whether to use PPE, due allowance should be made for its limitations and the practicalities of its use.

18.4.6 Health Surveillance

The purposes of health surveillance are essentially the protection of the health of employees and the evaluation of health hazards and control measures. The surveillance should lead to action and, before it is undertaken, the options and criteria should be established.

The health surveillance procedures include: (1) biological monitoring; (2) biological effect monitoring; (3) medical surveillance; (4) enquiries about symptoms, inspection, or examination by a suitably qualified person; (5) inspection by a responsible person; and (6) a review of the records and occupational history during and after exposure.

18.4.7 Schedule 5

Frequent reference has been made in the foregoing to Schedule 5 of the COSHH Regulations 1988. This contains in Column 1 the names of various substances and in Column 2 processes in which they are involved. One substance listed in this schedule is vinyl chloride monomer (VCM). The processes given in Column 2 opposite this substance are manufacture, production, reclamation, storage, discharge, transport, use, or polymerization.

18.4.8 Carcinogenic Substances

Exposure to a carcinogen involves risk of cancer. The risk increases with the exposure but there may be no short-term manifestation of adverse effects. The dose–response relation is quantal rather than graded. In the case of carcinogens, particular importance is attached to prevention. The use of carcinogenic substances should be kept to a minimum, these should be clearly labeled, and the areas where they are used should be delineated and non-essential personnel excluded. The first choice for control should be a totally enclosed system. Where this is not practical, alternative effective engineering measures should be taken including, where appropriate, the use of local exhaust ventilation (LEV). Care needs to be taken, however, that control measures do not aggravate the risk in the workplace or the outside environment.

18.5. DUST HAZARDS

In general industry, dust is one of the most serious hazards to health. Even in the chemical industry, where there are many toxic vapors, dust remains a relatively important airborne contaminant.

Some dusts have harmful effects, and some of the most harmful airborne contaminants are dusts. There are many dusts, however, which are inert, although they are unpleasant to breathe. Environmental Hygiene (EH 40) series requires that, in the absence of a specific exposure limit for a particular dust, exposure should be kept below both $10 \, \text{mg/m}^3$ 8-h time-weighted average (TWA) total inhalable dust and $5 \, \text{mg/m}^3$ 8-h TWA total respirable dust. Respirable dust is the fraction of inhalable dust which penetrates the lung. Much of the most important harmful effect from dusts is the various forms of fibrosis

such as silicosis from silica or asbestosis from asbestos dust. Metals which are essentially harmless in bulk form may have dusts which are to some degree toxic and cause severe lung inflammation. Some dusts are associated with lung cancer.

If the concentration of the dust is to be controlled in accordance with exposure limits, then a quantitative measurement of the total particle concentration is needed. It may also be necessary to measure the particle size distribution. The measurement of particle concentration and size distribution is a specialist matter. Good housekeeping and hygiene measures assume particular importance for dusts, especially highly toxic dusts.

18.6. LOCAL EXHAUST VENTILATION

A Local Exhaust Ventilation (LEV) system may be considered in terms of the following elements: (1) contaminant source, (2) hood, (3) ducting, (4) air filtration plant, and (5) fan.

The main problem in effecting LEV is the dispersion of the contaminant from the source. Once it has escaped from the immediate area of the source, it is very difficult to bring it under control. It is necessary, therefore, to obtain information on the behavior of the contaminant leaving the source, such as its direction and velocity.

It is usual to distinguish between two basic types of enclosure or hood systems used in LEV: (1) enclosures and receptor hoods and (2) captor hoods. Enclosures may be: (1) total enclosures or (2) partial enclosures.

The other main type of hood used in LEV is the captor hood. Whereas a receptor hood receives a flow of contaminated air carried into it, a captor hood draws the air flow in. The two types of hood may sometimes be similar in shape, but the principle of operation is different. A receptor hood operates on the push and a captor hood on the pull principle.

A particular type of captor hood arrangement for LEV is the low volume, high velocity (LVHV) system. This system utilizes a small hood (typically $<6 \, \text{cm}^2$) placed very close to the contaminant source and a high capture velocity ($\approx 50{-}100 \, \text{m/s}$). It may be used for vapor sources such as welding and for dust sources such as grinding. However, it is not easy to use and tends to be limited in application. Successful use requires that it be unobtrusive, correctly adjusted, and properly maintained.

18.7. SKIN DISEASE

Thus far, the discussion has dealt with toxic substances which are airborne. It is also necessary to prevent corrosive and toxic chemicals from causing irritation of the skin leading to skin disease, or dermatitis, and from entering the body through the skin.

The types of chemical which cause skin disease are classified as: (1) contact irritants, (2) contact sensitizers, and (3) photosensitizers. Contact irritants cause direct damage to the skin. Contact sensitizers cause a specific allergic response to develop over a period. Photosensitizers may be either irritants or sensitizers, but require the additional stimulus of sunlight or ultraviolet (UV) light.

Prevention of skin disease should be based on engineering measures to eliminate the causal agent, supplemented where necessary by personal protection. Where chemicals have to be handled, it is essential to use the proper methods and to wear the appropriate protective clothing, to observe the highest standards of personal hygiene, to change working clothes, including underwear, frequently and to take a thorough shower each day. Any break in the skin should be covered. Skin rashes should be reported, and wounds treated, promptly.

18.8. PHYSICO-CHEMICAL HAZARDS

There are a number of physico-chemical effects which are very simple and familiar to the engineer, but which have implications often not fully appreciated, particularly by process and maintenance personnel.

An account of some of these effects has been given by Jennings (1974a,b). They include: (1) density differences: gases and vapors and liquids, (2) phase changes, (3) vapor pressure, (4) flashing liquids, (5) nearly immiscible liquids, (6) heat of mixing, (7) exothermic reactions, and (8) impurities.

18.9. IONIZING RADIATION HAZARDS

Materials and machines which give off ionizing radiation are widely used in industry. They are potentially very hazardous, but they may be handled safely by taking proper precautions.

There are three main types of radioactive particle or ray: (1) α-particles, (2) β-rays, and (3) γ- and X-rays. α-Particles destroy bone marrow so that the red corpuscles are not replaced. β-Rays give rise to skin burns, skin cancer, dermatitis, and eye disease. γ-Rays and X-rays cause skin burns and cancer and aging effects. Fairly severe exposure is required to cause sterility, but genetic effects are believed to be proportional to exposure.

Occupational exposure to radiation should always be kept as low as practicable. The established limits should be observed, but since any level of radiation may involve some risk, every effort should be used to minimize exposure. Recommendations on limits for exposure are made by the International Committee on Radiological Protection (ICRP). These recommendations are widely applied internationally.

18.10. NON-IONIZING RADIATION

Artificial sources of non-ionizing radiation include: in the UV region, welding; in the IR region, welding, furnaces, and lasers; and in the microwave and radio frequency (RF) regions, microwave ovens, communications equipment, and radar.

The health effects of exposure to non-ionizing radiation are felt mainly on the eyes and skin. UV, IR, and microwave and RF radiations can each cause damage to both these parts of the body.

Thermal injury occurs when the rise in the temperature of the tissue is more than a few degrees. Thermomechanical damage ensues when thermal expansion of tissue is at such a rate that it results in explosive disruption. Photochemical injury is caused by absorption of a greater dose of energy than the body's repair mechanism can handle.

The principles of protection from non-ionizing radiations have many similarities with those for protection against the hazards already considered. The first line of defense is engineering measures to eliminate or reduce exposure, the second is systems of work for the control of exposure, and the third is PPE.

18.11. MACHINERY HAZARDS

Accidents caused by moving machinery have always been a serious problem in industry. The reduction of this type of accident has been one of the main concerns of industrial safety legislation.

The law on the safeguarding of machinery has become progressively stricter. The prime movers and their flywheels should be securely fenced, as should electric motors and generators, transmission machinery, and other dangerous machinery unless safe by construction or position, that devices for promptly cutting off the power from transmission machinery should be provided in every place where work is carried out, that efficient mechanical appliances should be provided to move driving belts to and from fast and loose pulleys, and that driving belts should not rest or ride on revolving shafts when the belts are not in use.

A large proportion of the hazards on moving machinery arise from running nips. A running nip occurs when material runs onto or over a rotating cylinder. Running nips exist, for example, where a belt runs over a pulley wheel, a chain over a sprocket wheel, a conveyor over a roller or material over a drum, or where two rolls create an in-running nip. Running nips cause many accidents and the accidents are frequently serious. Most of the methods of protection described for machinery in general are applicable to running nips.

Some accidents occur due to the inadvertent starting of machinery. Also accidents with abrasive wheels used for grinding are usually due to wheel failure. The number of accidents is small, but this is so only because great care is usually taken with grinding operations.

18.12. ELECTRICITY HAZARDS

Accidents in factories due to electricity are frequently associated with a failure to isolate or earth electrical equipment or to temporary or defective equipment. Electrical equipment should be installed, inspected, tested, and maintained only by competent electricians. Equipment should be regularly inspected and serviced. Equipment on which repair is to be done should be disconnected from the supply until the work is complete.

Temporary wiring should be avoided as far as possible, but if used should be to a safe standard and properly earthed. Circuits should not be overloaded, as this increases the risk of fire. Loading should be carefully supervised and circuits protected by fuses or circuit-breakers.

Leads should have wires of the standard colors.

18.13. PERSONAL PROTECTIVE EQUIPMENT

PPE includes respiratory equipment, protective clothing, footwear, and eye protection. Some tasks which may require PPE are (1) working with corrosive chemicals, (2) working with toxic chemicals, (3) working with dusts, (4) working in hot environments, (5) working in noisy environments, (6) welding, (7) fire fighting, and (8) rescue.

Protective clothing should be regarded as a last line of defense. It is much preferable to remove or control the hazard, if this is practicable. It is the responsibility of the management to specify when protective clothing and equipment are required and to provide them. The worker then has the duty to use them. If he is reluctant to do so, management is responsible for enforcing their use.

18.13.1 Eye Protection

The eye is a vulnerable organ and the avoidance of eye injury is particularly important. In the United Kingdom, the Factories Act 1961, Section 65, states that eye protection may be specified in certain processes, and eye protection has been covered by the Protection of the Eyes Regulations 1974 which give a schedule of processes for which appropriate eye protection is required. These processes include various cleaning and blasting operations using shot, water jets, and compressed air, various operations involving particular tools, especially power tools, and various processes with hot metal or molten salt.

The most appropriate eye protection depends, in principle, on the task. A view which is not only widely held but is also supported by successful company systems is that the most effective approach is to issue safety spectacles to all personnel who need eye protection and to require their use.

18.13.2 Respiratory Protective Equipment (RPE)

The use of RPE should not be a substitute for measures to maintain a breathable atmosphere. The regulations require that it be used only as a last line of defense in situations where it is not reasonably practicable to control exposure by other means.

There are four main types of situation in which it is used. These are (1) temporary situations, (2) short duration situations, (3) maintenance work, and (4) escape. Temporary situations are those which arise in non-routine circumstances such as commissioning. An important particular case is for activities in emergencies. Short duration situations may arise routinely. It is not normal for a worker to wear RPE all the time, but he may wear it routinely for certain tasks, for example, sampling or vessel entry.

18.13.3 Selection of RPE

The first stage of selection involves selection on the basis of the contaminant concentration, and the second involves selection on the basis of work-related and personal factors.

In selecting a respirator, account should be taken of the suitability of the filter for the particular gas or vapor and of the time until 'breakthrough,' when the filter becomes ineffective. Work-related factors which need to be taken into account in the second stage of selection include the length of time for which the RPE is to be worn, the physical work rate, and the needs for mobility, visibility, and communication. Personal factors include medical fitness, face shape and size, facial hair, and the use of spectacles. A worker should not be subjected to additional health risks by virtue of wearing an RPE. Persons with a respiratory disorder may find difficulty with respirators which rely on lung power to draw air through the filter.

18.13.4 Training for RPE

Persons involved in the use of RPE should be given suitable training. Management should have an appreciation of RPE, its applications, selection, use, maintenance, and limitations. Wearers should be trained in the wearing and use of the equipment and its limitations. Maintenance personnel should receive training in the maintenance, repair, and testing of the equipment.   •

Training should be both theoretical and practical. The former should include the hazards of contaminants and asphyxiation and the use of RPE to control exposure; the operation, performance, and limitations of the equipment; the operating procedures and permit systems which govern its use; the storage and maintenance of the equipment; and the factors which may reduce the protection which it provides.

Practical training should cover practice in inspecting the equipment before use; obtaining a good fit to the face, where applicable; putting the equipment on, wearing it and removing it; cleaning it after use; storing it; and replacing parts such as filters and cartridges.

Training is particularly important for the types of RPE which are more complex and are used in atmospheres immediately dangerous to life or health such as self-contained breathing apparatus (SCBA).

18.13.5 PPE and Training

Workers exposed to excessive noise must use appropriate PPE, including ear muffs, plugs, or both, when engineering or administrative controls are not feasible in reducing exposure to acceptable levels (Federal Register 29 CFR 1910.95). Training and education sessions must be offered at least annually to workers exposed to above 85 dB(A). These sessions must include information on the purposes and procedures of audiometric tests, the effects of noise on hearing, and the proper selection, fitting, use, and care of PPE.

18.14. RESCUE AND FIRST AID

Effective rescue and first aid arrangements are essential in the process industries. In the United Kingdom, the Factories Act 1961, Section 61, and formerly The Chemical Works Regulations 1922, Sections 6–14, contain various provisions for rescue and first aid. The responsibility for the provision of these services should be clearly defined. There are some features, such as rescue equipment, which might, in principle, come under the safety, the fire, or the medical department. All rescue and first aid equipment should be located with care and properly maintained. Personnel should be fully trained in its use.

18.15. ERGONOMICS

Ergonomics is the science of fitting job demands and working conditions to the capabilities of the working population. Although the scope of ergonomics is much broader, the term here refers to evaluation of those factors that may impose a risk of musculoskeletal disorders and

measures to alleviate them. Examples of ergonomic risk factors are found in requiring repetitive, forceful, or prolonged exertions of the hands; frequent or heavy pulling, lifting, pushing, or carrying of heavy objects; and prolonged awkward postures.

18.16. NOISE

The environment may be degraded not only by chemical pollution but also by noise. Process plant contains a variety of noise sources covering rotating equipment, piping, combustion equipment and vents to atmosphere. Noise from process plant sources may affect workers from the perspective of noise-induced hearing loss and the general public through environmental or community noise impacts. The control of noise from process plants needs to be controlled, and noise control design needs to be an integral part of the engineering and planning for any process facility.

18.16.1 Regulatory Controls

Regulations applicable to noise levels within an industrial facility are intended to protect employees from noise-induced hearing loss. The regulations are usually based on an employee noise exposure limits (e.g., 85 dB(A) averaged over 8 h). The regulations normally allow for exposure to higher noise levels for reduced time periods and mandate the use of hearing protection if necessary. This allows the employees' averaged noise exposure level to remain below the safe limits.

Typically, these regulations set a series of action levels based on employee noise exposure levels as follows:

1. First Action Level—daily personal noise exposure exceeds 85 dB(A)
   - The employer is responsible for initiating a hearing conservation program and having a noise assessment made by a competent person.
   - The employer is required to reduce the risk of hearing damage from exposure to noise to the lowest level reasonably possible.
   - Appropriate hearing protection should be available to the employee on request.
2. Second Action Level—daily personal noise exposure of 90 dB(A)
   - Appropriate hearing protection must be made available to all persons exposed.
   - The employer must reduce noise as far as reasonably possible.
   - Other requirements include assessment records and audiograms, designated ear protection zones, training on maintenance, and use of equipment and

provision of information to employees and information to employees.

In OSHA 1910.95 (Federal Register 29 CFR 1910.95), employee noise exposure threshold is defined for different duration periods. For 8 h shift per day, the noise exposure threshold is defined as 90 dB. If the noise level exceeds the defined threshold value, feasible administrative or engineering controls shall be utilized, and personal protection equipment may be required.

18.16.2 Noise Control Terminology

A summary of widely used noise control terminology is given below:

- Audible Frequency Range: The range of sound frequencies normally heard by the human ear. The audible range spans from 20 to 20 kHz, but for most engineering investigations only frequencies between 40 and 11 kHz (i.e., from 63 to 8 kHz octave bands) are considered.
- Decibels (dB): Ten times the logarithm (to the base 10) of the ratio of two mean square values of sound pressure, voltage, or current.
- Sound Pressure: A dynamic variation in atmospheric pressure. The pressure at a point minus the static atmospheric pressure at that point (N/m^2).
- Sound Pressure Level (L_p): The ratio, expressed in decibels, of the mean square pressure to a reference mean square pressure, which by convention has been selected to be equal to the assumed threshold of hearing.

$$L_p = 10 \log_{10} \left(\frac{P}{P_{ref}} \right)^2$$

where P = RMS sound pressure (N/m^2); P_{ref} = 2×10^{-5} N/m^2.

- Sound Intensity: The rate of sound energy transmission per unit area in a specified direction (W/m^2).
- Sound Intensity Level (L_I): The rate of sound energy transmission per unit area in a specified direction related to a reference intensity level.

$$L_I = 10 \log_{10} \left(\frac{I}{I_{ref}} \right)$$

where I = RMS value sound intensity (W/m^2); I_{ref} = 1×10^{-12} W/m^2.

- Sound Power: The total sound energy radiated by a source per unit time (W).

- Sound Power Level (L_w): The acoustic power radiated from a given sound source as related to a reference power level, expressed in decibels.

$$L_w = 10 \log_{10} \left(\frac{W}{W_{ref}} \right)$$

where W = RMS value sound power (W); W_{ref} = 1×10^{-12} W.

- Octave Bands: Frequency ranges in which the upper limit of each band is twice the lower limit. Octave bands are identified by their geometric mean frequency (center frequency).
- 1/3 Octave Bands Frequency: Ranges where each octave is divided into one-third octaves with the upper frequency limits being 1.26 times the lower frequency. They are identified by the center frequency of each band.
- A-Weighted Sound Pressure Level (dBA): The sound pressure level measured using an electronic filter that approximates the frequency response of the human ear.
- L_{Aeq}—Equivalent Continuous Sound Level: The A-weighted energy mean of the noise level averaged over the measurement period. It can be considered as the continuous steady noise level which would have the same total A-weighted acoustic energy as the real fluctuating noise measured over the same time period.

$$L_{Aeq} = 10 \log_{10} \frac{1}{T} \int_0^T \left(\frac{p_A(t)}{P_{ref}} \right)^2 dt$$

where T = total measurement time; $P_A(t)$ = A-weighted instantaneous sound pressure; P_{ref} = 2×10^{-5} N/m^2. L_{Aeq} is, therefore, an important number for the evaluation of a fluctuating noise level, because it reflects the actual energy content of the time-varying noise. L_{Aeq} is used in its own right in many national and international standards for rating some forms of community noise.

- L_N—Percentile Level: The A-weighted sound level equaled or exceeded by a fluctuating sound level $\times$ percent of the measurement time period. For example: L_{A10}—the A-weighted sound pressure level exceeded for 10% of the measurement period. L_{A50}—the A-weighted sound pressure level exceeded for 50% of the measurement period. L_{A90}—the A-weighted sound pressure level exceeded for 90% of the measurement period. The various L_N levels over a measurement period are important in the evaluation of the temporal nature of the sound. A widely varying sound level over a measurement period will tend to have large differences between the various L_N levels than a more constant steady noise.

- Daily Personal Exposure: The daily personal exposure of an employee is defined in the schedule to the HSE *Noise Guide 1* as follows:

$$L_{\text{EP,d}} = 10 \log_{10} \left[\frac{1}{T_0} \int_0^{T_e} \left(\frac{p_A(t)}{P_{\text{ref}}} \right)^2 \text{dt} \right]$$

where $L_{\text{EP,d}}$ is the daily personal exposure (dB(A)); $p_A(t)$ is the time-varying value of the A-weighted instantaneous sound pressure in the undisturbed field (Pa); t is the time (h); p_0 is the base value of the sound pressure (Pa); T_e is the duration of the exposure (h); T_0 a period of 8 h. As before, the value of P_{ref} is 2×10^{-5} Pa.

18.16.3 Noise Control

Noise control is generally considered under four heads: (1) at source, (2) in transmission, (3) on emission, and (4) on receipt.

- Control at Source: Control of noise at the source should be the first option considered. The approach is analogous to inherently safer design. The approach to noise control at source is to identify potential sources and to either eliminate them altogether or to reduce their level of vibration by one or more of the following: the surface, the amplitude, the frequency, and, in the case of a fluid, the flow velocity.
- Control in Transmission: In principle, noise associated with transmission through pipes and ducts may be structure-born or fluid-born. For gas or vapor pipes, the sound is predominantly fluid-born as evidenced by the effectiveness of control by the use of in-line silencers.
- Control on Emission: Control on emission involves modification of the route by which the noise reaches the worker. Another approach is to increase the distance between the worker and the noise source. Distance may be increased by the segregation of noisy machines, the use of remote controls, and the siting of exhausts.
- Control at Receiver: Control at the receiver involves the use of ear protection.

The importance of addressing noise control issues as early as possible in a process plant engineering project cannot be over emphasized. The earlier noise control design issues are identified and resolved on a project, the more effective is the noise control design.

The following sections highlight the key noise control design activities from the initial front end engineering design (FEED) phase of the project through Detailed Design to Commissioning and Start-up:

- Front End Engineering Design (FEED): Noise control must be executed as early as possible in a project to minimize the effect on project costs layout and schedule. The major noise control activities on a project, therefore, tend to focus toward the FEED phase of the project. During these initial stages of a project, baseline noise surveys may be required to establish existing noise levels at the plant site. The plant noise modeling must be based on international and industry accepted standards (e.g., EEMUA (2009), CONCAWE (1984), and ISO 9613). Noise model input data must be based on reliable and verifiable equipment and field noise data. The findings of the FEED Noise Study will provide the data to define definite and unambiguous noise limits and requirements for the Detailed Design phase of the project.
- Detailed Design: Noise control for the detailed design phase of a project will involve the engineering contractor, or contractors, executing the noise control design for the project in compliance with the noise limits defined during the FEED phase of the project.
  Commissioning and Start-up: The main noise control activities during commissioning and start-up will be noise surveys to demonstrate/verify compliance with the project noise limits. Depending on the results of the noise surveys corrective action may need to be initiated to reduce operation noise levels.

REFERENCES

CONCAWE, 1984. Methodologies for hazard analysis and risk assessment in the petroleum refining and storage industry. Fire Technol. 20 (3), 23.

EEMUA, Engineering Equipment Material Users Association, 2009. EEMUA 222: Guide to the Application of IEC 61511 to safety instrumented systems in the UK process industries. <https://www.eemua.co.uk/acatalog/INSTRUMENTATION_and_CONTROL.html>.

Jennings, A.J.D., 1974a. The physical chemistry of safety. Chem. Eng. (London). 290, 637.

Jennings, A.J.D., 1974b. The physical chemistry of safety. Course on Process Safety—Theory and Practice. Department of Chemical Engineering, Teesside Polytechnic. Middlesbrough, UK.

OSHA. Occupational Safety and Health Administration. Occupational Noise Exposure. USA Code of Federal Regulations Title 29 Part 1910.95.

Accident Research and Investigation

19.1. ACCIDENT RESEARCH

It has been considered by Suchman (1961), who distinguishes three defining characteristics, what actually constitutes an accident: (1) degree of expectedness, (2) degree of avoidability, and (3) degree of intention. Other characteristics are degree of warning, duration of occurrence, degree of negligence, and degree of misjudgment.

An event is more likely to be classed as an accident if it is unexpected, unavoidable, and unintended, or if it gives little warning and happens quickly and if there is a large element of negligence and misjudgment. A number of standard accident classifications are used in the world. These include the British classification used by the HSE in its annual *Health and Safety Statistics* (HSE, 1992), published in the *Employment Gazette* and that used in the annual *Yearbook of Labour Statistics* (International Labour Office, 1992).

19.1.1 Accident Causation

The main aim of accident research is to understand accidents so that they can be prevented. The attempt to understand an accident is often equated with the search for its cause. Root cause analysis is the main model by which accident causation is carried out.

The most widely used definition of a root cause of an event is an initial cause that allows an engineering failure to take place, while a direct (or immediate) cause is the event that directly leads to the failure. After completing the root cause analysis, recommendations must be assembled to learn from the accident in a reactive manner. This is perhaps the most important step, because in examining the accident, if a lesson cannot be learned, then the accident is doomed to repeat itself.

19.1.2 Accident Models

Recognition of the complexity of accidents has led to the development of a number of accident models. A particular model of the accident situation is that derived from epidemiology, following the work of Gordon (1949). The factors involved in an accident are defined as follows: (1) host, (2) agent, and (3) environment.

A second group of models is based on the concept that it is possible to define for a system a 'normal operating state', and hence deviations from that state, and that such deviations are associated with danger.

A third group of models is based on human information processing. They include those of Surry (1969), A.R. Hale and M. Hale (1970), and A.R. Hale and Glendon (1987).

19.1.3 Human Error

Work on the psychology of human error is exemplified by the work of Reason, described in *Human Error* (Reason, 1990) and in other work (Reason, 1977, 1979, 1986, 1987a,b, 1991, 2000). One aspect of this work is the recognition that many accidents occur due to absent-mindedness and that, while in the vast majority of cases the consequences of such absent-mindedness are not serious, in a proportion of cases they are. In the 2000 work by Reason, he asserted that there are two approaches to the identification of why human error exists. The first is the person approach, the approach that is normally associated with human error, which states that error is a problem of absent-mindedness, inattention, and moral weakness. The second is the systems approach, which acknowledges that there are systematic and exterior reasons that humans fail. Errors are seen as consequences of these exterior factors, and because they are seen as inevitable and difficult to reduce, countermeasures are taken to change the working environment and mitigate effects of errors. The focus of the systems approach is normally this kind of 'error management' that attempts to create a system that is better equipped to handle human error.

19.1.4 Impact of Safety Culture

Safety culture has proven a difficult term to define, despite having been studied extensively. Unfortunately, there is still no total consensus as to what it means and how it should be differentiated from safety climate.

Safety climate reflects the attitudes and perception that are present within an organization with regard to safety practices, regulations, and innovation, while safety culture is the overarching comprehensive pattern within the organization that supersedes and creates the climate (Guldenmund, 2000). Thus, safety culture is of utmost importance because it, in effect, helps determine the attitudes that individuals have toward safety.

Experimental methods in the field of safety culture normally involve a sociological and psychological approach, with a hope of quantifying factors that reflect the attitudes of individuals and therefore provide insight to the culture of the organization.

There are many obstacles to overcome when attempting to create a positive safety culture, and there is no universally accepted way to implement a healthy safety culture. An important consideration in safety culture is the fact that it is not necessarily effective in changing the basic personality of an individual. Rather, implementation

of culture has to be aimed at what can be changed, such as procedure and trust.

Safety culture is a hotbed of current research and industrial work. Current research focuses on the relationship between a measurable definition of safety culture and the number of injuries experienced in a given organization, such as the study done by Payne et al. (2009). Indeed, much of the focus of current research is on the development and validation of tools for the measurement of safety culture and climate. A tool developed by Butler et al. (2010) uses much the same procedure as the study by Payne to attempt to define a measure of safety climate and incorporates an evidence base to justify the inclusion of factors in the model.

19.1.5 Safety Training

Most prescriptions for the prevention of accidents lay much emphasis on training, and rightly so. It is necessary, however, that the training be effective. A study of the effectiveness of safety training by A.R. Hale (1984) found relatively little previous work on the topic. The essential finding of the study is that, unless effort is invested in analyzing the problem and defining the training requirements, the training is liable to be ineffective and to get the activity a bad name. Newer studies show that the effectiveness of safety training is highly dependent upon the workers' belief in its effectiveness. They also propose that management does not put enough emphasis on training, particularly in regard to protective clothing and equipment safety.

19.2. GENERAL INCIDENT INVESTIGATION CONCEPTS

Successful investigation is an iterative process based on scientific principles with the ultimate purpose being prevention of a repeat event. There are four stages that must be successfully completed in order to prevent a repeat event of a similar nature:

1. Identify what happened and how it happened (identifying and understanding the scenario).
2. Determine why it happened (identify the specific underlying and contributing causes).
3. Identify preventive remedies (recommendations and action items).
4. Implement changes to existing practices and systems (accompanied by sharing lessons learned to all those who could benefit).

19.2.1 Definitions

When an incident results in actual undesirable consequences, such as injury to people, the environment, damage to equipment, or adverse impact to operating profits, the event is most commonly labeled as an accident. The term *near-miss* is used in this chapter to describe an event that generated no actual adverse consequences, but with a slight change in circumstances, could have produced actual adverse consequences.

The term *root cause* in this chapter is used to represent: *the underlying reasons that allow defects in systems (physical equipment and systems and/or administrative systems) to exist, that the organization can correct them or not.*

An understanding of management system perspective is useful in effective root cause identification. The term *management system* is used to represent the total administrative activities and aspects associated with a dedicated task or an objective.

Management systems have common generic components such as: written procedures, training, performance expectations, required competencies, and assignment of different responsibilities. When searching for root causes, the investigation team should examine the components of all management systems involved in the incident (CCPS, 2003). In this context, management systems have common generic components as shown in Figure 19.1.

19.2.2 Investigation General Concepts

For effective investigations, an honest, thorough, and systematic approach is needed. The investigation team should objectively consider all facts, information, and evidence and use scientific principles to identify and evaluate all plausible cause scenarios.

Prematurely stopping before reaching the root cause level is a major and recurring challenge in most process incident investigations. If the investigation stops before the root cause level is reached, fundamental system weaknesses and defects remain in place pending another set of similar circumstances that will allow a repeat incident.

Figure 19.2 provides a graphical representation of the root cause investigation process. The first phase is determining 'What happened'. This is followed by efforts to identify 'Why it happened'. The final stage is the 'Prevention' stage.

Although investigations of process incidents share many common elements with traditional industrial incident investigation, there are a few relatively unique features. Many process incidents involve outside parties such as regulatory agencies, insurance companies, and in many

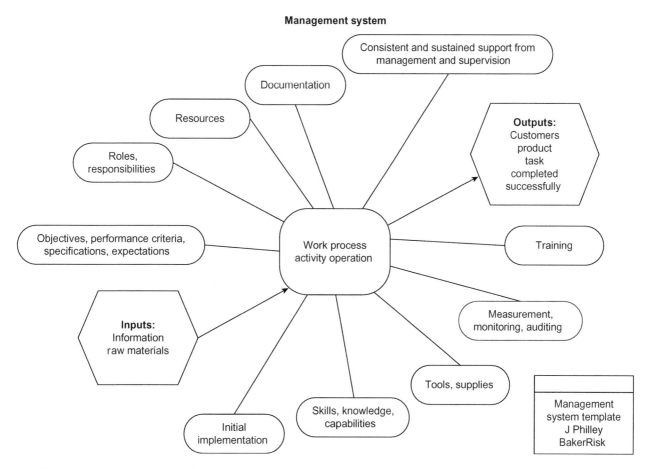

FIGURE 19.1   Management system template.

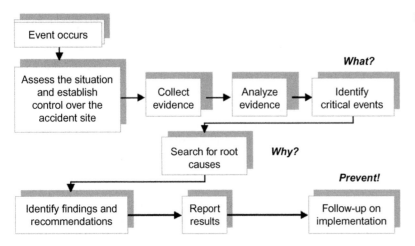

FIGURE 19.2 Root cause investigation process.

instances litigation issues. Process incident investigations by nature carry additional challenges. Actual or potential consequences are significant and can be catastrophic in terms of loss of life, impact to the environment, and financial impact on the organization. Specific design features of the manufacturing process system and sophisticated controls are provided to prevent adverse consequences.

19.2.3 Management's Role in Incident Investigation

Responsibility for prevention of incidents and repeat incidents is a responsibility shared at every level of the organization from the board of directors down to the entry-level employee. However, the ultimate responsibility rests with those in management positions who establish policy and allocate resources. Incident investigation should include:

- forming and providing training for an incident investigation team to be ready to quickly act in the case of an incident;
- personal involvement by line management in the review and approval of draft incident reports;
- rigorous and robust follow-up and implementation of recommendations and action items generated from incident investigations;
- proactive sharing of results, findings, and lessons learned from investigations;
- strong and sustained emphasis on reporting, investigating, and taking action on near-miss events;
- establishing priorities and implementation schedules for incident investigation action items or recommendations.

Each level of the management structure must have an adequate and appropriate understanding of the incident investigation policy, procedure, and responsibilities (DNV, 1996).

19.2.4 Investigation Team's Role in Incident Investigation

Depending on the technical issues of the incident, the likelihood of litigation, and the incident consequences, the investigation team should include someone knowledgeable of the work processes involved, experienced in accident causation, and investigative techniques. Moreover, in most cases, members of the investigation team can also include supervisor, safety officer, health and safety committee, 'outside' expert, and representative from government.

The investigating team has to decide whom to interview; prepare a list of questions, general and specific, for each witness; and carry out the interviews.

19.2.5 Human Factors in Incident Investigation

Well-designed management systems take human reliability and human performance into consideration when designing safeguards. Well-designed human factor safeguards are error tolerant and allow some opportunity to detect and correct deviations before catastrophic consequences are reached.

Two excellent examples of applied human factors are the emphasis on inherent safety in system design and the concept of alarm management. Modern distributed control system (DCS) (Shaw, 1991) instrumentation allows almost infinite capability to add computer-generated alarms and the result can be alarm overload/flood during the initial periods of process deviations and process upsets.

The underlying causes related to human factors can be improved in the incident investigation by a Human Factors Investigation Tool (HFIT). Based on the dominant psychological theories of accident causation, HFIT is developed and especially used for the offshore oil and gas industry.

19.3. EVIDENCE ISSUES

19.3.1 Evidence Collection and Management

For major chemical process incidents, evidence preservation, storage, and management are required. Effective collection and analysis of physical evidence should be conducted in a systematic fashion. All potentially important fragments, debris, and other physical items should be documented in place (photography, which should be used extensively, or other) before being moved or disturbed in any manner, noting the location, orientation, and the exact time when the piece of evidence is collected. It is standard practice to assign individual evidence numbers to each piece of physical evidence collected.

Investigatory photography (and video) has multiple purposes. It is most often used to document the 'as-found' position, location, configuration arrangement, damage pattern, and layering of physical evidence. In addition, photographic evidence is often useful in presenting the results of the investigation (reports) and in distributing the lessons learned (training).

19.3.2 Investigation Witness Interviews (Gathering Evidence from People)

Gathering verbal information from witnesses is one of the more challenging aspects of incident investigation. The first step is the collection of 'first reports' from witnesses. These reports should be completed as soon as possible but in any case within the first 24 h after the incident, when memory of the events is still fresh and uncolored by other evidence or the narratives of other witnesses.

Incident investigation interviews are conducted in a variety of circumstances. There are several types of investigatory interviews. The most common incident investigation interview is the initial screening interview, conducted for all possible witnesses who might be able to contribute to understanding the cause and event scenario.

There are four distinct stages in every effective interview:

1. opening;
2. witness statement;
3. interactive dialog;
4. closing.

Initial screening interviews are often conducted in series. Therefore, it is important to document thoughts, observations, questions, and follow-up issues immediately after each interview. If the interviewer waits until the end of the day, the interviewer will not remember everything and may not be able to recall which witness supplied which information. It is a good practice to allow additional time between interviews, for documentation or an unexpected extension.

In order to get optimum results, investigation should follow these recommendations. Promptness is critical and information would not be accepted selectively. An avoidable mistake is to disclose investigation team preliminary findings to the witness during the interview. An effective interviewer operates in a neutral mode to cause minimum impact on the witness. Another avoidable mistake is to make promises or commitments to the witness during the interview, especially if the interviewer is not in a position to ensure that the promise is met. Statements written by the individual witness are a potentially strong augment for enhancing the effectiveness of interviews. Also it is important to categorize witnesses. Interview room configuration can impact the quality and quantity of results. Non-verbal communication such as eye contact and body posture can sometimes provide additional clues and insights as to the information being offered by the witness.

19.3.3 Evidence Analysis

Evidence analysis can provide objective and scientific independent confirmation of the cause scenario speculated by the investigation team. Damage patterns provide information related to the origin and sequence. Investigators can also make useful determinations based on anomalies and by analyzing what remains undamaged. There are numerous publicly available resources for evidence analysis, including physical property data for melting temperatures, auto-ignition temperatures, and chemical incompatibilities. Some methods are nondestructive (Non-destructive Evaluation, NDE), while others require permanent modification of the evidence. Visual examination is the most common and one of the most powerful evidence analysis techniques.

19.4. THE INVESTIGATION TEAM

The incident investigation team for process-related events is typically composed of a cross-section of skill sets and competencies related to the process and nature of the event under investigation. Team size can vary from as few as 3 full-time members to as many as 15, with 6−8 being most common. One member of the investigation team will normally be devoted full-time to coordinating the witness interviews.

Team training occurs in two stages. There is normally an established formal training program for all potential investigation team members that is conducted prior to the event to maintain a pool of qualified potential team members. Some required training is extensive; US Hazardous Waste and Emergency Operations (US OSHA 1910.120) and respiratory protection (OSHA 1910.134) are examples and may require periodic refresher training and certification.

The first order of business is usually formal introductions of team members and their individual backgrounds, experiences, and skill sets that may have application to the investigation. The next order of business will address any legal and confidentiality issues. Communication protocols and expectations (both internal and external to the team) should be clearly explained with confirmation from each member that they understand and accept the communication controls. For major incidents, another item for the initial team meeting is coordination and liaison with other entities (inside or outside the organization). Evidence management should be discussed and preliminary arrangements made for collection, identification, documentation, chain-of-custody, storage, and access to evidence by team members and others.

19.5. IDENTIFYING ROOT CAUSES

Root cause determination is an iterative process which systematically seeks to identify and understand the underlying causes that allowed the event sequence to progress to its ultimate consequences. The investigators need a clear understanding of the term root cause in order to determine the proper stopping point for the investigation. Chemical process incidents are most often complex events with multiple root causes (Philley, 1992a,b; CCPS, 2003).

Basically, there are four major steps in root cause analysis as follows:

1. data collection;
2. causal factor charting;
3. root cause identification;
4. recommendation generation and implementation.

Most root cause methodologies incorporate use of a chronology tool to establish the sequence of events, activities, and conditions. Development of a timeline chronology is an iterative activity and allows the investigator to better understand the evidence and witness statements.

19.5.1 Logic Diagrams

Logic diagrams have several applications in investigations and are most often developed in an iterative fashion. As shown in the event tree logic diagram in Figure 19.3, in the early stages of an investigation they can be used to illustrate credibly possible reasons, conditions, and events to assist in determining the cause scenario. As shown in Figure 19.4, they can point the investigators to what specific additional information or evidence might be gathered in order to confirm or refute a postulated cause scenario. In the middle and late stages of an investigation, logic diagrams can be refined and used as a quality control tool to ensure the team is systematically addressing the information and that individual branches of the logic tree are consistent. Another use of logic diagrams is in presenting the findings in a formal written report. A simplified logic tree can be used to present the cause scenario and illustrate conditions and events so that the reader can follow a complex incident.

19.5.2 Truth Table Matrix

Truth table matrices are common tools in logic and troubleshooting diagnostic activities. This concept has useful application in incident investigations. This tool can be applied effectively to help the investigation team determine the most likely scenario among a set of speculated credible cause scenarios, as well as assist the investigation team in resolving inconsistencies generated during

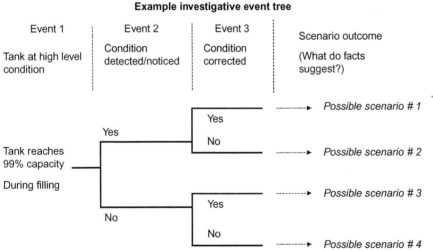

FIGURE 19.3 Example: investigative event tree.

witness interviews. The fact–hypothesis matrix is a special version of the logic truth tables used in developing control system logic and diagnosing/troubleshooting activities (CCPS, 2003). The concept is simple: known (or suspected) facts are placed on one axis of the matrix and potential cause scenarios are placed on the other axis. Then the facts are analyzed in conjunction with the speculated cause scenario for consistency (Table 19.1).

Facts fall into at least one of four categories:

1. + Yes—The fact *supports* the hypothesis (speculated cause scenario);
2. − No—Fact *contradicts* (refutes) the hypothesis;
3. Neutral—The fact neither supports nor contradicts the hypothesis (is neutral);
4. ? Insufficient information—The team does *not have enough information* to make a determination regarding category 1, 2, or 3. In this case, the team can then determine what specific information might be gathered

that would allow the team to confirm or refute the relationship between the fact and the speculated hypothesis cause scenario.

19.6. REPORTS

The written report is an accident prevention communications tool to share findings and lessons learned from the investigation. There is a wide variety of report style and format. Some reports are highly technical and provide extensive detail on complex systems, failure modes, and process operation. Other reports are as brief as one or two paragraphs and function as a heads-up alert communication. The report begins with a summary. This summary is most effective if limited to a single page highlighting the major items of what happened, why, and what the suggested action items are. For many process incident reports, it is beneficial to prepare a background section to allow the reader to fully understand the process, the associated potential hazards and properties of the materials, control measures, intended safeguards, and the nature of the incident scenario. The third section is often a narrative of the scenario, providing an explanation of the incident sequence, events, and enabling conditions. This section presents the WHO, WHEN, HOW, and WHAT HAPPENED information. The next section is often the 'findings and root causes' section that presents root causes and contributing causes (the WHY information).

19.7. DATABASES

As part of these programs and regulations, accident history data is often collected. There are two basic types

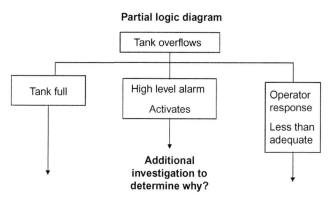

Partial logic diagram

FIGURE 19.4   Logic diagram for tank overflow incident.

TABLE 19.1 Fact–Hypothesis Matrix/Truth Table

Possible Cause Scenario	Fact #1	Fact #2	Fact #3
	Lab Sample at 06:00 showed elevated moisture content	High level alarm and high level condition noted in process sump	Vent compressor tripped twice on the shift before the explosion
A	Yes	?	0
Event triggered by contaminated feedstock raw material	Supports the scenario	Do not have enough information to decide	Neutral effect
B	0	Yes	No
Block valve left in incorrect position allowed backflow contamination	Neutral effect	Supports the scenario	Refutes the scenario
C	?	Yes	Yes
Leak in heat exchanger created contamination and overpressure condition	Do not have enough information to decide	Supports the scenario	Supports the scenario

of information. One is a database consisting of standard-ized fields of data usually for a large number of incidents. The second are more detailed reports of individual incidents.

Analysis of these accident history databases can provide insight into accident prevention needs. Statistical knowledge of the likelihood of the release of certain types of chemicals could help emergency responders determine the chemical releases in their areas and plan appropriate responses.

Incident Databases

- National Response Center (NRC) Incident Reporting Information System (IRIS)
- EPA Risk Management Plan (RMP)—5-year accident history
- Agency for Toxic Substances Disease Registry (ATSDR)—Hazardous Substances Emergency Events Surveillance (HSEES)
- EPA Accidental Release Information Program (ARIP)
- Accident Investigation System (AIS)—Occupational Safety and Health Administration, US Department of Labor
- US Department of Transportation (DOT)—Hazardous Material Incident Reporting System (HMIRS)
- US DOT Integrated Pipeline Information System (IPIS)
- Center for Chemical Process Safety (CCPS)
- Major Accident Hazards Bureau (MAHB), Major Accident Reporting System (MARS)
- Ionising Radiations Incident Database
- CBRN Incident Database (CID)
- National Toxic Substance Incidents Program (NTSIP)

The following are sources of statistical information regarding injuries and fatalities due to chemicals. They do not reveal the details of specific incidents but do provide statistical information about injuries due to chemicals in the United States.

**Injury Fatality Databases
(Not Tied to Specific Incidents)**

- Occupational Illness and Injury Statistics (OIIS)—Occupational Safety and Health Administration, US Department of Labor
- Census of Fatal Occupational Injuries (CFOI)—Occupational Safety and Health Administration, US Department of Labor
- Mortality Database, Centers for Disease Control, United States
- US Coast Guard (USCG)
- National Electronic Injury Surveillance System—Consumer Product Safety Commission, USA

Incident Investigation Reports in the United States

- United States Chemical Safety and Hazard Investigation Board (CSB)
- United States Department of Energy (DOE)
- United States Environmental Protection Agency (EPA)
- United States Fire Administration (USFA)
- National Institute for Occupational Safety and Health (NIOSH)
- Occupational Safety and Health Administration (OSHA)
- National Transportation Safety Board (NTSB)
- Minerals Management Service (MMS)
- United States Department of Labor—Mine Safety and Health Administration (MSHA)
- National Fire Protection Association (NFPA)
- United States Coast Guard (USCG)

REFERENCES

Butler, C., Lekka, C., Sugden, C., 2010. Development of a Process Safety Climate Tool. Mary Kay O'Connor Process Safety Center International Symposium, College Station, TX.

CCPS, 2003. Guidelines for Investigating Chemical Process Incidents, second ed. American Institute of Chemical Engineers, New York, NY.

DNV (Det Norske Veritas), 1996. Practical Loss Control Leadership. Revised edition. 088061-054-9, Chapter 4, DNV, Atlanta, GA.

Gordon, J.E., 1949. The epidemiology of accidents. Am. J. Pub. Health. 39, 504.

Guldenmund, F.W., 2000. The nature of safety culture: a review of theory and research. Saf. Sci. 34, 215–257.

Hale, A.R., 1984. Is safety training worthwhile? J. Occup. Accidents. 6, 17.

Hale, A.R., Glendon, A.I., 1987. Individual Behaviour and the Control of Danger. Elsevier, Amsterdam.

Hale, A.R., Hale, M., 1970. Accidents in perspective. Occup. Psychol. 44, 115.

HSE (Health and Safety Executive), 1992. Health and safety statistics 1990/91 (London). Issue Employ. Gazette. 110, 9.

International Labour Office, 1992. Yearbook of Labour Statistics 1991. Geneva: International Labour Office.

Payne, S.C., Bergman, M.E., Beus, J.M., Rodriguez, J.M., Henning, J.B., 2009. Leading and Lagging: The Safety Climate–Injury Relationship at One Year. Mary Kay O'Connor Process Safety Center International Symposium. College Station, TX.

Philley, J.O., 1992a. Investigate incidents with MRC. Hydrocarbon Process. 71 (9), 77.

Philley, J.O., 1992b. Acceptable risk: an overview. Plant/Operations Prog. 11, 218.

Reason, J., 1986. Recurrent errors in process environments: some implications for the design of intelligent support systems. In: Hollnagel, E., et al., op. cit., p. 255.

Reason, J., 1987a. The Chernobyl errors. Bull. Br. Psychol. Soc. 40, 201.

Reason, J., 1987b. A framework for classifying errors. In: Rasmussen, J., Duncan, K., Leplat, J., op. cit., p. 5.

Reason, J., 1990. Human Error. Cambridge University Press, Cambridge.

Reason, J., 1991. Too little and too late: a commentary on accident and incident reporting schemes. In: Van der Schaaf, T.W., Lucas, D.A., Hale A.R., op. cit., p. 9.

Reason, J., 2000. Human error: models and management. Br. Med. J. 320, 768–770.

Reason, J.T., 1977. Skill and error in everyday life. In: Howe, M.J.A. (Ed.), Adult Learning. Wiley, London, 21.

Reason, J.T., 1979. Actions not as planned: the price of automation. In: Underwood, G, Stevens, R. (Eds.), Academic Press, London, p. 67.

Shaw, J.A., 1991. Design your DCs to reduce operator error. Chem. Eng. Prog. 87 (2), 61.

Suchman, E.A., 1961. A conceptual analysis of the accident problem. Soc. Prob. 8 (3), 241.

Surry, J., 1969. Industrial Accident Research: A Human Engineering Appraisal. Report. Department of Industrial Engineering, University of Toronto.

Computer Aids and Expert Systems

The use of computers in the design and control of process plants is now widespread. The use of safety-related computer aids is described briefly in this chapter. The following sections deal with artificial intelligence, expert systems, and certain other advanced developments in computer aiding. The aim is to provide an overview of the field and an appreciation of the sorts of problems that are tackled, the types of technique developed, the terminology used, and the applications made. The first part of this chapter deals with common concepts regarding computer aids. The second part comprises an enumeration of application in the process industries.

A more general discussion of chemical engineering computer software is contained in the *Lees' Loss Prevention in Process Industries*, fourth edition; however, the availability of software is changing too fast to simply develop a list, which would be outdated even before

publication. Rather than provide exhaustive lists, the generic type of available software, as it applies to loss prevention will be described herein. References are also available in a number of chapters in the *Lees' Loss Prevention for Process Industries* fourth edition to various specific computer codes pertinent to the technology under discussion.

20.1. KNOWLEDGE REPRESENTATION

Knowledge is of many kinds and this is reflected in the forms of knowledge representation used. One distinction commonly made is between data and information. Other distinctions made about knowledge are domain-independent vs domain-specific, exact vs *fuzzy*, and procedural vs declarative knowledge. Some types of knowledge include: facts, models, distinctions, relationships, constraints, procedures and plans, and rules-of-thumb or heuristics.

The extent of the knowledge required varies greatly between problems. In certain limited worlds, such as the 'blocks world' extensively studied in Artificial Intelligent (AI), the world knowledge required is quite limited. In real-life, or mundane, situations on the other hand, the human comes to the problem possessed of a massive store of knowledge. He draws from this information not only about facts but also about other aspects such as constraints.

Examples of knowledge representation include: natural language, databases, classical logics, non-classical logics, probabilistic reasoning, structured knowledge, graphs, trees and networks, and programming languages.

20.1.1 Prepositional and Predicate Logic

Prepositional logic deals with the relationships between propositions. Typical propositions are (1) the valve is open and (2) the valve is open and the pipe is full of liquid.

The first of these is a simple proposition and the second a compound proposition. Propositions are also referred to as statements, sentences, or formulas. Statements are represented by prepositional symbols, which are capital letters, commonly P, Q, R, etc. The main feature of the syntax of prepositional logic is the connectives. Use is also made of parentheses ( ) as delimiters, but where there is no chance of ambiguity, parentheses may be omitted, in which case the order of precedence of the connectives is from highest to lowest: $\neg$, $\wedge$, $\_$, $\Rightarrow$, $\Leftrightarrow$.

Some principal features of the semantics of propositional logic are the properties of statements and the rules of inference. The semantics of a statement is the truth value assigned to it, the process of assignment being known as interpretation. The truth values assigned are true or false. The means to perform deductions is provided by the rules of inference.

The basic entities handled in First Order Predicate Logic (FOPL) are constants, variables, functions, and predicates. The first three are referred to as terms, and predicates are referred to as atomic formulas, or atoms. An atom or its negation is referred to as a literal.

A constant is a term that has a fixed value and a variable is a term that can take different values. A function maps elements of the domain to further elements of the domain. A predicate maps elements of the domain to truth values. Commonly used constant symbols are: a, b, c, ...; function symbols f, g, h, ...; variable symbols x, y, z; and predicate symbols P, Q, R. There are rules for the construction of syntactically correct formulas. Formulas derived by correct application of these rules are termed well-formed formulas, or WFFs (pronounced woofs). An atomic formula is a WFF. WFFs are obtained from predicates, or atomic formulas, by application of quantifiers and connectives. Use is also made of quantifiers. These are for existential quantification and $\forall$ for universal quantification. The meanings are:

$$\exists x \text{ For some } x; \text{ there is an}$$
$$\forall x \text{ For all } x$$

The expression $\exists x \, P(x)$ means that there exists at least one element of the domain which makes $P(x)$ true.

In respect of syntax, the logical connectives used in FOPL are the same as those for prepositional logic. On the other hand, the semantics of FOPL are somewhat more complex than those for prepositional logic. They largely revolve around the conversion of WFFs to clausal form and the application of the rules of inference and the principle of resolution. Accounts of the semantics for FOPL, including equivalence and truth tables are given by Frost (1986) and Patterson (1990). Interpretation of the WFFs is effected by application of the rules of inference for FOPL.

The form of inference just described is deductive inference, but this is not the only form of inference in use. Other forms are abductive inference, inductive inference, and analogical inference. These forms of inference are not strictly valid, but they are nevertheless in common use.

20.1.2 Production Rules

As already mentioned, one form in which knowledge may be represented is as a set of rules, also called productions. Such rules may be regarded as a special case of prepositional logic.

The rules are of the type IF...THEN.... They have a Left-Hand Side (LHS), which is variously known as the antecedent, premise, condition or situation and a Right-Hand Side (RHS), which is known as the consequent, conclusion, action, or response. The proposition on the LHS may be a compound one with a number of propositions ANDed together. A suitable set of rules, or productions, can be used to form the basis of a production system. Production systems are one of the main methods of implementing expert systems. A production system has three main features: the rule base, a working memory, and the inference engine. The rule base holds the set of rules that embody the expertise of the system. The working memory is supplied with the input data, or facts, on the problem to which the rules are to be applied. The inference engine controls the operation of the rules to deduce conclusions from these data. The rules in the rule base are laid out as a set of logic statements. The rule base may be inspected by an expert to check for correctness. Rules may be added, removed, or modified with relative ease.

20.1.3 Non-Classical Logics

A classical prepositional logic is not sufficiently expressive. The creation of another, now classical, logic, the predicate calculus, was a response to this deficiency. The same need has given rise to the development of further, non-classical logics. These newer logics extend the range of the knowledge that can be represented and are thus more expressive. They make it possible to handle knowledge that is incomplete, uncertain, vague and/or inconsistent, or associated with time. They are often more efficient in handling the knowledge.

The properties of a logic are that it has a well-defined method for representing knowledge, for obtaining proofs by syntactical manipulation and for extracting meaning using a semantic model. Examples of non-classical logics include: many-sorted logics, situational logics, non-monotonic logic, many-valued logics, fuzzy logic, modal logics, temporal logic, and epistemic logic.

20.1.4 Uncertainty and Inconsistency

The foregoing discussion has brought out the need to be able to handle uncertainty and inconsistency. There are a number of sources of uncertainty. Uncertainty arises due to the randomness of events and to incompleteness, vagueness, unreliability, and unbelievability of and contradictions in knowledge.

The traditional approach to handle uncertainty is the use of probabilistic reasoning. The treatment was in the context of reliability engineering, but the basic probability theory is of wide applicability. However, it also has a number of limitations, which have led to the development of alternative approaches. These approaches include: Certainty theory, Dempster/Shafer theory, possibility theory and fuzzy logic, incidence calculus, plausibility theory, truth maintenance systems, default reasoning, and heuristics. Limitations are perhaps best understood by considering these other techniques.

Much use is made in AI of Bayesian inference. Another formulation of probability that is often used in AI is odds. Odds concepts are used, for example, in the expert system PROSPECTOR, where the two likelihood ratios are termed Logical Sufficiency (LS) and Logical Necessity (LN). Another useful concept associated with odds is that of the weight of evidence.

Many accounts of knowledge representation and of inferential reasoning give examples cast either in the form of the predicate calculus or of a specific programming language such as LISP or Prolog.

20.1.5 Object-Oriented Programming

Another type of language that is now widely used in AI is Object-Oriented Programming (OOP). In OOP, the program consists entirely of objects. An object is a frame-like entity with slots that hold not only data but also procedures. Procedures may perform a wide range of functions such as symbolic manipulation, numerical calculation, and inferencing. Objects are arranged in hierarchies. Computation involves the sending of messages between objects. Two main features of OOP are data encapsulation and class hierarchy. OOP is an extremely powerful tool. A whole manufacturing process may be represented as an object, covering not only the physical plant but also its operation, documentation, hazards, and so on.

Work in AI makes use of a number of languages, and there are major differences between them. An account is given by Frost (1986). A distinction is often made between procedural languages and declarative languages, with conventional languages such as FORTRAN being described as procedural and those such as LISP and Prolog as declarative. However, the term "procedural" is not well defined.

The essential distinction is whether or not the 'flow of control' is specified. The flow of control is specified in conventional languages, using such commands as GO TO statements and DO loops, but not in languages such as LISP and Prolog. In these latter declarative languages, there are no explicit commands that determine the flow of control. This does not mean, however, that the programmer has no influence on control. In a language such as Prolog the flow of control is affected by the order in which the statements are written.

Languages used in AI work include: the functional language LISP (symbolic processing language); the logic programming language Prolog (based on the predicate calculus); the frame languages AM and KRL; and the script language SAM. The flexibility of a language may be increased by the use of an add-on facility. Thus, for example, the program FLEX provides Prolog with an object-oriented enhancement. A language such as Prolog is used particularly as a research tool. A commercial implementation may well be rewritten in a conventional language such as C++. Java, which was released in 1995 as a core component of Sun Microsystems's java platform, has been applied broadly as another object-oriented language in AI field.

20.2. STRUCTURED KNOWLEDGE

The foregoing sections have described some ways of representing knowledge. Several options are available in order to deal with the representation of knowledge in structured form, and these options include: indexing, associative networks, ISA hierarchy, inheritance, semantic networks, slot and filler systems, conceptual dependency, frames, scripts, constraints, relational databases, production systems, and objects.

The application of the tools described in problem-solving typically involves some form of search. Search is one of the fundamental activities in AI. Search algorithms found in the specialized literature include: search procedures, uninformed search, breadth-first search, depth-first search, informed search, heuristic search, evaluation function, hill-climbing, beam search, best-first search, British Museum method, branch-and-bound search, A* algorithm, games search, move generation and evaluation, minimax method, and alpha beta method.

Another major activity in AI, and one closely related to search, is matching. Some form of matching is involved in virtually all AI work but the nature of the matching process depends on the particular field of application. Areas of AI in which matching is required include logical reasoning, learning, planning, expert systems, vision, and natural language comprehension. Matching is required for a number of purposes. It may serve to identify and classify objects, retrieve objects from a knowledge base, establish the eligibility of objects for inheritance, select alternatives, control a sequence of operations, and so on.

Pattern recognition comprehends, but is somewhat broader than, matching. The term is applied particularly to vision. The process of pattern recognition involves feature selection, matching, and classification. Pattern recognition is often closely associated with learning by the development of classification rules. Pattern recognition is one of the strengths of neural networks.

20.3. PROBLEM-SOLVING, GAMES, AND VISION

An important part in the development of AI has been played by problem-solving of various kinds, particularly puzzles and games. An influential early program was General Problem Solver (GPS) by Newell and Simon (1963). It is described in *GPS: A Case Study in Generality and Problem-Solving* (Ernst and Newell, 1969).

GPS may be regarded as the first AI system to make a clear separation between problem-solving and task knowledge. It was designed to tackle the class of problems that can be formulated in terms of a set of objects, which include states, and of operators that are applied to these objects to transform them into goal objects, or states.

A quite different area of work is that of vision. Most applications of visual image processing lie outside the process field, but its use in robotics may have some bearing. There are also applications in accident investigation, as described below.

Visual image processing is generally described as involving an image acquisition stage followed by low-level,

intermediate, and high-level processing stages. The image acquisition stage yields a digitization of the image. The various processes utilize a number of specialized techniques such as the determination of the edge distance and surface direction, region growing, shape analysis, and so on. More advanced topics include three-dimensional images and motion. Work on aspects of visual image processing, notably on blocks world problems, has had influence on AI generally.

Image enhancement may sometimes be used in accident investigation. The Piper Alpha Inquiry heard evidence on image enhancement of photographs taken of the west side of the platform in the early stages of the fire.

20.4. LEARNING

Two principal categories of task in machine learning are (1) classification and (2) problem-solving. Also, learning may be classified in a number of ways. The learning modes considered here may be described as (1) learning by instruction, (2) learning by classification, (3) learning by exploration and conceptualization, (4) learning by experience and analogy, (5) learning from failure, and (6) learning by problem-solving. In addition to these learning modes, there are also particular devices that have a learning capability. Two that figure prominently in AI are (1) neural networks and (2) genetic algorithms. Of particular importance in learning is the use of inference by induction.

A large proportion of learning in AI proceeds by inference through induction. One of the principal uses of induction is the classification of objects, events, and situations, but applications of inductive reasoning are much wider than this and extend to concepts, rules, and so on. An account of induction in AI is given by Patterson (1990).

The concept underlying inductive learning is the basic logic of inductive inference:

$$
\begin{array}{ll}
\text{Assertion} & P(a_1) \rightarrow Q(b_1) \\
& P(a_1) \rightarrow Q(b_1) \\
\text{Conclusion} & \forall xy\, P(x) \rightarrow Q(y)
\end{array}
$$

Induction may be regarded as a process of class formation. The task is to partition the universe U of objects into classes. The minimum partition is into a single class C and the rest $U - C$. To this end use is made of concepts, a concept being a description or rule which subdivides a set. The target concept is the concept that classifies all the objects in the universe. The positive instances are those objects that fit the target concept and the negative instances are those that do not. A consistent classification rule is one that is true for all positive instances and false for all negative instances. A principal application of inductive learning is classification, in which the program is presented with a learning set of objects each of which is assigned to a class and described by a number of attributes. The task is then to induce a discrimination hierarchy, or a set of rules, for the assignment of the objects to the classes. In the more straightforward cases, the learning set is consistent in that it does not contain counterexamples. More advanced work addresses the problem of such noisy learning sets. One of the principal classification methods is the ID3 algorithm of Quinlan (1983a,b). Another program for inductive learning is INDUCE developed by Michalski and co-workers (Larson and Michalski, 1977; Dietterich and Michalski, 1981).

There are a number of recognized techniques of induction involving generalization and its converse, specialization. The basic process in induction is generalization. However, generalization may become too sweeping, so that the class created includes negative as well as positive instances. Such overgeneralization may be corrected by specialization.

20.5. NEURAL NETWORKS

Neural networks exemplify a quite different kind of learning device. Work on neural networks was originally inspired by research on modeling of the human brain. Another input comes from studies of associationist psychology, particularly behaviorism. A neural network is a large network of nodes connected by links. These nodes are information processing elements based on a simple model of the neuron in the human brain. A node sums the stimuli entering it through the links and has a threshold above which it will execute, or fire, and pass a stimulus to the connected nodes. A link has a weighting that is applied to the stimulus passing along it. The link weights enhance or inhibit the stimuli. The knowledge held in the network is distributed throughout it in the form of features such as these thresholds and weights. The network has an input layer of nodes, an output layer, and a number of intermediate layers in between. Learning in, or 'programming' of, a neural network proceeds broadly as follows. The network is set to its initial condition, typically with thresholds and weights set at random. It is presented with a learning set of pairs of inputs and outputs that it is then taken through. For each input the output is observed and adjustments are made to the thresholds and weights according to some strategy. The process is repeated until the learning set is exhausted.

20.6. GRAPHS, TREES, AND NETWORKS

AI makes use of a large number of graphs, trees, and networks, some of which are now briefly described. Some of these graphical forms constitute forms of knowledge representation, while others relate to the reasoning process. The graphical forms can be represented by suitable program structures in the programming languages used.

A graphical representation of some importance in engineering is the directed graph, or digraph. Digraphs have been widely used in process modeling.

Directed graphs are also known as directed linear graphs, directed networks, or signal flow graphs. A digraph consists of a set of vertices, or nodes, joined by branches, which are also termed links, arcs, or edges.

A digraph is said to be connected if it has no pair of vertices that are not connected by branches. It is strongly connected, or strong, if for each vertex, there exists a directed path from that vertex to another vertex. Otherwise it is said to be weakly connected or weak.

A path is a set of at least two connected branches. A closed path is called a cycle or loop. A digraph that contains no cycles, or loops, is referred to as an acyclic graph. A tree is an acyclic graph.

Three principal rules of flow graph algebra are (1) the addition rule, (2) the transmission rule, and (3) the product rule. There are a number of techniques for the reduction of SFGs. In addition to the product rule already mentioned, they include elimination of 'self-loops', various elementary transformations, and Mason's rule. An account of these techniques is given by Henley and Williams (1973).

Extensive use has been made of digraphs in the modeling of process plants, particularly in the modeling of fault propagation. In such work, the nodes represent process variables such as flow, pressure, and temperature, and the edges represent the effects of one variable on the other. In qualitative modeling, using digraphs a convention commonly employed is to use a transmittance function of +1, 0, and −1 for a positive influence, zero influence, and a negative influence, respectively. Some workers also use +10 and −10 to denote, respectively, strong positive and negative influences.

Generally, it is only certain variables which are of interest. In such cases use may be made of the techniques for the reduction of graphs mentioned earlier.

20.7. DATABASES, BIBLIOGRAPHIES, AND INDEXES

The safety of the chemical industry may dramatically improve by studying previous incidents. This data could provide efficient solutions to potential hazards based on incident occurrences and known consequences to chemical industries. But even though many different databases have been established, no certain criteria have been developed to set-up effective database development. This not only makes it hard to report incidents, but makes it harder to retrieve previous incident data. It is necessary to set-up a well-designed database system that could effectively allow access to the incident database. A well-managed database system could also be applied to process safety management and equipment reliability. Some tools used with data management systems, such as data mining, are being applied to the management of previous incident database developments.

A database stores and provides ready access to data. It imposes a discipline on the classification and storage of data. It is a natural form for the storage of knowledge about certain types of relationship between data. A rich variety of data structures are available, including: arrays, lists, strings, queues, stacks, bit maps, and records; direct access, sequential, inverted, and transposed files; hash tables; graphs and trees; and multi-lists, inverted lists, and bit lists. Aids to the design and implementation of databases exist in the form of database management systems (DBMSs) that provide a collection of procedures, documentation aids, languages, and programs. These tools allow for the application of traditional methods of statistics and other methods as well as new methods, such as decision trees and rules.

There are a number of databases, or data banks, for the physical, chemical, and toxic properties of materials. The databases are often associated with government agencies or technical societies, such as the following:

- Registry of Toxic Effects of Chemical Substances (RTECS), supplied by the National Institute for Occupational Safety and Health (NIOSH);
- HSDE created by the NLM as part of its Toxnet data system;
- Oil and Hazardous Materials Technical Assistance Data System (OHMTADS), supplied by the Environmental Protection Agency (EPA);
- Chemical Hazard Response Information System (CHRIS) supplied by the US Coast Guard;
- NIOSHTIC, supplied by the NIOSH;
- HSELINE, supplied by the Health and Safety Executive (HSE);
- CISDOC, supplied by the International Occupational Safety and Health Information Centre of the International Labor Office (ILO);
- MHIDAS, covering hazardous incidents, supplied by the Safety and Reliability Directorate;
- TOXLINE, supplied by NLM;
- RISKLINE, supplied by the National Chemical Inspectorate of Sweden;
- BIOSIS, CAS, IPA, and TOXBIB. PEST-BANK, supplied by the National Pesticide Information System, gives the key registration data of all US registered pesticides;
- European Inventory of Existing Commercial Chemical Substances (EINECS) is an inventory of European commercial chemicals;

- International Nuclear Information System (INIS) is supplied by the IAEA Medical databases, which includes MEDLINE and CANCER;
- Another set of databases are those from Microinfo. They include HAZARDTEXT (safe handling and incident response), INFOTEXT (regulatory listings), MEDITEXT (evaluation of exposed individuals), and SARATEXT (toxicity, Sara Title III);
- The Bretherick Reactive Chemical Hazards Database is the database version of the *Handbook of Reactive Chemical Hazards* (Bretherick, 1990b);
- CHEMSAFE and CHEMTOX are, respectively, compendia of the flammability and toxic properties of a large number of substances;
- A number of specialty databases on medical subjects are available from Elsevier Science Publishers, including the *Excerpta Medica* database EMBASE;
- IRIS is a database of monographs in support of the assessment of risk from a large number of chemicals.

There are three incident data banks operated by SRD. MHIDAS is the Major Hazard Incident Data Service. The other two are the Explosion Incidents Data Service (EIDAS), and the Environmental Incidents Data Service (EnvIDAS). In addition to MHIDAS, AEA Technology offers the AIRS code for recording incidents in the workplace. Another computer-based incident data bank is FACTS, of TNO.

20.8. PROCESS SAFETY WITH DESIGN AND OPTIMIZATION

In any chemical process, engineers are mostly interested in optimizing the unit design for the most effective, safe production. Safety systems are necessary to prevent and mitigate the consequence of inherent hazards. Although most of the safety systems are set to meet existing regulations, in most cases an adequate level of safety is required.

Traditional optimization would consider the system performance within a certain tolerable risk limit. However, traditional design optimization shows limited performance as the plant size grows larger due to the lack of proper knowledge. Finding the optimal point for economic, environmental, social, and safety issues requires better tools. Traditional mathematical tools are inappropriate and ineffective due to their lack of objective function to cover whole portions or whole time periods.

One of the alternative methods proposed is the fault tree analysis method. Because they integrate many different types of programs and techniques, fault tree solutions can provide potential designs the ability to create optimal systems with proper safety requirements. For each Safety

Instrumented System (SIS), models and algorithms are proposed to locate an optimal schedule for safety testing activities. This method finds the optimal point of cost and maintains sufficient safety measures and data. The Probability of Failure on Demand (PFD), another indicator for the integrity of a safety system, can also be used as a scale for optimizing safety systems. Locating optimal testing for SIS based on the improved PFD model allows for an increase in the safety level without changing the safety design in the process.

20.9. EXPERT SYSTEMS

One of the AI tools, which have been most widely taken up with a view to industrial application, is expert systems. The vast majority of expert systems are rule-based production systems, though other methods can be used such as neural networks. An expert system can aid in removing limitations on information that only the experts in the specific field could provide. The system can be programmed to provide knowledge in a particular specialized field in industry when required. If the number of input variables is limited, then an expert system could easily draw a conclusion from the possible solutions of the given data. The system could provide general solutions by selecting valves according to their specifications and standards, finding optimal solvents based on the specifications of the reaction and plant conditions, or choosing the proper specifications of a chemical reactor for a specific reaction.

Some expert systems on specific processes and units have been developed. A knowledge-based expert system that works with operating variables has been developed to optimize and maximize the production of a Crude oil Distillation Unit (CDU), as well as a Fluid Catalytic Cracking Unit (FCCU). Also, an expert system has been developed that can be applied to more complex systems, such as dealing with an entire power generating plant, to provide various designs for processes.

The strategic aspects of expert systems are considered in *Expert Systems: Strategic Implications and Applications* (Beerel, 1987). The management and organization of an expert system project depend very much on the nature and scale of the project. Most industrial applications are relatively modest. For a more substantial project, however, the management and organization may well determine the success.

The selection of an appropriate application is crucial. There are two essential conditions for undertaking the development of an expert system: that it be feasible and that, once created, it will be used. Knowledge acquisition in a project of any size is the responsibility of a knowledge engineer. The knowledge engineer is not a domain expert, but should be skilled both in knowledge acquisition and knowledge representation and in the building of expert systems. The basic team for the creation of an expert system is thus the project manager, the expert, the knowledge engineer, and the user.

The development of an expert system proceeds through the following typical stages: (1) system concept, (2) feasibility study, (3) outline specification, (4) preliminary knowledge acquisition, (5) knowledge representation, (6) tool selection, (7) prototype development, (8) main knowledge acquisition, (9) revised specification, (10) system development, (11) testing and evaluation, and (12) handover. The process is an iterative one, with looping back between some of these stages.

Prototyping covers the creation of the prototype, the writing of the documentation, the induction of the user, and use and evaluation by the user. The user should be involved from the start. This involvement should cover not only the specification and evaluation of the final system, but also evaluation at intermediate stages. An account of the development of a large expert system is given by J. Martin and Oxman (1988).

A minimal architecture for an expert system is a knowledge base, an inference engine, and a user interface. The knowledge base contains the set of facts and rules that constitutes the expertise. A distinction is sometimes made between a database of facts and a knowledge base containing rules. The knowledge base itself is frequently spoken of in terms of a set of knowledge bases. The inference engine performs the process of inference from the facts and rules in the knowledge base. It is a basic principle of the architecture that the knowledge base and inference engine are separate.

The user interface has several functions. It is the means whereby the user supplies additional information to the system, either in response to demand or by volunteering it, and the means whereby he receives the output.

To these three basic features may be added a number of others. They include a blackboard, a knowledge acquisition facility, a learning facility, and an explanation facility. A blackboard is a device for communicating with the knowledge bases and holding intermediate hypotheses and decisions. Only a proportion of systems incorporate a blackboard. A knowledge acquisition facility provides the means for the knowledge engineer to input knowledge elicited from the domain expert, allowing him to input and edit facts and rules, and so on. An inductive learning facility may also be provided. The user interface may contain an explanation facility.

The inference engine may incorporate any of a number of different inferencing, search and problem-solving methods. It may have forward or backward chaining, corresponding respectively to data- or goal-driven inference, or a mixture of both. It may use depth-first or breadth-first search, hill-climbing, means-ends analysis, generate-and-test, pattern-matching, and so on. Or it may use some hybrid scheme.

20.9.1 Blackboard, Knowledge Acquisition Facility, and Explanation Facility

A blackboard is a device for holding a set of intermediate hypotheses and decisions, in other words partial solutions. It communicates with and draws knowledge from the set of knowledge bases in an opportunistic way. In some cases it can be regarded as being split into three parts: plan, agenda, and solution. The plan part contains the goals, plans, states, and contexts; the agenda the potential actions awaiting execution; and the solution the hypotheses and decisions. The blackboard can be altered only by the knowledge bases. Such forms of control as the blackboard has acted to focus attention on a problem and the knowledge bases then indicate what contribution they can make.

It is helpful to the creation of an expert system if there is a Knowledge Acquisition Facility (KAF) that provides for the editing of facts and rules and for their entry into the knowledge base.

There may also be provided a facility for inductive learning. This may be regarded as part of the knowledge acquisition facility or as a separate feature. In any event, the purpose of this facility is to allow knowledge to be entered in the form of a learning set of examples provided by the expert and rules to be induced from these examples. This is an alternative to the provision of explicit rules by the expert.

Most expert systems are provided with an explanation facility which can provide some form of justification for the conclusion reached, but the sophistication of such facilities varies. Accounts of explanation facilities are given in *Expert Knowledge and Explanation* (Ellis, 1989) and by Sell (1985) and J. Martin and Oxman (1988).

There are two main types of users for an explanation facility, the knowledge engineer and the client user. There are several types of explanation that a user requires. One is elaboration of the questions that he is asked in order to assist him to answer them correctly. Another is justification for these questions in order to motivate him to provide answers to them. The third is an explanation of the reasoning process.

One of the simplest forms of explanation facility is rule-tracing. The system keeps track of the path by which it has reached the current point and can display this to the user on demand. A rule trace is relatively easy for the system to provide. A detailed example is given by Ellis (1989).

However, an explanation facility that utilizes a simple rule trace is somewhat crude. It is preferable that the line of reasoning be explained rather in terms of the principles that govern the domain, using domain knowledge in a structured way.

Generally, an expert system is used by a quite small number of people. There tend to be considerable differences in individual preferences for output from an explanation facility. One way of accommodating this is to provide options which individual users can select. In more advanced work on this topic, user models are utilized to improve the design of the explanation facility.

20.9.2 Knowledge Elicitation and Acquisition

Crucial activities in building an expert system are knowledge elicitation and knowledge acquisition. The two terms are often used almost interchangeably, but whereas in knowledge elicitation the emphasis is on eliciting the domain knowledge from the expert, in knowledge acquisition it is on providing the program with this knowledge. Accounts are given in *Knowledge Acquisition for Expert Systems* (Kidd, 1987) and *Knowledge Elicitation* (Diaper, 1989) and by Beerel (1987) and J. Martin and Oxman (1988).

The expertise of the domain expert has a number of facets. Typically, he or she has a wide experience and knowledge of case histories, problems, and solutions, both successful and failed. This background provides him with an understanding of the structure of the domain, of the crucial distinctions which have to be made, of the constraints which apply, of the problems to be solved, of the decompositions which can be made, of problem-solving and search strategies, of heuristics and rules-of-thumb, of facts and rules, and of exceptions and refinements. The most common method of knowledge elicitation is interviewing of the expert by the knowledge engineer. The aim is to make explicit the expertise. There is a large literature on interviewing, especially in the social sciences. Another approach is close observation of the expert as he performs the task. Typically this utilizes some form of verbal protocol in which the expert explains what he is doing and why he is doing it. A third approach is for the expert to provide sets of problems and solutions and for the knowledge engineer to induce from these the internal rules which the expert is evidently using to obtain the solutions. A fourth method is prototyping, in which the expert and knowledge engineer cooperate to build a system. The expert furnishes knowledge for, and provides tests of, the system, while the knowledge engineer tries to create a suitable system structure.

It is clearly necessary that at some stage an expert system should be evaluated, but such evaluation raises a number of issues that need to be addressed if it is to be done satisfactorily. The evaluation and validation of expert systems are discussed by Hayes-Roth et al. (1983), Sell (1985), Beerel (1987), and Ellis (1989).

20.9.3 Expert Systems: Some Systems and Tools

Some examples of experts systems and their applications include the following:

DENDRAL—Discovers molecular structures.

MYCIN—Diagnosis of infectious blood diseases.

CASNET—Diagnosis and treatment of glaucoma. Is a research program.

PROSPECTOR—Advice on mineral deposits.

CADUCEUS, originally called INTERNIST—Diagnosis of diseases of the internal organs. Is a research program.

XCON, originally called Rl—Configuration of VAX computer systems. Is in industrial use.

ART (Automated Reasoning Tool)—Building large systems and integrates a number of problem-solving techniques.

EXPERT—Diagnosis- or classification-type consultation systems.

EXTRAN, and the earlier EXPERTEASE—Classification tools based on the IDS algorithm for inductive learning.

20.10. QUALITATIVE MODELING

In engineering, the conventional method of modeling a physical situation is to create a mathematical model consisting of algebraic and differential equations based on classical physics. In recent years, there has been growing interest in an alternative approach that goes by the names of naive physics, and of qualitative modeling, qualitative physics, or qualitative simulation. Qualitative modeling has developed in three main directions. These are component-based, or device-based, modeling, as in the work of de Kleer and Brown (de Kleer, 1975, 1985; J.S. Brown and de Kleer, 1981; de Kleer and Brown, 1983, 1984, 1986); process-based modeling, as in the work of Forbus (1983, 1984, 1990); and constraint-based modeling, as in the work of Kuipers (1984, 1986). Of these three methods, component-based modeling appears the most suitable for the qualitative modeling of process plant.

20.11. ENGINEERING DESIGN

Applications have been mainly in the fields already described, but there is an increasing amount of work in engineering. The material of prime interest here is that on process plant design, which is described below, but first some consideration is given to design expertise and to AI applications in engineering design generally.

Turning now to the use of AI, expert system and other advanced techniques in the process industries, a number

of overviews have been published. They include those of Stephanopoulos and Townsend (1986), Westerberg (1989), Hutton et al. (1990), and Stephanopoulos (1990). Various computer aids fall into four broad categories, as aids for (1) the design project, (2) the design synthesis, (3) the design analysis, and (4) the plant operation.

Computer aids for process and plant synthesis may be termed front-end aids, while aids for plant analysis are back-end aids. Another way of looking at the matter is in terms of the generate-and-test concept: the synthesis aids represent the generate phase and the analysis aids the test phase. These various aids are considered in the sections that now follow.

It is appropriate to begin the account of AI aids for process plant design by considering those which address design problems at the highest level.

20.11.1 Aids for Representing Design Issues and History

Design involves the identification of a succession of issues, discussion of these issues and decisions to select particular options. There are a number of benefits to be had by adopting a more formal approach to the handling of design issues. Such an approach imparts greater structure to the decision-making process, assists communication and cooperation between the different design disciplines, and provides a record of design information. This information is of various kinds and includes the design intent, the chronological development of the design, the issues raised, the constraints recognized, the rules applied, and the decisions made.

This record is of particular value when a design is undertaken which draws on that for an existing plant. Unless such information can be retrieved, the designers of the new plant may be unaware of significant issues and constraints and may not appreciate the reasons for certain design features. In the more serious cases, some assumption or feature critical to safety is not appreciated.

An intelligent information system for safe plant design is described by Chung and Goodwin (1994). The system is issue-based, allowing the issues arising in the design and the arguments related to the issues to be represented and captured. It provides a historical record of the development of the design and of the issues affecting this development. Elements of the system are (1) the viewpoint mechanism, (2) the issue base, and (3) the rule base.

The viewpoint mechanism is used to represent the design hierarchy. The core of the system is the issue base. This provides a record of the issues considered. An issue is identified, positions are taken, and arguments adduced for or against each position. In due course a decision is made to select one of the positions, or options. The rule base constitutes the third main element of the system. This contains both general rules and rules specific to the particular design.

20.11.2 Aids for Representing Regulatory and Code Requirements

Another generic problem in design is the handling of the requirements of legislation, standards and codes, and company policies. The number of documents containing requirements and the number of individual requirements is generally very large, and this poses a major information handling problem. An account of the development of the aids to assist with such information handling has been given by Chung and Stone (1994). The problem has several different aspects, including the retrieval of the requirements relevant to a particular issue and the comprehension of these requirements.

Considering first an individual document such as a standard, three distinct approaches to the problem may be identified. The first is to accept the document as it is and provide aids to the user in finding his way around it. The second is to perform some kind of processing on the document. The third, and most radical, is to intervene at the stage when the document is being written and provide aids to assist the author in writing it.

The retrieval problem has two aspects: retrieval of whole documents and retrieval of requirements from within a document. For both types of retrieval, widespread use is made of keywords. For retrieval of whole documents, these may be words in the title, keywords provided by the author, or keywords searched in the text. The use of combinations of keywords narrows the search space.

20.11.3 Process Modeling

Before considering process and plant design as such, it is necessary at this point to say something about modeling. It has become clear from attempts to apply AI to a number of problems in this area that a common lesson is the need for modeling of the process and the plant. Such models are often referred to as deep level knowledge in contrast to compiled knowledge such as rules. Some areas in which modeling are required are process and plant design, planning of plant operations, and the diagnosis of faults on operating plant.

Such modeling draws particularly on the fundamental work on qualitative modeling. Some principal developments in qualitative modeling include modeling languages and environments such as OMOLA (Andersson, 1989; Nilsson, 1989), MODEL.LA (Stephanopoulos et al., 1990a,b), and ASCEND (Piela et al., 1991).

Another development is automatic model generation. Work on this topic has been described by Catino et al. (1991). These authors' work utilizes Forbus' Qualitative Process Theory (QPT). The approach taken is based not so much on process units such as reactors and heat exchangers as on process phenomena such as reactions and heat exchange. The authors describe the use of an adaptation of QPT to configure such models, and the creation of a prototype model library.

Another area of modeling is mathematical programming, specifically Mixed Integer Linear Programming (MILP) and Mixed Integer Non-Linear Programming (MINLP). Mathematical programming modeling process is quantitative in nature. An account of the relationship between MILP and logical inference for process synthesis is given by Raman and Grossman (1991), who show that qualitative knowledge on process synthesis that can be expressed in prepositional logic form can also be represented in the form of equivalent linear equations and inequalities.

20.11.4 Process Synthesis and Plant Design

Turning now to the actual design process, the most creative part of the design is the synthesis of the process and of the plant. The work of Rudd and Watson (1968) deals with in the first part, the creation and assessment of alternatives under the headings of (1) the synthesis of plausible alternatives, (2) the structure of systems, (3) economic design criteria, and (4) cost estimation; in the second part, optimization under the headings (5) the search for optimum conditions, (6) linear programming, (7) the suboptimization of systems with acyclic structure, (8) macrosystem optimization strategies, and (9) multilevel attack on very large problems; and in the third part, engineering in the presence of uncertainty under the headings (10) accommodation to future developments, (11) accounting for uncertainty in data, (12) failure tolerance, (13) engineering around variations, and (14) simulation.

Process synthesis is of great importance. In particular, it is relevant to (1) inherently safer design, (2) inherently cleaner design, and (3) inherently low energy design, or process integration.

Moving on to the design of the plant as opposed to that of the process, this is also an important area. Here some modest progress has been made in the development of aids for certain types of task, but there remains much scope.

One of the prime aims in the design of process plants is to promote the practice of inherently safer design right from the conceptual design stage. Apart from developments such as those in process synthesis and the selection of a reaction route, which are certainly important aspects of inherently safer design, it is not easy to identify any specific synthesis aid in this area.

An overview of the expert design of plant handling hazardous materials has been given by Bunn and Lees (1988). What these authors do is to describe a series of design problems, involving various kinds of design expertise, ranging from very simple, even trivial, design activities to designs with deep structure and to designs apparently better suited to solution by machine rather than by man.

The problems considered in this work are (1) the choice of whether or not to fit an emergency isolation valve, (2) the design of a flare system, (3) the design of a pressure relief system, and (4) the design of a valve sequencing system; together in each case with the nature of the appropriate AI aid.

Turning now from the structure of plant design synthesis problems to applications of AI to such synthesis, an account of AI applications that deals particularly with plant synthesis and analysis as well as process synthesis is given by Hutton et al. (1990).

20.11.5 Incremental Design and Safety Constraints

A development of some interest is the concept of incremental design as applied to safety. In conventional plant design, the designers proceed with the design, adhering to good design practice, and the design is subjected to safety reviews at prescribed stages. An alternative approach, which emphasizes the application of the safety constraints in an incremental way, has been championed by Ponton and co-workers. The application of this philosophy has been described in respect of the safety constraints on design by Waters et al. (1989) and Waters and Ponton (1992) and in respect of hazard studies by Black and Ponton (1992).

20.11.6 Plant Design: Analysis

The synthesis of plant designs is complemented by their analysis. The analysis constitutes the test element in the overall generate-and-test activity. There now exist a large number of computer programs for the analysis of various aspects of plant design. One category is those programs that compute an index and perform a ranking of the hazards of the plant. Another category is those which assist with hazard assessment. There are now beginning to appear aids for design analysis that have an AI flavor. Applications include: inherently safer design, fault propagation, hazard identification, and fault tree analysis.

20.11.7 Expert Systems: Some Process Systems

Expert systems were introduced in the process industries in the mid-1980s and are in widespread use. The typical expert system in the process industries is much more modest than the classic expert systems described in the previous section. Most have been created using shells. An appreciable proportion has been developed by a single engineer acting as the expert. The development time has generally been of the order of months rather than years.

Some principal applications are (1) physical, chemical, and thermodynamic properties; (2) selection of equipment, materials of construction, and processes and plants; (3) design of processes and plants; (4) process control; (5) process monitoring; and (6) fault administration.

The bulk of these applications are the handling of fault conditions, either on an individual item of equipment or on a process plant. Before considering these different application areas, it is of interest to consider a statistical profile of process industry applications.

A survey of expert system applications in the process industries has been described by Sangiovanni and Romans (1987). These authors found about 200 applications in some 30 companies. They classified these applications as (1) selection, (2) design, (3) planning and scheduling, (4) control, (5) situation analysis, (6) diagnosis, (7) prediction, (8) prescription, and (9) instruction. They define situation analysis as monitoring available data and information and inferring the system's state; diagnosis as inferring the cause of a malfunction or deviation; prediction as inferring the likely consequences of an action or set of actions; and prescription as recommending cures for a system malfunction or deviation.

Of these functions, selection accounted for some 15%. Situation analysis accounted for some 8%, diagnosis for 33%, prediction for 10%, and prescription for 21%. Thus one-third of the applications were for diagnosis and the group of four functions related to fault administration (situation analysis, diagnosis, prediction, and prescription) accounted between them for 66%. Only about 2% of applications were to design.

Of these expert systems, 30% had been developed by engineers, 29% by a team without the vendor, 29% by a team with the vendor, and 12% by an internal AI group. The development time was <3 months for 30%, 3–12 months for 40%, and >12 months for 27%.

Some two-thirds of the systems were in daily use, including cases where the system was used by different people at a number of locations. Some 90% of applications made use of an expert system shell. Over 60% were implemented on a PC. The dominant contributor to software costs was maintenance of the software, at some 67%, compared with design and coding and with testing and debugging, which each accounted for about 13%.

Expert systems are in use in the process industries for a variety of purposes. Reviews of applications and potential applications are given by M. Henry (1985), Coulsey (1986), and Barnwell and Ertl (1987).

20.12. FAULT PROPAGATION

Many of the methods used to identify and assess hazards at the design stage, such as hazard and operability studies, failure modes and effects analysis, fault trees and event trees, involve tracing the paths by which faults propagate through the plant. Fault propagation is thus a common feature of these techniques. Fault propagation is also a feature of methods for the diagnosis of alarms in the real-time computer control of processes.

Fault propagation is therefore a generic feature of work in this field. It follows that the computer aids for many of these methods may, in principle, draw on a methodology that is to some extent common. This concept has been elaborated by Andow et al. (1980).

The propagation of a fault through a plant may be represented in terms of the initiation of a fault in a unit that is unhealthy, the passage of the faults through units that are otherwise healthy, and the termination of the fault in a unit that is thereby rendered unhealthy. A unit may be modeled using a number of representations. Those considered here are (1) functional, or propagation, equations; (2) event statements, and (3) decision tables.

The propagation equations describe the propagation of faults through a healthy unit, while the event statements describe the initiation of a fault in an unhealthy unit or the termination of a fault in a unit that thus becomes unhealthy.

A propagation equation is a functional equation that describes the relation between an output parameter of a unit and the input and other output parameters. A typical propagation equation is:

$$L = f(Q_1 - Q_2) \tag{20.1}$$

which signifies that the level L increases if the inlet flow Q_1 increases or the outlet flow Q_2 decreases, and vice versa.

20.12.1 Fault Initiation and Termination

The propagation equations describe how a fault propagates but not how it is initiated or terminated. This information is provided by event statements. An initial event statement takes the form:

Initial fault: parameter deviation

The initial faults are usually mechanical failures. A terminal event statement takes the form:

Parameter deviation: terminal event

The terminal events are usually undesired events or hazards. A typical initial event statement is:

$$\text{F PART.BLK:} Q_2 \text{ LO} \tag{20.2}$$

which signifies that outlet flow Q_2 is low if there is a partial blockage. The letter F denotes a failure event. A typical terminal event statement is:

$$P_2 \text{ HI: OVERPRES} \tag{20.3}$$

which signifies that if the pressure P_2 is high, there is overpressure of the unit.

The third form is a decision table. A typical decision table is:

$$V Q_2 \text{REV} \quad V U_2 \text{HI} \quad T U_1 \text{HI} \tag{20.4}$$

This relation refers to conditions in reverse flow, U being the temperature in reverse flow and subscripts 1 and 2 denote the inlet and outlet ports under normal flow conditions, respectively. The relation signifies that the outlet temperature U_1 is high if there is reverse flow Q_2 and the inlet temperature U_2 is high. The letters T and V denote the top event of a mini-fault tree and a process variable, respectively.

20.12.2 Flow Propagation

In fault propagation modeling it is necessary to have a suitable method of modeling flow propagation. There are several features that need to be allowed for. One is that the representation should model correctly the two-way propagation of a fault. Thus, for example, if a valve is shut, a disturbance travels from it not only downstream but also upstream. The method for flow deviations should be compatible with that used for pressure deviations. A flow model that is based on pressure differences can run into difficulty because in some applications the deviations of pressure are not sufficiently well defined to use differences between them to determine flow. A third requirement in some cases is that the flow propagation should be able to handle a large leak from a component such that continuity of flow through that component is not maintained.

In cases where this latter aspect does not need to be taken into account, flow is often modeled by the simple relation:

$$Q_2 = f(Q_1) \tag{20.5}$$

For the case where a large internal leak needs to be allowed for, Kelly and Lees (1986a) have modeled flow propagation using, by convention, the following pair of relations:

$$Q_2 = f(G_1, G_2) \tag{20.6}$$

$$G_1 = f(Q_1, Q_2) \tag{20.7}$$

where G is the pressure gradient, and Q is the flow and subscripts 1 and 2 denote the inlet and outlet ports, respectively. Here, the pressure gradient G is essentially a surrogate for flow.

20.12.3 Fault Propagation Modeling

Fault propagation models may be represented in various ways, which may well be equivalent. Lees and co-workers (B.E. Kelly and Lees, 1986a,b; Parmar and Lees, 1987a,b) have represented their models initially in terms of propagation equations, event statements, and decision tables. They have utilized a proforma that contains slots for a description of the unit and for these relationships. Most workers, however, have used the digraph representation. The two representations are in principle, equivalent. In digraph terms, the propagation equations yield those nodes in the digraph that represent deviations of the process variables and the initial event statements those nodes that represent failures. The essential requirements for a model format are that (1) the modeling process be straightforward, so that it can be applied by practicing engineers as well as research workers, with as little effort and error as possible, and that (2) it supports the automation of model creation using model archetypes.

Models already created are stored in a model library. An essential feature of such a library is a taxonomy that helps the user to distinguish between different models and to identify those required or, alternatively, to confirm that some are not available in the library. The library needs to be governed by a discipline which ensures that appropriate information is provided about each model, such as the person creating it and the extent of its use, and that an incorrect model is not entered.

Generally, the user will find that some models are not in the library and need to be generated. In this case, it is necessary to configure the models and enter them into the library. Since this is liable to be the main and most difficult input that the user has to provide, it is highly desirable that the process of model generation be kept as simple as possible. The rules governing the manual generation of models should be formulated explicitly and documented.

In many cases, the new model to be created differs only slightly from an existing one. There is therefore scope for the use of model archetypes and templates and for core models or model sections. Core model elements may be provided for common types of unit such as heat exchangers and reactors.

As already indicated, fault propagation relations may be represented in a number of different ways. To some extent at least it is possible to map between them. An account of alternative representations and of mapping has been given by Aldersey et al. (1991).

Some forms of representation available include: (1) functional equations, (2) program rules, (3) digraphs, (4) block diagrams, (5) logical expressions, (6) truth tables, (7) cut sets, (8) fault trees, and (9) event trees.

Fault propagation modeling may be used to support a range of safety-related techniques such as hazard identification, fault tree synthesis, operating procedure synthesis and alarm diagnosis. There is clearly an advantage to be gained by formulating a type of generic model that supports all these methodologies and others. On the other hand, the development of a particular technique should not be unduly constrained by the requirement for commonality. This is illustrated in the work of Lees, Rushton, Chung, and co-workers.

A code QUEEN, designed to provide a front-end for fault propagation modeling in activities such as hazard identification, fault tree synthesis, operating procedure synthesis, and alarm diagnosis has been described by Chung (1993).

20.13. HAZARD IDENTIFICATION AND RISK EVALUATION

There are a number of codes available to assist in the conduct of the various techniques for hazard identification. Aids for HAZard and OPerability analysis (HAZOP) are particularly valuable for their housekeeping and follow-up aspects. Computer codes exist for a number of hazard models, and for the more complex models this is the only practical method. Qualitative modeling, and particularly fault propagation modeling, is an important feature in many of these aids. Some of the principal targets of study are jets; vaporization from pools; pool fires; fire engulfment; flame impingement; explosion in, and explosion relief of, large enclosures; and vapor cloud explosions.

An intelligent system is implemented in HAZOP to aid in identifying the hazards in process design. The knowledge engineering framework allows the program to process large quantities of data from safety analysis. This data is reduced in the way the user defines and to allow easy access when necessary. The data is processed so that it can be reused as a solution to a new problem. These systems are also equipped with functions to import and export data to other programs. This flexibility allows the system to be easily applicable to current industry.

The applicability of AI techniques in this area is discussed by Ferguson and Andow (1986). The identification of hazards is in many ways one of the least promising fields for computer methods. It is very difficult to devise a technique that can compete with man's ability to think laterally and to make apparently obscure connections with those from a conventional HAZOP study. They also give an account of an application session.

A number of works have described systems which perform functions broadly similar to those carried out in a hazard and operability, or hazop, study or which serve as an aid in the conduct of a study of this general type. Systems include: HAZID by Parmar and Lees (1987a,b); a system described by Weatherill and Cameron (1988, 1989); HAZOPEX by Heino et al. (1989a,b) and Suokas et al. (1990); HAZEXPERT by Goring and Schecker (1992); COMHAZOP by Rootsaert and Harrington (1992); QHI by Cartino and Ungar (1995); PHAzer by Srinivasan and Venkatasubramanian (1998); PHASuite by Zhao et al. (2005); IRIS by Bragatto et al. (2007); and PetroHAZOP, also named as HAZOPSuite by Zhao et al. (2009).

- COMHAZOP: The COMHAZOP system is described by Rootsaert and Harrington (1992). The plant configuration is entered and unit models are assigned to the units. The program examines deviations associated with the units and uses rules to identify the causes of these deviations.
- HAZEXPERT: Goring and Schecker (1992) describe the hazard identification system HAZEXPERT. This aid eschews a HAZOP-style approach. The plant configuration is entered. A set of pre-defined consequences such as overpressure, explosion is used. A search, limited in scope around the set of units under examination, is conducted to discover the causes of these events. The program contains a generic hazard knowledge base in which a key concept is disturbance of the mass or energy balance. HAZEXPERT is implemented using an expert system shell.
- PSAIS: PSAIS (Plant Safety AI System) is described by Schöneburg (1992). The program utilizes only a limited number of rules but holds a large collection of design cases. Examination of a plant design is based on accessing 'similar' cases in the design case database. If no similar case is found, a new case is created. Use is made of fuzzy matching. In this way PSAIS benefits from a continuous process of learning.
- HAZExpert: The HAZExpert system is described by Venkatasubramanian and Vaidhyanathan (1994). The general approach described appears broadly similar to that used in HAZID. The authors identify as distinguishing characteristics an emphasis on consequences and the resolution of ambiguities. HAZExpert is implemented in G2 by Gensim (1992).
- QHI: The QHI system, which is described by Catino and Ungar (1995), is a prototype hazard identification system. The system can exhaustively posit possible faults, build qualitative process models, simulate and obtain the hazard, using the qualitative differential equation (QDE)-based model.
- PHAzer: PHAzer is a HAZOP expert system which integrated SDG, Petri-Net, and fault tree synthesis. It has been used to analyze continuous as well as batch

chemical process (Srinivasan and Venkatasubramanian, 1998).

- PHASuite: PHASuite was developed by Zhao et al. (2005). This HAZOP expert system is designed mainly for analyzing the batch chemical process, including data acquirement, model construction, inference engine, report generator, integrating several technologies including SDG modeling, Petri-Net modeling, Ontology, Database, etc.
- IRIS: IRIS is a HAZOP assistant system (Bragatto et al., 2007), had the ability to interact with CAD system. The system can deduce the HAZOP result on an interactive way, which can display the hazard information directly in the graphical CAD system screen.
- PetroHAZOP platform: In the PetroHAZOP platform developed by Zhao et al. (2009), the existing hazard analysis reports produced by human experts were utilized to generate a case base. Depended on the case-based knowledge base and the related case-based reasoning technology, the system can provide the most suitable knowledge for the human experts. The case-based system PetroHAZOP, now renamed as HAZOPSuite, has been installed at Dushanzi PetroChemical company, a Petrochina company in Xinjiang province, China, for more than 2 years, and about 90 HAZOP study projects have been done at Dushanzi based on the expert system platform. Thousands of potential hazards have been identified and hundreds of human HAZOP experts have been trained by using this software.
- HASILT: For efficient management of process safety risks, HASILT, an intelligent software platform combining HAZOP, layer of protection analysis (LOPA), safety requirements specifications (SRS), and SIL validation together has been with its roots in HAZOPSuite. By using this HASILT, complex hazard scenarios identified through HAZOP analysis can be automatically loaded into a LOPA module to quantitatively assess their likelihoods by using the built-in initiating event database and the Probability of Failure on Demand (PFD) database of independent protection layers. If a new safety instrument function is needed to meet the tolerable risk criteria, its SRS can be generated by using HASILT and stored in its database. To ensure the reliability of LOPA, the validated SILs of existing safety instrument systems have to be used for LOPA. If the SILs of existing safety instrument systems have not been validated before LOPA is started, the SIL validation can be done within the same platform HASILT. HASILT is currently installed in a server of the Petrochina Corporation for large-scale tests.
- Hazard Identification (HAZID): The HAZID code, described by Parmar and Lees (1987a,b), follows closely the general approach taken in a HAZOP study,

but draws on generic fault propagation technology developed for fault tree synthesis.

20.13.1 Computer Codes

The main programs in the HAZID package are: MASTER, which does the housekeeping; CONFIGURATOR, which handles the configuration; IDENTIFIER, which is the core program and generates the cause and consequence lists for the parameter deviations; CONSOLIDATOR, which turns these lists into a table similar in form to that produced by a conventional HAZOP; and MODGEN, which generates the unit models. There is a unit model library. The task of IDENTIFIER is to handle rules and the program is written in Prolog. The other programs are written in Fortran.

Some other computer codes for hazard assessment systems are as follows:

- ARCHIE: Developed by Hazmat America for the Federal Emergency Management Agency (FEMA, 1989). This software contains a set of models for a variety of scenarios cast in a tree structure.
- SAFETI: The SAFETI code was developed for the Dutch Ministry of Housing, Physical Planning and Environment. This is a major code for the conduct of a complete probabilistic risk assessment (PRA). The program performs a complete PRA for a fixed installation. It is designed for ease of input, speed of processing, and transparency of results, both intermediate and final. It is also designed for minimum loss of detail, and allows the level of detail to be varied.
- WHAZAN: A rather simple code is WHAZAN, developed by Technica for the World Bank. WHAZAN provides a facility for the user to explore the consequences of a set of release scenarios.
- RISKAT: The HSE has developed its own hazard assessment program initially entitled the Risk Assessment Tool (RAT), and latterly RISKAT.

One of the most important topics covered by such codes is gas dispersion, particularly dense gas dispersion. Different models' capabilities include passive gas dispersion, dense gas dispersion, jets, vaporization from pools, fire events, condensed phase explosions, and vapor cloud explosions.

Codes for the simulation of explosions in large enclosures play an important role in the design of fire and explosion protection of offshore platforms. These programs solve the fundamental equations of fluid flow, taking into account turbulence and combustion. The three-dimensional Navier–Stokes equations, suitably augmented to include the effects of turbulence and combustion, are cast in discrete form, employing a finite volume technique, and are solved implicitly. Turbulence is

modeled in terms of eddy viscosity, and combustion is modeled in terms of a turbulent, mixing-limited reaction. The space modeled is divided into a grid of "boxes" of 1 m^3 volume. Normal assumptions are that the gas cloud is a stoichiometric homogeneous, quiescent mixture. Ignition is modeled by assuming that at time zero, half the flammable mixture in one of the boxes has undergone combustion.

Codes for Computational Fluid Dynamics (CFD) find widespread use both in design and in hazard assessment. The CFD tools can be used to extract potential, temperature, velocity, or gas volume fraction distribution from the given information in a two- or three-dimensional domain. These tools can provide information on consequence modeling or complex physical phenomena depending on the codes used.

For reactor venting, the SAFIRE code for the venting of reactors implements the design methods developed by the Design Institute for Emergency Relief Systems (DIERS) as part of its design package for emergency relief. There are also programs that calculate the emission flows from vessels, pipes, and pipelines, such as BLOWDOWN (Imperial College) and BLOWSIM (UCL).

20.14. FAULT TREE ANALYSIS

A large number of computer programs have been developed for the analysis of fault trees, which are often very large. The tasks carried out by these programs fall mainly into four categories. These are determination of (1) minimum cut sets, (2) common cause failures, (3) reliability and availability, and (4) uncertainty.

The synthesis of fault trees is a more difficult problem, but some progress has been made in developing aids for this. Again the codes utilize some form of fault propagation modeling. Some codes for fault tree construction include TREDRA and TREE.

There are a number of codes that have been developed to affect the automatic synthesis of fault trees. The different methods take different starting points, including: functional equations, graphical methods, reliability graphs or block diagrams, logic models, and tabular methods, such as decision or transition tables, and mini-fault tress.

20.15. OPERATING PROCEDURE SYNTHESIS

There are relatively few computer aids for the examination of plant operations such as start-up and shut-down, although it is often in these phases that incidents occur, but progress is now apparent in the synthesis of operating procedures through the application of Artificial Intelligence (AI) planning techniques. Pioneering work in this area was the study of Rivas and Rudd (1974) on the synthesis of operating sequences for valves.

Since the operating procedures (OPs) are used to instruct an operator to safely and optimally manage the batch process, safety issues related to each operation should be explicitly flagged in the corresponding operation instructions. To automate the information flow from iTOPS to Batch HAZOPExpert (BHE), an integration system of iTOPS and BHE was developed by Zhao et al. (2000).

20.16. PROCESS MONITORING

Process monitoring has several aspects. One is the detection of a disturbance or abnormality which falls short of an identifiable fault and which does not necessarily result in any process alarm. The second is the handling of alarms caused by some fault or operator action, the diagnosis of the fault and the response to it. The third is the detection of an incipient malfunction which has not yet resulted in a fault or an alarm. There are advanced methods, including Artificial Intelligence techniques, which address all three of these. An overview of some of the methods available for these functions is given in "A Review of Process Fault Detection and Diagnosis: Part I to Part III" (Venkatasubramanian et al., 2003a−c).

There is a considerable monitoring element in the expert systems for process control. In some expert systems, however, assistance to the process operator in assimilating process information is the prime aim. An example is ESCORT, described by Sachs et al. (1985).

Another development is the use of multimedia aids that exploit the potential of combinations of audio, visual, and other forms of information presentation. Accounts are given by Alty and McCartney (1991) and Alty and Bergan (1992). Alty and Bergan (1992) have described a laboratory study involving the Grossman water bath experiment on manual control with multimedia aiding.

20.17. ADVISORY SYSTEM

A real-time advisory system provides operators with the ability to supervise a process within the safety boundary. The system aims to resolve issues when the process goes out of bounds. The real-time advisory system can analyze the problem through the collected data and provide solutions. All the systems, from collecting to analyzing, are integrated into one system so that only the necessary data can be processed. All the alarm interruptions are taken into account in the optimized solutions.

For the safety application, the real-time advisory system identifies risks in real-time based on the different variables. Detailed plant conditions, such as weather and other plant situations, are taken into account and analyzed. The system recommends specific solutions based on the collected data. This data is collected by monitoring systems throughout the plants. Such systems could be applied in decision-making steps, like the optimal solvent section and in units, like distillation units. Also, safety measures could be added using the real-time advisory system on specific steps, such as start-up procedures or feeding system.

20.18. INFORMATION FEEDBACK

It is clearly essential that the industry should learn from its mistakes so that as far as possible it avoids repeating them. This need to learn has long been recognized and is dealt with by a system of accident reporting and investigation. On the basis of this information, accident statistics are compiled and an attempt is made to determine the trend in safety performance. Case histories are obtained and general lessons are derived. These lessons give rise to changes of practice, which are often embodied in standards and codes of practice or training programs.

There are statutory requirements for the reporting and investigation of accidents. Company requirements may be confined to these or may include other incidents. Accident statistics, safety performance trends, and case histories are produced, changes in practice are made and publicity is given to these at both company and national level. The effectiveness of this approach depends very much on the way in which it is implemented. There are wide differences, for example, in the quality of incident investigation and consequently in the benefit derived from it. There is also considerable variation in the extent to which the real lessons of case histories are appreciated and applied.

It is not enough; however, to discover the facts about accidents. It is even more necessary to make sure this information is used. A high proportion of accidents constitute repetitions, with only minor variations, of accidents which have already occurred elsewhere and which are well understood by the engineering profession as a whole. It is not to be expected that the individual engineer should familiarize himself with all these. But he should make it his business to become familiar with some of the exemplary case histories and their lessons. In addition, he should make use of aids such as checklists and codes of practice that embody much of this experience in the most readily usable form.

If learning is to take place, then, it is necessary to have both a means of obtaining information, for example, accident reporting and investigation, and a means of ensuring that this information is utilized, for example, checklists and codes of practice. It is desirable for the learning process to take place at both company and national levels and the measures described are, therefore, necessary at both levels.

20.19. EDUCATION AND TEACHING AIDS

The education of engineers normally includes some material on safety. In chemical engineering, this has traditionally tended to cover aspects such as legal requirements and personal safety. There has also generally been some coverage of the engineering aspects, particularly in design projects, but overall the treatment has been variable.

The occurrence of disasters such as Flixborough, Seveso, and Bhopal has highlighted the need for greater awareness of and knowledge about SLP. These three accidents have had particular impact in the United Kingdom, Continental Europe, and the United States of America, respectively. The trend has been for industry to urge universities to teach SLP and for professional bodies to introduce a requirement that SLP be part of the curriculum.

There are available a number of teaching aids of various kinds, suitable for use in the education of engineers at universities and/or of engineers and other personnel in industry. The IChemE has been active in the production of a variety of aids. Other aids include the information exchange scheme of the Loss Prevention Bulletin, and many of the books by Trevor Kletz. Another source of material is the CCPS of the American Institute of Chemical Engineers (AIChE). Specifically, educational material from this source includes *Safety Health and Loss Prevention in Chemical Processes: Problems for Undergraduate Engineering Curricula—Instructor's Guide* (CCPS, 1990).

20.19.1 Virtual Training

Virtual training covers simulations with applications and case studies. It provides more realistic challenges with hands-on experience and helps learners experience the control room during a real operating environment within a reasonable timescale. Virtual training aims to help understand a wider scope of the chemical process. Some field virtual training, such as unit operations and separations, chemical reactions, process systems engineering, and the biological processes, could be useful.

20.19.2 Emergency Response Simulation

An aid to training personnel in the handling of emergencies is available in the form of the handling emergencies simulation system of the IChemE. The trainee assumes the role of the main incident controller and is responsible for controlling all personnel, and issuing operational commands. One such system is SAFER (Dupont Safer Emergency Systems). The core of the code is a gas dispersion program together with data on the chemicals handled in the plant and their properties. The computer is provided with a network of instruments that measure meteorological conditions such as wind direction and speed. In an emergency, data are furnished to the computer both by these instruments and by operator inputs. SAFER is more than a computer code. It is essentially an emergency response system built around a computer package.

ACRONYMS

AI	Artificial Intelligent
FOPL	First Order Predicate Logic
WFFs	Well-Formed Formulas
LHS	Left-Hand Side
RHS	Right-Hand Side
LS	Logical Sufficiency
LN	Logical Necessity
OOP	Object-Oriented Programming
GPS	General Problem Solver
DBMSs	Database Management Systems
SIS	Safety Instrumented System
PFD	Probability of Failure on Demand
CDU	Crude oil Distillation Unit
FCCU	Fluid Catalytic Cracking Unit
KAF	Knowledge Acquisition Facility
QPT	Qualitative Process Theory
MILP	Mixed Integer Linear Programming
MINLP	Mixed Integer Non-Linear Programming
HAZOP	HAZard and OPerability analysis
HAZID	Hazard Identification
PRA	Probabilistic Risk Assessment
CFD	Computational Fluid Dynamics
OPs	Operating Procedures

REFERENCES

Aldersey, M.L., Lees, F.P., Rushton, A.G., 1991. Knowledge elicitation and representation for diagnostic tanks in the process industries: forms of representation and translation between these forms. Process Saf. Environ. 69B, 187.

Andersson, M., 1989. An Object-Oriented Language for Model Representation. Paper Presented at CACSD 89, Tampa, FL.

Alty, J.L., Mccartney, M.D., 1991. Managing multimedia resources in process control: problems and solutions. In: Kjelldahl, L., Multimedia: Systems, Interactions and Applications. Springer, Berlin, 293.

Alty, J.L., Bergan, M., 1992. Guidelines for multimedia design in a process control application. In: Oka, Y., Koshizuka, S., Atomic Energy Society of Japan, 43IV-1

Andow, P.K., Lees, F.P., Murphy, C.P., 1980. The propagation of faults in process plants: a state of the art review. Chem. Process Hazard. 7, 225.

Barnwell, J., Ertl, B., 1987. Expert systems and the chemical engineer. Chem. Eng. (London). 440, 41.

Beerel, A.C., 1987. Expert Systems: Strategic Implications and Applications. Ellis Horwood, Chichester.

Black, J.M., Ponton, J.W., 1992. A hierarchical method for line-by-line hazard and operability studies. Workshop on Interactions Between Process Design and Process Control, London, p. 227.

Bragatto, P., Monti, M., Giannini, F., Ansaldi, S., 2007. Exploiting process plant digital representation for risk analysis. J. Loss Prev. Process Ind. 20 (1), 69–78.

Bretherick, L., 1990b. Bretherick's Handbook of Reactive Chemical Hazards. Schoneburg (1992).

Brown, J.S., De Kleer, J., 1981. Towards a theory of qualitative reasoning about mechanisms and its role in troubleshooting. In: Rasmussen, J., Rouse, W.B. (Eds.), op. cit., p. 317.

Bunn, A.R., Lees, F.P., 1988. Expert design of plant handling hazardous materials. Design expertise and computer aided design methods with illustrative examples. Chem. Eng. Res. Des. 66, 419.

Catino, C.A., Ungar, L.H., 1995. Model-based approach to automated hazard identification of chemical plants. AIChE J. 41 (1), 97–109.

Catino, C.A., Grantham, S.D., Ungar, L.H., 1991. Automatic generation of qualitative models of chemical process units. Comput. Chem. Eng. 15, 583.

Chung, P.W.H., 1993. Qualitative analysis of process plant behaviour. In: Proceedings of the Sixth International Conference on Industrial and Engineering Applications of Artificial Intelligence and Expert Systems, Gordon & Breach, p. 277.

Chung, P.W.H., Goodwin, R., 1994. Representing design history. In: Gero, J. (Ed.), Artificial Intelligence in Design '94. Kluwer, Amsterdam.

Chung, P.W.H., Stone, D., 1994. Approaches to representing and reasoning with technical regulatory information. Knowledge Eng. Rev. 9 (2), 147.

Coulsey, H., 1986. How expert was my system? Chem. Eng. (London). 425, 55.

De Kleer, J., 1975. Qualitative and Quantitative Knowledge in Classical Mechanics. Report TR-352. AI Laboratory, MIT, Cambridge, MA.

De Kleer, J., Brown, J.S., 1983. Assumptions and ambiguities in mechanistic mental models. In: Gentner, D., Stevens, A.L. (Eds.), op. cit., p. 155.

De Kleer, J., Brown, J.S., 1984. A qualitative physics based on confluences. Artif. Intell. 24, 7.

De Kleer, J., Brown, J.S., 1986. Theories of causal ordering. Artif. Intell. 29, 33.

Diaper, D., 1989. Knowledge Elicitation. Ellis Horwood, Chichester.

Dietterich, T.C., Michalski, R.S., 1981. Inductive learning of structural descriptions: evaluation criteria and comparative review of selected methods. Artif. Intell. 16, 257.

Ellis, C., 1989. Expert Knowledge and Explanation: The Knowledge–Language Interface. Ellis Horwood, Chichester.

Ernst, G.W., Newell, A., 1969. GPS: A Case Study in Generality and Problem-Solving. Academic Press, New York, NY.

Federal Emergency Management Agency, 1989. Handbook of Chemical Hazard Analysis Procedures. Washington, DC.

Ferguson, G., Andow, P.K., 1986. Process plant safety and artificial intelligence. In: World Congress of Chemical Engineering, Tokyo, p. 1092.

Forbus, K.D., 1983. Qualitative reasoning about space and motion. In: Gentner, D., Stevens, A.L. (Eds.), op. cit., p, 53.

Forbus, K.D., 1984. Qualitative process theory. Artif. Intell. 24, 85.

Forbus, K.D., 1990. The qualitative process engine. In: Weld, D.S., de Kleer, J. (Eds.), op. cit.

Frost, R.A., 1986. Introduction to Knowledge Based Systems. Collins, London.

Gensym Corporation, 1992. G2 Reference Manual. Cambridge, MA.

Goring, M.H., Schecker, H.G., 1992. HAZEXPERT—an expert system for supporting hazard identification for process plants. Loss Prev. Saf. Promot. 7 (3), 150.1.

Hayes-Roth, F., Waterman, D.A., Lenat, D.B., 1983. Building Expert Systems. Addison-Wesley, Reading, MA.

Heino, P., Suokas, J., Karvonen, I., 1989a. Computer aided safety analysis of process systems. In: Safety Engineering Anniversary Seminar on Automation Safety in Design in Process and Manufacturing Industry, Espoo, Finland.

Heino, P., Suokas, J., Karvonen, I., 1989b. Expert system for hazop-studies. Loss Prev. Saf. Promot. 6, 112.1.

Henley, E.J., Williams, R.A., 1973. Graph Theory in Modern Engineering. Academic Press, New York, NY.

Henry, M., 1985. Expert systems and all that. Chem. Eng. (London). 420, 32.

Hutton, D., Ponton, J.W., Waters, A., 1990. AI applications in process design, operation and safety. Knowledge Eng. Rev. 5, 69.

Kelly, B.E., Lees, F.P., 1986a. The propagation of faults in process plants: 3, an interactive, computer-based facility. Reliab. Eng. 16, 63.

Kelly, B.E., Lees, F.P., 1986b. The propagation of faults in process plants: 4, fault tree synthesis of a pump system changeover sequence. Reliab. Eng. 16, 87.

Kidd, A.L., 1987. Knowledge Acquisition for Expert Systems. Plenum Press, New York, NY.

Kuipers, B., 1984. Commonsense reasoning about causality: deriving behaviour from structure. Artif. Intell. 24, 169.

Kuipers, B., 1986. Qualitative simulation. Artif. Intell. 29, 289.

Larson, J., Michalski, R.S., 1977. Inductive Inference in the Variable Value Predicate Logic System VL2: Methodology and Computer Implementation. Report 869. Department of Computer Science, University of Illinois, Urban, IL.

Martin, J., Oxman, S., 1988. Building Expert Systems: A Tutorial. Prentice Hall, Englewood Cliffs, NJ.

Newell, A., Simon, H.A., 1963. GPS, a program that simulates human thought. In: Feigenbaum, E.A., Feldman, J. (Eds.), 1963, op. cit., p. 279.

Nilsson, B., 1989. Structured modelling of chemical processes with control systems. American Institute of Chemical Engineers, Annual Meeting, San Francisco (New York), paper 27g.

Parmar, J.C., Lees, F.P., 1987a. The propagation of faults in process plants: hazard identification. Reliab. Eng. 17 (4), 277.

Parmar, J.C., Lees, F.P., 1987b. The propagation of faults in process plants: hazard identification for a water separator system. Reliab. Eng. 17 (4), 303.

Patterson, D.W., 1990. Introduction to Artificial Intelligence and Expert Systems. Prentice Hall, Englewood Cliffs, NJ.

Piela, P.C., Epperly, T.G., Westerberg, K.M., Westerberg, A.W., 1991. ASCEND: an object-oriented computer environment for modeling and analysis: the modeling language. Comput. Chem. Eng. 15, 53.

Quinlan, J.R., 1983a. Fundamentals of the knowledge engineering problem. In: Hayes, J.E., Michie, D. (Eds.), op. cit., p. 33.

Quinlan, J.R., 1983b. Learning efficient classification procedures and their application to chess end games. In: Michalski, R.S., Carbonell, J.G., Mitchell, T.M. (Eds.), op. cit., p. 463.

Raman, R., Grossman, I.E., 1991. Relation between MILP modelling and logical inference for chemical process synthesis. Comput. Chem. Eng. 15, 73.

Rivas, R.J., Rudd, D.F., 1974. Synthesis of failure-safe operations. AIChE J. 20, 320.

Rootsaert, T., Harrington, J., 1992. A knowledge based computer program "COMHAZOP" as aid for hazard and operability studies in process plants. Loss Prev. Saf. Promot. 7, 3156-1.

Rudd, D.F., Watson, C.C., 1968. Strategy of Process Design. Wiley, New York, NY.

Sachs, P.A., Paterson, A.M., Turner, M.H.M., 1985. ESCORT—An Expert System for Complex Operations. PA Computers and Telecommunications, London (Paper).

Sangiovanni, J., Romans, H.C., 1987. Expert systems in industry: a survey. Chem. Eng. Prog. 83 (9), 52.

Sell, P.S., 1985. Expert Systems—A Practical Introduction. Macmillan, London.

Srinivasan, R., Venkatasubramanian, V., 1998. Multi-perspective models for process hazards analysis of large scale chemical processes. Comput. Chem. Eng. 22 (1), S961–S964.

Stephanopoulos, G., 1990. Artificial intelligence in process engineering—current state and future trends. Comput. Chem. Eng. 14, 1259.

Stephanopoulos, G., Townsend, D.W., 1986. Synthesis in process development. Chem. Eng. Res. Des. 64, 160.

Stephanopoulos, G., Hanning, G., Leone, H., 1990a. MODEL.LA. A modeling language for process engineering. I, the formal framework. Comput. Chem. Eng. 14, 813.

Stephanopoulos, G., Hanning, G., Leone, H., 1990b. MODEL.LA. A modeling language for process engineering. II, multifaceted modeling of processing systems. Comput. Chem. Eng. 14, 847.

Suokas, J., Heino, P., Karvonen, I., 1990. Expert systems in safety management. J. Occup. Accid. 12 (1–3), 63.

Venkatasubramanian, V., Vaidhyanathan, R., 1994. A knowledge-based framework for automating HAZOP analysis. AIChE J. 40, 496.

Venkatasubramanian, V., Rengaswamy, R., Kavuri, S.N., 2003a. A review of process fault detection and diagnosis: Part II. Qual. Models Search Strat. 27, 313–326.

Venkatasubramanian, V., Rengaswamy, R., Kavuri, S.N., Yin, K., 2003b. A review of process fault detection and diagnosis: Part III: process history based methods. Comput. Chem. Eng. 27 (3), 327–346.

Venkatasubramanian, V., Rengaswamy, R., Yin, K., Kavuri, S.N., 2003c. A review of process fault detection and diagnosis: Part I: quantitative model-based methods. Comput. Chem. Eng. 27 (3), 293–311.

Waters, A., Ponton, J.W., 1992. Managing constraints in design: using an AI toolkit as a DBMS. Comput. Chem. Eng. 16, 987.

Waters, A., Chung, P.W.H.C., Ponton, J., 1989. Representing safety constraints upon chemical plant designs. In: Vadera, S. (Ed.), op. cit., p. 199.

Weatherill, T., Cameron, I., 1988. Preliminary hazop studies using expert systems. In: Third International Symposium on Process Systems Engineering, Sydney.

Weatherill, T., Cameron, I.T., 1989. A prototype expert system for hazard and operability studies. Comput. Chem. Eng. 13, 1229.

Westerberg, A.W., 1989. Synthesis in engineering design. Comput. Chem. Eng. 13, 365.

Zhao, C., Bhushan, M., Venkatasubramanian, V., 2005. PHASuite: an automated HAZOP analysis tool for chemical processes: Part I. Knowledge Eng. Framework. 83, 509–532.

Zhao, J., Viswanathan, S., Venkatasubramanian, V., Sauro, P., 2000. Industrial applications of intelligent systems for operating procedure synthesis and hazards analysis for batch process plants. In: Computer Aided Chemical Engineering. Elsevier.

Zhao, J., Cui, L., Zhao, L., Qiu, T., Chen, B., 2009. Learning HAZOP expert system by case based reasoning and ontology. Comput. Chem. Eng. 33 (1), 371–378.

Chapter 21

Inherently Safer Design

Chapter Outline

21.1. INTRODUCTION

Inherently safer design is a philosophy which focuses on elimination of hazards or reduction of the magnitude of hazards rather than the control of hazards. Kletz (1978) suggested that in many cases, a simpler, cheaper, and safer plant could be designed by focusing on the basic technology, eliminating or significantly reducing hazards and therefore the need to manage them. A process or plant is best described as 'inherently safer' with respect to a specific hazard or set of hazards, and with respect to other alternative designs.

21.2. DEFINITIONS

An inherently safer process refers the process where hazards are eliminated or significantly reduced, rather than controlled with safety equipment and procedures.

According to the CCPS (1992)—a hazard is an inherent physical or chemical characteristic that has the potential for causing harm to people, the environment, or property. A hazard is an inherent characteristic of a material or of its condition of use.

Inherently safer design is a philosophy for risk management for a chemical process. While inherently safer design is generally considered to be applicable to the consequence portion of the risk equation, it also applies to the likelihood of occurrence. An inherently safer design can either reduce the magnitude of a potential incident arising from a particular hazard, or make the occurrence of the accident highly unlikely, or perhaps impossible.

21.3. HISTORY OF INHERENTLY SAFER DESIGN

Over the centuries, engineers have invented and developed inherently safer designs in many technologies. However, the common philosophy of hazard elimination or reduction has not been recognized or identified as a generally applicable design strategy. Trevor Kletz of ICI introduced the concept of 'inherently safer design' in 1977 (Kletz, 1978). Over the years since 1977, Kletz developed a set of principles to assist engineers in identifying inherently safer design options for chemical processes (most recently, Kletz, 1998). Others picked up the concept and have further developed it. The philosophy has become an integral part of process design and process safety for many companies in the chemical industry.

21.4. STRATEGIES FOR PROCESS RISK MANAGEMENT

Process risk management strategies can be considered to fall into four categories (CCPS, 1996): inherent, passive, active, and procedural.

A complete risk management program for a process will include elements from all of these strategies. This is particularly true when one considers all of the multiple hazards in a process. These chemical process safety strategies should not be considered discrete categories with clear boundaries.

Consider a process which requires pneumatic conveying of a combustible powder from a storage silo to the processing equipment. The potential hazard is a dust explosion that might result in maximum pressure of 10 bar. The risk arising from this hazard might be managed by any of the following alternatives.

- Inherent Risk Management
  - Identify a non-combustible solid which can replace the combustible dust in the process.
  - Increase the particle size of the combustible solid so that a dust explosion cannot occur.
- Passive Risk Management
  - Build a conveying system with a pressure rating of 15 bar.
- Active Risk Management
  - Dust explosion venting panels to relieve the dust explosion pressure to a safe place.
  - Explosion suppression systems which detect the incipient explosion and inject a fire suppressant to reduce the explosion pressure to less than the failure pressure of the conveying system.
- Procedural Risk Management
  - Procedures to keep metal objects out of the conveying system. These objects could cause sparks in the conveying system which could ignite the powder.
  - Procedures to ensure that equipment is correctly assembled and that all metal parts of the conveying system are properly electrically bonded and grounded.

Some of the risk management strategies suggested to reduce the risk of dust explosion may introduce other hazards, increase the magnitude of other existing hazards, or inhibit the ability of the process to produce the required product. A complete risk management program for a complex chemical process will require elements from all of the risk management strategies to properly manage all hazards.

The application of various process risk management strategies to hazard control in the chemical process industries has been described as providing multiple layers of protection to the process (CCPS, 1993, 1996). A process that is inherently safer will require fewer, and less robust, layers of protection. The layers of protection concept have been developed as a methodology for quantitative risk analysis of a process, called Layers of Protection Analysis (LOPA) (CCPS, 2001).

21.5. INHERENTLY SAFER DESIGN STRATEGIES

Kletz (1998) has categorized inherently safer design strategies into four major categories: intensification, substitution, attenuation, and limitation of effects. CCPS (2009) suggests the same basic inherently safer design ideas as Kletz, but describes the four categories as—minimize, substitute, moderate, and simplify.

Minimize (Intensify): The quantity of hazardous material and energy in a process should be minimized. This applies to all unit operations in a plant as well as to raw material, intermediate, and hazardous product storage and to piping which connects equipment. Some areas where significant benefits have been realized include reaction, distillation, extraction, heat exchange, raw material storage, intermediate storage, and piping.

Substitute: In the chemical industry, the two major areas where substitution enhances inherently safer design are substitution of a less hazardous chemical synthesis route for a desired product and substitution of a less hazardous material for a specific application.

Moderate (Attenuate): Inherent safety can be enhanced by designing the plant to moderate the impact of an incident arising from the hazard. It includes moderating hazards of materials and moderating hazards of processes. Hazards of materials can be moderated by dilution, refrigeration, and changing physical properties. Substitution of alternate chemistry or development of improved catalysts can often allow a process to operate at lower temperature and pressure. This not only improves inherent safety, but is also almost certain to make the construction of the plant cheaper. There are many approaches to limiting the effects (moderating the effects) of an incident arising from a hazard of a process. They do not reduce the magnitude of the hazard, or eliminate the hazard, but they are very robust and reliable methods for minimizing the potential impact of an incident.

Simplify: Simplification is critical to a 'user friendly' plant. While most chemical plants are highly complex of necessity because of the complexity of the technology, designers should always be striving to eliminate all unnecessary complexity. In some cases, simplification will eliminate hazards by making it impossible to conduct operations which result in those hazards. In other cases, simplification will make it more difficult to make errors, perhaps not truly an example of inherently safer design, but still highly desirable.

21.6. INHERENTLY SAFER DESIGN CONFLICTS

It is almost always impossible to simultaneously maximize all desired characteristics of any engineered system. The central problem of all engineering design is to find the optimum combination of characteristics that best meets the overall objectives. The inherent safety characteristics of a process become one of the many characteristics which the design engineer must evaluate as he determines his optimum design for a particular facility.

Many decision-making tools have been developed to aid decision-makers in understanding the many conflicting characteristics of a multi-dimensional decision, allowing them to make logical and consistent decisions. Decision-making tools can also be applied to resolving conflicts between the inherent safety characteristics of a process with respect to different kinds of hazard, as well as conflicts between inherent safety and other important process characteristics.

21.7. MEASURING INHERENT SAFETY CHARACTERISTICS OF A PROCESS

In a recent global survey on the use and interest of people in inherently safer design, Gupta and Edwards (2002) reported at length that the survey responders were generally familiar with inherently safer design and desired to use it to their advantage. They desired a simple way to determine or measure the inherent safety of a process at the research and development stage before too much time and resources have been invested in the process development. Therefore, it is essential that the inherent safety index (ISI) be very simple.

There are no hard and fast rules or methods as to how to make a process inherently safer. One could use any one of several approaches possible. In inherently safer design measurement, we have come to a stage where several approaches are available. As experience is gained in using the existing indices, a universally accepted index will develop.

Dow Fire and Explosion Index (Dow Chemical Company, 1994) and Mond Index (ICI, 1993) are used for determining the hazard level of a process plant. However, to use these, the process design has to be fully in place since a lot of information required pertains to that. Hence, these are unsuitable for use in conceptual design and preliminary process development stages.

There are a variety of inherent safety indices. Edwards and Lawrence index (Edwards and Lawrence, 1993) was the first ever index that was developed and has been modified by later workers. Heikkila, Hurme, and Järveläinen argued that the safety is affected both by the properties of the chemicals and by the equipment used. They have, therefore, included the type of equipment and process structure and developed Heikkila and Hurme index (Heikkila et al., 1996). Palaniappan et al. have come up with an i-Safe index (Palaniappan, 2001). They expanded the Heikkilä and Hurme's chemical and process safety indices to include five other supplementary indices, viz, Hazardous Chemical Index (HCI), Hazardous Reaction Index (HRI), Total Chemical Index (TCI), Worst Chemical Index (WCI), and Worst Reaction Index (WRI). Khan and Amyotte Index (Khan and Amyotte, 2003) has

proposed an Integrated Inherent Safety Index (I2SI). It comprises of two main indices: a hazard index (HI) and an inherent safety potential index (ISPI). In all the above indices, except Khan and Amyotte's, there are sudden jumps in the score values at the extreme ends of each subdivision. Also, the scoring tables proposed by different authors for the same parameter do not always match. This aspect was addressed by Gentile et al. (2001). Experts from several corporations in the United Kingdom and Europe worked on an EU sponsored project for an index for inherent safety, health, and environment (ISHE). They called it an INSET toolkit (INSIDE Project Team, 1997). Gupta and Edwards (2003) established a method to express three concerns about the additive indices discussed above:

- Addition of different types of hazards or parameters is not justifiable.
- Arbitrary assignment of scores to different parameters without establishing the equality of hazard for the same numerical value.
- The total score will get biased by the number of steps or by one major score.

They proposed that the parameters of interest should be plotted individually for each step in a process route without carrying out any mathematical operation and then be compared with each other.

The indices developed thus far are expertise intensive. They need to be simplified so that a scientist or an engineer in the process industry can understand and use them easily without having to hunt for a lot of data and information. Most of the work has been done in academic institutions (except INSET Toolkit). Several practical examples from industry need to be evaluated using the proposed indices and modifications made to them to suit the needs of the industry. An alliance between the researchers, industry personnel, and the regulators would prove beneficial to all.

Inherent safety is an exciting field that has caught the attention of researchers, plant designers, management, and regulators worldwide. This method should work at the research stage itself and not require too much data. Researchers in several countries have produced indices toward this end. These need to be simplified. The intensive work put in by researchers needs to be consolidated and unified. The ultimate proof will be in the thorough testing of the indices by the industry and sharing of their experiences. Some industries have developed their own proprietary methods to test for inherent safety but these are not available in open literature.

Once the company personnel starting from research chemists, process engineers, and the rest start using inherently safer design, they would actually see the advantages and want to use it more and more. With process plants thus becoming significantly safer, the regulators are likely to gradually relax on process safety protocols. Once inherently safer design is successfully applied to process industries, it can be adapted by other accident-prone industries such as mining, construction, and transportation.

21.8. INHERENTLY SAFER DESIGN AND THE PROCESS LIFE CYCLE

Inherently safer design applies to a process at all phases in its life cycle. However, the designer will be applying inherently safer design principles in different ways at different stages of the life cycle. Early on, he will be considering the basic technology and synthesis routes; and in the later stages of the life cycle, the designers and operators will be looking at more specific opportunities for applying inherently safer design concepts to specific pieces of equipment. The greatest opportunities for making major improvements in inherent safety occur early in process development, when the designer may have many choices of basic technology and chemistry available, and may be free to choose less hazardous alternatives. In the conceptual process research and development stage, the selection of chemicals and materials plays a big role. It is crucial at this stage to assess many types of hazards. However, it is never too late to apply inherently safer design concepts and major improvements to the inherent safety characteristics of plants which have been in operation for many years. At the preliminary process design stage, the designers are encouraged to incorporate inherently safer design principles. In order to do so, one can evaluate the inherent safety indices as described previously, and integrate safety concepts into the process design by using optimization-based techniques.

21.9. IMPLEMENTING INHERENTLY SAFER DESIGN

Inherently safer design applies at all levels of chemical plant conception, research, design, and operation. It is more of a design philosophy or way of thinking than a specific set of tools, review meetings, or other specific activities. The initial focus of the designer should be on the elimination and minimization of hazards, rather than the control of hazards.

Some companies incorporate inherently safer design into the process hazard analysis (PHA) activities which are already included in their process safety programs. In the course of a safety review, HAZOP, or other process safety review activity, the review team and leader are charged with considering inherently safer design options.

Other companies have used separate inherent safety reviews.

There have been some regulatory proposals to require that the chemical industry consider inherently safer design options. For example, in the early 1990s, the United States Environmental Protection Agency considered requiring a 'Technical Options Analysis (TOA)' for plants covered by its Risk Management Program (RMP), as required by the US Clean Air Act Amendments of 1990.

However, the best way to invent and build inherently safer processes is to make it a part of the thinking of all engineers and chemists. If they recognize the importance and benefits of hazard elimination, their creativity in inventing new ways to eliminate or reduce hazards will go further toward enhancing inherent safety.

21.10. INHERENT SAFETY AND CHEMICAL PLANT SECURITY

With the advent of larger and more complex chemical plants, it is also necessary to thoroughly assess chemical plant security. Deliberate acts in such facilities can result in major impacts to the health and safety of workers, the public, and damage to the environment.

The security risk in a chemical plant consists of the components including Consequence, Threat, Vulnerability, and Attractiveness (CCPS, 2009). Alternatively, the team can use judgment to assess the ability of the adversary to achieve success by analyzing the threat, vulnerabilities, and countermeasures.

Some of the limitations to implement inherently safer measures in chemical plant security are as follows:

- Balancing competing security or other concerns;
- Creating new security concerns;
- Shifting security risk from one concern or location to another (CCPS, 2009).

The Chemical Facility Anti-terrorism Standard CFATS (2007) has been released as an interim rule by the US Department of Homeland Security. The rule imposes comprehensive federal security regulations for high-risk facilities. Prior to implementing any regulation and standards, it is important for science to precede them.

REFERENCES

CCPS, 1992. Guidelines for Hazard Evaluation Procedures, Second Edition with Worked Examples. New York, NY.

CCPS, 1993. Guidelines for Engineering Design for Process Safety. American Institute of Chemical Engineers, New York, NY.

CCPS, 1996. In: Crowl, D.A. (Ed.), Inherently Safer Chemical Processes: A Life Cycle Approach. American Institute of Chemical Engineers, New York, NY.

CCPS, 2001. Layer of Protection Analysis. American Institute of Chemical Engineers, New York, NY.

CCPS, 2009. Inherently Safer Process: A Life Cycle Approach. Wiley, Hoboken, NJ.

Dow Chemical Company, 1994. Dow's Fire and Explosion Index Hazard Classification Guide, Seventh Edition. American Institute of Chemical Engineers, New York, NY.

Edwards, D.W., Lawrence, D., 1993. Assessing the inherent safety of chemical process routes: is there a relation between plant costs and inherent safety? Process Saf. Environ. 71B, 252.

Gentile, M., Rogers, W.J., Mannan, M.S., 2001. Inherent safety index for transportation of chemicals. Fourth Annual Symposium: Beyond Regulatory Compliance, Making Safety Second Nature, 509. Mary Kay O'Connor Process Safety Center, College Station, TX.

Gupta, J.P., Edwards, D.W., 2002. Some thoughts on measuring inherent safety. Fifth Annual Symposium: Beyond Regulatory Compliance, Making Safety Second Nature. 29−30 October, 2002, College Station, TX, Track I: Inherent Safety Session (College Station, TX: Mary Kay O'Connor Process Safety Center).

Gupta, J.P., Edwards, D.W., 2003. A simple graphical method for measuring inherent safety. J. Hazard. Mater. 104 (1−3), 15−30.

Heikkila, A.-M., Hurme, A.M., Järveläinen, M., 1996. Safety considerations in process synthesis. Comp. Chem. Eng. 20 (Suppl. A), S115−S120.

ICI, 1993. Mond Index. second ed. ICI, Northwich.

Inside Project Team, 1997. The INSET Toolkit. Available from AEA Technology, Chesire, UK.

Khan, F.I., Amyotte, P.R., 2003. Integrated inherent safety index (I2SI): a tool for inherent safety evaluation. Proceedings of 37th Annual Loss Prevention Symposium. New Orleans, LA.

Kletz, T.A., 1978. Safety—the springs of action and innovation. Occup. Saf. Health. 8 (3), 10.

Kletz, T.A., 1998. Process Plants: A Handbook for Inherently Safer Design. Taylor and Francis, Philadelphia, PA.

Palaniappan, C., 2001. Expert System for Design of Inherently Safer Chemical Processes. National University of Singapore, Singapore.

Chapter 22

Reactive Chemicals

Chapter Outline

22.1. BACKGROUND

22.1.1 Motivation

In recent decades, greater recognition and resources have been directed toward preventing and mitigating incidents arising from uncontrolled reactivity. A sampling of incidents that have substantially heightened concerns regarding reactive hazards in the general public, in governmental agencies, and in industry includes:

1. The 1984 Union Carbide incident in Bhopal, India, in which methyl isocyanate was contacted with water, generating highly toxic cyanide gas and leading to thousands of fatalities.
2. The 1994 Napp Technology incident in Lodi, New Jersey, in which an uncontrolled reaction involving gold ore processing led to the deaths of five fire fighters.

3. The 2001 TotalFinaElf incident in Toulouse, France, in which ammonium nitrate being processed for nitrogen fertilizers exploded, leading to 30 fatalities.

22.1.2 Definition of Reactive Hazard

The US Chemical Safety Board (CSB) (Chemical Safety Board, 2002) has defined a reactive hazard as 'A *sudden* event involving an *uncontrolled* chemical reaction with significant increases in temperature, pressure, *or* gas evolution that has the potential to, or has caused *serious harm* to people, property or the environment.'

A common pathway for an uncontrolled, or runaway reaction involves overpressurization of equipment, either through buildup of pressure or through weakening of the equipment structural integrity, leading to possible generation of a blast wave or generation of a flammable vapor cloud that can explode upon finding an ignition source.

22.2. STRATEGIES FOR IDENTIFYING AND CHARACTERIZING REACTIVE HAZARDS

22.2.1 Overall Evaluation Strategy

The stages in the assessment and control of hazards of chemical reactors are shown in Figure 22.1. The strategy for reaction hazard evaluation described in the IChemE *Guide* involves the following: desk screening tests, explosibility tests, preliminary screening tests, characterization of normal reaction, and characterization of runaway reaction. These are now discussed under the categories of: desk screening, preliminary screening tests, and detailed tests.

22.2.2 Categorizing Reactivity Types

22.2.2.1 Stability vs Reactivity with Other Species

Stability issues are related to degradation or decomposition of a substance when subjected to elevated

1. Initial chemistry

　a. Characterization of
　　materials/process
　b. Suitability of production

2. Pilot plant

　a. Chemical reaction hazards
　b. Influence of plant on
　　hazard
　c. Definition of safe
　　procedures

3. Full-scale production

　a. Re-evaluation of chemical
　　reaction hazards
　b. Effect of expected
　　variations in process
　　conditions
　c. Hazards from plant
　　operations
　d. Definition of safe
　　procedures
　e. Interaction of technical
　　safety with engineering,
　　production, economic and
　　commercial aspects of
　　process

FIGURE 22.1 Stages in control of hazards of chemical reactions (N. Gibson et al., 1987). *Source: Courtesy of the Institution of Chemical Engineers.*

temperature, mechanical impact, or other disturbances. The term *reactivity* refers to the ability of the substance to react when placed in contact with certain other species.

22.2.2.2 Intended vs Unintended

Intended reactions are clearly those reactions targeted in a process. Unintended reactions are the side reactions that are typically avoided. Usually, there are no intended reactions in blending, distribution, transportation, and storage operations. An intended reaction occurring outside of its associated equipment can in that context be considered as unintended.

22.2.2.3 Vapor vs Gassy vs Hybrid

Generally, a 'vapor' refers to a non-condensed species that is in thermodynamic equilibrium with the corresponding species in a condensed phase. A 'gas' describes a non-condensable species that is not in thermodynamic equilibrium. The term 'hybrid' is applied to systems that exhibit both vapor and gassy features.

22.2.3 Characterizing Reactions

Thermodynamic schemes such as CHETAH (CHEmical Thermodynamics And energy Hazard evaluation) may be used to evaluate the thermodynamic characteristics of the reaction.

To characterize a chemical reaction adequately, however, it is desirable not only to obtain a qualitative understanding of the behavior of the reaction, but also to obtain quantitative data on various parameters needed for the specification of the process. The following approaches can be applied as well to intended chemistry as to unintended chemistry.

22.2.3.1 Desk Screening

Initial desk screening will provide information on the physical and chemical properties and the reactivity of the chemical. Where specific data are lacking, features such as group structure, thermodynamic data, oxygen balance, and reactivity of analogous substances provide pointers.

The starting point should always be a thorough literature search to determine the data already available on the properties, chemistry, and thermokinetics of the chemical. Material Safety Data Sheets (MSDSs) should also be an initial source for qualitative and limited quantitative reactivity information.

22.2.3.2 NFPA Classification

The NFPA 704 system (*Standard System for the Identification of the Hazards of Materials for Emergency*

TABLE 22.1 Degrees of Reactivity (Instability) Hazard As Defined in NFPA 49 (NFPA, 1994), currently NFPA 704 (NFPA, 2012).

Degree	Description
4	Materials that in themselves are readily capable of detonation or explosive decomposition or explosive reaction at normal temperatures and pressures. This degree usually includes materials that are sensitive to localized thermal or mechanical shock at normal temperatures and pressures.
3	Materials that in themselves are capable of detonation or explosive decomposition or explosive reaction, but that require a strong initiating source or that must be heated under confinement before initiation. This degree usually includes: materials that are sensitive to thermal or mechanical shock at elevated temperatures and pressures; materials that react explosively with water without requiring heat or confinement.
2	Materials that readily undergo violent chemical change at elevated temperatures and pressures. This degree usually includes: materials that exhibit an exotherm at temperatures less than or equal to 150°C when tested by differential scanning calorimetry; materials that may react violently with water or form potentially explosive mixtures with water.
1	Materials that in themselves are normally stable, but that can become unstable at elevated temperatures and pressure. This degree usually includes: materials that change or decompose on exposure to air, light, or moisture; materials that exhibit an exotherm at temperatures greater than 150°C but less than or equal to 300°C, when tested by differential scanning calorimetry.
0	Materials that in themselves are normally stable, even under fire conditions. This degree usually includes: materials that do not react with water; materials that exhibit an exotherm at temperatures greater than 300°C but <500°C, when tested by differential scanning calorimetry; materials that do not exhibit an exotherm at temperatures <500°C when tested by differential scanning calorimetry.

Source: Reproduced with permission from NFPA 49 Hazardous Chemicals Data, Copyright, 1994, National Fire Protection Association, Batterymarch Park, Quincy, MA.

Response, 2012) uses a diamond-shaped diagram of symbols and numbers to indicate the specific hazards and degree of hazards associated with a particular material or chemical. It addresses the health, flammability, reactivity/instability, and other related hazards that may be presented during a fire, spill, or similar emergency. Chemical Reactivity/Instability Ratings given in the NFPA 704 are given in Table 22.1.

22.2.3.3 Chemical Sensitivity to Decomposition (Reactive Conditions)

Oxygen Balance: Rapid energy release occurs if the constitution of a substance is such that its carbon and hydrogen are able to react with its own oxygen without needing to obtain oxygen from the surrounding air. If there is just enough oxygen to give a stoichiometric reaction of all the carbon to carbon dioxide and hydrogen to water, there is said to be a zero oxygen balance. The more reactive substances, such as explosives, typically contain enough oxygen to give such decomposition.

For an organic compound of formula $C_xH_yO_z$, the oxygen balance has been defined by Lothrop and Handrick (1949) as:

$$OB = -1600(2x + y/2 - z)M \qquad (22.1)$$

where M is the molecular weight. Thus, an unstable, or explosive material having perfect balance to yield carbon dioxide and water has a zero OB value—one lacking sufficient oxygen has a negative OB value, and one containing excess oxygen has a positive OB value.

Chemical Incompatibility with Other Materials: The US National Oceanic and Atmospheric Administration (NOAA) has posted on its website a program called the Chemical Reactivity Worksheet, which provides guidelines to facilitate chemical incompatibility assessment.

Related Previous Incidents: Knowledge and analysis of previous incidents is extremely valuable to help identify reactivity hazards including those caused by incompatibilities of different materials within the process.

22.2.3.4 Determination of Energy Hazard Potentials

Computer programs have been developed to calculate certain parameters that indicate energy hazard potentials. The ASTM interactive PC program, named CHETAH, calculates four quantities (Treweek et al., 1973): heat of decomposition ΔH_d, difference between heat of combustion and heat of decomposition $\Delta H_0 - \Delta H_d$, oxygen balance, and y criterion.

The y criterion is given by:

$$y = \frac{10(\Delta H_d)^2 MW}{N} \qquad (22.2)$$

MW = molecular weight; N = number of atoms in the molecule.

The energy hazard potentials are related to criteria (1−4) as given in Table 22.2 (Figure 22.2). These criteria correlate reasonably well with data on shock sensitivity.

TABLE 22.2 Energy Hazard Potential Criteria in CHETAH

Criterion	Energy Hazard Potential		
	Low	Medium	High
1	$\Delta H_d > -0.3$	$-0.7 < \Delta H_d < -0.3$	$\Delta H_d < -0.7$
2	See Figure 22.2		
3	OB > 240	120 < OB < 240	−80 < OB < 120
	OB < −160	−160 > OB < −80	
4	$y < 30$	$30 < y < 110$	$y > 110$

Source: After Treweek et al. (1973).

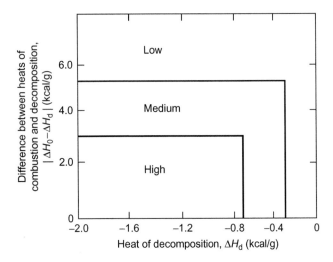

FIGURE 22.2 Energy hazard potential as a function of the heats of combustion and decomposition. *Source: After Traweek et al. (1973); Courtesy of the American Institute of Chemical Engineers.*

22.2.4 Preliminary Tests

The starting point for preliminary tests is usually to test for the occurrence of an exothermic reaction when the substance is heated.

22.2.4.1 Melting Point Tube Test

A simple test is the capillary melting point tube test in which the sample is heated up at a constant rate of about 10°C/min in a standard melting point apparatus. A result such as that shown in Figure 22.3(a), in which the sample temperature has exceeded that of the heating medium, indicates that there is an exotherm.

22.2.4.2 Dewar Test

A number of tests involve the use of a Dewar flask. These are of three types: (1) ramped temperature tests, (2) differential temperature tests, and (3) adiabatic tests.

In the first type, the sample is held in a tube in a heat transfer medium in a Dewar flask and the temperature of the medium is either ramped at a constant rate (ramped heating, Figure 22.3(a)) or increased in a series of steps (stepped heating, Figure 22.3(b)). In the latter, the temperature is held constant for a period at each step. If, at a given medium temperature, the sample temperature rises above that of the medium, an exotherm is present. In the second type of Dewar flask test, the sample and a reference material are put in two tubes in a Dewar flask filled with heat transfer medium, the medium is heated up at a constant rate and the temperature difference between the sample and the reference material is measured.

This temperature difference, ΔT, is plotted vs the temperature, T, of the heat transfer medium. A typical result for a material is illustrated in Figure 22.3(c).

The third type of Dewar flask test is the 'adiabatic test'. The sample is held in a pressure Dewar flask in an oven. The temperature of the oven is controlled to follow closely the sample temperature allowing measurement of any exotherm that occurs.

22.2.4.3 Delayed Onset Detection Test

The sample, held in a lagged Carius tube placed in an isothermal oven, is maintained isothermally at about 10°C above the oven temperature using a small electric heater. An exotherm is detected by a fall in the electrical power requirement.

22.2.5 Experimental Test Methods and Equipment for Detailed Evaluations

There are a number of more elaborate techniques, such as calorimetry, and special instrumentation has been developed that may be employed to detect and quantify exotherms. The principal techniques are differential thermal analysis (DTA), differential scanning calorimetry (DSC), Automatic Pressure Tracking Adiabatic Calorimeter (APTAC), and accelerating rate calorimetry (ARC). Among these calorimeters, some of them are used for screening reactive chemicals, and some are for detailed evaluation. Table 22.3 gives the differences among some of these calorimeters.

22.2.6 Screening Calorimeters

22.2.6.1 DTA

In DTA, the sample is held in a tube in a vessel surrounded by a heat transfer medium and is heated at a constant rate. A plot is made of the temperature difference, ΔT, between the sample and the heat transfer medium vs the temperature, T, of the latter. DTA may be carried out

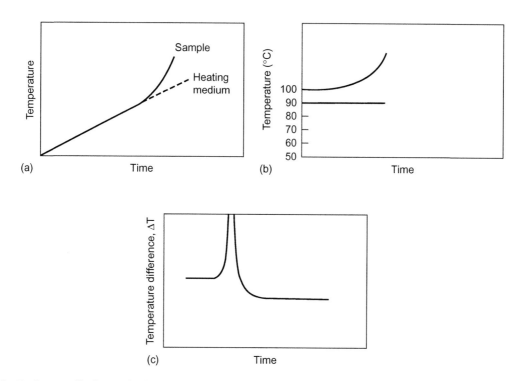

FIGURE 22.3   Exotherm profiles in tests for thermal stability of individual substances: (a) sample and heating medium temperatures in melting point tube test or Dewar flask ramped temperature test; (b) sample temperature in Dewar flask stepped temperature test; (c) temperature difference between sample and reference material in Dewar flask dynamic heating test. *Source: Courtesy of the American Chemical Society.*

TABLE 22.3 Comparison of Different Calorimeters

Calorimeter	Size	Temperature Range	Pressure Range	Operation Model	Sensitivity
DSC	0.1 g	300°C	N/A	Temperature-scanning	1 μW
RSST	10 mL	400°C	500 psig	Temperature-scanning	0.1°C/min
ARC	5 mL	500°C	200 bar	Isothermal, Heat-wait-search, Heat ramp	0.04°C/min
APTAC	500 mL	500°C	2000 psig	Heat-soak-search, Isothermal, Heat-wait-search, Heat ramp	0.05°C/min
VSP2	120 mL	400°C	130 bar	Adiabatic	0.05°C/min
RC-1	50 mL–5 L	−70 to 300°C	300 bar	Isothermal, Isoperibolic, and Adiabatic	0.02°C/min

Source: N. Gibson et al. (1987).

with open or closed tubes, the latter being more appropriate where gas evolution occurs.

In conventional DTA, the temperature is programmed to rise at a constant rate. Where the thermal stability of a substance may be a function of time, DTA can be performed isothermally.

22.2.6.2 DSC

In DSC, the sample and a reference material are held in pans in a vessel and are heated at a constant rate of typically 5–10°C/min. The heating is carried out by a control system that maintains the sample and the reference material at the same temperature. The variation in the heat that

must be supplied to the sample to keep it at the same temperature as the reference material gives a quantitative measure of any exotherm in the sample. A plot is made of the rate of change of heat input to the sample with time, dQ/dt, vs the temperature, T, of the reference sample. Again, open or closed pan methods may be used. Figure 22.4 (Duval, 1985) shows a typical DSC plot.

22.2.6.3 ARSST

The Advanced Reactive System Screening Tool (ARSST) is designed to provide the rapid screening of chemical reactivity and characterize the reactive nature of chemicals. The substance is placed in an open 10 mL glass test cell and then inserted in a 0.45 L containment cell and heated at a constant temperature rate by external heaters. The containment cell with the open test cell is pressurized with nitrogen to help retain sample liquid in a condensed state. The ARSST can detect exotherms to as low as 0.1°C/min using a heat-wait-search mode of operation.

22.2.7 Detailed Evaluation Calorimeters

22.2.7.1 Heat Flow Calorimeters

In heat flow calorimetry, the sample is held in the calorimeter (a jacketed mini-reactor) and a heat transfer medium is circulated around the jacket to eliminate the heat generated. The RC1 of Mettler-Toledo is an example of a heat flow calorimeter.

One important test carried out using a reaction calorimeter is the determination of the adiabatic temperature

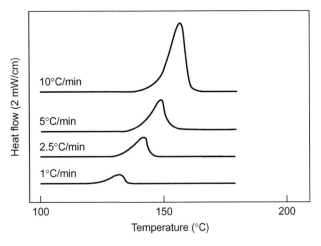

FIGURE 22.4 Exotherm profile in test for thermal stability of individual substances using differential scanning calorimetry (DSC) (Duval, 1985). *Source: Courtesy of the American Chemical Society.*

rise. This single test provides a large amount of information on the thermodynamic and kinetic characteristics of the reaction such as the heat capacity of the reaction mass, heat of reaction, and the rate of heat production.

22.2.7.2 Adiabatic Calorimeters

Dewar Calorimetry: In the adiabatic heating test, the oven temperature is held constant and any exotherm is measured. The test period is at least 8 h and preferably 24 h. This test is also known as the 'hot storage' or 'heat accumulation' test. The adiabatic storage test may be used to determine a suitable temperature limit. The most useful information yielded is the adiabatic induction time $T_a d$, which usually correlates with temperature according to the relation log $T_a d$ vs $1/T$. This correlation can be used to determine the temperature for an adiabatic induction time of 24 h, which is known as AZT24.

ARC: In the Accelerating Rate Calorimeter (ARC), the sample is held in a bomb calorimeter under adiabatic conditions. Adiabaticity is achieved by heating the surroundings of the sample cell so that the temperature outside the cell matches the temperature in the cell. ARC testing may be done in different ways, but one of the most common is the 'heat, wait, and search' method. The temperature is increased in steps. Then, the sample is held for a period of time at that temperature. The rate of any temperature increase is observed to see if it exceeds a set value, typically 0.02°C/min. If it does not, the temperature is raised by the incremental amount and the procedure is repeated. If the temperature rises at a high rate, there is an exotherm. A typical result, given by N. Gibson et al. (1987), for a material giving an exotherm at about 80°C is shown in Figure 22.5.

The ARC is also equipped to accommodate and measure large increases in sample pressure, which requires a relatively thick-walled cell consistent with the amount of

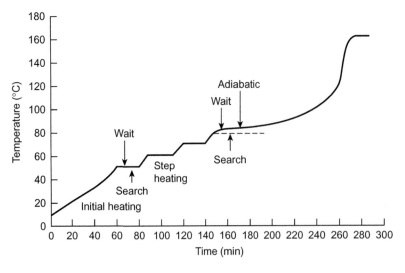

FIGURE 22.5 Exotherm profile in test for thermal stability of individual substances using accelerating rate calorimetry (ARC) (N. Gibson et al., 1987). *Source: Courtesy of the Institution of Chemical Engineers.*

sample contained. This requirement leads into the concept of the thermal inertia factor, ϕ, reflecting the amount of heat absorbed by the sample container relative to that retained in the sample itself. The thermal inertia is defined as

$$\phi = 1 + \frac{m_c Cp_c}{m_s Cp_s}$$

where m is the mass, Cp is the heat capacity, subscript c is the cell + stir bar, and s is the sample. The ϕ factor of industrial reactors approaches a value of unity, whereas a value of 1.5−5 is typical of ARC tests.

VSP: The Vent Sizing Package or VSP (originally the DIERS Bench-Scale Apparatus) was the first of the adiabatic calorimeters to incorporate pressure-balancing in its design. This feature allows use of a thin-walled sample while still enabling the sample to reach elevated pressures during a test. The VSP also operates with a 120 cm^3 sample cell. The use of a larger sample with a thin-walled cell yields thermal inertia factors typically 1.05−1.20.

PHI-TEC: Another adiabatic calorimeter equipped with pressure balancing, and thus operating with thin-walled cells and larger sample sizes. It has many of the newer features described above for the VSP.

APTAC: The Automatic Pressure Tracking Adiabatic Calorimeter (APTAC) is also a VSP-like instrument, but more fully instrumented and automated. The sample resides in a spherical cell and exhibits greater sensitivity for exotherms (0.04°C/min) than the VSP. The instrument can also be purchased with various integrated sample injection and venting options.

22.2.7.3 Deflagration Tests

Certain liquids such as peroxides and hydroperoxides are susceptible to violent decomposition, even in the absence of air. These decompositions show no simple temperature dependence, but occur at about 200°C and are characterized as deflagrations (subsonic propagating reactions). Liquid-phase deflagration tests are carried out in a small cylindrical vessel fitted with rupture disks in which the liquid sample is ignited by a heated wire. The maximum pressure resulting from a liquid decomposition would typically rule out simple containment as a design option.

Vapor phase deflagration tests are carried out in a spherical vessel in which the vapor sample is ignited by a fused wire. In this case, containment may be an option, although venting is also widely used.

22.2.8 Interpretation and Comparison of Experimental Test Methods

22.2.8.1 Reaction Characterization

The reaction system may be characterized in terms of the physical and chemical properties of the components:

reaction stoichiometry, heat of reaction, quantity of gas evolved, kinetics of reaction.

22.2.8.2 Exotherm Characterization

Some parameters characterizing an exotherm include: exotherm onset temperature; adiabatic temperature rise; heat of reaction; rate of temperature rise (also known as the self-heat rate); rate of heat generation; rate of pressure rise; velocity constant, activation energy, and pre-exponential constant; and adiabatic induction time.

22.2.9 Interpretation of Exotherm Onset Temperature Test

Although the 'onset temperature' is a convenient concept, it needs to be borne in mind that even at lower temperatures there is still a non-zero reaction rate. It is important to note that the onset temperature is a measure linked to the detection limit of the instrument utilized sample size, degree of adiabaticity, instrument sensitivity, etc. It is not a value unique to the reaction chemistry alone.

A traditional approach to reactor safety has been to set the temperature of reactor operation at some fixed temperature interval below the exotherm onset temperature measured in the laboratory test. A widely quoted value of this safety margin is 100°C, the '100 degree rule'.

Any material with an exotherm onset temperature within 100°C of the maximum operating temperature as determined by the simple exotherm test is considered a potential hazard. In such a case, the material is subjected to the adiabatic exotherm test. If the exotherm onset temperature in this test is 50°C or more above the operating temperature, it is not considered a hazard. If it is within this 50°C safety margin, it is subjected to a long-term hot storage test at 20°C above the operating temperature.

22.2.10 Test Schemes

A systematic approach to testing requires that it be performed within the framework of a scheme for the acquisition of the complete set of information necessary for reactor design. There are a number of test schemes described in the literature. The *Guidelines for Chemical Reaction Hazard Evaluation* (ABPI, 1989) address material stability as well as chemical reactivity. The overall scheme is illustrated in Figure 22.6. This scheme for individual substances covers: (1) explosive properties, (2) unexpected decomposition, and (3) maximum safe temperature for storage. The scheme also utilizes the adiabatic Dewar test, ARC, the heat flow calorimeter test, and the two gas evolution measurement tests to characterize the intended reaction.

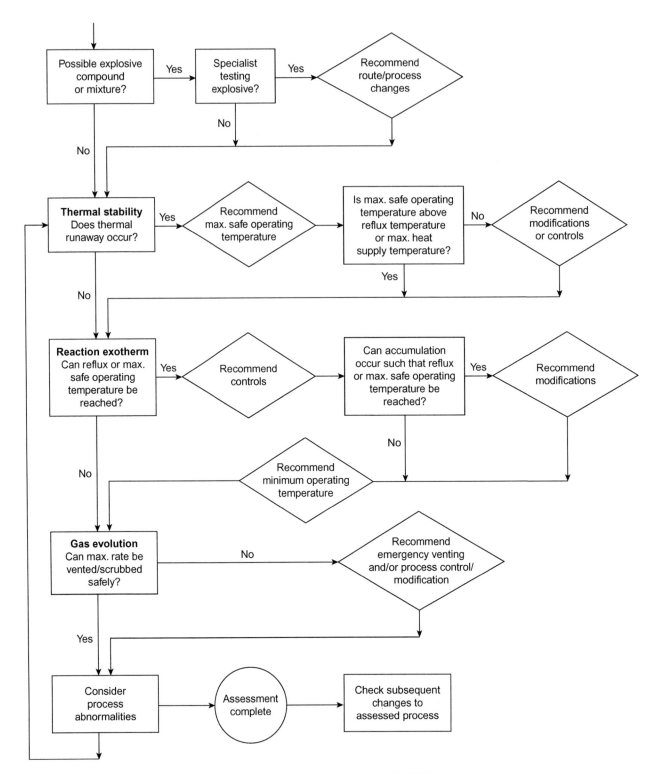

FIGURE 22.6 Test scheme for assessment of thermal stability of reactant materials ABPI (1989).

For explosive decomposition, the scheme involves assessment of explosive properties from: (1) chemical structure, (2) oxygen balance, (3) energy of decomposition, and (4) preliminary small-scale tests (e.g., DSC), and if necessary, larger-scale Koenen tube or Trauzl lead block tests.

For unexpected decomposition, DSC and closed tube tests are given as small-scale tests using ramped temperatures. Additionally, isothermal tests for delayed onset are given, including delayed onset detection test, ARC, and adiabatic Dewar test.

The tests for maximum safe storage temperature apply only to liquids. For solids, the temperature may not be uniform throughout the material, making it necessary to treat this situation as a self-heating problem.

22.3. IDENTIFICATION OF REACTIVE HAZARDS SCENARIOS

A review should be conducted to determine credible pathways by which the identified reactive hazards can potentially pose significant threats to the process or equipment. The review should focus exclusively on reactive hazards. All credible scenarios should be described in detail indicating clearly what the cause of the scenario is, how the unmitigated scenario develops, and what the ultimate consequence might be. Then, the risk of each event can be evaluated.

Emphasis in the review should focus on potential events that could lead to 'high consequence' events. This will encourage resources to be focused on the more significant scenarios.

Given the potential for a high consequence incident arising from a reactivity event, it may be prudent to consider selected multiple jeopardy scenarios (a 'jeopardy' being an initiating deviation, not including a failed response by people or instrumentation to a deviation). Note that a runaway reaction itself is not a jeopardy but a consequence of operating under upset conditions.

22.4. REACTIVE HAZARDS RISK ASSESSMENT

22.4.1 Consequence Assessment

The intent of consequence assessment is to determine which scenarios need prevention or mitigation action. To ascertain whether a high consequence event could occur, the extent of temperature and pressure rise associated with each scenario should be evaluated.

In evaluating the consequences of a scenario, it may be prudent to take no credit for operator procedures and training, instrumentation and mechanical relief (these can be examined during likelihood assessment). However, inherent constraints to the development of the consequence, such as maximum heating medium temperature, equilibrium (thermodynamic)-limited reactant conversion or catalyst deactivation, etc., can be accounted for in the calculations.

Consequence assessment results should be documented for each corresponding scenario generated from the

review along with further actions (if any) needed to mitigate the consequences.

22.5. BATCH REACTORS: BASIC DESIGN

22.5.1 Inherently Safer Design

A common application of inherent safety to batch reactors consists on avoiding high concentrations of reactants. This can be done for instance, by using semi-batch reactors instead of 'all-in' batch reactors. In the latter, all the reactants are added in the initial charge; in the former, one of the reactants is added continuously. The hazard in the all-in design is that conditions may occur in which there is a sudden massive reaction of the unreacted reactants. In the semi-batch reactor, the continuous feed arrangement, combined with a suitable reactor temperature, keeps the concentration of one of the reactants relatively low, and it is possible to effect prompt shut-off of the feed if a potentially hazardous operating deviation occurs.

The full benefit of the semi-batch reaction mode does depend, as just indicated, on the adoption of a suitable reaction temperature. If this temperature, and therefore the reaction rate are too low, accumulation of the fed reactant can occur.

22.5.2 Secondary Reactions

In addition to the primary reaction, there may be one or more secondary reactions. If such a secondary, or side, reaction is exothermic, it may constitute a hazard.

Whether a secondary reaction occurs may be established using DTA. If it does, the temperature T_s at which the secondary reaction takes place is estimated by subtracting 100 K from the start of the reaction as given by the DTA. This estimate of T_s is compared with the reaction temperature T_r or the maximum possible reactor temperature T_{max}:

$$T_{max} = T_r + \Delta T_{ad} \qquad (22.3)$$

where ΔT_{ad} is the adiabatic temperature rise (K), T_{max} is the maximum absolute temperature (K), T_r is the reactor temperature (K), and T_s is the absolute temperature of the secondary reaction. If the estimate of T_s is greater than T_{max}, there is no hazard and further investigation is not required.

If T_s lies between T_r and T_{max}, there may be an appreciable hazard, but it may still be possible to carry out the reaction provided it can be ensured that the reactor temperature does not reach T_s during the primary reaction. Alternatively, if this cannot be guaranteed, use may be made of appropriate protective measures. Operation in this region is a specialist matter.

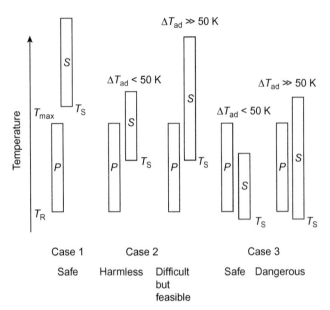

FIGURE 22.7 Hazards of side reactions (Grewer et al., 1989): P, primary reaction; S, secondary reaction; T_{max}, maximum reactor temperature; T_R, temperature of reactor; T_s, temperature at which secondary reaction occurs; ΔT_{ad}, adiabatic temperature rise.

If T_s for a highly exothermic reaction is equal to or less than T_r, it is rarely possible to carry out the primary reaction safely. These guidelines are summarized in Figure 22.7.

22.5.3 Decompositions

For single phase decomposition of organic substances, if the decomposition energy based on total reaction mass is $<100-200$ kJ/kg, which corresponds roughly to $\Delta T_{ad} \approx 50-100$ K, the effect of the decomposition is generally not critical (Grewer et al., 1989). Even for a decomposition energy above the range just quoted, the effect may not be critical if there are other compensating effects such as endothermic reactions or high heat removal capacity. For inorganic substances, the criterion decomposition energy should be lower, due to the lower specific heat, while for substances containing water it should be higher.

22.6. LIKELIHOOD ASSESSMENT

For scenarios that are ranked as 'high' consequence, the likelihood can be evaluated in semi-quantitative terms; semi-quantitative terms, as determined through a Layers of Protection Analysis (LOPA); or in fully quantitative terms, through a fully executed Quantitative Risk Assessment (QRA) such as discussed by Modarres (2006).

The likelihood assessment should take into account the availability and reliability of relevant factors in the possible development of a scenario. For example,

probability of a deviation, process instrument failure, human error, and human intervention should all be considered. Availability of components to respond to system demand includes a continuous test and maintenance schedule, which should be optimized for cost effective reduction of upset scenarios.

Once the likelihood has been obtained, the composite of the likelihood with the consequence gives the risk of the scenario reactive hazard.

22.7. PREVENTION MEASURES

Prevention measures can be divided in two categories: changes in process conditions (process changes) and changes to the hardware (design changes).

22.7.1 Process Changes (Inherent Safety)

Process changes make the process inherently safer as they are inherent to the chemistry and therefore reduce the consequences of a scenario or eliminate (or reduce) the likelihood of a scenario. Some examples include: limiting the concentrations of the reactants (e.g., feedstock specification) to reduce the maximum adiabatic end temperature and pressure to within the design limits of the unit; changing to a non-reactive or less combustible solvent; and using a different catalyst resulting in a lower temperature. However, the effects produced by a different catalyst should be thoroughly tested in order to avoid unintended consequences, such as undesirable reactions occurring at lower temperatures.

22.7.2 Hardware Changes

Hardware changes comprise any changes in equipment/hardware, excluding mechanical relief systems. Some examples include: removing heating coil from storage tank to prevent product decomposition, removing linework (pipes and fittings) to prevent mixing of incompatible materials, reducing flow rates by restricting orifices or smaller control valves, eliminating or substantially reducing storage of reactive intermediate species.

22.8. MITIGATION MEASURES

The safeguarding philosophy employed can be dependent on regulatory requirements as well as the amount and type of risk posed by the scenarios. For example, relief venting is frequently a more cost-effective means than high integrity instrumentation to limit pressure build-up in equipment. However, in several countries, relief venting is not the preferred option, particularly when potentially toxic or corrosive materials could be released. Addition of any single prevention or mitigation option or a combination thereof can be considered. Once intolerable risk levels are precluded, alternative mitigations can be evaluated through a cost/benefit analysis.

22.8.1 Instrumentation and Control

Safe operation of a batch reactor requires close monitoring and control of reactant and additive flows to the reactor, of the operating temperature, and of the agitation. For the temperature, a safety margin is set between the operating temperature and the exotherm temperature. The operating temperature is displayed and provided with an alarm. The instrument should not allow a hot spot to develop undetected and it should not have a large time lag or a high failure rate. It is good practice to have separate measurement devices for the temperature control loop and the temperature alarm.

Actions that can be taken to counter an increase in operating temperature such as shut-off of feed or use of full cooling are identified, and appropriate actions are selected to be activated by either operator intervention or trip. Actions to be taken in case of a loss of agitation depend on the reaction. In semi-batch reactors, it is common to provide a trip to shut off the feed.

22.8.2 Emergency Safety Measures

The prime measures are inhibition of reaction, quenching of reaction, and dumping.

The reaction may be stopped by the addition of an inhibitor, or a short stop, which involves the use of an inhibitor specific to the reaction in question. Quenching involves adding a quenching agent, usually water under gravity, to the reaction mass to cool and dilute it. The third method is dumping, which involves dropping the reactor charge under gravity into a quench vessel beneath which contains a quench liquid.

22.9. CHEMICAL SECURITY

22.9.1 Introduction

The world of safety and security in the chemical process industries has certainly changed since the terrorist attack on the United States on September 11, 2001. Security management is now required more than ever to protect the assets (including employees) of the facility, maintain the ongoing integrity of the operation, and to preserve the value of the investment.

The foundation of the security management system is a security vulnerability assessment (SVA), which is intended to identify security vulnerabilities from a wide range of threats ranging from vandalism to terrorism. With the recognition of threats, consequences,

vulnerabilities, and the evaluation of the risk of security events, a security management system can be organized that will effectively mitigate the risks.

22.9.2 Security Management System

The purpose of the management system is to ensure the ongoing, professional, and systematic application of security principles and programs to achieve a level of specified security. A comprehensive process security management system must include management program elements that integrate and work in concert to control security risks.

The *Security Code* published by the American Chemistry Council (2001) is designed to help companies achieve continuous improvement in security performance using a risk-based approach. Each company must implement a risk-based security management system for people, property, products, processes, information, and information systems throughout the chemical industry value chain. The corresponding security management system must include the following 13 management practices:

1. *Leadership Commitment.* Senior leadership commitment to continuous improvement through published policies, provision of sufficient and qualified resources and established accountability.
2. *Analysis of Threats, Vulnerabilities, and Consequences.* Prioritization and periodic analysis of potential security threats, vulnerabilities, and consequences using accepted methodologies.
3. *Implementation of Security Measures.* Development and implementation of security measures commensurate with risks and taking into account inherently safer approaches to process design, engineering and administrative controls, and prevention and mitigation measures.
4. *Information and Cyber-Security.* Recognition that protecting information and information systems is a critical component of a sound security management system.
5. *Documentation.* Documentation of security management programs, processes, and procedures.
6. *Training, Drills, and Guidance.* Training, drills, and guidance for employees, contractors, service providers, value chain partners, and others, as appropriate, to enhance awareness and capability.
7. *Communications, Dialogue, and Information Exchange.* Communications, dialogue, and information exchange on appropriate security issues with stakeholders, balanced with safeguards for sensitive information.
8. *Response to Security Threats.* Evaluation, response, reporting, and communication of security threats as appropriate.

9. *Response to Security Incidents*. Evaluation, response, investigation, reporting, communication, and corrective action for security incidents.
10. *Audits*. Audits to assess security programs, and processes and implementation of corrective actions.
11. *Third-Party Verification*. Third-party verification that chemical operating facilities with potential for off-site impacts have implemented the physical site security measures to which they have committed.
12. *Management of Change*. Evaluation and management of security issues associated with changes involving people, property, products, processes, information, or information systems.
13. *Continuous Improvement*. Continuous performance improvement processes entailing planning, establishment of goals and objectives, monitoring of progress and performance, analysis of trends and development, and implementation of corrective actions.

22.9.3 Security Strategies

It is difficult to prescribe security measures that apply to all facilities in all industries. The specific situations must be evaluated individually by local management using best judgment of applicable practices. It is suggested to use SVA as a means of identifying, analyzing, and reducing vulnerabilities. Appropriate security risk management decisions must be made commensurate with the risks. Resources are best applied to mitigate high risk situations primarily.

Security strategies for the process industries are generally based on the application of four key concepts against each threat (CCPS, 2003):

> *Deter*: A security strategy to prevent or discourage the occurrence of a breach of security by means of fear or doubt. Physical security systems such as warning signs, lights, uniformed guards, cameras, and bars are examples of systems that provide deterrence.
> *Detect*: A security strategy to identify an adversary attempting to commit a malicious act or other criminal activity in order to provide real-time observation, interception, and post-incident analysis of the activities and identity of the adversary.
> *Delay:* A security strategy to provide various barriers to slow the progress of an adversary in penetrating a site to prevent an attack or a theft, or in leaving a restricted area to assist in apprehension and prevention of theft.
> *Response:* The act of reacting to detected criminal activity either immediately following detection or post-incident via surveillance tapes or logs.

A complete security design includes these four concepts in 'Layers of Protection' or a 'Defense in Depth' arrangement. Examples of these physical protection

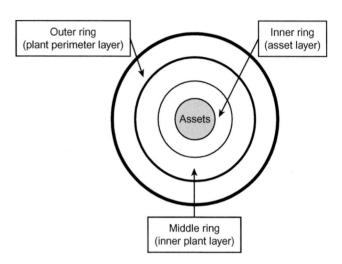

FIGURE 22.8 Defense in depth concept for security (layers of protection). *Source: Adapted from American Chemistry Council (2001).*

concepts and their associated countermeasures are graphically depicted in Figure 22.8 and 22.9.

22.9.4 Countermeasures and Security Risk Management Concepts

Countermeasures are actions taken to reduce or eliminate one or more vulnerabilities. Security risk reduction at a site can include the following strategies: Physical Security, Cyber Security, Crisis Management and Emergency Response Plans, Policies and Procedures, Information Security, Intelligence, and Inherent Safety.

22.9.5 Defining the Risk to be Managed

For the purposes of an SVA, the definition of risk is shown in Figure 22.10. The risk that is being analyzed for the SVA is defined as an expression of the likelihood that a defined threat will target and successfully attack a specific security vulnerability of a particular target or combination of targets to cause a given set of consequences. This is contrasted with the usual accidental risk definitions. The risk variables are defined as shown in Figure 22.11.

22.9.6 Overview of an SVA Methodology

The SVA process is a risk-based and performance-based methodology. The user can choose different means of accomplishing the general SVA method so long as the end result meets the same performance criteria.

There are several SVA techniques and methods available to the industry, all of which share common elements. One approach to conducting an SVA is shown in Figure 22.12.

Security countermeasure	Outer perimeter (plant boundary layer)				Middle perimeter (inner plant/asset level)				Inner perimeter (asset/subsystem level)			
	Deter	Detect	Delay	Respond	Deter	Detect	Delay	Respond	Deter	Detect	Delay	Respond
Access control	■	■	■		■	■	■		■	■	■	
Background checks	■	■			■	■			■	■		
Bollards	■		■		■		■		■		■	
CCTV		■		■		■		■		■		■
Counter surveillance		■				■				■		
Door/gate locks	■		■		■		■		■		■	
Emergency planning				■				■				■
Emergency shutdown				■				■				■
Fences	■		■		■		■		■		■	
Guard force	■	■	■	■	■	■	■	■	■	■	■	■
ID system	■	■			■	■			■	■		
Intelligence gathering		■			■					■		
Intrusion detection		■		■		■		■		■		■
Jersey barriers	■		■		■		■		■		■	
Lighting	■	■		■	■	■		■	■	■		■
Network firewall	■	■	■		■	■	■		■	■	■	
Target hardening	■		■		■		■		■		■	
Trenches	■		■		■		■		■		■	
Vehicle checks	■	■			■	■			■	■		
Vigilance training		■		■		■		■		■		■
Key	■	Countermeasure may be applicable to the security concept										

FIGURE 22.9   Defense in depth/layers of protection concept for a chemical process facility. *Source: Moore (2004).*

Intentional release risk is a function of:	Accidental release risk is a function of:
Consequences of a successful attack against an asset and likelihood of a successful attack against an asset.	Consequences of an accidental event and likelihood of the occurrence of the event.
Likelihood is a function of: The attractiveness to the adversary of the asset, the degree of threat posed by the adversary and the degree of vulnerability of the asset.	Likelihood is a function of: The probability of an event cascading from initiating event to the consequences of interest and the frequency of the events over a given period.

FIGURE 22.10   Intentional release vs accidental release—risk definitions.

Ultimately, it is the responsibility of the owner/operator to choose the SVA method and depth of analysis that best meets the needs of his specific location. The overall five-step approach of the SVA methodology proposed by the American Petroleum Institute (API) and the National Petrochemical and Refineries Association (NPRA) (2004), is described as follows.

22.9.6.1 Step 1: Asset Characterization

Essentially, this step identifies the assets (people, facilities, information, reputation, business) of value, analyzes why it is of value and identifies its importance, determines the interaction of the assets with other neighboring facilities, suppliers, or customers or other economic interdependencies. Assumptions are made on the worst credible security event consequences to determine the impacts. The estimate of severity of the consequences is one of the four risk factors.

22.9.6.2 Step 2: Threat Assessment

The intentional release risk includes possible attacks by outsiders or insiders, or a combination of the two

Consequences	The potential impacts of the event
Likelihood	Likelihood which is a function of the chance of being targeted for attack, and the conditional chance of mounting a successful attack (both planning and executing) given the threat and existing security measures. This is a function of three variables below.
Threat	Threat, which is a function of the adversary existence, intent, motivation, capabilities and known patterns of potential adversaries. Different adversaries may pose different threats to various assets within a given facility.
Vulnerability	Any weakness that can be exploited by an adversary to gain access and damage or steal an asset or disrupt a critical function. This is a variable that indicates the likelihood of a successful attack given the intent to attack an asset.
Target attractiveness	Target attractiveness, which is surrogate measure for likelihood of attack. This factor is a composite estimate of the perceived value of a target to the adversary and their degree of interest in attacking the target.

FIGURE 22.11 API/NPRA (2004) SVA methodology risk variables.

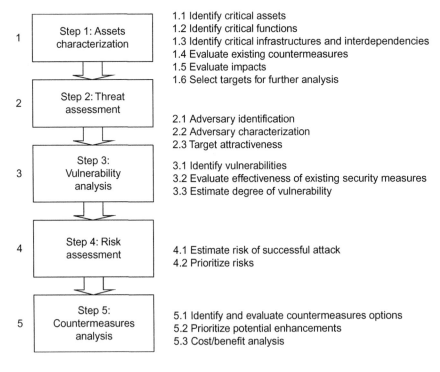

FIGURE 22.12 API/NPRA (2004) security vulnerability assessment methodology.

adversaries. It may be perpetrated by a number of different adversaries with varying intents, motivations, weapons, tactics, and capabilities.

The threat definition results in a determination of the design basis threat for the facility. The threat assessment results in a 'fixed' and 'variable' design basis threat. The fixed threat forms the basis for the design and is the baseline threat estimate. The variable design basis threat assessment is an estimate of the change in threat levels given certain possible future conditions.

The concept of fixed and variable design basis threats is useful for making decisions on facility design and operation. If the threat to insiders is considered significant, countermeasures designed to limit those risks are imperative. The fixed threat may determine the need for background screening, limiting the span of control of individuals, strong supervision, monitoring of activities, audits, surveillance, password controls, and other measures. Threat to a particular asset varies with several factors including attractiveness and potential impact if the asset was attacked. For this reason, the threat assessment includes a step whereby each asset is analyzed from the perspective of each potential adversary to determine the degree of attractiveness of the asset to the adversary.

22.9.6.3 Step 3: Vulnerability Analysis

The vulnerability analysis includes the relative pairing of each target asset and threat to identify potential vulnerabilities related to process security events. This involves the identification of existing countermeasures and their level of effectiveness in reducing those vulnerabilities. The degree of vulnerability of each valued asset and threat pairing is evaluated by the formulation of security-related scenarios or by an asset protection basis. If certain criteria are met such as higher consequence and attractiveness ranking values, then it may be useful to apply a scenario-based approach to conduct the Vulnerability Analysis. This approach option is very similar to the process hazard analysis (PHA) techniques employed to analyze accidental releases. It includes the assignment of risk rankings to the security-related scenarios developed.

22.9.6.4 Step 4: Risk Assessment

Using the assets identified during Step 1 (Asset Characterization), the risks are prioritized based on the likelihood of a successful attack which is a function of the threats assessed under Step 2 and the degree of vulnerability identified under Step 3.

A risk assessment matrix can be used for the purpose of assessing risk. Risk assessment is only possible when there is some frame of reference. Since the events in question are extremely rare events, it is necessary to (1) use surrogate factors such as attractiveness and threat to determine the likelihood of interest of attack of any particular asset and (2) use vulnerability as a measure of the likelihood of a successful attack given the desire to attack. Then the analyst can use performance criteria to set risk goals.

22.9.6.5 Step 5: Countermeasures Analysis

Countermeasure options will be identified to further reduce vulnerability at the facility. These include improved countermeasures that follow the process security doctrines of deter, detect, delay, respond, mitigate, and possibly prevent. Some of the factors to be considered are as follows:

1. Reduced probability of successful attack;
2. Degree of risk reduction by the options;

3. Reliability and maintainability of the options;
4. Capabilities and effectiveness of mitigation options;
5. Costs of mitigation options;
6. Feasibility of the options.

The countermeasure options should be re-ranked to evaluate effectiveness and prioritized to assist management decision making for implementing security program enhancements.

22.9.7 Chemical Facility Anti-Terrorism Standards (CFATS)

Considering the extreme consequences of events like 9/11, the United States Congress enabled the Department of Homeland Security (DHS) to create and promulgate the Chemical Facility Anti-Terrorism Standards (CFATS).

CFATS represent a set of federal regulations issued by the DHS on April 9, 2007. The objective of this measurement is to identify, assess, and ensure effective security at high-risk chemical facilities by following the Interim Final rule (IFR) under the 6 CFR part 27. Based on IFR, the DHS requires all chemical facilities to develop and implement an adequate Site Security Plan (SSP) considering their individual Security Vulnerability Assessment (SVA).

Appendix A of the 6 CFR part 27 (published on November 20, 2007) provides a list of chemicals of interest and their specific Screening Threshold Quantity (STQ). For this, the DHS considered three different pieces of security criteria which can affect human life and health. These include:

1. Release of toxic, flammable, or explosive materials.
2. Theft or diversion of any material that can be easily employed for weapons fabrication or any other technology.
3. Sabotage or contamination of any material that can be mixed with an accessible material and create unfavorable consequences.

Based on this, all facilities covered by CFATS are required to use the Chemical Security Assessment Tool (CSAT) developed by DHS. This web computer system appeals to an information protection regime called Chemical-Terrorism Vulnerability Information (CVI). This means only CVI certified individuals are authorized to access the CSAT system.

22.9.8 Inherently Safer Technology (IST)

Lately, this approach has also received considerable attention for chemical security applications, since it can reduce potential consequences of terrorist attacks. IST provides chemical security to facilities based on their use of technology and chemicals whose intrinsic properties can be reduced by the following methods and minimize hazards: intensification, substitution, moderation, and simplification.

REFERENCES

American Chemistry Council, 2001. Site Security Guidelines for the U.S. Chemical Industry. <http://www.socma.com/assets/File/socma1/PDFfiles/securityworkshop/SecurityGuideFinal10-22.pdf>.

American Petroleum Institute/National Petrochemical and Refiners Association (2004). Security Vulnerability Assessment Methodology for the Petroleum and Petrochemical Industries. Second Edition.

Association of the British Pharmaceutical Industry (ABPI), 1989. Guidelines for Chemical Reaction Hazard Evaluation, second ed. London.

Center for Chemical Process Safety (CCPS), 2003. Guidelines for Analyzing and Managing the Security Vulnerabilities of Fixed Chemical Sites. New York: John Wiley & Sons, Inc.

Chemical Safety Board (CSB), 2002. Hazard investigation—improving reactive hazard management. In: U.S. Chemical Safety and Hazard Investigation Board. <http://www.csb.gov/investigations/detail.aspx?SID=61>.

Duval, R.C., 1985. Thermochemical hazard evaluation. In: Chemical Process Hazard Review. ACS Symposium Series 274, American Chemical Society Washington, DC, 1985, pp. 57–68.

Gibson, N., Rogers, R. L., Wright, T. K., 1987. Chemical Reaction Hazards—An Integrated Approach. In IChemE Symposium, Series (No. 102).

Grewer, T., Klusacek, H., Loüffler, U., Rogers, R.L., Steinbach, J., 1989. Determination and assessment of the characteristics values for the evaluation of the thermal safety of chemical processes. J. Loss Prev. Process Ind. 2 (4), 215.

Lothrop, W.C., Handrick, G.R., 1949. The relationship between performance and constitution of pure organic explosive compounds. Chem. Rev. 44, 419.

Modarres, M., 2006. Risk Analysis in Engineering: Techniques, Tools, and Trends. Taylor & Francis.

Moore, D., Applying inherently safer technologies for security of chemical facilities. 38th Annual Loss Prevention Symposium, AIChE 2004 Spring National Meeting, New Orleans, LA, April 26–29, 2004.

National Fire Protection Association, 1994. NFPA 49, Hazardous Chemicals Data. National Fire Protection Association, Batterymarch Park, Quincy, MA.

National Fire Protection Association, 2012. NFPA 704, Standard System for the Identification of the Hazards of Materials for Emergency Response. National Fire Protection Association, Batterymarch Park, Quincy, MA.

Treweek, D.N., Claydon, C.R., Seaton, W.H., 1973. Appraising energy hazard potentials. Loss Prev. 7, 21.

Benchmarking in the Process Industry

23.1. INTRODUCTION

According to Webster's Dictionary (2013), benchmarking is defined as 'a systematic process for evaluating and comparing services, products, and work processes in order to facilitate improvement or strategic advantage.' It is imperative to understand that benchmarking exercises help to investigate reasons behind safety performance gaps and then develop best practices to improve safety performance (Mearns and Thom, 2000). Motivations for benchmarking in the chemical industry include avoidance of undesired incident occurrences, comparison of existing safety-related practices, a desire to become a best-practice model among industry competitors, a desire to accomplish a goal of zero incidents, and company resource pressures (Arendt, 2006).

Comparative benchmarking is often used to identify specific areas of weakness in an organization. In the process safety sphere, the use of benchmarking techniques can assist in (1) identifying gaps in the safety performance and finding improvements, (2) avoiding repeating incidents caused by the same or similar causes, (3) creating an active discussion platform to focus on current practices, (4) sharing lessons learned from individual personal or corporate experiences, (5) determining an industry consensus on Recognized and Generally Accepted Good Engineering Practices (RAGAGEP), and (6) optimizing regulatory compliance. Specifically, benchmarking exercises enable participants to reach a better understanding of the key elements that improve process safety performance.

23.2. BENCHMARKING OUTLINE

Camp (1989) suggests a generic benchmarking process with three stages: planning, analysis, and integration. In the planning phase, the objectives should be identified, as well as the safety performance indicators (e.g., lost time incident rate), the peer companies or organizations, and the data collection method. It will lead to a clear understanding of the jobs and tasks for the benchmarking project. Afterwards, an analysis phase should be initiated to identify current safety performance gaps and determine the levels of future safety performance. The benchmark findings should be communicated within all organizational levels to gain intra-organizational acceptance and used to set organizational goals that will be achieved in the integration phase. Finally, if the organization decides on implementing the benchmarking findings, operational procedures should be established and implemented at all levels of the organization and then continuously monitored to check the implementation progress. Additionally, a recalibration process should take place to ensure the effectiveness of the implementation of the benchmarking findings.

23.2.1 Types of Benchmarking

Benchmarking can be conducted in several different ways depending on the objectives, e.g., internal and external benchmarking. Typically, internal benchmarking is exercised to compare internal operations and upgrade to higher standard levels in most large or globally operated companies. Each division under one company may not have similar practices or procedures due to different personnel, circumstances, regulations, and cultures. Thus, internal benchmarking can help identify strengths and weaknesses of each division. In addition, this internal benchmarking exercise may assist in validating the necessity of external benchmarking. External benchmarking can be subdivided into competitive, functional, and generic benchmarking. Competitive benchmarking is to directly compare a product or service with competitors in the same business area by investigating comparative advantages and disadvantages. The functional benchmarking approach focuses on identifying functional competitors, industry leaders, or other practices in different industries to improve their operations and safety practices instead of concentrating on direct product competitors. Generic benchmarking, the most pure form of benchmarking, is usually conducted to identify differences and similarities among organizations with similar functions and procedures. This benchmarking approach can discover the best practices and provide the highest long-term benefits. However, it often requires more objectivity from the benchmarking researchers as well as thorough, careful understanding of conceptualization and generic processes.

23.2.2 Data Collection Methods

In order to conduct a benchmarking study, it is necessary to gather data. The amount and quality of data gathered may significantly impact the accuracy and validity of the study results. The data and information can be collected through intra-organizational resources. External resources might also be utilized and will have a wider range of knowledge and information. These resources include publications or databases created by professional associations (e.g., CCPS, API, NFPA, ISO, ASME, and ANSI), publications from journals or conferences, seminars, special industry technical reports, industry data firms, independent research organizations such as universities, research institutes (e.g., the Mary Kay O'Connor Process Safety Center), newsletters, and external consulting experts and firms.

In order to gather data effectively, Camp (1989) recommends categorizing the available data into logical groups. The author suggests that benchmarking researchers investigate internal data and public information and then ultimately conduct benchmarking research and investigations.

The questionnaire survey, administered by mail, email, Internet, and telephone, is one widely used method because it is relatively low cost and can ensure full anonymity. Sometimes, during the safety performance benchmarking process, managerial and psychological questions may need to be included along with technical queries to focus on issues related to management systems and personnel behaviors. Direct-site visits can also be employed to follow up the questionnaires and to focus on specific issues. If the information gathered by questionnaires, and direct site visits is not sufficient enough, then investigators might use more advanced approaches, such as panel discussions with benchmarking participants. The use of focus groups (e.g., panels) involves direct observation sharing between invitees from benchmarking organizations through discussions. However, it may be better to have the panel discussion hosted by a third party to maintain confidentiality.

23.2.3 Benchmarking Parameters

Before going into detail, it is necessary for researchers to determine what should be benchmarked and what parameters are needed. From the safety system point of view, OSHA/PSM and EPA/RMP requirements can be chosen as the parameters of interest. From the safety performance indicator point of view, benchmarking parameters of interest can be determined by leading indicators of safety, such as leadership and auditing, because they can significantly impact safety performance. Another source for safety performance parameters can be lagging indicators such as incidents, lost time incident rates, and financial loss. More detailed examples for oil and gas industries can be found in Tables 23.1 and 23.2. In addition, benchmarking parameters for the evaluation of quantitative risk assessment for onshore plants are described in Table 23.3. However, special care should be taken when determining the parameters because they can influence the success or failure of benchmarking exercises.

The collected parameters can be assessed in a qualitative or quantitative manner. Qualitative assessment is more often used when identifying strengths and weaknesses of organizations in management and operational systems. For comparative benchmarking studies, quantitative assessments may be more user-friendly since they enable users to compare estimated safety performances for other organizations. In fact, many companies have been adopting both methods to increase the quality of the results (Gomm et al., 2000; Mearns and Thom, 2000).

To quantitatively compare the overall performance of organizations, mathematical models are often used. The fuzzy logic model has been frequently used within the sugar industries (Krajnc et al., 2007). In Krajnc's paper, the environmental performances of three sugar plants were compared with the industry's best practices, and their performances were estimated in terms of the consumption of energy, that is, water, and the production of wastes, that is, wastewater. The paper concludes that the developed fuzzy model is able to assist in combining various production indicators and resolving difficult issues that arise in the benchmarking of environmental performance.

23.3. POSSIBLE BARRIERS AND RESOLUTIONS FOR BENCHMARKING

23.3.1 Unwillingness to Change and Skepticism

It is critical to make personnel accept new practices administered through benchmarking studies since there will always be skeptics (Camp, 1989). Reluctance to accept new findings from the benchmarking investigations may be overcome through a carefully designed communication approach that provides rationale for validating the findings and expected benefits in both individual and company aspects. Top management should approve the new practices quickly and then show active commitment ahead of lower level personnel. In addition, it is useful to provide initiatives for change, operational acceptance, and validation from multiple sources to target personnel with a clear interpretation of the benchmarking outcomes. Reward and recognition programs that compensate employees who adopt the new practices may be introduced to increase the acceptance rate. With the long-term perspective focused on acceptance, management should put effort into creating an organizational culture where employees are willing to accept new and creative ideas.

23.3.2 Low Rate Participation Against Benchmarking

Poor response rates adversely impact the validity and reliability of the collected data and analysis, thus resulting in

---

TABLE 23.1 Benchmarking Parameters of Health and Safety Performance

Parameters in Safety Management	Parameters in Workforce Perspective
• Installation/business unit information • Accident and incident data • Health and safety policies • Organizing for health and safety • Management commitment • Workforce involvement • Health surveillance and promotion • Health and safety audits and follow-up	• Employer, occupation, supervisory status, experience, and accident involvement over past year • Degree of exposure and understanding of the company health and safety policy • Level of involvement in discussing the effectiveness of the safety management system, discussing procedures and risk assessments, setting safety objectives and safety improvement plans, safety auditing • Communication about safety issues • Job satisfaction (including contractors) • Self-reported safety behavior • Satisfaction with safety activities • Willingness to report accidents and incidents • Perceptions of management commitment to safety • Perceptions of supervisor commitment to safety • Attitudes toward rules and regulations

Source: Adapted and modified from *Mearns and Thom* (2000).

TABLE 23.2 Parameters of Upstream Benchmarking

Sections	Company Descriptors	Result Measures	Activity Indicators
Parameters	• To HSE policy document • Assurance of policy compliance • Organization and governance • Number of HSE people • Employee and contractor hours worked • Operational performance • Joint venture/subsidiary policy • Stakeholder communication • Linkage of HSE into the business	• Safety performance statistics (number of fatalities, cause of fatalities, number of Lost Workday Cases, number of recordable injuries and illnesses, etc.) • Major environmental incidents • Oil discharged to the environment (number of oil spills, quantity of oil spills, mass of oil discharges in produced water and oil-based mud and cuttings) • Air emissions (gas flared, fugitive hydrocarbon emissions)	• Number and types of audits • Number of bus with management systems implemented • Emergency response plans and drills • Description of training • Major incident reporting process • Reward and recognition programs • Goals, objectives, and targets

Source: Adapted and modified from Gomm et al. (2000).

TABLE 23.3 Benchmarking Parameters on Risk Assessment Evaluation

Category	Specific Parameters
Hazard identification analysis	• Credibility and quality of the incident scenarios • Completeness of the list of incident scenarios • Proper selection of PHA methodology and its completeness (e.g., HAZOP, FMEA, Check-list, What-If) • Appropriate composition of the PHA team (e.g., facilitators, process engineers, safety and fire engineers, operators, instrument engineers, external experts)
Frequency assessment	• Proper usage of failure data (e.g., OREDA, CCPS, plant specific data) • Appropriate assumptions and calculations • Completeness and suitability of frequency assessment (e.g., FTA, ETA) • Appropriate address on uncertainty in calculation
Consequence assessment	• Proper choice of methodology, assumptions and calculations in source term analysis (e.g., models, release rate, orifice size, discharge coefficient) • Proper usage of consequence models (e.g., dispersion models, fire, and explosion models) • Appropriate assumptions and calculations in consequence analysis • Appropriate understanding of uncertainty in calculations
Interpretation of estimated risks and recommendations	• Suitable interpretation of estimated risks against company and industry criteria (e.g., individual risk criteria, societal risk criteria) • Appropriate recommendations for improving safety performance

uncertainty in adopting the benchmarks. Keren and West in conjunction with the Mary Kay O'Connor Process Safety Center, an independent research institute, executed a study on the benchmarking of MOC practices (Keren et al., 2002), and questionnaire was created, but only half of facilities responded. In order to address this problem, all responses should be completely anonymous and confidential. A well reviewed and specific plan on what is going to be shared and how it is going to be shared will encourage facilities to share more information. In addition, statistical techniques may be associated with benchmarking studies because they can improve effectiveness (Mearns and Thom, 2000). Another resolution may be to create a benchmarking consortium of similar companies that assign a third party to perform a benchmarking study that is confidential and independent.

23.3.3 Range of Responses to Benchmarking

Participants can produce a wide range of responses to comparative benchmarking exercises for specific topics (e.g., evaluation of risk assessment). These variations may make it difficult to combine benchmarked results into a common consensus and may create uncertainties in the conclusions. To resolve these shortcomings, benchmarking teams can provide standardized procedures, methodology, tools, and reference databases for participants to use. Additionally, the benchmarking team can provide specific

outcomes to the attendees to narrow down the response spectrum.

23.3.4 Time Consuming and Expensive Activities

Benchmarking is a time consuming and expensive activity; therefore, there should be an obvious need before getting started. Important questions such as 'Where are we with respect to our own company and with respect to other companies?', 'Are we already in an excellent position?', and 'Does the cost/benefit ratio warrant further improvement?' should be answered by the management.

23.4. EXAMPLES OF BENCHMARKING ACTIVITIES

A number of benchmarking activities have been conducted in the process industry to improve safety performance and develop best practices for the petrochemical industry; for instance, MOC practices, safety cultures of upstream businesses, emergency response planning practices, and incident investigation.

23.4.1 MOC Practices

A benchmarking study of MOC practices in the process industries has been described by Keren et al. (2002).

Chemical companies under the jurisdiction of OSHA PSM and EPA RMP regulations should develop and put into practice documented procedures to manage changes in processes, equipment, and procedures. This research provides benchmarking results from 26 out of 50 plants aimed at identifying the diversity in MOC implementation practices and developing a benchmark for MOC practices in the process industries. To do this, the researchers prepared a questionnaire to survey topics such as policy development, the size of MOC programs, emergency and temporary changes, MOC record management, and audit exercises. This study identified that local personnel developed most organization's MOC policies and procedures without using external resources. This benchmarking exercise was able to identify common difficulties and insufficiencies in implementing MOC practices that should be focused to improve the performance and the efficiency of MOC elements in PSM and RMP.

23.4.2 Emergency Response Planning

Emergency response planning is a crucial element in PSM and RMP requirements and deals with controlling hazardous events and minimizing their severity. Keren et al. (2005) found benchmarking study results on emergency preparedness and response practices gathered from 15 facilities in the process industry. A questionnaire based on the 'Guidelines for Technical Planning for On-Site Emergencies' was developed and distributed to more than 50 plants, 15 of which responded. This paper addresses key aspects for preparing emergency response plans as follows: identification methods of credible scenarios, assessment of resources and capabilities (e.g., shelters, medical facilities, fire brigades, and emergency operating centers), communication systems with local agencies, and training statuses. This study found that the most widely used methods of scenario identification are process hazard analysis (PHA) studies, such as HAZOP. However, there are numerous types of emergency planning practices used in the industry based on the size and risk involved. It would be beneficial to reach a consensus on generally accepted best practices in industries of a similar nature.

23.4.3 Safety Culture (or Safety Climate) on Offshore Installations of Oil and Gas Industry

A benchmarking study on the safety culture of the UK oil and gas industry was presented (Mearns and Thom, 2000) in 13 offshore companies to understand what forms a UK

safety culture and, eventually, promotes the sharing of the best practices. An informal benchmarking exercise was conducted from four perspectives: workforce opinions, safety management, loss costing and safety investment, and identification of best practices. The paper suggests that in order to improve the safety culture of offshore businesses, management should improve communication, policy awareness, workforce involvement, and job satisfaction.

Another benchmarking study on safety climate in hazardous environments has been conducted (Mearns et al., 2001). This paper introduces a benchmarking exercise that identifies and compares key parameters of safety climates against nine North Sea oil and gas installations. For this purpose, two types of offshore safety questionnaires were developed and distributed by the installation manager and returned by mail. It reports that a willingness to report incidents leads to a lower probability of accident involvement.

23.4.4 Incident Investigation

The second edition of CCPS on *Guidelines for Investigating Chemical Process Incidents* (2003) presents the best practices used by the leading practitioners in incident investigation. CCPS conducted survey to benchmark the methodologies used by its members in investigating incidents. The responses of the survey suggested that the companies use either one of the following combination of tools to investigate minor or major incidents:

1. The time line construction followed by logic tree development.
2. The time line construction, identification of causal factors, followed by the use of predefined trees or checklists.

REFERENCES

Arendt, S., 2006. Continuously improving PSM effectiveness—a practical roadmap. Process Saf. Prog. 25 (2), 86–93.
Camp, R.C., 1989. Benchmarking: The Search for Industry Best Practices That Lead to Superior Performance. Quality Press (American Society for Quality Control), Milwaukee, WI.
Gomm, C., Brosnan, C., Grundt, H. J., Hartog, J., Thoem., T., 2000. Global upstream HSE benchmarking. SPE International Conference on Health, Safety and Environment in Oil and Gas Exploration and Production, 26–28 June 2000, Stavanger, Norway.
Keren, N., West, H.H., Mannan, M.S., 2002. Benchmarking MOC practices in the process industries. Process Saf. Prog. 21 (2), 103–112.
Keren, N., West, H.H., Mannan, S.M., 2005. Benchmarking emergency preparedness and response practices in the process industry. J. Emergency Manage. 3 (3), 25–32.

Krajnc, D., Mele, M., Glavic, P., 2007. Fuzzy logic model for the perfor-
    mance benchmarking of sugar plants by considering best available
    techniques. Comput. Aided Chem. Eng. 24, 111−116.
Mearns, K., Thom, G., 2000. Benchmarking safety culture on the
    UKCS. SPE International Conference on Health, Safety and
    Environment in Oil and Gas Exploration and Production, 26−28
    June 2000, Stavanger, Norway.
Mearns, K., Whitaker, S.M., Flin, R., 2001. Benchmarking safety cli-
    mate in hazardous environments: a longitudinal, interorganizational
    approach. Risk Anal. 21 (4), 771−786.
Webster's Dictionary, 2013. <http://www.websters-online-dictionary.
    org/definition/benchmarking> (accessed in May, 2013.).

Liquefied Natural Gas

24.1. LNG PROPERTIES AND SUPPLY CHAIN

LNG is natural gas that is refrigerated to its liquid state at approximately $-162°C$ ($-260°F$) under atmospheric pressure. LNG consists mainly of methane and a small portion of ethane, propane, and other heavier hydrocarbons. Table 24.1 provides the properties and flammable limits of LNG (Mary Kay O'Connor Process Safety Center, MKOPSC, 2008). It is colorless, odorless, nontoxic, noncorrosive, and weighs almost 45% of the weight of water. The liquefaction process reduces the volume of natural gas by 600 times, which makes it easier and economically feasible to store and transport by vessels (Institution of Chemical Engineers, IChemE, 2007). Table 24.2 summarizes the LNG codes, standards, and regulations.

The LNG supply chain consists of four interlinked and independently operated parts. The first part of the LNG supply chain is exploration and production. Exploration activities involve seismic measurements, drilling, and well completions. The composition of natural gas varies depending on where the gas reservoir is located (Wang and Economides, 2009). Liquefaction is another key part of the LNG supply chain. The natural gas first passes through pretreatment to remove contaminants or heavier hydrocarbons to meet quality specifications at the delivery point. The residue gas mainly composed of methane is further cooled until completely liquefied and the volume of gas is reduced by a factor of 600. The LNG is stored in double-walled tanks at atmospheric pressure. The inner wall is in contact with the LNG and is made of materials suitable for cryogenic service, which includes 9% nickel steel, aluminum, or other cryogenic alloys. LNG tankers are specially designed ships to transport LNG across the seas. These tankers are constructed with double hulls to increase the integrity of the containment system and prevent leakage or rupture in an accident. Three types of cargo containment systems have evolved with industry standards, which are the spherical (Moss) design, the membrane design, and the structural prismatic design. LNG receiving terminals may include facilities to directly load LNG into tanker trucks for road distribution or, power stations, where natural gas is burned for electricity generation. The revaporized natural gas is then regulated for pressure and is transported to residential and commercial customers via a pipeline system (Foss, 2007; Wang and Economides, 2009).

24.2. LNG HAZARDS

The hazards related to handling of LNG should be addressed and fully understood in design, construction, and operations of every part of the LNG supply chain to ensure a safe operation. The two most quoted accidents related to LNG facilities and transportation are summarized in Table 24.3.

When LNG comes into contact with skin or other living tissues, it vaporizes rapidly and causes "cold burn" or frostbite. Exposure of people to a large amount of vapor accumulated near the ground could lead to asphyxiation. Many materials such as rubber, plastic, and carbon steel can become brittle at extremely cold temperatures. The vaporization of LNG can produce enormous pressures that could rupture a container or a vessel without sufficient venting or pressure relief devices. As the LNG vapor disperses downwind and the flammable portion (5–15% by volume) encounters an ignition source, the vapor cloud will burn and flame can travel through the

TABLE 24.1 LNG Properties and Flammable Limits

Property	Value	Units
Molecular weight	16.043	kmol/kg
Critical temperature	190.6	K
Critical pressure	4.64E + 06	Pa
Atmospheric boiling temperature	111.6	K
Freezing temperature	91.0	K
Liquid density at boiling point (for pure methane)	422.6	kg/m^3
Liquid density at boiling point (commercial LNG)	450.0	kg/m^3
Vapor density at boiling point	1.82	kg/m^3
Density of gas at NTP (1 atm, 20°C)	0.651	kg/m^3
Heat of vaporization	510	kJ/kg
Heat of combustion (lower)—LHC	50.0	MJ/kg
Heat of combustion (higher)—HHC	55.5	MJ/kg
Specific heat of vapor at constant pressure	2200	J/kg K
Ratio of specific heats	1.30815	
Stoichiometric air–fuel mass ratio	17.17	
Stoichiometric methane vapor concentration in air (volumetric)	9.5	%
Upper flammability limit in air (volumetric concentration)	15	%
Lower flammability limit in air (volumetric concentration)	5	%

TABLE 24.2 LNG Codes, Standards, and Regulations

Type	Title
49 CFR Part 193	Liquefied Natural Gas Facilities: Federal Safety Standards
33 CFR Part 127	Waterfront Facilities Handling Liquefied Natural Gas and Liquefied Hazardous Gas
NFPA 59A	Standard for the Production, Storage, and Handling of Liquefied Natural Gas (LNG)
NFPA 57	Standard for Liquefied Natural Gas (LNG) Vehicular Fuel Systems
EN 1473	Installation and Equipment for Liquefied Natural Gas—Design of Onshore Installations
EN 1160	Installations and Equipment for Liquefied Natural Gas—General Characteristics of Liquefied Natural Gas
EN 14620	Design and Manufacture of Site Built, Vertical, Cylindrical, Flat-Bottomed Steel Tanks for the storage of refrigerated, liquefied gases with operating temperatures between 0°C and −165°C

heat transfer from large volumes of water causes the LNG to vaporize violently, which is known as a rapid phase transition (RPT). Since there is no combustion involved, it is also described as a physical or flameless explosion (Alderman, 2005; Zinn, 2005; Qiao et al., 2006; Cormier et al., 2009).

24.3. LNG HAZARD ASSESSMENT

Consequence analysis of LNG hazards usually involves three phases: source term, dispersion, and hazardous effect. Table 24.4 summarizes some LNG spill experiments that had been conducted to understand the complex LNG vapor behavior in different scenarios. The source term describes the behavior of LNG immediately after the loss of containment, which is dominated by storage conditions and local ambient conditions. The outputs from the source term model, including the release rate (MKOPSC, 2008), pool spread (Briscoe and Shaw, 1980), and vaporization rate (Webber et al., 2010; Crowl and Louvar, 2002; Briscoe and Shaw, 1980), are used as input to subsequent dispersion modeling or fire modeling. The density of methane at its boiling point (1.82 kg/m^3) is much higher than that of air at ambient temperature, thus LNG vapor clouds behave like dense gas in the atmosphere in the event of accidental release. A broad range of consequence models have been developed to model LNG vapor dispersion and to determine exclusion zones (Ivings et al.,

cloud back to the release source and result in LNG pool fire. The typical surface emissive power of an LNG pool fire lies in the range of $220 \pm 50 \text{ kW/m}^2$, but for large fires increased production of smoke will significantly reduce this heat output, which can cause severe damages to surrounding equipment and burns to people caught within the cloud. If the release occurs in a confined space or in a heavily congested plant area, LNG vapor cloud can explode and produce damaging overpressures. Rollover refers to the rapid release of large amounts of vapor due to sudden mixing of stratified LNG layers in a storage tank, which results from either introducing LNG of a different composition to a partially filled tank or auto-stratification if sufficient nitrogen is present. As a result, the liquid from the lower layer will be superheated and give off a large amount of vapor to overpressurize the tank. When LNG is released on or under water, rapid

TABLE 24.3 Significant Accidents in LNG Industry

Date	Location	Cause	Consequence	Reference
October 20, 1944	Cleveland, OH	Primary containment fail (brittle fracture)	LNG spill, LNG fire (128 fatalities, 225 injuries)	US Bureau of Mines (1946)
October 6, 1979	Cove Point, MD	Primary containment fail (seal leak)	LNG spill, LNG explosion/fire (1 facility, 1 injury)	CH-IV International (2009)
				US Government Accountability Office (GAO) (2007)

TABLE 24.4 LNG Spill Experiments

Test Name	Organization	Type	Reference
Burro	Naval Weapons Center, Lawrence Livermore National Laboratory	1 liquid nitrogen spill, 8 LNG spills	Koopman et al. (1982, 1989)
Maplin Sands	National Maritime Institute	34 spills (LNG/refrigerated liquid propane)	Puttock et al. (1982), Koopman et al. (1989)
Coyote series	Naval Weapons Center, Lawrence Livermore National Laboratory (LLNL)	7 LNG spills, 2 liquid methane spills, 1 liquid nitrogen spill	Goldwire et al. (1983), Rodean (1984), Koopman et al. (1989)
Falcon series	Lawrence Livermore National Laboratory	5 large-scale LNG spills	Koopman et al. (1989), Brown et al. (1990)

2007). The modified Gaussian puff/plume models are some types of the empirical correlations, which are based upon the general observation that the concentration profiles downwind can be represented by a Gaussian distribution (Havens, 1980). Integral models such as SLAB, HEGADAS, and DEGADIS use similarity profiles that assume a specific shape for the crosswind profile of concentration and other properties. Shallow-layer models are based on the assumption that the lateral dimensions are much greater than the vertical dimension, which is representative of dense gas releases where low wide clouds result. These models are a compromise between Navier–Stokes equations-based models and one-dimensional integral models (Luketa-Hanlin, 2006). The most notable CFD codes for LNG vapor dispersion is FEM3 developed by LLNL and it is subsequently upgraded to FEM3C. FEM3 uses a Galerkin finite element scheme in space and a finite difference scheme in time. The latest version, FEM3C, models flow over variable terrain and objects, as well as complex cloud structures such as vortices and bifurcation (Chan, 1992, 1994; Luketa-Hanlin, 2006).

Two common approaches to estimate radiant heat from a pool fire include (1) solid flame model and (2) point source model. The solid flame model is based on some semi-empirical correlations and represents fire by a geometrical shape and its orientation due to wind effects (Raj, 2007a). Point source model assumes the overall radiation

TABLE 24.5 Mean Surface Emissive Power (MSEP) Values of Different LNG Fire Experiments

Fire Diameter (m)	Substrate Under LNG Pool	MSEP over the Visible Fire Plume Height (kW/m²)	Remarks
15	Water	185–224	China Lake tests
20	Land	140–180	Maplin Sands tests
35	Land	175 ± 30	Montoir tests GDF

energy released from a ground level point in a spherical space. Some published values for mean surface emissive power measured from LNG fire experiments are shown in Table 24.5 (MKOPSC, 2008). As the diameter of an LNG pool fire increases, smoke or soot production was observed in the field experiments. Soot resulting from pyrolysis of fuel and incomplete combustion forms in some parts of a fire where the temperature gradient is of the order of 1000 K/mm. A radiation heat model developed by Raj takes into account the soot effect (Raj, 2007b).

24.4. SAFETY MEASURES IN LNG FACILITY

Layers of protections are the safety barriers which are used to prevent, control, or mitigate undesired events or accidents. A variety of protection layers or safeguards have been used by process designers to provide an indepth defense against catastrophic accidents. Several layers of protections are currently in use in the LNG industry for controlling and minimizing the consequences associated with LNG spills, vapor dispersions, and subsequent fires or explosions.

LNG facilities are generally constructed with spill prevention as a primary goal. A method to reduce the probability of LNG release is to ensure materials in contact with the cryogenic temperatures will maintain their integrity by proper material selection, high-quality fabrication methods, proper design of components, testing of completed assemblies, etc. Even in the most carefully designed and constructed plants, there is still some chance that a spill can occur due to factors such as mechanical failure, human error, or external events. Lom (1974) discussed some of these inherent features that could be incorporated at the design stages to reduce hazards.

Controlling the vapor cloud and fire hazards of LNG spills also can be done by minimizing the area over which the spill spreads. Curbs or dikes are used to confine the spill to a predetermined location and divert the spill from the leak source to a more remote location.

Several techniques exist to enhance the dispersion of vapors created by liquefied gas spills, thus reducing the size of the resulting flammable vapor cloud. These techniques are based on reducing the effects of the variables that lead to vapor cloud formation. Other techniques are based on enhancing those variables that promote the cloud to disperse. One of the most effective ways of controlling LNG vaporization is to decrease the surface area of the LNG pool, because a larger pool surface area results in faster LNG evaporation. This can be accomplished by installing insulated dikes or natural barriers. Dikes should be able to contain 105–110% of the full capacity of an LNG tank, which assists in minimizing the surface area and directs the fluid away from the facility to a sump or a deeper pool to handle the LNG spill (Prugh and Johnson, 1987). Water curtain has been used in industry as one of the most economic and promising LNG vapor mitigation techniques. Water curtains enhance LNG vapor cloud dispersion and reduce the 'vapor cloud exclusion zone' effectively, if properly designed (IChemE, 2007). The effectiveness of water curtains depends on its own characteristics and extrinsic parameters, such as gas properties, cloud features, wind speed, and atmospheric stability.

Mitigating techniques are implemented when a vapor cloud and/or fire occurs as a result of inadequacy or failure of preventive and controlling methods. Historically, water curtains, dry chemicals, and foam have been used in controlling LNG fires. Insulating equipment and structural supports can minimize the damage in the immediate area of a fire. Facility design should always minimize the chance of a fire propagating from one part of the process area to another, especially in an LNG plant where flammable materials are present. Three main types of thermal coatings exist for exposure protection: concrete, often applied as gunite (i.e., sprayed-in-place), refractory materials, such as firebrick, tumescent, ablative, and subliming coatings (Vervalin, 1981).

In case the LNG vapors ignite and a fire results, high expansion foam application is one of the effective ways presently known to mitigate LNG vapors and suppress an LNG pool fire. Dry chemicals suitable for extinguishing LNG fires in exposed locations include sodium bicarbonate ($NaHCO_3$, ordinary), potassium bicarbonate ($KHCO_3$), or mono-ammonium phosphate (multipurpose). Common strategy for dry chemical application is to apply it on a high expansion foam-controlled LNG pool fire, where the fire size is smaller, radiant heat is reduced, and firefighters can approach the fire and apply dry chemical effectively. Dry chemical (potassium bicarbonates type) is usually combined with an Aqueous Film Forming Foam (AFFF) to create dual agent systems, which creates greater firefighting capabilities (Nohan, 1996). Some limitations of dry chemical use are: reduction of visibility after discharge, posing breathing hazards, clogging ventilation filters, inducing corrosions of exposed metals, etc.

Foamglas®PSF is the trade name for cellular glass and is widely used as a nonflammable, loading-bearing insulating material. A layer of granules or blocks of nonflammable low-density solid material is placed in the dike or bund area around LNG storage tanks. In the event of an accidental release or leak of LNG from a storage tank, the solid material placed in the bund area will float and act as a shield covering the bund area from back radiation if LNG is ignited.

REFERENCES

Alderman, J.A., 2005. Introduction to LNG safety. Process Saf. Prog. 24, 144–151.

Briscoe, F., Shaw, P., 1980. Spread and evaporation of liquid. Prog. Energ. Combust. Sci. 6, 127–140.

Brown, T.C., Cederwall, R.T., Ermak, D.L., Kooperman, R.P., McClure, J.W., Morris, L.K., 1990. Falcon Series Data Report: 1987 LNG Vapor Barrier Verification Field Trials. Gas Research Institute.

CH-IV International, 2009. Safety History of International LNG Operations.

Chan, S.T., 1992. Numerical simulations of LNG vapor dispersion from a fenced area. J. Hazard. Mater. 30, 195–224.

Chan, S.T., 1994. FEM3C: An Improved Three-Dimensional Heavy-Gas Dispersion Model: User's Manual, UCRL-MA-116567. Lawrence Livermore National Laboratory, Livermore, CA.

Cormier, B.R., Suardin, J., Rana, M., Zhang, Y., 2009. Development of design and safety specifications for LNG facilities based on experimental and theoretical research. In: Pitt, E.R., Leung, C.N. (Eds.), OPEC, Oil Prices and LNG. Nova Science Publishers.

Crowl, D.A., Louvar, J.F., 2002. Chemical Process Safety: Fundamentals with Applications.

Foss, M.M., 2007. Introduction to LNG: An Overview on Liquefied Natural Gas (LNG), its Properties, Organization of the LNG Industry and Safety Considerations. Center for Energy Economics, University of Texas at Austin, Houston, TX.

Goldwire, H.C., Mcrae, T.G., Johnson, G.W., Hipple, D.L., Koopman, R.P., Mcclure, J.W., 1983. Coyote Series Data Report: LLNL/NWC 1981 LNG Spill Tests, Dispersion, Vapour Burn, and Rapid Phase Transition—Rep. UCID-19953. Lawrence Livermore Laboratory, Livermore, CA.

Havens, J.A., 1980. An assessment of predictability of LNG vapor dispersion from catastrophic spills onto water. J. Hazard. Mater. 3 (3), 267–278.

Institution of Chemical Engineers (IChemE), 2007. BP Process Safety Series—LNG Fire Protection & Emergency Response. IChemEIChemE), 2007, UK.

Ivings, M.J., Jagger,, S.F., Lea, C.J., Webber, D.M., 2007. Evaluating Vapor Dispersion Models for Safety Analysis of LNG Facilities. Health and Safety Laboratory, UK.

Koopman, R.P., Cederwall, R.T., Ermak, D.L., Goldwire, H.C., Hogan, W.J., Mcclure, J.W., 1982. Analysis of Burro series 40 m³ LNG spills. J. Hazard. Mater. 6 (1–2), 43.

Koopman, R.P., Ermak, D.L., Chan, S.T., 1989. A review of recent field tests and mathematical modelling of atmospheric dispersion of large spills of denser-than air gases. Atmos. Environ. 23, 731.

Lom, W.L., 1974. Liquefied Natural Gas. Applied Science Barking. John Wiley & Sons, New York, NY.

Luketa-Hanlin, A., 2006. A review of large-scale LNG spills: experiments and modeling. J. Hazard. Mater. 132 (2–3), 119–140.

Mary Kay O'Connor Process Safety Center (MKOPSC), 2008. White paper: LNG pool fire modeling. In: Mannan, M.S. (Ed.), MKOPSC, Texas A & M University, College Station, TX.

Nohan, D.P., 1996. Handbook of Fire and Explosion Protection Engineering Principles for Oil, Gas, Chemical, and Related Facilities. Noyes Publications, Saddle River, NJ.

Prugh, R.W., 1987. Evaluation of unconfined vapor cloud explosion hazards. Vapor Cloud Modeling, p. 712.

Prugh, R.W., Johnson, R.W., 1987. Guidelines for Vapor Release Mitigation. Center for Chemical Process Safety.

Puttock, J.S., Blackmore, D.R., Colenbrander, G.W., 1982. Field experiments on dense gas dispersion. J. Hazard. Mater. 6, 13–41.

Qiao, Y., West, H.H., Mannan, M.S., Johnson, D.W., Cornwell, J.B., 2006. Assessment of the effects of release variables on the consequences of LNG spillage onto water using FERC models. J. Hazard. Mater. 130, 155–162.

Raj, P.K., 2007a. LNG fires: a review of experimental results, models and hazard prediction challenges. J. Hazard. Mater. 140, 444–464.

Raj, P.K., 2007b. Large hydrocarbon fuel pool fires: physical characteristics and thermal emission variations with height. J. Hazard. Mater. 140, 280–292.

Rodean, H.C., 1984. Effects of a spill of LNG on mean flow and turbulence under low wind speed, slightly stable atmospheric conditions. In: Ooms, G., Tennekes, H. (Eds.), op. cit., pp. 157.

US Bureau of Mines, 1946. Report on the Investigation of the Fire at the Liquefaction, Storage, and Regasification Plant of the East Ohio Gas Co., Cleveland, OH.

US Government Accountability Office (GAO), 2007. Maritime Security: Public Safety Consequences of a Terrorist Attack on a Tanker Carrying Liquefied Natural Gas Need Clarification. GAO.

Vervalin, C.H., 1981. Know loss-prevention information sources. Hydrocarbon Process. 60 (3), 221.

Wang, X., Economides, M., 2009. Advanced Natural Gas Engineering. Gulf Publishing Company.

Webber, D.M., Grant, S.E., Ivings, M.J., Jagger, S.F., 2010. LNG Source Term Models for Hazard Analysis: A review of the State-of-the-Art and an Approach to Model Assessment. Health and Safety Laboratory, UK.

Zinn, C.D., 2005. LNG codes and process safety. Process Saf. Prog. 24, 158–167.

Sustainable Development

Since the beginning of the twentieth century, the prosperity of the industry has promoted the quality of human life dramatically. Industries have transformed various types of resource and energy to higher value products, no matter sustainable or non-sustainable, which has led to environmental pollution from the release of toxic substances and produced a large amount of non-reusable waste. When the danger of the environmental pollution and the crisis of energy were realized in the 1980s, sustainability started getting more and more attention. The increased interest in sustainability was followed by continuous development of disciplines, technologies, and government strategies towards creating sustainable processes and products. The tenet of sustainability has been studied a lot in the past decade. This chapter provides a brief introduction to the concept of sustainable development (SD), the disciplines of SD, the measurement and analytical tools to realize SD.

25.1. SUSTAINABLE DEVELOPMENT CONCEPTS

The environmental issue was first mentioned in Silent Spring (Carson, 1962). At that time, the concept was still in the initial stage and only non-formally described in literatures. The concept of sustainable development originated with the environmental scope in 1980s (World Conservation Strategy for Conservation of Nature and Natural Resources, 1980), which proposed three basic factors—social, ecological, and economic—which have been continuously developed until today. The formulation of sustainable development was defined as:

For development to be sustainable, it must take account of social and ecological factors, as well as economic ones; of the living and non-living resource base; and of the long-term as well as the short-term advantages and disadvantages of alternative actions.

The concept of sustainable development gained wide recognition in the international scientific community after the famous report 'Our common future' (Brundtland, 1987) was published by the World Commission on Environment and Development in 1987 (Azapagic et al., 2004). Sustainable development was defined by the Commission as, 'development that meets the needs of the present without compromising the ability of future generations to meet own needs'.

After the 'Our common future' report was published, the discussions of sustainable development were conducted in different perspectives, while amplifying the concept comprehensively. One of the most important events was the Summit held by the United Nations Conference on Environment and Development in Rio de Janeiro in 1992 (UNCED, 1992), also called 'The Earth Summit'.

The message from the summit was the complexity of the problem to the world: excessive consumption by affluent populations damaging the environment, as well as the poverty issues at the same time. Different governments were required to redirect plans and policies to follow the decisions of the summit. The content of this sustainable development was defined much more precisely. After this conference, sustainable development attracted much broader attention by most of the countries of the world, and it has been greatly developed through a wide range of agreements, national legislations, and scientific studies.

Three concepts arising from the interchangeable usage and description are Green Chemistry, Green Engineering, and Sustainable Development. There is overlap and similarity among the concepts, but each of them has unique characteristics. Green chemistry deals with the development of chemical reactions using more environmentally-friendly chemicals producing less hazardous chemicals as waste. Green engineering identifies the overall environmental impact of a process using life cycle concepts and improves the process design. However, only sustainable development normally places the focus within the societal and social impacts.

Abraham (2004) did a brief discussion of the different contexts involved, 'while green chemistry addresses issues of natural capital, and green engineering addresses both natural capital and economic viability, sustainability also addresses the human condition and implores the individual to improve the quality of life for all inhabitants'.

Sustainability requires considering the social implications of the production, which is not relative to technology. The constraints are presented by economics, society, and the environment. Sustainable engineering seeks solutions that are broader than those of green engineering, by considering the system as one part of the global ecosystem including all of humanity.

25.2. SUSTAINABLE DEVELOPMENT PRINCIPLES FOR ENGINEERING

A comprehensive review of sustainable development principles for engineering was provided by Gagnon et al. (2009). Some important principles are discussed in the following sections.

25.2.1 Twelve Principles of Green Chemistry

Green Chemistry's objective is to reduce the use of hazardous substances in chemical processes, ranging from the reaction and process design, manufacture management to the products usage. Compared to Green Engineering, Green Chemistry is inclined more towards the greener pathways, reaction conditions, and chemicals.

In 1998, Anastas and Warner summarized a set of 12 principles to design environmentally beneficial products and processes or to evaluate the chemical processes (Anastas and Warner, 1998). The principles are listed below:

1. Prevention is better than cure;
2. Maximize the incorporation of all the materials used in the synthesis into the final product;
3. Use of synthetic methods to produce substances that reduces harm to human health and environment, as much as possible;
4. Reduce the toxicity of chemical products while preserving their effectiveness;
5. Reduction in the use of solvents and separating agents when possible;
6. Reduction in energy requirements;
7. Use of renewable raw materials;
8. Unnecessary chemical processes or steps should be avoided whenever possible;
9. Prefer the use of catalytic reagents over stoichiometric reagents;
10. The chemical products should degrade into harmless products at the end of their function;
11. Development of real-time, in-process monitoring, and control of processes;
12. Substances used in the process should minimize the potential for chemical incidents, releases, explosions, and fires.

25.2.2 Twelve Principles of Green Engineering

Anastas and Zimmerman (2006) proposed the 12 principles of green engineering as follows:

1. Designers need to strive to ensure that all material and energy inputs and outputs are as inherently non-hazardous as possible;
2. Prevention of waste is better than cleaning up after it is formed;
3. Separation and purification should be incorporated into the process design;
4. Design system components to maximize mass, energy, and temporal efficiency;
5. System should be output pulled rather than input pushed;
6. Embedded entropy and complexity must be viewed as an investment when making design choices on recycle, reuse, or beneficial disposition;
7. Emphasis on durability, not immortality;
8. Design for unnecessary capacity should be treated as a design flaw;
9. Minimize material diversity in multi-component products;
10. Design of processes and systems must include integration and interconnectivity with available energy and material flows;

11. Performance metrics include designing for performance in commercial 'after-life';
12. Design should be based on renewable and readily available inputs throughout life cycle.

25.2.3 The Natural Step Principle

Founded in 1989, the Natural Step (TNS) is a not-for-profit organization which is dedicated to sustainable development. The objective of TNS is to improve the production towards sustainability that leads to new creation, reduced costs, and reduced social and environmental impacts. TNS proposed a framework of four System Conditions for sustainability. The principle is based on the scientific foundation that the earth should be a sustainable system itself and the materials and energy flow between the crust and the human society should be balanced.

The scientific principles are: (1) Matter and energy cannot be created or destroyed (first law of Thermodynamics), (2) Matter and energy tend to disperse (second law of Thermodynamics), (3) Society consumes the quality, purity, or structure of matter, not its molecules, and (4) Sun-driven processes are increasing net material quality on earth.

25.2.4 Cradle-to-Cradle Principle

The Cradle-to-Cradle principle (McDonough and Braungart, 2002) depends on observing one issue in a novel perspective. The waste or byproducts of one process may be the input for another process. So, from a different viewpoint, the efficiency of the material and energy will be improved dramatically. The three aspects of C2C are:

1. *Waste equals food.*
2. *Use current solar income.*
3. *Respect biodiversity.*

25.3. SUSTAINABILITY MEASUREMENT

Measurement and accounting of sustainability is the second step after the principles are established. A problem will be difficult to solve without measuring the magnitude of it. The 1992 United Nations Conference on Environment and Development (UNCED, 1992) held in Rio de Janeiro clarified the way to develop sustainable development: 'develop and identify indicators of sustainable development in order to improve the information basis for decision making at all levels' (UNCED, 1992; Agenda 21, Chapter 40). Measuring sustainability is a complicated task because the concept of sustainability is too broad and difficult to be exactly defined. Until now there has not been uniform metrics of sustainability (Sikdar, 2007) and the argument is continuing in various

government agencies, non-profit institutions, industries, and universities. A review of sustainability metrics for process industries are given in 'Transforming Sustainability Strategy into Action' (Beloff et al., 2005). Recently, Powell (2010) and Gavani et al. (2009) provided in-depth reviews of sustainability indicators in their books.

The requirements for selecting appropriate indicators include (1) the connection to the definitions and principles of sustainability (Pezzey, 1992), (2) the representation of holistic fields (Custance and Hiller, 1998), (3) reliability and availability of data (Ramachandran, 2000; Barrios and Komoto, 2006), (4) indicator selection for specific process (Radke, 1999), (5) political objectives (Esty et al., 2006), and (6) adequate normalization, aggregation, and weighting of variables (Böhringer and Jochem, 2007).

In this section, we look at indicators of sustainability in a hierarchy of different levels, followed by several commonly used sets of metrics relevant to the process industries.

25.3.1 Public Policy-Level Indicators

Public policy-level indicators are more used by international organizations, national governments, and global non-profit organizations to evaluate the sustainability and constitute relative policies. The scale of these indicators is very broad. Böhringer and Jochem (2007) identified the most widely used indices from more than 500 reported public policy-level indicators of sustainability (Parris and Kates, 2003).

25.3.1.1 Living Planet Index

The Living Planet Index (LPI) indicates the biological diversity state of the earth, measuring changes in 7953 populations of 2554 species of vertebrates in terrene, freshwater, and seawater ecosystems. The *World Wide Fund for Nature* (WWF) developed LPI in 1997 and the results are biennially presented in the *WWF Living Planet Report*.

LPI firstly calculates the ratio of populations between consecutive years for every species, and then aggregates them together. LPI can provide the information that which habitats or ecosystems are losing species most rapidly. People can use this indicator to figure out the impact source and take effective actions to prevent the biodiversity loss.

In Living Planet Report 2010 (WWF, 2010), the index shows that species declines by 28% between 1970 and 2007 (as shown in Figure 40.1), which indicates that the natural ecosystem are being degraded because of human activities.

25.3.1.2 Ecological Footprint

The Ecological Footprint (EF) is used to judge if human's demand is larger than the Earth's tolerance extent. The first publication about EF was developed by William Rees (Rees, 1992). In 1996, the book 'Our Ecological Footprint: Reducing Human Impact on the Earth' introduced this methodology comprehensively (Wackernagel and Rees, 1998).

EF is calculated on the basis of land and water requirements to ensure the sustainability of a nation's living standard for a long time. EF is the quotient of required resource and available resource, which means if EF is larger than one then the development is unsustainable. Therefore, EF compares human demand with the ecological ability to generate resources. The survey concludes the footprint values with carbon, food, housing, goods, and services as well as the total footprint number. Currently, EF is widely used as an index of environmental sustainability around the whole world.

25.3.1.3 City Development Index

The City Development Index (CDI) was developed in 1996 and is used for measuring the sustainability level of cities. The Urban Indicators Program of the United Nations Human Settlements Program (UN-Habitat) developed this index to rank development levels of different cities.

25.3.1.4 Human Development Index

The Human Development Index (HDI) is reported annually to rank countries by 'human development' levels, which is categorized as 'very high', 'high', 'medium', and 'low' (UNDP, 2005). HDI (2010) is composed of three sub-indices as: Life Expectancy Index, Education Index and Income index. Figure 40.3 indicates the HDI map of the whole world, showing that Northern America and Europe gain the highest score.

25.3.1.5 Environmental Sustainability Index/ Environmental Performance Index

Esty et al. (2005) states 'Environmental Sustainability Index (ESI) score quantifies the likelihood that a country will be able to preserve valuable environmental resources effectively over the period of several decades'. The ESI was published by Yale University's Center for Environmental Law and Policy and developed to evaluate environmental sustainability with the data from 1999 to 2005. After 2005, a new index was developed as the 'Environmental Performance Index' (EPI), which has been published in 2006, 2008, and 2010. The structure of

EPI including Index, Objectives, Policy Categories, and Indicators is shown in Figure 25.1.

25.3.1.6 Environmental Vulnerability Index

The Environmental Vulnerability Index (EVI) was developed to recognize the severity of various types of environmental issues across the globe. This index was developed by the South Pacific Applied Geoscience Commission (SOPAC), the United Nations Environment Program, and other partners. EVI provides insights of economic and social vulnerability that can influence sustainable development of countries.

25.3.1.7 Index of Sustainable Economic Welfare/Genuine Progress Indicator

The Index of Sustainable Economic Welfare (ISEW) is used to account national welfare accounting by integrating environmental and social factors (Cobb, 1989). In 1995, ISEW was modified and renamed as the Genuine Progress Indicator (GPI) (Cobb et al., 1995). ISEW uses the following five categories to calculate the GDP which is proper to measure social welfare: distribution of income, economic activities not counted in the conventional gross national income, time adjustments, damage caused by economic activity, and the consideration of net capital endowment of foreign investors.

25.3.1.8 Well Being Index

The Well Being Index (Prescott-Allen, 2001) is the mean of the Human Well Being Index (HWI) and the Ecosystem Well Being Index (EWI). The HWI consists of five sub-indices: Health and Population, Welfare, Knowledge, Culture and Society, and Equity. The EWI also includes five sub-indices: land, water, air, species and genes, and resources deployment.

25.3.1.9 Genuine Savings

Hamilton et al. (1997) defined the Genuine Savings Index (GS) to assure that the societal capital stock never declines. The societal capital stock is composed of produced capital, human capital (e.g., knowledge) and natural capital (e.g., natural resources).

25.3.2 Corporate-Level Indicators

Unlike public policy-level indicators designed for national scale, corporate-level indicators focus on characteristics of specific industries. These indicators can provide quantitative evaluation about the sustainability level of the supply chain of the corporate or a manufacturing plant. In

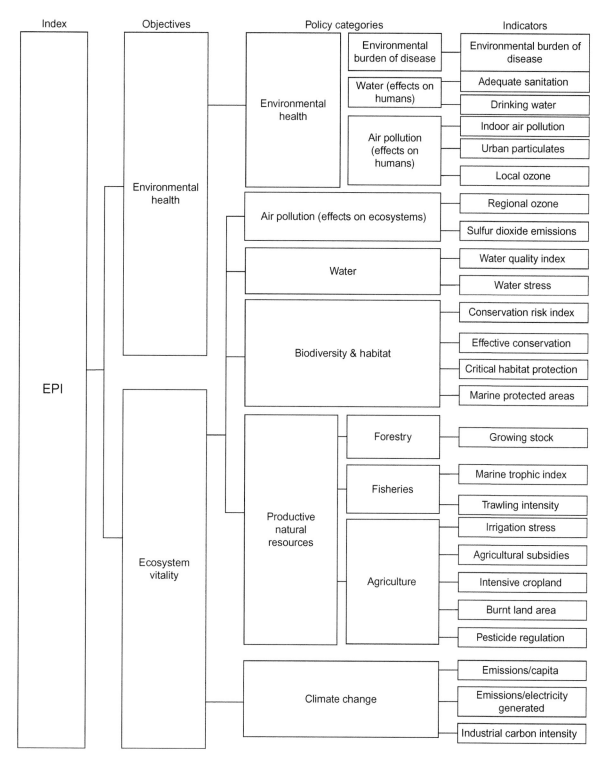

FIGURE 25.1   Construction of the EPI (Yale Center for Environmental Law & Policy, 2008).

this section, the diverse indicators are categorized into groups as:

1. Economic indices
2. Environmental indices
3. Social indices
4. Thermodynamics indices
5. Health and Safety indices.

25.3.3 Widely Used Sustainable Development Metrics for Process Industries

25.3.3.1 GRI Metrics

The Global Reporting Initiative's (GRI's) first draft was released in 1999, entitled Sustainability Reporting Guidelines. GRI is the first global framework for comprehensively reporting sustainable development and encompasses the three pillars: economic, environmental, and social issues. Over 60 countries have used GRI as their reporting basis. Figure 25.2 shows the structure of GRI.

25.3.3.2 IChemE Sustainability Metrics

The Institution of Chemical Engineers (IChemE) developed a spreadsheet of metrics of sustainability for process industries (IChemE, 2002):

1. Environmental Indicators
   A. Material Intensity
   B. Energy
   C. Water
   D. Land
   E. Emissions to atmospheric, aquatic, and land
2. Economic Indicators
   A. Profit, value, taxes
   B. Investments
3. Social Indicators
   A. Work place (employment situation, health and safety)
   B. Society.

Except for the specific indicators, weighing factors are also provided for environmental impacts.

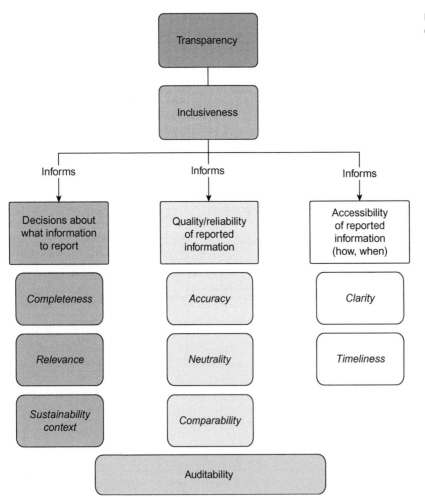

FIGURE 25.2 GRI framework guidelines (Garcia-Serna et al., 2007).

25.3.3.3 BASF Eco-Efficiency Metrics

BASF developed the eco-efficiency analysis tool to address the issues from marketplace, politics, and research (Saling et al., 2002) on the entire life cycle. The eco-analysis is based on the following aspects:

1. Total cost from customer viewpoint
2. Life cycle analysis (LCA)
3. Impacts on the health of people
4. Dangers for the environment
5. Risk potentials
6. Weighting of LCA
7. Relationship between ecology and economy
8. Weakness analysis
9. Scenarios assessment
10. Sensitivity analysis
11. Business options.

25.3.3.4 CWRT Metrics

The Center for Waste Reduction Technologies (CWRT) developed an approach using life cycle assessment to evaluate environmental and social costs on the same basis as economic costs (AIChE, 2000), so that all of the three pillars can be estimated with the same standard (Powell, 2010). Costs are categorized as follows:

1. Direct Costs
   A. Capital
   B. Fixed (labor)
   C. Feedstock
   D. Waste disposal
   E. Operating and Maintenance
2. Indirect Costs
   A. Non allocated and overhead costs
3. Future Liabilities
   A. Fines, penalties, legal fees
   B. Business interruptions
   C. Cost of environmental cleanup
4. Intangible Internal Costs
   A. Reputation
   B. Customer acceptance and loyalty
   C. Employee morale and safety
5. Intangible External Costs
   A. Costs borne by society
   B. Environmental costs within compliance.

25.4. ANALYTICAL TOOLS: LCA

25.4.1 Introduction of LCA

LCA stands for Life Cycle Analysis and is also known as 'cradle-to-grave' analysis. LCA refers to the assessment of environmental impacts of a product throughout its life cycle from its raw materials to disposal or recycling. The various stages considered in assessing the environmental impacts start from the extraction of raw materials and move towards materials processing, manufacture, distribution, use, repair, maintenance, and disposal or recycling (Life Cycle Assessment, 2011). According to ISO 14040, LCA is a 'compilation and evaluation of the inputs and outputs and the potential environmental impacts of a product system throughout its life cycle' (Life Cycle Assessment, 2005). LCA provides a clear picture of the energy and material required for a process and the environmental releases, and also identifies the potential environmental effects associated with the required energy and materials as well as the environmental releases. LCA interprets the results regarding the environmental impact, thus helps us to make better decisions. One aspect of LCA that needs to be kept in mind is that it only deals with the environmental impacts of a product or a process, not any financial, political, or social issues (Life Cycle Assessment, 2005). LCA provides us the choices of different pathways we want to use for a process or a product with least environmental impacts as possible. Even though LCA has four major steps, the steps are iterative in nature. Rather than being just steps, these parts of LCA form a circle which we need to go around several times for a single system (Life Cycle Assessment, 2005).

25.4.2 Steps to Develop LCA

ISO states four different phases towards the development of an LCA (LCA 101, 2011):

1. Goal and Scope Definition
2. Inventory Analysis
3. Impact Assessment
4. Interpretation

25.4.2.1 Goal and Scope Definition

This step defines the goal of the intended study, the functional unit, the reference flows, the system under study, assumptions, limitations, impact categories, and target audience (Life Cycle Assessment, 2011).

Functional unit refers to what exactly is being studied. The reference flow refers to the measurement of the needed inputs from a process in a system that is needed to fulfill the functionality of the defined functional unit (Life Cycle Assessment, 2005). This step also defines the depth of the study required to attain the goal and explains how and to whom the results are to be directed. The goal and scope of LCA should be defined and consistent with intended application.

25.4.2.2 Inventory Analysis

The basis of inventory analysis is the unit process. A number of unit processes linked together form a system or a process which is assessed by LCA. This step primarily deals with all inputs and outputs in a product's life cycle, beginning from product composition, origin of the materials, where they go, and the inputs and outputs related to the materials during their lifetime (Life Cycle Analysis, 2008).

The inventory of inputs includes water, energy, and raw materials whereas the outputs include the releases to water, air, and land (Life Cycle Assessment, 2011). The purpose of the inventory analysis is to determine the quantity of what comes in and what goes out of a system, including the energy and material associated with extraction, manufacture, assembly, distribution, use, and disposal at the end lifetime (Life Cycle Analysis, 2008). This is the most time-consuming phase of LCA.

25.4.2.3 Impact Assessment

In this stage, the environmental impacts identified in the previous stage are evaluated, such as the environmental impacts of energy generation for the process and the hazardous wastes generated through the process lifetime (Life Cycle Analysis, 2008). The environmental impacts of all the inputs and outputs of a product's life cycle are analyzed in this stage. The impacts associated with the materials or processes defined in the previous step are evaluated in order to get the best results in terms of environmental impact (Life Cycle Assessment, 2005).

According to ISO 14040, impact assessment is a 'phase of LCA aimed at understanding and evaluating the magnitude and significance of the potential environmental impacts of the product system' (Life Cycle Assessment, 2005). There are four mandatory steps to this phase and three optional steps that ISO has provided for effective impact assessment. The first four steps are mandatory whereas the last three steps are optional (Life Cycle Assessment, 2005):

1. Selection of impact categories
2. Selection of characterization methods: category indicators, characterization models, and factors
3. Classification (assignment of inventory results to impact categories)
4. Characterization
5. Normalization
6. Grouping
7. Weighting.

25.4.3 Software Applications

Some often-used LCA databases are listed in Table 25.1.

25.4.4 Application of LCA

LCA can be used at operational level to improve the product development, design as well as for comparison of different products, like which product is greener to the environment. LCA can also be used at strategic level by companies or authorities in product development, research, investment decisions, and waste management (Life Cycle Assessment, 2005). Increasing concerns over environmental impacts has increased the strategic use of LCA by companies for different processes and products. Also, proper use of LCA by companies is always regarded positive for their image in the society, as it reflects their seriousness regarding environmental problems. LCA has

TABLE 25.1 References to Some Often-Used LCA Databases

Name	Domain	Website
Ecoinvent 2000	Energy, transport	www.ecoinvent.ch
Buwal 250	Packaging materials	http://www.umwelt-schweiz.ch/buwal/eng/fachge-biete/fg_produkte/umsetzung/oekobilanzen/index.html
APME	Plastics	http://www/apme.org/
The Boustead Model	Fuels and materials	http://www/boustead-consultinweg.co.uk/products.htm
Japanese Database	All kinds of materials and fuels	http://www.japanfs.org/db/database.cgi?cmd = dp&num = 411&UserNum = &Pass = &AdminPass = &dp = data_e.html
FEFCO	Corrugated board	http://www.fefco.org/index.php?id = 62
IVAM LCA DATA 4	Building materials, plastics, metals, agriculture, electro-technical industry, waste management	http://www.ivam.uva.nl/nl/
Eco-Indicator 99	All kinds of materials, fuels, and waste processes	http://www.pre.nl/eco-indicator99/ei99-reports.htm

Source: Life Cycle Assessment (2005).

been applied in numerous products ranging from simple ones as shopping bags and packaging materials to complex ones like computers and automobiles (Life Cycle Assessment, 2005). LCA is not only limited to products, but can also be applied to processes, services, and activities. Examples are LCAs on hazardous waste site cleanup, waste management strategies, railroad transport (Life Cycle Assessment, 2005). LCA provides a better picture of the interactions of an activity with the environment, thus serving as a tool for environmental management. LCA helps to determine the environmental performance of a product or a process providing better view of alternatives. LCA provides a basis for assessing potential improvements in the environmental performance of the system (Life Cycle Assessment, 2005). According to Azapagic, 'These applications have mainly included the following uses, but are not limited to (Azapagic, 1999):

- strategic planning or environmental strategy development,
- product and process optimization, design, and innovation,
- identification of environmental improvements opportunities,
- environmental reporting and marketing,
- creating a framework for environmental audits.'

REFERENCES

Abraham, M.A., 2004. Sustainable Engineering: An initiative for chemical engineers. Environmental Progress. 23 (4), 261–263.

Anastas, P.T., Warner, J.C., 1998. Green Chemistry: Theory and Practice. Oxford University Press, New York, NY.

Anastas, P.T., Zimmerman, J.B., 2006. The twelve principles of green engineering as a foundation for sustainability. In: Abraham, M.A. (Ed.), Sustainability Science and Engineering Defining Principles. Elsevier.

AIChE (American Institute of Chemical Engineers), 2000. Total Cost Assessment Methodology Manual. AIChE Center for Waste Reduction Technology, Institute for SustainabilityAmerican Institute of Chemical Engineers), 2000, New York, NY.

Azapagic, A., 1999. Life cycle assessment and its application to process selection, design and optimisation. Chem. Eng. J. 73, 21.

Azapagic, A., Perdan, S., Clift, R., 2004. Sustainable Development in Practice: Case Studies for Engineers and Scientists. John Wiley & Sons.

Barrios, E., Komoto, K., 2006. Some approaches in the construction of a sustainable development index for the Philippines. Int. J. Sustain. Develop. World Ecol. 13, 277–288.

Beloff, B., Lines, M., Tanzil, D., 2005. Transforming Sustainability Strategy into Action. Wiley, Hoboken, NJ.

Böhringer, C., Jochem, P.E.P., 2007. Measuring the immeasurable: a survey of sustainability indices. Ecol. Econ. 63 (1), 1.

Brundtland, G.H., 1987. For World Commission on Environment and Development—Our Common Future. Oxford University Press, Oxford.

Carson, R., 1962. Silent Spring. Houghton Mifflin, Boston, MA.

Cobb, C.W., 1989. The index for sustainable economic welfare. In: Daly, H., Cobb, J.B. (Eds.), For the Common Good-Redirecting the Economy Toward Community, the Environment, and a Sustainable Future. Beacon Press, Boston, MA, pp. 401–457.

Cobb, C.W., Halstead, T., Rowe, J., 1995. The Genuine Progress Indicator: Summary of Data and Methodology. Redefining Progress, Washington, DC.

Custance, J., Hiller, H., 1998. Statistical issues in developing indicators of sustainable development. J. R. Stat. Soc. A. 161 (3), 281–290.

Database IVAM LCA Data 4.04. European Commission - DG Joint Research Centre – Institute for Environment and Sustainability, 2010. <http://lca.jrc.ec.europa.eu/lcainfohub/database2.vm?dbid = 126>.

Eco-Indicator 99 Impact Assessment Method for LCA, PRé Consultants, n.d. <http://www.pre-sustainability.com/content/eco-indicator-99>.

Ecoinvent 2000: Project Description, 2006. <http://gabe.web.psi.ch/projects/ecoinvent2000/index.html>.

Environmental Performance Index Report, 2008. Yale Center for Environmental Law & Policy/Center for International Earth Science Information Network at Columbia University. <http://www.yale.edu/epi/files/2008EPI_Text.pdf>.

Esty, D.C., Levy, M.A., Srebotnjak, T., de Sherbinin, A., 2005. Environmental Sustainability Index: Benchmarking National Environmental Stewardship. Yale Center for Environmental Law & Policy, New Haven, CT.

Esty, D.C., Levy, M.A., Srebotnjak, T., De Sherbinin, A., Kim, C.H., Anderson, B., 2006. Pilot Environmental Performance Index. Yale Center for Environmental Law & Policy, New Haven, CT.

Gagnon, B., Leduc, R., Savard, L., 2009. Sustainable development in engineering: a review of principles and definition of a conceptual framework. Environ. Eng. Sci. 26, 1459–1472.

Garcia-Serna, J., Perez-Barrigon, L., Cocero, M.J., 2007. New trends for design towards sustainability in chemical engineering: green engineering. Chem. Eng. J. 133, 7–30.

Gavani, F., Genti, G., Perathoner, S., Trifiro, F., 2009. Sustainable Industrial Process. Wiley-VCH Verlag GmbH&Co. KGaA, Weinheim.

Hamilton, K., Atkinson, G., Pearce, D.W., 1997. Genuine Savings as an Indicator of Sustainability. CSERGE Working Paper GEC97–03, Norwich.

HDI 2010 Index, 2010. Human Development Index. <http://hdr.undp.org/en/media/Lets-Talk-HD-HDI_2010.pdf>.

IICHE (2002). "Sustainable Development Progress Metrics: Recommended for Use in the Process Industries.": 1–29.

LCA 101, 2011. <http://www.epa.gov/nrmrl/lcaccess/lca101.html>.

Life Cycle Analysis, 2008. <http://www.enviroliteracy.org/subcategory.php/334.html>.

Life Cycle Assessment, 2005. Kirk–Othmer Encyclopedia of Chemical Technology. John Wiley & Sons.

Life Cycle Assessment, 2011. <http://en.wikipedia.org/wiki/Life_cycle_assessment>.

McDonough, W., Braungart, M., 2002. Cradle to Cradle: Remaking the Way We Make Things. North Point Press.

Parris, T.M., Kates, R.W., 2003. Characterizing and measuring sustainable development. Annu. Rev. Environ. Resour. 28 (13), 1–28.

Pezzey, J., 1992. Sustainable Development Concepts—An Economic Analysis. World Bank, Washington, DC.

Powell, J.B., 2010. Sustainable Development in the Process Industries: Cases and Impact. John Wiley & Sons.

Prescott-Allen, R., 2001. The Wellbeing of Nations. Island Press, Washington, DC.

Radke, V., 1999. Nachhaltige Entwicklung. Springer, Heidelberg.

Ramachandran, N., 2000. Monitoring Sustainability: Indices and Techniques of Analysis. Concept Publishing Company, New Delhi.

Rees, W.E., 1992. Ecological footprints and appropriated carrying capacity: what urban economics leaves out. Environ. Urban. 4 (2), 121−130.

Saling, P., Kicherer, A., Dittrich-Kramer, B., Wittlinger, R., Zombik, W., Schmidt, I., et al., 2002. Eco-efficiency analysis by BASF: the method. Int. J. Life Cycle Assess. 7 (4), 203−218.

Sikdar, S.K., 2007. Sustainability perspective and chemistry-based technologies. Ind. Eng. Chem. Res. 46 (14), 4727−4733.

The Boustead Model, 2011. <http://www.boustead-consulting.co.uk/products.htm>.

UNDP (United Nations Development Programme), 2005. Human Development Report 2005. Oxford University PressUnited Nations Development Programme), 2005, Oxford.

UNEP/WWF/IUCNNR, IUCN, 1980. World Conservation Strategy. International Union for Conservation of Nature and Natural Resources, Gland, <www.iucn.org>.

United Nations Conference on Environment and Development (UNCED), 1992. Information for Decision making, Agenda 21, Chapter 40 Rio de Janeiro. June 3−14. <http://www.un.org/earth-watch/about/docs/a21ch40.htm>.

Wackernagel, M., Rees, W., 1998. Our Ecological Footprint: Reducing Human Impact on the Earth. New Society Press.

WWF, 2010. Living Planet Report 2010. <http://assets.panda.org/downloads/lpr2010.pdf>.

Case Histories

26.1. INTRODUCTION

An essential feature of the learning process in safety and loss prevention is the study of case histories.

26.1.1 Incident Sources

The powerful impact of the internet has changed the way we obtain our information on process safety, as well as everything else. In recent years, various governmental agencies and private organizations have improved our access to details of recent and past case histories. A number of websites provide a narrative of the event, the technical reasons for the accident, as well as the management system shortcomings and valuable lessons learned. Written reports sometimes supported by graphics and photos are often published by various agencies and provided on governmental websites. There are also university and corporate organizations that collect incidents and publish the information on their websites.

Since about 1996, the US Environmental Protection Agency (EPA), under the Chemical Emergency Preparedness and Prevention, has investigated and reported on many high profile US chemical plant and refinery case histories. As a result of these investigations, some Alerts focusing on certain type of incidents have been developed. EPA reports are available at the following internet website http://www.epa.gov/emergencies/publications.htm.

The US Chemical Safety and Hazard Investigation Board (CSB) was created by an Act of the US Congress in the late 1990s and their mission is to promote the prevention of chemical accidents. The focus of the CSB is on-site and off-site chemical safety, determining causes and preventing chemical-related incidents, fatalities, injuries, property damage and enhancing environmental protection. The CSB started investigations on selected important chemical plant and refinery incidents in about 1998, and also provide reports on the internet at the following web address: http://www.csb.gov/investigations/default.aspx

The Center for Chemical Process Safety (CCPS) of the American Institute of Chemical Engineers started providing brief one-page summaries of types of hazards in late 2001. These high-impact messages were designed with the chemical process operator and others involved with the manufacturing processes to focus quickly on an important topic. This publication is entitled the *Beacon* and can be found at: www.aiche.org/ccps/safetybeacon.htm.

The Mary Kay O'Connor Process Safety Center at Texas A&M University in College Station, Texas, is an international leader in process safety and is increasing in stature each year. The Center is a rich source of process safety information as well as documented case histories. The Center's web address is: http://process-safety.tamu.edu.

The British Institution of Chemical Engineers (IChemE) offers links to a comprehensive list of well over 150 other websites of incident reports. The IChemE site mainly describes incidents of interest to chemical or process engineers, and includes a time span of incidents from historical to today's events. The collection includes both information and viewpoint. The address of this site is: http://slp.icheme.org/incidents.html.

The Delft University of Technology of the Netherlands offers a vivid array of about two dozen outstanding fire and explosion photos, an accident database, and more. The university's main research effort is focused on measuring and modeling dust and gas explosions. The incident photos can be found at: http://www.dct.tudelft.nl/part/explosion/gallery.html.

Another internet source of process safety information, including case histories, is www.acusafe.com. The AcuSafe website provides a comprehensive electronic newsletter entitled AcuSafe News. Their newsletter is devoted to serving industry, government, and interested members of the public by providing process safety practices, incident news, lessons learned, and regulatory developments.

A booklet entitled *Large Property Damage Losses in the Hydrocarbon-Chemical Industry—A Thirty Year Review* is a source of information in print form. This publication presents summaries of large property losses of incidents from around the globe. The information is from a property insurance viewpoint and is presented in an easy-to-understand form. Most of the damages are the result of fires and explosions occurring in refineries, petrochemical plants, gas-processing plants, terminals, and offshore, with a miscellaneous category also provided. It is updated every few years. The 21st edition of Marsh's *The 100 Largest Property Losses 1972–2009 (in the Hydrocarbon Industry)* reviews the 100 largest property damage losses that have occurred in the hydrocarbon processing industries since 1972.

Collections of case histories have been published in *Case Histories of Accidents in the Chemical Industry* by the Manufacturing Chemists Association (1962– <c1975>) and by Kier and Muller (1983). Specialized collections have been published on particular topics such as major hazards (Harvey, 1979, 1984; Carson and Mumford, 1979; Lees, 1980; Marshall, 1987), fire and explosions (Vervalin, 1964, 1973; Doyle, 1969), vapor cloud explosions (Gugan,

1979; Slater, 1978; Lewis, 1980; Davenport, 1987; Lenoir and Davenport, 1993), LPG (L.N. Davis, 1979), instrumentation (Doyle 1972a,b; Lees, 1976), transport (Haastrup and Brockhoff, 1990), and pipelines (Riley, 1979). The series *Safety Digest of Lessons Learned* by the API gives case histories with accompanying analyses. Another such series is published by the American Oil Company (Amoco) which includes *Hazards of Water* and *Hazards of Air*. The Canvey Reports and the Rijnmond Report gives information on certain incidents. The reports of the HSE provide case histories of investigations by a regulator, as do those of the NTSB, which deal with rail, road, pipeline, and marine accidents. Incidents involving large economic loss are described in the periodic review *100 Large Losses* by Marsh and McLennan.

Case histories are also given in the Annual Report of HM Chief Inspector of Factories, The Chemical Safety Summary of the Chemical Industry Safety and Health Council (CISHC), and in journals such as Petroleum Review and NFPA Journal. In addition, the Loss Prevention Bulletin issued by the IChemE is a major source of safety case studies for the process industries. There are also numerous case histories described in much greater detail in various papers and reports. The analysis of case histories to draw the relevant lessons is exemplified pre-eminently in the work of Trevor Kletz's many books and numerous technical articles:

- *Critical Aspects of Safety and Loss Prevention* (1990),
- *Lessons from Disaster—How Organisations Have No Memory and Accidents Recur* (1993)
- *What Went Wrong?—Case Histories of Process Plant Disasters,* 5th edn (2009)
- *An Engineer's View of Human Error*, third edition (2001a)
- *Learning from Accidents*, third edition (2001b) and numerous technical articles.
- *Still Going Wrong!: Case Histories of Process Plant Disasters and How They Could Have Been Avoided* (2003)

26.1.2 Incident Databases

There are a number of databases specifically dealing with case histories. They include the following:

- The incident databases MHIDAS (the Major Hazard Incident Data Service) is a database of incidents involving hazardous materials that had an off-site impact, or had the potential to have an off-site impact. MHIDAS is maintained by AEA Technology plc, on behalf of the UK Health and Safety Executive and EIDAS (Explosives Incidents Database Advisory Service), which contains readily accessible and searchable information about the causes and effects of explosives and firework incidents in the United Kingdom, from the 1850s up to the present day, maintained by HSE. Website: http://www.hse.gov.uk/explosives/eidas.htm.

- TNO has developed the FACTS incident database covering 24,000 (industrial) accidents (incidents) involving hazardous materials or dangerous goods that have happened worldwide over the past 90 years.
- Major Accident Reporting System (MARS) is a distribution information network, consisting of 15 local databases from the European Union and the European Commission's Joint Research Centre in Ispra (MAHB). The database allows complex text retrieval and pattern analysis.
- Hydrocarbon Releases Database System contains information, dating from October 1992, concerning offshore releases of hydrocarbons reported to the HSE Offshore Division (OSD) under the Reporting of Injuries, Diseases and Dangerous Occurrences Regulations 1995 (RIDDOR), and prior offshore legislation. Website: https://www.hse.gov.uk/hcr3/.
- HSEES: US federal database. The activities of the HSEES database program are now administered by the National Toxic Substance Incidents Program (NTSIP) Website: http://www.atsdr.cdc.gov/ntsip/National_Database.html.
- ARIA (France) database records accidents which had or could have had an adverse effect on public safety or health, agriculture, nature, and the environment. Website: http://www.aria.developpement-durable.gouv.fr/The-ARIA-Database--5425.html.
- Fire & Explosion in the Canadian Upstream Oil and Gas Industry. Website: http://www.firesandexplosions.ca/index.php.

26.1.3 Reporting of Incidents

The extent and accuracy of the reporting of incidents and injuries is variable and this creates problems, particularly for attempts to perform statistical analysis of incident data.

Three distinct problems may be identified: (1) occurrence of an incident, (2) injuries associated with an incident, and (3) national injury statistics. These three cases are considered in this section and in the next two sections, respectively.

The awareness of incidents in the engineering community worldwide varies according to the country in which the incident has occurred and the size and impact of the incident. For example, in the recent past, incidents in the United States have generally been reported in the technical press, but reports of comparable incidents in the USSR have been relatively few.

With regard to the effects of scale and impact, the probability of worldwide reporting of an incident clearly increases with these factors. The probability that an accident of the magnitude of Bhopal is not reported in countries with a free press is negligible. However, as the size of the incident decreases, the probability that it is not reported or at least is not picked up in incident collections and databases increases.

OSHA (Regulations) *Part 1094: recording and reporting occupational injuries and illness* was developed to require employers to report and record work-related fatalities, injuries, and illness. The standard also provides guidelines about the process that should be performed. For more information, refer to OSHA's website: http://www.osha.gov/pls/oshaweb/owasrch.search_form?p_doc_type=STANDARDS&p_toc_level=1&p_keyvalue=1904.

Key highlights of Standard part 1904:

- All employers covered by the OSH Act must report to OSHA any workplace incident that results in a fatality or the hospitalization of three or more employees [1904.39].
- Each employer is required by Part 1904.4 to keep records of fatalities, injuries, and illnesses, and must record each fatality, injury, and illness that is work-related, is a new case, and meets one or more of the general recording criteria of part 1904.7, or if it applies to specific cases in part 1904.8 through part 1904.12.
- Employers must save the OSHA 300 Log, the privacy case list (if one exists), the annual summary, and the OSHA 301 Incident Report forms for 5 years following the end of the calendar year that these records cover [Part 1904.33].

Within 8 h after the death of any employee from a work-related incident or the in-patient hospitalization of three or more employees as a result of a work-related incident, employers must orally report the fatality/multiple hospitalization by telephone or in person to the Area Office for OSHA, and the US Department of Labor that is nearest to the site of the incident [part 1904.39] (OSHA).

26.1.4 Reporting of Injuries in Incidents

For various reasons, accounts of incidents tend to differ in the number of injuries and, to a lesser extent, fatalities that are reported. A discussion of this problem has been given by Haastrup and Brockhoff (1991).

There are a number of reasons for differences in the numbers quoted. One is that early reports of an incident tend not be very accurate, but are sometimes quoted without sufficient qualification and the numbers then receive currency.

With regard to fatalities, a difference arises between immediate and delayed deaths. A case where a proportion of delayed deaths is fairly common is burn casualties.

The most frequent and large differences, however, are in 'injuries'. Here, much of the difference can usually be accounted for by differences in definition.

As an illustration, consider the injuries in the explosion at Laurel, Mississippi, on January 25, 1969. The NTSB report on this incident (NTSB 1969 RAR) states that 2 persons died, 33 received treatment in hospital, and numerous others were given first aid. Some authors have therefore quoted this as 2 dead, 33 injured. Eisenberg et al. (1975) refer to the NTSB report and also to a private communication from a railroad source and state that 976 persons were injured, 17 being in hospital for more than a month. This incident is one of those quoted by Haastrup and Brockhoff (1991) as an example of the problem.

26.1.5 Reporting of Injuries at National Level

It is normal for there to be a regulatory requirement for the reporting, as a minimum, of deaths and injuries. In the United Kingdom, the relevant regulations are RIDDOR 1985. The information gathered in this way is published by the HSE in the series *Health and Safety Statistics.*

The reporting is incomplete. A study supplementary to the household-based Labour Force Survey 1990 in the United Kingdom showed that only approximately 30% of non-fatal injuries were reported to the HSE, but that the level of reporting varied significantly across industries (Kiernan, 1992). For the energy industry, the proportion of such accidents reported was 75%, and it seems probable that the level of reporting in process industries such as oil refining, petrochemical, and chemicals is similar.

In the United States, the Hazardous Substances Emergency Events Surveillance (HSEES) system, maintained by the Agency for Toxic Substances and Disease Registry (ATSDR), is a key component in monitoring the acute health effects, causes, and circumstances of chemical, biological, radiological, and nuclear release in the United States (MKOPSC, 2009). For all incidents, other than petroleum-only releases, the HSEES system details the substances involved, causes of the incident, associated equipment items, the type of location, victim demographics, the type of emergency response, injury details, personnel protective equipment in use, nearby vulnerable populations, and other pertinent information. Currently, 14 state health departments have cooperative agreements with ATSDR to participate in HSEES: Colorado, Florida, Iowa, Louisiana, Michigan, Minnesota, New Jersey, New York, North Carolina, Oregon, Texas, Utah, Washington, and Wisconsin.

26.1.6 Incident Diagrams, Plans, and Maps

Many accounts of incidents give diagrams, plans, or maps showing features such as derailed tank cars, location of missiles, and location of victims. In particular, such diagrams are a feature of the accident reports by the HSE and the NTSB and of case histories described in the Loss Prevention Bulletin.

26.1.7 Incidents Involving Fire Fighting

A feature of some interest in incidents is the experience gained in fire fighting. Further information may be found in the following accounts:

Brindisi, 1977 (Mahoney, 1990)
Milford Haven, 1983 (Dyfed County Fire Brigade, 1983)
Thessalonika, 1986 (Browning and Searson, 1989)
Grangemouth, 1987 (HSE, 1989)
Port Heriot, 1987 (Mansot, 1989)

26.1.8 Incidents Involving Condensed Phase Explosives

A number of the incidents given in the following sections involve explosives. There have been, however, a large number of other explosives and munitions incidents that are of only marginal interest here. An account of explosions up to 1930 is given in *History of Explosions* by Assheton (1930), later treatments are in *Explosions in History* by Wilkinson (1966), and *Darkest Hours* by Nash (1976). Some principal explosions are listed by Nash. His list gives death tolls that in some cases differ from those given elsewhere.

26.1.9 Incidents Involving Spontaneously Combustible Substances

Spontaneous combustible substances are materials which can ignite without any flame, spark, heat, or other ignition source. Hence, the definition of 'spontaneous combustion' is 'combustion that results when materials undergo atmospheric oxidation at such a rate that the heat generation exceeds heat dissipation and the heat gradually builds up to a sufficient degree to cause the mass of material to inflame' (Carson and Mumford, 2002). A number of materials can be classified as spontaneous combustibles including linseed oil, alkyd enamel resins, and drying oils. A more comprehensive list of spontaneous combustibles can be found in the 'National Fire Protection Handbook (18th edition), Table A.10: Materials subject to spontaneous heating'. The worst consequences of spontaneous combustion include fatalities through blast trauma and asphyxiation from CO and mental disorders in survivors of disaster/accidents.

A large number of incidents involving spontaneous combustion happened in the underground mining environment. For example, there has been an increase in the number of spontaneous combustion incidents in the past years, culminating in the closure of Southland Colliery in December 2003. Since 1972, spontaneous combustion has resulted in three underground mine explosions which killed 41 workers in Queensland and some pit closures in New South Wales (Ham, 2005). A report by Richardson and Ham (1996) outlines several incidents of spontaneous combustion in Queensland underground coal mines. A total of 51 spontaneous combustion incidents were reported in Queensland between 1972 and 2004. The most damaging of these incidents were Box Flat in 1972, Kianga in 1975, and Moura No 2 Mine in 1994. These incidents involved explosions that resulted in 17, 13, and 11 fatalities, respectively, as well as closures of the mines.

26.1.10 Case Histories

One of the principal sources of case histories is the MCA collection. There are a number of themes which recur repeatedly in these case histories. They include:

Failure of communications
Failure to provide adequate procedures and instructions
Failure to follow specified procedures and instructions
Failure to follow permit-to-work systems
Failure to wear adequate protective clothing
Failure to identify correctly plant on which work is to be done
Failure to isolate plant, to isolate machinery and secure equipment
Failure to release pressure from plant on which work is to be done
Failure to remove flammable or toxic materials from plant on which work is to be done
Failure of instrumentation
Failure of rotameters and sight glasses
Failure of hoses
Failure of, and problems with, valves
Incidents involving exothermic mixing and reaction processes
Incidents involving static electricity
Incidents involving inert gas.

In the 2108 case histories described, the most frequently mentioned chemicals and the corresponding number of entries are given in Table 26.1.

The most frequently mentioned operations are given in Table 26.2.

The most frequently mentioned kinds of equipment are given in Table 26.3.

It is emphasized, however, that in many instances it is not appropriate to assign a single cause. For more details, refer to MCA collections.

TABLE 26.1 The Most Frequently Mentioned Chemicals

Ammonia	48	Chlorine	37
Caustic soda	88	Sulfuric acid	46

TABLE 26.2 The Most Frequently Mentioned Operations

Loading	38	Steaming	32
Maintenance	82	Tank entry	39
Pipe fitting	27	Transfer	46
Process reaction	66	Unloading	64
Sampling	24	Welding	36

TABLE 26.3 The Most Frequently Mentioned Equipment

Centrifuge	24	Machine	27
Cylinder	26	Pump	60
Drum	76	Rotameter, sight glass	27
Hose	57	Tank	46
Industrial truck	31	Tank car	59
Laboratory	85[a]	Tank truck	35
Line	105	Valve	86

[a]In the first 1623 case histories.

26.2. FLIXBOROUGH

At about 4:53 p.m. on Saturday, June 1, 1974, the Flixborough Works of Nypro (UK) Ltd (Nypro) were virtually demolished by an explosion of war-like dimensions. Of those working on the site at the time, 28 were killed and 36 others suffered injuries. If the explosion had occurred on an ordinary working day, the number of casualties would have been much greater. Outside the Works, injuries and damage were widespread, but no one was killed. Fifty-three people were recorded as casualties by the casualty bureau, which was set up by the police; hundreds more suffered relatively minor injuries, which were not recorded. Property damage extended over a wide area, and a preliminary survey showed that 1821 houses and 167 shops and factories had suffered to a greater or lesser degree (Parker, 1975—the *Flixborough Report*, para. 1).

The Flixborough explosion was by far the most serious accident that had occurred in the chemical industry in the United Kingdom for many years.

Within a month of the disaster, a Court of Inquiry under the chairmanship of Mr. Parker was set up under Section 84 of the Factories Act 1961 to establish the causes and circumstances of the disaster and to point out any lessons which might be learned.

The Court's report *The Flixborough Disaster, Report of the Court of Inquiry* (Parker, 1975) (the *Flixborough Report*), was the most comprehensive inquiry conducted in the United Kingdom at that time of a disaster in the chemical industry.

The Flixborough disaster was of crucial importance in the development of safety and loss prevention in the United Kingdom. It made both the industry and the public much more aware of the potential hazard of large chemical plants and led to an intensification of both the efforts within industry to ensure the safety of major hazard plants and of the demands for public controls on such plants.

The setting up of the Advisory Committee on Major Hazards (ACMH) at the end of 1974 was a direct result of the Flixborough disaster.

26.2.1 The Site and the Works

The nearest villages are Flixborough itself and Amcotts, both of which are about half a mile away. The town of Scunthorpe lies at a distance of approximately 3 miles. The works is surrounded by fields, and the population density in the neighborhood beyond is very low.

Other plants on the site included an acid plant and a hydrogen plant. There was also a large ammonia storage sphere.

26.2.2 The Process and the Plant

The cyclohexane plant consisted of a train of six reactors in series in which cyclohexane was oxidized to cyclohexanone and cyclohexanol by air injection in the presence of a catalyst. The feed to the reactors was a mixture of fresh cyclohexane and recycled material. The product from the reactors still contained approximately 94% of cyclohexane. The liquid reactants flowed from one reactor to the next by gravity. In subsequent stages, the reaction product was distilled to separate the unreacted cyclohexane, which was recycled to the reactors, and the cyclohexanone and cyclohexanol, which were converted to caprolactam. The design operating conditions in the

reactor were a pressure of 8.8 kg/cm^2 and a temperature of 155°C. The reaction is exothermic.

The heat required for initial warm-up and for supplementation of the heat of reaction during normal operation was provided by a steam-heated heat exchanger on the reactor feed. The steam flow to the exchanger was controlled by an automatic control valve. There was a bypass around this valve, which was needed to pass the larger quantities of steam required during start-up.

Removal of the heat of reaction from the reactors during normal operation was effected by vaporizing part of the cyclohexane liquid. The vaporized cyclohexane passed out in the off-gas from the reactors. The rest of this off-gas was mainly nitrogen with some unreacted oxygen.

The off-gas passed through the feed heat exchanger and then through a cooling scrubber and an absorber, in which the cyclohexane was condensed out, and thence via an automatic control valve to a flare stack.

The atmosphere in the reactor was controlled using nitrogen from high-pressure nitrogen storage tanks. The nitrogen was brought into the works by tankers.

The reactor pressure was controlled by manipulating the control valve on the off-gas line. Safety valves venting into the relief header to the flare stack were set to open at $11 \text{ kg}_f/\text{cm}^2$.

A trip system was provided which shut off air to, and injected nitrogen into, the reactors in the event of either a high oxygen content in the off-gas or a low liquid level in the nitrogen supply tank. This trip could be disarmed; however, by setting the timer to zero, fixing the duration of the purge.

26.2.3 Events Prior to the Explosion

On the evening of March 27, 1974, it was discovered that Reactor No. 5 was leaking cyclohexane. The reactor was constructed of 1/2 in. mild steel plate with 1/8 in. stainless steel bonded to it on the inside. A vertical crack was found in the mild steel outer layer of the reactor. The leakage of cyclohexane from the crack indicated that the inner stainless steel layer was also defective. It was decided that the plant be shut down for a full investigation. The following morning's inspection revealed that the crack had extended some 6 ft. This was a serious state of affairs and a meeting was called to decide on a plan of action. The decision was made to remove Reactor No. 5 and to install a bypass assembly to connect Reactors No. 4 and 6 so that the plant could continue production.

The openings to be connected on these reactors were 28 in. diameter, with bellows on the nozzle stubs, but the largest pipe which was available on site and which might be suitable for the bypass was 20 in. diameter. The two flanges were at different heights so that the connection had to take the form of a dog-leg of three lengths of

20 in. pipe welded together with flanges at each end bolted to the existing flanges on the stub pipes on the reactors.

Calculations were done to check (1) that the pipe was large enough for the required flow and (2) that it was capable of withstanding the pressure as a straight pipe. No calculations were made which took into account the forces arising from the dog-leg shape of the pipe.

No drawing of the bypass pipe was made other than in chalk on the workshop floor.

The bypass assembly was supported by a scaffolding structure. This scaffolding was intended to support the pipe and to avoid straining of the bellows during construction of the bypass. It was not suitable as a permanent support for the bypass assembly during normal operation.

No pressure testing was carried out either on the pipe or on the complete assembly before it was fitted. A pressure test was performed on the plant, however, after installation of the bypass. The equipment was tested to a pressure of $9 \ kg_f/cm^2$, but not up to the safety valve pressure of $11 \ kg_f/cm^2$. The test was pneumatic, not hydraulic.

Following these modifications, the plant was started up again. The bypass assembly gave no trouble. There did appear, however, to be an unusually large usage of nitrogen on the plant, and this was being investigated at the time of the accident.

On May 29, the bottom isolating valve on a sight glass on one of the vessels was found to be leaking. It was decided to shut the plant down to repair the leak.

On the morning of June 1, start-up began. The precise sequence of events is complex and uncertain. The crucial feature, however, is that the reactors were subjected to a pressure somewhat greater than the normal operating pressure of $8.8 \ kg_f/cm^2$.

A sudden rise in pressure up to $8.5 \ kg_f/cm^2$ occurred early in the morning when the temperature in Reactor No. 1 was still only $110°C$ and that in the other reactors was less, while later in the morning, when the temperature in the reactors was closer to the normal operating value, the pressure reached $9.1-9.2 \ kg_f/cm^2$.

The control of pressure in the reactors could normally be affected by venting the off-gas, but this procedure involved the loss of considerable quantities of nitrogen. Shortly after warm-up began, it was found that there was insufficient nitrogen to begin oxidation and that further supplies would not arrive until after midnight. Under these circumstances, the need to conserve nitrogen would tend to inhibit reduction of pressure by venting.

26.2.4 The Explosion

During the late afternoon, an event occurred which resulted in the escape of a large quantity of cyclohexane. This event was the rupture of the 20 in. bypass system,

with or without contribution from a fire on a nearby 8 in. pipe. The cyclohexane formed a vapor cloud and the flammable mixture found a source of ignition.

At about 4:53 p.m., there was a massive vapor cloud explosion. The explosion caused extensive damage and started numerous fires. The blast of the explosion shattered the windows of the control room and caused the control room roof to collapse. Of the 28 people who died in the explosion, 18 were in the control room. Some of the bodies had suffered severe injuries from flying glass. Others were crushed by the roof. No one escaped from the control room.

The main office block was also demolished by the blast of the explosion. Since the accident occurred on a Saturday afternoon, the offices were not occupied. If they had been, the death toll would have been much higher.

The fires on the site burned for many days. Even after 10 days the fires were hindering rescue work on the site. The large ammonia sphere was lifted up a few inches. It leaked slightly at a flange, but the leak was not serious.

26.2.5 Some Lessons of Flixborough

There are numerous lessons to be learned from the Flixborough disaster. The lessons include both public controls on major hazard installations and the management of such installations by industry. In the latter area, there are lessons on both management systems and technological matters and on both design and operational aspects.

The fact that alternative hypotheses were advanced concerning the cause of the explosion does not detract from these lessons, but rather means that a greater number of lessons can be drawn. Some of these lessons are now considered.

26.2.5.1 Public Controls on Major Hazard Installations

The effect of the Flixborough disaster was to raise the general level of awareness of the hazard from chemical plants and to make the existing arrangements for the control of major hazard installations appears inadequate.

The government therefore set up the ACMH to advise means of control for such installations. The committee issued three reports (Harvey, 1976, 1979, 1984).

This work was a major input to the development of the EC Major Accident Hazards Directive, which was implemented in the United Kingdom as the CIMAH Regulations 1984. These require the operator of a major hazard installation to produce a safety case. Major hazard installations receive a greater degree of supervision by the local Factory Inspectorate. The CIMAH safety case plays an important part in this.

26.2.5.2 Siting of Major Hazard Installations

As the *Flixborough Report* (para. 11) points out, the number of casualties from the explosion might have been much greater if the site had not been on open land.

The question of the siting of major *hazard* installations, or more generally land use planning in relation to such installations, became a principal concern of the ACMH, and is treated in its three reports.

26.2.5.3 Licensing of Storage of Hazardous Materials

The situation at Flixborough revealed the need for better methods of notification of major hazard installations to the local planning authorities and for greater guidance to these authorities by the HSE. The notification of major hazard installations is a requirement of the 1982 NIHHS Regulations and the 1984 CIMAH Regulations.

26.2.5.4 Regulations for Pressure Vessels and Systems

The escape of cyclohexane at Flixborough was caused by a failure of the integrity of a pressure system.

At the time of the Flixborough incident, legislation on pressure systems in the United Kingdom consisted of regulations for steam boilers and receivers and air receivers. It failed to cover the Flixborough situation in two crucial respects. It applied only to steam and air, not to hazardous materials such as cyclohexane, and it dealt only with pressure vessels, not with pressure systems. This latter point is relevant, because at Flixborough, the failure occurred in a pipe not a vessel.

The *Flixborough Report* (para. 209) recommends that existing regulations relating to the modification of steam boilers should be extended to apply to pressure systems containing hazardous materials.

26.2.5.5 The Management System for Major Hazard Installations

The works did not have a sufficient complement of qualified and experienced people. Consequently, management was not able to observe the requirement that persons given responsibilities should be competent to carry them out. In particular, there was no works engineer in post and no adequately qualified mechanical engineer on site. Moreover, individuals tended to be overworked and thus more liable to make errors.

The management system, however, is more than the individuals. It includes the whole structure which supports them. Thus the system should provide, for example, for the coverage of absence due to resignation, illness, and so on.

The use of a comprehensive set of procedures is another important aspect of the management system. A crucial procedure which was deficient at Flixborough was that for the control of plant modifications.

The role of the safety officer at Flixborough was not well defined.

The importance for major hazard installations of the management and the management system was the single most prominent theme in the work of the ACMH. Emphasis by the HSE on management aspects has steadily grown.

26.2.5.6 Relative Priority of Safety and Production

The *Flixborough Report* (paras 57, 206) drew attention to the conflict of priorities between safety and production. It states:

We entirely absolve all persons from any suggestion that their desire to resume production caused them knowingly to embark on a hazardous course in disregard of the safety of those operating the Works. We have no doubt, however, that it was this desire which led them to overlook the fact that it was potentially hazardous to resume production without examining the remaining reactors and ascertaining the cause of the failure of the fifth reactor. We have equally no doubt that the failure to appreciate that the connection of Reactor No. 4 to Reactor No. 6 involved engineering problems was largely due to the same desire.

26.2.5.7 Use of Standards and Codes of Practice

As the *Flixborough Report* (paras 61–73) describes, the 20 in. bypass assembly was not constructed and installed in accordance with the relevant standards and codes of practice. The bellows manufacturer, Teddington Bellows Ltd, produced a *Designer's Guide* which made it clear that two bellows should not be used out of line in the same pipe without adequate support for the pipe.

26.2.5.8 Limitation of Inventory in the Plant

The *Flixborough Report* (para. 14) makes it clear that the large inventory of flammable material in the plant contributed to the scale of the disaster.

The *Second Report* of the ACMH proposes that limitation of inventory should be taken as a specific design objective in major hazard installations.

The limitation of inventory is a particular aspect of the more general principle of inherently safer design, which is now widely recognized.

26.2.5.9 Engineering of Plants for High Reliability

The explosion at Flixborough occurred during a plant start-up. The *Flixborough Report* (para. 206) suggests that special attention should be given to factors which necessitate the shut-down of the chemical plant so as to minimize the number of shut-down/start-up sequences and to reduce the frequency of critical management decisions.

26.2.5.10 Dependability of Utilities

The high-pressure nitrogen required for the blanketing of the reactors at Flixborough was brought into the works by tankers. There was insufficient nitrogen available during the start-up when the explosion occurred. The *Flixborough Report* (para. 211) emphasizes the importance of assuring dependable supplies of nitrogen where these are necessary for safety.

26.2.5.11 Limitation of Exposure of Personnel

The *Flixborough Report* (para. 1) states that the number of casualties would have been much greater if the explosion had occurred on a weekday instead of on a Saturday.

The *First Report* of the ACMH (para. 68) suggests that limitation of exposure of personnel be made a specific design objective.

Aspects of limitation of exposure are controls on access to hazardous areas and design and location of buildings in or near such areas.

26.2.5.12 Design and Location of Control Rooms and Other Buildings

Of the 28 deaths at Flixborough 18 occurred in the control room. The *Flixborough Report* (para. 218) refers to various suggestions made to the inquiry concerning the siting of control rooms, laboratories, offices, etc., and the construction of control rooms on blockhouse principles.

The construction of buildings for chemical plant is considered in the *First Report* of the ACMH (para. 69).

26.2.5.13 Control and Instrumentation of Plant

The control and instrumentation system was not a prominent feature in the Flixborough inquiry. The *Flixborough Report* (para. 204) considered that the controls in the control room followed normal practice. But it also states 'Nevertheless we conclude from the evidence that greater attention to the ergonomics of plant design could provide rewarding results'.

26.2.5.14 Decision Making Under Operational Stress

The *Flixborough Report* (para. 205) draws attention to the problem of decision making under operational stress and emphasizes the desirability of reducing the number of critical management decisions which have to be made under these conditions.

Such critical decisions are not necessarily confined to management, however. Process operators may also be required to take important decisions in emergency conditions.

26.2.5.15 Restart of Plant After Discovery of a Defect

Following the discovery of the serious defect in Reactor No. 5 at Flixborough, the reactor was removed, a bypass assembly was installed, and the plant was started up again. The *Flixborough Report* (para. 57) is critical of the fact that the remaining reactors were not examined, and the cause of the failure in the fifth reactor was not ascertained before plant start-up.

26.2.5.16 Control of Plant and Process Modifications

The *Flixborough Report* (para. 209) states:

The disaster was caused by the introduction into a well-designed and constructed plant of a modification which destroyed its integrity. The immediate lesson to be learned is that measures must be taken to ensure that the technical integrity of plant is not violated.

As it happens, there was also a process modification at Flixborough, although this is not emphasized in the report. The agitator in Reactor No. 4 was not in use at the time of the disaster. The absence of agitation in the reactor could have allowed a water layer to accumulate more easily in the bottom of the vessel.

26.2.5.17 Security of and Control of Access to Plant

The *Flixborough Report* (para. 194) draws attention to the fact that there were two unguarded gates through which it was possible for anyone at any time to gain access, although this fact did not contribute to the disaster.

26.2.5.18 Planning for Emergencies

The *Flixborough Report* (para. 222) calls for a disaster plan for major hazard installations.

26.2.5.19 The Metallurgical Phenomena

The Flixborough disaster drew attention to several important metallurgical phenomena. Thus, the *Flixborough Report* describes the nitrate stress corrosion cracking of mild steel (paras 53, 212); the creep cavitation of stainless steel (para. 214); the zinc embrittlement of stainless steel (para. 213); and the use of clad mild steel vessels (para. 224).

The HSE subsequently issued *Technical Data Notes* (1976 TON 53/1, 53/2; 1977 TON 53/3) on the first three of these problems.

It was apparent from the discussion in the engineering profession following publication of the report that although metallurgical specialists were aware of failure due to zinc embrittlement, the phenomenon was not well known among engineers generally.

26.2.5.20 Vapor Cloud Explosions

The explosion at Flixborough was a large vapor cloud explosion. Although such explosions had become more common in the preceding years, none compared with Flixborough in scale and impact. The *Flixborough Report* (para. 215) draws attention to the marked lack of information on the conditions under which a vapor cloud can explode.

26.2.5.21 Investigation of Disasters and Feedback of Information on Technical Incidents

The Flixborough disaster was investigated by a Court of Inquiry. There was some feeling in the engineering profession that a legal inquiry of this kind is not a satisfactory means of establishing the facts concerning technical incidents.

The *Flixborough Report* (para. 216) states that the inquiry would have been greatly assisted if the essential instrument records in the control room had not been destroyed in the explosion, and recommends that consideration be given to systems that record and preserve vital plant information, such as the 'black box' used in aircrafts.

The points just outlined by no means exhaust the lessons to be learned from the Flixborough disaster. In particular, there are many instructive aspects of the 8 in. pipe hypothesis relating to such features as lagging fires, directed flames, and sprinkler sensor performance.

26.3. SEVESO

At 12:37 p.m. on Saturday July 9, 1976, a bursting disc ruptured on a chemical reactor at the works of the ICMESA Chemical Company at Meda near Seveso, a town of about 17,000 inhabitants some 15 miles from Milan. A white cloud drifted from the works and material from it settled out downwind. Among the substances deposited was a very small amount of TCDD, one of the most toxic chemicals known. There followed a period of great confusion due to lack of communication between the company and the authorities and the latter's inexperience in dealing with this kind of situation. Over the next few days in the contaminated area, animals died and people fell ill. A partial and delayed evacuation was carried out. In the immediate aftermath, there were no deaths directly attributable to TCDD, but a number of pregnant women who had been exposed had abortions.

A Parliamentary Commission of Inquiry, drawn equally from the Chamber of Deputies and the Senate and chaired by Deputy B. Orsini, was set up. The Commission's report (the Seveso Report) (Orsini, 1977, 1980) is a far-ranging inquiry not only into the disaster but also into controls over the chemical industry in Italy.

The impact of the Seveso disaster in Continental Europe has in some ways exceeded that of Flixborough and has led to much greater awareness of process industry hazards on the part of the public and demands for more effective controls.

The EC Directive on Major Accident Hazards of 1982 was a direct result of the Seveso disaster, and indeed this Directive was initially often referred to as the 'Seveso Directive'.

In addition to the Seveso Report, there are other accounts of the Seveso disaster and of TCDD in *The Superpoison* by Margerison et al. (1980) and *The Chemical Scythe* by Hay (1976a−e, 1981, 1982), Bolton (1978), Marshall (1980), Theofanous (1981, 1983), Rice (1982), Sambeth (1983), and Howard (1985).

26.3.1 The Site and the Works

When ICMESA built its works at Meda, the site was surrounded by fields and woods. Over the years, however, the area near the site was developed.

The reactor, Reactor A101, in which the runaway occurred, was in Department B of the works.

26.3.2 The Process and the Plant

The process which gave rise to the accident was the production of 2,4,5-trichlorophenol (TCP) in a batch reactor.

TCP is used for herbicides and antiseptics. Givaudan (parent company of ICMESA) required it for making the bacteriostatic agent hexachlorophene. It manufactured its own because the herbicide grades contained impurities unacceptable in this application. Between 1970 and 1976, some 370 te of the chemical were produced.

The reaction was carried out in two stages. Stage 1 involved the alkaline hydrolysis of 1,2,4,5-tetrachlorobenzene (TCB) using sodium hydroxide in the presence of a solvent ethylene glycol at a temperature of $170 - 180°C$ to form sodium 2,4,5-trichlorophenate. The reaction mixture also contained xylene, which was used to remove the water by azeotropic distillation. In Stage 2, the sodium trichlorophenate was acidified with hydrochloric acid to TCP and purified by distillation.

On completion of the Stage 1 reaction, some 50% of the ethylene glycol would be distilled off and the temperature of the reaction mixture lowered to $50 - 60°C$ by the addition of water.

The process was a modification by Givaudan of a process widely used in the industry. The conventional process used methanol rather than ethylene glycol and operated at some 20 bar pressure.

In this reaction, the formation of small quantities of TCDD as a byproduct is unavoidable. At a reaction temperature below $180°C$, the amount formed would be unlikely to exceed 1 ppm of TCP, but with prolonged heating in the temperature range $230-260°C$, it could increase a thousand fold.

During manufacture, nearly all (99.7%) of the TCDD formed concentrated in the distillation residues from which it was collected and incinerated. Only 0.3% found its way into the TCP, giving a maximum concentration of 10 ppb.

The reactor was a 138751 vessel with an agitator and with a steam jacket supplied with steam at 12 bar. The saturation temperature of steam at this pressure is $188°C$. The controls on the reactor were relatively primitive. There was no automatic control of the heating.

The reactor was provided with a bursting disc set at 3.5 bar and venting direct to atmosphere. The prime purpose of this disc was to prevent overpressure when compressed air was being used on the reactor.

The works had an incinerator for the destruction of hazardous plant residues at temperatures of $800-1000°C$.

26.3.3 TCDD and its Properties

The properties of TCDD are given in the *Seveso Report* and in *Dioxin, Toxicological and Chemical Aspects* by Cattabeni et al. (1978) and by Rice (1982). 2,3,7,8 Tetrachlorodibenzo-p-dioxin is also known as TCDD or dioxin. It is generated by the elimination of two molecules of HCl from 2,4,5 trichlorophenol.

TCDD is one of the most toxic substances known. The lowest LD_{50} quoted is for guinea pigs and is $0.6 \mu g/kg$, which is a dose of 0.6×10^{-9} per unit of body weight.

TCDD can be taken into the body by ingestion, inhalation, or skin contact. A leading symptom of TCDD poisoning is chloracne, which is an acne-like skin effect

caused by chemicals. A mild case of chloracne usually clears within a year, but a severe case can last many years. Other effects of TCDD include skin burns and rashes and damage to liver, kidney, and urinary systems and to the nervous system. It appears to have an unusual ability to interfere with the metabolic processes. There are varying degrees of evidence for carcinogenic, mutagenic, and teratogenic properties. These are reviewed in the report.

TCDD is a stable solid which is almost insoluble in water and resistant to destruction by incineration except at very high temperatures.

26.3.4 Events at Seveso

The start of the batch began at 16:00 on Friday, July 9 (Table 26.4). The reactor was charged with 2000 kg TCB, 1050 kg sodium hydroxide, 3300 kg ethylene glycol, and 600 kg xylene.

After the reaction had taken place, part of the ethylene glycol was distilled off, but the fraction removed was only 15% instead of the usual 50% so that most of the solvent was left in the vessel. Distillation was interrupted at 5:00 on July 10 and heating was discontinued, but water was not added to cool the reaction mass. The reactor was not brought down to the $50 - 60°C$ temperature range specified. The temperature recorder was switched off, with $158°C$ being the last temperature recorded.

The shift ended at 6:00. This time coincided with the closure of the plant for the weekend. The reactor was left with the agitation turned off but without any action to reduce the temperature of the charge.

During the weekend, with the steam turbine on reduced load, the steam supply to the reactor jacket became superheated at a temperature of about $300°C$.

26.3.5 The Release − 1

The release contaminated the vicinity of the plant with TCDD. Figure 26.1 is a map of the area, showing the zones later established, and Table 26.5 gives the concentrations later measured in these zones.

The area of Zone A was 108 ha ($1.08 km^2$) with concentrations of TCDD averaging $240 \mu g/m^2$ and rising to over $5000 \mu g/m^2$. The area of Zone B was 269 ha ($2.69 km^2$) with concentrations averaging $3 \mu g/m^2$ but rising to $43 \mu g/m^2$. In the Zone of Respect, Zone R, which had an area of 1430 ha ($14.3 km^2$), the concentrations varied from indeterminable to $5 \mu g/m^2$.

26.3.6 The Release − 2

It is known that above $230°C$, such a reactive mixture will undergo an exothermic decomposition reaction. However,

TABLE 26.4 Timetable of Events at Seveso

July	09 Friday	16:00 Final Reactor Batch Started
	10 Saturday	5:00 Final reactor batch interrupted
		12:37 Bursting disc on reactor ruptures
		Barni visits houses near plant to warn against eating garden produce. He asks carabinieri to repeat warning but they refuse
	11 Sunday	Barni and Paoletti inform von Zwehl. They are unable to contact Ghetti or his deputy. They inform Rocca and Malgrati. Von Zwehl informs Sambeth
	12 Monday	Rocca and Uberti visit ICMESA. Letter of von Zwehl to Health Officer
	13 Tuesday	Report of Uberti to Rocca and Malgrati
	14 Wednesday	Sambeth and Vaterlaus inspect Dept B and contaminated area. Uberti writes to Provincial Health Officer in Milan
	15 Thursday	Dubendorf Laboratories give first analyses showing high TCDD content. Sambeth telegraphs Waldvogel in Turkey and asks von Zwehl to inform local authorities. Sambeth informs Hoffmann La Roche and seeks clinical advice and permission to close plant. Rumours grow in Seveso. Citizens invade Uberti's office. Rocca and Uberti meet von Zwehl. They decide to declare polluted zone and to post warning notices
	16 Friday	Workforce goes on strike. Efforts are made to contact Ghetti in remote holiday farmhouse. Warning notices are erected. Rocca, Uberti, and others meet von Zwehl again. Uberti insists on evacuation. Rocca requests and obtains permission for evacuation from Deputy Prefect of Milan. Rocca contacts journalist friend
	17 Saturday	Il Giorno carries front-page headlines on 'poison gas' at Seveso. Press descends on town. Uberti contacts Cavallaro
	18 Sunday	Cavallaro's team of health inspectors investigate contaminated area. Rocca, Ghetti, Cavallaro, and Adamo meet von Zwehl. Cavallaro presses to know identity of poison released. Adamo threatens to arrest ICMESA management
	19 Monday	Waldvogel arrives in Seveso and offers local authority financial compensation, which is refused
	20 Tuesday	Cavallaro and Ghetti meet Vaterlaus in Zurich. Vaterlaus reveals that poison released was TCDD
		Uberti's letter to Provincial Health Officer in Milan arrives
		16:00 Ghetti telephones Rocca to say substance was TCDD. Prefect in Milan is informed
	21 Wednesday	9:30 Meeting at prefecture. Rivolta and Carreri take over responsibility. Carabinieri arrest von Zwehl and Paoletti
	23 Friday	9:00 Meeting at prefecture. Rivolta decides against evacuation. Reggiani visits Rocca and urges evacuation. Reggiani asked to leave Deputy Prefect's meeting unheard
		14:00 Rivolta reassures meeting in Seveso town hall. Reggiani argues with Rivolta and calls for evacuation. Carabinieri try unsuccessfully to arrest Reggiani, who returns to Switzerland
	24 Saturday	9:30 Meeting of regional health council. Vaterlaus presents map of locations affected. Council decides to recommend evacuation and defines evacuation zone (part of Zone A)
	26 Monday	179 people evacuated
	29 Thursday	Zone A extended and further 550 evacuated

accidents have been known to occur involving a reaction runaway above this temperature.

Due to the interruption of the batch, the usual sharp reduction in the temperature of the charge at the termination of the reaction did not take place, and after 7.5 h, the explosion occurred.

Investigations carried out after the accidents using differential thermal analysis showed that there exist two slow exotherms (Theofanous, 1981, 1983). One starts at about 185°C, peaking at 235°C and giving a 57°C adiabatic temperature rise. The other starts at about 255°C, peaking at 265°C and giving an estimated 114°C temperature rise.

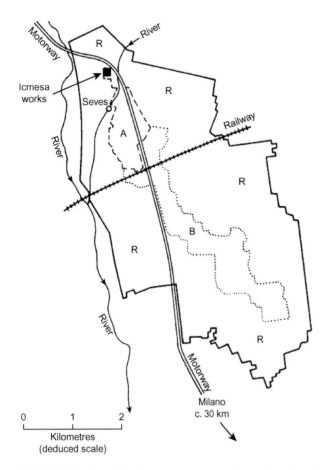

FIGURE 26.1 Plan of the Seveso area, showing Zones A and B and Zone of Respect. *Source: After Orsini (1980).*

TABLE 26.5 Concentrations of TCDD in Zones A and B at Seveso

Zone	Concentration (mg/m²)		
	Mean	Maximum	Minimum
A1	580.4	5477	n.v.d
A2	521.1	1700	6.1
A3 (north)	453.0	2015	1.7
A3 (south)	93.0	441	n.v.d.
A4	139.9	902	n.v.d.
A5	62.8	427	n.v.d.
A6	29.9	270	n.v.d.
A7	15.5	91.7	n.v.d.
B	3	43.8	n.v.d.

n.v.d. = no value determined.
Source: After Orsini (1980).

The adiabatic induction times for the two exotherms are 2.1 and 0.5 h, respectively. The decomposition exotherm starts at about $280 - 290°C$ and shows a rapid pressure rise at about $300°C$.

A mechanism for the reaction runaway has been proposed by Theofanous. It states that due to layering of the reaction mix, the amount of residual heat in the upper section of the reactor wall was sufficient to raise the temperature of the top layer of liquid to $200-220°C$, a temperature high enough to initiate exotherms leading to decomposition.

In conducting the final batch, there were failures of adherence to operating procedures. Howard (1985) instances (1) failure to distil off 50% of the glycol; (2) failure to add water; (3) failure to continue agitation; (4) switching off of the reactor temperature recorder; and (5) failure to bring the reactor temperature down from its value of $158°C$ to the normal value of $50-60°C$.

26.3.7 The Later Aftermath, Contamination, and Decontamination

There was a prolonged period in which the authorities tried to assess the situation and determine measures for dealing with it. They were advised by a team from Cremer and Warner (C&W), and an account has been given by Rice (1982).

The release was modeled using fluid jet and gas dispersion models, and predictions were made on the probable ground level concentrations of TCDD. Accounts have been given by Comer (1977) and Rice (1982). In this work, the reaction mass was taken as 2800 kg ethylene glycol, 2030 kg trichlorophenol 2030, 542 kg sodium chloride, and 562 kg sodium hydroxide.

Estimates of the amount of TCDD generated in the reactor and dispersed over the countryside vary. Cattabeni et al. (1978) give an estimated range of 0.45–3 kg released. The amount assumed in the C&W modeling work was 2 kg.

Other parameters used in the modeling were bursting disc rupture pressure at 376 kPa; a vent pipe diameter of 127 mm; and a discharge height above ground of 8 m. The bursting disc rupture was assumed to a first approximation to be due to the vapor pressure of ethylene glycol at $250°C$. From these data, the value obtained for the vapor exit velocity was 274 m/s. Two limiting cases were considered for the discharge: Case A, pure vapor (density 1.61 kg/m³) and Case B, a two-phase vapor–liquid mixture (density 8.99 kg/m³). Eyewitness accounts indicated that the actual event was intermediate between these two extremes. For Case A, the total plume rise was estimated at 83 m with the downwind distance to the maximum height as 103 m, and the corresponding figures for Case B were 55 and 95 m, respectively.

The modeling work proved useful in defining the problem and planning the decontamination. The results showed that beyond about 1.5 km, there was little difference in the concentration estimates for the two cases. There was much agreement between the predicted and the measured concentrations. This gave some confidence that the amount released was indeed about 2 kg. There had been some speculation by scientists that an amount as high as 130 kg had been released.

Various methods of decontamination were put forward by various parties, but were rejected by various reasons. TCDD is virtually insoluble in water, which reduces the threat to the water supply, but means that the effect of rain is to transfer it from the vegetation to the soil rather than to remove it.

The decontamination measures actually carried out are described by Rice. They include collection of vegetation and cleaning of buildings by high intensity vacuum cleaning followed by washing with high-pressure water jets.

It is worth mentioning that people may also have been affected by materials other than TCDD in the reactor charge. Marshall (1980) indicates that on the basis of the ground concentrations found in the contaminated area, a minimum of 0.25 kg was released. He points out that there were some 5−10 te of charge in the reactor and that several tons of material, including sodium hydroxide, must have been ejected. He suggests that burns around the hands and other parts of the body suffered by 413 persons who did not develop chloracne were probably caused by sodium hydroxide.

Data from the official report on ground concentrations in zones A and B are given in Table 26.5.

Kimbrough et al. (2010) provides a short account of the extent to which humans actually were exposed to the chemicals from soil, food, or air. The large air borne release of TCDD may have provided an acute inhalation dose compared to the one ingested from locally grown produce after the release.

26.3.8 Some Lessons from Seveso

Some of the numerous lessons to be learnt from the Seveso disaster, many of which were brought out in the Commission's report, are considered.

26.3.8.1 Public Control of Major Hazard Installations

The Seveso disaster had the effect of raising the general level of awareness of the hazards from chemical plants in Italy and in Europe, and highlighted the deficiencies in the existing arrangements for the control of major hazards.

The Italian government set up the Commission of Inquiry to investigate the cause of the disaster and to make recommendations. It also gave strong support to the development of the EC Directive.

26.3.8.2 Siting of Major Hazard Installations

The release at Seveso affected the public because in the period since the site was first occupied, housing development had encroached on the area around the plant. The accident underlined the need for separation between hazards and the public.

26.3.8.3 Acquisition of Companies Operating Hazardous Processes

A lesson which has received relatively little attention is the problem faced by a company which becomes owner by acquisition of another company operating a hazardous process. At Seveso, the problem was compounded by the fact that ICMESA was owned by Givaudan and the latter by Hoffmann La Roche, so that the company ultimately responsible was twice removed. As a result, the directors of the latter were not familiar with the hazards.

26.3.8.4 Hazard of Ultra Toxic Substances

Seveso threw into sharp relief the hazard of ultra toxic substances. The toxicity of TCDD is closer to that of a chemical warfare agent than to that of the typical toxic substance which the chemical industry is used to handling.

As it happened, the British Advisory Committee on Major Hazards was presenting its *First Report*, giving a scheme for the notification of hazardous installations, when Seveso occurred. It contained no notification proposals for ultra toxics. In the *Second Report*, this deficiency was rectified. The EC Directive places great emphasis on toxic and ultra toxic materials.

Research on the toxicity of dioxins continues to shed light on the effects that were observed after the incident. Population-based studies (Pesatori et al., 2008) and short/long term morbidity and mortality (Pesatori et al., 2003) have been investigated. During the 20 year period, mortality and morbidity findings indicate that there is an increased risk from lymphoemopoietic neoplasm, digestive system cancer, and respiratory system cancer.

26.3.8.5 Hazard of Undetected Exotherms

The characteristics of the reaction used had been investigated, and the company believed it had sufficient information. It was well aware of the hazard from the principal exotherm. Subsequent studies showed, however, that other, weaker exotherms existed which, given time, could also cause a runaway.

The complete identification of the characteristics of the reaction being operated is particularly important. If these are not known, or only partially known, there is always the danger that the design will not be as inherently safe as intended, that operating modifications will have unforeseen results, and/or that protective measures will be inadequate.

After the incident, studies were carried out to better understand and investigate the trichlorophenol synthesis, such as the research conducted by Braun et al. (1999).

26.3.8.6 Hazard of Prolonged Holding of Reaction Mass

The interruption of the reaction and the holding of the reaction mass for a prolonged period after the main reaction was complete without reducing the temperature gave time for the weak exotherms to occur and lead to runaway.

26.3.8.7 Inherently Safer Design of Chemical Processes

The reactor was intended to be inherently safe to the extent that the use of steam at 12 bar set a temperature limit of 188°C so that the contents could not be heated above this by the heating medium. Unfortunately, this feature was defeated by allowing the reaction charge to stand hot for too long so that weak exotherms occurred, which the company did not know about.

The bursting disc was not intended for reactor venting, but, as the Commission pointed out, if the set pressure of the disc had been lower, the reactor would have vented at a lower temperature and with less dioxin in the charge.

26.3.8.8 Control and Protection of Chemical Reactors

The control and protection system on the reactor was primitive. Operation of the reactor was largely manual. There was no automatic control of the cooling and no high temperature trip.

The reactor was not designed to withstand a runaway reaction. It was not rated as a pressure vessel to withstand pressure build-up prior to and during venting, the disc was not designed for reaction relief, and there was no tank to take the ultra toxic contents of the reactor.

26.3.8.9 Adherence to Operating Procedures

The conduct of the final batch involved a series of failures to adhere to the operating procedures.

26.3.8.10 Planning for Emergencies

As the account given above indicates, the handling of the emergency was a disaster in its own right. Information on the chemical released and its hazards was not immediately available from the company. There was failure of communication between the company and the local and regulatory authorities, and within those authorities. Consequently, there was lack of action and failure to protect and communicate with the public. These deficiencies might in large part have been overcome by emergency planning.

26.3.8.11 Difficulties of Decontamination

The incident illustrates the difficulties of decontamination of land where the contaminant is both ultra toxic and insoluble in water.

26.4. MEXICO CITY

At about 5:35 a.m. on the morning of November 19, 1984, a major fire and a series of explosions occurred at the terminal at San Juan Ixhuatepec (commonly known as San Juanico), Mexico City. As a consequence of this incident, more than 500 people were killed, around 7000 sustained injuries, and the terminal was destroyed.

The accident was investigated by a team from TNO who visited the site about 2 weeks after the disaster and has published its findings (Pietersen, 1985, 1986a,b, 1988). A further account has been given by Skandia International (1985).

26.4.1 The Site and the Plant

The oldest part of the plant dated from 1961 to 1962, and was thus over 20 years old. In the intervening period, residential development had crept up to the site. By 1984, the housing was within 200 m of the installation, with some houses within 130 m.

The terminal was used for the distribution of LPG, which came by pipeline from three different refineries. The main LPG storage capacity of 16,000 m^3 consisted of 6 spheres and 48 horizontal cylinders. The daily throughput was 5000 m^3. The layout of the terminal with storage tank capacities is shown in Figure 26.2. The two larger storage spheres had individual capacities of 2400 m, and the four smaller spheres had capacities of 1600 m^3. The site covered an area of 13,000 m^2.

The plant was said to have been built to API standards, and much of it to have been manufactured in the United States.

A ground level flare was used to burn off excess gas (see 7 in Figure 26.2). The flare was submerged in the

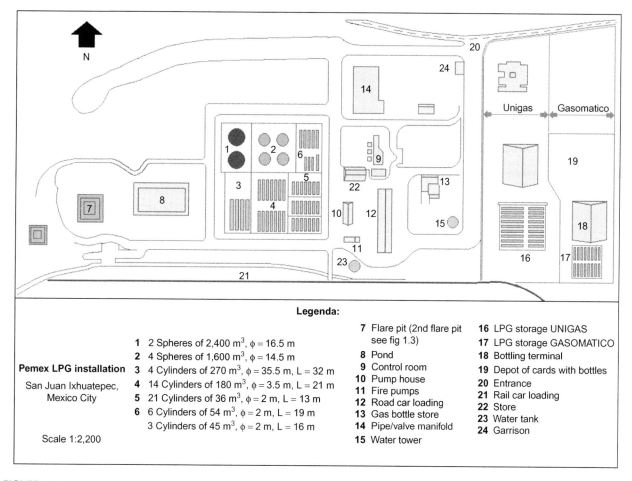

FIGURE 26.2   Layout of the PEMEX site at Mexico City (Pietersen, 1985). *Source: Courtesy of TNO.*

Legenda:

Pemex LPG installation

San Juan Ixhuatepec,
Mexico City

Scale 1:2,200

1 2 Spheres of 2,400 m³, φ = 16.5 m
2 4 Spheres of 1,600 m³, φ = 14.5 m
3 4 Cylinders of 270 m³, φ = 35.5 m, L = 32 m
4 14 Cylinders of 180 m³, φ = 3.5 m, L = 21 m
5 21 Cylinders of 36 m³, φ = 2 m, L = 13 m
6 6 Cylinders of 54 m³, φ = 2 m, L = 19 m
   3 Cylinders of 45 m³, φ = 2 m, L = 16 m

7 Flare pit (2nd flare pit see fig 1.3)
8 Pond
9 Control room
10 Pump house
11 Fire pumps
12 Road car loading
13 Gas bottle store
14 Pipe/valve manifold
15 Water tower

16 LPG storage UNIGAS
17 LPG storage GASOMATICO
18 Bottling terminal
19 Depot of cards with bottles
20 Entrance
21 Rail car loading
22 Store
23 Water tank
24 Garrison

ground to prevent the flame being extinguished by the strong local winds.

Adjoining the PEMEX plant, there were distribution depots owned by other companies. The Unigas site was some 100–200 m to the north and contained 67 tank trucks at the time of the accident. Further away was the Gasomatico site, with large numbers of domestic gas cylinders.

26.4.2 The Fire and Explosion – 1

Early on the morning of November 18, the plant was being filled from a refinery 400 km away through 8 in. feed pipeline. The previous day, the plant had become almost empty, and refilling started during the afternoon. The two larger spheres and the 48 cylindrical vessels had been filled to 90% full, and the four smaller spheres to about 50% full, so that the inventory on site was about 11,000 m³, when the incident occurred.

About 5:30 a.m., a fall in pressure was registered in the control room and at a pipeline pumping station 40 km away. The 8 in. pipe between sphere F4 and the Series G

cylinders had ruptured. The control room personnel tried to identify the cause of the pressure fall but were not successful.

The release of LPG continued for some 5–10 min. There was a slight wind of 0.4 m/s. The wind and the sloping terrain carried the gas toward the south-west. People in nearby housing heard the noise of the escape and smelled the gas.

When the gas cloud had grown to cover an area which eyewitnesses put at 200 × 150 m with a height of 2 m, it found a flare and ignited at 5:40 a.m. The cloud caught fire over a large area, causing a high flame and a violent ground shock. When this general fire had subsided, there remained a ground fire, a flame at the rupture, and fires in some 10 houses.

Workers at the plant now tried to deal with the escape. One drove off to another depot to summon help. Five others who may have been on their way to the control room or to man fire pumps were found dead and badly burned. At a late stage, someone evidently pressed the emergency shut-down button.

TABLE 26.6 Timetable of Events at Mexico City

A Seismograph Readings

1	5 h	44 min	52 s	6	6 h	49 min	38 s
2	5 h	46 min	01 s	7	6 h	54 min	29 s
3	6 h	15 min	53 s	8	6 h	59 min	22 s
4	6 h	31 min	5 s	9	7 h	01 min	27 s
5	6 h	47 min	56 s				

B General Timetable

5:30	Rupture of 8 in. pipe. Fall of pressure in control room
5:40	Ignition of gas cloud. Violent combustion and high flame
5:45	First explosion on seismograph, a BLEVE
	Fire department called
5:46	Second BLEVE, one of most violent
6:00	Police alerted and civilian traffic stopped
6:30	Traffic chaos
7:01	Last explosion on seismograph, a BLEVE
7:30	Continuing tank explosions[a]
11:00	Last tank explosion
8:00–10:00	Rescue work at its height
12:00–18:00	Rescue work continues
23:00	Flames extinguished on last large sphere

Disturbances 2 and 7 were the most intense with a Richter scale intensity of 5. Skandia suggests that the first violent combustion may not have been recorded.
[a]*Explosions of cylindrical vessels.*

In the neighboring houses, some people rushed out into the street, but most stayed indoors. Many thought it was an earthquake.

At 5:45 a.m., the first BLEVE occurred. About a minute later, another explosion occurred, one of the two most violent during the whole incident. One or two of the smaller spherical BLEVEs produced a fireball 300 m in diameter.

A rain of LPG droplets fell on the area. Surfaces covered in the liquid were set on fire by the heat from the fireballs. People also caught on fire.

There followed a series of explosions as vessels suffered BLEVEs. There were some 15 explosions over a period of an hour and a half. BLEVEs occurred in the four smaller spheres and many of the cylindrical vessels.

The explosions during the incident were recorded on a seismograph at the University of Mexico. The timing of the readings is given in Table 26.6 Section A. As the

footnote indicates, it is suggested by Skandia (1985) that the initial explosion, or violent deflagration, was probably not recorded.

Numerous missiles were generated by the bursting of the vessels. Many of these were large and travelled far. Twenty-five large fragments from the four smaller spheres weighing 10 − 40 te were found 100 − 890 m away. Fifteen of the 48 cylindrical vessels weighing 20 te became missiles and rocketed over 100 m, one travelling 1200 m. Four cylinders were not found at all. The missiles caused damage both by impact and by their temperature, which was high enough to set houses on fire.

A timetable of events during the disaster is given in Table 26.6, Section B.

26.4.3 The Emergency

Accounts of the emergency give little information about the response of the on-site management in the emergency, and deal mainly with the rescue and firefighting.

The site became the scene of a major rescue operation which reached a climax in the period 8:00 − 10:00 a.m. Some 4000 people participated in rescue and medical activities, including 985 medics, 1780 paramedics, and 1332 volunteers. At one point, there were some 3000 people in the area. There were 363 ambulances and five helicopters involved.

The rescuers were at risk from a large BLEVE. The Skandia report states, 'If a BLEVE had occurred during the later morning, a large number of those 3000 people who were engaged in rescue and guarding would have been killed'.

The fire services were called by surrounding plants and by individual members of the public at about 5:45 a.m. They went into the plant area only 3 h after the start of the incident. Initially, they moved toward the Gasomatico site where a sphere fragment had landed and started a fire that caused the domestic gas cylinders to explode.

The fire brigade also fought the fire on the two larger spheres, which had not exploded. They were at an appreciable risk from a BLEVE in these two spheres. In this event, however, these burned themselves out. The last flames on the spheres went out at about 11:00 p.m. Some 200 firemen were at the site.

26.4.4 The Fire and Explosion − 2

The TNO report gives technical information on the course of the disaster and on the fire and explosion phenomena which occurred during it.

The report discusses the effects of explosions, including vapor cloud explosions (VCEs), BLEVEs, and physical explosions; the effects of fire engulfment and heat radiation; and the effects of missiles, including fragments from bullet tanks and spheres.

The report gives estimates of the overpressure from the BLEVEs of the principal vessels. It states that the degree of blast damage to the housing was not great, that the VCE effects were not responsible for major damage, that the second explosion, a BLEVE, was the most violent and did damage houses, that the worst explosion damage was probably from gases which had accumulated in houses, and that much of the damage was caused by fire.

Films were available for many of the BLEVEs, though not for the second, violent explosion. From this evidence, the BLEVEs had diameters of 200 − 300 m and durations of some 20 s. Heavy direct fire damage was found at distances up to about 300 m, which agrees reasonably well with the estimates of fireball size.

A very large fire burned on the site for about an hour and a half, punctuated by BLEVEs. Details are given on the number, size, and range of fragments from spheres and bullet tanks.

26.4.5 Some Lessons of Mexico City

Mexico City incident is one of largest incidents in process safety and involving catastrophic BLEVEs. Things to be learned from the incident may be as described here.

26.4.5.1 Siting of Major Hazard Installations

The high death toll at Mexico City occurred because housing was too near the plant. At the time the plant was constructed the area was undeveloped, but over the years the built-up area had gradually crept closer to the site.

26.4.5.2 Layout and Protection of Large LPG Storages

The total destruction of the facility occurred because there was a failure in the overall system of protection, including layout, emergency isolation, and water spray systems.

26.4.5.3 Gas Detection and Emergency Isolation

One feature which might have averted the disaster is more effective gas detection and emergency isolation. The plant had no gas detector system and, probably as a consequence, emergency isolation happened too late.

26.4.5.4 Planning for Emergencies

One particularly unsatisfactory aspect of the emergency was the traffic chaos which built up as residents sought to flee the area and the emergency services tried to get in.

Another risk was due to the large number of rescuers who came on site due to the possibility of a BLEVE of one of the larger spheres.

26.4.5.5 Fire Fighting in BLEVE Hazard Situations

The fire services appear to have taken a considerable risk in trying to fight the fire on the two larger spheres. The potential death toll if a BLEVE had occurred was high.

26.4.5.6 Boiling Liquid Expanding Vapor Explosions

After Flixborough the problem of vapor cloud explosions received much attention. Mexico City demonstrates that boiling liquid expanding vapor explosions are an equally important hazard.

The Mexico City incident represents the largest series of major BLEVEs which have occurred, and provides much information on BLEVEs. However, the TNO report remains as the major source of information about this accident. More than two decades have passed after the explosion at the San Juanico LPG terminal, and no official investigation report has been released to the public by PEMEX or Mexican authorities.

26.5. BHOPAL

Early in the morning of December 3, 1984, a relief valve lifted on a storage tank containing highly toxic methyl isocyanate (MIC) at the Union Carbide India Ltd (UCIL) works at Bhopal, India. A cloud of MIC gas was released onto housing, including shantytowns, adjoining the site. Close to 2000 people died within a short period and tens of thousands were injured.

The accident at Bhopal is by far the worst disaster which has ever occurred in the chemical industry. Its impact has been felt worldwide, but particularly in India and the United States. There are a lot of lessons we should learn.

26.5.1 The Site and the Works

The works was in a heavily populated area. Much of the housing development closest to the works had occurred since the site began operations in 1969, including the growth of the J.P. Nagar shantytown. Although these settlements were originally illegal, in 1984 the government gave the squatters rights of ownership on the land to avoid having to evict them. Other residential areas which were affected by the gas cloud had been inhabited for over 100 years.

26.5.2 The Process and the Plant

Monomethylamine (MMA) is reacted with excess phosgene in the vapor phase to produce methylcarbamoyl chloride (MCC) and hydrogen chloride, and the reaction products are quenched in chloroform. The unreacted phosgene is separated by distillation from the quench liquid and recycled to the reactor. The liquid from the still is fed to the pyrolysis section where MIC is formed. The stream from the pyrolyser condenser passes as feed to the MIC refining still (MRS). MIC is obtained as the top product from the still. The MIC is then run to storage.

The MIC storage system (MSS) consisted of three storage tanks, two for normal use and one for emergency use. The tanks were 8 ft diameter by 40 ft long with a nominal capacity of 15,000 USgal. They were made out of 304 stainless steel with a design pressure of 40 psig at 121°C and with a hydrostatic test pressure of 60 psig. A diagram of the storage tank system is shown in Figure 26.3.

A 30 ton refrigeration system was provided to keep the tank contents at 0°C by circulating the liquid through an external heat exchanger.

On each storage tank, there was a pressure controller which controlled the pressure in the tank by manipulating two diaphragm motor valves (DMVs), a make-up valve to admit nitrogen and a blowdown valve to vent vapor. Each tank had a safety relief valve (SRV) protected by a bursting disc. They also had high temperature alarms and low and high level alarms.

A vent gas scrubber (VGS) and a flare were provided to handle vented gases. The VGS was a packed column 5 ft 6 in. diameter in which the vent gases were scrubbed with caustic soda. There were two vent headers going into the column: the process vent header (PVH), which collected the MIC system vents, and the relief valve vent header (RVVH), which collected the safety valve discharges. Each vent header was connected both to the VGS and the flare and could be routed to either. The vent stack after the VGS was 100 ft (33 m) high.

The VGS had the function of handling process vents from the PVH and of receiving contaminated MIC, in either vapor or liquid form, and destroying it in a controlled manner.

The function of the flare was to handle vent gases from the carbon monoxide unit and the MMA vaporizer

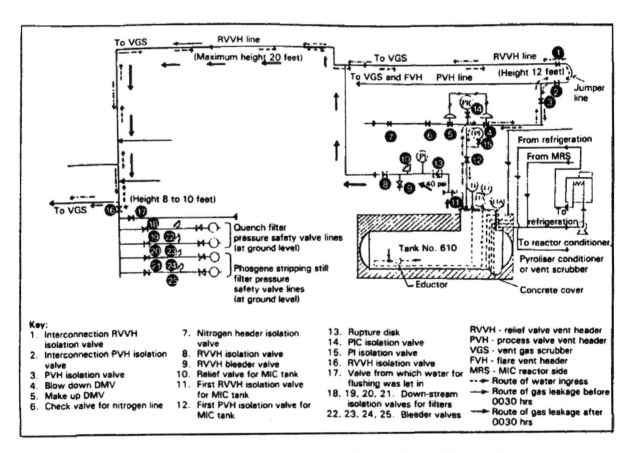

FIGURE 26.3 Flow diagram of Tank 610 system (Bhushan and Subramanian, 1985) *Source: Courtesy of Business India.*

safety valve and also vent gas from the MIC storage tanks, the MRS and the VGS.

In the 2 years preceding, the number of personnel on site were reduced, 300 temporary workers were laid off and 150 permanent workers pooled and assigned as needed to jobs, some of which they said they felt unqualified to do when interviewed. The production team at the MIC facility was cut from 12 to 6.

26.5.3 MIC and its Properties

MIC is a colorless liquid with a normal boiling point of 39°C. It has a low solubility in water. It is relatively stable when dry, but is highly reactive and in particular can polymerize and will react with water. It is flammable and has a flashpoint of -18°C and a lower flammability limit of 6% v/v. It is biologically active and highly toxic.

The high toxicity of MIC is indicated by the fact that its TLV at the time was 0:02 ppm. This is very low, relative to most typical compounds handled in industry.

MIC is an irritant gas and can cause lung edema, but it also breaks down in the body to form cyanide. The cyanide suppresses the cytochrome oxidase necessary for oxygenation of the cells and causes cellular asphyxiation.

Information on the inhalation toxicity of MIC is given by Kimmerle and Eben (1964) and ten Berge (1985).

MIC can undergo exothermic polymerization to the trimer, the reaction being catalyzed by hydrochloric acid and inhibited by phosgene. It also reacts with water, iron being a catalyst for this reaction. This reaction is strongly exothermic.

26.5.4 Events Prior to the Release

In 1982, a UCC safety team visited the Bhopal plant. Their report gave a generally favorable summary of the visit, but listed ten safety concerns.

Following this visit, valves on the MIC plant were replaced, but degraded again. At the time of the accident, the instruments on Tank 610 had been malfunctioning for over a year.

Between 1981 and 1984, there were several serious accidents on the plant. In December 1981, three workers were gassed by phosgene and one died. Two weeks later, 24 workers were overcome by another phosgene leak. In February 1982, 18 people were affected by an MIC leak. In October 1982, three workers were injured and nearby residents affected by a leak of hydrochloric acid and chloroform.

Following this latter accident, workers from the plant posted a notice in Hindi which read, 'Beware of fatal accidents ... lives of thousands of workers and citizens in danger because of poison gas ... Spurt of accidents in the factory, safety measures deficient'. These posters were also distributed in the community.

About a year before the accident, a 'jumper line' was connected between the PVH and the RVVH. Figure 26.3 shows the MIC storage tank and pipework arrangements. The jumper line is between valves 1 and 2. The object of the modification was to allow gas to be routed to the VGS if repairs had to be made to one of the vent headers.

In June 1984, the 30 ton refrigeration unit cooling the MIC storage tanks was shut down. The charge of Freon refrigerant was drained from the system.

In October, the VGS was turned off, apparently because it was thought unnecessary when MIC was only being stored not manufactured. In the same month, the flare tower was taken out of service, a section of corroded pipe leading to it was removed so that it could be replaced.

Another difficulty was experienced in pressurizing MIC storage Tank 610. It appeared that since nitrogen was passing through the make-up valve satisfactorily, the blowdown valve was leaking and preventing pressurization.

According to plant workers, there were other instrumentation faults. The high temperature alarm had long been faulty. There were also faults on the pressure controller and the level indicator.

The plant had a toxic gas alarm system. This consisted of a loud siren to warn the public and a muted siren to warn the plant. These two sirens were linked and could be activated from a plant toxic alarm box. The loud siren could be stopped from the control room by delinking the two. A procedure had been introduced according to which after delinking, the loud siren could be turned on only by the plant superintendent.

Plant workers stated that on the morning of December 2, washing operations were undertaken. Orders were given to flush out the downstream sections of four filter pressure safety valves lines. These lines are shown in Figure 26.3. In order to carry out this operation, Valve 16 on the diagram was shut, Valves $18-21$ and $22-25$ were opened and then Valve 17 was opened to admit water.

It was suggested water might have entered MIC storage Tank 610 as a result of this operation—the water washing theory. In this hypothesis, water evidently leaked through Valve 16, into the RVVH and passed through the jumper line into the PVH and then into Tank 610. This would require that Valves 3 and 12 were open to connect the tank to the PVH and Valves 1 and 2 open to connect the RVVH to the PVH via the jumper line.

26.5.5 The Release

On the evening of December 2, a shift change took place at the plant at 22:45. At 23:00, the control room operator noticed that the pressure in Tank 610 was

10 psig. This was higher than normal but within the $2-25$ psig operating pressure of the tank. At the same time, the field operator reported a leak of MIC near the VGS. At 00:15, the field operator reported an MIC release in the process area, and the control room operator saw that the pressure on Tank 610 was now at 30 psig and rising rapidly. He called the supervisor and ran outside to the tank. He heard rumbling sounds coming from the tank and a screeching noise from the safety valve and felt heat from the tank. He returned to the control room and turned the switch to activate the VGS, but this was not in operational mode because the circulating pump was not on.

At 00:20, the production supervisor informed the plant superintendent of the release. At 00:45 operations in the derivative unit were suspended due to the high concentration of MIC.

At 01:00, an operator in this unit turned on the toxic gas alarm siren. After 5 min the loud siren was switched off leaving only the muted siren on.

At about the same time, the plant superintendent and control room operator verified that MIC was being emitted from the VGS stack into the atmosphere, and turned on and directed water monitors at the stack fixed fire to knock down the vapor.

Water was also directed at the MIC tank mound and at the vent header for the VGS. Steam issued from the cracks in the concrete showing that the tank was hot.

One plant supervisor tried to climb the structure to plug the gas leak but was overcome, falling and breaking both legs.

Some time between 01:30 and 02:30, the safety valve on Tank 610 reseated and the release of MIC ceased.

At about 02:30, the loud siren was switched on again.

The cloud of MIC gas spread from the plant toward the populated areas to the south. There was a light wind and inversion conditions.

People in the housing around the plant felt the irritant effect of the gas. Many ran out of their houses, some toward the plant. Within a short period, animals and people began to die.

At Railway Colony some 2 km from the plant, where nearly 10,000 people lived, it was reported that within 4 min 150 died, 200 were paralyzed, and 600 rendered unconscious, with 5000 people severely affected.

People tried to telephone the plant but were unable to get through. At 01:45, a magistrate contacted the plant superintendent.

The cloud of toxic gas hung around the area for the entire day of December 3. During the day, it stopped moving toward the city, but resumed its movement in that direction during the night.

26.5.6 Some Lessons Learned from the Bhopal Incident

The Bhopal disaster was a wakeup call for the chemical industry around the world. Concerns from the public before the incident were mainly health issues along with environmental losses. The Bhopal incident brought up questions of public confidence in the chemical industries. There are numerous lessons learned from the Bhopal incident. Some of these lessons are now considered.

26.5.6.1 Public Control of Major Hazard Installations

The disaster at Bhopal received intense publicity for an extended period and put major hazards on the public agenda worldwide, but particularly in India and the United States, who had not reacted as strongly to Flixborough and Seveso as Europe had. The incident was a stimulus to question the practices and attitudes toward the safety of chemical plants. Extensive legislative and industrial improvements were brought up worldwide throughout decades.

26.5.6.2 Siting of and Development Control at Major Hazard Installations

Very large numbers of people were at risk from the plant at Bhopal. This situation was due in large part to the encroachment of the shantytowns, which came up to the site boundary. Although these settlements were illegal, the Indian authorities had acquiesced to them.

In this instance, however, this was not the whole story. The accident showed that the site was built close enough to populated areas to present a hazard, when used for the production of a chemical as toxic as MIC. If the manufacture of such a chemical was envisaged from the start, the problem may be regarded as one of siting. If not, it may be viewed as one of intensification of the siting hazard.

26.5.6.3 Management of Major Hazard Installations

The plant at Bhopal was by any standards a major hazard and needed to be operated by a suitable competent management. The standards of operation and maintenance do not give confidence that this was so.

There had been recent changes in the responsibility for the plant which suggest that the new management may not have been familiar with the demands of a major hazardous operation. However, many of the problems at the plant appear to have predated these changes.

26.5.6.4 Highly Toxic Substances

MIC is a highly toxic substance, much more toxic than substances such as chlorine which are routinely handled in the chemical industry. The hazard from such highly toxic substances has perhaps been insufficiently appreciated.

This hazard will only be realized if there is a mechanism for dispersion. At Bhopal, this mechanism was the occurrence of exothermic reactions in the storage tank.

26.5.6.5 Runaway Reaction in Storage

The hazard of a runaway reaction in a chemical reactor is well understood, but such a reaction in a storage tank had received very little attention. At Bhopal, this occurred due to ingress of water. When such a reaction could act as the mechanism of dispersion for a large inventory of a hazardous substance, the possibility of its occurrence should be carefully reviewed.

26.5.6.6 Water Hazard in Plants

In general terms, the hazard of water ingress into plants is well known. In particular, water may contact hot oil and vaporize with explosive force or may cause a frothover, it may corrode the equipment and it may cause a blockage by freezing. Bhopal illustrates the hazard of an exothermic reaction between a process fluid and water.

26.5.6.7 Relative Hazard of Materials in Process and in Storage

There has been a tendency to argue that the risks from materials in storage are less than from materials in process, since, although usually the inventories in storage are larger, the probability of a release is much less. The release at Bhopal was from a storage tank, albeit from one associated with a process.

26.5.6.8 Relative Priority of Safety and Production

The features which led to the accident have been described above. As indicated, the *Union Carbide Report* Union Carbide Corporation (1985) itself refers to a number of these.

The ICFTU-ICEF report states that at the time of the accident the plant was losing money and lists a number of measures which had been taken, apparently to cut costs. These include the manning cuts and the cessation of refrigeration.

26.5.6.9 Limitation of Inventory in the Plant

The hazard at Bhopal was the large inventory of highly toxic MIC. The process was the same as that used at UCC's West Virginia plant. UCIL had stated that it regarded this inventory as undesirable, but was overruled by the parent company, which wished to operate the same process at both plants.

Processes are available for the manufacture of MIC which requires only small inventories of the material. Moreover, carbaryl can be made by a route which does not involve MIC. The alternatives to the use of MIC are discussed by Kletz (1988).

26.5.6.10 Set Pressure of Relief Devices

It is preferred from an operational viewpoint that the set pressure of a relief valve be such that the valve opens when a pressure rise threatens the integrity of the vessel, but not when normal, minor operating pressure deviations occur. However, when the cause of a potential pressure rise is a runaway reaction, there is a penalty for a high set pressure as it may allow the reaction to reach a higher temperature and to proceed more rapidly before venting starts. There is a need to balance these two factors.

26.5.6.11 Disabling of Protective Systems and Alarms

It was evident that the flare system was not viewed as a critical component for the protection of the plant, since it was allowed to remain out of commission for the 3 months prior to the accident. The community alarm, which was activated 5 min after the leak, was then deactivated for nearly an hour. It is essential that there be strict procedures for the disabling of any item which is critical for protection and alarm systems. It is also very important that the time for which an item is out of action be kept to a minimum.

26.5.6.12 Maintenance of Plant Equipment and Instrumentation

The 1982 UCC safety team drew attention to the problems in the maintenance of the plant. The *Union Carbide Report* Union Carbide Corporation (1985) gives several examples of poor maintenance of plant equipment and instrumentation, and the ICFTU-ICEF report gives further details. Workers stated that leaking valves and malfunctioning instruments were common throughout the plant.

Maintenance was very slow. Although the crucial safety system was not operational, no action was conducted to fix the problem. The flare system, which was a critical protective system, had been out of commission for 3 months before the accident. The vent gas scrubber that was supposed to neutralize MIC gas with caustic soda was not fully operational. Temperature and pressure gauges were not giving reliable information and alarms failed to work. The refrigeration system to cool

MIC to around 0°C was shut down. Maintenance to keep a certain level of safety in the process should never be overlooked.

26.5.6.13 Isolation Procedures for Maintenance

A particular deficiency in the maintenance procedures was the failure to properly isolate the section of the plant being flushed out by positive isolation using a slip plate or equivalent means. The fact that the water may not have entered in this way does not detract from this lesson.

26.5.6.14 Control of Plant and Process Modifications

A principal hypothesis to explain the entry of water into Tank 610 is that the water passed through the jumper line. The installation of this jumper line was a plant modification. In a chemical plant, even the smallest change must go through reviews to verify any potential hazards before restarting the process. Company procedures called for plant modifications to be checked by the main office engineers, but they were evidently disregarded. The management of change (MOC) program is meant to prevent any unforeseen consequences that may result in disastrous outcomes.

Insufficient MOC in Bhopal was one of the reasons why several safety measures at the plant failed to operate at the time of the incident. The spare tank which was designed to be empty for emergency situations was not empty, and other tanks were filled over the recommended level with MIC. A properly managed MOC program is a crucial part of safety procedures in running a chemical plant.

26.5.6.15 Information for Authorities and Public

UCIL failed to provide full information on the substances on site to the authorities, emergency services, workers, and members of the public exposed to the hazards. Budget cuts in the company resulted in insufficient training for many plant personnel, causing a lack of general safety awareness. Many workers interviewed said they had had no information or training about the chemicals. The training, although done within the lower administrative level, can never be neglected, as it can play a significant part in performing safety measures. This is more crucial where process safety management principles and measures are insufficiently provided.

26.5.6.16 Planning for Emergencies

The response of the company and the authorities to the emergency suggests that there were no effective emergency plans, procedures, or actions in place.

Within the works, defects revealed by the emergency include the hesitation about the use of the siren system and the lack of escape routes.

The preliminary condition for emergency planning to protect the public outside the works is provision to the authorities of full information about the hazards. This was not done. In consequence, the people exposed did not know what the siren meant or what action to take, the hospitals did not know what they might be called on to handle, and so on.

Likewise, the essential action in an actual emergency is to inform the authorities what has happened and what the hazards are. On the morning of the accident, the hospitals were in the dark about the nature and effects of the toxic chemical whose victims they were trying to treat.

Legislation to frame the outline of chemical emergency responses, covering specification on emergency planning, emergency release notification, hazardous chemical storage reporting requirements, and toxic chemical release inventory, is required at the community level.

26.5.6.17 Hazard Evaluations

The MIC process was running without fully functional safety systems. The refrigeration system was shut down and gauges and alarms were not reliable. The vent gas scrubber was not fully operative and the flare tower and water curtain system were not designed to prevent any unexpected release of MIC. If any proper hazard evaluations had been done, the management level would question the safety of running the process without the factors mentioned above.

Hazard evaluations help to identify any inherent hazard or potential critical scenario when running the process. It is recommended to evaluate the hazards that may arise from the product and communicate the results to the plant workers as well.

26.5.7 The Worldwide Impact of the Bhopal Incident

The Bhopal incident was a tragedy that attracted widespread attention to the concept of process safety. Although process safety had been generally slighted in the chemical industry, this incident provided the momentum to implement process safety into the industry's standard practices.

The US Congress passed the Clean Air Act Amendments (CAAA) in 1990 after Bhopal and several other domestic incidents. It included some major provisions toward the chemical industry and enlisted the Occupational Safety and Health Administration (OHSA) to enforce the Process Safety Management of Highly Hazardous Chemicals (29 CFR 1910.119). The legislation also directed the

Environmental Protection Agency (EPA) to establish the Risk Management Program (40 CFR 68) to target the safety of personnel and communities in case of an accidental release.

Despite the complex political and geographical background, the European Union has tried to set a common regulatory frame work toward the requirements of the European Commission and the United Nations Economic Commission for Europe. Before the Bhopal incident, the Seveso incident in 1976 initiated the 'Council Directive 82/501/EEC on the major-accident hazards of certain industrial actives' known as 'Seveso I' to prevent serious consequences to the environments and public. After the Bhopal incident, the regulation was amended and the Directive 88/610/EEC 'Seveso II' related to the Control of Major Accident Hazards (COMAH) came into effect.

Before the Bhopal incident, Indian regulations lacked a sufficient system to protect the public from industry hazards. The Environment Protection Act of 1986 was established and, along with the Indian Factories Act of 1948, the Air Act was amended to regulate the facilities dealing with hazardous materials. The Hazardous Waste Rules of 1989 and the Public Liability Insurance Act of 1991 followed, and the Environment Protection Rules were amended in 1992.

26.6. PASADENA

Shortly after 1:00 p.m. on October 23, 1989, a release occurred in a polyethylene plant at the Phillips 66 Company's chemical complex at Pasadena, near Houston, Texas. A vapor cloud formed and ignited, giving rise to a massive vapor cloud explosion. There followed a series of further explosions and a fire. Twenty-two people on the site were killed and one later died from injuries, making a death toll of 23. The number injured is variously given as 130 and 300.

A report on the investigation of the accident has been issued by OSHA (1990). Other accounts include those of Mahoney (1990), T. Richardson (1991), and Scott (1992).

26.6.1 The Site and the Plant

The Phillips works was sited in the Houston Chemical Complex along the Ship Channel, the location of a number of process companies.

The plant in which the release occurred was Plant V, one of the two active polyethylene plants in the complex. The plant operated at high pressure (700 psi) and high temperature. The process involved the polymerization of ethylene in isobutane, the catalyst carrier. Particles of polyethylene settled out and were removed from settling legs.

26.6.2 Events Prior to the Explosion

On the previous day work began to clear three of the six settling legs on Reactor No. 6, which were plugged. The three legs were prepared by a company operator and were handed over to the specialist maintenance contractors, Fish Engineering.

At 8:00 a.m. on Monday, October 23, work began on the second of the three blocked legs, Leg No. 4. The isolation procedure was to close the DEMCO ball valve and disconnect the air lines to it.

The maintenance team partially disassembled the leg and were able to remove part of the plug, but part remained lodged in the pipe 12−18 in. below the ball valve. One of the teams was sent to the control room to seek assistance. Shortly after, at 1:00 p.m., the release occurred.

Although both industry practice and Phillips corporate safety procedures require isolation by means of a double block system or a blind flange, at local plant level a procedure had been adopted which did not conform to this.

It was subsequently established that the DEMCO ball valve was open at the time of the release. The air hoses to the valve had been cross-connected so that the supply which should have closed it actually opened it. The hose connectors for the 'open' and 'close' sides of the valve were identical, thus allowing this cross-connection to be made. Although procedures laid down that the air hoses should not be connected during maintenance, there was no physical barrier to the making of such a connection. The ball valve had a lockout system, but it was inadequate to prevent the valve being inadvertently or intentionally opened during maintenance.

26.6.3 The Explosion

The mass of gas released was estimated as some 85,200 lb of a mixture of ethylene, isobutane, hexene, and hydrogen, which escaped within seconds. The release was observed by five eyewitnesses. A massive vapor cloud formed and moved rapidly downwind.

Within 90 − 120 s, the vapor cloud found a source of ignition. Possible ignition sources were a gas-fired catalyst activator with an open flame; welding and cutting operations; an operating forklift truck; electrical gear in the control building and the finishing building; 11 vehicles parked near the polyethylene plant office; and a small diesel crane, although this was not operating.

The TNT equivalent of the explosion was estimated in the OSHA report as 2.4 tons. An alternative estimate from seismograph records is 10 tons.

There followed two other major explosions, one when two 20,000 gallon isobutane storage tanks exploded and the other when another polyethylene plant reactor failed catastrophically, the timings being some 10–15 min and some 25–45 min, respectively, after the initial explosion. One witness reported hearing 10 separate explosions over a 2 h period.

Debris from the explosion was found 6 miles from the site.

All 22 of those who died at the scene were within 250 ft of the point of release and 15 of them were within 150 ft.

Injuries which occurred outside the site were mainly due to debris from the explosion.

The explosion resulted in the destruction of two HOPE plants.

26.6.4 The Emergency and the Aftermath

People in the immediate area of the release began running away as soon as they realized that gas was escaping. The alarm siren was activated, but the level of noise in the finishing building was such that there was a question whether some employees there failed to hear it.

The immediate response to the emergency was provided by the site fire brigade, which undertook rescue and care of the injured and began fighting the fire. Twenty-three people were unaccounted for, but for an extended period, the area of the explosion remained dangerous to enter.

Severe difficulties were experienced in fighting the fires resulting from the explosion. There was no dedicated fire water system, water for firefighting being drawn from the process water system. The latter suffered severe rupture in the explosion so that water pressure was too low for firefighting purposes. Fire hydrants were sheared off by the blast. Fire water had to be brought by hose from remote sources such as settling ponds, a cooling tower, a water treatment plant, and a water main on a neighboring plant. These difficulties were compounded by failures of the fire pumps. The electrical cables supplying power to the regular fire pumps were damaged by the fire so that these pumps were put out of action. Further, of the three backup diesel fire pumps, one was down for maintenance and one quickly ran out of fuel. Despite these problems, the fire was brought under control within some 10 h.

The handling of the emergency was handicapped by the facts that the intended command center had been damaged and that telephone communications were disrupted. Telephone lines were jammed for some hours following the accident.

The emergency response was coordinated by the site chief fire officer and involved the local Channel Industries Mutual Aid (CIMA) organization, a cooperative of some 106 members in the Houston area.

More than 100 people were evacuated from the administration building across the Houston Ship Channel by the US Coast Guard and by Houston fireboats; they

would otherwise have had to cross the area of the explosion to reach safety.

The media were quickly aware of the explosion, and within an hour, there were on-site 150 media personnel from 40 different organizations.

The financial loss in this accident is comparable with, and may exceed, that of the Piper Alpha disaster. Redmond (1990) has quoted a figure of $1400 million, divided almost equally between property damage and business interruption losses.

On the basis of a review of company reports and of the defects found during the investigation of the disaster, OSHA issued a citation to the company for wilful violations of the 'general duty' clause. The citation covered the lack of hazard analysis; plant layout and separation distances; flammable gas detection; ignition sources; building ventilation intakes; and the fire water system; the permit system; and isolation for maintenance.

26.6.5 Some Lessons of Pasadena

Some of the lessons to be learned from Pasadena are described here:

26.6.5.1 Management of Major Hazard Installations

The OSHA report details numerous defects in the management of the installation. Some of these are described below.

26.6.5.2 Hazard Assessment of Major Hazard Installations

According to the report, the company had made no use of hazard analysis or an equivalent method to identify and assess the hazards of the installation.

26.6.5.3 Plant Layout and Separation Distances

The report was critical of the separation distances in the plant in several respects. It stated that the separation distances between process equipment plants did not accord with accepted engineering practice and did not allow time for personnel to leave the polyethylene plant safely during the initial vapor release; and that the separation distance between the control room and the reactors was insufficient to allow emergency shut-down procedures to be carried out.

26.6.5.4 Location of Control Room

As just mentioned, the control room was too close to the plant. In addition, the control room was not built up to the expected explosion over protection necessary for their location. It was destroyed in the initial explosion.

26.6.5.5 Building Ventilation Intakes

The ventilation intakes of buildings close to or downwind of the hydrocarbon processing plants were not arranged so as to prevent intake of gas in the event of a release.

26.6.5.6 Minimization of Exposure of Personnel

Closely related to this, there was a failure to minimize the exposure of personnel. Not only the control room but the finishing building had relatively high occupancy.

26.6.5.7 Escape and Escape Routes

As already stated, the separation distances were not such as to allow personnel on the polyethylene plant to escape safely. Further, the only escape route available to people in the administration block (other than across the ship channel) was across the area of the explosion.

26.6.5.8 Gas Detection System

Despite the fact that the plant had a large inventory of flammable materials held at high pressure and temperature, there was no fixed flammable gas detection system.

26.6.5.9 Control of Ignition Sources

The control of ignition sources around the plant was another feature criticized in the OSHA report.

26.6.5.10 Permit-to-Work Systems

The OSHA report stated that an effective permit system was not enforced for the control of the maintenance activities either of the company's employees or of contractors.

26.6.5.11 Isolation Procedures for Maintenance

In this incident, the sole isolation was a ball valve, which was meant to be closed but was in fact open. There was no double block system or blind flange.

The practice of not providing positive isolation was a local one and violated corporate procedures. The implication is that it had not been brought to light by any safety audits conducted.

26.6.5.12 Integrity of Fire Water System

The practice of relying for fire water on the process water system and the failure to provide a dedicated fire water system meant that the fire water system was vulnerable to an explosion.

26.6.5.13 Dependability of Fire Pumps

The electrical cables to the regular fire pumps were not laid underground and were therefore vulnerable to damage by explosion and fire. One of the back-up diesel pumps had insufficient fuel and one had been taken out for maintenance without informing the chief fire officer.

26.6.5.14 Audibility of Emergency Alarm

As described, the level of noise in some areas was such that the employees might not have been able to hear the siren.

26.6.5.15 Follow-Up of Audits

The OSHA report criticized the company's failure to act upon reports issued previously by the company's own safety personnel and by external consultants, which drew attention to unsafe conditions.

26.6.5.16 Planning for Emergencies

The disaster highlighted a number of features of emergency planning. The company had put a good deal of effort into planning and creating personal relationships with the emergency services, by means such as joint exercises, and these paid off. The value of planning, training, and personal relations was one of the most positive lessons drawn.

Another area in which a proactive approach proved beneficial was in relations with the media. Senior personnel made themselves available, and the company evidently felt it received fair treatment.

One weakness of the emergency planning identified was that it had not envisaged a disaster of the scale which actually occurred.

The incident brought out the need to be able to respond clearly to calls from those liable to be affected about the toxicity of the fumes and smoke generated in such an event.

The behavior of rescue helicopters posed a problem. Personnel on the ground had no means of communication with them and the craft tended to come in low, creating the danger of blowing flames or toxic fume onto those below. A need was identified for altitude and distance guidelines for helicopters.

26.7. CANVEY REPORTS

The most comprehensive hazard assessment of non-nuclear installations in the United Kingdom is the Canvey study, carried out for the HSE by SRD.

The first phase of the work is described in *Canvey: An Investigation of Potential Hazards from Operations in the*

Canvey Island/Thurrock Area (the First *Canvey Report*) (HSE, 1978). The report is in two parts: Part 1 is an introduction by the HSE and Part 2 is the SRD study.

The origin of the investigation was a proposal to withdraw planning permission for the construction of an additional refinery in the area. Two oil companies, Occidental Refineries Ltd and United Refineries Ltd, had been granted planning permission for the construction of oil refineries. The construction of the Occidental refinery was begun in 1972, but was halted in 1973 pending a major design study review. United Refineries had valid planning consents, but had not started construction. It was a public inquiry into the possible revocation of the planning permission for the United Refineries development which gave rise to the investigation.

Responses to the First *Canvey Report* centered mainly on two aspects: the methodology used and the magnitude of the assessed risks. The HSE commissioned further work, leading to the Second *Canvey Report* (HSE, 1981), in which the methodology used is revised and the assessed risks are rather lower.

26.7.1 First *Canvey Report*

The terms of reference of the investigation were:

In the light of the proposal by United Refineries Limited to construct an additional refinery on Canvey Island, to investigate and determine the overall risks to health and safety arising from any possible major interactions between existing or proposed installations in the area, where significant quantities of dangerous substances are manufactured, stored, handled, processed and transported or used, including the loading and unloading of such substances to and from vessels moored at jetties; to assess the risk; and to report to the Commission.

The members of the investigating team were appointed as inspectors of the HSE under the provisions of the Health and Safety at Work etc. Act of 1974 and were given specified powers to enable them to make the necessary inquiries. The overall approach taken in the investigation was

1. To identify any potentially hazardous materials, their location and the quantities stored and in process.
2. To obtain and review the relevant material properties such as flammability and toxicity.
3. To identify the possible ways in which failure of plants might present a hazard to the community.
4. To identify possible routes leading to selected failures. Typically, the factors examined included operator errors, fatigue or aging of the plant, corrosion, loss of process control, overfilling, impurities, fire, explosion, missiles and flooding.

5. To quantify the probability of the selected failures occurring and their consequences.

The investigation involved the identification of the principal hazards of the installations and activities in the area, the assessment of the associated risks to society and to individuals, and the proposal of modifications intended to reduce these risks.

Some 30 engineers were engaged in the investigation, which cost about £400,000.

26.7.2 First *Canvey Report*: Installations and Activities

The principal hazardous installations and activities identified in the investigation are summarized in Table 26.7.

26.7.3 First *Canvey Report*: Identified Hazards

The investigation identified several principal hazards in the area. These are:

1. oil spillage over bund;
2. LNG vapor cloud release (1000 t);
3. LPG vapor cloud release (1000 t);
4. ammonium nitrate explosion (4500 t of 92% solution);
5. ammonia vapor cloud release (1000 t);
6. hydrogen fluoride cloud release (1000 t).

The figures in parentheses indicate standard cases considered in the study.

A severe fire might occur if there is an escape of flammable liquids from storage so that large quantities flow down into the residential area. This hazard is presented by the large storages of flammable liquids.

A severe vapor cloud fire and/or explosion might occur if there is a spillage of LNG so that a vapor cloud forms and ignites. This hazard is presented by the large LNG terminal and storage at British Gas. The spillage might occur at sea or on land.

Similarly, a severe vapor cloud fire and/or explosion might occur if there is a spillage of LPG so that a vapor cloud forms and ignites.

A severe explosion might occur if there is a rupture of an ammonium nitrate storage tank.

A severe toxic release might occur if there is a spillage of ammonia. The spillage might occur at sea or on land.

A severe toxic release might also occur if there is a rupture of storage or process plant containing hydrogen fluoride. This hazard is presented by the alkylation facilities at Shell, at the Mobil extension and at the proposed Occidental refinery.

TABLE 26.7 First Canvey Report: Principal Hazardous Installations and Activities at Canvey

Location	Company	Installation or Activity	Storage	Employees	Transport In	Transport Out
Canvey Island	British Gas Corporation	LNG terminal	Fully refrigerated storage of LNG (atmospheric pressure, $<-162°C$) 6×4000 t above-ground tanks 2×1000 t above ground tanks $4 \times 20,000$ t in ground tanks. Fully refrigerated storage of butane (atmospheric pressure, $<10°C$) $1 \times 10,000$ t tank 2×5000 t tanks	200	Sea	Mainly pipeline (as vapor) but some road
	Texaco Ltd	Petroleum products storage	Atmospheric storage of petroleum products >80,000 t total capacity	130	Sea	Pipeline, road, sea
	London and Coastal Wharves Ltd	Flammable and toxic liquids storage	Atmospheric storage of liquids >300,000 t total capacity	50	Mainly sea but some road	Pipeline (Texaco oil). Rest mainly road but some sea
	Occidental Refineries Ltd	Oil refinery (proposed)	Pressure storage of LPG (atmospheric temperature) 2×750 t propane spheres 2×400 t butane spheres			
	United Refineries Ltd	Oil refinery (proposed)	Pressure storage of LPG (atmospheric temperature) 4×200 t propane spheres 3×900 t butane spheres. Process and storage containing hydrogen fluoride			
Coryton	Mobil Oil Co. Ltd	Oil refinery	Pressure storage of LPG (atmospheric temperature) 1×1000 t + 14 other vessels, giving 4000 t total capacity	800	LPG produced on site	Pipeline, road, rail, sea
		Oil refinery (extension)	Pressure storage of LPG (atmospheric temperature) 4×1000 t LPG spheres. Fully refrigerated storage of LPG (atmospheric pressure) 1×5000 t tank. Process and storage containing hydrogen fluoride		LPG produced on site	Pipeline, road, rail, sea
	Calor Gas Ltd	LPG terminal	Pressure storage of LPG (atmospheric temperature) 3×60 t propane vessels 2×60 t butane vessels + cylinders, giving 500 t total inventory	100	Pipeline (from oil refineries)	Road (cylinders, bulk tankers)
Shell Haven	Shell Oil U.K. Ltd	Oil refinery	Pressure storage of LPG (atmospheric temperature) 1×1700 t butane sphere + 3 other butane spheres, giving 3200; t total butane capacity 4×400 t propane spheres 3×135 t LPG horizontal vessels	1900	LPG produced on site	Pipeline, road, rail, sea
			Fully refrigerated storage of liquid anhydrous ammonia (atmospheric pressure, $-33°C$) $1 \times 14,000$ t tank. Process and storage containing hydrogen fluoride 2×40 t vessels		Sea	Sea, road, occasionally rail
Stanford-le-Hope	Fisons Ltd	Ammonium nitrate plant	Storage of 92% aqueous ammonium nitrate solution 1×5000 t tank 1×2000 t tank	80	Ammonium nitrate produced on site	Road
		Ammonia storage	Semi-refrigerated pressure storage of liquid anhydrous ammonia (pressure above atmospheric, $6°C$) 1×2000 t sphere		Rail	Ammonia used on site
Canvey/Thurrock area	General	Transport of hazardous materials by river, road, rail, and pipeline				

26.7.4 First *Canvey Report:* Failure and Event Data

The investigation required the estimation of the probabilities of various occurrences and of their consequences.

Some of the sources of information on such probabilities used in the study were:

1. UK industries, including oil, chemical and other process industries and transport;
2. government organizations such as those concerned with fire, road, rail, sea, and air transport;
3. professional institutions, for example, Institution of Chemical Engineers, Institution of Civil Engineers, American Institute of Chemical Engineers;
4. international safety conference proceedings, for example, loss prevention in the process industries, ammonia plant safety, and hazardous materials spills;
5. industry-based associations, for example, Chemical Industries Association, Institute of Petroleum, American Petroleum Institute, Liquefied Petroleum Gas Industry Technical Association;
6. international insurance interests, for example, Lloyds, Det Norsk Veritas, and industrial risk insurers, Fire Protection Association;
7. overseas government and international agencies, for example, US Coast Guard, US Department of Transportation, OECD, and EEC;
8. specialized research laboratories;
9. individual subject specialists known or recommended to the investigating team.

The degree of uncertainty associated with the probability estimates is indicated by the following code:

a. assessed statistically from historical data—this method is analogous to the use of aggregate estimates in economic forecasting;
b. based on statistics as far as possible but with some missing figures supplied by judgment;
c. estimated by comparison with previous cases for which fault tree assessments have been made;
d. 'dummy' figures—likely always to be uncertain, a subjective judgment must be made;
e. not used;
f. fault tree synthesis, an analytically based figure which can be independently arrived at by others.

26.7.5 First *Canvey Report*: Hazard Models and Risk Estimates

The investigation involved the study of a wide range of hazards and scenarios.

The projects that were initiated as part of the investigation were:

1. consideration of known history of identified storage tanks and their possibility of failure;
2. probability of particular storage tanks or process vessels being hit by missiles caused by fires or explosions on site or adjacent sites, by fragmentation of rotating machines or pressure vessels, or transport accidents;
3. effects of vapor cloud explosions on people, houses, engineering structures, etc.;
4. evaporation of LNG from within a containment area on land or from a spill on water;
5. special problems of frozen earth storage tanks for LNG and the effect of flooding;
6. study of possible failures in handling operations;
7. consideration of the possible benefits and practicality of evacuation;
8. civil engineering aspects of the sea-wall—the chance of it being breached by subsidence, explosion or impact of ships, consideration of the timing of improved defenses, consideration of the time for floods to rise;
9. statistics of ship collisions and their severity, groundings, etc., applying extensive world experience to the Canvey Island area;
10. reliability and analysis of fluid handling practices, ship to shore, and store to road vehicles and pipelines;
11. toxicology of identified hazardous substances;
12. studies to determine the lethal ranges for various releases of toxic or explosive materials leading to a number of special studies such as
    a. the behavior of ammonia spilt on water or land, and
    b. an assessment of the relative importance of explosion or conflagration from a cloud of methane or liquefied petroleum gas.

The subjects that are considered in appendices to the report are:

1. a review of current information on the causes and effects of explosions of unconfined vapor clouds (F. Briscoe);
2. fires in bunds—calculations of plume rise and position of downwind concentration maximum (R. Griffiths);
3. a quantitative study of factors tending to reduce the hazards from airborne toxic clouds (Q.R. Beattie);
4. the dispersal of ammonia vapor in the atmosphere with particular reference to the dependence on the conditions of emission (F. Abbey, R.F. Griffiths, S.R. Haddock, G.D. Kaiser, R.J. Williams, and B.C. Walker);
5. statistics on fires and explosions at refineries (Q.H. Bowen);
6. missiles—penetration capability (E.A. White);
7. discussion of data base for pressure vessel failure rate (T.A. Smith);

8. risk of aircraft impacts on industrial installations in the vicinity of Canvey Island (L.S. Fryer);
9. the dispersion of gases that are denser than air, with LNG vapor as a particular example (G.D. Kaiser);
10. not used;
11. the risk of a liquefied gas spill to the estuary (D.F. Norsworthy);
12. the toxic and airborne dispersal characteristics of hydrogen fluoride (Q.R. Beattie, F. Abbey, S.R. Haddock, G.D. Kaiser);
13. transient variation of the wall temperature of an LNG above-ground storage tank during exposure to an LNG fire in an adjacent bund (I.R. Fothergill);
14. the escape of 1000 t of anhydrous ammonia from a pressurized storage tank (L.S. Fryer, G.D. Kaiser, and B.C. Walker);
15. graphical calculation of toxic ranges for a release of 1000 t of ammonia vapor (J.H. Bowen);
16. effect of unbunded spill of hydrocarbon liquid from refinery at Canvey Island (A.N. Kinkead);
17. estimated risk of missile damage causing a vapor cloud release from existing and proposed LPG storage vessels at the Mobil refinery, Coryton (D.F. Norsworthy);
18. risks of accidents involving road tankers carrying hazardous materials (L.S. Fryer);
19. toxicology of lead additives (S.R. Haddock);
20. compatibility of materials stored at London and Coastal Wharves Ltd (S.R. Haddock);
21. blast loading on a spherical storage vessel (Q. Wall);
22. statistical comment on data on distribution of cracks found on inspection of steel vessel (Q.C. Moore);
23. calculation of resistance of ship hull to collision (A.N. Kinkead);
24. reduction of apparent risk by shared experience (Q.H. Bowen).

The treatment of some of the topics which is given in the report is now described as follows:

1. failure of pressure vessels;
2. failure of pressure piping;
3. failure of pipelines;
4. generation of and rupture by missiles;
5. crash of and rupture by aircraft;
6. ship collision and other accidents;
7. flow of a large release of oil;
8. temperature of the wall of an LNG tank exposed to an LNG fire in an adjacent bund;
9. evaporation of LNG and ammonia on water;
10. dispersion of an LNG vapor cloud;
11. unconfined vapor cloud fire and explosion;
12. ammonium nitrate explosion;
13. toxicity of chlorine, ammonia, hydrogen fluoride, and lead additives;

14. dispersion of ammonia and hydrogen fluoride vapor clouds;
15. factors mitigating casualties from a toxic release;
16. road tanker hazards;
17. evacuation.

For a more complete discussion of the methods and of the background to and application, please consult the report.

26.7.6 First *Canvey Report*: Assessed Risks and Actions

The hazards described in Section 26.7.3 were assessed using appropriate failure and event data as described in Section 26.7.4. Hazard models and risk estimates to be used are listed in Section 26.7.5.

The results of the risk assessments are presented by the investigators as risks of causing casualties, that is, severe hospitalized casualties or worse. This is in accordance with established practice (e.g., Department of Defense, n.d.; Glasstone, 1964). It was considered misleading to attempt to distinguish between severe injury and death.

These results have several interesting features. The hazards may be ranked for societal risks in order of descending frequency for accidents of different magnitude.

The hazard arising from the very large quantities of LNG stored is a serious one, but is no worse than that from the considerably smaller quantities of LPG.

The obvious hazards of LNG, LPG, and ammonia are equaled by others, such as oil and hydrogen fluoride, which are perhaps less well appreciated.

The relative importance of the hazards changes with the scale of the accident. For the smaller scale accidents, oil spillage, flammable vapor clouds, and toxic gas clouds are all important. As the scale increases, it is the toxic gas clouds that dominate.

There are a number of interactions identified both within sites and between sites. These include the threat to LPG storage at Mobil from the Calor Gas site, to the oil storage at Texaco from explosives barges, to the ammonia sphere at Fisons from rotating machinery, and from the ammonium nitrate plant at Fisons, to various installations from process and jetty explosions, and possibly to the ammonia storage tank at Shell from explosion in the Shell refinery.

The relative hazard of the pressure storage of anhydrous ammonia at Fisons is much greater than that of the refrigerated storage of the same chemical at Shell. The risk for the latter was assessed as negligible with the possible exception of rupture by an explosion.

The assessed societal risks are shown in Figure 26.4. Figure 26.4(a) gives the societal risks for all the existing

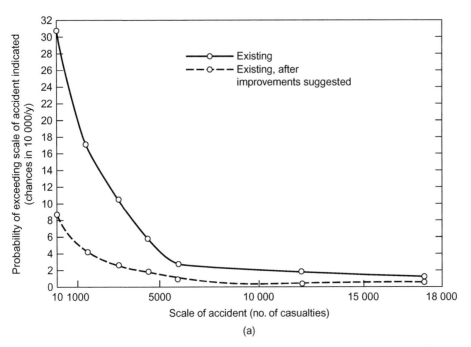

(a)

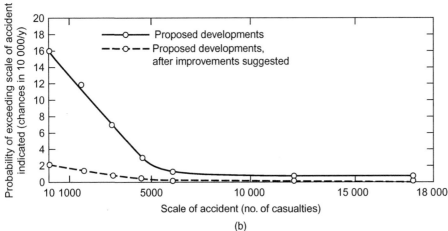

(b)

FIGURE 26.4 First Canvey Report: societal risks for existing installations and proposed developments (Health and Safety Executive, 1978). (a) All existing installations. (b) All proposed developments. *Source: Courtesy of HM Stationery Office.*

installations. It shows that the risk of an accident causing more than 10 casualties is 31×10^{-4}/year. Figure 26.4(b) gives the societal risks for all the proposed developments. It shows that the risk of an accident causing more than 10 casualties is 16×10^{-4}/year.

The assessed individual risks are given for all the existing installations as given in Table 26.8.

The investigators made a number of recommendations for the reduction of the hazards. These included

1. oil spillage—construction of a simple containing wall around the London and Coastal Wharves and Texaco sites and the proposed Occidental and UR refinery sites;

2. LNG tank flooding—construction of a dike;
3. spontaneous failure of LPG spheres—high standard of inspection;
4. LPG vessel rupture by missile (Mobil)—measures including fitting of pressure relief valves on cylinders at Calor Gas depot;
5. LPG tank failure at BG jetty—improvement of bund;
6. LPG pipeline failure (BG)—removal of pipeline;
7. spontaneous failure of ammonia sphere—high standard of inspection, control of ammonia purity;
8. ammonia release at Shell jetty—provision of water sprays at jetty;
9. HF plant rupture (Mobil, Occidental)—provision of water sprays;

TABLE 26.8 First Canvey Report: Rank Order of Societal Risks for Some Principal Hazards of Existing and Proposed Installations at Canvey

Frequencies in Units of 10^{-6}/year								
>10 Casualties			>4500 Casualties			>18,000 Casualties		
1	Oil spillage	1366	1	Ammonia vapor cloud	258	1	HF vapor cloud	100
2	LPG vapor cloud (BG)	970	2	HF vapor cloud	246	2	Ammonia vapor cloud	73
3	Ammonia vapor cloud	735	3	Oil spillage	150	3	LNG vapor cloud	7
4	LPG vapor cloud (others)	637	4	LPG vapor cloud (other)	96			
5	LNG vapor cloud	497	=5	LNG vapor cloud	80			
6	HF vapor cloud	464	=5	LPG vapor cloud (BG)	80			
7	AN explosion	85	7	AN explosion	17			

Source: After Health and Safety Executive, (1978).

TABLE 26.9 First Canvey Report: Rank Order of Societal Risks for Some Principal Hazards of Existing and Proposed Installations at Canvey, After Improvements Suggested

Frequencies in Units of 10^{-6}/year								
>10 Casualties			>4500 Casualties			>18,000 Casualties		
1	LPG vapor cloud (other)	421	1	Ammonia vapor cloud	71	1	HF vapor cloud	24
2	LNG vapor cloud	396	2	HF vapor cloud	67	2	Ammonia vapor cloud	18
3	Ammonia vapor cloud	240	3	LNG vapor cloud	66	3	LNG vapor cloud	7
4	HF vapor cloud	115	4	LPG vapor cloud (other)	63			
5	AN explosion	8	5	AN explosion	2			

Source: After Health and Safety Executive (1978).

10. ship collision—strict enforcement of the speed restriction of 8 knots;
11. road tanker hazards—road tanker traffic restriction to new road only (if road built).

The assessed effect of the proposed modifications is to eliminate the hazards from oil spillage, from LPG vapor cloud due to explosion at the jetty and pipeline failure at British Gas, and from vessel rupture at Mobil due to missiles from Calor Gas, and to reduce greatly many of the other hazards.

The rank order for societal risk of the hazards assuming that the proposed modifications are carried out is then as given in Table 26.9. Oil spillage and LPG vapor cloud (BG) are eliminated.

26.7.7 Second *Canvey Report*

In 1981, the Second Canvey Report was issued, entitled Canvey: A Review of Potential Hazards from Operations in the Canvey Island/Thurrock Area Three Years after Publication of the Canvey Report (HSE, 1981). The report was again in two parts: Part 1, an introduction by the HSE and Part 2, the SRD study. In the next two sections, accounts are given of the reassessed risks and actions and of some technical aspects of the second report.

26.7.8 Second *Canvey Report:* Reassessed Risks and Actions

The risks assessed in the Second *Canvey Report* are generally less than those in the first report. Thus, for example, at Stanford-le-Hope the individual fatality risk is given as 0.6×10^{-4}/year, as opposed to 5×10^{-4}/year in the first report, a reduction by a factor of 8.

The HSE in Part 1 adduce the following reasons for this: (1) physical improvements already carried out; (2) changes in operation; (3) detailed studies by companies; (4) changes in assessment techniques; (5) correction of errors; and (6) further changes firmly agreed but yet to be made.

One major change in operation was the cessation of ammonia storage at Shell.

In a number of cases in the first report, estimates of the risks had to be made without benefit of detailed studies. Subsequent studies by the companies involved showed that in some cases the risks had been overestimated. A case in point was the limited ammonia spill at Fisons.

In a large proportion of cases, the effect of the use of improved models for heavy gas dispersion was to reduce the travel distance of the cloud and hence the risks.

The HSE describes a number of further actions taken or pending to reduce the risks.

26.7.9 Second *Canvey Report:* Technical Aspects

The Second *Canvey Report* revisits some of the frequency estimates and gives a revised set of hazard models and injury relations, covering the following aspects: (1) emission; (2) gas dispersion; (3) ignition; (4) fire events; (5) vapor cloud explosions; and (6) toxic release. With these improvements, the second report may give more reasonable assessment.

Another comprehensive hazard assessment is the Rijnmond study carried out for the Rijnmond Public Authority by Cremer and Warner.

This work is described in *Risk Analysis of Six Potentially Hazardous Industrial Objects in the Rijnmond Area*, a Pilot Study issued by the Rijnmond Public Authority (1982) (the Rijnmond Report).

The Rijnmond is the part of the Rhine delta between Rotterdam and the North Sea. Within the Rijnmond area of 15×40 km^2, about one million people are located, as well as a vast complex of oil, petrochemical, and chemical industries and the largest harbor in the world. A description of the industrial complex at Rijnmond has been given by Molle and Wever (1984). There are five oil refineries with a refining capacity of some 81 Mt/year, including the large refineries of Shell at Pernis and BP near Rozenburg, and major petrochemicals complexes operated by Shell, Gulf, and Esso.

26.8. RIJNMOND REPORT

26.8.1 The Investigation

The aim of the work was to evaluate the methods of risk assessment for industrial installations and to obtain experience in the practical application of these methods. Such evaluation was considered to be essential before any decision could be made on the role of such methods in the formulation of safety policy.

The COVO steering committee was reinforced with experts from industry. The overall approach taken in the investigation was:

1. to collect basic data and define the boundary limits of the installation;
2. to identify the potential failure scenarios;
3. to select and apply the best available calculation models for physical phenomena;
4. to collect and apply the best available basic data and models to calculate the probabilities of such events;
5. to choose and develop different forms of presentation of the final results;
6. to investigate the sensitivity of the results for variations in the assumptions used and to estimate the accuracy and reliability of these results;
7. to investigate the influence of risk reducing measures.

26.8.2 Installations and Activities

The six installations studied were Acrylonitrile storage (Paktank); Ammonia storage (UKF); Chlorine storage (AKZO); LNG storage (Gasunie); Propylene storage (Oxirane); Hydrodesulfurizer (Shell).

26.8.3 Event Data

The event data used in the study were a mixture of generic data, plant data, and estimates.

26.8.4 Hazard Models

The principal hazard models used are given in Table 26.10.

26.8.5 Population Characteristics

The study covered the risks both to employees and to the public. For the off-site population, a grid was used showing the number of people over an area of some 75 km^2, which covered the whole of the Rijnmond. Populations were estimated for each 500 m^2 square using data from the 1971 census, updated in 1975.

TABLE 26.10 Rijnmond Report: Some Hazard Models Used

Event	Model
A Emission	
Flow from	
A. Vessel containing liquid at atmospheric pressure	
B. Vessel containing liquid above atmospheric boiling point	
C. Vessel containing gas only	
Two-phase flow	Fauske–Cude (Fauske, 1964; Cude, 1975)
B Vaporization	
Spreading of liquid spill	Shaw and Briscoe (1978 SRD R100)
Vaporization of cryogenic liquid	
On water	Shaw and Briscoe (1978 SRD R100)
On land	Shaw and Briscoe (1978 SRD R100); AGA (1974)
Combined spreading and vaporization of cryogenic liquid on land	
Vaporization from spill into complex bunds	
Evaporation of volatile liquid	Pasquill (1943), Opschoor (1978)
C Gas dispersion	
Neutral density gas dispersion	Pasquill–Gifford
Dense gas dispersion	R.A. Cox and Roe (1977)
Dispersion of jet of dense gas	Ooms, Mahieu and Zelis (1974)
D Vapor cloud combustion	
Initial dilution with air	
Continuous release	Ooms et al. (1974)
Instantaneous release	Empirical relation
Ignition sources	
Fire versus explosion	
E Vapor cloud explosion	
Vapor cloud explosion	TNO (Wiekema, 1980)
F Pool fire	
Flame emissivity	
Liquid burning rate	Burgess et al. (1961)
Flame length	P.H. Thomas (1963)
Flame tilt	P.H. Thomas (1965b)
Fraction of heat radiated	
View factor	Rein, Sliepcevich and Welker (1970)
G Jet flame	
Flame length	API (1969 RP 521)
Deflection effect	API (1969 RP 521)
Buoyancy and liquid effects	MITI (1976)
View factor	Rein et al. (1970)
H Fireball	
Fireball dimensions	
Radius	R.W. High (1968)
Duration	R.W. High (1968)
View factor	McGuire (1953)

Outside the population grid, average population densities were used. These were required only in cases where an exceptionally large cloud occurred.

For the particular installations studied, data was obtained on the numbers of employees present during day and night and on their locations.

For toxic gas hazard, it was assumed that 1% of the population is outdoors. This estimate includes an allowance for people taking shelter. For explosion hazard, it was assumed that 10% are outdoors.

26.8.6 Mitigation of Exposure

The principal mitigations of exposure considered were evacuation and shelter. For evacuation, the estimate of the proportion evacuated is taken as a function of the warning time. In some cases, the proportion is estimated as 90−100%.

For the effect of shelter from toxic gas, the single-exponential stage model of ventilation is used.

26.8.7 Individual Assessments

The individual assessments included:

- *Acrylonitrile storage (Paktank)*: The acrylonitrile (AN) storage is one of a number of storages at the Botlek terminal of Paktank, a tank storage company providing bulk storage and associated services to the oil and chemical industries.
- *Ammonia storage (UKF)*: The ammonia storage is part of a complex of plants producing ammonia, nitric acid, ammonium nitrate, urea, and other fertilizers at the Pernis site of UKF.
- *Chlorine storage (AKZO)*: The chlorine storage consists of five spheres associated with chlorine cell rooms at the Botlek site of AKZO.
- *LNG storage (Gasunie)*: The LNG storage consists of two tanks and associated equipment at the peak shaving plant of Gasunie.
- *Propylene storage (Oxirane)*: The propylene storage is associated with the propylene oxide plant of Oxirane at Seinehaven. There are two spheres, each with a capacity of 600 te. The ambient temperature was taken in the study as 15°C, that corresponds to a gauge pressure of 8 bar.
- *Hydrodesulfurizer (Shell)*: The section of the hydrodesulfurizer studied is the diethanolamine (DEA) regenerator. The latter is part of an absorber−regenerator unit which removes hydrogen sulfide from sour gas. The hydrogen sulfide is absorbed into DEA in the absorber and the fat DEA passes to the regenerator where a reversible reaction occurs releasing the hydrogen sulfide again.

26.8.8 Assessed Risks

The results of the assessment include risk contours, risk to employees, risk to public (individual risk, societal risk, FN tables, FN curves), and average annual fatalities.

26.8.9 Remedial Measures

For each installation, recommendations were made for remedial measures to reduce the risks. In accordance with the remit of the study, the proposals made were illustrative rather than exhaustive and are suggestions for further evaluation.

For the AN storage, the main measure proposed is nitrogen blanketing to prevent internal explosion.

For the ammonia storage, the principal measure suggested is the provision of a bund. Another proposal made is the provision of an emergency air supply at the control room to allow the operators time to close emergency isolation valves.

For the chlorine storage, the main recommendations are directed to reduction of the size of escapes from pipework by the use of excess flow valves.

For the LNG storage, the main measure proposed is directed toward the prevention of escalation due to the brittle fracture of carbon steel from a small release of cold liquefied gas.

For the propylene storage, the installation of remote isolation valves is the principal measure proposed. It was estimated that this would reduce the total risk by half. Proposals are also made for modifications to the vapor return system.

For the hydrodesulfurizer the measures proposed are minor. It is suggested that there should be monitoring of hydrogen sulfide in the control room and that the hazard of spread of DEA through the drain system be investigated.

26.8.10 Critiques

Rijnmond Report itself contains its own critique in the form of a review by outside consultants and industrial comment. Much of this relates to the failure and event data, the hazard models, and the injury relations and particularly to the models.

26.9. SAN CARLOS DE LA RAPITA DISASTER

At about 14:30 in the afternoon of July 11, 1978, a road tanker carrying propylene caught on fire, eventually leading to a BLEVE. The incident occurred just in front of Los Alfaques camp site, The Sand Dunes, located in the municipality of Alcanar (Tarragona, Spain) between

Barcelona and Valencia, only 3 km from the village of San Carlos de la Rapita. As a result, 217 people lost their lives and approximately 600 were injured (Manich, 2008).

26.9.1 The Camp Site

The camp site was a triangular shaped piece of land between the coastal road and the sea, some 3 km from the nearest township, San Carlos de la Rapita, to the north. It was 200 m in length and 10,000 m^2 in area, tapering from about 100 m wide at the north to 30 m at the south end and 60 m wide at the point of the accident. There was a brick wall on a concrete foundation between the road and the camp.

The legal capacity of the camp site was 260 people; however, on the day of the accident, the camp site was fully booked with some 800 people staying on the original grounds as well as on the two new undeclared extensions.

26.9.2 The Road Tanker

The road tanker was manufactured in 1973, designed with a maximum capacity of 45 m^3, and a maximum allowed mass at 8 bars and 4°C of 19.35 tons. It had no pressure relief valve and no pressure test certificate.

At 12:05 the tanker took on a load of propylene at the ENPETROL Tarragona refinery. At the moment when the vehicle left the refinery (12:35), it was plated as suitable for a maximum load of 22 tons, and it was actually weighed out at 23.5 tons. The driver had a recommended route, which was the motorway, and was provided with the toll money, but actually took the N340 coast road which went past the Los Alfaques camp.

26.9.3 The Fire and Explosion

The sequence of events at the camp site is not entirely clear, but several eyewitness accounts are available. Marshall (1987) quotes the following. A young man serving a customer in the camp shop off the main site heard a bang. He got into his car to investigate and 2 min after the first sound heard a severe explosion. A 'fireball' appeared on the site. A tourist in the camp restaurant heard a 'pop' from a tanker on the main road and saw a milky cloud drifting toward him. He ran to move his car and seconds later the cloud ignited.

It is reported that until the main 'fireball' event, people stood around watching the ball of smoke from the fire. When it did occur, large numbers of people, many scantily clad, were burned, some running into the sea to escape or douse the flames.

Over 90% of the camp site was gutted by fire.

The tanker vehicle was torn into four main pieces. Some two-thirds of the tank, landed on the ground about 150 m, about 300 m from the starting point. The middle section traveled about 100 m to the north-east, into the camp. The cab unit traveled some 60 m along the line of the road to the south. The front end cap fetched up about 100 m beyond the cab.

26.9.4 The Emergency and the Aftermath

For about half an hour after the fire, there were some 200–300 people milling around the camp, many seriously burned and calling for help. Private cars and taxis began to ferry the injured to a hospital at Amposta 13 km away. The first ambulance came at 14v45 from the Shell oil drilling site at San Carlos. The municipal ambulances came at 15:05 and the fire brigade at about 15:30.

The road was still blocked by the burning cab; victims had to be rescued from both sides of the camp and taken north to Barcelona or south to Valencia as circumstances permitted. Those taken north to Barcelona received primary medical care at points en route, while those taken south to Valencia, 165 km away, did not.

Over 100 people died outright. Others died later from burns. Out of 148 cases, 122 had third degree burns on 50% or more of the body. Either there were fewer people only slightly burned or they received treatment elsewhere (Arturson, 1981).

26.9.5 Lessons of San Carlos De La Rapita Disaster

The most important lessons of Los Alfaques Disaster are:

1. Education on equipment, procedures, supervision, and training to prevent overfilling of vehicles carrying hazardous materials.
2. Provision for pressure relief on vehicles carrying flammable materials.
3. Routing of vehicles carrying hazardous materials away from populated areas and vulnerable targets.
4. Prompt treatment of burn victims.

26.10. PIPER ALPHA

At 10:00 p.m. on July 6, 1988 an explosion occurred in the gas compression module of the Piper Alpha oil production platform in the North Sea. A large pool fire took hold in the adjacent oil separation module, and a massive plume of black smoke enveloped the platform at and above the production deck, including the accommodation. The pool fire extended to the deck below, where after 20 min it burned through a gas riser from the pipeline connection between the Piper and Tartan platforms. The gas from the riser burned as a huge jet flame. The lifeboats were inaccessible due to the smoke. Some 62 men escaped, mainly by climbing down knotted ropes or by jumping from a height, but 167 died, the majority in the quarters.

The Piper Alpha Inquiry has been of crucial importance in the development of the offshore safety regime in the UK sector of the North Sea.

26.10.1 The Company

The Piper Alpha oil platform was owned by a consortium consisting of Occidental Petroleum (Caledonia) Ltd, Texaco Britain Ltd, International Thomson plc, and Texas Petroleum Ltd and was operated by Occidental.

26.10.2 The Field and the Platform

The Piper Alpha platform was located in the Piper field some 110 miles north-east of Aberdeen. The Piper platform separated the fluid produced by the wells into oil, gas, and condensate. The oil was pumped by pipeline to the Flotta oil terminal in the Orkneys, the condensate being injected back into the oil for transport to shore. The gas was transmitted by pipeline to the manifold compression platform MCP-01, where it joined the major gas pipeline from the Frigg Field to St Fergus.

There were two other platforms connected to Piper Alpha. Oil from the Claymore platform, also operated by Occidental, was piped to join the Piper oil line at the 'Claymore T.' Claymore was short of gas and was therefore connected to Piper Alpha by a gas pipeline so that it could import Piper gas. Oil from Tartan was piped to Claymore and then to Flotta and gas from Tartan was piped to Piper and thence to MCP-01.

The production deck level consisted of four modules, Modules A–D. Module A was the wellhead, Module B the oil separation module, Module C the gas compression module, and Module D the power generation and utilities module.

Module A was about 150 ft long east to west, 50 ft wide north to south, and 24 ft high. The other modules were of approximately similar size. There were firewalls between Modules A and B, between Modules B and C, and between Modules C and D (the A/B, B/C and C/D firewalls, respectively); these firewalls were not designed to resist blasts.

The pig traps for the three gas risers from Tartan and to MCP-01 and Claymore were on the 68 ft level. Also on this level were the dive complex and the JT flash drum, the condensate suction vessel and the condensate injection pumps.

There were four accommodation modules: the East Replacement Quarters (ERQ), the main quarters module; the Additional Accommodation East (AAE); the Living Quarters West (LQW); and the Additional Accommodation West (AAW).

The control room was in a mezzanine level in the upper part of D Module. It was located about one quarter of the way along the C/D firewall from the west face.

There were two flares on the south end of the platform, the east and west flares, and there was a heat shield around Module A to provide protection against the heat from the flares.

The main production areas were equipped with a fire and gas detection system. In C Module the gas detection system was divided into five zones: C1 and C2 in the west and east halves of the module and C3, C4, and C5 at the three compressors, respectively.

The fire water deluge system consisted of ring mains which delivered foam to Modules A–C and part of Module D and at the Tartan and MCP-01 pig traps and water at the condensate injection pumps. The fire pumps were supplied from the main electrical supply, but there were backup diesel-driven pumps.

26.10.3 Events Prior to the Explosion

On July 6, there was a major work program on the platform. This included the installation of a new riser for the Chanter field and work on a prover and metering loop.

The GCM was also out of service for changeout of the molecular sieve driers. In consequence, the plant operation had reverted to the phase 1 mode so that the gas was relatively wet.

The resulting increased potential for hydrate formation was recognized by management onshore. The increased methanol injection rates required were calculated, the methanol injection rates needed were some 12 times greater than for normal phase 2 operation.

There was an interruption of the methanol supply to the most critical point, at the JT valve, between 4:00 and 8:00 p.m. that evening.

The operating condensate injection pump was pump B. Pump A was down for maintenance. There were three maintenance jobs to be done on this pump: (1) a full 24 month preventive maintenance (PM), (2) repair of the pump coupling, and (3) recertification of PSV 504. In order to carry out the 24 month PM, the pump had been isolated by closing the gas operated valves (GOVs) on the suction and delivery lines but slip plates had not been inserted.

With the pump in this state, with the GOVs closed but without slip plate isolation, access was given to remove PSV 504 for testing. It was taken off in the morning of July 6. They were not able to restore the PSV that evening. The supervisor in this team came back to the control room sometime before 6:00 p.m. to suspend the permit-to-work (PTW).

At about 4:50 p.m. that day, just at shift changeover, the maintenance status of the pump underwent a change. The maintenance superintendent decided that the 24 month PM would not be carried out and that work on the pump should be restricted to the repair of the pump coupling.

About 9:50 p.m. that evening pump B tripped out on the 68 ft level. The lead production operator and the phase 1 operator attempted to restart it but without success. The loss of this pump meant that with pump A also down, condensate would back up in the JT flash drum and within some 30 min would force a shut-down of the gas plant.

The lead operator came up to the control room. It was agreed to attempt to start pump A. The lead maintenance hand came down to the control room to organize the electricians to deisolate the pump. They were observed at the pumps by the phase 2 operator and an instrument fitter, but the evidence of these witnesses was inconclusive. There was no doubt that the lead operator intended to start pump A.

About 9:55 p.m., the signals for the tripping of two of the centrifugal compressors in Module C came up in the control room. This was followed by a low gas alarm in zone C3 on centrifugal compressor C. Then, the third centrifugal compressor tripped. Before the control room operator could take any action a further group of alarms came up: three low gas alarms in zones C2, C4, and C5 and a high gas alarm. The operator had his hand out to cancel the alarms when he was blown across the room by the explosion.

Just prior to the explosion, personnel in workshops in Module D heard a loud screeching sound which lasted for about 30 s.

26.10.4 The Explosion, the Escalation, and the Rescue

The initial explosion occurred at 10:00 p.m. It destroyed most of the B/C and C/D firewalls and blew across the room the two occupants of the control room, the control room operator, and the lead maintenance hand.

The explosion was followed almost immediately by a large fireball which issued from the west side of Module B and a large oil pool fire at the west end of that module.

The large oil pool fire gave rise to a massive smoke plume which enveloped the platform from the production deck at the 84 ft level up.

The offshore installation manager made his way to the radio room and had a Mayday signal sent.

The Tharos effectively took on the role of On-Scene Commander. Rescue helicopters and a Nimrod aircraft for aerial on-scene command were dispatched.

Most of the personnel on the platform were in the accommodation, the majority in the ERQ. Within the first minute, flames also appeared on the north face of the module and the module was enveloped in the smoke plume coming from the south. The escape routes from were impassable.

At the 68 ft level, divers were working with one man under water. They followed procedure, got the man up and briefly through the decompression chamber. They were unable to reach the lifeboats, which were inaccessible due to the smoke. They therefore launched life rafts and climbed down by knotted rope to the lowest level, the 20 ft level.

The drill crew also followed procedure and secured the wellhead.

The oil pool in Module B began to spill over onto the 68 ft level where another fire took hold. There were drums of rigwash stored on that level which may have fed the fire.

The fire water drench system did not operate. There was only a trickle of water from the sprinkler heads.

The explosion disabled the main communications system which was centered on Piper. The other platforms were unable to communicate with Piper. They became aware that there was a fire on Piper, but did not appreciate its scale. They continued for some time in production and pumping oil.

Some 20 min after the initial explosion, the fire on the 68 ft level led to the rupture of the Tartan riser on the side outboard of the ESV. This resulted in a massive jet flame.

The emergency procedure was for personnel to report to their lifeboat, but in practice, most evacuations would be by helicopter and personnel would be directed from the lifeboats to the dining area on the upper deck of the ERQ and then to the helideck. Personnel in the ERQ found the escape routes to the lifeboats blocked and waited in the dining area. The OIM told them that a Mayday signal had been sent and that he expected helicopters to be sent to effect the evacuation. In fact, the helideck was already inaccessible to helicopters.

Some 33 min into the incident, the Tharos picked up the signal. 'People majority in galley. Tharos come. Gangway. Hoses. Getting bad.'

No escape from the ERQ to the sea was organized by the senior management. However, as the quarters began to fill with smoke, people filtered out by various routes and tried to make their escape.

Some men climbed down knotted ropes to the sea. Others jumped from various levels, including the helideck at 174 ft.

The vessels around the platform launched their fast rescue craft (FRCs). The first man rescued, by the FRC of the Silver Pit, was the oil laboratory chemist, who, on experiencing the explosion, simply walked down to the 20 ft level. Most survivors, however, were rescued from the sea.

The FRC of another vessel, the Sandhaven, was destroyed with only one survivor.

At about 10:50 p.m. the MCP-01 riser ruptured, and about 11:18 p.m. the Claymore riser ruptured. The pipe deck collapsed and the ERQ tipped. By 12:15 a.m. on July 7, the north end of the platform had disappeared. By the morning only Module A, the wellhead, remained standing.

26.10.5 Some Lessons of Piper Alpha

A list of some of the lessons is given in Table 26.11.

26.10.6 Recommendations on the Offshore Safety Regime

The *Piper Alpha Report* makes recommendations for fundamental changes in the offshore safety regime.

The basis of the recommendations is that the responsibility for safety should lie with the operator of the installation and that nothing in the regime should detract from this.

A central feature of the regime proposed is the safety case for the installation. In the offshore safety case, it is required that the operator should demonstrate that the installation has a TSR in which the personnel on the

---

TABLE 26.11 Some Lessons of Piper Alpha

Regulatory control of offshore installations Safety Management System
Documentation of plant
Fallback states in plant operation
Permit-to-work systems
Isolation of plant for maintenance
Training of contractors personnel
Disabling of protective equipment by explosion itself
Offshore installations: control of pressure systems for hydrocarbons at high pressure, limitation of inventory on installation, emergency shut-down system, fire and explosion protection, temporary safe refuge, formal safety, use of wind tunnel test and explosion simulations in design
The explosion and fire phenomena: explosions in semi-confined modules, pool fires and jet flames
Publication of reports of accident investigations

---

installation may shelter while the emergency is brought under control and evacuation organized.

Further, it is recommended that this demonstration should be by QRA.

The recommendation on the safety case includes a requirement that the operator should demonstrate that it has a safety management system (SMS).

The report considers that the then current regulatory body, the DoEn, is unsuitable as the body to be charged with implementing the new regime and recommends the transfer of responsibility for offshore safety to the HSE.

These recommendations were accepted immediately by the government and the new regime under the HSE began in April 1991.

26.11. THREE MILE ISLAND

At 4:00 on March 28, 1979, a transient occurred on Reactor No. 2 at Three Mile Island, near Harrisburg, Pennsylvania. A turbine tripped and caused a plant upset. The operators tried to restore conditions, but misinterpreting the instrument signals, misjudged the situation, and took actions which resulted in the loss of much of the water in the reactor and the partial uncovering of the core. Radioactivity escaped into the containment building. Site and general emergencies were declared.

The accident at Three Mile Island (TMI) (also referred to initially as Harrisburg) was the most serious accident which had occurred in the US nuclear industry.

26.11.1 The Company

The company which operated TMI was Metropolitan Edison (Met Ed).

26.11.2 The Site and the Works

TMI is situated on the Susquehanna River, some 10 miles south-east of Harrisburg, Pennsylvania.

26.11.3 The Process and the Plant

The plant consisted of two pressurized water reactors (PWRs) of approximately 900 MWe each.

The reactor itself is a pressure vessel containing the reactor core. There are some 37,000 of these fuel pins arranged in clusters in the core. The core is cooled by the primary cooling system containing water at a high pressure ($<<15.2$ MPa) to prevent it from boiling.

Heat is removed from the reactor in this primary cooling circuit, the reactor cooling system (RCS), and used in the steam generators to generate steam to drive a pair of turbogenerators. The primary cooling system consists of two independent loops containing water at high pressure.

Since the volume of the cooling water changes with temperature, it is necessary to have some free space or 'bubble' in the primary circuit to control the system pressure. This is provided by the pressurizer, which is equipped with electrical heating to raise steam and thus increase the pressure, and cold water sprays to condense steam and thus lower the pressure. Steam can be blown off through a pilot-operated relief valve (PORV). The tail pipe from the PORV is run to the reactor coolant drain tank.

An emergency core cooling system (ECCS) is provided to give backup cooling. This consists of putting coolant into the system at a pressure greater than the normal coolant pressure. Another is the core flooding system, which automatically injects cold borated water into the system when the coolant pressure falls below about 4.1 MPa. The third is the low pressure injection system which operates at 2.8 MPa. This takes water either from a large tank of borated water or from the dump of the containment building and is thus capable of keeping up recirculation indefinitely.

The reactor, the primary cooling circuit, and the steam generators are housed in a containment building which is leak-tight. The containment is designed to retain the contents of the primary circuit in the event of a major leak.

The turbogenerator set is housed in a separate building which also contains the condensate and water treatment plants. Feedwater for the secondary circuit is from the condenser. The condensate water is purified by ion exchange in a condensate polishing system, a package unit.

The water converted to steam on the secondary side in the steam generators is supplied to them by the main feedwater pumps. These are backed up by auxiliary, or emergency, feedwater pumps. Both sets of feedwater pumps are housed in the turbine building.

26.11.4 Events Prior to the Excursion

Some 2 days prior to the incident, routine testing was done on the valves on the auxiliary feedwater pumps. Two of the valves, the 'twelve valves', had inadvertently been left shut.

For some 11 h before the incident, the operators had been trying to clear a blockage in the condensate polishing system. In order to do this, they made use of compressed air. The service air line used for this purpose was cross-connected to the instrument air line. The pressure of the air in the instrument line was less than that of the water in the units and some water got into the instrument air line, despite the presence of a non-return valve on that line. This water found its way to some of the plant instruments.

Another feature was a persistent slight leak from either the regular safety valves or the PORV.

26.11.5 The Excursion – 1

At the time of the incident, the reactor was operating under automatic control at 97% of its rated output. The incident began when water which had entered the instrument air line caused the isolation valves on the condensate polishing system to drift shut and the condensate booster pump to lose suction pressure and trip out. The main feedwater pumps in the secondary circuit then tripped and almost immediately the main turbine tripped. The time was 4:00:37.

Valves allowing steam to be dumped to the condenser opened, and the auxiliary feedwater pumps started up. The removal of heat from the RCS fell, and the pressure of the RCS rose. Within 3–6 s the PORV set pressure of 15.5 MPa was reached and the valve opened, but this was insufficient to relieve the pressure, and at 8 s in, pressure rose to 16.6 MPa, the set point for reactor trip. The control rods were driven into the core to stop the reaction.

The RCS pressure now fell below the PORV set pressure, but the valve failed to shut. However, the valve position indicator in the control room, which actually indicated not the position of the valve but the signal sent to it, showed the valve as closed.

In the secondary circuit condensate was no longer being returned to the steam generators. The auxiliary feedwater pumps were running, but the valves between the pumps and the steam generators were also, inadvertently, closed. At 1 min 45 s into the incident the steam generators boiled dry.

Meanwhile the RCS pressure was dropping. At 2 min the pressure fell to 11 MPa at which point the first of the ECCS systems, the HPI system, came on and injected high pressure water. The liquid level in the pressurizer, however, was rising.

The display of liquid level was an indicator of the level of fluid in the pressurizer, but no longer of the mass of water; the liquid now contained steam bubbles and had thus been rendered less dense. To this extent the level display had become misleading.

The operators became concerned that the HPI system was increasing the inventory of water in the RCS, and that the circuit would become full of water and thus go 'solid'. At 4 min 38 s they therefore shut down one of the HPI pumps and throttled back the other.

With water passing through the PORV, the pressure in the reactor coolant drain tank built up and at 7 min 45 s the reactor building sump pump came on, transferring water to the waste tanks in the auxiliary building.

At 8 min the operators realized that the steam generators were dry. They found that the auxiliary feedwater system valves were closed and opened them, thus restoring feedwater to the steam generators.

At 60 min the operators found that the RCS pumps were vibrating. This was due to the presence of steam, though they did not realize this. At 1 h 14 min they shut off the two pumps in Loop B to prevent damage to the pumps and pipework, with possible loss of coolant, and at 1 h 40 min shut off the two pumps in Loop A.

At 2 h 18 min the PORV block valve was at last shut off, thus stopping the loss of water from the RCS. The RCS pressure then began to rise.

In the 2 h since the turbine trip periodic alarms had warned of low level radiation in the containment building. At about 2 h in, there was a marked increase in the radiation readings. By 2 h 48 min in, high radiation levels existed in several areas. At 2 h 55 min a site emergency was declared.

Attempts were made to restart the RCS pumps. One pump in Loop B operated for some 19 min but was shut down again by vibration trips at 3 h 13 min. At 3 h 30 min a general emergency was declared.

At some point, perhaps about 4 h in, the station manager appears to have realized that the reactor core had suffered damage.

From 4 h 30 min attempts were made to collapse the steam bubbles in the RCS loops, but without success and at 7 h these efforts were abandoned.

The operators then tried to bring in the lower pressure cooling system by reducing pressure in the RCS. They began at 7 h 38 min by opening the PORV block valve. At 8 h 41 min the pressure fell to 4.1 MPa at which point the core flooding system operated.

One consequence of the exposure of the core was that a steam-zirconium reaction occurred, generating hydrogen. During the depressurization, hydrogen from the RCS was vented into the containment building. At 9 h 50 min, there occurred in the containment building what was apparently a hydrogen explosion. The sprays came on for some 6 min.

When the significance of the event in the containment building became appreciated, concern grew about a possible explosion of a hydrogen bubble there.

26.11.6 The Emergency and the Aftermath

The site emergency was declared at 7:00 on March 28. At 7:24 a general emergency was declared.

The plant was contacted by the local WKBO radio station. The radio station asked what kind of emergency it was. They were told that it was a general emergency, a 'red-tape' sort of thing required by the NRC when certain conditions exist, but that there was no danger to the public.

The emergency did not end with the establishment of stable operation. The following day, the core damage was found to be more serious than first thought. The hydrogen bubble problem persisted over the next 8 days and caused

particular concern. There were also a number of other problems, mainly associated with the large quantities of water contaminated with low level radioactivity.

On March 30, a radiation reading outside the plant, later found to be erroneous, led NRC officials to recommend evacuation. The local director of emergency preparedness was notified to stand by for an evacuation; he notified fire departments and had a warning broadcast that an evacuation might be called. Then the NRC chairman assured the Governor that an evacuation was not necessary. Later, the Governor decided that pregnant women and pre-school children should be evacuated. By April 1, the NRC had concluded that the hydrogen bubble posed no threat but failed to announce this properly.

26.11.7 The Excursion − 2

The operators in the TMI-2 control room made a number of errors. Some of these were failures to make a correct diagnosis of the situation, while others were undesirable acts of intervention.

The first was the failure to realize that the PORV had stuck open. There were several central-indications, however, which pointed to the fact that the PORV was open.

Closely connected with this first diagnostic failure was the failure to recognize that the mass of primary coolant water had fallen. Here the operators were misled by the level measurement.

Another counterproductive action was the shutting down of the primary coolant pumps. About 60 min into the incident, the pumps began vibrating.

These failures of diagnosis and the fear of the system going solid have frequently been characterized as illustrations of mindset.

The operators were poorly served by the displays and alarms in the control room.

26.11.8 Some Lessons

The Three Mile Island incident yields numerous lessons. Some of these are listed in Table 26.12.

26.12. CHERNOBYL

On Monday, April 28, 1986, a worker at the Forsmark nuclear power station in Sweden put his foot in a radiation detector for a routine check and registered a high reading. The station staff thought they had had a radioactive release from their plant and the alarm was raised. However, as reports came in of high radioactivity in Stockholm and Helsinki, the source of the release was identified as the Soviet Union. In fact, an accident had occurred on Unit 4 at Chernobyl in the Ukraine some 800 miles from Sweden on Saturday April 26.

TABLE 26.12 Some Lessons of Three Mile Island

Regulatory control of major hazard installations
People-related versus equipment-related problems
Formal safety assurance
Relative importance of large and small failures
Influence of operator in an emergency
Fault diagnosis by the process operator
Accident management for major hazard installations
Follow-up of safety issues
Learning from precursor events
Package and other ancillary units
Display of variable of interest
Plant status display and disturbance analysis systems
Abuse of instrument air
Limitations of non-return valves

At 1:24 on that day an experiment to check the use of the turbine during rundown as an emergency power supply for the reactor went catastrophically wrong. There was a power surge in the reactor, the coolant tubes burst and a series of explosions rent the concrete containment. The graphite caught fire and burnt, sending out a plume of radioactive material. Emergency measures to put out the fire and stop the release were not effective until May 6.

26.12.1 The Site and the Works

The Chernobyl nuclear power station is situated in a region which at the time was relatively sparsely populated. There were some 135,000 people within a 30 km radius. Of these, 49,000 lived in Pripyat to the west of the plant's 3 km safety zone and 12,500 in Chernobyl 15 km to the south-east of the plant.

There were four nuclear reactors on the site.

26.12.2 The Process and the Plant

Unit 4 at Chernobyl was designed to supply steam to two turbines each with an output of 500 MWe. The reactor was therefore rated at 1000 MWe or 3200 MWt.

Unit 4 was an RBMK-1000 reactor, a boiling water pressure tube, graphite moderated reactor. The reactor is cooled by boiling water, but the water does not double as the moderator, which is graphite.

The design of the reactor avoided the use of a large pressure vessel.

The reactor contained 192 te of uranium enriched to 2%. The subdivision of the uranium into a large number of separately cooled fuel channels greatly reduced the risk of total core meltdown.

The reactor was cooled by water passing through the pressure tubes. The water was heated to boiling point and partially vaporized. The steam—water mixture with a mass steam quality of 14% passed to two separators where steam was flashed off and sent to the turbines, while water was mixed with the steam condensate and fed through downcomers to pumps which pumped it back to the reactor. There were two separate loops each with four pumps, three operating and one standby. This system constituted the multiple forced circulation circuit (MFCC).

The reactor contained a stack of 2488 graphite blocks with a total mass of 1700 te.

An emergency core cooling system (ECCS) was provided to remove residual heat from the core in the event of loss of coolant from the MFCC.

There was a control system which controlled the power output of the reactor by moving the control rods, in and out.

The reactor was housed in a containment built to withstand a pressure of 0.45 MPa. There was a 'bubble' condenser designed to condense steam entering the containment from relief valves or from rupture of the MFCC.

An important operating feature of the reactor was its reactivity margin, or excess reactivity. Below a certain power level the reaction would be insufficient to avoid xenon poisoning. It was therefore necessary to operate with a certain excess reactivity. The operating instructions stated that a certain excess reactivity was to be maintained; the equivalent number of control rods was 30.

Another important operating characteristic was the reactor's 'positive void coefficient'. An increase in heat from the fuel elements would cause increased vaporization of water in the fuel channels and this in turn would cause increased reaction and heat output. There was therefore an inherent instability, a positive feedback, which could be controlled only by manipulation of the control rods.

26.12.3 Events Prior to the Release

The origin of the accident was the decision to carry out a test on the reactor. Electrical power for the water pumps and other auxiliary equipment on the reactor was supplied by the grid with diesel generators as back-up. However, in an emergency, there would be a delay of about a minute before power became available from these generators. The objective of the test was to determine whether during this period the turbine could be used as it ran down to provide emergency power to the reactor. A test had already been carried out without success, so modifications had been made to the system, and the fresh test was to check whether these had had the desired effect.

The program for the test included provision to switch off the ECCS to prevent its being triggered during the test.

At 1:00 on April 25 the reduction of power began. At 13:05 Turbogenerator 7 was switched off the reactor. At 14:00 the ECCS was disconnected. However, a request was received to delay shutting down since the power output was needed. It was not until 23:10 that power reduction was resumed.

At some stage, the trip causing reactor shut-down on loss of steam to Turbogenerator 8 was disarmed, apparently so that if the test did not work the first time, it could be repeated. This action was not in the experimental program.

The test program specified that rundown of the turbine and provision of unit power requirements was to be carried out at a power output of 700−1000 MWt, or 20% of rated output. The operator switched off the local automatic control, but was unable to eliminate the resultant imbalance in the measurement function of the overall automatic controls and had difficulty in controlling the power output, which fell to 30 MWt. Only at 1:00 on April 26 was the reactor stabilized at 200 MWt, or 6% of rated output. Meanwhile the excess reactivity available had been reduced as a result of xenon poisoning.

It was nevertheless decided to continue with the test. At 1:03 the fourth, standby pump in one of the loops of the MFCC was switched on and at 1:07 the standby pump in the other loop was switched on.

26.12.4 The Release

With the reactor operating at low power, the hydraulic resistance of the core was less, and this combined with the use of additional pumps resulted in a high flow of water through the core. This condition was forbidden by the operating instructions, because of the danger of cavitation and vibration. The steam pressure and water level in the separators fell. In order to avoid triggering, the trips which would shut-down the reactor, the trips were disarmed. The reactivity continued to fall, and the operator saw from a printout of the reactivity evaluation program that the available excess reactivity had fallen below a level requiring immediate reactor shut-down. Despite this, the test was continued.

At 1:23:04 the emergency valves on Turbo generator 8 were closed. The reactor continued to operate at about 200 MWt, but the power soon began to rise. At 1:23:40 the unit shift foreman gave the order to press the scram button. The rods went down into the core, but within a few seconds shocks were felt and the operator saw that the rods had not gone fully in. He cut off the current to the servo drives to allow the rods to fall in under their own weight. Within 4 s, by 1:23:44, the reactor power had risen, according to Soviet estimates, to 100 times the nominal value.

At about 1:24, there were two explosions, the second within some 3 s of the first.

It is thought that the fuel fragmented, causing a rapid rise in steam pressure as the water quenched the fuel elements, so that there was extensive failure of the pressure tubes. The explosive release of steam lifted the reactor top shield, exposing the core. Conditions were created for a reaction between the zirconium and steam, producing hydrogen. An explosion involving hydrogen occurred in and ruptured the containment building, and ejecting debris and sparks were seen.

The accident was aggravated by the fact that the 200 te loading crane fell onto the core and caused further bursts of the pressure tubes.

26.12.5 The Emergency and the Immediate Aftermath

The fire brigades from Pripyat and Chernobyl set out at 1:30. The fires in the machine hall over Turbogenerator 7 were particularly serious, because they threatened Unit 3 also. These therefore received priority. By 5:00 the fires in the machine hall roof and in the reactor roof had been extinguished.

However, the heating up of the core and its exposure to the air caused the graphite to burn. The residual activity of the radioactive fuel provided another source of heat. The core therefore became very hot and the site of raging fire.

During the next few days the fire raged and a radioactive plume rose from the reactor. The accident had become a major disaster. It was necessary to evacuate the population from a 30 km radius round the plant and deal with the casualties and to take a whole range of measures to dampen and extinguish the fire, to cover and enclose the core, and to deal with the radioactivity in the surrounding area.

Three evacuation zones were established: a special zone, a 10 km zone and 30 km zone. A total of 135,000 people were evacuated.

Accounts of the evacuation are incomplete and sometimes contradictory, but what appears to have happened is as follows. Within a few hours of the accident, an emergency headquarters was set up in Pripyat. About 14:00 on April 26, there was an evacuation of some 1000 people from within a 1.6 km zone around the plant. A much larger evacuation took place about 14:00 the next day, when about 1000 buses were brought in and within 2 h had evacuated some 40,000 people. However, it was not until some 9 days later that the authorities in Moscow ordered complete evacuation from a 30 km zone around the site.

A medical team was set up within 4 h of the accident, and within 24 h triage of the 100 most serious cases had been affected.

A system was set up to control movements between the zones. People crossing the zone boundaries had to change clothing and vehicles were decontaminated.

Decontamination of the area outside the site was complicated by the fact that the distribution of the radionuclides changed with time according to the terrain, particularly during the first 3–4 months. Hence the full value of measures to decontaminate a particular area is generally obtained only temporarily.

The death toll from Chernobyl cannot be known with certainty, since most deaths are due to excess cancers. Two men died dealing with the accident itself. By the end of September the toll of dead in the USSR was 31 with 203 injured. Other European countries were also affected. Radioactive contamination was measured across Europe.

26.12.6 The Later Aftermath

Of those on site at the time, two died on site and a further 29 in hospital over the next few weeks. A further 17 are permanent invalids and 57 returned to work but with seriously affected capacity. The others on site, some 200, were affected to varying degrees. There is apparently no detailed information on the effects on the military and civilian personnel brought in to deal with the accident.

A detailed discussion of such effects from Chernobyl is given in the Second *Watt Committee Report*. For EC countries the estimates of the NRPB in 1986 were 100 fatal thyroid cancers and 2000 fatal general cancers, or 2100 fatal cancers. For the USSR the situation is much more complex, but the report quotes an estimate by Ryin et al. (1990) of 1240 fatal leukemia cases and 38,000 fatal general cancers.

26.12.7 Some Lessons of Chernobyl

A list of some of the lessons learned from Chernobyl, that are applicable for process industry is given in Table 26.13.

26.13. HURRICANES KATRINA AND RITA

26.13.1 Introduction

Despite the high hurricane activity throughout the years in the Gulf of Mexico, 2005 was predominantly active in the Gulf Coast with 28 tropical storms and 15 hurricanes, of which seven were category 3 or higher (Blake et al., 2007). 2005 also registered for the first time two category-five hurricanes, hurricanes Katrina and Rita,

TABLE 26.13 Some Lessons of Chernobyl

Management of, and safety culture in, major hazard installations

Adherence to safety-related instructions

Inherent safer design of plants

Sensitivity and operability of plants

Design of plant to minimize effect of violations

Disarming of protective systems

Planning and conduct of experimental work on plants

Accidents involving human error and their assessment

Emergency planning for large accidents

Mitigating features of accidents

with disastrous consequences (NOAA, Hurricane Rita—National Climatic Data Center). Katrina is the costliest hurricane in the US history.

26.13.2 Hurricane Katrina

On August 23, 2005, a tropical depression developed in the southeastern Bahamas, upgrading to tropical storm Katrina the next day. On August 25, Katrina became a category 1 hurricane and passed by the Southeastern coast of Florida. Continuing its southwest movement, Katrina reached its highest intensity on August 28. Katrina made landfall near Buras, LA on August 29 at 7:10 a.m. EDT as a category 3 hurricane with winds of 125 mph. Inland, the hurricane moved near the Louisiana–Mississippi border and through Mississippi, maintaining hurricane intensity for almost 100 miles. Katrina continue to move north through Tennessee and on August 30, it became a tropical depression. Katrina developed 62 tornados and produced rainfalls of up to 15 inches (Johnson, 2006).

Shortly after the hurricane hit Louisiana, the flooding problems in New Orleans worsened when four levees breached resulting in floodwaters that covered about 80% of the city up to 20 feet high. The New Orleans floodwaters were completely removed 43 days after Katrina made landfall (Johnson, 2006).

Highways and roads were damaged or flooded, and many businesses were destroyed causing the loss of tens of thousands of jobs. Louisiana and Mississippi were significantly affected by Hurricane Katrina; Florida, Georgia, and Alabama to a lesser degree (Knabb et al., 2006).

During Katrina, the evacuation was not very effective, Federal emergency officials failed to provide ground transportation, and neither buses nor trains were deployed (Litman, 2006).

Because of the approximately 1800 direct and indirect fatalities (Knabb et al., 2006), the 275,000 homes affected, and the nearly 81 billion dollars of property damage, hurricane Katrina is the costliest and one of the deadliest hurricanes in the United States (Blake et al., 2007; Johnson, 2006).

26.13.3 Hurricane Rita

Twenty-six days after Katrina, Hurricane Rita impacted the Texas−Louisiana border. Hurricane Rita developed as a tropical depression on September 18, 2005. On September 20, Rita became a category-two hurricane and passed by the Florida Keys. Rita intensified to category five the next day where a maximum wind speed of 175 mph was reached. Rita weakened to a category three and made landfall on September 24 at 3:40 a.m. EDT with winds that reached 120 mph. Inland, the winds impacted more than 150 miles causing significant damage. The winds continued until the Louisiana−Texas−Arkansas border. Rita became a tropical depression on September 25. Approximately 90 tornados and rainfall of five to nine inches were reported in Louisiana, Arkansas, Mississippi, and Alabama (NOAA, Hurricane Rita—National Climatic Data Center; Knabb et al., 2006). Texas and the Florida Keys were also affected by the floods and winds (Knabb et al., 2006).

The damage caused by Rita is estimated at approximately $10 billion in the United States (Knabb et al., 2006; Carpenter et al., 2006).

Of the approximately 120 deaths reported during Rita, only 7 were directly caused by the hurricane (Knabb et al., 2006; Carpenter et al., 2006). Most of the deaths occurred during the evacuations (Carpenter et al., 2006). Due to the massive evacuation, the roadways were highly congested preventing rapid medical responses to emergencies (Carpenter et al., 2006). Additionally, fuel, food, and water shortages as well as cars out of gas worsened the traffic congestion. The evacuation of patients was inefficient (Carpenter et al., 2006), and Hobby Airport and Bush Intercontinental Airport suffered delays because many employees evacuated the area (Litman, 2006). In Texas, a bus exploded killing 23 passengers, all of them nursing home evacuees (Carpender, 2006). Because of the extreme heat and humidity (Sanders, 2008), many people died due to heat exhaustion (Knabb et al., 2006).

26.13.4 Effect on the Industry

In 2005, hurricanes Katrina and Rita destroyed 109 facilities, damaged 53 others, and caused a significant amount of oil and hazardous materials spills, both onshore and offshore (Kaiser and Pulsipher, 2007; Cruz and Krausmann, 2009).

Over 1070 onshore and offshore releases were reported to the Incident Reporting Information System (IRIS) between 2005 and 2008 due to Hurricane Katrina, and even 2 years later releases were reported due to damaged or sunk structures and leaks in pipelines (Kessler et al., 2008; Santella et al., 2010).

26.13.4.1 Offshore Industry

During Katrina and Rita, 113 platforms were destroyed, 457 pipelines were damaged (BOEMRE, 2006a, 2006b), and 611 releases at offshore facilities were recorded between 2005 and 2006 (Cruz and Krausmann, 2009). Additionally, the damage, evacuation and loss of production affected the price of oil (Cruz and Drausmann, 2008).

During Katrina, more than 700 platforms and rigs were evacuated, 44 platforms were destroyed (mostly in shallow waters), and 21 were severely damaged (BOEMRE 2006a, 2006b). The major damage occurred in the Shell platform Mars. Four drilling rigs were destroyed, nine were severely affected, and six were set adrift (Krausmann and Cruz, 2008); The most severely damage rigs were the Rowan New Orleans and the Diamond's Ocean Warwick.

During hurricane Rita, 69 platforms were destroyed and 32 were severely damaged, mostly in shallow water. One rig was destroyed, 10 others were severely damaged, and 13 were set adrift (Krausmann and Cruz, 2008); MOEMRE 2006a, 2006b. For Mobile Offshore Drilling Units (MODUs), both hurricanes combined destroyed 5 drilling rigs and damaged 19 (Krausmann and Cruz, 2008).

More releases occurred on platforms during hurricane Katrina than during Rita, probably due to the preparedness for Rita or because some of the facilities from Katrina remained shut-down (Cruz and Krausmann, 2009).

26.13.4.2 Onshore Industry

Numerous onshore facilities were also affected by the strong winds and the surges of hurricane Katrina and Rita. Companies with no major damage still required approximately 2 months to return to normal operating conditions. Other facilities were severely damage and remained shut down for long periods of time (Sanders, 2008; Santella et al., 2010). Prior to Katrina, four refineries reduced runs and nine refineries were shutdown, four of which remained closed for some time due to the damage from the hurricane (Johnson, 2006; Santella et al., 2010). Three natural gas facilities, the Dynegy Plant of Yschoskey, LA, the Dynegy Plant of Venice, LA, and the BP Plant of Pascagoula, MS, were damaged by hurricane Katrina, but no releases were reported (Santella et al., 2010). The most significant spill during Katrina occurred in the Murphy Oil refinery in the

St. Bernard Parish (Meraux) where 25,100 barrels of mixed crude oil were released when a partially filled storage tank was dislodged and ruptured. The spill affected approximately 1800 homes (Santella et al., 2010). During Rita, 15 refineries were taken offline in advance in Texas and Louisiana, and one of them remained shut down until 1 month after the hurricane made landfall (Godoy, 2007).

26.13.5 Lessons Learned

It is important to recognize the devastating power that hurricanes can have and, even more important, learn from others that experienced the destruction of hurricanes Katrina and Rita (Blake et al., 2007; Sanders, 2008; Keel et al., 2008; Krausmann and Cruz, 2008). Two key aspects should be considered to avoid the loss of lives and property and to prevent potential releases: design to withstand hurricanes and plan in advance to minimize the damage.

26.13.6 Recommendations

Key factors should be implemented to avoid loss of life and minimize potential damage caused by hurricanes. The following lists some practices known to provide effective protection and response.

Planning: Develop hurricane preparedness plans that consider the probability of occurrence, the damage expected from winds and surges, and the vulnerability of the facility. Then, identify suitable mitigations options. Provide detailed shutdown and start-up procedures closely related to evacuation plans. Include evacuation plans that ensure the transportation of workers and their families. Ensure the availability of hurricane shelters and harden buildings, supplies, backup communication system, and storage of gasoline (Santella et al., 2010).

Evacuation: Have an accurate weather information system. For offshore operations, the shutdown and evacuation should start 5–7 days before the hurricane. Decrease inventory of hazardous materials. Transfer the material from storage tanks and fill the tanks with water. Inspect the foundation of tanks to determine the potential for a storage tank to be moved by the storm (Keel et al., 2008).

Post-storm: Ensure that post-storm plans are followed. Assess the damage and report to management, government agencies and property insurance carriers accordingly. Share experiences with other industries to develop best practices and determine the aspects that need to be improved (Cruz and Krausmann, 2009).

26.14. BP AMERICA REFINERY EXPLOSION, TEXAS CITY, TEXAS, USA

On March 23, 2005, a massive explosion occurred at the BP Texas City Refinery, in Texas City, Texas, USA. This accident resulted in 15 fatalities and 180 injuries and was considered as one of the most catastrophic industrial accident in the US history. The explosion happened at approximately 1:20 p.m. during the startup of a hydrocarbon isomerization unit when a distillation tower flooded with hydrocarbons, caused over pressurization and relief devices to open resulting in a geyser-like release of flammable liquid from the vent stack that was not equipped with a flare. Most victims at the time of the incident were located in office trailers located near to the blowdown vessel. In addition to casualties, houses and buildings located as far as three quarters of a mile away from the refinery were significantly damaged. The financial losses from this accident exceed $1.5 billion.

26.14.1 The Company

Texas City refinery is the company's largest refinery worldwide with capacity up to 10 million gallons of gasoline per day. There are 29 oil refining units and 4 chemical units within its 1200 acre site with approximately 1800 BP workers. At the time of the incident, BP was performing a turnaround and there were about 800 contractor workers onsite to support the work.

26.14.2 Incident Description

The incident at the BP refinery in Texas City, Texas, occurred on March 23, 2005. The incident occurred during the restarting of the refinery's ISOM process unit after turnaround maintenance. The ISOM unit process consists of an Ultrafiner desulfurizer, a Penex reactor, a vapor recovery/liquid recycle unit and a raffinate splitter. The incident started at the raffinate splitter section when the raffinate splitter tower was overfilled with liquid during the startup. Heavy raffinate was pumped from the bottom of the splitter tower and circulated through the reboiler furnace and back to the bottom tray. Heavy raffinate product was taken of as a side stream and sent to storage. A level transmitter was installed in the splitter tower and provided the reading of the liquid level to the control room. The tower was also equipped with alarms that indicated high liquid level.

The raffinate splitter tower was protected by three parallel safety relief valves located in the overhead vapor line. The relief valves were designed to open and release primarily vapor when their set pressures were exceeded. The vapor was then sent to the disposal header collection system. The header collection system consisted of a

14-inch NPS elevated pipe approximately 270 m away from the raffinate splitter tower. This section and two other additional headers from other ISOM unit sections were connected to the blowdown drum.

Around the time of the incident, two major turnarounds were underway in the ultracracker unit (ULC) and the aromatics recovery unit (ARU).

A month before the incident, the raffinate splitter section was shut down and the raffinate splitter tower was drained, purged, and steamed-out to remove residual hydrocarbons. At the time of the incident, almost all scheduled maintenance work had been completed except for the Penex reactor which was awaiting a gasket; however, BP decided to start up the raffinate splitter section.

During the incident, no pre-startup safety review (PSSR) procedure were conducted to verify the adequacy of the safety system and equipment in the ISOM process unit although the company policy required a formal PSSR prior to startup following a turnaround. The process safety coordinator responsible for an area of the refinery including the ISOM unit was not familiar with the applicability of PSSR, and therefore, the PSSR was not performed.

During pre-startup equipment checks, the instrumentation and equipment which was identified as malfunctioning were not repaired. During the ISOM turnaround, operations personnel informed the turnaround supervisor that the splitter level transmitter and level sight glass were malfunctioning. When the ISOM unit was operating, this equipment could not be repaired because the block valves needed to isolate the level transmitter were leaking. The isolation valves were replaced during the turnaround but the BP supervisors decided not to repair the level transmitter due to insufficient time to complete the job in the existing turnaround schedule. The supervisors planned to repair the level transmitter after the startup.

In the days prior to the raffinate splitter startup, one of the operations personnel found that a pressure control valve was malfunctioning both during the unit shutdown and the equipment checks. In addition, the functionality check of all alarms and instruments required prior to startup were not completed.

The raffinate section in the ISOM unit was started during the night shift on March 22, 2005; then it was stopped to be re-started during the next shift. The Night Lead Operator did not record the completed steps for the filling process of the raffinate section, and therefore, the operators on the next shift did not have any records on what had been done.

In the early morning on March 23, 2005, the raffinate splitter tower was filled beyond the set points of both alarms to a level reading of 99% on the transmitter; however, only one alarm was activated and sounded at 3:09 a.m.

The splitter tower feed and bottoms pumps were shut off and the raffinate section was filled and the circulation was shut down. The level control valve remained in the 'closed' position for the following shift to resume the startup.

At 6:00 a.m. in the morning shift change took place. The Night Lead Operator had left approximately an hour before the shift change and left the Night Board Operator with brief explanation of the actions he had taken. During the shift change, the Night Board Operator provided only few details about the night shift's raffinate section startup activities to the Day Board Operator. The Day Board Operator read the logbook, interpreted that the liquid was only added to the tower and was unaware that the heat exchangers, the piping, and associated equipment had also been filled during the previous shift.

On the morning of March 23, a series of miscommunications took place with regard to whether the startup could proceed or not. The Day Supervisor did not have a solid knowledge of what steps the night crew had completed and did not review the applicable startup procedure with the crew.

Prior to sending additional liquid hydrocarbons to the raffinate section, the 8-inch NPS chain-operated valve was open by outside operators to remove nitrogen from the splitter tower. A mis-communication occurred regarding how the feed and product would be connected to the tower. The Day Board Operator believed that the heavy raffinate product should not be sent to storage; however, the outside operators believed that they were instructed not to send the light raffinate product to storage.

At 9:40 a.m. prior to the restart of the splitter tower, the Day Board Operator opened the level control valve to 70% output and approximately 12,000 barrel per day (bpd) of heavy raffinate flowed out of the tower for about 3 min. He closed the level control valve to 0% output, but the flow indication showed that the flow dropped only to 4300 bpd rather than 0 bpd although no heavy raffinate flowed out of the tower. Then the Day Board Operator resumed the startup, restarted the circulation and introduced the feed on the splitter tower without considering the written procedure which indicated the exact stage of the startup.

The Day Board Operator believed that the verbal instruction required his action to maintain higher level in the tower to protect downstream equipment although the startup procedure called for level control valve to be set in 'automatic' and 50% to establish heavy raffinate flow to the storage. The Day Board Operator noticed the level transmitter indicated 97% and thought it was normal. At around 10:00 a.m., as much as 20,000 bpd of raffinate feed was pumped into the tower and the output flow showed to be 4100 bpd although the Day Board Operator was aware that the level control valve was shut. CSB investigation post-incident concluded that there was likely no flow leaving the tower at that period.

As the tower was filled, the tower level transmitter showed the liquid level less than 100%. The level sight glass was reported nonfunctional for several years.

When the unit was being heated, the Day Supervisor who had ISOM-experience left the plant due to a family emergency. The second Day Supervisor had very little experience with the ISOM unit and therefore did not get involved in the ISOM startup. At 9:55 a.m. two burners which pre-heated the feed flowing into the heater were lit heating the liquid at the bottom of the tower. About an hour and a half later, two additional burners were lit. Post-incident analysis by the CSB determined that the level in the tower was actually 20 ft when the level reading was 93% and when the level decreased to 88% the level in the tower was actually 30 m.

The pressure inside the tower rose to 33 psig (228 kPa) because of the compression of the nitrogen by the liquid in the raffinate section. As a result of this high pressure, the outside operators opened the 8-inch NPS chain-operated valve that vented directly to the blowdown drum in order to reduce the pressure in the tower. During the startup, the temperature of the reboiler return to the tower was as high as 307°F (153°C), the temperature increased at a rate of 73°F (23°C) per hour during the period 10 a.m. to 1 p.m. The heat to furnace was then reduced by decreasing the fuel gas flow to the furnace. At this time, the level transmitter at the raffinate splitter showed 80% but the tower level was actually 43 m. At 12:42 p.m., the splitter level control was opened to 15% output. The valve was opened five times but the heavy raffinate flow had only begun at 12:59 p.m. The heavy raffinate flowed out the tower at 20,500 bpd at 1:02 p.m and 2 min later it spiked to 27,500 bpd. The level of liquid at that time was 48 m but the level transmitter reading continued to decrease up to 78%. The total liquid in the 52-m tower was decreased due to the opening of the heavy raffinate control valve; however, the level of the liquid at the top of the column kept increasing.

Heating from the furnace caused contact between the bubbles of hot vapor rising through the column and the cold liquid. The raffinate leaving the bottom of the column circulated onto the heat exchanger, heated the feed and resulted in rapid heating of the column section above the feed inlet. The increase in the overall column temperature gave rise to the decrease of the liquid hydrocarbon density and therefore significantly increased the level of the liquid inside the tower. The entire column was eventually heated approaching the boiling point of the liquid so that vapor was accumulated and no longer condensed.

The hydrostatic head in the line increased due to the entrance of the liquid to the overhead line and exceeded the set pressures of the safety relief valves. The valves opened and liquid raffinate flowed to the raffinate splitter disposal header collection system.

The Day Board Operator and the outside operators noticed that the splitter tower pressure had spiked dramatically so the day board operator tried to troubleshoot the problem by reducing the fuel gas to feed preheater and fully opened the level control valve to raffinate storage. However, data showed that the incoming feed to the raffinate splitter tower had not been stopped. The data also indicated that the three safety relief valves remained fully open for about 6 min. The amount of material and pressure in the tower overhead reduced due to the heavy raffinate rundown and raffinate splitter reflux flow and thus the pressure dropped inside the tower overhead and the safety relief valves closed. Approximately 196,000 liters of flammable liquid flowed from the overhead vapor line through the safety relief valves to the collection header and then discharged to the blowdown drum during the opening of the safety relief valves. Because the blowdown drum completely filled, the flammable liquid flowed into the ISOM unit sewer system.

The geyser-like flammable liquid was discharged from the stack and fell to the ground. The first notification that the blowdown was overflowing was sent via radio. In response to this message, the Board Operator and Lead Operator used the computerized control system to shut the flow of the fuel to the heater. Other operators left the satellite room and ran toward an adjacent road to redirect the traffic away from the blowdown drum. These actions were required by the company emergency response procedure. The ISOM operators did not have sufficient time to sound the emergency alarm before the explosion. When the explosion occurred, hundreds of alarms registered in the control room sounded but the operations personnel did not have sufficient time to assess the situation. A number of witnesses reported seeing geyser-like liquid flowing down the 6.1-m stack.

The liquid hydrocarbon was released and a flammable vapor cloud was formed and reached an estimated area of 200,000 ft^2 (18,581 m^2). The vapor cloud covered this area in a very short interval due to the direct dispersion from evaporation prior to ignition as well as the burning by the subsonic flames through the flammable cloud. The impact of the falling liquid through the stack onto process equipment, structural components, and piping enhanced the liquid fragmentation into relatively small droplets and promoted the evaporation and formation of the vapor cloud.

The wind also contributed in enhancing the mixing between the flammable vapors with air. The wind at the time of the incident was reported to be at 5 miles (8 km) per hour in to the southeast direction.

Several potential ignition sources were identified in this incident, but the idling diesel pickup truck was suspected to be the most likely ignition point. Once ignited, the flame rapidly spread through the flammable vapor cloud pushing the gas ahead of it to create a blast pressure wave. The flame was accelerated by the combination of congestion/confinement and the flammable mix so that intensified the blast pressure in several areas.

26.14.3 Aftermath and Emergency

The explosion killed 15 contract employees who worked in or near the trailer site between the ISOM and NDU unit. Sixty-six workers were seriously injured and 114 others required medical attention. Of those who were seriously injured, 14 were BP employees and the rest were contractor employees from 13 different organizations.

The blast caused severe damage within the ISOM unit and the surrounding parking areas. About 70 vehicles and 40 trailers were damaged; 13 trailers were completely destroyed. Buildings surrounding the ISOM unit were also impacted by the blast and were characterized by broken windows, cracked wall, damaged door, bent metal structure and dispersal of interior contents.

The actions of the emergency response teams were fast and effective in helping the victims. Member companies of Texas City Industrial Aid System (IMAS) assisted with fire hose line and search-and-rescue of the victims.

26.15. BUNCEFIELD

26.15.1 Description of the Incident

Buncefield site was collectively operated by a number of different companies. However, the incident was initiated in one of the tanks of Hertfordshire Oil Storage Ltd. which was a COMAH site housing about 200,000 metric tons of cold gasoline inventory (BMIIB initial report, 2006).

On December 10, 2005 around 7 p.m. in the evening, one of the tanks entitled Tank 912 began receiving about 296,644 ft^3 unleaded gasoline fuel. The tank gauge records at this point indicated that there was 38,105 ft^3 of unleaded motor fuel. The initial flow rate of the tank was 19,423 ft^3/h. Around 3:00 a.m., the Automated Tank Gauging System showed that the level of the tank was two-thirds full. By 5:35 a.m. on the same day, the fuel in Tank 912 began to overflow with an increase in the temperature. Within the following 20 min, the flow rate to Tank 912 had also increased to 34,430 ft^3/h. Around 6 a.m. about 300 metric tons of unleaded gasoline had overflowed from Tank 912 resulting in a vapor cloud explosion. The initial explosion was followed by several other explosions which eventually engulfed 23 tanks in the tank farm causing a very large scale fire.

While there were no serious injuries or fatalities resulting from this incident, 43 people were directly affected and about 2000 people were evacuated from their houses (Mannan, 2009; BMIIB, 2006).

The Buncefield incident resulted in significantly damaging many commercial as well as residential properties close to its vicinity. The Buncefield Depot was close to the Maylands Industrial Estate having around 630 firms with 16,500 people. The estimated losses in the Maylands businesses ranged between \$207−270 million.

Significant environmental damages were reported as an aftermath of the incident in terms of the water pollution and the decreased air quality in the affected areas to very high cost of clean-up and environmental remediation.

26.15.2 Causes of the Incident

Lack of learning culture: Following the incident, it was claimed that cold gasoline had never resulted in a vapor cloud explosion previously. However, a review of literature and past incidents (Kletz, 1986) indicate otherwise that there are many incidents with significant similarities with the Buncefield incident. According to Kletz (2001c), similar types of incidents keep occurring. Hence, he concludes that organizations have no memory and there is a greater need for digging into the past by performing incident investigations, root cause analysis of near-misses, and estimating probabilities of major incident occurrences (Mannan, 2011).

Inadequate and non-functional equipment: The fuel level in the tank was monitored by using the servo gauge technique, which is based on level and temperature measurements. Servo gauges are utilized because they require less maintenance and it is possible to also monitor the tank level manually. Additionally, the temperature of the fuel was measured using a temperature sensor (Powell, 2006).

Human factors: In this incident, actively assessing the process hazards and analyzing the risks were not prioritized (DNV report, 2008). Similarly, the possibility of the formation of vapor clouds because of overfilling was never investigated and modeling of the resulting vapor pressure, heat radiation, and domino effects were also not considered (DNV report, 2006).

Siting of occupied vehicles and running engines: Mannan (2009) suggests that based on the accident investigation, the average height of the vapor cloud would have been at least 6.6 ft. The footage received from CCTV and the description of eye witnesses corroborate that the vapor cloud could be given a conservative estimate of about 1,610,000 ft^2 which would indicate that the cloud volume could range between 5,650,000 ft^3 and 10,590,000 ft^3 (Mannan, 2009; Powell, 2006; Gant and Atkinson Report, 2006).

26.15.3 Lessons Learned from the Incident

Lessons from past incidents: In case of the Buncefield explosion, there was an underlying belief that cold gasoline and its vapors could not explode in the open air. Unfortunately, this was a belief shared by many oil and gas companies owning tank farms. Authorities and companies alike were unaware of other similar incidents such as Newark, New Jersey in 1983, Naples, Italy in 1995, and St. Herblain, France in 1991.

Conducting proper hazard and risk assessments: The study of the Buncefield incident has indicated that generally insufficient attention to incident precursors and leading indicators result in devastating explosions (DNV, 2009; Mannan et al., 2010). Emphasis is lacking in properly understanding different processes and their hazards. There is also a lack of awareness of the properties of chemical compounds utilized in process operations.

Vapor cloud formation and explosion mechanism: In the case of the Buncefield incident, the BMIIB Third Progress Report discusses the spread of the vapor cloud and its depth of 3.3 ft (1 m) in the area between bund A and the loading gantry, and 16.4−23 ft down Three Cherry Trees Lane. Eye witnesses also reported the presence of a gasoline mist and odor along with automatic revving of cars despite their ignitions being turned off. Powell (2006) in the third BMIIB reports that smaller explosions followed one major explosion, but only the major explosion was recorded seismically. The delay in subsequent explosions after the initial explosion could indicate that they were because of an internal tank explosion and fuel leakage from the pipework and tank damage.

It is safe to assume that generally, tank farms have plenty of ignition sources in their surroundings. Ignition sources might result due to improper procedures in hot works such as welding and cutting during maintenance operations. Sometimes, ignition sources are also present because of smoking near tanks.

High overpressures generation in open air settings: The preliminary assessment of the blast damage by the investigation team suggested the generation of large overpressures. They found that the overpressure values were of the range of 10.2−14.5 psi in the car parks of the buildings in the facility. The car park location most likely was the area where the gasoline vapors would have ignited.

26.16. SPACE SHUTTLE COLUMBIA DISASTER

26.16.1 Development of the Space Shuttle Program

On July 20, 1969, Apollo 11 made the first lunar landing by a manned space craft. The landing was the culmination of 8 year of preparation following President John F. Kennedy's mission statement which ignited the space race with the Soviet Union. Apollo 11 used a lunar orbit rendezvous (LOR) to achieve the lunar landing. This required a rocket to achieve orbit with the moon and a separate lunar module which would make the trip to the surface and return the astronauts to the orbiting spacecraft. Early in the space program, most engineers at the National Aeronautics and Space Administration (NASA) initially preferred a direct ascent mode of travel. This would utilize one craft to both reach the moon and land on its surface. Subsequently, the same craft could return the astronauts to Earth. This was set aside in favor of LOR, but would be revived later as the Space Shuttle Program.

26.16.2 *Columbia's* Final Flight

Columbia began its 113th (STS-107) flight on January 16, 2003. There were seven crew members aboard. The mission consisted of 16 days in orbit around Earth for scientific research in areas including microgravity and various Earth science disciplines.

During the shuttle's ascent, a piece of thermal insulation foam separated from the external fuel tank and struck the left wing. Shuttle fuel tanks contain liquid hydrogen at −423°F (−253°C) (also oxygen) for propulsion and are fitted with insulation to prevent ice formation and melting of the external tanks. Previous shuttle launches had also shown foam separation, an event that came to be termed 'foam shedding'. In these previous missions, the results had been primarily benign, although the resulting damage to *Atlantis* during mission STS-27 was of enough concern that the mission's commander, Robert Gibson, expected the shuttle to be destroyed upon reentry.

Columbia began its scheduled reentry on February 1, 2003 with the expectation that the damage sustained during takeoff would not affect the landing. As the shuttle entered the atmosphere, excessive heating and strain were observed on the left wing's leading edge. Heating and hydraulic sensors began to fail and TPS tiles and other debris were shed as the shuttle began to break apart over the southwest United States. Just before 9:00 a.m., Mission Control lost contact with the crew. Witnesses to the crash notified NASA and a recovery team was sent to find the crew and shuttle remains (CAIB, 2003).

26.16.3 Accident Analysis

Foam shedding became an accepted phenomenon of shuttle launches by NASA. It was observed on at least 6 prior missions and yielded no significant accidents, despite the concerns raised during STS-27. Normalization of deviance is a term applied to a safety environment in which events initially considered outside the accepted safety standards occur without serious incident. The events are then viewed as acceptable and the bounds of safety are expanded to include this new risk. This type of mentality was applied to NASA after both the *Challenger* and *Columbia* disasters.

After *Columbia* entered orbit, NASA engineers began issuing requests to the Department of Defense (DOD) for imaging to assess the extent of the wing damage as well as information obtained by the astronauts' own inspections. These requests were not met and NASA management chose to focus on reentry scenarios rather than a complete analysis and potential repair of the damage.

In actuality, the foam had significantly damaged the TPS, leaving the left wing susceptible to the extreme heat it experiences entering the Earth's atmosphere (>2800°F). Hot atmospheric gases were able to penetrate the leading edge of the wing and damage the internal structure leading to a loss of hydraulics and failure of the wing's temperature sensors. This led directly to the breakup of the shuttle as it shed various components across Texas and Louisiana.

26.16.4 Other Factors Considered

The CAIB's report included the immediate foam strike as the cause of the accident, a detailed analysis of the safety environment at NASA and the decision-making process leading up to and during STS-107. The safety culture at NASA was cited as a significant root cause.

Budgetary problems were also noted. US competition with the Soviet Union for space supremacy largely ended with the Cold War, making access to funding and support for NASA more difficult.

The overall risk management strategy of NASA was criticized. Reliance on simulation software not programmed for this specific situation was determined to be an ineffective method to assess risk.

Many recommendations were generated as a result of the accident. Some of the most important include:

- Proper camera placement upon shuttle launch.
- Reduction of schedule pressures focusing on available resources.
- Creation of two independent entities, one for technical engineering and one for safety.
- Development of the ability to analyze and repair shuttle wings in space. In *Columbia*'s case, the crew had enough oxygen and supplies to last for 30 days, giving another shuttle time to rendezvous for a repair attempt.

26.16.5 The Accident's Organizational Causes

The investigations that followed both the *Challenger* and *Columbia* disasters identified physical (technical) as well as organizational causes as the foundation for what went wrong during these missions. Adams (2004) refers to the 'destructive organizational dynamics' as root causes that led to the inevitable accident.

The organizational structure of NASA at the time of the accident was basically a dependent structure that left room for potential conflicts of interest. The overlapping jurisdiction of the safety programs at different levels by the Division Chief blurred the lines of responsibility and authority, and instead, produced faulty channels of communications, lack of corporate memory, flawed management and hierarchy, and a generally poor safety culture (CAIB, 2003).

One of the major recommendations by the Rogers commission was the creation of a Headquarters Office of Safety, Reliability, and Quality Assurance, which would fulfill the most significant gaps in the safety culture of the organization. This shift in activities was believed to involve an impartial party that would re-establish effective communication channels and would have authority over safety, reliability, and quality issues involved in shuttle missions. However, NASA neglected full implementation of this change from the beginning as evidenced when independent power was not provided as specified. This created a false sense of safety that compromised the perception of management having a fully operational program with a low probability of failure, rather than a program in the development stages (CAIB, 2003).

After the *Columbia* disaster, the commission recommended the establishment of an independent Technical Engineering Authority to ensure a shift in the organization's pathway to destruction. The objective of this measure was to guarantee the unbiased supervision of the division by removing their dependency on program resources by providing guidance on technical standards, hazard analysis, and risk assessment (CAIB, 2003).

26.16.6 Implications for the Future of Human Space Flight

The CAIB noted several faults with actions taken by NASA officials prior to and during the *Columbia* accident. However, it acknowledged that the shuttle is not inherently unsafe in its current form and sufficient changes to the safety practices could maintain the program as safe for flight. Potential causes for future disasters should be constantly sought out while never settling for a culture that begins to accept anomalies as acceptable risk.

26.16.7 Recommendations

The CAIB report (CAIB, 2003) laid out 29 specific recommendations that addressed many areas, from shuttle operation to safety culture at NASA. In relation to shuttle operation, they specify the development and implementation of an inspection plan to be conducted before the shuttle's return flight and also call for the ability to image the shuttle's exterior in orbit to determine its ability for safe return.

Specific safety environment recommendations were also included. Shuttle flight schedules should be realistic and appropriate with available resources and the ability to alter the schedule as the need arises.

26.17. DEEPWATER HORIZON

At approximately 9:49 p.m. on April 20, 2010, a cloud of flammable hydrocarbons released from a US Gulf of Mexico deepwater well encountered an unknown source of ignition and caused two explosions in rapid succession and a subsequent fire. The events that even now continue to unfold uncover many underlying problems in the process safety considerations for all involved parties.

The incident occurred on the *Deepwater Horizon* mobile offshore drilling unit owned and operated by Transocean located about 48 miles off the coast of Louisiana. The lease was held by BP who contracted Transocean to drill the well for their Macondo Prospect. Eleven workers were killed in the explosion and a further 17 were injured. The fire that was caused by the explosions burned for 36 h before the rig sank on April 22, 2010. The oil spill that resulted from the uncontrolled wellbore flow incident developed into a matter of worldwide interest as it became the largest oil spill to originate in US waters in history.

26.17.1 The Companies

The companies generally considered to be directly involved in the operations of the well are BP, the lessee of the drilling area, and Transocean, the owner of Deepwater Horizon and drilling contractor. Related parties responsible for various parts of the drilling operations include Halliburton, who was the cementer for the job, and Cameron, the manufacturer of the blowout preventer. Numerous other contractors in diverse fields had employees aboard at the time of the accident, including companies supporting other drilling operations, administrative support, and housekeeping (USCG).

The drilling operations were designed by BP, but the physical work of actually running the well was contracted to Transocean, as is common practice (Presidential Commission). These employees worked under the direction of two 'Well Site Leaders' provided by BP for guidance and direction, as well as for communication with onshore BP engineers.

26.17.2 The Site and the Works

The Macondo well was located in Mississippi Canyon Block 252, about 48 miles off the coast of Louisiana. BP jointly acquired the 10 year lease to this area in the central Gulf of Mexico on March 19, 2008 from the Minerals Management Service (MMS). BP owned 65% of the lease, while Anadarko owned 25%, and MOEX Offshore owned 10%. BP was the lease operator.

The Macondo well was an exploration well, as there was promise of oil to be found below the surface; however, BP had limited information about the geology, as it was its first well on Block 252. The well was designed to be completed into a production well in the case that sufficient hydrocarbons were found to justify production.

The exploration plan for the lease was accepted by MMS on April 6, 2009 and approved in revised form 10 days later. The drilling permit was approved by the MMS on May 22, 2009. The semi-submersible drilling unit *Marianas* was used to begin drilling the well in October of 2009. In November of 2009, Hurricane Ida caused extensive damage to *Marianas* and caused it to go offline.

Deepwater Horizon was used as the substitute to the damaged *Marianas* after the MMS approved a revised drilling application for the well on January 14, 2010. *Deepwater Horizon* had been a familiar drilling unit to BP, who had leased it from Transocean for 9 years at the beginning of the Macondo drilling project. It had been used for approximately 30 drilling projects by BP, the most recent before Macondo being the Kodiak appraisal well, also off the coast of Louisiana in Mississippi Block 771 (SubseaIQ).[1]

26.17.3 Events Prior to the Explosions

BP initially estimated the cost of completion of the Macondo well to be $96.2 million and the expected time of completion was 51 days. At the time of the incident, BP was $58 million over budget and 6 weeks behind schedule. This situation likely contributed to decisions made up to the day of the blowout that served to save time and money perhaps in lieu of safety considerations.

26.17.4 Events Prior to April 19

Events that led to or foreshadowed the blowout began at the time that *Deepwater Horizon* arrived at the site to replace the damaged *Marianas* unit.

---

1 Information in sections 26.17.6 and 26.17.7 is primarily taken from the various reports for the *Deepwater Horizon* incident; specifically, BP (2010), Presidential Commission (2011), DNV (2011), and Transocean (2011).

Yellow pod of BOP leaks: Problems with the blowout preventer (BOP) itself began within the first month of operation when on February 23 a pilot valve leak of 1 gpm was noticed on the yellow pod of the blowout preventer. Operation was switched to the blue pod, and leaking was reduced.

Original drill pipe sheared: A well control event occurred on March 8 that caused the drill pipe to become stuck in the formation and subsequently abandoned.

Lost circulation event: On April 9, at 18,198 ft below sea level, much shallower than the objective depth of 20,200 ft, the pressure exerted by the drilling mud began to damage the fragile formation.

Selection of long string casing: With the lost circulation event, it was obvious that the original design objectives for the casing had to be reevaluated to ensure that the formation would be able to withstand future production operations.

Selection of adequate number of centralizers: The original design of the long-string casing was determined to require 16 or more centralizers, while only 6 were available from the supplier at the time. Further calculations and simulations after the decision between long-string and liner showed that the well needed more than 6 centralizers and could even need as many as 21.

26.17.5 Events on April 19 and 20

The events beginning with the final casing run, completed on April 19, and continuing through the cementing process, pressure tests, and preparations for temporary abandonment had a direct effect on the incident itself, as problems began to compound.

Final casing run: The final casing run actually started early on April 18, and was not completed until mid-afternoon on April 19. The number of stabilizers used was the aforementioned 6 as opposed to the recommended 21.

Float collar: The final casing run was completed to a terminal depth of 18,304 ft and the shoe track was fitted with the float collar installed at the top and a reamer shoe at the bottom. After the casing run was installed, nine attempts were made to establish circulation. The final attempt at establishing circulation was deemed successful at a peak pressure of 3142 psi; however, this pressure was far greater than the expected pressure from the manufacturer's guidelines and BP's procedures, which specified 500–700 psi at a fluid circulation rate of 5–8 barrels per minute, which was also not attained.

The cement job: Cementing is figuratively regarded as a combination of science and engineering with an element of art. The engineering tells the engineer what properties are needed for the cement to complete the objective, while the science allows one to design the cement toward

that end. However, in the end, the process of cementing comes down to the art of running a blind operation that depends on the prior science and engineering to succeed. In cementing, the crew is not able to see the results of the job, so it must rely upon well-designed procedures, and testing to ascertain whether the job has been done properly.

Preparing for temporary abandonment: The final plan for temporary abandonment was not delivered to the rig until the morning of the procedure, after it had already begun. It included an abnormally large amount (425 barrels, or twice the normal amount) of spacer to be used to complete the displacement, which was formed from two highly viscous water-based pills so that the pill material could be discharged overboard legally instead of taken to shore for disposal. It differed from the previous plans in the ordering of events and the presence of a negative-pressure test. As noted by BP, it did not include a formal risk assessment of the annulus cement barriers, which did not conform to their own guidelines and may have identified problems with the cement job.

Pressure tests: Both positive- and negative-pressure tests were carried out as a part of the procedure for temporary abandonment. The positive-pressure test is meant to examine the mechanical integrity of the casing under pressure, while the negative-pressure test is designed to ensure that the bottom-hole cement job can keep the hydrocarbons from flowing out of the well.

Between the positive pressure test and the negative pressure test, the sequence for temporary abandonment was followed and the drill pipe was run into the ground and the mud was displaced with seawater to a point above the BOP. At about 5 p.m. the negative-pressure test began.

The crew opened the bleed valve to initialize the pressure to 0 psi, measured on the drill pipe. The procedure is to bleed the line to 0 psi, close the valve, and check that it stays at zero. However, for the first trial the pressure normalized at between 240 and 266 psi and, furthermore, when the crew closed the valve, it shot up to between 1250 and 1262 psi. The annular preventer was tightened to stop a suspected leak that was causing the fluid level in the riser to drop. When the line was bled again, 0 psi was reached; however, when the well was shut in, once again the pressure spiked, this time to 773 psi. One more attempt was made, and once again the pressure fell to zero and spiked when the valve was closed.

The results of these tests may have been explained incorrectly (though Transocean disputes that this was ever used as an explanation) by attributing the pressure to the 'bladder effect', where mud in the riser exerts pressure on the annular preventer, which transfers to the drill pipe, thereby giving incorrect readings. The crew decided to carry out another negative-pressure test, but this time on the kill-line where, they reasoned, there would be no bladder effect.

The test was carried out on the kill line. After bleeding the pressure down to 0 psi, the crew closed the valve and waited for 30 min without observing any flow or pressure change, though the pressure on the drill pipe was still 1400 psi. The negative-pressure test was deemed a success even though the difference in the pressures was never explained. The validity of using the kill line as the pressure monitor point is a point of dispute in the reports, with Transocean saying that it was not valid and BP saying that it was valid, but the kill line may have been plugged or a valve may have been left closed.

Mud displacement: When the negative-pressure test was accepted, the next step was to displace the rest of the drilling mud with seawater and then add the cement plug to block off the well temporarily. The flow of hydrocarbons into the wellbore was first restricted by the hydrostatic pressure of the dense drilling mud on the formation, but would be less restricted by the less dense seawater. Theoretically, if the negative pressure test is correct, there should be no flow from the formation into the well at this point; if there is flow, it is called a kick and it indicates that there is a serious problem with the integrity of the well.

When displacement activities commenced at about 8 p.m., all signs looked normal. The pressure in the drill pipe was decreasing, as expected, and the mud was being redirected to different holding tanks as necessary to accommodate the volume coming in from the wellbore. However, according to the Presidential Commission report, as the operation went on, mud was also directed from the sand traps and trip tank (though not simultaneously) to the holding tanks used for the displaced mud. Because the holding tanks were being filled from several locations at the same time, the measurement of the volume of displaced mud to check for a kick would have been much more complicated or even impossible.

The well became underbalanced at about 8:52 p.m., meaning that the pressure exerted by the seawater and mud could not balance the pressure of the fluids in the formation, causing flow of hydrocarbons into the wellbore. It was estimated that by 9:08, 39 barrels of fluid from the formation had found its way into the mud being displaced, but there was no indication that this was recognized by the well crew. This has widely been attributed to the distraction of preparing the next operation, the surface plug. But before the addition of the surface plug, a sheen test was performed.

Sheen test: At 9:08 p.m., the spacer fluid between the drilling mud and the seawater displacing it was considered to have come to the top of the riser, and pumping was shut down to conduct a sheen test, or an inspection of the fluid to detect any free oil in the spacer before discharging it to sea. The test itself only took about

6 min, but in the time it took to test the fluid, the pressure in the drill pipe rose by 246 psi, indicating flow from the underbalanced well. This is because the seawater initially surrounding the drill pipe was being pushed up from below by the remaining drilling mud below the pipe, which, in turn, was being pushed up by the flowing hydrocarbons from the pay zone.

It is not known why the pressure was not monitored during the sheen test. At any rate, the pumps were restarted at 9:14 p.m., obscuring the pressure data from the well.

Displacement continues: The displacement was continued after the sheen test was deemed to be acceptable. Flow was diverted overboard to discard the spacer, and four pumps were turned on to displace the rest of the fluid. However, one of the pumps was connected to the kill line and this happened before the kill line valve was open; at 9:18 p.m., pressure in the line rose to over 7000 psi and the emergency relief valve opened.

Three of the four pumps were taken off-line in order to determine which had failed and to subsequently deal with the repair. When the pump in need of repair had been ascertained, it began to be repaired while the other two came back on-line. This caused a normal and expected increase in pressure, but within the next 5 min an unexpected increase in kill line pressure occurred.

Before the blowout: The increase in pressure was not noticed until about 9:30 p.m., when it had reached 833 psi. The pumps were shut down to investigate this anomaly, and the pressure fell accordingly. However, after the initial fall, the pressure began to build back up.

Between 9:31 and 9:34, the pressure on the drill pipe increased by approximately 560 psi and began to stabilize. At this point, in an attempt to bleed the excess pressure, a floorhand opened a drill pipe valve. The pressure was relieved partially, but only momentarily before it crept up again and stabilized, though at a slightly lower pressure than initially. At this point, the well was producing between 60 and 70 barrels per minute.

At roughly 9:38, the pressure in the drill pipe began to drop at a constant rate, indicating that the hydrocarbons were pushing through the annulus past the drilling mud toward the top of the riser.

Influx of hydrocarbons: After the pressure drop, the influx of hydrocarbon was just a matter of a short delay. Between 9:40 and 9:43, the drilling mud displaced by the hydrocarbons spewed onto the drilling floor. At 9:41 the annular preventer of the BOP was activated and, though it closed, failed to fully seal the well. The crew concurrently decided to reroute the flow to the mud-gas separator to prevent it from spilling on the drilling floor. Another option, rerouting the fluid overboard, was not chosen.

Unfortunately, the mud-gas separator (MGS) was not designed for such high flow rates, and was quickly overwhelmed. The fluid being diverted soon began to pour from its outlet lines. One minute after this event, at about 9:47, natural gas from the well began to be vented from the MGS through goosenecked vents that angled flow back toward the surface of the platform. Gas alarms sounded, but were not equipped to trip any equipment that could be a source of ignition. Also at this time, the crew activated the variable bore ram mechanism of the BOP, which temporarily sealed the annulus.

This evasive action was too late. At 9:49, main power generation engines started to overspeed. The rig lost power. About 5 s later flammable gas ignited and the first explosion occurred; only 10 s later another explosion rocked the platform. 11 workers lost their lives in direct conjunction with the explosions and subsequent fire.

26.17.6 Lessons from the Deepwater Horizon Incident

Summarized below are a series of Lessons Learned based on the four incident investigation reports. The incident reports, in general, found very similar lessons to be learned from *Deepwater Horizon*. Management and procedural problems were commonly cited as underlying causes, but there were also many circumstantial causes of the accident to learn from as well. No accident report attributes the incident to a single root cause. Some causes from which lessons can be learned developed over extended amounts of time, such as the failure to update engineering drawings and maintenance failures, while other causes developed during a short duration leading up to incident—for example, the decisions made based upon test data (Table 26.14).

26.17.7 Impact

The Deepwater Horizon blowout, explosions and fire, sinking, and oil spill gathered worldwide attention, not only for the disaster itself, but for the sheer size and duration of the incident. It was the largest accidental marine oil spill in the US history with impacts on the environment, economy, and regulations. Short term environmental impacts are easily inferred from the many reports and images of shores and animals covered in oil from the gulf coast, while various initiatives are assessing the long-term impact. President Obama insisted that BP set up a $20 billion compensation fund, which was managed by a government official, to compensate those economically impacted by the spill.

One of the early impacts on regulation was the reorganization of the Minerals Management Service. In the aftermath of the disaster, this reorganization was put into

TABLE 26.14 Some Lessons from the Deepwater Horizon Incident

Relative priority of safety and production
Management of change in drilling procedures
Review process
Lost time versus higher risk
Frequent changes
Management splitting and hierarchy in operations with more than one company
Adequacy of and adherence to company procedures and guidelines
Planning for an emergency
Training
Lessons learned from similar near-misses
Procedures for maintenance
Update of engineering drawings
Design of blowout preventer and automatic blowout prevention system
Use of simulation data and other data-at-hand
Interpretation of test results
Identification of hazardous situations in a timely fashion
Design of offshore gas detection systems
Decision-making under stress

action and the MMS was rebranded as BOEMRE. This reorganization would also lead to the establishment of the Bureau of Ocean Energy Management, for the management of leasing and permitting for energy resources, and the Bureau of Safety and Environmental Enforcement, which would be in charge of safety activities such as inspections. The US government also put a temporary ban on deepwater drilling, and issued new requirements for all offshore drillwells. As part of this the voluntary API SEMP program, is now required for offshore operations.

Yet to be determined are the long-term impacts of *Deepwater Horizon* on the regulatory atmosphere of the gulf and the exact impact that the disaster will have on the parties in question.

ACRONYMS

ACMH   Advisory Committee on Major Hazards

Amoco	American Oil Company
API	American Petroleum Institute
ATSDR	Agency for Toxic Substances and Disease Registry
BG	British Gas
BLEVE	boiling liquid expanding vapor explosion
BOEMRE	Bureau of Ocean Energy Management, Regulation and Enforcement
CAAA	Clean Air Act Amendments
CCPS	Center for Chemical Process Safety
CIMAH	Control of Industrial Major Accident Hazards Regulations 1984
CISHC	Chemical Industry Safety and Health Council
CSB	Chemical Safety and Hazard Investigation Board
DMV	diaphragm motor valve
DNV	Det Norske Veritas
EC	European Community
EIDAS	Explosion Incidents Data Service
EPA	Environmental Protection Agency
HSE	Health and Safety Executive
HSEES	Hazardous Substances Emergency Events Surveillance
ICFTU	International Confederation of Free Trade Unions
IChemE	Institution of Chemical Engineers
ICMESA	Industrie Chimiche Meda Societa Azionara
LNG	liquefied natural gas
LPG	liquefied petroleum gas
MARS	Major Accident Reporting System
MCA	Manufacturing Chemists Association
MCC	methylcarbamoyl chloride
MHIDAS	Major Hazard Incidents Data Service
MIC	methyl isocyanate
MMA	Methyl Methacrylate
MOC	Management of change
MRS	MIC refining still (Bhopal)
MSS	MIC storage system
NIHHS	Notification of Installations Handling Hazardous Substances Regulations 1982
NTSB	National Transportation Safety Board
NTSIP	National Toxic Substance Incidents Program
OECD	Organization for Economic Cooperation and Development
OSD	Offshore Safety Division
OSHA	Occupational Safety and Health Administration
PHA	process hazard analysis
PVC	polyvinyl chloride
PVH	process vent header
RIDDOR	Reporting of Injuries, Diseases and Dangerous Occurrences Regulations 1985
RVVH	relief valve vent header
SRD	Safety and Reliability Directorate
SRV	safety relief valve
TCB	1,2,4,5-tetrachlorobenzene
TCDD	TCDD 2,3,7,8-tetrachlorodibenzo-*p*-dioxin
TCP	2,4,5-trichlorophenol
TNO	Toegepast-Natuurwetenschappelijk Onderzoek
TNT	trinitrotoluene
UCC	Union Carbide Corporation
UCIL	Union Carbide India Ltd
VCE	vapor cloud explosion
VGS	vent gas scrubber

REFERENCES

Assheton R., 1930. History of Explosions. Charles L. Storey Company for Institute of Makers of Explosives, Wilmington, DL.

Assini, J., 1974. Choosing welding fittings and flanges. Chem. Eng. 81 (September), 90

ten Berge, W.F., 1985. The toxicity of methylisocyanate for rats. J. Hazard. Mater. 12 (3), 309.

Bhushan, B., Subramanian, A., 1985. Bhopal: what really happened? Business India February 25—March 10, 102.

Blake, E.S., Rappaport, E.N., Landsea, C.W., 2007. The Deadliest, Costliest and Most Intense United States Tropical Cyclones from 1851—2006 (and Other Frequently Requested Hurricane Facts). NOAA, Technical Memorandum NWS-TPC-5, 43 pp.

Bolton, L., 1978. What happened in Seveso. Chem. Eng. 85 (September), 104C.

Braun, R., Frilling, M., Schönbucher, A., 1999. Simulation of a reaction network in a semibatch reactor. Chem. Eng. Technol. 22 (11), 919—923.

Browning, B., Searson, A.H., 1989. The lessons of the Thessaloniki oil terminal fire. Loss Prevention and Safety Promotion 6, 39.1.

Buncefield major incident investigation board explosion mechanism advisory group report, 2006.

Buncefield MIIB—Final Report, 2006. <http://www.buncefieldinvestigation.gov.uk/reports/index.htm#final/>. Access date: October 3, 2013.

Burgess, D.S., Strasser, A., Grumer, J., 1961. Diffusive burning of liquid fuels in open trays. Fire Res. Abs. Rev. 3, 177.

Carpenter, S, Bennett, E, Peterson, G., 2006. Scenarios for ecosystem services: an overview. Ecol. Soc.29.

Carson, P.A., Mumford, C.J., 1979. Analysis of incidents involving major hazards in the chemical industry. J. Hazards Mater. 3, 149—165.

Carson, P.A., Mumford, C.J., 2002. Hazardous Chemical Handbook. second ed. Elsevier, London.

Cattabeni, F., Cavallaro, A., Galli, G. (Eds.), 1978. Dioxin, Toxicological and Chemical Aspects. Halsted Press, New York, NY.

Comer, P.J., 1977. The dispersion of large-scale accidental releases such as Seveso. Royal Meteorological Society Meeting on Atmospheric Surface Exchanges of Pollution, London.

Cox, R. A., Roe, D. R., 1977. A model of the dispersion of dense vapour clouds. Loss Prevention and Safety Promotion 2, 359.

Cruz, A.M., Krausmann, E, 2009. Hazardous-materials releases from offshore oil and gas facilities and emergency response following hurricanes Katrina and Rita. J. Loss Prev. Process Ind. 22, 59—65.

Cude, A.L., 1975. The generation, spread and decay of flammable vapour clouds. In: Course on Process Safety Theory and Practice, Department of Chemical Engineering, Teesside Polytechnic, Middlesbrough.

Cullen, J., 1985. What can we learn from Pemex? Chem. Eng. (London). 418, 21.

Davenport, J.A., 1987. Gas plant and fuel handling facilities: an insurer's view. Plant/Operations Prog. 6, 199.

Davis, L.N., 1979. Frozen Fire. Friends of the Earth, San Francisco, CA.

Doyle, W.H., 1969. Industrial explosions and insurance. Loss Prev. 3, 11.

Doyle, W.H., 1972a. Instrument connected losses in the CPI. Instrum. Technol. 19 (10), 38.

Doyle, W.H., 1972b. Some major instrument-connected CPI losses. Instrumentation in the Chemical and Process Industries 8. Instrument Society of America, Pittsburgh, PA.

Dyfed County Fire Brigade, 1983. Report of the Investigation into the Fire at Amoco Refinery August 30, Carmarthen, Dyfed.

Eisenberg, N.A., Lynch, C.J., Breeding, R.J., 1975. Vulnerability Model: A Simulation System for Assessing Damage Resulting from Marine Spills. Rep. CG-D-136 75. Enviro Control Inc., Rockville, MD.

Fauske, H.K., 1964. The discharge of saturated water through tubes. Seventh National Heat Transfer Conference. American Institute of Chemical Engineers, New York, p. 210.

Gant, S., Atkinson, G., 2006. Buncefield Investigation: Dispersion of the Vapour Cloud Health and Safety Laboratory Report, CM/06/13.

Glasstone, S., 1964. The Effects of Nuclear Weapons. Atomic Energy Commission, Washington, DC.

Gugan, K., 1979. Unconfined Vapour Cloud Explosions. Institution of Chemical Engineers, Rugby.

Haastrup, P., Brockhoff, L., 1990. Severity of accidents with hazardous materials. A comparison between transportation and fixed installations. J. Loss Prev. Process Ind. 3 (4), 395.

Haastrup, P., Brockhoff, L.H., 1991. Reliability of accident case histories concerning hazardous chemicals. An analysis of uncertainty and quality aspects. J. Hazard. Mater. 27, 339.

Hagon, D.O., 1986. Pemex (letter). Chem. Eng. (London). 421, 3.

Ham, B., 2005, A review of spontaneous combustion incidents. In: Aziz, N. (Ed.), Coal 2005: Coal Operators' Conference, University of Wollongong & the Australasian Institute of Mining and Metallurgy, pp. 237–242.

Harvey, B.H., 1976. First Report of the Advisory Committee on Major Hazards. HM Stationery Office, London.

Harvey, B.H., 1979. Second Report of the Advisory Committee on Major Hazards. HM Stationery Office, London.

Harvey, B.H., 1984. Third Report of the Advisory Committee on Major Hazards. HM Stationery Office, London.

Hay, A., 1976a. Seveso: the aftermath. Nature. 263, 537.

Hay, A., 1976b. Seveso: dioxin damage. Nature. 266, 7.

Hay, A., 1976c. Seveso: seven months on. Nature. 265, 490.

Hay, A., 1976d. Seveso: solicitude. Nature. 266, 384.

Hay, A., 1976e. Toxic cloud over Seveso. Nature. 262 (August), 636.

Hay, A., 1981. Seveso: the intrigue and the infighting. Nature. 290 (March), 71.

Hay, A., 1982. The Chemical Scythe. Lessons of 2,4,5-T and Dioxin. Plenum Press, New York, NY.

Health and Safety Executive (HSE), 1978. Canvey: An Investigation of Potential Hazards from Operations in the Canvey Island/Thurrock Area. HM Stationery OfficeHSE), 1978, London.

Health and Safety Executive (HSE), 1981. Canvey: A Second Report. A Review of the Potential Hazards from Operations in the Canvey Island/Thurrock Area Three Years After Publication of the Canvey Report. HM Stationery OfficeHSE), 1981, London.

Health and Safety Executive (HSE), 1989. The Fires and Explosion at BP Oil (Grangemouth) Refinery Ltd. HM Stationery OfficeHSE), 1989, London.

High, R.W., 1968. The Saturn fireball. Ann. N.Y. Acad. Sci. 152 (Art. 1), 441.

Howard, W.B., 1985. Seveso: cause-prevention. Plant/Operations Prog. 4, 103.

Kaiser, M.J., Pulsipher, A.G., 2007. Generalized functional models for drilling cost estimation. SPE Drill Compl. 22 (2), 67–73, SPE-98401-PA.

Kessler, R.C., Galea, S., Gruber, M.J., Sampson, N.A., Ursano, R.J., Wessely, S., 2008. Trends in mental illness and suicidality after Hurricane Katrina. Mol. Psychiatry. 13, 374–384. Available from: http://dx.doi.org/10.1038/sj.mp.4002119.

Kier, B., Muller, G., 1983. Handbuch Störfalle. Erich Schmidt, Berlin.

Kimbrough, R., Krouskas, C.A., Carson, M.L., Long, T.F., Bevan, C., Tardiff, R.G., 2010. Human uptake of persistent chemicals from contaminated soil: PCDD/Fs and PCBs. Regul. Toxicol. Pharmacol. 57, 43–54.

Kimmerle, G., Eben, A., 1964. On the toxicity of methyl isocyanate and its quantitative determination in air. Arch. Toxikol. 29, 235.

Kletz, T., 2003. Still Going Wrong!: Case Histories of Process Plant Disasters and How They Could Have Been Avoided. Butterworth-Heinemann publications, Burlington MA 01803, USA.

Kletz, T., 2009. What Went Wrong: Case Histories of Process Plant Disasters and How They Could Have Been Avoided. fifth ed Butterworth-Heinemann publications, Burlington MA 01803, USA.

Kletz, T.A., 1986. Accident reports and missing recommendations. Loss Prev. Safety Promot. 5, 201.

Kletz, T.A., 1988. Learning from Accidents in Industry. Butterworths, London.

Kletz, T.A., 1990. Critical Aspects of Safety and Loss Prevention. Butterworth, London.

Kletz, T.A., 1993. Lessons from Disaster: How Organizations Have No Memory and Accidents Recur. Gulf Professional Publishing, UK.

Kletz, T.A., 2001a. An Engineer's View of Human Error. third ed. Institution of Chemical Engineers, Rugby.

Kletz, T.A., 2001b. Learning from Accidents. third ed. Butterworth-Heinemann, Oxford, UK and Woburn, MA.

Kletz, T.A. 2001c. Some problems and opportunities that have been overlooked. Proceedings of the Mary Kay O'Connor Process Saf. Cent. Annu. Symp. College Station,Texas.

Knabb, R.D., Rhome, J.R., Brown, D.P., 2006. Tropical cyclone report, Hurricane Katrina, 23–30, August 2005. National Hurricane Center, 43 pp. Available online at http://www.nhc.noaa.gov/pdf/TCRAL122005_Katrina.pdf.

Krausmann, E., Cruz, A.M., 2008. Natech disaster: when natural hazards trigger technological accidents. Spec. Issue Nat. Hazards. 46 (2).

Lees, F.P., 1976. The reliability of instrumentation. Chem. Ind. March, 195.

Lees, F.P., 1980. Loss Prevention in the Process Industries, vol. 12. Butterworths, London.

Lenoir, E.M., Davenport, J.A., 1993. A survey of vapor cloud explosions: second update. Process Saf. Prog. 12 (1), 12-33.

Lewis, D.J., 1980. Unconfined vapour cloud explosions: historical perspective and predictive method based on incident record. Prog. Energy Combust. Sci. 6, 151.

Litman, T., 2006. Lessons from Katrina and Rita: what major disasters can teach transportation planners. J. Trans. Eng. 132, 11–18. Available online at: http://scitation.aip.org/teo; http://www.vtpi.org/katrina.pdf.

Mcguire, J.H., 1953. Heat Transfer by Radiation. Fire Research Special Report 2. HM Stationery Office, London.

Mahoney, D.G., 1990. Large Property Damage Losses in the Hydrocarbon-Chemical Industries: A Thirty-Year Review. thirteenth ed. M&M Protection Consultants, New York, NY.

Manich, O., 2008. Diseñö de las bases de una logística aplicada a desastres y catástrofes en el ámbito de la Provincia de Barcelona. Anexo D. Department de Projected d'Enginyeria. Universitat Politècnica de Catalunya. Spain.

Mannan, M.S., et al., 2009. Analysis of the Buncefield oil depot explosion: explosion modeling and process safety perspective. Proceedings of the Mary Kay O'Connor 2009 International Symposium.

Mannan, M.S., 2011. The Buncefield explosion and fire lessons learned. Proc. Safety Prog. 30, 138–142. Available from: http://dx.doi.org/10.1002/prs.10444.

Mansot, J., 1989. Incendie du Depot Shell du Port Edouard Herriot a' Lyon les 2 et 3 Juin 1987. Loss Prevention and Safety Promotion 6, 41.1.

Margerison, T., Wallace, M., Hallenstein, D., 1980. The Superpoison. Macmillan, London.

Marshall, V.C., 1980. Seveso, an analysis of the official report. Chem. Eng. (London). 358, 499.

Marshall, V.C., 1987. Major Chemical Hazards. Ellis Horwood, Chichester.

Nash, J.R., 1976. Darkest Hours. Nelson Hall, Chicago, IL.

Occupational Safety and Health Administration (OSHA), 1990. Phillips 66 Company Houston Chemical Complex Explosion and Fire. OHSAOSHA), 1990, Washington, DC.

Ooms, G., Mahieu, A.P., Zelis, F., 1974. The plume path of vent gases heavier than air. Loss Prevention and Safety Promotion 1, 211.

Opschoor, G., 1978. Rep. 78-0834. TNO, Apeldoorn.

Orsini, B., 1977. Parliamentary Commission of Inquiry on the Escape of Toxic Substances on July 10, 1976 at the ICMESA Establishment and the Consequent Potential Dangers to Health and the Environment due to Industrial Activity. Final Report, Rome.

Orsini, B., 1980. Seveso (Translation of Official Italian Report by Health and Safety Executive). Health and Safety Executive, London.

Parker, R.J., 1975. The Flixborough Disaster. Report of the Court of Inquiry. HM Stationery Office, London.

Pasquill, F., 1943. Evaporation from a plane free-liquid surface into a turbulent air stream. Proc. R. Soc. Ser. A 182, 75.

Pesatori, A.C., Baccarelli, A., Consonni, D., Lania, A., Beck-Peccoz, P., Bertazzi, P.A., et al., 2008. Aryl hydrocarbon receptor-interacting protein and pituitary adenomas: a population-based study on subjects exposed to dioxin after the Seveso, Italy, accident. Eur. J. Endocrinol. 159, 699–703.

Pesatori, A.C., Consonni, D., Bachetti, S., Zocchetti, C., Bonzini, M., Baccarelli, A., et al., 2003. Short- and long-term morbidity and mortality in the population exposed to dioxin after the Seveso accident. Industrial Health. 41, 127–138.

Pietersen, C.M., 1985. Analysis of the LPG Incident in San Juan Ixhuatepec, Mexico City, November 19, 1984. Rep. 85-0222. TNO, Apeldoorn, the Netherlands.

Pietersen, C.M., 1986a. Analysis of the LPG disaster in Mexico City, November 19, 1984. Gastech. 85, 112.

Pietersen, C.M., 1986b. Analysis of the LPG disaster in Mexico City. Loss Prevention and Safety Promotion. 5, 21–31.

Pietersen, C.M., 1988. Analysis of the LPG-disaster in Mexico City. J. Hazard. Mater. 20, 85.

Powell, T. 2006. The Buncefield Investigation: Third Progress Report. Report prepared for the Buncefield Major Incident Investigation by the Health and Safety Executive (HSE) and the Environmental Agency (EA).

Redmond, T.C., 1990. Piper Alpha AU: 176 the cost of the lessons. Piper Alpha: Lessons for Life Cycle Management. Institution of Chemical Engineers, Ruby, 113 p.

Rein, R.G., Sliepcevich, C.M., Welker, J.R., 1970. Radiation view factors for tilted cylinders. J. Fire Flammability 1, 140.

Rice, A.P., 1982. Seveso accident: dioxin. In: Hazardous Materials Spills Handbook, G. F. Bennett, F. S. Feates, and I. Wilder, eds., McGraw-Hill, New York, 11–44.

Richardson, J., Ham, B.W., 1996. Incidence of spontaneous combustion in Queensland underground coal mines. Department of Mines and Energy Internal Report.

Richardson, T., 1991. Learn from the Phillips explosion. Hydrocarbon Process. 70 (3), 83.

Rijnmond Public Authority, 1982. Risk Analysis of Six Potentially Hazardous Industrial Objects in the Rijnmond Area: A Pilot Study. Springer.

Riley, R.V., 1979. Accidents with pipelines in the USA: selected case histories and a review of the activities of the National Transportation Safety Board. Third International Conference on Internal and External Protection of Pipes. British Hydromechanics Research Association, Cranfield, Bedfordshire.

Sambeth, J., 1983. What really happened at Seveso. Chem. Eng. 90 (May), 44.

Sanders, R.E., 2008. Hurricane Rita: an unwelcome visitor to PPG industries in Lake Charles, Louisiana. J. Hazard. Mater. 159 (1), 58–60.

Santella, N., Steinberg, L.J., Sengul, H., 2010. Petroleum and hazardous material releases from industrial facilities associated with hurricane Katrina. Risk Analyis. 30 (4), 635–649, Published Online: March 16, 2010.

Scott, J.N., 1992. Succeeding at emergency response. Chem. Eng. Prog. 88 (12), 62.

Skandia International, 1985. BLEVE! The Tragedy of San Juanico. First 2 (November).

Shaw, P. and Briscoe, F. (1978) Evaporation from spills of hazardous liquids. UKAEA, SRD R100, Risley, UK, May 1978, p333. In Marshall, V.C. (1987) Major Chemical Hazards. Ellis Horwood Ltd, ISBN 085312969X.

Slater, D.H., 1978. Vapour clouds. Chem. Ind. 6 (May), 295.

Theofanous, T.G., 1981. A physicochemical mechanism for the ignition of the Seveso accident. Nature. 291 (June), 640.

Theofanous, T.G., 1983. The physicochemical origins of the Seveso accident. Chem. Eng. Sci. 38 (1615), 1631.

Thomas, P.H., 1963. The size of flames from natural fires. Combustion 9, 844.

Thomas, P.H., 1965. Fire Spread in Wooden Cribs, Part III: the Effect of Wind. Fire Res. Note 600. Fire Research Station, Borehamwood.

Union Carbide Corporation, 1985. Bhopal methyl isocyanate incident. Investigation Team Report. Danbury, CT.

University Engineers Inc., 1974. An Experimental Study on the Mitigation of Flammable Vapor Dispersion and Fire Hazards Immediately Following LNG Spills on Land. AGA Proj. IS-100-1

Vervalin, C.H. (Ed.), 1964. Fire Protection Manual for Hydrocarbon Processing Plants. Gulf, Houston, TX.

Vervalin, C.H., 1973. Fire Protection Manual for Hydrocarbon Processing Plants. second ed. Gulf, Houston, TX.

Wilkinson, N.B., 1966. Explosions in History. The Hagley Museum, Wilmington, DE.

Wiekema, B.J., 1980. Vapour cloud explosion model. J. Haz. Materials 3 (3), 221.

Laboratories and Pilot Plants

Laboratories and pilot plants are an integral part of the activities of the process industries. Even though they are smaller in scale compared to the full-scale plants, they have a same set of potential hazards; thus the approach to hazard control used for full-scale plants is applicable in large part to activities in laboratories and pilot plants. More detailed discussion on laboratories and pilot plants and other related topics can be found in *Lees' Loss Prevention in the Process Industries*, fourth edition, including: laboratories (Appendix 9), pilot plants (Appendix 10), legislation (Chapter 3), management and management systems (Chapter 6), hazard identification (Chapter 8), reactive hazards (Chapter 33), process, pressure system, and control system design (Chapters 11−13), fire, explosion, and toxic release (Chapters 16−18), personal safety (Chapter 25), plant operation and maintenance (Chapters 20 and 21), and safety systems (Chapter 28). The process reaction(s) should be screened to identify any exothermic reaction which might lead to a runaway reaction as described in Chapter 33.

27.1. LABORATORIES

Laboratories are an integral part of the activities of the process industries. They include not only analytical laboratories and laboratories carrying out research and development work in industry but also university teaching laboratories where future industrial managers are trained.

Laboratory safety management is required by regulations. In the United Kingdom, the Factories Act 1961 covers industrial workplaces, including laboratories, while the HSWA 1974 applies to all places of work and thus extends coverage to university as well as industrial laboratories. In the United States, OSHA 1910.1450 on 'Occupational exposure to hazardous chemicals in laboratories' (OSHA, 1990) provides regulatory guidelines for all employers engaged in the laboratory use of hazardous chemicals.

Accounts of laboratories and of laboratory safety are given in many publications such as *Code of Practice for Chemical Laboratories* (the RIC Laboratories Code) has been published by the Royal Institute of Chemistry (RIC, 1976). *Chemical Laboratory Safety and Security, A Guide to Prudent Chemical Management* was published by the National Research Council of the National Research Council (2013). A review of legislation applicable to chemical engineering laboratories is given in the IChemE Laboratories Guide. AIHA Laboratory health and safety committee also provides a significant source of information on the practice of industrial hygiene and safety in the laboratory and associated research settings (AIHA, 2013). It provides accounts of various laboratory safety incidents (such as asphyxiation, autoclave incidents, centrifuge explosions, chemical exposure, electrical incidents, fires, ultraviolet burns), as well as information on various lab health and safety technical topics, such as chemical safety/MSDS, security/emergency preparedness and response, hazardous waste, ergonomics, fire safety ventilation, management systems.

Accounts of laboratory incidents have been provided on the AIHA website. Several notable incidents occurred in university laboratories are:

1. University of California, Los Angeles California on December 29, 2008, in which a graduate student died from injuries sustained in a chemical fire involving *tert*-butyllithium (tBuLi);
2. Texas Tech, Lubbock, Texas, in which a graduate student was severely injured in an explosion during the handling of a high-energy metal compound which suddenly detonated (Kemsley, 2010);
3. Carnegie Mellon, Pittsburgh, Pennsylvania, on January 8, 1999, in which a graduate student was injured because of an explosion involving azobisisobutyronitrile, several members of EH&S team also experienced temporary symptoms from exposure; and
4. Texas A&M University, College Station, Texas, on January 12, 2006, involving catastrophic rupture of a nitrogen tank in the Chemistry building (Mattox, 2006).

27.1.1 Laboratory Management Systems and Personnel

The laboratory should have a management system with suitable organization and competent people, systems and procedures, standards and codes of practice, and documentation. All this information should be incorporated in laboratory safety manuals. There should be a clear chain of command and separation of executive from advisory, or staff functions. Some principal systems in laboratory management systems include hazard review, permit systems, incident reporting, and safety audits.

Laboratory codes provide systematic methodologies to manage laboratories. Examples of those codes are the RIC Laboratories Code, the NFPA 45, and the IChemE Laboratories Guide.

There tend to be wide variations in the capability of the personnel who work in laboratories, ranging from well-trained and experienced permanent staff to relatively inexperienced research workers and students, indicating that training assumes particular importance. This training needs to cover the hazards, the equipment, the procedures, and the systems. As with all training, it should aim to motivate as well as to inform. Accounts of laboratory training are given by Bretherick (1981).

27.1.2 Laboratory Hazards

Reviews of hazards in chemical laboratories are given in the RIC Code and the IChemE Guide and in most texts on laboratory safety such as those of Bretherick (1981). Recognized hazards in chemical laboratories include reactive substances, flammable substances, toxic substances, radiation hazards, electrical hazards, mechanical hazards, operating condition hazards, and water release hazards.

— *Reactive substances*: Where reactive substances are handled or processed, every effort should be made to obtain the fullest information on their behavior.
— *Flammable substances*: Many of the liquids and gases handled in laboratories are flammable. Accounts of the hazards of flammable and combustible materials are given in the various NFPA codes, including for laboratories NFPA 45.
— *Toxic substances*: Where toxic substances are handled, it is necessary to consider all three modes of entry into the body and both short- and long-term toxic effects. Where a toxic hazard exists, the requirements of the Control of Substance Hazardous to Health (COSHH) Regulations 1988 apply. Hazards associated with nanomaterials and nanotechnology must also be taken in consideration (Savolainen et al., 2010).
— *Radiation hazards*: There are a variety of radiation hazards, most of which may be found on occasion in laboratories, such as (1) devices containing radioactive sources: liquid level gauges, gas chromatograph detectors, leakage detectors, anti-static devices on balances, and fire detectors; (2) apparatus producing voltages above 5 kV may be a source of X-rays; and (3) non-ionizing radiation sources include lasers, microwaves, and UV and IR devices.
— *Electrical hazards*: Personnel may be at risk of electrocution from temporary hook-ups and repairs, defective and damaged cabling, unearthed components, and poor grounding. The electrical hazards are little different from those in industrial situations generally, but unless good practice is observed a laboratory tends to be especially vulnerable to them.
— *Mechanical hazards*: These mechanical hazards include those associated with rig equipment, workshop machinery, hand and power tools, lifting equipment, rotating equipment, and pinching equipment. Accidents are liable to occur where laboratory personnel use equipment with which they are unfamiliar in order to progress the job.
— *Operating condition hazards*: Hazards associated with high or very low temperature surface; cryogenic fluid; high pressure sources (steam, air, compressed gas cylinders, and water); vacuum.
— *Water release hazards*: Accidental release of water, particularly in the form of a jet, can result in short circuiting, thermal shock, extinction of gas jets, and reaction with water-reactive chemicals.

27.1.3 Laboratory Design

Laboratory design and layout is discussed in the IChemE Guide and Baum and Diberardinis (1987). The factors needed to be considered in laboratory design include layout, toxic chemicals inventory, ventilation, fume hoods, and support facilities such as workshop, stores, receipt bays, analytical services, and staff facilities.

- *Laboratory layout*: The design and layout of a laboratory should proceed on principles broadly by analyzing the needs of the experimental activities using a flow diagram showing the flow of materials between the experimental rigs, the workshops, stores, analytical services, waste disposal facilities, and developing layout diagrams in which the low and high hazard features are separated, with minimization of exposure near the latter. There should be adequate staff facilities provided.
- *Toxic chemicals*: Where toxic chemicals are handled, the design should aim to keep the concentrations in the laboratory environment below the relevant exposure limits. For toxic chemicals the COSHH Regulations 1988, including monitoring workplace atmosphere and keeping down the concentration of contaminants through the use of ventilation and fume hoods.
- *Ventilation*: The most common method of controlling the concentration of contaminants in the workplace is ventilation. The exhaust from the ventilation should pass to a safe place (e.g., away from air intakes).
- *Fume hoods*: For conducting experiment involving some toxic or noxious chemicals.
- *Laboratory support*: The support includes: (1) workshop, (2) stores, (3) receipt bays, (4) analytical services, and (5) staff facilities. Particular attention is needed for storage space to achieve required segregation, for example, that separate stores are normally required for solvents, different classes of chemicals, explosives, gas cylinders, and cryogenic materials.

The laboratory should be designed for fire protection, in accordance with building and fire protection codes and with advice from the fire authorities. Certain NFPA codes which can be applied to laboratory safety for fire hazards are NFPA 10, NFPA 30, NFPA 45, NFPA 101, NFPA 704, and NFPA Code 45. Some basic elements of design for fire protection include the fire resistance of the laboratory envelope, including doors, internal layout, hazardous area classification, mechanical ventilation, and a fire alarm system.

27.1.4 Laboratory Operation

There are multiple factors involving laboratory operation such as chemical information, design of experiments, hazard assessment, control of substances hazardous to health assessment (COSHH Regulations of 1988), operating procedures, emergency procedures, equipment maintenance, permit systems, housekeeping, out-of-hours working, unattended operation, and access to laboratory. All these factors must be considered properly and adequately to ensure safe working environment.

Laboratory generally contains a large range of equipment, each of which should only be used in suitable applications; its specification and manufacturers' guidance should be consulted. Examples of typical equipment that need attention when being used in laboratory include glassware, hotplates, ovens and furnaces, and centrifuges.

Laboratories also use a wide range of services such as water, steam, compressed air, fuel gas, and electrical power. Others include refrigerated coolants, vacuum, oxygen, and other piped gases. Measure should be taken to ensure the proper use of those services and preventing hazards associated with those services.

The storage of materials for the laboratory is discussed in the IChemE Guide and by Bretherick (1981). The main inventory of hazardous chemicals should be kept in the chemical storage locations. General principles governing such storage are as follows: (1) segregation of incompatible materials, (2) right types of containers, (3) receipt points, (4) acquisition, stock-taking and disposal, (5) minimization of inventory, and (6) identification, ownership, safety and health information, and labeling.

Emergency planning should be undertaken to identify the potential causes and types of emergencies and their consequences, and the countermeasures required. The emergency planning should cover all part of laboratory including storage and services. A detailed account of laboratory emergency planning is given in the IChemE Guide.

27.2. PILOT PLANTS

Pilot plants are intermediate in scale between the laboratory bench and the full-scale plant. They are used for a variety of purposes, essentially to obtain information on process design and operation and on products, including information relevant to safety. Some hazards are liable to show up first at the pilot plant stage and the pilot plant is therefore an important tool for hazard identification. Here consideration is given to the more general aspects of pilot plants and particularly to their characteristic features and hazards, and to safety in pilot plants.

Accounts of pilot plants are given in publications such as Pilot Plant Design, Construction and Operation by Palluzi (1992) and by Constan (1984), and Carr (1988). Other authors provide more discussion on pilot plants regarding aspects such as justification for use; overall program chemical and process information; scale-up issues; project review; flow sheet review; hazards; hazard identification—what-if review; plant construction; staffing requirements, operating and safety aspects; operating manual, operator training manual, and computer control systems (J. Jones et al., 1993a,b).

27.2.1 Pilot Plant Uses, Types, and Strategies

The purposes for which pilot plants are built are the development of a new product, a new process, or an existing process. For a new product, small quantities may be required for testing and for market development. For a new process, information is required to prove feasibility, specify operating conditions, resolve scale-up issues, provide design data, identify problems, develop operating procedures, and provide experience and training. For an existing process, work may be required to check the suitability of different raw materials, improve product quality, explore modified operating conditions, improve the treatment of the effluents, and effect cost reductions and other optimizations. A principal reason for building a pilot plant is to fill the gaps in the information necessary for the design and operation of the full-scale plant, so that pilot plants can be differentiated in both type and scale such as general-purpose pilot plant, specific-purpose pilot plant, and multi-purpose pilot plant (Palluzi, 1992). Pilot plants could be designed and located in-house or located in-house but be designed by a design contractor. Other options available for using pilot plant include hiring of outside pilot plant facilities and contracting out of the pilot plant works (Palluzi, 1991).

Accounts of pilot plant programs emphasize a number of themes, including keeping attention focused on the basic chemistry; keeping the program simple, first establishing feasibility, before getting involved in matters such as yield and effluent, applying lateral thinking and seeking alternative approaches; being prepared to take the calculated risk that the process may not work, although taking no risks with safety; and making use of expert advice from other sources. It is also important to recognize the handover and decommissioning stage.

27.2.2 Pilot Plant Features and Hazards

While the hazards encountered in pilot plants are essentially similar to those in full-scale plants, the characteristics of pilot plants are such that the hazard profile is somewhat different. The specific features and hazards of pilot plants are discussed by a number of authors, including Carr (1988) and Capraro and Strickland (1989). Some characteristic features of pilot plants include gaps in knowledge; novelty of chemicals, process, equipment, and operations; scale effects; extent of manual activities; frequency of modification; multiplicity of tasks; materials storage and transfer; flow features; recycles; utilities

features; frangible elements; plant layout features; location in a building; and research staff involvement.

A pilot plant is on a scale intermediate between those of the laboratory and of the full-size plant. Relative to the laboratory, the larger scale of the pilot plant means that hazards which were obscured at the laboratory scale may become apparent. Operation on the laboratory scale has some of the features of an inherently safer design. The most obvious is the limited inventory, but there are also others such as good ventilation. In some cases, the problem may be that the change in scale is accompanied by other changes, such as the use of less pure raw materials. In others it may simply be that features always present become more obvious with scale, such as the need to dispose of noxious effluents. Moreover, the pilot plant is the stage at which the process is first carried out in process plant as opposed to laboratory equipment and thus the stage at which difficulties of processing, of equipment or of measurement will show up.

In contrast to most large-scale plants, which are in the open, many pilot plants are in buildings, and this has a number of implications. There is often a more congested layout. There is an increased hazard from accumulation of gases or vapors. These may be flammable or toxic vapors, whether from open vessels or from leaks and spillages. Or they may be asphyxiant gas, such as nitrogen, or flammability-enhancing gas, such as oxygen. Likewise, there may be occupational hygiene problems because fugitive emissions disperse less readily.

Research staff involved in pilot plant work are likely to be less attuned than those familiar with plant operation to the disciplines necessary, and specific steps may need to be taken to rectify this.

27.2.3 Pilot Plant Design

There are a number of features which are characteristic of the design of pilot plants, including design objectives, standards and codes, usage, layout, safety systems and control systems. There are a number of features which are characteristic of the design of pilot plants, including design objectives, standard and codes, usage, layout, equipment, and protective and control systems. Some of these are listed below:

— *Design objectives*: The purpose of the pilot plant is to provide information for the design of the full-scale plant. The plant should be designed and operated so as to yield information in which confidence can be placed and which is conformable to established design methods.
— *Design standards and codes*: The standards to be applied should be declared at an early stage and any potential conflict identified, whether between

different standards or between a standard and the design.

— *Flexibility and multiple uses*: Flexibility is important in the design of a pilot plant, because it is to extend knowledge and deal with novel issues.
— *Plant layout*: The basic principles of plant layout apply to pilot plants also such as minimizing the length of pipe work and ensuring a convenient arrangement. Hazards analysis should be conducted.
— *Safety systems*: Providing protection and mitigation against hazards. Particular attention should be on protective barriers, pressure systems (especially in high-pressure plants), pressure relief, frangible elements, gas cylinders, and ignition source control.
— *Plant classification*: Classifying pilot plants for the purposes of design and operation and of personnel protection. There are four classes based on potential hazards: 1—detonation reactions; 2—rupture and fires; 3—leaks and small fires; and 4—low hazard operations.
— *Control systems*: The pilot plant should be provided with instrumentation sufficient not only to obtain design data but to ensure safety with the same principles applied for full-scale plants.

Cost estimation for pilot plants is also an essential part of design process. Estimating the capital cost for pilot plants is difficult in some circumstances because it is the most challenging topic in the process of plant design. Different from cost estimations for full-scale process plants, pilot plant cost estimates are usually made before the design is complete because technology, equipment, and process that are not completely understood and chemistry that could still be under investigation will commonly be used (Palluzi, 2005). Accounts of cost estimating for pilot plants are given by Palluzi (2005) and by Dysert (2003).

The cost for pilot plants can be estimated based on cost estimation for small sized, full-scale plants, neglecting some of the cost components that are not directly related to the process operation, such as process buildings, administrative buildings, laboratories, shipping, transportation, utility, shops, and other permanent parts of the full-scale plant. The cost could be divided into directed cost of a pilot plant involving the purchase of equipment, equipment installation, instrumentation and control systems, piping, electrical equipment, service facilities, and land; and indirect costs are engineering and supervision, construction expenses, contractor's fees, startup expense, and contingency.

27.2.4 Pilot Plant Operation

There are multiple factors involving pilot plant operation. The following factors are considered: suitability of plant, personnel and training, operating and emergency procedures, specifications and documentation, and mothballing.

— *Suitability of plant*: If the pilot plant is a multipurpose one and is thus used to investigate a number of different processes, a particularly careful check should be made that it is fully suitable for the proposed process.
— *Personnel and training*: The magnitude of the hazard in a pilot plant is less than in the full scale, but in other respects the operations tend to be more demanding. The materials, the process, the equipment, the plant, and the procedures are all relatively unfamiliar. The operating team needs strong leadership and experienced personnel.

As with plant operation generally, training is critical. Many of the topics related to pilot plan operation imply the need for training. Training is required for management and research personnel also. The latter may well be used for laboratory rather than plant situations, and need to become familiar with plant disciplines.

— *Operating and emergency procedures*: The operations to be performed should be identified and for each a suitable operating procedure should be developed, and examined from the safety viewpoint. Operations which are likely to be dominant in pilot plant work include manual operations, reactor operations, sampling, and measurement activities. There also needs to be suitable emergency procedures.
— *Specifications and documentation*: Despite the small scale, it is desirable to adhere to a certain formality in pilot plant operation such as formal specification for all chemicals presented, record keeping rules for data, and procedures.
— *Mothballing*: The possibility of mothballing should be taken into account in the design, including all aspects during shut-down and start-up to protect equipment and identify hazards.

27.2.5 Pilot Plant Safety

An accident in a pilot plant, like one in a laboratory, is generally on a much smaller scale than one in a full-scale plant, but again it can give rise to considerable direct and consequential loss. Safety issues in pilot plants includes procedures, hazard identification, process design, mechanical design, and design review; scale-up from pilot to full scale; engineering standards for flammable, explosive, corrosive, and toxic materials and for radiation protection; maintenance procedures; and high-pressure and high-temperature processes. Thus, it is important to have a system of project safety reviews adapted to pilot plant design and operation.

There should be a system of project safety reviews adapted to pilot plant design and operation. These should be the subject of a formal requirement and should be documented. The general methods of hazard identification should be used to discover potential hazards in plant design and operation. In addition, there are hazard identification procedures which are particularly relevant to pilot plants. The information on the chemicals handled, the reactions involved, and the materials of construction for the plant should be as complete and as well documented as practical. The transfer of information from the chemist to the engineer should be regulated by formal procedures. The chemist should give a full description of the process, including the reaction kinetics and heats of reaction, limits of operating parameters such as pressure and temperature, and procedures and precautions adopted. The engineer should study the research reports to envision problems which may arise in scale-up to the pilot plant.

ACRONYMS

HSWA   Health and Safe at Work etc. Act
OSHA   Occupational Safety and Health Administration
AIHA   American Industrial Hygiene Association
IChemE Institution of Chemical Engineers
RIC   Royal Institute of Chemistry
MSDS   Material Safety Data Sheet
COSHH Control of Substance Hazardous to Health
NFPA   National Fire Protection Association

REFERENCES

AIHA, 2013. < http://www.aiha.org/get-involved/VolunteerGroups/ LabHSCommittee/Pages/default.aspx > .

Baum, J., Diberardinis, L., 1987. Designing safety into the laboratory. In: Young, J.A. (Eds.), op. cit., p. 275.

Bretherick, L., 1981. Hazards in the Chemical Laboratory. 3rd ed. Royal Society of Chemistry, London.

Capraro, M.A., Strickland, J.H., 1989. Preventing fires and explosions in pilot. Plant/Operations Prog. 8, 189.

Carr, J.W., 1988. Taking the guesswork out of pilot plant safety. Chem. Eng. Prog. 84 (9), 52.

Constan, G.L., 1984. Pilot plants for medium-sized companies. Chem. Eng. Prog. 80 (2), 56.

Dysert, L.R., 2003. Sharpen your cost estimating skills. Cost Eng. 45.

Jones, J., Asher, W., Bomben, J., Bomberger, D., Marynowski, C., Murray, R., et al., 1993a. Keep pilot plants on the fast track. Chem. Eng. 100 (11), 98.

Jones, J., Asher, W., Bomben, J., Bomberger, D., Marynowski, C., Murray, R., et al., 1993b. Tips for justifying pilot plants. Chem. Eng. 100 (4), 138.

Kemsley, J.N., 2010. Texas tech lessons. Chem. Eng. News. 88 (34), 34–37.

Mattox, B.S., 2006. Investigative Report on Chemistry 301A Cylinder Explosion. Environmental Health and Safety Department, Texas A&M University, College Station, TX.

National Fire Protection Association, 1996. NFPA 30, Flammable and Combustible Liquids Code.

National Fire Protection Association, 1996. NFPA 45, Standard on Fire Protection Using Chemicals.

National Fire Protection Association, 2007. NFPA 10, Standard for Portable Fire Extinguishers.

National Fire Protection Association, 2009. NFPA 101, Life Safety Code.

National Fire Protection Association, 2011. NFPA 704, Standard System for the Identification of the Hazards of Materials for Emergency Response.

National Research Council, 2013. Chemical Laboratory and Safety: A Guide to Prudent Chemical Management. National Academy of Sciences.

Occupational Safety and Health Administration (OSHA), 1990. 29CFR 1910.1450, Occupational Exposure to Hazardous Chemicals in Laboratories (Laboratory Standard). US Government OSHA), 1990, Washington, DC.

Palluzi, R.P., 1991. Understand your pilot plant options. Chem. Eng. Prog. 87 (1), 21.

Palluzi, R.P., 1992. Pilot Plant Design, Construction and Operation. McGraw-Hill, New York, NY.

Palluzi, R.P., 2005. But what will it cost? The keys to success in pilot plant cost estimating. Chem. Eng. 112 (12), 6.

Savolainen, K., Pylkkänen, L., Norppa, H., Falck, G., Lindberg, H., Tuomi, T., et al., 2010. Nanotechnologies, engineered nanomaterials and occupational health and safety—a review. Saf. Sci. 48 (8), 957–963.

The Royal Institute of Chemistry (RIC), 1976. Code of Practice for Chemical Laboratories. Royal Institute of Chemistry (RIC), London.

Earthquakes

An earthquake is one of the principal natural hazards from which process plants world-wide are at risk. Some account is therefore necessary of the seismic design of plants and the assessment of seismic hazard to plants.

Accounts of earthquakes are given in Seismicity of the Earth and Associated Phenomena by Gutenberg and Richter (1954), Introduction to Seismology by Bath (1979), Earthquakes and the Urban Environment by Berlin (1980), Earthquakes by Eiby (1980), An Introduction to the Theory of Seismology by Bullen and Bolt (1985), and Earthquakes by Bolt (1978). Treatments of earthquake engineering, most of which give some coverage of seismology, include Dynamics of Bases and Foundations by Barkan (1962), Earthquake Engineering by Wiegel (1970), Fundamentals of Earthquake Engineering by Newmark and Rosenbleuth (1971), Earthquake Resistant Design by Dowrick (1977, 1987), Ground Motion and Engineering Seismology by Cakmak (1987), and Manual of Seismic Design by Stratta (1987). UK conditions are dealt with in Earthquake Engineering in Britain by the Institution of Civil Engineers (1985) and by Lilwall (1976) and Alderson (1982 SRD R246, 1985). It is proper to recognize also the large amount of work done on seismicity and seismic engineering in Japan.

28.1. EARTHQUAKE GEOPHYSICS

The structure of the earth is approximately as follows: a crust 30 km thick, a mantle 2900 km thick, and a core of 3470 km radius, giving a total radius of 6370 km. The core has an inner core of 1400 km radius.

The earth's crust has a degree of elasticity and when subject to stress due to the earth's forces, it undergoes crustal strain. This property is the basis of the elastic rebound theory of Reid. Reid suggests that 'the crust, in many parts of the earth, is being slowly displaced, and the difference between displacements in neighboring regions sets up elastic strains, which may become greater than the rock can endure. A rupture then takes place, and the strained rock rebounds under its own elastic stresses, until the strain is largely or wholly relieved'.

The origin of an earthquake is termed the focus, or hypocentre, and the point on the earth's surface directly above the focus the epicenter. Earthquakes are classified as shallow focus (focus depth <70 km), intermediate focus (depth 70−300 km), and deep focus (depth >300 km).

During an earthquake, waves pass through the earth and impart motion to the ground. There are two broad types of wave: body waves and surface waves. Body waves are classified as primary, or P, waves and secondary, or S, waves. The other main type of wave is surface waves. Surface waves, long waves or L waves.

An instrument for the recording of the ground motion caused by an earthquake is known as a seismometer and the record that it produces a seismogram.

Measurements may be made of any of the three main time-domain parameters: displacement, velocity, and acceleration. The measurement of ground motion is well developed and records have been obtained for the amplitude and for the acceleration of a large number of earthquakes.

28.2. EARTHQUAKE CHARACTERIZATION

The quantitative characterization of earthquakes is largely in terms of magnitude and intensity scales and of empirical correlations. Earthquakes are a geographical phenomenon and hence in many cases the original correlations have been derived for a specific region, often California. This should be borne in mind in respect of the relationships quoted. The relation of focus and epicenter, magnitude and magnitude scales, wave energy and the surface wave magnitude, frequency and return period, intensity and intensity scales are demonstrated in the full version Lees' book (fourth edition).

28.3. EARTHQUAKE EFFECTS

The effects of an earthquake may be classified as direct or indirect. The direct effects include the following:

1. ground shaking;
2. ground lateral displacement;
3. ground up-lift and subsidence and the indirect effects;
4. ground settlement;
5. soil liquefaction;
6. slope failure:
   a. avalanches, landslides,
   b. mud slides;
7. floods;
8. tsunamis and seiches;
9. fires.

Of the direct effects, ground shaking usually occurs over a wide area. At faults, fault displacement results in lateral movement of the ground and surface breaks. The distance over which the surface breaks occur is variable. There may be tectonic up-lift and subsidence.

Of the indirect effects, ground settlement may occur due to compaction of unconsolidated deposits. There may be soil liquefaction, giving a 'quick condition' failure. Slope failures may occur in the form of avalanches, landslides, and mud slides.

28.4. EARTHQUAKE INCIDENTS

Tabulations of earthquakes are given in a number of texts. Eiby (1980) gives a chronological list; this includes the values of the magnitudes. Berlin gives tables showing earthquakes that have caused major fatalities world-wide, major fatalities in the United States, and major property damage in the United States. Bolt (1978) gives tables of earthquakes world-wide, in the United States and Canada and in Central and South America; the North American list gives MM intensities. Berlin (1980) also gives lists of earthquakes with special features such as shattered earth, vertical displacement and bad ground, and soil liquefaction effects.

Table 28.1 gives some earthquakes in this century that have caused major loss of life world-wide or property damage in the United States. Accounts of these earthquakes are given in the references shown in the table.

28.5. EARTHQUAKE DAMAGE

There are numerous accounts available of earthquake damage. Most texts on seismology contain descriptions and illustrations of the damage caused by particular earthquakes. These accounts deal largely, often exclusively, with damage to buildings, which is not the prime concern here. There is rather less information available on damage to plants.

A review of seismic damage to process plant and utilities has been given by Alderson (1982 SRD R246). Information is also given by Berlin (1980). Information about damage of the earthquake in Kern County near Bakersfleld in 1952 (Case History A20), earthquake at the Paloma Cycling Plant, earthquake at San Fernando in 1971, and other earthquakes are described in the full version of Lees' book (fourth edition).

Over the past several decades, earthquakes have hit China severely, causing dreadful fatalities and damages especially to Chinese chemical plants and subsequently catastrophes to people, property, and the environment. A recent earthquake shook Wenchuan County on May 12, 2008, causing a fault rupture exceeding 200 km in length and affected a total area of about 500,000 km^2. With the vast number of damaged and broken flanges, pipes, and vessels, there was a large amount of chemicals leaked. The collapse of so many chemical facilities leads to a higher chance of fatality. The extensive damages of electric-power, gas, and water-supply systems forced many industrial plants to interrupt production. It was reported that natural gas production reduced as much as 13.2 million cubic meters daily after the devastation.

Japan has suffered more crippling earthquakes than just about any place on Earth. In addition to September 1 (1923 Great Kanto earthquake) and January 17 (1995 Kobe earthquake), 11 March, 2011 Heisei Tohoku earthquake tsunami disaster devastated the Pacific coast of northeastern parts of Japan, Sanriku, Miyagi, Fukushima, and Ibaragi, where the earthquake followed by tsunamis took place (One Year after the 2011 Great East Japan Earthquake, 2012). The magnitude of the 2011 Tohoku earthquake was estimated to be 9.0 or 9.1 and occurring at N38.1, E142.9 with a magnitude of 9.0 and a depth of 24 km (JMA, 2011), which is the largest in Japanese

TABLE 28.1 Some Earthquakes Causing Major Loss of Life or Property Damage

A. Earthquakes Causing Major Loss of Life[a]

Year	Location	Number of Deaths
1970	Peru	66,000
1972	Managua, Nicaragua	12,000[b]
1976	Guatemala	22,000
1976	Tangshan, China	≈250,000[a]
1976	Hopei, China	655,000
1978	Iran	15,000
1988	Spitak, Armenia	25,000
1990	Western Iran	50,000
1999	Turkey	17,118
2001	Gujarat, India	20,085
2003	Southeastern	31,000
2004	Sumatra	227,898
2005	Pakistan	86,000
2008	Sichuan, China	87,587
2010	Haiti region	316,000
2011	Japan	19,500

B. Earthquakes Causing Major Property Damage in the United States[b]

		Property Damage ($ million)
1971	San Fernando, CA	553
1987	Whittier Narrows, CA	358
1989	Loma Prieta, CA	6000
1992	Landers, CA	92
1994	Northridge, CA	15,000

[a]For China in 1976, this is Bolt's entry; Berlin gives Hopei with 655,000 deaths.
[b]Earthquakes since 1900 causing $40 million or more property damage.
Sources: Berlin (1980); Bolt (1985); U.S. Geological Survey (2011).

history. The earthquake was ranked as the fourth largest in the world following the 1960 Chile (M9.5), 2004 Sumatra (M9.3), and 1964 Alaska (M9.2) earthquakes. The earthquake had a long period (about 3 min with the largest slip found to be approximately 30 m) (USGS, 1975). The maximum recorded earthquake intensity was 7, the maximum level on the Japanese scale (JMA, 2011). The earthquake early warning system was issued 8 s after the detection of a first P-wave. A tsunami warning was issued 3 min after the earthquake and revised several times after getting real time seismic and tsunami data. The after shocks in the source region off the Pacific region have still continued to happen around the earthquake source. The devastating tsunami with a maximum height of 39 m, the large number of casualties more than 19,500, which is the worst in Japan after World War II, and several types of tsunami impact such as inundation in a large area more than 500 km^2, destructive force damaging houses, buildings, infrastructures, road, and railways, and change of topography due to the erosion and deposition are also reported to be severe. Fires also resulted in extensive damage. The Great East Japan earthquake revealed the fact that an infrequent and high consequence event can cause unforeseeable damage once it happened.

A serious accident at Fukushima Dai-ichi Nuclear Power Station made the problem more complex (see Chapter 30).

28.6. GROUND MOTION CHARACTERIZATION

The ground motion caused by an earthquake may be characterized in terms of the peak values of the acceleration, velocity, and displacement; of the full time-domain responses of these quantities, particularly the accelerogram, or of the response spectrum. The latter is described in the full version of Lees' book (fourth edition).

The information required for design depends on the method used. Some methods utilize the peak acceleration. The trend is, however, to use the full response in the form of an accelerogram or a response spectrum.

For a single parameter, such as the PGA, methods of prediction are available. For a full response, such as an accelerogram or a response spectrum, it is necessary to select one that is judged appropriate for the site.

28.7. GROUND, SOILS, AND FOUNDATIONS

The characteristics of the ground have profound effects on virtually all aspects of earthquake assessment and earthquake-resistant design.

Experience with earthquakes shows that differences in the ground can result in large differences in the damage done. According to Eiby (1980), in extreme cases, this can amount to up to four degrees of intensity on the MM scale. It is, therefore, important to avoid siting structures on bad ground.

Some of the worst earthquake damage has been due to a quick condition failure in which liquefaction of the soil has occurred. It is important, therefore, to be able to assess susceptibility to soil liquefaction. Damage may also occur due to soil compaction and settlement short of liquefaction.

It is practice to carry out a site investigation to determine the characteristics of the ground. An account of the laboratory and field tests and of the information which they yield is given by Dowrick (1977). Laboratory tests on relative density and particle size distribution and field tests on groundwater conditions and penetration resistance bear on the problems of soil liquefaction and settlement, while field tests on soil distribution, layer depth, and depth to bedrock, groundwater conditions, and the natural period of the soil provide data for response calculations.

Where dynamic analysis is to be performed, it may be necessary to modify the accelerogram to take account of the nature of the ground. The case of bedrock would seem at first sight to be the most straightforward, but this is not necessarily so, since most accelerograms are for softer soils, and there are very few for bedrock. The modification of accelerograms to take account of ground conditions is a specialist matter. An account is given by Dowrick (1977). The ground motion is also affected by any structure placed on it. The accelerogram relevant to the site as a whole is, therefore, the free field accelerogram.

It has been found by experience that there is an appreciable interaction between the soil and the structure and that it is necessary to take this into account. The effects of soil—structure interaction (SSI) are to modify the dynamics of the structure and to dissipate its vibrational energy.

28.8. EARTHQUAKE-RESISTANT DESIGN

The design of an earthquake-resistant structure includes the following principal steps:

1. seismic assessment;
2. ground assessment;
3. selection of structural form;
4. selection of the materials;
5. overall design of the structure.

It has been found by experience that certain structural forms resist earthquakes well while others do not. An account of the features of structural form that give earthquake resistance is given by Dowrick (1977). In broad terms, the structure should be simple and symmetrical and not excessively elongated in plan or elevation. It should have a uniform and continuous distribution of strength. It should be designed so that horizontal members fail before vertical members.

The properties of materials of construction that confer good resistance to earthquakes include homogeneity, ductility, and high strength/weight ratio. Materials generally quoted as being suitable for earthquake-resistant structures include steel, *in situ* reinforced concrete, and prestressed concrete.

The design methods described give information on the forces to be expected at different points in the structure. The detailed design of the various elements of the structure can then be performed. Discussions of the behavior of structural elements and of detailed design using different materials are given by Newmark and Rosenblueth (1971) and Dowrick (1977).

28.9. EARTHQUAKE DESIGN CODES

Guidance on the design of earthquake-resistant structures is available in earthquake design codes. These codes have been developed particularly in the United States and Japan, and many countries have adopted or adapted the US codes.

Some earthquake design codes which have been developed in the United States include the Uniform

Building Code (UBC) of the International Conference of Building Officials (ICBO), the Basic Building Code of the Building Official and Code Administration (BOCA), the Standard Building Code (SBC) of the Southern Building Code Congress International, the recommendations of the Structural Engineers Association of California (SEAOC), and the ANSI code.

28.10. DYNAMIC ANALYSIS OF STRUCTURES

There are a variety of methods of dynamic analysis. Features of these methods are the unsteady-state model of the structure, the use of decomposition techniques, the forcing function, and the domain in which the analysis is conducted.

In the account described in the full version Lees' book (fourth edition), dynamic analysis is described mainly in terms of a single design earthquake. It is generally recommended that the analysis be repeated for further earthquakes.

28.11. SEISMICITY ASSESSMENT AND EARTHQUAKE PREDICTION

The assessment of the seismic hazard to which an installation may be exposed is normally based on information about the seismicity of the area in question. In principle, this may be complemented by information from the monitoring of ground motions and prediction of any impending earthquake. However, such prediction is not sufficiently precise to be of much use for this purpose.

There are a number of methods for the prediction of earthquakes. Such methods are applicable mainly to active regions. Accounts are given by Berlin (1980), Eiby (1980), and Bolt (1978).

One important concept is that of the seismic gap. Essentially, a seismic gap is an area that has not had an earthquake for some time and in which strain is accumulating. The argument is that due to this strain accumulation, the probability of an earthquake increases with time.

28.12. DESIGN BASIS EARTHQUAKE

In earthquake-resistant design, an installation is designed to withstand seismic events so that the frequencies at which various degrees of damage are incurred do not exceed the specified values. The choice of the design basis earthquake is governed by these frequency specifications.

Essentially, there are three main approaches to characterize the ground motion of an earthquake. These are characterizations in terms of peak acceleration, of the response spectra, or of the acceleration–time profile. These three methods of characterization accord with the three main methods of analysis, i.e., static analysis, response spectra, and dynamic analysis.

28.13. NUCLEAR INSTALLATIONS

The seismic hazard is of particular importance for nuclear power plants, and a large amount of work has been done both on seismic design and on seismic hazard assessment.

The NRC has set earthquake-resistant design criteria and has undertaken a large amount of work on seismic engineering of nuclear plants. In 1973, the NRC issued *Regulatory Guide 1.60* (*RG 1.60*) containing its seismic design response spectra. The NRC has for some years had a major program of work on seismic safety margins. Accounts of this work include those of Cummings (1986) and Kennedy et al. (1989). A review of methods for the calculation of the seismic margin of items of equipment is given by Kennedy et al. (1989).

28.14. PROCESS INSTALLATIONS

There is rather less work published on the seismic design and seismic assessment of process plants, but there are some accounts available. Moreover, the methodologies developed in the nuclear industry are in large part applicable to the process industries.

The guide TID-7024 *Nuclear Reactors and Earthquakes* (NTIS, 1963) is quoted in the NPFA codes such as NFPA 59A: 1985 *Production, Storage and Handling of Liquefied Natural Gas* (*LNG*) for guidance on seismic loading.

A review of seismic design and assessment for process plants with special reference to major hazards is given in *Seismic Design Criteria and Their Applicability to Major Hazard Plant within the United Kingdom* by Alderson (1982 SRD R 246). Further accounts of seismic assessment of process plants include those of Ravindra (1992), and, for the United Kingdom, of Davies (1982), Alderson (1985), and Kunar (1985).

Methods for seismic design are given by the NRC (1979c). There is also a good deal available for particular items. For example, for liquid storage tanks, accounts have been published by Veletsos and co-workers (Veletsos and Yang, 1976; Veletsos, 1984; Veletsos and Tang, 1986), Haroun and Housner (1981), Kennedy (1979, 1989), and Priestly (1986). Guidance on analysis of individual failure modes is given in standard texts on structures such as Timoshenko (1936), Warburton (1976), Pilkey and Chang (1978) and, for wind loads, Sachs (1978).

REFERENCES

Alderson, M.A.H.G., 1985. Seismic engineering requirements in British industry. Earthquake Engineering in Britain. Institution of Civil Engineers, London.

Barkan, D.D., 1962. Dynamics of Bases and Foundations. McGraw-Hill, New York.

Bath, M., 1979. Introduction to Seismology. Birkhauser Verlag, Basel.

Berlin, G.L., 1980. Earthquakes and the Urban Environment, vols. 1–3. CRC Press, Boca Raton, FL.

Bolt, B.A., 1978. Earthquakes. 1988 (revised ed.), 1993. W.H. Freeman, New York, NY.

Bullen, K.E., Bolt, B.A., 1985. An Introduction to the Theory of Seismology. Cambridge University Press, Cambridge.

Cakmak, A.D. (Ed.), 1987. Ground Motion and Engineering Seismology. Elsevier, Amsterdam.

Cummings, G.E., 1986. Summary Report on the Seismic Safety Margins Research Program. Rep. NUREG/CR-4431. Nuclear Regulatory Commission, Washington, DC.

Davies, R., 1982. Seismic risk to liquefied gas storage plant in the United Kingdom. In: The Assessment of Major Hazards. Institution of Chemical Engineers, Rugby.

Dowrick, D.J., 1977. Earthquake Resistant Design. Wiley, Chichester.

Dowrick, D.J., 1987. Earthquake Resistant Design. second ed. Wiley, Chichester.

Eiby, G.A., 1980. Earthquakes. Heinemann, London.

Gutenberg, B., Richter, C.F., 1954. Seismicity of the Earth and Associated Phenomena. Princeton Univ. Press, Princeton, NJ.

Haroun, M.A., Housner, G.W., 1981. Seismic design of liquid storage tanks. J. Tech. Councils ASCE. 107, 191.

Institution of Civil Engineers, 1985. Earthquake Engineering in Britain. University of East Anglia, London, UK. April 18–19.

Japan Meteorological Society, 2011. Tokyo. < http://www.jma.go.jp/ jma/indexe.html > < www.jma.go.jp >.

Kennedy, R.P., 1979. Above ground vertical tanks. In: Nuclear Regulatory Commission 1979 NUREG/CR-1161, Sec. 2.2.

Kennedy, R.P., Murray, R.C., Ravindra, M .K., Reed, J.W., Stevenson, J.D., 1989. Assessment of Seismic Margin Calculation Methods. Nuclear Regulatory Comm., Washington, DC (Rep. NUREG/CR-5270).

Kennedy, R.P., Murray, R.C., Ravindra, M.K., Reed, J.W., Stevenson, J.D., 1989. Assessment of Seismic Margin Calculation Methods. Rep. NUREG/CR-5270. Nuclear Regulatory Commission, Washington, DC.

Kunar, R.R., 1985. Methods of seismic qualification for hazardous facilities in the UK. Earthquake Engineering in Britain. Institution of Civil Engineers, London.

Lilwall, R.C., 1976. Seismicity and Seismic Hazard in Britain. HM Stationery Office, London.

Newmark, N.M., Rosenblueth, E., 1971. Fundamentals of Earthquake Engineering. Prentice-Hall, Englewood Cliffs, NJ.

Nuclear Regulatory Commission, 1979. Recommended Revisions to Nuclear Regulatory Commission Seismic Design Criteria. Rep. NUREG/CR-1161, Washington, DC.

One Year After the 2011 Great East Japan Earthquake. In: Proceedings from the International Symposium on Engineering Lessons Learned from the Giant Earthquake, Tokyo, March 1–4, 2012.

Phillips, D.W., 1982. Seismic structural analysis of a typical plant item. In: Appendix II of Alderson, M.A.H.G. (Eds.), op. cit., p. 44.

Pilkey, W.D., Chang, P.Y., 1978. Modern Formulas for Statics and Dynamics. McGraw-Hill, New York, NY.

Priestly, M.J.N., 1986. Seismic design of storage tanks. Bull. N.Z. Nat. Soc. Earthquake Eng. 19, 4.

Ravindra, M.K., 1992. Seismic assessment of chemical facilities under California: risk management and prevention program. Hazard Identification and Risk Analysis, Human Factors and Human Reliability in Process Safety, p. 11.

Sachs, P., 1978. Wind Forces in Engineering. second ed. Pergamon Press, Oxford.

Stratta, J.L., 1987. Manual of Seismic Design. Prentice-Hall, Englewood Cliffs, NJ.

Timoshenko, S., 1936. Theory of Elastic Stability. McGraw-Hill, New York, NY.

US Geological Survey, 1975. The Interior of the Earth. US Government Printing Office, Washington, DC.

Veletsos, A.S., 1984. Seismic response and design of liquid storage tanks. Guidelines for the Seismic Design of Oil and Gas Pipeline Systems. American Society of Civil Engineers, Reston, VA.

Veletsos, A.S., Tang, Y.U., 1986. Dynamics of vertically excited liquid storage tanks. J. Struct. Eng. 112, 1228.

Veletsos, A.S., Yang, J.Y., 1976. Earthquake response of liquid storage tanks. In: Advances in Civil Engineering Through Earthquake Mechanics. American Society of Civil Engineers, New York, NY.

Wiegel, R.L., 1970. Earthquake Engineering. <http://www.google.com/ search?tbo = p&tbm = bks&q = inauthor:%22Robert + L. + Wiegel %22&source = gbs_metadata_r&cad = 3>

Warburton, G.B., 1976. The Dynamical Behaviour of Structures. Pergamon Press, Oxford.

Offshore Process Safety

Any account of loss prevention needs to include some mention of offshore oil and gas activities, even if this is necessarily brief. There is a continuous interaction between developments onshore and offshore. The treatment given here is confined to an outline of offshore safety activities.

Offshore installations operate in a difficult and often hostile environment. The problems, not only of structures but also of processing, are challenging. The solution to these problems often involves technological innovation. The significance of this for safety and loss prevention is clear.

29.1. NORTH SEA OFFSHORE REGULATORY ADMINISTRATION

Two principal elements of the administration are the systems of internal control and of risk assessment. The *Guidelines for the Licencee's Internal Control 1979* describe in effect an SMS. The *Regulations Related to the Licencee's Internal Control 1985* make this a regulatory requirement.

With regard to risk assessment, the *Regulations Concerning Safety Related to Production and Installation 1976* contained a requirement that if the living quarters were to be located on a platform where drilling, production, or processing of petroleum was taking place, a risk evaluation should be carried out. At this date the evaluation was largely qualitative. The move to a more quantitative approach came with the *Guidelines for Safety Evaluation of Platform Conceptual Design 1981*. These had as a central feature, the provision of a sheltered area, required the conduct of a concept safety evaluation (CSE), and specified numerical acceptance criteria.

The Guidelines defined a design accidental event as one that does not violate any of the following three criteria:

1. At least one escape way from central positions which may be subjected to an accident, shall normally be intact for at least an hour during a design accidental event.
2. The shelter area shall be intact during a calculated accidental event until safe evacuation is possible.
3. Depending on the platform type, function, and location, when exposed to the design accidental event, the main support structure must maintain its load carrying capacity for a specified time.

Risk assessment is now the subject of the *Regulations Relating to the Implementation and Use of Risk Assessment in the Petroleum Activities 1990*. Ognedal emphasized that the Norwegian attitude is flexible in its approach to risk assessment and tries to avoid its degenerating into a 'numbers game'.

The *Cullen Report* (Cullen, The Honourable Lord, 1990) recommended far-reaching changes in the regulatory administration in the British sector of the North Sea. The recommendations flow from the evidence given on the Piper

Alpha disaster and on the regulatory administration up to that date. The administration envisaged is one of goal-setting rather than prescriptive regulations and of the use of QRA to demonstrate compliance. The report also recommended that an operator should submit a safety case and that this should be given structure by a requirement to demonstrate by QRA the integrity of a temporary safe refuge (TSR). Another major recommendation is that the operator demonstrates, as part of the safety case, an appropriate safety management system (SMS). The recommendations of the *Cullen Report* (Cullen, The Honourable Lord, 1990) were accepted by the government and the new administration with the HSE as the regulatory body was put in place in 1991.

29.2. GULF OF MEXICO OFFSHORE REGULATORY ADMINISTRATION

The offshore petroleum industry began in the US Gulf of Mexico in the 1950s. The Coast Guard was the original regulatory agency, but the responsibility for offshore safety was eventually transferred to the Minerals Management Service (MMS). The safety regulations are found in the *US Code of Federal Regulations,* (Code of Federal Regulations, Title 250.30, Part 250).

MMS directly references many of the American Petroleum Institute standards and recommended practice documents. The API 14 series contain standards for sub-surface safety valves, platform safety systems, piping systems, electrical systems, and fire safety systems. The API t series recommended practices for offshore staff training are also codified by the MMS.

API 14C (API, 2008a), the recommended practice for analysis, design, and installation of basic surface safety systems on offshore production platforms was originally promulgated in 1972. API 14C embodies two very important process safety concepts:

1. Every identifiable failure mode requires two functionally independent safeguards.
2. Recommended safeguards for each type of offshore process unit (e.g., pressure vessel, pump, and heater) were developed using a generic failure mode and effects analysis.

The documentation showing the results of a 14C approach are presented in a Safety Analysis and Functional Evaluation diagram, known as a SAFE chart. The SAFE chart is a special type of cause and effect diagram.

29.3. OFFSHORE PROCESS SAFETY MANAGEMENT

An offshore platform has both marine and process characteristics. It shares with marine vessels the hazards of severe weather and of collision and *in extremis* those of escape to the sea. At the same time, it contains high pressure plant processing oil and gas. The platform is also a self-contained community with its own power plant, accommodation, and other facilities.

In the 1990s, MMS requested voluntary adherence to the API 75 (API, 2004) and API 14G (API, 2007) recommended practice documents, which present process safety concepts to the offshore environment. These standards are very similar to the OSHA process safety management regulation, containing virtually identical wording in many sections. Following the 2011 Macondo disaster, the newly reconstituted bureau of ocean energy management and regulatory enforcement (BOEMRE) incorporated API 75 by reference and thus requiring mandatory compliance with API 75 for offshore operations.

29.4. OFFSHORE INCIDENTS

The most common types of offshore incidents are fires, collisions, explosions, and loss of well control (blowout). Hurricanes can also cause significant damage to offshore platforms and rigs. The confined space on an offshore platform, combined with the presence of heavy machinery and hydrocarbons, make offshore processes potentially dangerous. Every year approximately 50 offshore workers will suffer significant injuries and approximately 10 will lose their lives.

Some of the deadliest accidents in the offshore history are Occidental's Piper Alpha platform fire and explosions in 1988 that led to 167 fatalities; the failure of the brace of Alexander L. Kielland, a semi-submersible in 1980, causing 123 deaths; the Seacrest drillship capsized in 1989 during Typhoon Gay in Gulf of Thailand, with the loss of 91 people; the Ocean Ranger capsized due to a ballast control malfunction during a ferocious storm in the North Atlantic in 1982 led to 84 fatalities, and the Macondo oil spill disaster (BP, 2010) that led to 11 fatalities and extensive environmental impact.

The most typical causes of these accidents include equipment failure, human errors, and extreme natural impacts (i.e., seismic activity, ice fields, and hurricanes). In drilling activities, accidents usually happen with unexpected blowouts of liquid and gaseous hydrocarbons from the well due to high pressure. These drilling accidents can be controlled by shutting down the well with the blowout preventers (BOP) and by changing the density of the drilling fluid. However, failure to control BOP could cause catastrophic spills, such as the BP Deepwater Horizon incident in the Gulf of Mexico in 2010. BOP is one layer of protection during the drilling process.

29.5. INHERENTLY SAFER DESIGN

Inherently safer design concepts are particularly useful for risk reduction and are highly recommended as a first choice in offshore process designs. Three examples follow.

One is a reduction of the oil inventory in the production separators. The separators constitute a major oil inventory on the platform and they fed the oil pool fire on Piper Alpha.

A second example is adoption of a layout that minimizes the possibility of an oil pool forming near a gas riser.

A third example concerns the potential for a jet flame from a gas riser to impinge on the accommodation. Following Piper Alpha, Shell reviewed their platforms to establish whether in each case this was a possibility (Chamberlain, 1989). In those cases where it was, action was taken.

29.5.1 Friendly Plant

Closely related is the design of plant that is friendly to the process operator. In oil and gas extraction, there are strong pressures to maintain production. In these circumstances, it is highly desirable that there be fallback states of plant operation to which the operator can resort, without facing the all-or-nothing choice of continuing normal production or effecting total shut-down.

Evidence at the *Piper Alpha Inquiry* appeared to indicate that in some systems fuel changeover on the generators from gas to diesel could not be fully relied on. If this situation exists, it puts a greater pressure on the operator to keep going.

29.5.2 Plant Layout

Offshore production platforms carry large amounts of equipment held in a small space. Space and weight are both at a premium, since they are difficult and expensive to provide. Spacing between equipment in a module has to be less generous than at onshore plants. Where a 15 m distance is widely used in the latter, the distance offshore is often half that. It is also necessary to ensure that the layout is such that the center of gravity of the total mass of equipment is at the center of the supporting structure. The decks of the platform are divided into modules. The modules are separated by fire walls which inevitably reduce the ventilation.

29.5.3 Platform Systems

There are a number of basic systems that are critical for safe operation of the platform. The account given here of these systems is limited to a brief overview. It is convenient to cast it as a description of the systems on Piper Alpha. The *Cullen Report* gives a detailed description of these systems and of their behavior on the night of the disaster.

29.5.3.1 Electrical Power Systems

The first of the systems is the electrical power supply system. Basic power is supplied by a pair of main turbine-driven generators with dual fuel firing, gas being the normal fuel with diesel fuel as standby. As backup, there is an emergency generator, turbine-driven and diesel fueled. Changeover of fuel on the main generator and start up of the emergency generator are automatic. Drilling is served by a separate power supply from a diesel-driven generator with its own emergency generators.

The emergency generator for the main supply is designed to provide supply to critical services, which include HVAC, instrumentation and valves, and emergency lighting. Further backup is provided by an uninterruptible power supply (UPS) drawn from batteries designed to provide power during the momentary interruption while the emergency generator starts up and, if necessary, for a period in the event of total failure of the main supply.

Electrical systems for offshore production platforms are the subject of API RP 14F (API, 2008b).

29.5.3.2 Fire Protection System

The fire protection system comprises a number of complementary elements including hazardous area classification, fire walls, and fire and gas detection systems. Fire prevention and control on offshore production platforms is covered in API RP 14G: 1986.

The first line of defense against fire is the use of hazardous area classification to reduce the risk of ignition of any flammable leak that may occur. This is covered by the general onshore and offshore code RP 500: 1991. The traditional form of active fire protection is the water deluge system. This is covered by the NFPA Fire codes, *Construction and Use Regulations 1974* and the associated guidance in the *Offshore Design Guide*. There is an extensive fire and gas detection system, with sensors in the main production modules and elsewhere, utilizing both combustible gas detectors and fire detectors.

A fixed water deluge system with distribution throughout the main modules and at the risers furnishes a basic level of active fire protection. Water for the deluge system is drawn from the sea by fire pumps operating off the main power supply but with diesel-driven pumps at standby.

Certain closed volumes may be provided with a halon total flooding system. Enclosures which may be protected in this way include the centrifugal compressor enclosures, the generator area, the electrical switchgear room, and the control room. Where, as in the latter case, operators may be present, a non-toxic agent is used and there is prior alarm. These fixed systems are supplemented by the fire fighting teams.

As just indicated, the traditional means of fire protection has been passive protection by firewalls and active protection by a uniform water deluge. Alternative approaches have tended to be inhibited by the need to comply with the two separate sets of regulations covering fire, one requiring passive and the other active fire protection measures. In addition to the trade-off between passive and active fire protection, there may also be a trade-off between ventilation and active fire protection. If a module is well ventilated, gas from a leak is less likely to accumulate in the first place.

There are a number of issues related to fire protection. One is the availability of standard fire tests for hydrocarbon fires, as opposed to building fires, such tests being needed for design of fire walls. Another is the use of passive fire protection on risers, which might involve a risk of corrosion beneath the lagging. Generally, there is a strong argument for the use of passive fire protection in that once installed it appears less subject to human failings.

There is now considerable activity in the investigation of fire events on platforms, covering oil pool fires and jet flames; impingement of flames on vulnerable targets such as risers and passive protection of these targets. Further details are given in Section 29.7.

29.5.4 Emergency Shut-Down System

The platform is provided with an emergency shut-down (ESD) system, the main functions of which are (1) to shut down the flow from the reservoir, (2) to shut off the flow through the pipelines entering and leaving the platform, (3) to shut down the main items of equipment, and (4) to initiate blowdown of the inventories to flare.

29.5.5 Process Blowdown System

The process here is basically a three-phase separation process. The well fluids enter the separation system and are separated into oil, gas, and water. Separated oil is then de-gassed, dehydrated, and compressed before being exported through the pipeline. The water may be disposed of into the sea after treatment. Usually on the platform, an automatic blowdown system in the process module is required. The blowdown system will be triggered based on the detector signals within the area. It is usually combined with an emergency shut-down (ESD) system and connected to a flare system.

29.5.6 Evacuation, Escape, and Rescue System

The evacuation, escape, and rescue (EER) system is built around the three main means of leaving the platform, which are (1) helicopter, (2) lifeboat, and (3) life raft. The emergency procedure is for personnel to collect at designated muster points. In the vast majority of cases, evacuation is by helicopter, personnel being summoned to the helideck from their muster points. However, there are a number of reasons why evacuation by helicopter may be impractical. They include high winds and smoke from the platform, which can prevent landing, and time taken to reach the platform.

If helicopter evacuation is not practical, the other main means of getting off the platform is by lifeboat. The use of lifeboats is also subject to limitations. Lifeboats may sometimes be difficult to reach and have limitations in high seas. If the lifeboats cannot be used, the resort is to escape to the sea by launching life rafts and climbing down knotted ropes. The use of knotted ropes requires a certain degree of fitness and is not without risk.

A variety of devices have become available for escape, ranging from individual packages which can be hooked onto a guard rail to chutes and slides.

Another aspect is the integrity of the escape routes from locations where personnel are likely to be to the means of escape. Escape routes therefore need to be designed against a variety of scenarios. Of particular importance are the routes to the lifeboats from the accommodation where at any given time a large proportion of the personnel will be. One approach is to locate lifeboats in a protected area integral with the accommodation.

29.5.7 Explosion Protection

In many earlier platform designs, there was little protection against explosion over and above that against fire, namely, partition walls being designed as firewalls rather than as blast walls. This situation no longer pertains and much effort is being devoted to explosion protection. CFD simulation is being used to study the overpressures generated by explosions in modules, with particular reference to the enhancing effect of obstacles and to the mitigating effects of venting and of water spray systems. A large amount of work has been done on the development and venting of explosions in modules and other obstructed spaces. Much of the work using CFD explosion simulation codes has been directed to the offshore module situation.

29.5.7.1 Blast Walls

An increasingly common method of explosion protection is the use of blast walls. The design of a blast wall is partly a matter of formulating suitable accident scenarios and predicting by simulation the resultant overpressure so that it fulfills its protective function even if it deforms somewhat.

29.5.7.2 Smoke Minimization and Protection

Another area of work is the investigation of the hazard presented by smoke, particularly that from an oil pool fire. Aspects of this are the generation and movement of smoke and the exclusion of smoke from the accommodation. Smoke engulfment of the Living Quarters and muster area will result in a panicked evacuation. It is necessary to install smoke detectors around the platform, especially in the Living Quarters and muster area.

29.5.8 Design Basis Accidents

The design of a plant to cope with accidental events requires that there be defined a set of design basis accidents. The concept is that the plant is then designed to withstand the design basis accidents in each category but does not have to be designed for events more severe than this. The selection of the design basis accidents is therefore closely linked with the estimates for the frequency of such events, so that overall risk in the design is an appropriate one.

29.5.8.1 Drilling

Drilling is a major activity on the platform that has no counterpart onshore. One aspect of this activity is the ever-present hazard of a well blowout. Drilling activities form the major contributor to blowout risk-based costs. A blowout preventer (BOP) is a very important layer to ensure the offshore operational safety and the pollution prevention. The BP Deepwater Horizon oil rig fire and oil spill have shown the significant challenges in operating deepwater BOPs.

Another aspect that can impinge on the operation of the platform as a whole is the need to avoid loss of power such that the drill becomes stuck.

29.5.8.2 Diving

Diving is another activity specific to an offshore platform. For the most part, it does not impinge to any major extent on other activities except in so far as precautions are necessary to ensure the safety of the divers. At the time of the explosion on Piper Alpha, the fire pumps had been turned to manual start to prevent sudden start-up of the pumps with consequent danger of a diver being drawn into the pump water inlet.

29.5.8.3 Contractors

The proportion of contractors on a platform may well be of the order of 70% or more. Typically, specialist teams such as drillers and divers are contractors with one person from the operating company acting as a liaison. Some contractors may remain on a platform for long periods, others come and go. The operating company attempts to ensure the quality of contractors admitted to the platform by means of a quality assurance system. Personnel from the company visit the contractor company onshore and make the usual quality assurance audit.

It is the responsibility of a contractor to ensure that its personnel are properly trained in offshore emergency procedures as well as in safe systems of work and PTW systems. However, there may well be features of the systems operated on a platform that are particular to that platform. The operator needs, therefore, to ensure that contractors within its system are familiar with them.

29.5.9 Bow-Tie Analysis

Bow-tie analysis displays and illustrates the relationship between hazards, controls, consequences, and mitigations. Bow-ties are qualitative expression or combination of Fault Tree and Event Tree methods. The bow-tie approach is an extremely powerful representation of hazard analyses and risk assessment.

In a typical case of applying the Bow-tie approach, the left hand side can show a number of possible threats that could potentially cause hazards and thus produce the top event. In the case of offshore operations, potential hazards could be blowout, jet fires or explosions from the release of oil and gas. Then some barriers can be put to illustrate typical protection measures. The right hand side illustrates the consequential outcomes of the top event, and some recovery measures can be illustrated here. The typical recovery measures for fire scenarios are blow-down systems, gas detection systems, and fire protection measures.

29.6. OFFSHORE EMERGENCY PLANNING

Offshore emergency planning has the same broad features as that for onshore. In principle, they include the detection of the incident, the assessment of its nature and seriousness and, if necessary, the declaration of the emergency, the assumption of command and control, and the implementation of the emergency plan.

29.6.1 Emergency Scenarios

The first step in emergency planning is the definition of the set of scenarios on which the plan is to be based. A wide range of scenarios need to be considered if the plan is to be robust. It is not uncommon for an incident to require the evacuation of the platform, but in the vast majority of cases this will be by helicopter and management may well feel it has not done as well as it might if even one person is injured.

At the other extreme is the sort of situation that arose on Piper Alpha where there was no prospect of an evacuation by any of the conventional means and where escape to the sea was the only option. This implies that the person in command explicitly instruct personnel to make their own escape.

29.6.2 Safety Case

The offshore safety case is largely modeled on the onshore case, but has three features not explicitly contained in the latter.

1. Safety Management System
2. Temporary refuge
3. Quantitative risk assessment.

29.7. OFFSHORE EVENT DATA

As for onshore installations, information on offshore events is of two main types: (1) incident data and (2) equipment failure data. The *World-wide Offshore Accident Databank (WOAD, 1988)* enumerates the principal accidents involving offshore structures involved in oil and gas activities. There is also a need, particularly in hazard assessment, for more detailed information on events, such as leaks, fires and explosions, dropped loads, and so on. The HSE has created the HCR database, as described by Bruce (1994).

Equipment failure data collection is the subject of a major co-operative exercise carried out under the aegis of OREDA and initiated some years before Piper Alpha.

REFERENCES

API, 2004. Recommended Practice for Development of a Safety and Environmental Management Program for Offshore Operations and Facilities, third ed. Recommended Practice 75. <http://www.api.org/~/media/Files/Publications/Catalog/Final-catalog.pdf>

API, 2007. Recommended Practice for Fire Prevention and Control on Fixed Open-type Offshore Production Platforms, fourth ed. Recommended Practice 14G. <http://www.api.org/~/media/Files/Publications/Catalog/Final-catalog.pdf>

API, 2008a. Recommended Practice for Analysis, Design, Installation and Testing of Basic Surface Safety Systems on Offshore Production Platforms, 7th ed. Recommended Practice 14C.

API, 2008b. Recommended Practice for the Design, Installation, and Maintenance of Electrical Systems for Fixed and Floating Offshore Petroleum Facilities for Unclassified and Class 1, Division 1 and Division 2 Locations, Recommended Practice 14F. <http://www.api.org/~/media/Files/Publications/Catalog/Final-catalog.pdf>

BP, 2010. Deepwater Horizon Accident Investigation Report. <http://www.bp.com/liveassets/bp_internet/globalbp/globalbp_uk_english/incident_response/STAGING/local_assets/downloads_pdfs/Deepwater_Horizon_Accident_Investigation_Report.pdf>

Bruce, R.A.P., 1994. The offshore hydrocarbon releases (HCR) database. Hazards. XII, 107.

Chamberlain, G.A., 1989. Piper Alpha Public Inquiry. Transcript, Days 134 and 138.

Cullen, The Honourable Lord, 1990. The Public Inquiry into the Piper Alpha Disaster. HM Stationery Office, London.

Norwegian Petroleum Directorate, 1981. Guidelines for safety evaluation of platform conceptual design, Stavanger.

Norwegian Petroleum Directorate, 1985. Regulations concerning the licensee's internal control in petroleum activities on the Norwegian Continental Shelf, Stavanger.

Norwegian Petroleum Directorate, 1990. Regulations concerning implementation and use of risk analyses in the petroleum activities with guidelines, Stavanger.

Oil and Gas and Sulphur Operations in the Outer Continental Shelf, Code of Federal Regulations Title 30, Pt. 250. <http://www.sba.gov/advocacy/815/753077>

World Offshore Accidents Databank, 1988. WOAD Statistical Report. Veritas Technology and Services A/S, Hovik, Norway.

Nuclear Energy and Safety

Nuclear energy is one of the major sources for mitigating the energy needs of humans. About 13.5% of world's electricity demand is provided from the nuclear power plants. With the use of very small amount of world's resource, the nuclear reactions generate enormous amount of energy. For example, the fission of Plutonium 289 generates 83.61 TJ/kg. As a result, the number of nuclear power plants increased significantly with time. In the middle of the 1950s, the commercial use of nuclear energy began. In 1979, there were 233 reactors being built. At the end of 1987, there were still 120 reactors in process. As of August 2012, worldwide there were 435 nuclear reactors operating and 66 new plants were under construction.

Early perceptions of the hazard from the nuclear industry were colored by the image of an explosion from a nuclear weapon. As a result of this fear, pressure has always been put to make nuclear industry much safer. Therefore, in comparison with the other types of industry, nuclear industry has evolved as the leader in the field of safety and risk analysis. In the context of accident of nuclear industry: the major accidents in the nuclear industry are core meltdown, which is quite different than its colored image. There would not be the initial intense blast and radiation effects, but rather a dispersion of radioactive material from the site by the wind. Such an accident occurred at Chernobyl. Despite having much different accident scenario, the consequences of the small nuclear accidents such as a small radiation leak may have significant effect on mankind. For example, the potential health effects of radiation at relatively low doses are: (1) birth defects from radiation received by the fetus; (2) thyroid cancers from radioactive iodine from 3 years onwards (3) leukemia over a period 5–40 years (4) cancers over a generation and (5) genetic effects.

30.1. REGULATION AND CONTROL OF NUCLEAR INDUSTRY

In the United Kingdom the regulatory body is the Nuclear Installations Inspectorate (NII), which is part of the HSE. Also there is a standing committee, the Advisory Committee on the Safety of Nuclear Installations (ACSNI) which advises on nuclear safety. Advice on radiological protection in the United Kingdom is the responsibility of the National Radiological Protection Board (NRPB). In the United States, the regulatory body changed in the mid-1970s from the Atomic Energy Commission (AEC) to the Nuclear Regulatory Commission (NRC). There is also an Advisory Committee on Reactor Safety (ACRS).

Bodies operating at the international level include the International Atomic Energy Agency (IAEA) in Vienna and the International Committee on Radiological Protection (ICRP). The IAEA was set up in 1957 as an intergovernmental organization linked to the UN.

Work for the nuclear industry is a main source of support for a number of major research laboratories and organizations whose output is also of interest to the process industries. In the United Kingdom, these include the UK Atomic Energy Authority (UKAEA), which operates laboratories at Harwell. The Systems Reliability Directorate (SRD) is part of the UKAEA.

In the United States, major laboratories with a nuclear interest include the Argonne National Laboratory (ANL), Brookhaven National Laboratory (BNL), Electric Power Research Institute (EPRI), Lawrence Livermore National Laboratory (LLNL), Oak Ridge National Laboratory (ORNL), and Sandia Laboratories (SL).

30.2. NUCLEAR REACTORS

The principal and exclusive unit of nuclear industry is the reactor. Some principal types of nuclear reactor are:

1. Gas-cooled reactors:
   a. Magnox reactor;
   b. Advanced gas-cooled reactor (AGR);
   c. High temperature gas-cooled reactor (HTGR);
2. Light water reactors (LWRs):
   a. Pressurized water reactor (PWR);
   b. Boiling water reactor (BWR);
3. Heavy water reactors:
   a. CANDU reactor;
   b. Steam generating heavy water reactor (SGHWR);
4. Fast breeder reactors.

30.2.1 Reactor Design Features

In a nuclear reactor, energy is released by a chain reaction involving neutrons. The reaction is sustained by those neutrons which are absorbed. Slower neutrons are absorbed in larger proportion than faster ones. Use is made, therefore, of a 'moderator' to slow down the neutrons. In some designs, the moderator is graphite, for example, the Magnox reactor, while in others it is water, for example, the PWR. The power output of the reactor is controlled by movement of control rods, containing materials which are strong absorbers of neutrons, and the reactor is shut down by their full insertion. Emergency shut down, or scram, is affected by dropping in the rods. Even when the reactor is shut down, it has a certain output of heat, due to radioactivity in the core. This decay heat needs to be removed if the core is not to overheat. The removal of the decay heat is one of the main problems which have to be handled in reactor design. The reactor core is cooled by a coolant which in the AGR is carbon dioxide and in the PWR is water.

The underlying concept of the advanced gas-cooled reactors (AGR) is that if forced convection circulation of the gas coolant is lost, natural convection will still affect a degree of cooling. To this extent, the reactor is an application of inherently safer design.

The majority of nuclear reactors installed world-wide are pressurized water reactors (PWRs). In the PWR design the core is cooled by water which also doubles as the moderator. A PWR is provided with multiple systems to inject cooling water into the reactor to keep the core cool and with numerous protective systems to ensure the operation of this backup cooling. The reactor is vulnerable in the event of the failure of these instrument and cooling systems. On a PWR an emergency can escalate within a relatively short time scale. In a PWR the fuel is inside three sets of containment. The first is the fuel rods. The second is the reactor pressure vessel (RPV) in which the fuel rods are held. The third is the containment building which encloses the reactor vessel. The reactor is vulnerable to loss of coolant and much attention has been focused on the large loss of coolant (LOCA) accident.

In view of the nature of the hazard from a nuclear reactor, there is interest in inherently safer design. Mention has already been made of the inherent safety aspects of the AGR. A review of approaches to inherently safer design of nuclear reactors is given in *Proposed and Existing Passive and Safety-related*

Structures, Systems and Components (Building Blocks) for Advanced Light Water Reactors by Forsberg et al. (1989). Some examples are discussed by Kletz (1988). The HTGR is a small reactor cooled by high pressure helium. The reactor is designed to withstand a high temperature and to have an adequate heat loss by radiation and natural convection if the coolant should fail. The Swedish process-inherent ultimate safety (PIUS) reactor is a water-cooled reactor immersed in boric acid solution, so that if the coolant pumps fail this solution is drawn through the core by natural convection. Another aspect of inherent safety is the reliability of the overall system for the control and protection of the reactor, which includes both the automatic control and protection and the process operator. As already mentioned, a PWR is vulnerable to equipment failure in the instrument and coolant systems which protect it. It is therefore vulnerable also to operator actions which render this protection ineffective.

30.3. NUCLEAR WASTE TREATMENT

The residue that comes from the use of radioactive materials is known as nuclear waste; the most radioactive are those that were used for weapon construction and from commercial nuclear factories. Nuclear waste can be divided into three main classes:

1. High level waste (HLW) is the most radioactive product of the nuclear fuel cycle. Can be spent fuel or liquid and solid products from the reprocessing of spent fuel.
2. Transuranic waste (TRU) means elements of atomic number larger than 92. This waste contains more than $0.1 \, \mu Ci$ of long-lived ($T > 20$ years) transuranic alpha-particles per gram of material.
3. Low level waste (LLW), the material left that comes from nuclear facilities.

Nuclear reprocessing uses chemical procedures to separate the useful components from the fission products and other radioactive waste in spent nuclear fuel obtained from nuclear reactors.

There are two acceptable storage methods for spent fuel after it is removed from the reactor core: spent fuel pools and dry cask storage (NRC, 2007).

The spent fuel assemblies still generate significant amounts of radiation and heat when removed from the reactor. It must be shipped in containers which can shield and contain the radioactivity and dissipate the heat.

There are several approaches for the final disposal of nuclear waste, which are given in David (2004). These include deep geologic disposal, deep-borehole disposal, rock melt, deep-seabed, and others.

30.4. NUCLEAR SYSTEM RELIABILITY

Some of the important features that are considered in nuclear system reliability are as follows.

30.4.1 Pressure Systems

The integrity of the pressure system is critical to the safety of a nuclear reactor such as the PWR. Therefore, much attention has been focused on the potential for pressure vessel failure. Pressure vessel failure is considered in the Rasmussen Report. It is also discussed by Bridenbaugh et al. (1976) and Cottrell (1977).

30.4.2 Protective Systems

From the start nuclear industry made extensive use of instrumentation particularly of instrumented protective systems. Accounts of this work are given in Eames (1965), and Green and Bourne (1966).

This work led to interest in reliability technologies, including fault trees and dependent failure. Since it was necessary to have a target to aim for in design of such protective systems, this also prompted the development of risk criteria.

30.4.3 Reliability Engineering

The nuclear industry joined the defense and aerospace industries in developing reliability techniques. A treatment of reliability engineering from the viewpoint of the nuclear industry and with special reference to protective systems is Reliability Technology by Green and Bourne (1972).

30.4.4 Failure and Event Databases

The practice of reliability engineering created the need for failure and event databases. This is illustrated by the development of the database of the UKAEA Systems Reliability Service (Ablitt, 1973).

30.4.5 Fault Trees

The need to identify failure paths and to estimate failure frequencies for systems with multiple layers of protection has led in the nuclear industry to extensive use of fault tree analysis. It is the basic method used for frequency estimation in the Rasmussen Report. The technique is described in Reliability and Fault Tree Analysis by Barlow et al. (1975) and in Fault Tree Handbook by NRC (1981).

30.4.6 Event Trees

Likewise, the nuclear industry has made much use of event trees, both to identify and quantify the effects of certain failures, such as those of the power supply, and the outcomes of a release to atmosphere.

30.4.7 Dependent Failures

Nuclear reactors rely on extensive protective systems to give a very high degree of reliability, and it is characteristic of such systems that they are liable to be defeated by dependent failure. In some cases different assumptions about dependent failure can result in a dramatic increase in the estimated frequency of an accident. The nuclear industry has therefore put much effort into the study of this problem. The problem of dependent failure bulks large in the analyses in the Rasmussen Report and in the critiques of that report.

30.5. NUCLEAR HAZARD ASSESSMENT

30.5.1 Probabilistic Risk Assessment

The nuclear industry has for long made use of probabilistic risk assessment (PRA), or probabilistic safety assessment (PSA), as a means of ensuring that its hazards are identified and under control and as a basis for dialogue on this with the regulatory authority. Guidance is given in *PRA Procedures Guide* by the NRC (1983). Another source of PRA is *Procedures for Conducting Probabilistic Safety Assessments of Nuclear Power Plants* by the IAEA (1992). More recent guidance is given in *Handbook of Parameter Estimation for Probabilistic Risk Assessment* by NRC (2003).

30.5.2 Accident Sequences and Scenarios

In a nuclear PRA considerable effort is devoted to the treatment of the accident sequences, or scenarios. Thus a large proportion of the IAEA document just referred to is devoted to procedures for the identification of accident initiators and of accident sequences, for the classification of accident sequences into plant damage states and for the determination of the frequency of the accident sequences.

30.5.3 Rare Events

There are a number of natural and man-made threats which may hazard a nuclear plant. Examples of these two types of events are, respectively, an earthquake and an aircraft crash. Such events, their consequences and frequency, have therefore been the subject of much study by the nuclear industry.

30.5.4 Explosions

One type of the event which may put a nuclear plant at risk is an explosion. The industry has examined various kinds of explosion, including vapor cloud explosions, and certain types of particular interest to it such as steam explosions and hydrogen explosions.

30.5.5 Hydrogen Explosions

The fuel elements in the reactor core are held in metal cladding tubes. At high temperatures the zirconium metal used for this cladding can react with steam to produce hydrogen, with the risk of a hydrogen explosion.

30.5.6 Missiles

Studies of some relevance to process plants are those on missiles such as the work reported by Baum (1984).

30.5.7 Earthquakes

Of rare natural events, the earthquake hazard bulks large both in plant design and in risk assessment. The nuclear regulator has put much effort into defining the type of earthquake which the plant should be required to withstand. Guidance is given in *Design Response Spectra for Seismic Design of Nuclear Power Plants* by the US AEC (1973) and *Seismic Considerations for the Transition Break Size* by the NRC (2008).

30.5.8 Expert Judgment

Where hard information is lacking, resort may be had to the use of expert judgment. Thus expert judgment may be used to obtain estimates of failure and event rates. It may also be applied to other aspects such as the formulation of accident sequences. Guidance on this approach is given in *Eliciting and Analysing Expert Judgement: a Practical Guide* by Meyer and Booker (1990).

30.5.9 Computer Error

A modern nuclear power station operates under computer control. The control computer, both hardware and software, is therefore a safety critical system. The quality assurance of the computer software is a particularly intractable problem. An account of this aspect of the Sizewell B computer system is given by Hunns and Wainwright (1991).

30.5.10 Human Error

The conduct of PRA for nuclear plants has led naturally to a requirement for methods of handling the human aspects, particularly in fault trees and in event trees. An overall methodology for this problem is described in *Handbook of Human Reliability Analysis with Emphasis on Nuclear Power Plant Applications* by Swain and Guttman (1983). A method of estimating error probabilities is addressed in *SLIM-MAUD: an Approach to Assessing Human Error Probabilities Using Structured Expert Judgement* by Embrey et al. (1984). More guidance is given in NUREG-1842 *Evaluation of Human Reliability Analysis Methods Against Good Practices* by NRC (2006).

30.5.11 Source Terms

Each accident scenario needs to be associated with a defined source term, so that the dispersion of radioactive materials may be modeled.

30.5.12 Emission Models

The nuclear industry was one of the first to develop a sustained interest in two-phase flow, both flow within the plant and flow of leaks from it.

30.5.13 Gas Dispersion Models

Much of the work on gas dispersion modeling has been in support of work on the consequences of a nuclear accident. This is illustrated by one of the early texts, *Meteorology and Atomic Energy 1968* by Slade (1968).

30.5.14 Severe Accident Risk

A focus to work on PRA has latterly been provided by the Severe Accident Risk study, for which the basic document is NUREG-1150 *Reactor Risk Reference Document* of the NRC (1987).

30.5.15 Risk Criteria

The quantification of risks from nuclear reactors has necessarily led to the need for risk criteria by which to judge these risks. Some of the first risk comparisons were those given in the *Rasmussen Report*. Since then, there has been a wide-ranging debate of risk criteria for man-made hazards of all kinds. This aspect is also described in *Development of Risk Criteria in Nuclear Power Plants—Problems and Solutions* (Cepin, 2006).

30.6. NUCLEAR REACTOR OPERATION

The *Kemeny Report* and, even more, the *Task Force Report* on TMI contain numerous recommendations on nuclear reactor operation, particularly on the design of display and alarm systems and of control rooms generally, operating and emergency procedures, and operator training. Some of the emphasized points are human factors, process operator, display and alarm, control room design, operating procedures, emergency procedures, operator training, nuclear emergency planning, accident management plans, and off-site emergency plans.

30.7. NUCLEAR INCIDENT REPORTING

In most regulatory regimes the nuclear industry is required to report to the regulator the occurrence of specified types of incidents. In the United States, the system is that described in the NUREG/CR-2000 *Licensee Event Report* (LER) compilation. The LERs record some events which are of such a nature that they could well have escalated into more severe events, even though in the particular case this has not occurred. These events are precursors of more serious incidents and serve as warnings. Periodic analyses are made of such precursors such as NUREG/CR-4674 *Precursors to Potential Severe Core Damage Accidents 1990: Status Report* by the NRC (1998).

30.8. NUCLEAR INCIDENTS

There have been a number of incidents on nuclear plants which are instructive for the process industries also. In addition to TMI and Chernobyl, they include Windscale, Browns Ferry, and Fukushima.

Accounts of incidents include those in *Nuclear Power* by Patterson (1976), and *Nuclear Lessons* by Curtis et al. (1980); later incidents are covered by May (1989).

30.8.1 Radioactive Sources

Although the incidents of prime interest here are those involving nuclear plant, other nuclear sources should not be neglected. The following case, believed to be the worst of its kind to date, illustrates the hazard from equipment containing a radioactive source.

In 1985, in Goiania, Brazil, a radiotherapy institute moved premises, leaving behind a unit containing a cesium-137 source in a stainless steel container and without notifying the licensing authorities. Subsequently, the building was partially demolished. Some 2 years after the institute's departure, on September 13, 1987, two men entered and tried to dismantle the machine for scrap, one taking home the steel cylinder; both fell ill. However, some 5 days later one broke open the steel cylinder and the cesium spilled out. A train of events ensued which resulted in 249 people being exposed to radiation, of whom four died.

30.8.2 Military Reactors and Weapons Plants

30.8.2.1 Hanford, the Green Run, 1949

On December 2, 1949, at Hanford, Washington, there occurred one of a series of releases from an installation of eight nuclear reactors producing plutonium for the Manhattan Project. This was the Green Run, an experiment conducted to investigate monitoring methods which would be useful in intelligence work on the nuclear capabilities of other countries. The release formed a plume 200 miles $\times$ 40 miles in dead calm conditions and involved 20,000 Ci of xenon and 7780 Ci of iodine-131.

30.8.2.2 Rocky Flats, Colorado, 1957

On September 11, 1957, at Rocky Flats, Colorado, a fire occurred in a glove box at an atomic weapons plant. Plutonium shavings ignited spontaneously, workers tried to put the fire out and explosions occurred which blew out the ventilation filters. For half a day there was a release of plutonium contaminated smoke. On May 11, 1969, there occurred another glove box fire. All new weapons production was shut down for 6 months.

30.8.2.3 Idaho Falls, Idaho, 1961

On January 3, 1961, at Idaho Falls, Idaho, an incident occurred on a prototype military nuclear power plant, the Stationary Low Power Reactor SL-1, one of 17 reactors at the AECs National Reactor Test Station (NRTS). The reactor was shut down and control rods had been disconnected to install additional instrumentation. The emergency started with the sounding of alarms. As it developed, three men working on the control rods were found to be missing. One was discovered dead, pinned to the ceiling by a control rod, another was also dead, and the third died soon after. The precise sequence of events is uncertain. One explanation given is that the central control rod was partially withdrawn and there was a power surge, generating steam which lifted the lid of the pressure vessel, causing it to rise 9 ft and then drop back.

30.8.3 Windscale

Accounts of the Windscale incident are given in *Windscale Fallout* by Breach (1978), The *Windscale Fire, 1957* by Arnold (1992) and by May (1989).

On October 10, 1957, at Windscale, United Kingdom, an incident occurred which led to a serious radioactive release. The installation consisted of two simple air-cooled, graphite-moderated, atomic piles used for the production of plutonium. The graphite was subject to deformation and build-up of energy due to neutron bombardment, the so-called Wigner effect. A procedure had been established to release the Wigner energy by turning off the fans, making the pile critical and letting heat build up. On October 7, 1957, this procedure had been followed, but it appeared that not all the Wigner energy had been released, so the pile was heated up again. The sensors then showed an abnormal temperature rise and the power was reduced. By October 9, conditions seemed normal except that there was a hot spot. However, on October 10, a rise in radioactivity was detected at the filters in the chimney. An attempt was made to inspect the core using a remote scanner, which jammed. Staff removed a charge plug and looked in. All the fuel channels which they could see were on fire and a fuel cartridge had burst. The air movement caused by the fans was now fanning the blaze. Despite fears that it might lead to a hydrogen–oxygen explosion, the decision was taken to use water to put out the fire, which it did. As a result of the incident, workers on the site were exposed to radiation and milk from cattle in the area had to be discarded. The release was mitigated by the presence of the filters, installed at the insistence of Sir John Cockcroft, prominent at the top of the chimney, and known locally as 'Cockcroft Folly'. The piles were decommissioned and set in concrete.

30.8.4 Civil Nuclear Reactors and Reprocessing Plants

30.8.4.1 Chalk River, Ontario, 1957

On December 12, 1957, at Chalk River, Ontario, there was a loss of control on a nuclear reactor leading to damage of the core so that it had to be removed and buried. In outline the initial sequence of events was as follows. Outside the control room, a supervisor found that an operator was opening valves which caused the control rods to withdraw. He immediately set about shutting the valves, but some of the rods did not drop back. He telephoned the operator in the control room, intending to ask him to push a button which would drop control rods in, but by a slip of the tongue actually referred to a button which would withdraw rods. In order to comply, the operator moved away from the phone so that the supervisor temporarily lost touch with him. The operator in the control room soon recognized from the rising reactor temperature that there was something wrong and pressed the scram button, but events had been set in train which led to core damage.

30.8.4.2 Detroit, Michigan, 1966

On October 5, 1966, near Detroit, Michigan, the Enrico Fermi reactor experienced an incident. The reactor was the first commercial fast breeder reactor in the United States. It was being started up for a series of tests when abnormalities were observed and some of the control rods were found not to be in their expected positions. Tests on the sodium coolant showed it to be contaminated, suggesting that part of the fuel core had melted, but the cause was unknown. The authorities were alerted and went on standby for evacuation. Subsequent investigation found that separation had occurred of parts from a steel cone at the bottom of the vessel, installed so that in a meltdown the fuel would spread out and not reach critical mass. In a last-minute modification, zirconium plates had been put onto the cone. One plate had worked loose, had been forced into the core by the sodium coolant and had blocked coolant flow to two of the fuel assemblies.

An account of this incident is given in *We Almost Lost Detroit* by Fuller (1984).

30.8.4.3 Browns Ferry, Alabama, 1975

Prior to TMI one of the US incidents which received particular attention was that on March 22, 1975, at Browns Ferry. Accounts are given in the *Rasmussen Report* and in *Browns Ferry: the Regulatory Failure* by Ford, Kendall and Tye (1976).

30.8.4.4 Beloyarsk, USSR, 1978

On December 30−31, 1978, at Beloyarsk, USSR, there occurred an incident which until Chernobyl was the most serious in the nuclear industry. The complex was 50 km from Sverdlovsk and housed two RMBK reactors and a BM600 fast breeder reactor. A serious fire broke out in the machine hall, causing steel girders and the concrete roof to collapse, opening up a huge hole above No. 2 Generator and disabling the fire protection system. Emergency procedures called for both RMBK reactors to be shut down. But as it was close to −50°C outside, it was feared the reactor cooling systems would freeze and the cores would overheat. Attempts were made to keep No. 1 Reactor and its turbine running, but despite the fire the turbine froze. There followed an extended emergency which eventually, with the arrival of fire fighters from outside, was brought under control.

30.8.4.5 Pennsylvania, US, 1979

On March 28, 1979, near Middletown, Pennsylvania, a major accident happened at Three Mile Island nuclear plant. The accident started from a series of human and mechanical failures. After that, cooling water was lost and temperatures soared above 5000°C and the top portion of the reactor's 150-ton core melted. More than 200,000 people needed to flee due to the contaminated coolant water that escaped and evaporated as radioactive gas. The accident led to a minimum of 430 infant deaths, as estimated in Dr. Ernest J. Sternglass's study.

30.8.4.6 Cap la Hague, France, 1981

On January 6, 1981 at Cap la Hague, Normandy, France, in a nuclear fuel reprocessing facility, there occurred the most serious of a series of incidents. Fire broke out in a spent fuel dry waste silo. Ignition occurred in cotton waste, soaked in solvent and in contact with uranium and magnesium. The cotton waste had been used in a decontamination operation there several weeks earlier. The fire was attacked first by water, which formed steam, and then by liquid nitrogen. Radioactivity was released inside and outside the plant.

30.8.4.7 New York State, 1982

On January 25, 1982, in New York State the Ginna reactor experienced tube rupture, due to corrosion, in the steam generator, causing contamination of the secondary, clean steam circuit by the radioactive water from the primary water circuit cooling the core. High pressure in the steam circuit caused a pressure relief valve to open, initially in 5 min bursts and then by sticking open for 50 min. Another PRV on the primary circuit was deliberately opened to reduce the pressure in that circuit but also stuck open. A steam bubble formed, but it proved possible to correct it by pumping in more water.

30.8.4.8 Frankfurt, FRG, 1987

On December 16−17, 1987, at Frankfurt, FRG, the Biblis A nuclear reactor experienced an incident involving the low pressure injection (LPI) system. It was a requirement that this system remain isolated to prevent (1) leakage of primary coolant out through it and (2) overpressure of the system itself by the primary coolant. During start up the main valve TH22 S006 on the pipe between the LPI system and the primary circuit had been left open; there were also two secondary valves on the line. The status of the main valve was shown by a red light on the control panel, but the operators assumed that the fault was on the light itself. There were two changes of shift before it was realized that valve TH22 S006 was open. The operators took 2 h to decide on a plant shut down and then 10 min later changed their minds and tried an alternative, and hazardous, procedure. They decided to initiate flow through valve TH22 S006, expecting that the valve would then be shut by its trip system. They controlled the flow by cracking open one of the secondary valves and primary coolant started to escape, but valve TH22 S0006 did not shut, and after a few seconds they desisted. The operators then shut the reactor down.

30.8.4.9 Waterford, Connecticut, 1996

On February 20, 1996, in Waterford, Millstone Nuclear Power Plant Units 1 and 2 shut down due to a leaking valve, which led to the loss of more than 254 million dollars.

30.8.4.10 Ibaraki Prefecture, Japan, 1999

On September 30, 1999, in Ibaraki Prefecture, workers at the Tokaimura nuclear fuel processing facility try to save time by mixing dangerously large amounts of treated uranium in metal buckets. Two of the workers later die from their injuries, and more than 40 others are treated for exposure to high levels of radiation.

30.8.4.11 Oak Harbor, Ohio, 2002

On February 16, 2002, in Oak Harbor, severe corrosion of control rod forces 24-month outage of Davis-Besse reactor, which cost more than 143 million dollars.

30.8.4.12 Fukui Prefecture, Japan, 2004

On August 9, 2004, in Fukui Prefecture, a steam explosion happened at Nihama Nuclear Power Plant, which killed five workers and injured dozens more.

30.8.4.13 Fukushima, Japan, 2011

The Fukushima Daiichi Nuclear Power Plant has been stupendously affected by the massive earthquake and tsunami of the Great East Japan Earthquake occurred on March 11, 2011 (see Chapter 28), where the cooling system for reactors and spent fuel storage pools became uncontrollable. As a result, reactor cores were damaged at unit 1−3. Furthermore, the cooling function of the spent fuel storage pools of unit 1−4 was also lost, and some fuels may have been damaged. The 'Cliff Edge' factors are DC power loss, the loss of cooling system, and the loss of the ultimate heat sink. The direct causes of the accident at Fukushima Dai-chi Nuclear Power Plant arise from the loss of multiple safety functions of systems/components vital to safety; the quake-hit plant was no longer able to withstand further crisis beyond design basis. The height of tsunami triggered by the Tohoku Pacific Ocean Earthquake yielded no allowance. It has resulted in large and long-lasting consequence to the modern society.

30.9. RASMUSSEN REPORT

A comprehensive hazard assessment of nuclear power plants in general is the *Reactor Safety Study: An Assessment of Accident Risks in US Commercial Nuclear Power Plants* by the Nuclear Regulatory Commission (NRC) (1975). The work was done by a team led by Professor N.C. Rasmussen and is often referred to as the *Rasmussen Report*. It is also known as the *Reactor Safety Study* (RSS) and also as *WASH 1400*. This study was a major exercise involving some 70 man years of work and costing $4 million. The report is a document of nine volumes some 15 cm thick.

The RSS constituted a watershed in probabilistic risk assessment. It not only brought the PRA approach to the forefront as an aid in decision-making, but created a framework and brought together the various techniques needed to carry out such a study. The work is of interest in respect of its methodology, its treatment of particular problems in risk assessment, its compilations of failure data, and its presentation and evaluation of risks. Also of interest are the critiques of the report. The methodology is based on the extensive use of fault trees and event trees and addresses the problem of uncertainty in the results obtained.

30.9.1 Risk Assessment Methodology

The report provides a framework for probabilistic risk assessment. The overall methodology described is based on selection of a set of release scenarios; estimation of the frequency of each scenario using fault trees and of the frequency of each associated set of outcomes using event trees; estimation of the consequences using hazard models for the physical phenomena and models of population exposure, mitigation of exposure by shelter and by escape and evacuation and injury from radioactivity and evaluation of these results using risk criteria.

30.9.2 Event Data

The *RSS* contains a compilation of a large amount of failure and event data. These are based primarily on nuclear industry experience supplemented by other sources. Most of the data refer to 2 years' experience at 17 nuclear plants which were operational in the United States in 1972. The average plant in the sample was some 4 years old. Data for equipment are given as failure rates or probabilities of failure on demand. It is assumed that the failures are random and that the exponential failure distribution is applicable. The possibility of failure due to aging was considered but excluded. The report states: It should be recognized that the study did not include extreme aging consideration since the applicability of its results is limited to only the next 5 years. The data on failure rates and probabilities are given not as point values but as ranges. Most of these data are presented in terms of the 90% confidence range and are correlated using the log–normal distribution with values quoted for the lower and upper bounds, median and error factor.

30.9.3 Fault Trees

The *RSS* makes extensive use of fault trees. The need to do this arises because the accident events considered can occur only if there are failures of a number of protective systems and because such failures are rare and insufficient historical data for them are available. The volume of the report which deals with the fault trees contains some 150 figures; most of them are fault tree diagrams.

30.9.4 Event Trees

The basic accident sequence considered is an initiating event followed by system failure and then containment failure. The *RSS* makes extensive use of event trees for the accident sequences.

30.9.5 Common Mode Failure

Common mode failures are recognized in the *RSS* as a factor which may greatly increase the frequency of the hazard. Several techniques are used to handle such common mode failures. For similar events which may in some way be coupled, such as human actions to calibrate an instrument or open a valve, lower and upper bound probabilities of failure are defined. The lower bound probability p_l is the probability calculated assuming complete independence of the events. The upper bound probability p_u is the probability calculated assuming complete coupling between the events. The actual probability p is then taken as the geometric mean, or log–normal median:

$$p = (p_l p_u)^{1/2} \qquad (30.1)$$

The example given is the reclosing of two valves. The probability of failing to close one valve is taken as 10^{-2}. Then, the lower bound is 10^{-4} ($10^{-2} \times 10^{-2}$) and the upper bound 10^{-2} ($10^{-2} \times 1$), while the probability given by Equation 30.1 is 10^{-3} (($10^{-4} \times 10^{-2}$)$^{1/2}$). The *RSS* concludes that common mode failures do not make a large contribution to the overall frequency of core melt failure.

30.9.6 Human Error

The report gives an extensive treatment of human error and estimates for error rates for a number of types of error. It also deals with problems such as non-independence of, or coupling between, errors.

30.9.7 Rare Events

An important rare event considered in the *RSS* is failure of the reactor pressure vessel. The report states: 'Potentially large ruptures in the vessel were considered that could prevent effective cooling of the core by the ECCS. Since certain of these ruptures appeared to be capable of causing missiles (such as the reactor vessel head) with sufficient momentum to rupture the containment, this area was explored with some care. There is some small probability that a large vessel missile could in fact impact directly on the containment and penetrate through the wall. This type of rupture could involve a core meltdown in a non-intact containment'.

A study by the Advisory Committee on Reactor Safeguards (ACRS) concluded that 'the disruptive failure probability of reactor vessels designed, constructed and operated according to code is even lower than 1×10^{-6} per vessel year'. The figure used in the *RSS* is 1×10^{-7}/ vessel year. The effect of this low estimate of reactor pressure vessel failure rate is to take this event out of consideration as a cause of release. The *RSS* states 'Gross vessel rupture would have to be at least about 100 times more likely than the value estimated in order to contribute to the PWR core melt probability'.

30.9.8 External Threats

The *RSS* considers external threats such as fire, earthquakes, and sabotage. The prediction of earthquakes is highly uncertain and the report emphasizes this. It nevertheless states that a 'reasonable estimate (of core melt due to earthquake) is 10^{-7} per reactor year'. The *RSS* acknowledges the threat posed by sabotage, but does not take this into account in the risk estimates made. These estimates apply to *bona fide* accidents and exclude those due to sabotage.

30.9.9 Release Scenarios

The consequences of a core degradation or meltdown depend on (1) core inventory; (2) fraction released; (3) dispersion and deposition; (4) population exposed; (5) mitigating features (evacuation, shelter); and (6) injury relations.

30.9.10 Population Characteristics

The population at risk was characterized by a composite model based on the 68 actual sites holding the first 100 reactors. Six composite sites of different types were defined. The population density was determined as follows. The first type of site was an Atlantic coastal site and 14 of the reactors were assigned to this composite site. The actual population around each of these reactors was determined from census data for a distance of 50 miles. For each actual site 16 sectors were defined, making 224 sectors in total. These 224 sectors were then ranked in order of population density and 16 representative sectors were determined for the composite site. Data from a time use study by Robinson and Converse (1966) were used to determine the fraction of time spent at different locations and hence the probability that a person was inside shelter.

30.9.11 Mitigation of Exposure

Factors which mitigate the exposure of the population include shelter and evacuation. The *RSS* uses a single exponential stage evacuation model. This model is based on a study of actual evacuations by Hans and Sell (1974). Measures to mitigate long-term exposure include interdiction of the contaminated zone and decontamination of this zone. The effect of shelter on concentration is modeled using a single exponential stage model for ventilation.

30.9.12 Injury Relations

The effects of radioactivity on people are complex and are both short-term, or prompt, and long-term. A general criticism is the relatively high uncertainty in injury relations. In order to determine these effects it is first necessary to estimate the doses of radioactivity received and then to apply a dose—response relation. The *RSS* gives a detailed treatment of all these aspects and the report contains a large number of tables and graphs showing early and late fatalities and injuries due to different modes of injury and the contribution to these of individual radionuclide. It also deals with the contamination of land.

30.9.13 Uncertainty in Results

An attempt is made in the report to put bounds on the errors of the estimates. A number of methods are used. Failure rates and probabilities are taken not as single values but as a range of values with a log—normal distribution, as described above. The inputs to the fault trees are therefore not single values of frequency or probability but distributions of these. The frequency or probability of the top event is then computed, also as a distribution, by sampling from the distributions of these individual events using Monte Carlo simulation. Such simulation may also be used to take account of common mode failures by

arranging coupling between the failure rates or probabilities in question.

30.9.14 Evaluation of Results

The *RSS* evaluates the risk from the 100 nuclear reactors by making comparisons with those from other hazards. The main report gives numerous tables showing risks from various natural and man-made hazards such as earthquakes, hurricanes, tornadoes, meteorites, fires, dams, explosions, toxic gas releases, and aircraft crashes. One of the principle comparisons made is the FN curve for the 100 reactors and for other hazards. The overall conclusion of the report is that the risks to the public from 100 nuclear reactors are very much less than those from other natural and man-made hazards. The report has a slim Executive Summary which describes the report as a whole and gives a presentation and evaluation of the results.

30.9.15 Critical Assumptions

The results obtained in the *RSS* depend on the validity of a number of assumptions.

Some of the more critical assumptions are:

- plant standards are not seriously below average;
- failures are random rather than wearout;
- reactor pressure vessel operating conditions are closely controlled;
- site is not subject to unusually severe earthquakes;
- population density around site is not unusually high;
- emergency plans exist and are exercised;
- medical facilities exist to handle large numbers of injured; and also
- sabotage is not considered.

30.9.16 Critiques

The *RSS* has been the subject of much comment and criticism. Three principal critiques are the comments included as Appendix XI of the report itself, the *Risk Assessment Review Group Report* (Lewis, 1978) and *The Risks of Nuclear Power Reactors* by the Union of Concerned Scientists (1977). There are also reviews by the Electric Power Research Institute (EPRI) (Leverenz and Erdmann, 1975, 1979).

REFERENCES

Barlow, R.E., Fussell, J.B., Singpurwalla, N.D., 1975. Reliability and Fault Tree Analysis. Society for Industrial and Applied Mathematics, Philadelphia, PA.

Bridenbaugh, D.G., Hubbard, R.B., Minor, G.C., 1976. Testimony of Dale G. Bridenbaugh, Richard B. Hubbard, Gregory C. Minor before the Joint Committee on Atomic Energy, February 18, 1976. Union of Concerned Scientists, Cambridge, MA.

Cepin, M., 2006. Development of risk criteria in nuclear power plants—problems and solutions. Int. J. Mater. Struct. Reliab. 4 (1), 53–63.

Cottrell, A. , 1977. Reactor pressure vessel failure. In: Union of Concerned Scientists, op. cit., p. 197.

Curtis, R., Hogan, E.R., Horowitz, S., 1980. Nuclear Lessons. Turnstone Press, Wellingborough.

David, B., 2004. Nuclear Energy—Principles, Practices, and Prospects. Springer, Washington, DC.

Eames, A.R., 1965. Reliability Assessment of Protective Equipment for Nuclear Installations. AHSB (S) R 99.

Embrey, D., Humphreys, P.C., Rose, E.A., Kirwan, B., Rea, K., 1984. SLIM-MAUD: An Approach to Assessing Human Error Probabilities Using Structured Expert Judgment. Rep. NUREG/CR-3518. Nuclear Regulatory Commission, Washington, DC.

Ford, D.F., Kendall, H.W., Tye, L.S., 1976. Browns Ferry: The Regulatory Failure. Union of Concerned Scientists, Cambridge, MA.

Forsberg, C.W., Moses, D.L., Lewis, E.B., Gibson, R., Pearson, R., Reich, W.J., et al., 1989. Proposed and Existing Passive and Inherent Safety-Related Structures, Systems, and Components (Building Blocks) for Advanced Light-Water Reactors, ORNL-6554. Oak Ridge National Laboratory, Oak Ridge, TN.

Fuller, J.G., 1984. We Almost Lost Detroit. Berkley Books, New York, NY.

Green, A.E., Bourne, A.J., 1966. Safety Assessment with Reference to Automatic Protective Systems for Nuclear Reactors. AHSB (S) R 117.

Green, A.E., Bourne, A.J., 1972. Reliability Technology. Wiley, New York, NY.

Hans, J.M., Sell, T.C., 1974. Evacuation Risks—An Evaluation. Rep. EPA-520/6-74-002. National. Environmental Research Center, Environmental Protection Agency, Las Vegas, NV.

Hunns, D.M., Wainwright, N., 1991. Software-based protection for Sizewell B: the regulator's perspective. Nucl. Eng. Int. 36 (436), 38–40.

Kletz, T.A., 1988. Learning from Accidents in Industry. Butterworths, London.

Leverenz, F.L., Erdmann, R.C., 1975. Critique of the AEC Reactor Safety Study (WASH 1400). Rep. EPRI 217-2-3. Electric Power Research Institute, Palo Alto, CA.

Leverenz, F.L., Erdmann, R.C., 1979. Comparison of the EPRI and Lewis Committee Review of the Reactor Safety Study. Rep. NP-1130. Electric Power Research Institute, Palo Alto, CA.

Lewis, H.W., 1978. Risk Assessment Review Group Report to the US Nuclear Regulatory Commission. Rep. NUREG/CR-0400. Nuclear Regulatory Commission, Washington, DC.

Meyer, M.A., Booker, J.M., 1990. Eliciting and Analyzing Expert Judgment: a Practical Guide. Rep. NUREG/CR-5424. Nuclear Regulatory Commission, Washington, DC.

Nuclear Regulatory Commission, 1987. Reactor Risk Reference Document (Draft). Rep. NUREG/CR-1150. Nuclear Regulatory Commission, Washington, DC.

Nuclear Regulatory Commission, 2006. Evaluation of Human Reliability Analysis Methods Against Good Practices. NUREG-1842. Nuclear Regulatory Commission, Washington, DC.

Nuclear Regulatory Commission, 2007. NRC Staff Guidance for Activities Related to US Department of Energy Waste Determinations. NUREG-1854. Nuclear Regulatory Commission, Washington, DC.

Patterson, W.C., 1976. Nuclear Power. Penguin, Harmondsworth.

Robinson, J.P., Converse, P.E., 1966. Summary of US Time Use Survey (Report). Institute for Social Research. University of Michigan, Ann Arbor, MI.

Slade, D.H., 1968. Meteorology and Atomic Energy 1968 (Report). Office of Information Services, Atomic Energy Commission, Washington, DC.

Swain, A.D., Guttman, H.E., 1983. Handbook of Human Reliability Analysis with Emphasis on Nuclear Power Plant Applications. Final Report. Rep. NUREG/CR-1278. Nuclear Regulatory Commission, Washington, DC.

Union of Concerned Scientists, 1977. Union of Concerned Scientists. The Risks of Nuclear Power Reactors, Cambridge, MA.

Index

Note: Page numbers followed by "*f*", "*t*" and "*b*" refer to figures, tables and box, respectively.

Printed and bound by CPI Group (UK) Ltd, Croydon, CR0 4YY

08/05/2025

01864922-0001